Global Climates since
the Last Glacial Maximum

Global Climates since the Last Glacial Maximum

H. E. Wright, Jr., J. E. Kutzbach, T. Webb III,
W. F. Ruddiman, F. A. Street-Perrott, and
P. J. Bartlein, editors

University of Minnesota Press
Minneapolis
London

This material is based upon work supported by the National Science Foundation, Climate Dynamics Program, under Grant Numbers ATM91-07750 (Brown University), ATM91-05839 (Columbia University/University of Virginia), ATM91-02452 (University of Minnesota), ATM91-03000 (University of Oregon), ATM91-01919 (University of Wisconsin-Madison), and ATM89-02849 (University of Wisconsin-Madison). Support from the U.S. Department of Commerce (NOAA) and the U.S. Nuclear Regulatory Commission (NRC) also supported this work. The National Science Foundation also provided funds to help support the publication costs of this book.

Published by the University of Minnesota Press
2037 University Avenue Southeast, Minneapolis, MN 55455-3092
Printed in the United States of America on acid-free paper

Library of Congress Cataloging-in-Publication Data

Global climates since the last glacial maximum / H. E. Wright, Jr. . . . [et al.], eds.
      p.   cm.
  Includes bibliographical references and index.
  ISBN 0-8166-2145-4 (alk. paper)
  1. Climatic changes. 2. Paleoclimatology. I. Wright, H. E.
(Herbert Edgar), 1917-
QC981.8.C5G663   1993
551.6--dc20   93-9332
                                                                                                              CIP

Contents

v

Preface

As Holocene paleoecological data began to accumulate at an increasing rate in the 1970s, and as supercomputers made possible the application of climate models to paleoclimatic problems, it became apparent that a combination and comparison of the two approaches could deepen our understanding of environmental changes during the past 18,000 yr. The well-known CLIMAP program (Climate: Mapping, Analysis, and Prediction), a multi-institutional consortium led by J. Imbrie, J. D. Hays, N. Shackleton, and A. McIntyre, took up this challenge, particularly with regard to oceanic data. The success of CLIMAP stimulated discussions among J. E. Kutzbach, T. Webb III, H. E. Wright, Jr., and E. J. Cushing of Wisconsin, Brown, and Minnesota that resulted in the organization of a formal group called COHMAP, with the objective of doing for the terrestrial scene what CLIMAP had done for the marine realm. The acronym COHMAP originally stood for "Climates of the Holocene—Mapping Based on Pollen Data."

The initial efforts of COHMAP focused on the history of the prairie-forest border of the midwestern United States and the history of the monsoonal regions, but early model simulations soon showed that a broader regional treatment was needed and that Holocene events should be placed in the context of the entire pattern of deglaciation and climatic change. At this point F. A. Street-Perrott, W. F. Ruddiman, W. L. Prell, and P. J. Bartlein became heavily involved in the program, and numerous cooperators were recruited to provide paleoclimatic data from the rest of North America and from most other areas as well. To acknowledge this expanded role, COHMAP members modified the definition of the acronym to "Cooperative Holocene Mapping Project."

Annual workshops in Madison, Wisconsin, brought together most of the COHMAP participants. Well-dated paleoclimatic data sets were compiled to provide synoptic views of past climates, hydrologic budget models and multivariate statistical methods were developed to calibrate the data in climatic terms, and simulations were generated from models based on past boundary conditions involving Milankovitch orbital variations in solar radiation, progressive reductions in ice-sheet size, changes in sea-surface temperatures (based on CLIMAP estimates), and inferred modifications in atmospheric chemistry.

The results from both the paleoclimatic data analysis and the model simulations were represented by a series of "snapshots" at 3000-yr intervals from 18,000 yr ago to the present. An article in the journal *Science* in 1988 summarized some of the results. The present volume provides detailed synopses for different parts of the globe and includes an atlas showing changes in the several climatic variables simulated by the model experiments for summer and winter for the time of the last glacial maximum (18,000 yr ago), the time of the maximum solar heating of the northern continents in summer (9000 yr ago), and the present.

Throughout its activities COHMAP has been supported by the Climate Dynamics Program of the National Science Foundation (NSF) and by the Carbon Dioxide Research Division of the Department of Energy (DOE). Additional support was provided by the Nuclear Regulatory Commission (NRC) to NSF. The model simulations were undertaken at the National Center for Atmospheric Research in Boulder, Colorado, which is sponsored by NSF. Program managers for these agencies who facilitated the funding and provided encouragement include Eugene Bierly, Alan Hecht, Thomas Crowley, Hassan Virji, and William Curry of NSF, Michael Riches of DOE, and Robert Kornasiewicz of NRC. Cooperating scientists in addition to the chapter authors of this book include R. A.

Bryson, R. Chervin, E. J. Cushing, K. Gajewski, R. Gallimore, D. G. Gaudreau, E. C. Grimm, S. Hastenrath, S. E. Howe, M. Kennedy, P. Klinkman, B. Molfino, P. Newby, B. Otto-Bliesner, J. T. Overpeck, S. H. Schneider, L. C. K. Shane, R. Steventon, A. M. Swain, W. Washington, R. S. Webb, and M. Woodworth, the administrative coordinator for COHMAP. The efforts of all these persons were important in the development of the COHMAP program. The production of the book itself has been facilitated by the particular interest of Barbara Coffin, natural science editor of the University of Minnesota Press.

CHAPTER *1*

Introduction

H. E. Wright, Jr.

Ever since the introduction and acceptance of the concept of continental glaciation, geologists and paleoecologists have been trying to decipher the sequence and chronology of glacial and interglacial climates—and the cause or causes of these conditions, which apparently affected the entire globe. Regarding chronology, the time elapsed since the last glaciation was calculated in Europe from the Swedish varve chronology and in America from rates of waterfall retreat. The duration of glacial and interglacial intervals was estimated generally from the thickness of soils formed on glacial tills. But regarding causes, numerous theories were introduced, but no consensus developed.

One theory was based on the cyclical variations in the earth's orbit about the sun. The roots of this concept go back to the 19th century (e.g. Croll, 1864), but it was the laborious computations of orbital variations by Milankovitch (1920) that prompted contemporary climatologists like Köppen and Wegener (1924) to apply the theory to the problems of multiple continental glaciation. At the same time Spitaler (1921) postulated that changes in tropical and monsoonal climates should be attributed to orbital variations. The early history of these ideas has been told with some elegance by Imbrie and Imbrie (1979) and more recently from a different viewpoint by Kutzbach and Webb (1991), who emphasize the early efforts to model climatic changes on the basis of astronomic factors.

My own exposure to the concept first came from reading the work of the British geographer C. E. P. Brooks (1926), who tried to develop a comprehensive picture of climate history, but I was most challenged by the books of the British geological archaeologist F. E. Zeuner (1945, 1946), who proposed a detailed scheme of correlation and dating of European climatic and prehistoric cultural events on the basis of the Milankovitch cycles.

In Zeuner's day little credence was given to the Milankovitch theory by glacial geologists. R. F. Flint in his authoritative textbooks (1947, 1957, 1971) listed many reasons why the theory should not be accepted. However, two critical developments formed the basis for serious reconsideration of this approach. One was the introduction of radiocarbon dating, which led to the development of consistent chronologies for the last 40,000 yr. The other was the demonstration that cores from the world's oceans contain stratigraphic sequences of microfossils that indirectly record systematic changes in sea-surface temperatures (Imbrie and Kipp, 1971) and in the magnitude of glacial ice stored on the continents (Emiliani, 1955; Shackleton and Opdyke, 1973). The stratigraphic pattern of multiple changes matched the orbital changes calculated by Milankovitch (Hays *et al.*, 1976; Imbrie and Imbrie, 1980).

The obvious potential of this approach, combined with the availability of hundreds of ocean cores, led to the formation of a consortium of paleoceanographers known as CLIMAP (Climate Mapping, Analysis and Prediction), whose goal was to develop a spatial and temporal framework from stratigraphic analyses to test the global application of the Milankovitch theory. The reports of the CLIMAP Project Members (1976, 1981) made a strong case that orbital variations were indeed the "pacemaker" of the ice ages (Hays *et al.*, 1976). Simultaneous development of paleoclimatic models of atmospheric general circulation, with boundary conditions based on the data from CLIMAP (1976, 1981), made it possible to estimate specific global climatic conditions for the last glacial maximum (Gates, 1976; Manabe and Hahn, 1977).

Although the ocean record indirectly provided a chronology for continental glaciation and global climatic change, it could say little directly about terrestrial conditions. Meanwhile, glacial geologists and paleoecologists had been steadily accumulating dated stratigraphic and geomorphic evidence to build local and regional syntheses that could be interpreted in climatic terms. The intensification of research that had been impelled by radiocarbon dating resulted in a significant buildup of data. Moraines were more and more carefully mapped and dated, and pollen studies were beginning to rough out the postglacial vegetational sequence in different parts of North America, as they had long ago done in Europe. Studies of packrat middens in the American Southwest were adding a new paleoecological dimension to climatic interpretations in that region, supplementing previous knowledge of the history of lake expansions, and studies of fossil insects in the northern United States and Canada provided further evidence for significant climatic changes.

Although certain regional paleoclimatic generalizations were well established on the basis of such field studies—for example, the postglacial "climatic optimum" in western Europe and the mid-Holocene shift in the prairie-forest border in the Minnesota area—many major puzzles remained. Why were the lakes in the American Southwest at high levels during the glacial period, whereas those in the Sahara and northern Africa were high during the early Holocene instead? Why was it difficult to find modern faunal and floral assemblages of glacial or late-glacial age? Was the northward progression of major tree types in the eastern United States in the Holocene a relatively immediate response to contemporaneous climatic change, or was it a delayed manifestation of differential rates of seed dispersal and range expansion?

At the same time as these questions were being raised in light of better data, the introduction of supercomputers made it possible to adapt general circulation models of the atmosphere for paleoclimatic reconstructions, using boundary conditions for time intervals that could be prescribed with some confidence. The COHMAP (Cooperative Holocene Mapping Project) group had its formal inception in 1977 with the support of grants from the Climate Dynamics Program of the National Science Foundation and later the U.S. Department of Energy. Its central strategy involved close interaction of "data people" and

paleoclimatic modelers, in the belief that the data could be used to test the accuracy of model simulations and that the models could lead to an understanding of the mechanisms that created the climatic conditions for certain times and places. Although many types of geologic and paleoecologic data have been used in late-Quaternary paleoclimatic reconstructions, the COHMAP efforts have concentrated on a broad distribution of pollen, lake-level, and marine microfossil records, all of which are amenable to quantitative representation and thus to comparisons with model results.

The pollen concentration of lake sediments and peat deposits reflects the local and regional vegetation in semiquantitative ways. The successful CLIMAP methodology of transfer functions was applied to the hundreds of pollen surface samples that had been collected in eastern North America: surface pollen assemblages were calibrated against local climate variables in much the same way as marine microfossil assemblages were calibrated against known sea-surface temperatures. The way was then clear to apply the calibrations to stratigraphic sequences in order to reconstruct the vegetational and climatic history at particular sites. The techniques have since been modified to produce response surfaces that depict graphically and in more detail the abundance of certain pollen types in the "climate space" defined by key climate variables.

Closed-basin lakes are sensitive to changes in the hydrologic balance, which is controlled directly or indirectly by climate. Lake levels higher than present can be recorded by raised shoreline features and lakeside archaeological sites, and levels lower than present by the buried stratigraphic record of littoral sediments and macrofossils. These lake-level records are particularly important in paleoclimatic reconstructions of arid and semiarid regions, where pollen records are generally sparse.

The species composition of marine plankton reflects the surface-water temperature and other controlling variables such as nutrients and salinity. The sequence of microfossil assemblages thus can record changes in water-mass boundaries or in environmental conditions.

The boundary conditions specified for the model simulations included orbitally determined insolation, ice-sheet and mountain orography, atmospheric trace-gas concentrations, sea-surface temperatures, sea-ice limits, snow cover, albedo, and effective moisture. The orbital variations that affect latitudinal and seasonal distributions of insolation include the 22,000-yr precession cycle, which controls the time when the earth is closest to the sun, the 40,000-yr tilt cycle, and the 100,000-yr eccentricity cycle.

The first successful model simulations by the COHMAP group were made at the University of Wisconsin on a model with relatively coarse spatial resolution and simplified treatment of physical processes (Kutzbach, 1981; Kutzbach and Otto-Bliesner, 1982). The experiment was designed to test the hypothesis that the early-Holocene lake-level rise in northern intertropical Africa, documented by extensive lake-level data accumulated at Oxford University, could be explained by the enhancement of monsoonal circulations over the northern continents. Although the maximum insolation was at 10-11 ka (thousands of years ago), the time for the experiment was set for 9 ka, when the lake levels were highest, to reduce possible effects of the still-existing North American ice sheet on the atmospheric general circulation. The results of this experiment clearly showed that the increased summer insolation warmed the Eurasian-African landmass sufficiently to increase the monsoonal circulations and raise lake levels in northern Africa.

Emboldened by these results, the COHMAP group then set a plan for using the Community Climate Model (CCM) of the National Center for Atmospheric Research (NCAR), a model with finer spatial resolution and better parameterization than the earlier model, to undertake experiments for every 3000 yr from 18 ka to the present, to examine not only the global effects of changing insolation but also the influence of the ice sheets on the general atmospheric circulation. The results of these experiments have helped to explain several persisting puzzles in late-Quaternary climatic history mentioned above. For example, the presence of the massive North American ice sheet displaced the jet stream and associated storm tracks well to the south, bringing the increased rains to the Southwest that raised lake levels. And the cooling that the waning ice sheet still provided in summer delayed the maximum insolational warming in the Middle West until about 7.5 ka.

Because the COHMAP model results were global, the geographic scope of data application was broadened from the original restricted interest in eastern North America, and cooperating investigators were brought into the effort to cover the rest of the North American continent as well as Eurasia, the Southern Hemisphere, and the world oceans. Annual workshops were organized to maintain an international, interdisciplinary approach, always with a focus on comparing field and analytical data with model simulations. Results to date were summarized in *Science* (COHMAP Members, 1988). The present book contains the details of the model output in a newly de-

signed set of maps for 18, 12, 9, 6, and 0 ka as well as regional summaries for most of the globe. The conditions for 9 and 6 ka are emphasized, but several chapters also provide an overview of the entire deglaciation from 18 ka to the middle Holocene and on to the present. Comparison of the field and analytical data with results of the model simulations is an objective of each of the regional summaries. Discrepancies (where they exist) point to gaps in or misinterpretation of the data or to inadequacies in the boundary conditions or parameterizations of the model simulations. The results as a whole are summarized by Webb, Ruddiman, *et al.* (this vol.), and the potential for further investigations is discussed in the Epilogue (Kutzbach, Bartlein, *et al.,* this vol.).

This introduction cannot close without a comment on the implications of this work for our understanding not only of the past but of the future. Perhaps the closing paragraph of the article in *Science* (COHMAP Members, 1988:1051) might be quoted:

> Climate has influenced human activities. Two great developments emerged at about the time of major environmental change between 12 and 10 ka—the earliest appearance of agriculture in the Old World, and the cultural changes accompanying extinction of the Pleistocene megafauna in the New World. Both developments may have been caused at least indirectly by the types of climatic change examined here. Now we may be faced with the reverse: human modification of climate and of related aspects of the physical environment. Application of a well-tested climate model is at present the only method for predicting these climatic and environmental changes and can help in planning a response to them. COHMAP research is contributing to the testing of these models and to the understanding of past climates.

References

Brooks, C. E. P. (1926). "Climate through the Ages." E. Benn, Ltd., London. (Second revised edition, 1970, Dover, New York)

CLIMAP Project Members (1976). The surface of the ice-age earth. *Science* 191, 1138-1144.

———. (1981). Seasonal reconstructions of the earth's surface at the last glacial maximum. *Geological Society of America Map and Chart Series* MC-36.

COHMAP Members (1988). Climatic changes of the last 18,000 years: Observations and model simulations. *Science* 241, 1043-1052.

Croll, J. (1864). On the physical cause of the change of climate during geological epochs. *Philosophical Magazine* 4, 28, 121-137.

Emiliani, C. (1955). Pleistocene temperatures. *Journal of Geology* 63, 538-578.

Flint, R. F. (1947). "Glacial Geology and the Pleistocene Epoch." Wiley, New York.

———. (1957). "Glacial and Pleistocene Geology." Wiley, New York.

———. (1971). "Glacial and Quaternary Geology." Wiley, New York.

Gates, W. L. (1976). Modeling the ice-age climate. *Science* 191, 1138-1144.

Hays, J. D., Imbrie, J., and Shackleton, N. J. (1976). Variations in the earth's orbit: Pacemaker of the ice ages. *Science* 194, 1121-1132.

Imbrie, J., and Imbrie, J. Z. (1980). Modeling the climatic response to orbital variations. *Science* 207, 943-953.

Imbrie, J., and Imbrie, K. P. (1979). "Ice Ages: Solving the Mystery." Enslow Publishers, Short Hills, N.J.

Imbrie, J., and Kipp, N. G. (1971). A new micropaleontological method for quantitative paleoclimatology: Application to a late Pleistocene Caribbean core. *In* "The Late Cenozoic Glacial Ages" (K. K. Turekian, Ed.), pp. 71-181. Yale University Press, New Haven.

Köppen, W., and Wegener, A. (1924). Die Klimate des Quartärs. *In* "Die Klimate der geologischen Vorzeit," chapter VII. Gebrüder Borntraeger, Berlin.

Kutzbach, J. E. (1981). Monsoon climate of the early Holocene: Climate experiment using the earth's orbital parameters for 9000 years ago. *Science* 214, 59-61.

Kutzbach, J. E., and Otto-Bliesner, B. L. (1982). The sensitivity of the African-Asian monsoonal climate to orbital parameter changes for 9000 yr B.P. in a low-resolution general circulation model. *Journal of the Atmospheric Sciences* 39, 1177-1188.

Kutzbach, J. E., and Webb, T. III (1991). Late Quaternary climatic and vegetational change in eastern North America: Concepts, models, and data. *In* "Quaternary Landscapes" (L. C. K. Shane and E. J. Cushing, Eds.), pp. 175-217. University of Minnesota Press, Minneapolis.

Manabe, S., and Hahn, D. G. (1977). Simulation of the tropical climate of an ice age. *Journal of Geophysical Research* 82, 3889-3911.

Milankovitch, M. (1920). "Théorie mathématique des phénomènes thermiques produits par la radiation solaire." Gauthier-Villars, Paris.

Shackleton, N. J., and Opdyke, N. D. (1973). Oxygen isotope and paleomagnetic stratigraphy of equatorial Pacific core V28-238. *Quaternary Research* 3, 39-55.

Spitaler, R. (1921). "Das Klima des Eiszeitalters." R. Spitaler, Prague.

Zeuner, F. E. (1945). "The Pleistocene Period: Its Climate, Chronology and Faunal Successions." Ray Society, Bernard Quaritch, London.

——. (1946). "Dating the Past: An Introduction to Geochronology." Methuen, London.

CHAPTER 2

Conceptual Basis for Understanding Late-Quaternary Climates

J. E. Kutzbach and T. Webb III

The geological records of past climates are stored in many localities across the earth's surface from the bottoms of the oceans to the tops of ice sheets and mountains, and the collection and correct interpretation of these data present a challenge to field and laboratory researchers. Paleoclimatologists face wide gaps in coverage and seemingly contradictory information from different types of data. A general understanding of the climate system is therefore required to help bridge the gaps and resolve the contradictions, and this understanding can be articulated as either qualitative concepts or formal mathematical models.

This book focuses on a series of paleoclimatic modeling experiments aimed at elucidating the causes, patterns, magnitudes, and mechanisms of late-Quaternary climatic changes. These detailed numerical experiments were motivated by qualitative, conceptual considerations of how orbitally induced changes in solar radiation and the lingering effects of the slowly melting ice sheets might have affected global and regional climates. This chapter (revised from Kutzbach and Webb, 1991) describes many of the concepts, new and old, that have guided us in planning the detailed experiments and in interpreting the results in relatively simple terms.

Orbital Geometry

Variations in the seasonal and latitudinal distribution of solar radiation caused by changes in earth-sun geometry are at the core of our understanding of late-Quaternary climate changes. These solar variations are expressed compactly in the form of equations,

tables, and charts (Milankovitch, 1920; Berger, 1978), but the underlying changes in earth-sun geometry can be illustrated graphically. Figure 2.1 shows the orbital configurations for 18, 15-12, 9, and 6-3 ka and the present. The earth-sun geometry at 18 ka, the approximate time of the last glacial maximum, was similar to that at present. Between 18 and 9 ka, the time of peri-

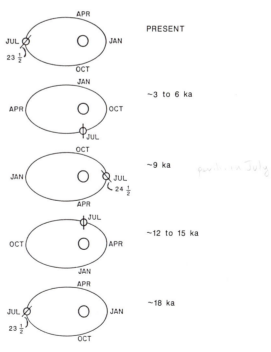

Fig. 2.1. Earth-sun geometry for about 18, 15-12, 9, and 6-3 ka and the present. Perihelion is in January at present and in July around 9 ka, and tilt is greater at 9 ka than at 18 ka and at present. From Kutzbach and Webb (1991) in "Quaternary Landscapes" (L. C. K. Shane and E. J. Cushing, Eds.).

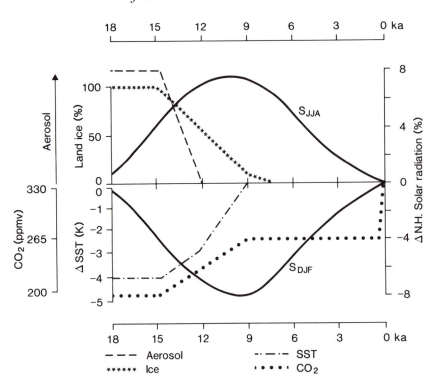

Fig. 2.2. Boundary conditions for the COHMAP simulation with the Community Climate Model (see Wright, this vol.) for the last 18 ka. External forcing is shown for Northern Hemisphere solar radiation in June through August (S_{JJA}) and December through February (S_{DJF}) as the percentage difference from the radiation at present. Internal boundary conditions include land ice (Ice) as a percentage of 18-ka ice volume, global mean annual sea-surface temperature (SST, K) as a difference from present, excess glacial-age aerosol (Aerosol) on an arbitrary scale, and atmospheric CO_2 concentration in parts per million by volume. The horizontal scale indicates the times of the seven sets of simulation experiments. The simulation for 18 ka included the lowered CO_2 concentration (200 ppmv, dots); the others had the CO_2 concentrations of the control case (330 ppmv) rather than the indicated stepwise increase. Experiments incorporating increased glacial-age aerosol loading have not been completed. Reprinted by permission from *Nature* vol. 317, pp. 130-134, copyright 1985 Macmillan Magazines Ltd.

helion (when the earth is closest to the sun) changed from January to July, advancing about 1 day every 60 yr, and the axial tilt increased. Since about 9 ka, axial tilt has decreased, and the time of perihelion has advanced through the remainder of the year, so that today it again occurs during the northern winter.

These orbital changes combined to produce the changes in the seasonal cycle of solar radiation portrayed in Figure 2.2, here summarized in terms of Northern Hemisphere averages. Compared to our present regime, the seasonality of solar radiation was considerably greater during the interval from around 12 to 6 ka, reaching a maximum around 9 ka, when solar radiation was greatest in northern summer. Summer insolation was about 30 W/m² (about 8%) more than today, and winter insolation was less (Fig. 2.3, top and middle). In the Southern Hemisphere the seasonality of solar radiation was less pronounced around 9 ka than at present.

A conceptual model of the response of climate to these insolation variations can be based on the differ-

ent heat capacities of land and water. The small heat capacity of land (North America, North Africa/Eurasia) compared to ocean (the North Pacific and Atlantic) causes the response of surface temperature to insolation changes to be much larger over land than over ocean (Fig. 2.3, bottom). Realistic values of the heat capacity of land and of a 100-m mixed-layer ocean (that part of the ocean that experiences the largest seasonal cycle) lead to estimates that northern continental interiors were as much as 5°C warmer than at present in summer around 9 ka (and generally 2-4°C warmer from 12 to 6 ka), whereas the northern oceans were only about 1°C warmer. The seasonal warming of the ocean would have lagged behind the warming of the land by a month or more (Fig. 2.3, bottom). Northern winters would have been correspondingly colder than at present. The warmer summers and colder winters would have increased the seasonality and continentality of Northern Hemisphere climate (with the reverse holding for southern continents), but with perhaps little if any change in

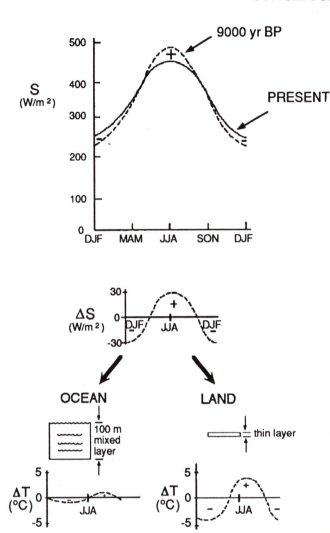

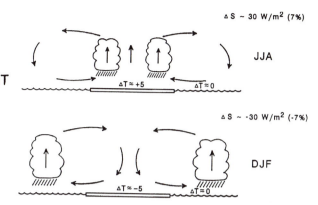

Fig. 2.4. Schematic vertical cross section along an east-west transect across tropical oceans (wavy line) and land (thin slab). The increased temperature over land in northern summer (JJA) at 9 ka causes increased rising motion, high-level outflow of air, and low-level inflow of moist air. Monsoon precipitation is enhanced along the coasts and to a certain distance inland. The interior is hotter but does not receive increased precipitation and therefore becomes drier. In northern winter (DJF) the reverse vertical circulation cell develops. These changes (at 9 ka) enhance the normal vertical (monsoonal) cells. From Kutzbach and Webb (1991) *in* "Quaternary Landscapes" (L. C. K. Shane and E. J. Cushing, Eds.).

Fig. 2.3. Top: The seasonal cycle of Northern Hemisphere average solar radiation (S, W/m²), calculated for the top of the earth's atmosphere, at present (solid line) and 9 ka (dashed line). Middle: The change in input of solar radiation (ΔS) at 9 ka compared to present. Bottom: Schematic temperature responses (ΔT, C) of a 100-m slab of ocean (left) and the insulative land surface (right) to the same change in solar radiation. It is this differential thermal response of land and ocean to changes in the seasonal cycle of solar radiation that drives the global monsoonal response to changes in earth's orbit. From Kutzbach and Webb (1991) *in* "Quaternary Landscapes" (L. C. K. Shane and E. J. Cushing, Eds.).

annual-average temperature. The magnitude of these temperature changes was confirmed in our modeling experiments (Kutzbach and Otto-Bliesner, 1982; Kutzbach and Guetter, 1986; Kutzbach and Gallimore, 1988), as detailed in chapters that follow.

On the basis of hydrostatic considerations, we expect that the relatively large increase in summer temperature over northern land (compared to ocean) would cause higher air temperatures and vertical ex-

pansion of the column of air over land, outflow of air from land to ocean at high levels, lower surface pressure over land (compared to over ocean), and inflow of air from ocean to land at low levels (Fig. 2.4); that is, an increased summer monsoon. The increased inflow would increase precipitation and create moister conditions along the coasts and some distance inland; but far inland, beyond the reach of the increased flux of moisture, the climatic conditions would be drier because of increased evaporative losses.

By analogous reasoning, monsoon circulations for northern winters would also be intensified (Fig. 2.4). However, net annual moisture (precipitation minus evaporation [P-E]) would tend to increase, because the increase in summer moisture would outweigh any decrease in winter moisture. This annual-average increase in moisture is another important concept. It results in part because of the nonlinearity between temperature and water-vapor saturation pressure. This nonlinearity, which is represented by the Clausius-Claperon relation, means that an increase in summer temperature raises the moisture-holding capacity of an airmass, and hence precipitation, more than they are lowered by a corresponding decrease in winter temperature.

If solar-radiation changes were the only causal factor to be considered in Holocene climatic change, then one would arrive at a simple model where the magnitude of the "monsoonal" change depended only on

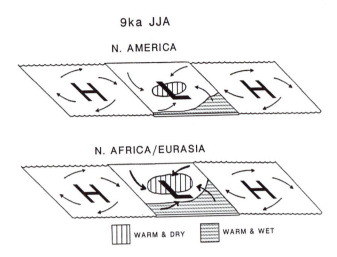

Fig. 2.5. Idealized changes in circulation and climate on small (North America) and large (North Africa/Eurasia) continents resulting from the enhanced northern summer (JJA) monsoons caused by orbitally induced changes in solar radiation. The temperature in the interior should increase most over the large continent (as shown by the size of the region of greatest warmth), and the region of increased moisture should extend farther into the interior on the south and east sides of the large continent (as shown by the size of the region of greatest precipitation). The monsoon circulation is also more intense on the large continent (as shown by the thickness of the wind arrows). Modified from Kutzbach and Webb (1991) *in* "Quaternary Landscapes" (L. C. K. Shane and E. J. Cushing, Eds.).

the size and perhaps the latitude of the continent. Continent size is a very important concept because the maximum temperature response mentioned earlier (about 5°C) occurs only if the continent is so large that the interior is isolated from the moderating (advective) influence of airstreams from the cooler ocean. Large continents amplify the seasonal extremes of temperature, as clearly demonstrated in experiments with energy-budget climate models (Crowley *et al.*, 1986). Thus a conceptual view of northern monsoon enhancement at 9 ka would anticipate a larger temperature, moisture, and circulation response for North Africa/Eurasia than for North America (Fig. 2.5).

The sketch of an enhanced summertime monsoon circulation at 9 ka over the idealized North African/ Eurasian continent (Fig. 2.5, bottom) is based on the above-mentioned concepts, and it agrees with the results of our detailed climate simulations for 12–6 ka for the large North African/Eurasian landmass. Observations of a moister climate at that time consist of raised lake levels compared to the preceding and following periods. Both the detailed model simulations and the observations are described in subsequent chapters

(e.g., Kutzbach, Guetter, *et al.*, this vol.; Street-Perrott and Perrott, this vol.). This conceptual model might not fully describe the climate of North America at 9 ka, however, because of the residual effects of the melting ice sheet on climate.

Ice Sheets

Ice sheets are a second major controlling influence on climate. Whereas orbitally caused insolation changes are external to the climate system, ice sheets are very much a part of it. Nevertheless, for simplicity, in our model the size and shape of ice sheets are prescribed as a lower boundary condition rather than calculated as a function of accumulation, flow, and ablation. In this simplified conceptual model the glacial and postglacial climates of North America, the North Atlantic, and Europe can be viewed as a combined response to solar-radiation changes caused by orbital changes and to the lingering effects of the slowly melting ice sheets (Fig. 2.2). However, it is difficult to assess the relative importance of these two climatic controls without the aid of numerical climate models. Several results that have emerged from our detailed modeling studies are useful in summarizing the flow patterns around ice sheets and the contrasting monsoonal circulations. These results can be illustrated by a vertically stacked set of "snapshots" of the climatic patterns of 18–15, 12, 9, and 6 ka and today for an idealized North America (Fig. 2.6).

The Laurentide ice sheet had a profound effect on glacial-age circulation because of its size, height, and high reflectivity. The air in contact with the ice was cold, of course, compared to the surrounding regions, which resulted in a sinking motion above the ice sheet, low-level outflow of air along the ice perimeter, and via earth-rotational effects, a clockwise, anticyclonic surface wind regime known as a glacial anticyclone (Brooks, 1926; Hobbs, 1926; Bryson and Wendland, 1967). A strong north-south temperature gradient developed near and south of the ice front and displaced the polar front south of its present position. A major jet stream was located above this polar front. Simultaneously, the northern flank of the ice sheet became a boundary between truly Arctic airmasses to the north and the sinking (and thus warmer) air over the ice sheet itself. This secondary north-south thermal gradient also resulted in a midtroposphere wind maximum. The glacial-age jet stream pattern over North America could therefore be described as split into two branches (northern and southern), in contrast to the dominant single jet core at present. A more detailed explanation of the split jet of glacial times in-

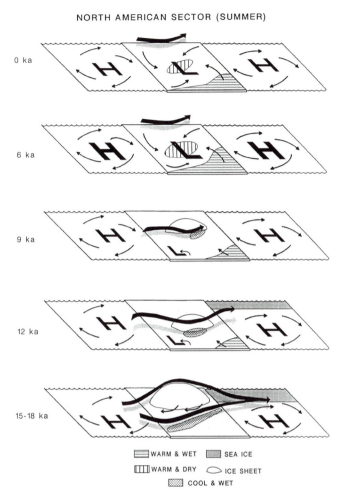

NORTH AMERICAN SECTOR (SUMMER)

0 ka

6 ka

9 ka

12 ka

15-18 ka

WARM & WET SEA ICE
WARM & DRY ICE SHEET
COOL & WET

Fig. 2.6. Conceptual changes in summertime circulation for North America, 18 ka to present, based on simulation results of COHMAP Members (1988) and concepts illustrated in Figures 2.1-2.5. At 15-18 ka the circulation was strongly influenced by the large North American ice sheet and by extensive sea ice in the North Atlantic. The jet stream split into two branches over North America and extended far downstream over the Atlantic. The southern branch was shifted far south and was accompanied by precipitation associated with frequent cyclonic storms. Along the southern ice-sheet margin, conditions were relatively dry and were dominated by a glacial anticyclone. By 12 ka the ice sheet was smaller and the sea ice less extensive. The jet stream was weaker than at 15-18 ka and was located along the southern margin; it was accompanied by precipitation associated with cyclonic storms. Increased summertime solar radiation associated with orbital changes helped establish warmer conditions west of the ice sheet, a low-pressure center in the southwest, and increased precipitation in the south and southeast. By 9 ka the ice sheet was very small, and the orbitally produced enhancement of summertime conditions was becoming more evident. By 6 ka northern summer monsoons over land and anticyclonic circulations over the ocean, caused by orbitally induced changes in solar radiation, were stronger than at 9 ka (when the presence of the small ice sheet still moderated the orbital effects) and also stronger than at present (see Figs. 2.1-2.5). The interior of the continent was warmer than at present, and the region of increased moisture extended farther into the continent on the south and east margins. At present (0 ka), these summertime circulation features are not as pronounced as they were at 6 ka. Modified from Kutzbach and Webb (1991) *in* "Quaternary Landscapes" (L. C. K. Shane and E. J. Cushing, Eds.).

volves dynamical processes related to flow around a barrier (like currents around a rock in a stream) as well as the thermal process described here.

To the climatologist, the conceptual sketch (Fig. 2.6) of upper and low-level flow patterns implies many different regional thermal and moisture regimes related to the ice sheet at its maximum: (1) storm track precipitation in the American Southwest and along the path of the southern jet; (2) cold and dry conditions in the Northwest and near the ice front; but (3) climates perhaps no colder than at present in parts of Alaska, where the flow would have had a more southerly component than at present. The cold northwesterly flow along the northeast flank of the ice sheet may have helped chill the North Atlantic and displace the sea-ice border southward (Manabe and Broccoli, 1985).

By 12 ka our modeling experiments point toward significant changes in the flow patterns from those experienced earlier (Fig. 2.6). The insolation increase during northern summer (Figs. 2.1 and 2.2) had significantly warmed the region well south and west of the ice sheet, but regions near and downwind of the ice remained cold. Significant wastage of the North Amer-

ican ice sheet (Fig. 2.2) had reduced its extent and height (and perhaps reflectivity) sufficiently so that the simulated glacial anticyclone had weakened, and the primary thermal gradient (and jet) hugged the southern boundary of the ice sheet. This change consolidated the split jet of full-glacial time into a single jet stream. In the east the climate was cool and moist along the ice front. West of the ice sheet in parts of northwestern North America, this time would have been warm and dry, coinciding with maximum insolation. The continuing cool conditions in central North America would have interfered with the establishment of the midcontinent heat low that developed in North Africa/Eurasia (Fig. 2.5). Indeed our model simulations for 12 ka show only a hint of this pure monsoonal response, expressed as a center of pressure lower than at present in the Southwest and increased inflow to that region. Thus the model suggests the possibility that in parts of the Southwest the cool, moist glacial conditions could have given way to continued moister conditions than at present, associated with an enhanced but regionally limited summer monsoon.

The trends discussed above for 12 ka would have been even more evident at 9 ka as the climatic effects of the ice sheet diminished further. By 6 ka the residual ice sheet was too small to fill even a single model grid square, and therefore our simulations for North America for 6 ka show the simple monsoonal response to increased insolation that occurred earlier in North Africa/Eurasia (Fig. 2.6). At 6 ka the maximum warming (compared to present) was centered in the continental interior. The enhanced monsoonal circulation corresponded closely to the pattern over North Africa/Eurasia (Fig. 2.5), except that it was subdued because of the smaller size of the continent. At this time the east and south would have experienced the maximum increase in summer precipitation due to the enhanced southerly flow. Some of the central interior, however, would have been drier because of the increased evaporative losses associated with increased insolation and temperature and possibly decreased precipitation if the flow became more westerly to the west and southwest of the low-pressure center. This simulated climatic sequence (and the conceptual sketch in Fig. 2.6) perhaps comes close to depicting the mid-Holocene climate and explaining why it was the warmest and driest period in the North American continental interior (i.e., the "Hypsithermal" prairie expansion [see Wright, 1976]). This topic is discussed in detail in subsequent chapters (e.g., Webb, Bartlein, et al., this vol.).

The simulated wintertime climate from around 9 to 6 ka, when insolation was less than at present, was generally colder than at present; however, this radiative effect on temperature would probably have been modified by advective effects from the oceans in western North America and western Europe (Kutzbach and Gallimore, 1988).

Conclusions

In summary, because of the presence of the large ice sheet, the glacial-age circulation and climate of North America were vastly different from today, as exemplified by the split jet and altered storm tracks (Fig. 2.6). Orbital changes increased the seasonality of northern climates and strengthened northern summer monsoons in the early to middle Holocene. Near the slowly melting ice sheets, the direct effect of the orbital changes was delayed or reduced. The climatic changes associated with orbital changes were greater in North Africa/Eurasia than in North America because the former continent is larger. The mid-Holocene warm and dry period in the North American Midwest, and simultaneously the moist phase in

HYDROLOGIC BUDGETS and LAKE LEVELS

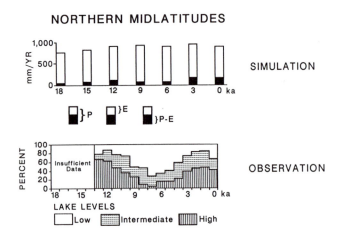

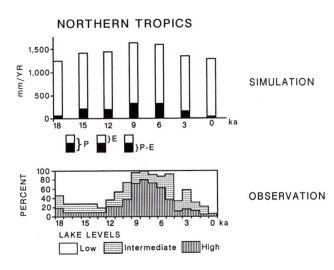

Fig. 2.7. Observed status of lake levels (low, intermediate, high) and model-simulated moisture budgets (precipitation [P], evaporation [E], and precipitation minus evaporation [P-E]) for each 1000-yr (observations) or 3000-yr (model) interval from 18 ka to the present. The temporal variation in percentage of lakes with low, intermediate, or high levels is taken from the Oxford lake-level data bank. In both northern midlatitudes (top) and the northern tropics (bottom), the shape of the high-lake-level curve corresponds roughly to the shape of the P-E curve. That is, lake levels and P-E were low around 9-6 ka in northern midlatitudes and high around 9-6 ka in the northern tropics. Modified from Kutzbach and Street-Perrott (1985) and Street-Perrott (1986); reprinted by permission from Nature vol. 317, pp. 130-134, copyright 1985 Macmillan Magazines Ltd., and from the Office for Interdisciplinary Earth Studies, University Corporation for Atmospheric Research.

the northern tropics, were both caused by the orbital change that increased summer warmth and reduced P-E in northern midlatitude continental interiors and increased P-E in the northern tropics. By examining the detailed climatic simulations in the light of these concepts, we have been able to compare the timing, location, and magnitude of simulated and observed events (e.g., Fig. 2.7).

These concepts and the idealized sketches of some of the results of the paleoclimatic simulations should set the stage for the more detailed results and comparisons profiled in the following chapters. Our goal is not only to document the major climatic changes since the last glacial maximum but also to understand the underlying causes of these changes and to describe both the causes and the changes in relatively simple terms.

References

Berger, A. L. (1978). Long-term variations of caloric insolation resulting from the earth's orbital elements. *Quaternary Research* 9, 139-167.

Brooks, C. E. P. (1926). "Climate through the Ages: a Study of the Climatic Factors and Their Variations." E. Benn, Ltd., London. (Second revised edition, 1970, Dover, New York)

Bryson, R. A., and Wendland, W. M. (1967). Tentative climatic patterns for some late-glacial and postglacial episodes in central North America. *In* "Life, Land, and Water" (W. J. Mayer-Oakes, Ed.), pp. 271-298. University of Manitoba Press, Winnipeg.

COHMAP Members (1988). Climatic changes of the last 18,000 years: Observations and model simulations. *Science* 241, 1043-1052.

Crowley, T. J., Short, D. A., Mengel, J. G., and North, G. R. (1986). Role of seasonality in the evolution of climate during the last 100 million years. *Science* 231, 579-584.

Hobbs, W. H. (1926). "The Glacial Anticyclone: The Poles of Atmospheric Circulation." Macmillan, New York.

Kutzbach, J. E., and Gallimore, R. G. (1988). Sensitivity of a coupled atmosphere/mixed-layer ocean model to changes in orbital forcing at 9000 years B.P. *Journal of Geophysical Research* 93(D1), 803-821.

Kutzbach, J. E., and Guetter, P. J. (1986). The influence of changing orbital parameters and surface boundary conditions on climate simulations for the past 18,000 years. *Journal of the Atmospheric Sciences* 43, 1726-1759.

Kutzbach, J. E., and Otto-Bliesner, B. (1982). The sensitivity of the African-Asian monsoonal climate to orbital parameter changes for 9000 years B.P. in a low-resolution general circulation model. *Journal of the Atmospheric Sciences* 39, 1177-1188.

Kutzbach, J. E., and Street-Perrott, F. A. (1985). Milankovitch forcing of fluctuations in the level of tropical lakes from 18 to 0 kyr B.P. *Nature* 317, 130-134.

Kutzbach, J. E., and Webb, T. III (1991). Late Quaternary climatic and vegetational change in eastern North America: Concepts, models, and data. *In* "Quaternary Landscapes" (L. C. K. Shane and E. J. Cushing, Eds.), pp. 175-217. University of Minnesota Press, Minneapolis.

Manabe, S., and Broccoli, A. J. (1985). The influence of continental ice sheets on the climate of an ice age. *Journal of Geophysical Research* 90, 2167-2190.

Milankovitch, M. (1920). "Théorie mathématique des phénomènes thermiques produits par la radiation solaire." Gauthier-Villars, Paris.

Street-Perrott, F. A. (1986). The response of lake levels to climatic change—Implications for the future. *In* "Climate-Vegetation Interactions" (C. Rosenzweig and R. Dickinson, Eds.), pp. 77-80. Office for Interdisciplinary Earth Studies (OIES), University Corporation for Atmospheric Research (UCAR), Boulder, Colo.

Wright, H. E., Jr. (1976). The environmental setting for plant domestication in the Near East. *Science* 194, 385-388.

CHAPTER *3*

Model Description, External Forcing, and Surface Boundary Conditions

J. E. Kutzbach and W. F. Ruddiman

Overview of the Design of the COHMAP Experiments

Atmospheric general circulation models (GCMs) are typically used in paleoclimatic studies for two purposes: (1) to generate simulations of atmospheric circulation during times in the past when geologic data provide reasonable constraints on critical boundary conditions at the earth's surface (such as ice-sheet size, land albedo, and sea-surface temperatures), and (2) to test the sensitivity of the simulations by altering key features of the climate system one at a time to isolate their unique impact on the atmospheric circulation and climate patterns. Sensitivity tests thus do not necessarily represent climatic configurations that actually existed at any time in the past.

Kutzbach and Guetter (1986) first ran a series of experiments to reconstruct atmospheric circulation every 3000 yr during the last 18,000 yr. The choice of 3000-yr intervals was based in part on the fact that boundary condition estimates of sea-surface temperature cannot be derived for shorter intervals from most bioturbated ocean records. It also resulted from an attempt to select intervals of relatively stable climate and avoid times of especially rapid change, such as 10 ka in North America and Europe and 13 and 11–10 ka in the North Atlantic.

These first experiments were simulations, in the sense that many key changes in boundary conditions were reasonably well known for each time level and could be used as input for the model (Fig. 3.1). These experiments and related studies showed that changes in insolation and in surface boundary conditions (particularly the ice sheets) can explain many of the cli-

matic responses recorded by changes in vegetation and lake levels on the continents over the last 18,000 yr (Kutzbach and Street-Perrott, 1985; Webb *et al.,* 1987).

In another sense, however, these experiments were only a first attempt to model the complexities of the deglaciation. Some of the boundary conditions used, such as ice-sheet height and sea-ice extent, had a considerable range of uncertainty. Others were set at modern values (atmospheric CO_2 and aerosols) even though the limited evidence available suggests that they were significantly different during the last 18,000 yr. As additional boundary conditions become sufficiently well defined in the geologic record to be entered into GCM runs, further refinements of these initial simulations will become possible.

The modeling experiments reported in this volume are largely the same as those of Kutzbach and Guetter (1986) and use the same 3000-yr time steps. For the 12-ka level, however, several boundary conditions have been modified on the basis of geologic evidence. These changes are discussed below and are summarized in Table 3.1.

We also report here initial results from three sensitivity tests of the model for 18 ka, one with lower ice sheets, one with less extensive winter sea ice in the North Atlantic, and one with reduced atmospheric carbon dioxide concentration. These experiments were undertaken to evaluate the climatic implications of the uncertainty in these boundary conditions, but they can also be regarded as alternative climate simulations for 18 ka.

GCMs have been well suited to the central purpose of COHMAP work, namely to compare model-simu-

12

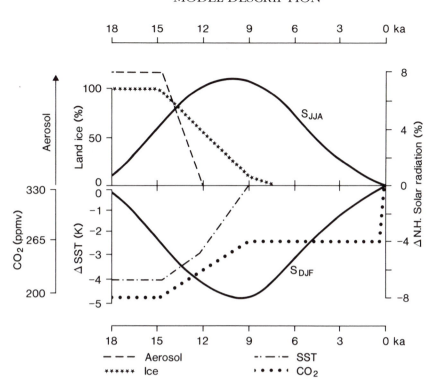

Fig. 3.1. Boundary conditions for the COHMAP simulation with the Community Climate Model for the last 18,000 yr. External forcing is shown for Northern Hemisphere solar radiation in June through August (S_{JJA}) and December through February (S_{DJF}) as the percentage difference from the present. Surface boundary conditions include land-ice volume (Ice) as a percentage of the 18-ka value (with the "maximum model" ice sheets of Denton and Hughes [1981]) and global mean annual sea-surface temperature (SST, K) as departure from present. The horizontal scale indicates the times of the seven sets of simulation experiments. An additional experiment for 18 ka included a lower glacial atmospheric CO_2 concentration of 200 ppmv (dots); the other deglacial experiments used the modern value of CO_2 concentration (330 ppmv). Experiments including glacial-age aerosols have not yet been completed. Reprinted by permission from *Nature* vol. 317, pp. 130-134, copyright 1985 Macmillan Magazines Ltd.

lated atmospheric circulation with quantitatively calibrated geologic data on land. Although these models reveal how atmospheric circulation may have differed at times in the past, they do not treat in an interactive way other parts of the climate system, including sea ice, surface-ocean and deep-ocean temperatures, land vegetation and soil moisture, and the long-term behavior of ice sheets. Other models now coming into use have begun to address some of these more complex linkages in the climate system. These new models will be able to use oceanic data to verify model predictions rather than simply to set boundary conditions.

Model Description

The COHMAP simulation experiments for the last deglaciation were made with the Community Climate Model (CCM) of the National Center for Atmospheric Research (NCAR). This model is fully described by Pitcher *et al.* (1983) and Ramanathan *et al.* (1983).

GCMs contain the most detailed physical representation available of the atmospheric general circulation. The NCAR CCM includes atmospheric dynamics based on the equations of fluid motion. It includes radiative and convective processes, condensation, and evaporation. It has nine vertical levels in the atmosphere, and the horizontal fields of wind, temperature, pressure, and moisture are represented in spectral form up to wavenumber 15. This equates to a horizontal resolution of about 4° of latitude and 7.5° of longitude.

The model was typically run for 450 simulated days for each July and January for each interval chosen in the deglaciation. The first 60 days of each segment were ignored, in order to allow the atmospheric circulation to approach equilibrium with the altered boundary conditions. Three 90-day averages were obtained, each separated by 60 days and thus assumed to be independent (Blackmon *et al.,* 1983).

The three 90-day averages were then combined to form a 90-day ensemble average that was compared

Table 3.1. Boundary conditions

Time level	Ice-sheet extent	Ice-sheet height	Surface albedo	Sea-ice extent	Sea-surface temperature	CO_2 aerosols
18 ka	CLIMAP (1981); Denton and Hughes (1981)	CLIMAP (1981)[a]	Ice, 0.8; land, modern[b]	CLIMAP (1981)[c]	CLIMAP (1981)	Modern[d]
15 ka[e]	Denton and Hughes (1981)	Same as 18 ka	Same as 18 ka	Same as 18 ka	Same as 18 ka	Modern
12 ka[e]	Denton and Hughes (1981)	50% of CLIMAP (1981)	Halfway between 18 ka and modern	Same as 18 ka in N. Pacific; modern in Antarctic[f]	Halfway between 18 ka and modern[g]	Modern
9 ka	Denton and Hughes (1981)	800 m in N. America	Modern, except 0.5 over ice in N. America	Modern	Modern	Modern
6 ka[e]	Modern	Modern	Modern	Modern	Modern	Modern
3 ka[e]	Modern	Modern	Modern	Modern	Modern	Modern

[a]See text, however, for results of sensitivity experiment with reduced North American ice-sheet height at 18 ka.
[b]Except for parts of South America and central Asia, which changed from 0.25 to 0.13.
[c]See text, however, for results of sensitivity experiment with reduced North Atlantic winter sea ice at 18 ka.
[d]See text, however, for results of sensitivity experiment with reduced atmospheric carbon dioxide concentration at 18 ka.
[e]These exepriments were run for longer time intervals than those reported by Kutzbach and Guetter (1986).
[f]Subpolar North Atlantic, same as modern in east during winter, same as 18 ka in west during winter and summer. This differs from Kutzbach and Guetter (1986).
[g]Except for subpolar North Atlantic: same as modern in east, same as 18 ka in west. This differs from Kutzbach and Guetter (1986).

to the control simulation of modern circulation represented by a similarly constructed 90-day ensemble average. Ensemble averages are constructed in order to determine the model's natural variability, much of which is quantitatively equivalent to the natural year-to-year variability observed in the climate system today. This comparison permits an assessment of the level of statistical significance of the climate changes simulated in the model experiments.

Input to the model consists of orbitally determined insolation, mountain and ice-sheet orography, atmospheric trace-gas concentrations, sea-surface temperatures, sea-ice limits, snow cover, land-surface albedo, and effective soil moisture. The solar constant is set at 1370 W/m².

As output, the CCM computes an array of surface and upper-air parameters. Atmospheric energy budgets are partitioned into kinetic and potential energy and into zonal, stationary-eddy, and transient-eddy components. Sea-level pressure and geopotential height fields are calculated. The model-simulated features of major paleoclimatic interest are those that can be directly compared with surface geologic evidence (such as temperature, winds, and moisture balances over land) and those that summarize the large-scale atmospheric circulation (such as the position of the jet streams).

External Forcing

Solar radiation at each 3000-yr interval during the deglaciation (Fig. 3.1) was prescribed according to changes computed by Berger (1978) from astronomical calculations of three orbital parameters: axial tilt, eccentricity, and precession (position of perihelion). Unlike the surface boundary conditions, these changes in external forcing are known with great accuracy over the last 18,000 yr.

At the glacial maximum 18,000 yr ago, the seasonal and latitudinal distributions of solar radiation in January and July were within 1% of their modern values. Simulated differences in atmospheric circulation for the glacial maximum must therefore have resulted mainly from the major differences in glacial-age boundary conditions (large ice sheets, extensive sea ice, and colder sea-surface temperatures).

Between 15 and 9 ka, as the ice sheets and sea ice retreated and the oceans warmed, the influence of these features on atmospheric circulation waned, and insolation changes became more important. During this interval, the seasonal and latitudinal distributions of solar radiation changed significantly in response to the increased tilt of the earth's orbit and the repositioning of perihelion in northern summer, in contrast to the modern (and glacial-maximum) position in northern winter. These changes increased the seasonality of radiation in the Northern Hemisphere and decreased it in the Southern Hemisphere.

At about 9 ka solar radiation averaged over the Northern Hemisphere was about 8% higher in July and about 8% lower in January than today. These differences were large enough to cause significant climatic changes. By 6 ka, with the disappearance of the glacial ice sheets, the still-large seasonal radiation extremes continued to be the dominant influence on Northern Hemisphere climate. Insolation subsequently returned gradually to modern values.

Surface Boundary Conditions

Boundary conditions set by CLIMAP Project Members (1981) were used for the glacial maximum at 18 ka. These values were adjusted either halfway or all the way to modern values during succeeding deglacial and Holocene intervals, except for cases described below and indicated in Table 3.1. Although CLIMAP's (1981) published boundary conditions were for February and August, Kutzbach and Guetter (1986) and the experiments reported in this volume used the values as input for January and July. Values for the NCAR CCM grid were derived from CLIMAP data by averaging values within 2° squares on the CLIMAP grid.

ICE-SHEET EXTENT

Ice-sheet extents at 18 ka were taken from CLIMAP (1981) and follow the limits of the "maximum model" of Denton and Hughes (1981). Limits at all subsequent intervals were taken from ice-retreat maps published by Denton and Hughes (1981).

The lateral extents of the ice sheets during deglaciation are relatively well known, particularly in relation to the size of the model grid boxes. Recent summaries of evidence for less extensive glacial-maximum ice limits (Andrews, 1987; Hughes, 1987) do not differ from the maximum model of Denton and Hughes (1981) by more than 100 km in any region, which is a small fraction of the CCM grid box. In general, the largest uncertainties relate to the northern sector of the Laurentide ice sheet early in the deglaciation.

Ice limits at 15 ka were little changed from those of 18 ka, but large-scale retreat is evident by 12 ka. By 9 ka European ice had shrunk to less than one grid box and was omitted from the model. Residual North American ice masses shrank below the size of the model grid box by 6 ka.

ICE-SHEET VOLUME AND HEIGHT

Estimates of the difference between the glacial-maximum global ice volume and today's ice volume,

hereinafter referred to as *excess ice volume,* range widely (Flint, 1971; Denton and Hughes, 1981). Differences in these estimates (as well as in estimates for individual ice sheets) mainly reflect uncertainties about ice elevation, particularly in North America and Antarctica. Second-order uncertainties include the existence and/or areal extent and height of several marine ice sheets and ice shelves.

Global sea-level lowering is potentially the best constraint on glacial-maximum ice volume, but interpretation of most sea-level records older than 11 ka is complicated by questionable ^{14}C dates, disagreements about whether the shells used for ^{14}C dating are in situ, and uncertain corrections for the form of local isostatic adjustments to ice and water loading. Recently, however, Fairbanks (1989) drilled a sequence of submerged coral reefs off Barbados and radiocarbon-dated the corals recovered to within the age range from 17 to 7 ka. He found a glacio-eustatic sea-level lowering of about 121 ± 5 m at 17 ka, which lies right in the middle of the very broad range of previous estimates of sea-level lowering (80-160 m). This value does not, however, include the volume of ice in floating marine ice shelves or in submerged portions of marine ice sheets that displace modern-day oceans, lakes, and seas. Neither of these kinds of ice would displace sea level. This effect might add a few cubic meters of ice volume to the amount indicated by a 121-m drop in eustatic sea level (Dodge *et al.,* 1983).

$\delta^{18}O$ signals from planktonic and benthic foraminifera are an independent constraint on excess ice volume. If local temperature overprints can be removed, the remaining $\delta^{18}O$ signal can be attributed to excess global ice volume. These values can then be compared to $\delta^{18}O$ values derived by multiplying the estimated volume of excess ice by the assumed isotopic composition of the ice.

Estimates of the fraction of the global $\delta^{18}O$ signal attributable to ice volume range widely. At the lower extreme, Labeyrie *et al.* (1987) estimated an ice-volume component of 1.1‰ from Norwegian Sea benthic foraminiferal $\delta^{18}O$ records inferred to lack any temperature overprint. In other benthic foraminiferal records, however, and in low-latitude planktonic foraminifera from cores with high deposition rates, the total amplitude of the observed $\delta^{18}O$ signal averages around 1.7‰ (Broecker, 1986; Mix, 1987), with considerable regional variation. This value can be taken as an upper limit on the excess ice-volume component (Mix, 1987).

The component of most $\delta^{18}O$ signals attributable to glacial-age cooling has been estimated at 0.35-0.45‰ for both benthic and low-latitude planktonic foraminifera (Shackleton and Opdyke, 1973; Chappell and

Shackleton, 1986). Subtracting this overprint from 1.7‰ would reduce the excess ice-volume component of the $\delta^{18}O$ signals to 1.3‰. Birchfield (1987), using a different technique, estimated an ice-volume component of 1.3-1.4‰. Given this range of estimates of the excess global ice-volume component of the $\delta^{18}O$ signal, 1.3‰ seems a reasonable middle estimate.

To constrain the volume of excess global ice, this estimate can be compared to the value obtained by summing the individual isotopic contributions of individual (excess) ice sheets. The main uncertainty left is the isotopic composition of the excess ice. Estimates of -30 to -35‰ have been published for North American ice (Dansgaard and Tauber, 1969; Broecker, 1978), but little effort has been made to estimate the composition of excess ice volume in other ice sheets.

Mix (1987) recently assigned isotopic values to all ice sheets in both the minimum- and maximum-model reconstructions of Denton and Hughes (1981). Choosing isotopic compositions for each ice sheet consistent with the glacial ice still in existence and/or with estimates from other sources, he calculated that the total amplitude of the excess ice-volume effect in the $\delta^{18}O$ signal would be 2.03‰ for the Denton-Hughes maximum model, 1.48‰ for the Denton-Hughes minimum model, and 1.33‰ for the Denton-Hughes minimum model with the still-smaller North American ice sheet of Paterson (1972). The isotopic data thus favor an excess ice volume close to the Denton-Hughes minimum model, but with the even-smaller North American ice sheet of Paterson (1972). The 121-m glacio-eustatic sea-level lowering from Fairbanks (1989) indicates an even smaller volume of global ice, which strongly implies even less Laurentide ice.

This research raises the question of whether the large ice sheets used as boundary conditions for 18 ka in previous studies (Kutzbach and Street-Perrott, 1985; Kutzbach and Guetter, 1986) give a reasonably valid picture of the atmospheric circulation at that time. The largest concern is the Laurentide ice sheet, whose surface, based on the maximum model of Denton and Hughes (1981), is so high that it causes major shifts in the planetary waves and splits the jet stream into two branches. The volume of North American ice in the Denton-Hughes minimum model (30.5×10^6 km³) is only 11% smaller than in the maximum model (34.2×10^6 km³), whereas the Paterson (1972) estimate (25×10^6 km³) is 27% smaller. Boulton et al. (1985) considered an even larger range of possible volumes for the Laurentide ice sheet ($22-44 \times 10^6$ km³).

We have retained the original experiment of Kutzbach and Guetter (1986) with large ice sheets (including North America) as the standard 18-ka experiment for this volume, but we have also run a sensitiv-ity test with a smaller North American ice sheet to assess the impacts of this choice on the atmospheric circulation. For this test we uniformly reduced the height of the North American ice sheet by 20%, which is equivalent to a 27% reduction in thickness for an ice sheet that is fully compensated isostatically by bedrock adjustment (using the 1:4 ratio of Denton and Hughes [1981]). The ice thickness for this sensitivity test thus equates to the 18-ka ice-volume estimate of Paterson (1972) and is in line with the $\delta^{18}O$ estimate of global excess ice volume. The sensitivity test is described briefly at the end of this chapter; a more detailed analysis is in preparation.

For the 15-ka model run the same maximum-model ice sheets of Denton and Hughes (1981) as in the 18-ka experiment were used as boundary conditions. This choice is based on the persistence until 15 ka of ice-sheet margins at or near their glacial-maximum limits and of $\delta^{18}O$ values at full-glacial values (Duplessy et al., 1981; Berger et al., 1985; Mix and Ruddiman, 1985; Mix, 1987). On the other hand, Fairbanks (1989) showed a rise in glacio-eustatic sea level of at least 10 m prior to 15 ka, indicating that appreciable ice had melted somewhere on the globe.

The ice sheets used in the model for 12 ka were reduced in area according to glacial-geologic constraints and were also substantially lowered in elevation. The mean reduction in ice-sheet thickness and height is 50% if the Denton-Hughes maximum model is used as a baseline and 33% relative to the Denton-Hughes minimum model.

The elevation of the central region of the North American ice sheet was set at about 1650 m, equating to a thickness of some 2200 m for full (1:4) isostatic compensation. In view of the roughly 30% decrease in Northern Hemisphere ice area by 12 ka (Bloom, 1971; MacDonald, 1971; Paterson, 1972), a 50% reduction in height relative to the glacial-maximum baseline of the Denton-Hughes maximum model is probably excessive. However, relative to the smaller 18-ka ice thickness used in the sensitivity test, the reduction in thickness (about 33%) is more reasonable. If bedrock rebound was delayed during rapid deglaciation, the ice sheet could have been several hundred meters thicker for this choice of ice height. Fairbanks (1989) found a rise in eustatic sea level of 25-45 m by 12 ka, which represents 20-36% of the glacial-interglacial difference.

By 9 ka only the Laurentide ice sheet still occupied a large area, at roughly 35-40% of its glacial-maximum limits (Bloom, 1971; MacDonald, 1971; Paterson, 1972). For the model the ice sheet was trimmed to a uniform height of 800 m, equivalent to a thickness of 1067 m in the central region if the Denton-Hughes 1:4 isostatic compensation is assumed. Delayed bedrock rebound during deglaciation would imply a thicker ice sheet for this choice of ice height. This choice of relatively thin ice at 9 ka is open to question. For example, Mitchell *et al.* (1988) used a model with a 9-ka North American ice sheet more than 2000 m high. Fairbanks (1989) found eustatic sea level near –40 m at 9 ka, which implies the persistence of a considerable volume of ice at that time (33% of the glacial-maximum total). His coral reef data, however, suggest that surprisingly low sea levels persisted until late in the deglaciation (e.g., –23 m at 7.4 ka). Ice-sheet margins were everywhere so reduced by this time that it is not clear where the remaining melting could have come from, or whether these values are plausible.

Marine shells dated by radiocarbon indicate that a large ice-free embayment penetrated deep into the southern part of Hudson Bay by 7.8 ka (Andrews and Falconer, 1969). If so, either the ice sheet was still thick at 9 ka and then collapsed catastrophically by 7.8 ka, or the ice sheet had already thinned considerably by 9 ka. No evidence for a catastrophic collapse between 9 and 7.8 ka has been found in the nearby marine record (Ruddiman and McIntyre, 1981), which suggests that the North American ice sheet over Hudson Bay had indeed thinned considerably by 9 ka (Ruddiman, 1987).

The actual configuration at 9 ka probably involved small ice-sheet domes centered over Labrador, Keewatin, and Foxe Basin, with lower ice over Hudson Bay. This complexity is greatly simplified in the thin, flat-surfaced ice sheet used in the model for 9 ka. By 6 ka (and thereafter) no excess ice was entered in the model.

SURFACE ALBEDO

Surface albedo varies with changes in glacial ice on land, sea ice over the ocean, and vegetation. Surfaces covered with glacial ice were assigned an albedo of 0.8, with one exception: the low-lying (800-m) North American ice sheet at 9 ka was assigned an albedo of 0.5 because its surface was assumed to consist of old ice, perhaps ponded with meltwater and partly covered with debris from both melting and deposition.

Land albedo for 18 ka was changed from today's value only in regions where the CLIMAP-estimated albedo difference (18 ka minus today) was more than half of the difference in albedo between the two types of land differentiated in the modern simulation: nondesert (0.13) and desert (0.25). Because albedo was altered only for very large changes in vegetation, the model's land albedo for 18 ka was the same as today's

except in parts of South America and central Asia, which were much drier than today.

Albedos at 15 ka were set identical to those at 18 ka. At 12 ka land albedos were set halfway between those of 18 ka and today in the two locations mentioned. At 9 ka and thereafter, all land albedos were set at modern values.

Snow cover is prescribed in the model, which excludes all snow-albedo-temperature feedbacks except those caused by prescribed changes in glacial ice and sea ice. For all experiments, Northern Hemisphere land is snow-covered north of 68.9° in July and 42.2° in January. In the Southern Hemisphere, Antarctica is always snow-covered.

SEA ICE

Sea-ice boundary conditions for 18 ka were taken from CLIMAP (1981). These limits are largely inferred; the available geologic indicators do not positively fix the seasonal limits of sea ice in any of the three polar oceans.

For the North Atlantic at 18 ka, the greater than 95% dominance of the planktonic foraminifer *Neogloboquadrina pachyderma* (left-coiling) in all latitudes above 45°N indicates winter temperatures below 0°C and summer temperatures below 6°C but does not prove that the ocean was frozen (–1.8°C) in either season. This contour thus sets an upper bound for winter sea ice but does not necessarily show its actual limits.

The North Atlantic sea-ice limits chosen by CLIMAP (1981) were based on changes in the concentration of this polar species in the sediments. A partial decrease in concentration was assumed to reflect seasonal suppression of productivity by winter sea ice and was thus used to set the winter ice limit. A much greater (near-total) drop in concentration was assumed to reflect suppression of foraminiferal flux to the sea floor by year-round ice cover and was thus used to position the summer ice limit.

The greatest source of uncertainty in these criteria is the assumption that foraminiferal concentrations (number per gram of sediment) directly reflect flux rates from the surface waters (number per square centimeter per 1000 yr). This assumption holds only if the time scale is accurately known, but dating glacial-age levels of North Atlantic cores is difficult because the oceanic carbonate is contaminated by older, ^{14}C-dead limestone and dolomite. Dissolution on the sea floor can also complicate interpretations by altering foraminiferal concentrations after deposition.

On the other hand, the CLIMAP choice of winter (and summer) sea ice in the North Atlantic has independent support from a sensitivity test. Manabe and Broccoli (1985) inserted glacial-maximum ice sheets into a model of the modern world and found that cold air advected from the ice sheets froze the North Atlantic sea surface southward to a point near (actually somewhat beyond) the CLIMAP winter and summer sea-ice limits. Because this experiment lacked ocean dynamics and used the large 18-ka ice sheets from the Denton-Hughes (1981) maximum model, it may have overpredicted the glacial extent of North Atlantic sea ice, but it does give first-order confirmation of the CLIMAP choices.

For modeling purposes, accurate placement of the winter sea-ice limit in the North Atlantic is important for several reasons. First, it anchors a particularly vigorous maximum of the winter jet stream. Second, it results in extremely cold surface-air temperatures over the sea ice. And third, this cold air is advected into Europe and northernmost Africa. To test the regional climatic impact of the choice of winter North Atlantic sea ice, we ran a sensitivity test in which 18-ka winter sea ice was reduced back to the summer limit, with all other surface boundary conditions unchanged. This reduced limit probably underestimates the average sea-ice limits at 18 ka but may approximate conditions in extreme (warm) individual years. Results of this sensitivity test are reported in a later section.

CLIMAP (1981) chose sea-ice limits for the North Pacific at 18 ka based on biotic data, mainly the more extensive past distributions of radiolarian and diatom species that today are abundant only in the Sea of Okhotsk, which is ice-covered in winter and ice-free in summer. Summer sea ice at 18 ka was absent south of the Bering land bridge in the CLIMAP (1981) reconstruction; winter sea ice was confined to a small portion of the northwestern Pacific. Because the marine geologic data do show clearly that there was relatively little sea ice in the North Pacific, errors in placement of the limits are unlikely to have a large effect on hemispheric climate.

CLIMAP (1981) fixed the winter sea-ice limit for the Southern Ocean at 18 ka at the northernmost position of abundant volcanic ash delivered by rafting on sea ice, although it is unclear how the limit of drifting and melting ice would relate to the limit of fixed pack. The summer sea-ice limit was placed at the southernmost occurrence of diatom-rich sediments (Hays, 1978). Burckle *et al.* (1982) questioned the criteria used to specify the summer sea-ice limit, arguing that the diatom-rich sediments better reflect the ice position during the spring bloom and that therefore the actual summer ice limit was probably less extensive than that shown by CLIMAP (1981).

Recent GCM experiments have lent some support to the CLIMAP (1981) sea-ice reconstruction. Broccoli and Manabe (1987) showed that the reduced glacial-age CO_2 concentrations in the atmosphere could have caused an annual mean Southern Ocean cooling and sea-ice reduction close to those estimated by CLIMAP (1981). Unfortunately, seasonal distributions of sea ice were not reported in these experiments, which also lacked ocean dynamics and deep-water and intermediate-water formation.

For the time levels during the deglaciation, sea-ice boundaries were assigned on the simplest possible basis, in view of the uncertainty surrounding the assignment of even the glacial-maximum limits. For each time level except 12 ka, sea-ice limits in all oceans were set either at the full-glacial position or at the modern position. The specific choices for each ocean were based on marine geologic evidence defining the timing of the major sea-surface warming.

For the 15-ka experiment, sea-ice limits were kept at the 18-ka positions. Because none of the high-latitude oceans show any evidence of a strong warming by that time, there was no reason to change the 18-ka limits.

For 12 ka the North Pacific sea-ice limits were left in the 18-ka position, although available geologic data do not constrain this choice closely (Morley and Dworetzky, this vol.). Winter sea ice was removed from the eastern sector of the North Atlantic, which had warmed to full interglacial conditions (Ruddiman and McIntyre, 1981), but it was kept at the 18-ka limits in the west along the still-cold margin of North America and Greenland. Kutzbach and Guetter (1986), however, kept winter sea ice at the 18-ka limits across the entire subpolar North Atlantic. Summer sea ice along the North American margin was kept at the 18-ka winter limits, both in the simulations reported in this volume and by Kutzbach and Guetter (1986).

By 12 ka the sub-Antarctic ocean had begun to warm, reflecting its long-term tendency to respond several thousand years ahead of ice-volume changes (Hays *et al.,* 1976). Hays (1978) proposed that circum-Antarctic sea ice retreated abruptly at 14 ka, although this choice of date depends on the correlation of a radiolarian signal from undated high-latitude cores with that of better-dated midlatitude cores. This evidence of an early Antarctic Ocean response to deglaciation was the basis for moving Antarctic sea ice back to its modern limits for the 12-ka model run reported in this volume. Kutzbach and Guetter (1986), in contrast, left Antarctic sea ice at 12 ka at the glacial-maximum limit.

By 9 ka most high-latitude oceans had warmed to values near or above modern temperatures (Ruddiman and Mix, this vol.; Morley and Dworetzky, this vol.). Sea-ice limits in all oceans were thus placed at the modern positions for the 9-ka run and kept there for the 6- and 3-ka experiments. Although the northern fringes of the Labrador Sea still remained colder than today at 9 ka (Ruddiman and McIntyre, 1981), setting sea-ice limits beyond those of today seems unjustified for any Labrador Sea grid boxes.

The extent of sea ice in the Arctic at 9 ka is somewhat uncertain. Modeling experiments by Kutzbach and Gallimore (1988) and Mitchell *et al.* (1988) suggest that Arctic sea ice at 9 ka may generally have been thinner and slightly less extensive than today. In addition, G. Jones (personal communication, 1989) has evidence for a slightly warmer and/or less ice-covered Arctic Ocean at 80°N near Svalbard. On the other hand, the reduced sea-ice limits in these experiments occurred mainly in the autumn; simulations for January and July showed differences in ice thickness only. In some regions the cooling effects of the residual Laurentide ice sheet may also have nullified this warming (Mitchell *et al.*, 1988). In any case, we used modern Arctic sea-ice limits for the 9-ka experiment as well as for the 6- and 3-ka runs.

SEA-SURFACE TEMPERATURES

CLIMAP (1981) estimated another key boundary condition, sea-surface temperature (SST). General circulation modeling has independently confirmed that the CLIMAP SST estimates at high and middle latitudes are probably of the correct magnitude. The ice-sheet experiment of Manabe and Broccoli (1985) yielded SST estimates very close to those predicted by CLIMAP for Atlantic and Pacific waters north of 30–40°N. The Broccoli and Manabe (1987) experiment adding realistic glacial-age CO_2 values predicted annual mean SST values close to the CLIMAP estimates at latitudes poleward of 40°S.

The validity of CLIMAP (and other) SST estimates for 18 ka at tropical and subtropical latitudes (30°S–30°N) is open to question (e.g., Webster and Streten, 1978). Rind and Peteet (1985) summarized the argument as follows: CLIMAP low-latitude SSTs appear too warm relative to the large lowering of tropical snow and ice lines and also relative to some pollen-based estimates of land temperatures at low altitudes. These estimates are based on modern observed lapse rates and use fairly crude conceptual ecological models of changes in vegetation (Peterson *et al.*, 1979).

Although the CLIMAP SST estimates could be too warm, other interpretations are possible. Tropical vegetation may respond more directly to the precipitation-evaporation balance than to temperature, and if so temperature estimates based on tropical pollen may be incorrect. Moreover, the specific climatic controls on modern snow and ice lines in the tropics are not clearly defined, and estimates of past temperature based on these data may also be in error. In several of the critical regions, the evidence for lowered snowlines is undated and cannot be attributed to 18 ka with certainty. Finally, most of the published model runs were based on simplified GCMs that lacked features that might affect the apparent disparity, such as aerosols, and that used a relatively coarse spatial resolution and a correspondingly smoothed and relatively low orography in the vicinity of tropical mountain glaciers.

Extensive testing of transfer function methodology using analogue measures (Overpeck *et al.*, 1985) has confirmed the CLIMAP SST estimates over most of the tropical ocean and even exacerbated the land-sea discrepancies somewhat in some regions (Prell, 1985). It appears that the CLIMAP estimates could not be several degrees too cold everywhere in the tropical oceans unless the plankton were not sensitive to temperature in the way suggested by modern distribution patterns. In addition, Broecker (1986) found that $\delta^{18}O$ data from planktonic and benthic foraminifera broadly support the CLIMAP view of a small glacial-age cooling of the tropical ocean.

The low-latitude SST controversy thus remains unresolved. In this book the CLIMAP SST estimates for 18 ka are used as boundary conditions for the model. If these estimates are incorrect, the most important consequence for data-model comparisons would be that the model would simulate temperature and precipitation incorrectly on oceanic islands such as New Zealand and along continental coasts also dominated by maritime climates (McGlone *et al.*, this vol.).

The SST values for 18 ka were retained for 15 ka, because no ocean shows a major warming during the intervening 3000 yr. This is consistent with the treatment by Kutzbach and Guetter (1986).

For 12 ka SST was brought halfway toward modern values in most regions, including all portions of the Antarctic and North Pacific from which sea ice was removed relative to the 18-ka limits. SST was set at modern values in the eastern North Atlantic but was kept at 18-ka values for the western North Atlantic, consistent with the pattern of polar-front retreat (Ruddiman and McIntyre, 1981).

SSTs were fixed at modern values in all regions for 9 ka and were held there for the 6- and 3-ka experiments. This is in accord with the observed pattern in most oceans, which warmed to modern or near-modern SSTs by 9 ka. It also agrees with results of GCM experiments with coupled mixed-layer oceans, which indicate that seasonal SST differences from today were less than 1°C in most regions at 9 ka (Kutzbach and Gallimore, 1988; Mitchell *et al.,* 1988). Fringes of the Labrador Sea along the North American coast near the residual Laurentide ice sheet still remained cool at 9 ka (Ruddiman and McIntyre, 1981), but these regions are only a fraction of a model grid box. By 6 (and 3) ka SSTs in all oceans were at modern values, at least within the uncertainty of the estimates.

SEA LEVEL

For 18 ka sea level was lowered by 100 m relative to today (Fairbanks [1989] inferred a value of –121 m). As a result, some grid points that are ocean today were land at 18 ka, particularly between Alaska and Siberia and in the Australasian region. Sea level at 15 ka was held at the 18-ka level. In view of rapid melting of the ice sheets during deglaciation, sea levels were set 40 m lower than today at 12 ka, 10 m lower than today at 9 ka, and at the modern value for 6 and 3 ka.

CARBON DIOXIDE

Evidence from ice cores (Oeschger *et al.,* 1983; Lorius *et al.,* 1984; Barnola *et al.,* 1987) indicates that the atmospheric CO_2 concentration was about 200 ppmv at the glacial maximum and until at least 15 ka. It rose to about 265-275 ppmv by about 9 ka and remained at that level until the last century (Neftel *et al.,* 1982; Lorius *et al.,* 1984; Stuiver *et al.,* 1984). The sensitivity of climate simulations to these changes in CO_2 is discussed in a later section.

ATMOSPHERIC AEROSOLS

The COHMAP experiments have not yet incorporated the effect of different tropospheric loadings of land and marine aerosols during the last 18,000 yr. This effect probably varied widely from region to region. The two largest sources of land aerosols—Africa and Asia—appear to have responded very differently during this time. Eolian fluxes from Africa to the subtropical and equatorial Atlantic appear to have been highest at or near the last glacial maximum, as indicated both by increased mass flux of terrigenous sediment (Gardner, 1975; Sarnthein *et al.,* 1982) and by GCM experiments with glacial-age boundary conditions (Joussaume, 1989). Large eolian influxes during the glacial maximum are also recorded in ice cores on Greenland (Hammer *et al.,* 1986) and Antarctica (Petit *et al.,* 1981). On the other hand, influxes from Asia to the northwestern Pacific may have peaked during the early Holocene (Leinen, 1989). Future modeling efforts should address the time-varying impact of aerosols on deglacial climate.

UNCERTAINTIES IN THE RADIOCARBON CHRONOLOGY

As this volume goes to press, new data are suggesting that the radiocarbon chronology before 9 ka, the previous limit of testing by tree-ring dating, may be systematically in error. Bard *et al.* (1990) obtained uranium-series mass spectrometer dates for the same sequence of submerged corals radiocarbon-dated by Fairbanks (1989). The U-series dates are close to the radiocarbon dates in the early-middle Holocene (7.4 ka) but smoothly and systematically diverge at older levels, becoming 1500 yr older by 11 ka and about 2500 yr older by 17 ka. One of the several possible explanations for this divergence is that the ^{14}C dates (these and all others) are systematically in error.

If the U-series dates prove to be correct, the calendar-year insolation values used in the COHMAP modeling experiments would still be correct, but all of the assigned boundary conditions, which in one way or another are ultimately linked to radiocarbon-dated geologic data, would be improperly positioned vis-à-vis the insolation values. For example, insolation values appropriate for 18 ka would have been combined with ice sheets characteristic of 20.5 ka, and insolation values correct for 11 ka would have been combined with ice sheets correct for 12.5 ka. Although the offsets are small late in the deglaciation, they are significant for the early and middle portions.

Sensitivity Tests

REDUCED NORTH AMERICAN ICE HEIGHT AT 18 ka

To evaluate the effects of smaller North American ice sheets on the model simulations, we ran a sensitivity test in which the Denton-Hughes maximum-model Laurentide ice sheet was everywhere reduced in thickness by 27%, which corresponds roughly to the ice volume suggested by Paterson (1972). We then incorporated the Denton-Hughes assumption of full (1:4) isostatic compensation of bedrock under the ice sheet, allowing the ice surface to come to a higher elevation. The net change was thus a 20% reduction in ice-sheet elevation relative to the boundary conditions in the standard model run. Other 18-ka boundary conditions were left unchanged.

The two simulations are very similar except in the vicinity of the northern ice sheets. The differences are greatest in January. A prominent feature of the maximum ice-sheet simulation is the pronounced split in the jet stream near the North American ice sheet (COHMAP Members, 1988): one branch flows north of the ice sheet, the other south. This split is still present but is less pronounced in the simulation with the minimum ice sheet. At considerable distances from the North American ice sheet, the amplitudes of the waves in the westerlies are less pronounced in the minimum ice-sheet case.

These circulation differences are associated with differences in surface temperature in the two cases, due primarily to differences in advection. The most extreme example of this effect is in northeastern Siberia/western Alaska, where January surface temperature is on the order of 5-10°C higher than at present in the maximum ice-sheet experiment because of increased southerly flow to the west of the North American ice sheet and 5-10°C lower than at present in the minimum ice-sheet experiment. Differences occur in other regions as well but are generally much smaller. Nevertheless, the sensitivity test demonstrates that simulations of glacial-age climates of the Northern Hemisphere midlatitudes are indeed sensitive to assumptions about the height of the North American ice sheet, especially in northern winter.

REDUCED NORTH ATLANTIC WINTER SEA ICE AT 18 ka

To evaluate the regional climatic effects of the choice of winter sea ice in the North Atlantic, we ran a sensitivity test in which the winter sea-ice limit at 18 ka was reduced to the position chosen by CLIMAP (1981) for summer, with all other 18-ka winter boundary conditions unchanged.

The two simulations are very similar except in the North Atlantic sector in midlatitudes between about 60°W and 30°E. The extended-ice experiment has continuous sea-ice cover to about 45°N, whereas the reduced-ice experiment has sea-ice cover to about 60°N. In the region that is alternatively ice-covered (extended) or ice-free (reduced), the surface temperature is about 30°C lower in the extended-ice experiment. The temperature differences away from the core region and along the continental margins of eastern North America and western Europe are smaller (5-10°C).

The circulation adjusts to this change in temperature in several ways. The core of the surface westerlies is displaced about 15° of latitude farther south in the extended-ice case than in the reduced-ice case. Over western Europe the winds are more westerly in the reduced-ice case than in the extended-ice case, and they advect warmer air. As a result, January surface temperatures in western Europe are about 5-10°C higher in the reduced-ice case. This experiment illustrates that simulations of glacial-age climates of the

North Atlantic sector, including the bordering lands of eastern North America and western Europe, are indeed sensitive to assumptions about the extent of North Atlantic sea ice.

REDUCED ATMOSPHERIC CARBON DIOXIDE CONCENTRATION AT 18 ka

The CO_2 concentrations in the series of experiments reported by Kutzbach and Guetter (1986) and in this volume were set at the modern value of 330 ppmv. However, Kutzbach and Guetter (1986) also reported an 18-ka glacial-maximum experiment run with a CO_2 concentration of 200 ppmv, with other boundary conditions unchanged from the basic 18-ka experiment. Because SST and sea-ice limits were prescribed, the effect of the lower CO_2 concentration on climate over the oceans was small. Over land, there was a small additional lowering of surface temperature of up to 1-2°C locally and 0.2°C on average and a decrease in precipitation and precipitation-minus-evaporation compared to the 18-ka experiment with modern CO_2 values.

Broccoli and Manabe (1987) used a GCM coupled to a mixed-layer ocean model to explore the effects of lower CO_2 on glacial climates. Compared to a simulation without lowered CO_2, their simulation had significantly lower surface temperature (greater than 2°C) around Antarctica and somewhat smaller reductions (1-1.5°C) in high and middle latitudes of the Northern Hemisphere.

References

Andrews, J. T. (1987). The Late Wisconsin glaciation and deglaciation of the Laurentide ice sheet. *In* "North America and Adjacent Oceans during the Last Deglaciation" (W. F. Ruddiman and H. E. Wright, Jr., Eds.), pp. 13-37. The Geology of North America, Vol. K-3. The Geological Society of America, Boulder, Colo.

Andrews, J. T., and Falconer, G. (1969). Late glacial and post-glacial history and emergence of the Ottawa Islands, Hudson Bay, Northwest Territories: Evidence on the deglaciation of Hudson Bay. *Canadian Journal of Earth Sciences* 6, 1263-1276.

Bard, E., Hamelin, B., Fairbanks, R. G., and Zindler, A. (1990). Calibration of the ^{14}C time scale over the past 30,000 years using mass spectrometric U-Th ages from Barbados corals. *Nature* 345, 405-410.

Barnola, J. M., Raynaud, D., Korotevich, Y. S., and Lorius, C. (1987). Vostok ice core provides 160,000 year record of atmospheric CO_2. *Nature* 329, 408-414.

Berger, A. L. (1978). Long-term variations of caloric insolation resulting from the earth's orbital elements. *Quaternary Research* 9, 139-167.

Berger, W. H., Killingley, J. S., Metzler, C. V., and Vincent, E. (1985). Two-step deglaciation: ^{14}C-dated high-resolution $\delta^{18}O$ records from the tropical Atlantic. *Quaternary Research* 23, 258-271.

Birchfield, G. E. (1987). Changes in deep-ocean water $\delta^{18}O$ and temperature from the last glacial maximum to the present. *Paleoceanography* 2, 431-442.

Blackmon, M. L., Giesler, J. E., and Pitcher, E. J. (1983). A general circulation model study of January climate anomaly patterns associated with interannual variation of equatorial Pacific sea surface temperatures. *Journal of the Atmospheric Sciences* 40, 1410-1425.

Bloom, A. L. (1971). Glacial-eustatic and isostatic controls of sea level since the last glaciation. *In* "The Late Cenozoic Glacial Ages" (K. K. Turekian, Ed.), pp. 355-380. Yale University Press, New Haven.

Boulton, G. S., Smith, G. D., Jones, A. S., and Newsome, J. (1985). Glacial geology and glaciology of the last mid-latitude ice sheets. *Journal of the Geological Society of London* 142, 447-474.

Broccoli, A. J., and Manabe, S. (1987). The influence of continental ice, atmospheric CO_2 and land albedo on the climate of the last glacial maximum. *Climate Dynamics* 1, 87-99.

Broecker, W. S. (1978). The cause of glacial to interglacial climatic change. *In* "Evolution of Planetary Atmospheres and Climatology of the Earth," pp. 165-177. Centre national d'études spatiales, Nice, France.

———. (1986). Oxygen isotope constraints on surface ocean temperatures. *Quaternary Research* 26, 121-134.

Burckle, L. H., Robinson, D., and Cooke, D. (1982). Reappraisal of sea-ice distribution in Atlantic and Pacific sectors of the Southern Ocean at 18,000 yr B.P. *Nature* 299, 435-437.

Chappell, J., and Shackleton, N. J. (1986). Oxygen isotopes and sea level. *Nature* 324, 137-138.

CLIMAP Project Members (1981). Seasonal reconstructions of the earth's surface at the last glacial maximum. *Geological Society of America Map and Chart Series* MC-36.

COHMAP Members (1988). Climatic changes of the last 18,000 years: Observations and model simulations. *Science* 241, 1043-1052.

Dansgaard, W., and Tauber, H. (1969). Glacier oxygen-18 content and Pleistocene ocean temperatures. *Science* 166, 499-502.

Denton, G. H., and Hughes, T. J. (1981). "The Last Great Ice Sheets." Wiley-Interscience, New York.

Dodge, R. E., Fairbanks, R. G., Benninger, L. K., and Maurrasse, F. (1983). Pleistocene sea levels from raised coral reefs of Haiti. *Science* 219, 1423-1425.

Duplessy, J. C., Delibrias, G., Turon, J. L., Pujol, C., and Duprat, J. (1981). Deglacial warming of the northeastern Atlantic Ocean: Correlation with the paleoclimatic evolution of the European continent. *Palaeogeography, Palaeoclimatology, Palaeoecology* 35, 121-144.

Fairbanks, R. G. (1989). A 17,000-year glacio-eustatic sea level record: Influence of glacial melting rates on the Younger Dryas event and deep-ocean circulation. *Nature* 342, 637-642.

Flint, R. F. (1971). "Glacial and Quaternary Geology." Wiley, New York.

Gardner, J. V. (1975). Late Pleistocene carbonate dissolution cycles in the eastern equatorial Atlantic. *In* "Dissolution of Deep-Sea Carbonates" (W. F. Sliter, A. W. H. Be, and W. H. Berger, Eds.), pp. 129-141. Cushman Foundation for Foraminiferal Research Special Publication 13.

Hammer, C. U., Clausen, H. B., and Tauber, H. (1986). Ice core dating of the Pleistocene/Holocene boundary applied to a calibration of the ^{14}C time scale. *Radiocarbon* 28, 284-291.

Hays, J. D. (1978). A review of the late Quaternary climatic history of Antarctic seas. *In* "Antarctic Glacial History and World Paleoenvironments" (E. M. van Zinderen Bakker, Ed.), pp. 57-71. A. A. Balkema, Rotterdam.

Hays, J. D., Imbrie, J., and Shackleton, N. J. (1976). Variations in the earth's orbit: Pacemaker of the ice ages. *Science* 194, 1121-1132.

Hughes, T. (1987). Ice dynamics and deglaciation models when ice sheets collapsed. *In* "North America and Adjacent Oceans during the Last Deglaciation" (W. F. Ruddiman and H. E. Wright, Jr., Eds.), pp. 183-220. The Geology of North America, Vol. K-3. The Geological Society of America, Boulder, Colo.

Joussaume, S. (1989). Desert dust and climate: An investigation using an atmospheric general circulation model. *In* "Paleoclimatology and Paleometeorology: Modern and Past Patterns of Global At-

mospheric Transport" (M. Leinen and M. Sarnthein, Eds.), pp. 253-263. NATO ASI Series, Vol. C282. Kluwer Academic, Boston.

Kutzbach, J. E., and Gallimore, R. G. (1988). Sensitivity of a coupled atmosphere/mixed-layer ocean model to changes in orbital forcing at 9000 yr B.P. *Journal of Geophysical Research* 93(D1), 803-821.

Kutzbach, J. E., and Guetter, P. (1986). The influence of changing orbital parameters and surface boundary conditions on climate simulations for the past 18,000 years. *Journal of the Atmospheric Sciences* 43, 1726-1759.

Kutzbach, J. E., and Street-Perrott, F. A. (1985). Milankovitch forcing of fluctuations in the level of tropical lakes from 18 to 0 kyr B.P. *Nature* 317, 130-134.

Labeyrie, L., Duplessy, J. C., and Blanc, P. L. (1987). Variations in mode of formation and temperature of oceanic deep waters over the last 125,000 years. *Nature* 327, 477-482.

Leinen, M. (1989). The late Quaternary record of atmospheric transport to the Northwest Pacific from Asia. *In* "Paleoclimatology and Paleometeorology: Modern and Past Patterns of Global Atmospheric Transport" (M. Leinen and M. Sarnthein, Eds.), pp. 693-732. NATO ASI Series, Vol. C282. Kluwer Academic, Boston.

Lorius, C., Raynaud, D., Petit, J.-R., Jousel, J., and Merlivat, L. (1984). Late-glacial maximum-Holocene atmospheric and ice-thickness changes from Antarctic ice-core studies. *Annals of Glaciology* 5, 88-94.

MacDonald, B. C. (1971). Late Quaternary stratigraphy and deglaciation in eastern Canada. *In* "The Late Cenozoic Glacial Ages" (K. K. Turekian, Ed.), pp. 331-354. Yale University Press, New Haven.

Manabe, S., and Broccoli, A. J. (1985). The influence of continental ice sheets on the climate of an ice age. *Journal of Geophysical Research* 90, 2167-2190.

Mitchell, J. F. B., Grahame, N. S., and Needham, K. J. (1988). Climate simulations for 9000 years before present: Seasonal variations and effect of the Laurentide ice sheet. *Journal of Geophysical Research* 93, 8283-8303.

Mix, A. C. (1987). The oxygen-isotope record of glaciation. *In* "North America and Adjacent Oceans during the Last Deglaciation" (W. F. Ruddiman and H. E. Wright, Jr., Eds.), pp. 111-135. The Geology of North America, Vol. K-3. The Geological Society of America, Boulder, Colo.

Mix, A. C., and Ruddiman, W. F. (1985). Structure and timing of the last deglaciation. *Quaternary Science Reviews* 4, 59-108.

Neftel, A., Oeschger, H., Schwander, J., Stauffer, B., and Zumbrunn, R. (1982). Ice core sample measurements give atmospheric CO_2 content during the past 40,000 yr. *Nature* 295, 220-223.

Oeschger, H., Beer, J., Siegenthaler, U., Stauffer, B., Dansgaard, W., and Langway, C. C. (1983). Late-glacial climate history from ice cores. *In* "Paleoclimatic Models and Research" (A. Ghazi, Ed.), pp. 95-107. Reidel, Dordrecht/Boston.

Overpeck, J. T., Webb, T. III, and Prentice, I. C. (1985). Quantitative interpretation of fossil pollen spectra: Dissimilarity coefficients and the method of modern analogs for pollen data. *Quaternary Research* 23, 87-108.

Paterson, W. S. B. (1972). Laurentide ice sheet: Estimated volumes during Late Wisconsin. *Reviews of Geophysics and Space Physics* 10, 885-917.

Peterson, G. M., Webb, T. III, Kutzbach, J. E., van der Hammen, T., Wijmstra, T., and Street, F. A. (1979). The continental record of environmental conditions at 18,000 yr B.P.: An initial evaluation. *Quaternary Research* 12, 47-82.

Petit, J.-R., Briat, M., and Royer, A. (1981). Ice age aerosol content from East Antarctic ice core samples and past wind strength. *Nature* 293, 391-394.

Pitcher, E. J., Malone, R. C., Ramanathan, V., Blackmon, M. L., Puri, K., and Bourke, W. (1983). January and July simulations with a spectral general circulation model. *Journal of the Atmospheric Sciences* 40, 580-604.

Prell, W. L. (1985). "The Stability of Low-Latitude Sea-Surface Temperatures: An Evaluation of the CLIMAP Reconstruction with Emphasis on the Positive SST Anomalies." U.S. Department of Energy Report TR025. U.S. Government Printing Office, Washington, D.C.

Ramanathan, V., Pitcher, E. J., Malone, R. C., and Blackmon, M. L. (1983). The response of a spectral general circulation model to refinements in radiative processes. *Journal of the Atmospheric Sciences* 40, 605-630.

Rind, D., and Peteet, D. (1985). Terrestrial conditions at the last glacial maximum and CLIMAP sea-surface temperature estimates: Are they consistent? *Quaternary Research* 24, 1-22.

Ruddiman, W. F. (1987). Synthesis: The ocean/ice sheet record. *In* "North America and Adjacent Oceans during the Last Deglaciation" (W. F. Ruddiman and H. E. Wright, Jr., Eds.), pp. 463-478. The Geology of North America, Vol. K-3. The Geological Society of America, Boulder, Colo.

Ruddiman, W. F., and McIntyre, A. (1981). The North Atlantic Ocean during the last deglaciation. *Palaeogeography, Palaeoclimatology, Palaeoecology* 35, 145-214.

Sarnthein, M., Thiede, J., Pflaumann, U., Erlenkeuser, H., Futterer, D., Koopman, B., Lange, H., and Seibold, E. (1982). Atmospheric and oceanic circulation patterns off northwest Africa during the past 25 million years. *In* "Geology of the Northwest African Continental Margin" (U. von Rad, K. Hinz, M. Sarnthein, and E. Seibold, Eds.), pp. 545-604. Springer-Verlag, Berlin/New York.

Shackleton, N. J., and Opdyke, N. D. (1973). Oxygen isotope and paleomagnetic stratigraphy of equatorial Pacific core V28-238: Oxygen isotope temperatures and ice volumes on a 10^5 and 10^6 year scale. *Quaternary Research* 3, 39-55.

Stuiver, M., Burk, R. L., and Quay, P. D. (1984). $^{13}C/^{12}C$ ratios in tree rings and the transfer of biospheric carbon to the atmosphere. *Journal of Geophysical Research* 89, 11734-11748.

Webb, T. III, Bartlein, P. J., and Kutzbach, J. E. (1987). Climatic change in eastern North America during the past 18,000 years: Comparisons of pollen data with model results. *In* "North America and Adjacent Oceans during the Last Deglaciation" (W. F. Ruddiman and H. E. Wright, Jr., Eds.), pp. 447-462. The Geology of North America, Vol. K-3. The Geological Society of America, Boulder, Colo.

Webster, P. J., and Streten, N. A. (1978). Late Quaternary ice age climates of tropical Australasia: Interpretations and reconstructions. *Quaternary Research* 10, 279-309.

CHAPTER 4

Simulated Climatic Changes: Results of the COHMAP Climate-Model Experiments

J. E. Kutzbach, P. J. Guetter, P. J. Behling, and R. Selin

Until about 15 yr ago, most theoretical work concerning past climates was derived from conceptual models of climatic change and involved either qualitative reasoning from simple models or interpretation of the geologic record. The recent development of general circulation models (GCMs) for the atmosphere and oceans has permitted quantitative paleoclimatic modeling and thus has led to a major advance in theoretical studies. With the advent of supercomputers, climate modelers gained the ability to make the large number of calculations required for paleoclimatic studies.

The first experiments with GCMs for the climate of 18 ka were those of Williams *et al.* (1974), Gates (1976a,b), and Manabe and Hahn (1977). Gates and Manabe and Hahn used the initial CLIMAP estimates of glacial-age boundary conditions (CLIMAP Project Members, 1976), whereas subsequent modelers used the revised CLIMAP estimates (CLIMAP Project Members, 1981). The climate of 18 ka has also been simulated with a coupled atmosphere-ocean model (Manabe and Broccoli, 1985a,b).

The COHMAP experiments were the first to simulate a sequence of paleoclimatic conditions covering the period from 18 ka to the present (COHMAP Members, 1988). The experiments were made with the Community Climate Model (CCM) of the National Center for Atmospheric Research (NCAR) (Pitcher *et al.,* 1983). Orbitally determined insolation, mountain and ice-sheet orography, sea-surface temperatures (SSTs), sea-ice limits, snow cover, land albedo, and effective soil moisture were prescribed as input to the model. The choice of values for these boundary conditions is discussed in the previous chapter (Kutzbach

and Ruddiman, this vol.). The numerical experiments produced "snapshots" of the climate, including such variables as temperature, precipitation, sea-level pressure, and winds, for January and July at 3000-yr intervals during the past 18,000 yr. Experiments covering several seasonal cycles for the present and for 9 ka were used to check that the results from January and July simulations (Kutzbach and Guetter, 1986) were similar to the results from full-year simulations (Kutzbach, 1981; Kutzbach and Otto-Bliesner, 1982).

An atmospheric GCM is a tool for illustrating in quantitative terms some of the climatic consequences of changes in solar radiation and surface boundary conditions. It can be used to demonstrate links or associations between climatic changes in widely separated places and to identify the physical mechanisms through which changes in boundary conditions alter climate. In these ways the model simulations provide possible explanations for climatic changes evident at a continental and global scale. At the same time, discrepancies between the model simulations and the paleoclimatic data reveal the need for new research to improve the model physics, the boundary conditions, the geologic data, and the methods used to infer climatic information from the data.

Because the seasonal and latitudinal distributions of solar radiation at 18 ka were similar to those of today, differences in the atmospheric conditions simulated for glacial times must be attributed primarily to altered surface boundary conditions or to altered atmospheric composition (CO_2, methane, aerosols), which may in turn be related to climate and surface conditions. Between 15 and 9 ka, the distance of the earth from the sun decreased in northern summer and the axial tilt increased, enhancing the seasonality of climate in the Northern Hemisphere and diminishing it in the Southern Hemisphere.

The COHMAP experiments also adjusted the surface boundary conditions between 15 and 9 ka to reflect in a broad way the retreat and melting of continental ice sheets and sea ice, the warming of the oceans, and changes in land-surface albedo (see Kutzbach and Ruddiman, this vol.). Therefore, the atmospheric conditions simulated for the period between 15 and 9 ka must be attributed to the combined effects of the orbitally caused changes in solar radiation and the changes in surface boundary conditions and atmospheric composition. At about 9 ka average solar radiation over the Northern Hemisphere was about 8% higher in July and about 8% lower in January than it is today. After 9 ka these seasonal radiation extremes decreased toward modern values. At 6 ka and thereafter the only changes in our experiments were the externally imposed solar-radiation changes. Thus the simulated climatic changes at 6 and 3 ka must be attributed exclusively to orbitally induced solar-radiation changes.

In this chapter we present maps of key climatic variables covering almost the full set of COHMAP experiments: 18, 12, 9, 6, and 0 ka (the control simulation). The maps for 15 ka (not shown) are nearly identical to those for 18 ka. The maps for 3 ka (not shown) illustrate the same climatic differences from the present that are simulated at 6 ka, but in subdued form. The pattern of change is similar but the amplitude is reduced at 3 ka relative to 6 ka because the orbitally induced solar-radiation differences for 3 ka compared to the present are approximately one-half those for 6 ka. We also present some results of simulations for 18 ka with somewhat different surface and atmospheric boundary conditions than those used in the standard experiment (see Kutzbach and Ruddiman, this vol.).

In this chapter we first describe the format of the paleoclimatic charts, then summarize related experiments and discuss acknowledged limitations in the model and the boundary conditions that should be kept in mind when the charts are used. Some other studies that have used the results of the COHMAP experiments are listed in the Appendix.

Description of Charts

The results from the paleoclimatic experiments are summarized in maps of individual climatic variables as a function of time and in maps of many climatic variables for individual times. The first format is convenient for studying changes in individual climatic variables, such as surface temperatures, at specific locations over time. The second format is convenient for studying the linkages among climatic variables (winds, pressure, temperature, precipitation) at specific times, namely 18, 12, 9, 6, and 0 ka (January and July). In all the charts, ice sheets are indicated by a solid light blue-gray coloring, sea ice by light blue-gray hatching, and continents by a darker blue-gray border.

We first present the maps of individual climatic variables (January and July) as a function of time (18, 12, 9, 6, and 0 ka). Figures 4.1 and 4.3 show simulated surface temperature for January and July, respectively, and Figures 4.2 and 4.4 show the differences between the simulated past and present temperatures. The difference maps also indicate, with fine hatching, regions where departures are statistically significant (two-sided t-test) at or above the 95% confidence level compared to the model's inherent variability (see Kutzbach and Guetter [1986] for a full description of the statistical test).

Figures 4.5 and 4.7 show simulated sea-level pressure and surface winds in January and July, respectively, and Figures 4.6 and 4.8 show the differences in these variables compared to modern values. Simulated precipitation (smoothed with a two-dimensional, nine-point binomial filter) for January and July is shown in Figures 4.9 and 4.11, respectively, and the differences between simulated past and present precipitation are shown in Figures 4.10 and 4.12.

Figure 4.13 shows the differences between simulated past and present annual-average precipitation-minus-evaporation. This climatic variable is the only one for which we present estimated annual averages rather than January and July estimates. We use the term *estimated annual average* because the values are derived from averages of the perpetual simulations of January and July.

Simulated winds at a level of 500 mb (about 5.5 km) in January and July are presented in Figures 4.14 and 4.15, respectively. The simulated storm tracks shown in Figures 4.16 and 4.17 were estimated from standard deviations of time series of surface pressure. These time series consisted of the pressure variations between periods of 2.5 and 6 days, obtained with the aid of a band-pass filter as described by Blackmon and Lau (1980). The regions with maxima in this index are those where centers of high and low pressure alternate with periods of several days.

We next present the maps of many climatic variables for individual times. These maps show sea-level pressure, surface winds, winds and wind speed contours at 500 mb, surface storm tracks, and differences in sea-level pressure, surface temperature, and

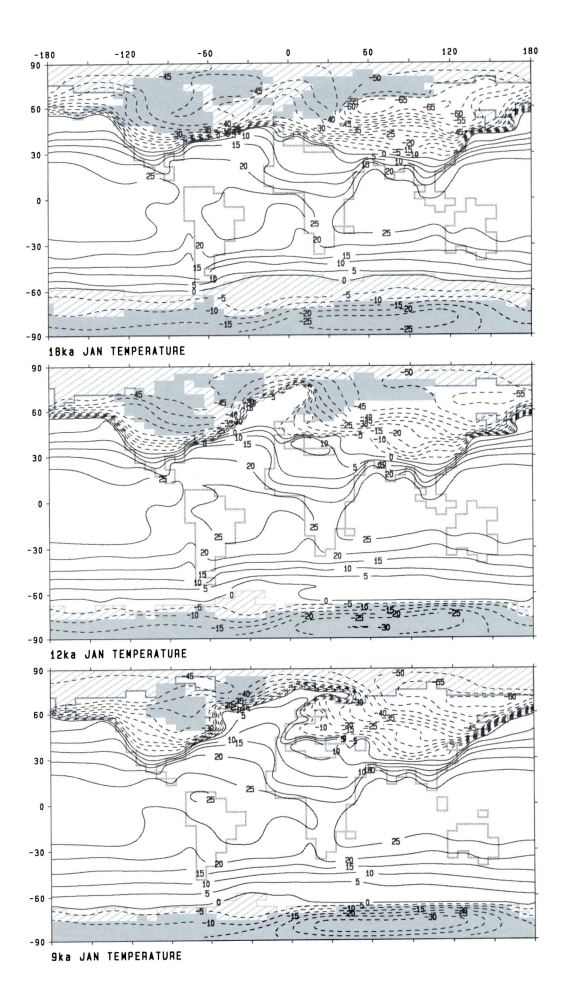

18ka JAN TEMPERATURE

12ka JAN TEMPERATURE

9ka JAN TEMPERATURE

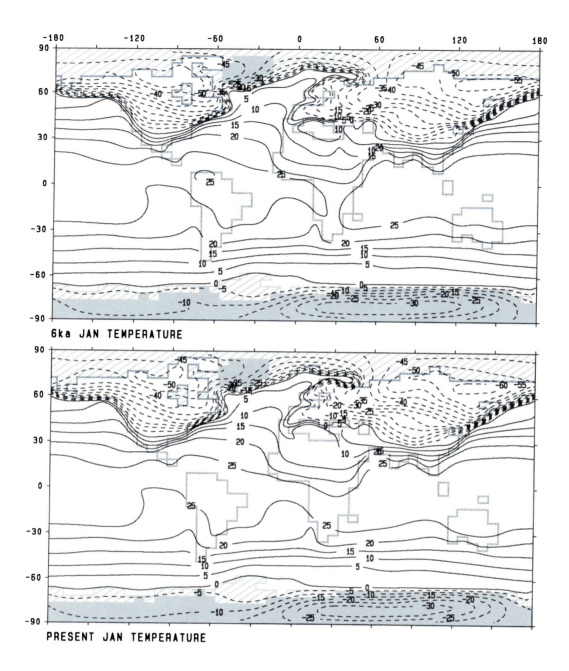

6ka JAN TEMPERATURE

PRESENT JAN TEMPERATURE

Fig. 4.1. Simulated January surface temperature (°C) for 18, 12, 9, 6, and 0 ka. Temperatures are low over the northern ice sheets and over the northern and southern sea ice at 18 ka, with strong gradients of temperature along the land- and sea-ice margins and south of the northern ice sheets. Temperatures increase, particularly over North America, the North Atlantic, and Europe, with the withdrawal of the ice sheets and sea ice and the northward advance of the warm water associated with the Gulf Stream.

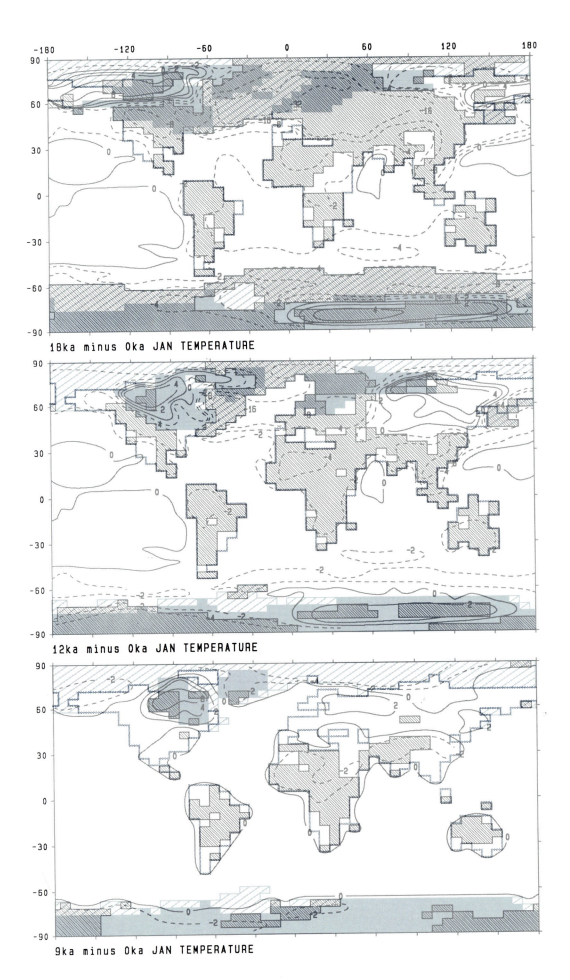

18ka minus 0ka JAN TEMPERATURE

12ka minus 0ka JAN TEMPERATURE

9ka minus 0ka JAN TEMPERATURE

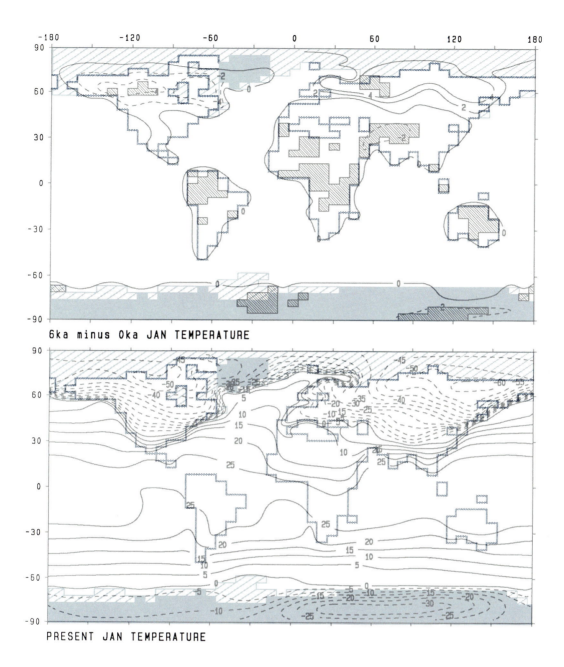

Fig. 4.2. Differences (°C) between simulated past (18, 12, 9, and 6 ka) and present (0-ka control) January surface temperature. Temperatures are much lower than at present at 18 and 12 ka, especially in much of the northern middle and high latitudes. At high latitudes in the Northern (winter) Hemisphere the inherent (year-to-year) variability of temperature is high, so that even relatively large temperature differences are not statistically significant. For example, the higher-than-present temperatures in eastern Siberia at 12 ka are generally not statistically significant. Beginning at 12 ka, and especially at 9 and 6 ka, the colder-than-present conditions in the tropics are due primarily to the solar-radiation changes associated with orbital changes.

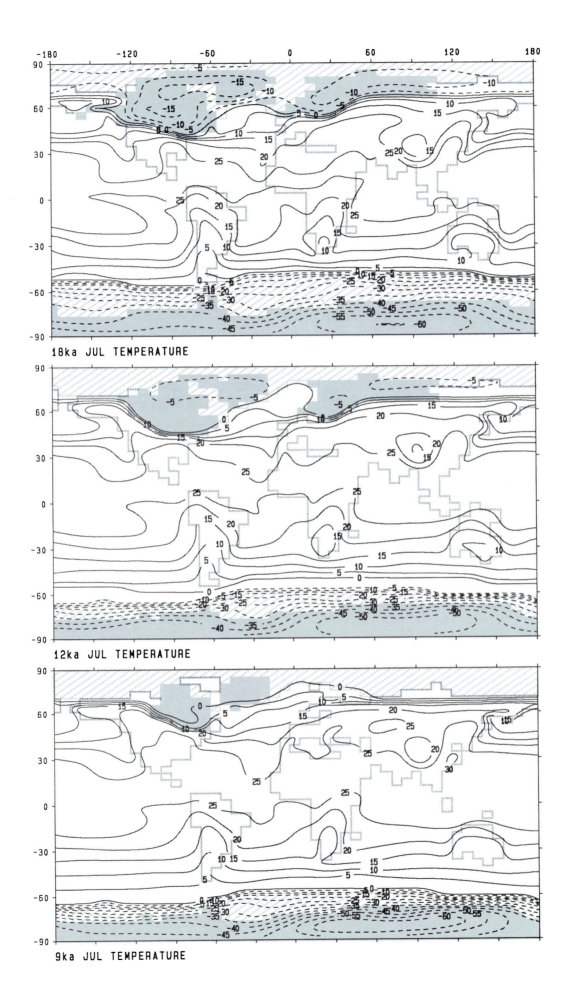

18ka JUL TEMPERATURE

12ka JUL TEMPERATURE

9ka JUL TEMPERATURE

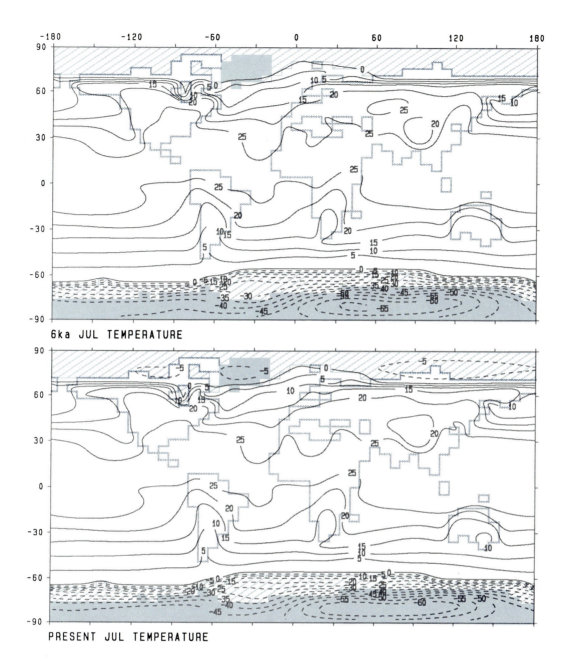

Fig. 43. Simulated July surface temperature (°C) for 18, 12, 9, 6, and 0 ka. Temperatures are low over the northern ice sheets and over the extended sea ice at 18 ka. Temperatures increase with the withdrawal of the ice, particularly over North America, the North Atlantic, and Europe.

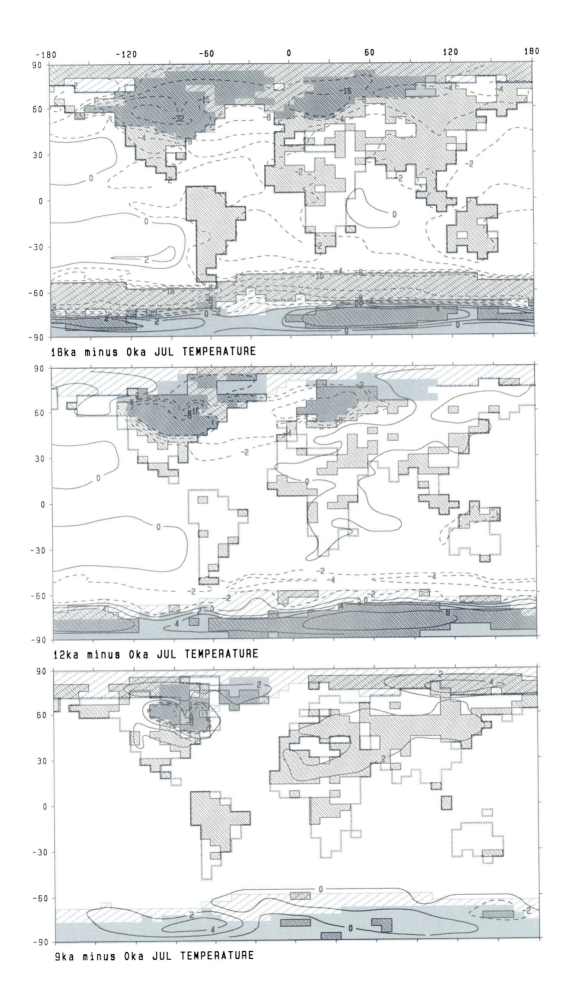

18ka minus 0ka JUL TEMPERATURE

12ka minus 0ka JUL TEMPERATURE

9ka minus 0ka JUL TEMPERATURE

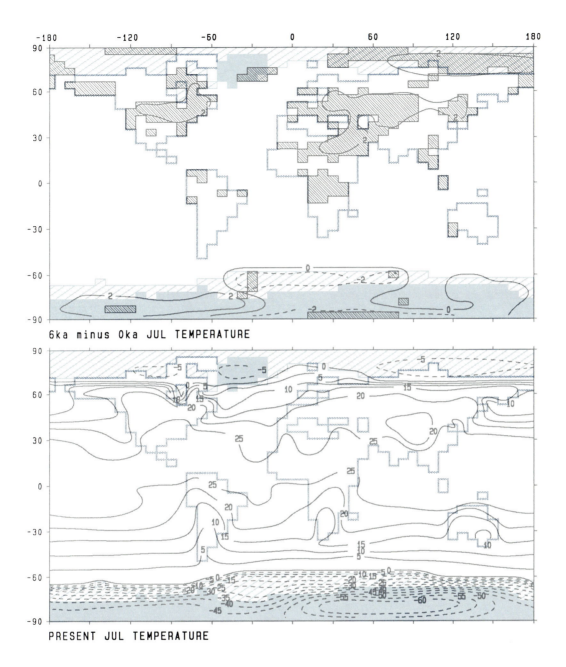

Fig. 4.4. Differences (°C) between simulated past (18, 12, 9, and 6 ka) and present (0-ka control) July surface temperature. Temperatures are lower than at present at 18 and 12 ka in much of the northern middle and high latitudes and continuing to 9 ka over the North American ice sheet. At high southern latitudes the temperature is lower than at present at 18 ka over the extended sea ice and higher than at present over parts of the Antarctic continent. It is colder than at present at 18 ka over South America and much of Southeast Asia and Australia. Beginning at 12 ka, and especially at 9 and 6 ka, the warmer-than-present continental conditions in the tropics and northern middle latitudes are due to solar-radiation changes associated with orbital changes. This trend is delayed in North America and Europe by the cooling effects of the retreating ice sheets.

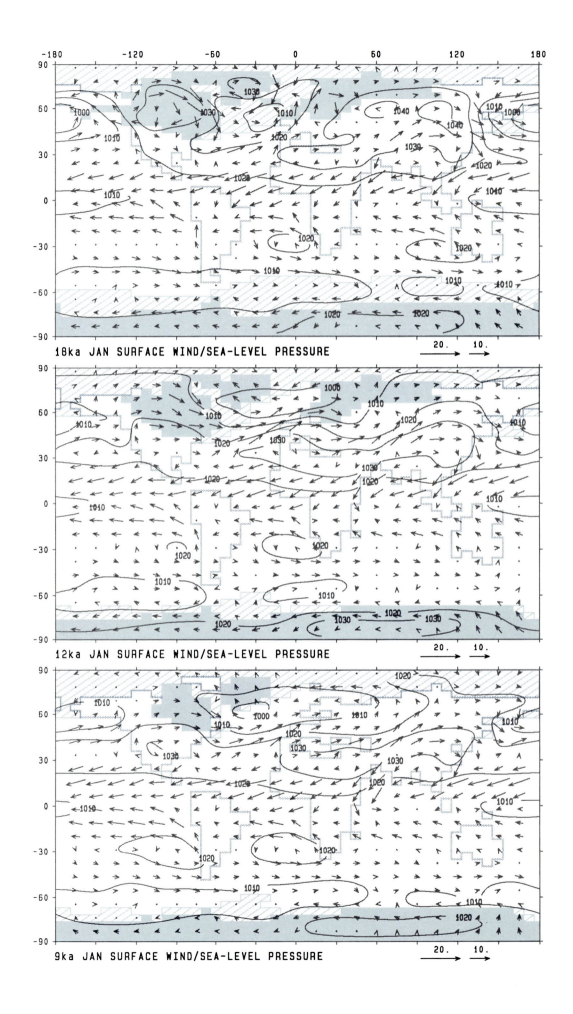

18ka JAN SURFACE WIND/SEA-LEVEL PRESSURE

12ka JAN SURFACE WIND/SEA-LEVEL PRESSURE

9ka JAN SURFACE WIND/SEA-LEVEL PRESSURE

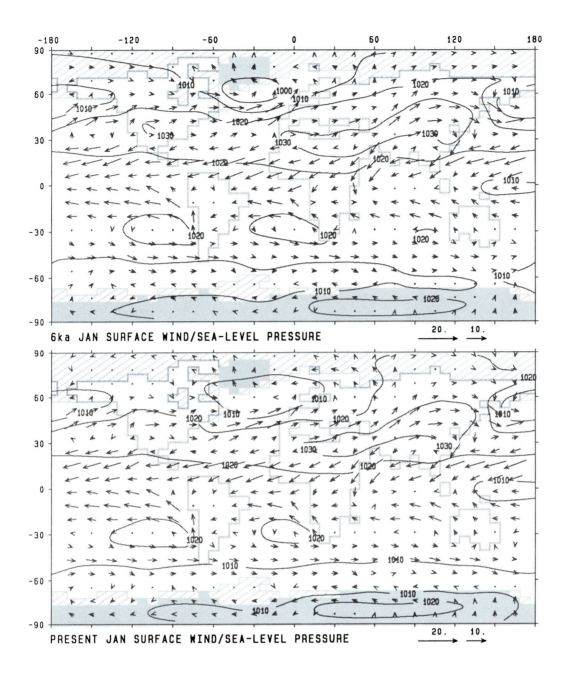

Fig. 4.5. Simulated January sea-level pressure (mb) and surface winds (m/s; see scale at bottom) for 18, 12, 9, 6, and 0 ka. Glacial anticyclones are located over the North American and European ice sheets and along the southern margin of the Asian ice sheet at 18 ka. Westerly flow is reestablished in these same areas by 12 ka. This westerly flow weakens to near-present intensity by 9 and 6 ka.

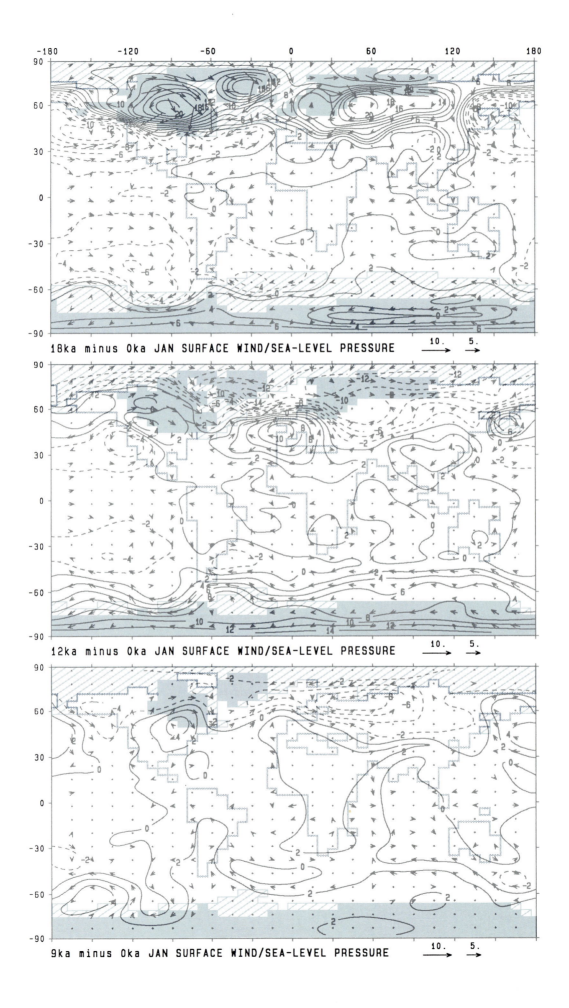

18ka minus 0ka JAN SURFACE WIND/SEA-LEVEL PRESSURE

12ka minus 0ka JAN SURFACE WIND/SEA-LEVEL PRESSURE

9ka minus 0ka JAN SURFACE WIND/SEA-LEVEL PRESSURE

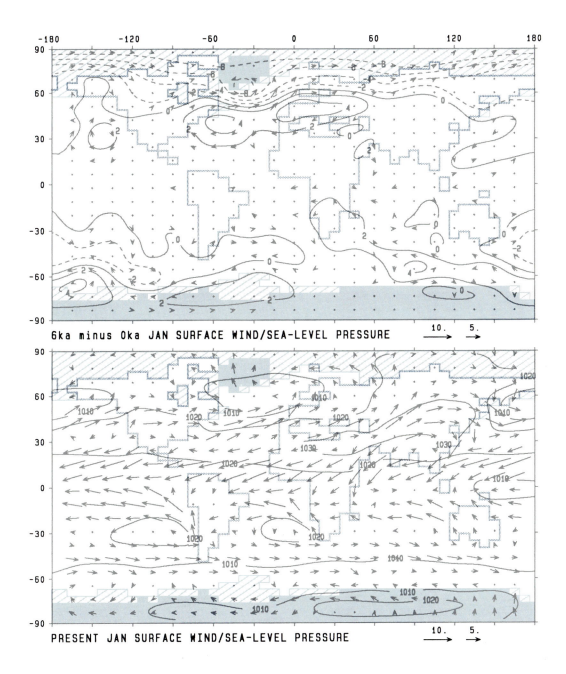

Fig. 4.6. Differences between simulated past (18, 12, 9, and 6 ka) and present (0-ka control) January sea-level pressure (mb) and surface winds (m/s; see scale at bottom). Sea-level pressure is greater than at present over the North American and Greenland ice sheets and the North Atlantic sea ice at 18 ka and also along the southern margin of the Eurasian ice sheet. Sea-level pressure is less than at present south of the ice sheets, particularly in North America and over the North Atlantic to the south of the sea-ice margin; this is the region of increased storminess. Sea-level pressure is also lower than at present over the North Pacific, where the Aleutian Low is intensified. At 12 ka the sea-level pressure pattern is less strongly influenced by the presence of the ice sheets, and at 9 ka sea-level pressure is greater than at present only over a small area of the North American ice sheet. Sea-level pressure is greater than at present over tropical and some northern midlatitude lands from 12 to 6 ka because of solar-radiation changes associated with orbital changes.

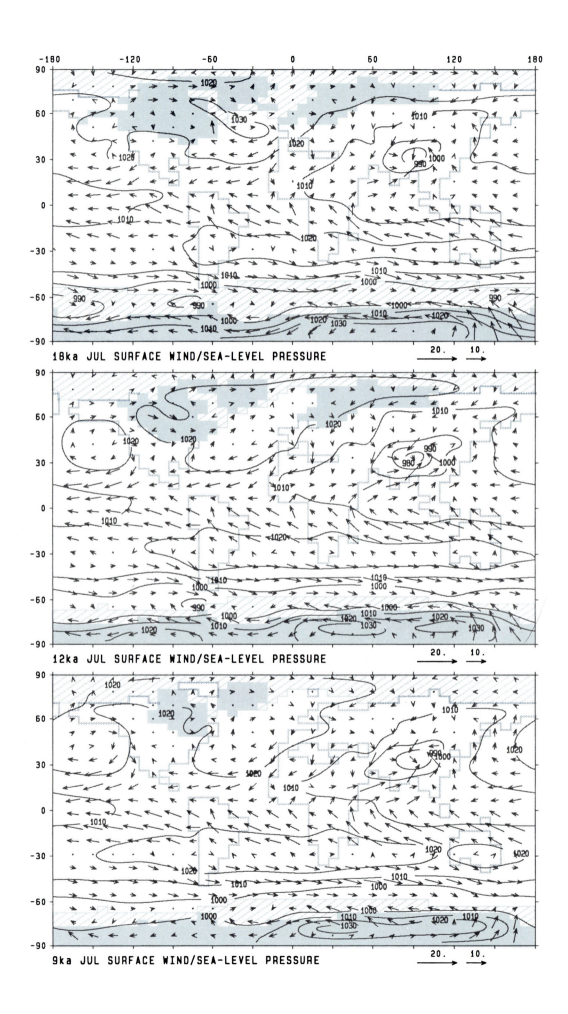

18ka JUL SURFACE WIND/SEA-LEVEL PRESSURE   20.   10.

12ka JUL SURFACE WIND/SEA-LEVEL PRESSURE   20.   10.

9ka JUL SURFACE WIND/SEA-LEVEL PRESSURE   20.   10.

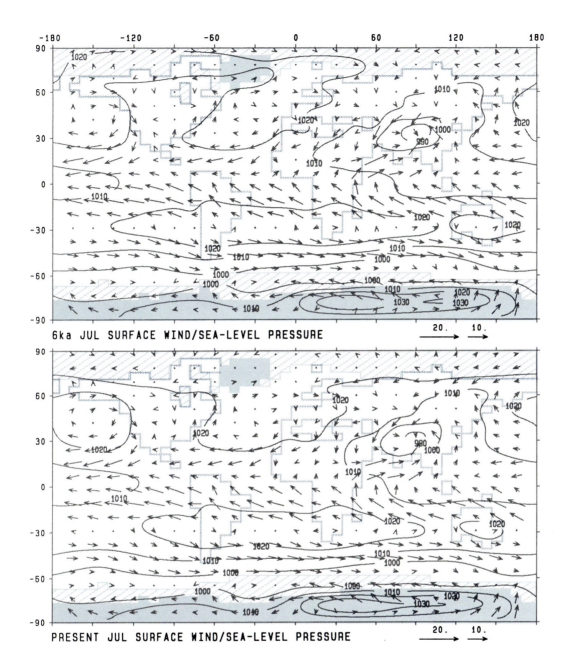

Fig. 4.7. Simulated July sea-level pressure (mb) and surface winds (m/s; see scale at bottom) for 18, 12, 9, 6, and 0 ka. Sea-level pressure is greater than at present in high northern latitudes over the North American and Eurasian ice sheets and the North Atlantic at 18 and 12 ka. The African-Asian monsoon is weak at 18 ka; for example, the southwesterly surface winds of the Arabian Sea are weak. The weak monsoon is caused by the glacial-age boundary conditions. Between 12 and 6 ka the African-Asian monsoon is strong because of solar-radiation changes associated with orbital changes.

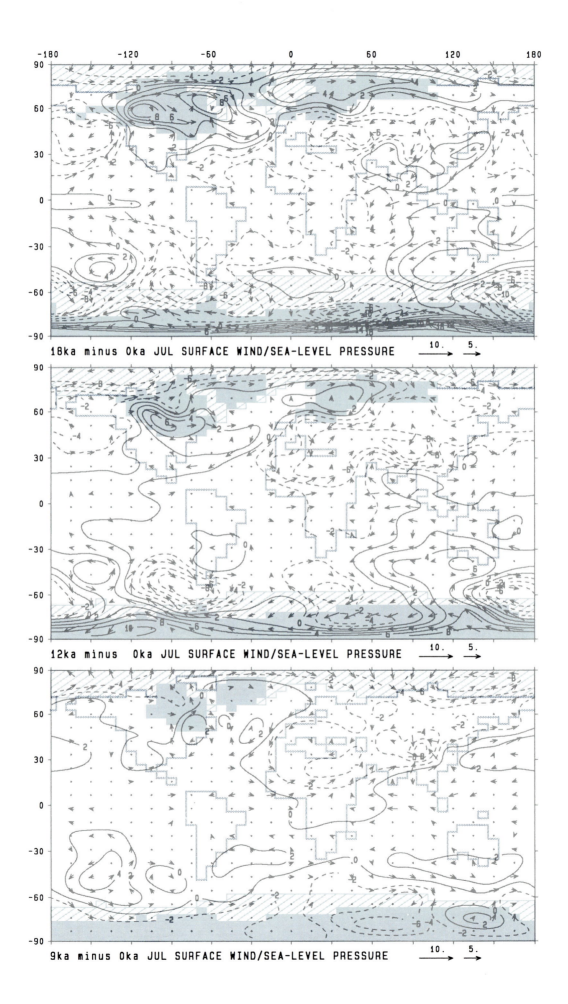

18ka minus 0ka JUL SURFACE WIND/SEA-LEVEL PRESSURE   10.   5.

12ka minus   0ka JUL SURFACE WIND/SEA-LEVEL PRESSURE   10.   5.

9ka minus 0ka JUL SURFACE WIND/SEA-LEVEL PRESSURE   10.   5.

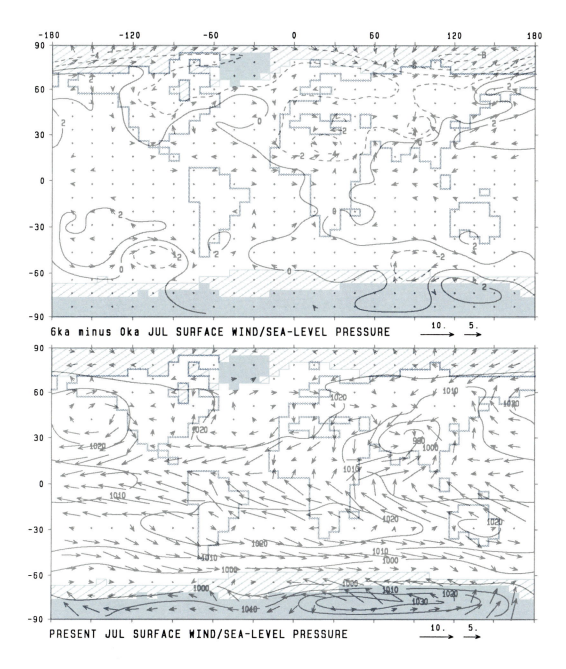

Fig. 4.8. Differences between simulated past (18, 12, 9, and 6 ka) and present (0-ka control) July sea-level pressure (mb) and surface winds (m/s; see scale at bottom). Sea-level pressure is greater than at present over the northern ice sheets and sea ice at 18 ka and over the North American ice sheet at 12 ka, and the associated wind differences are anticyclonic. Sea-level pressure is lower than at present south of the Eurasian ice sheet at 18 ka, and this is associated with increased storminess. At 12-6 ka over northern Africa-southern Asia and also at 9-6 ka over parts of North America, sea-level pressure is lower than at present because of solar-radiation changes associated with orbital changes. The wind differences show higher-than-present southwesterly (or southeasterly) flow in parts of northern Africa and southern Asia at 9-6 ka. The lower sea-level pressure over land is compensated for by higher sea-level pressure in the regions of the oceanic subtropical highs.

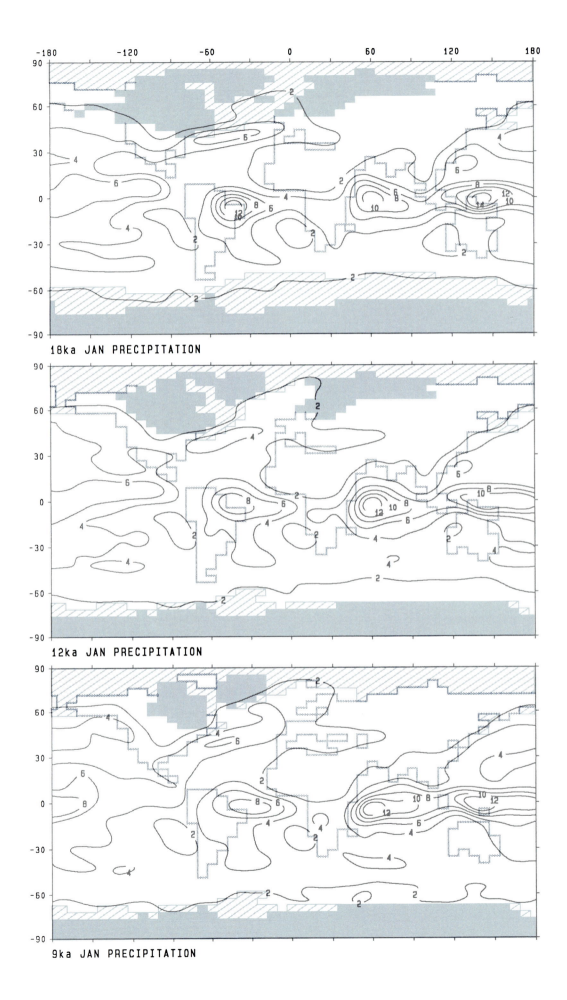

18ka JAN PRECIPITATION

12ka JAN PRECIPITATION

9ka JAN PRECIPITATION

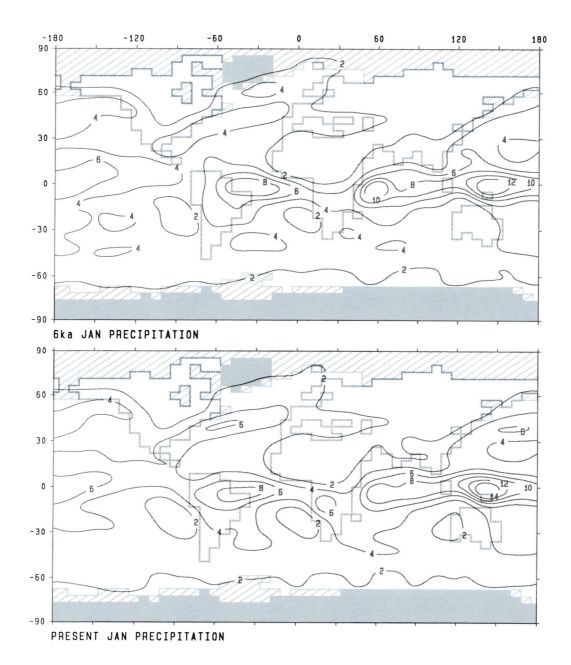

6ka JAN PRECIPITATION

PRESENT JAN PRECIPITATION

Fig. 4.9. Simulated January precipitation (mm/day) for 18, 12, 9, 6, and 0 ka. Precipitation is at a maximum from southwestern North America to Europe at 18 ka. The maximum parallels the ice-sheet/sea-ice border over eastern North America and the North Atlantic. This feature is weaker at 12 ka and thereafter. A band of precipitation associated with the intertropical convergence zone occurs along and south of the equator on each map.

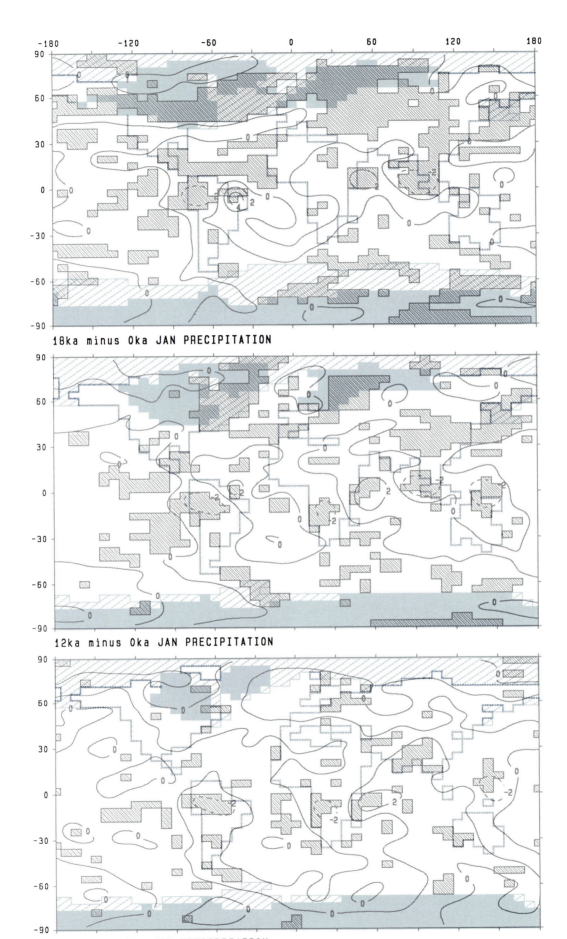

18ka minus 0ka JAN PRECIPITATION

12ka minus 0ka JAN PRECIPITATION

9ka minus 0ka JAN PRECIPITATION

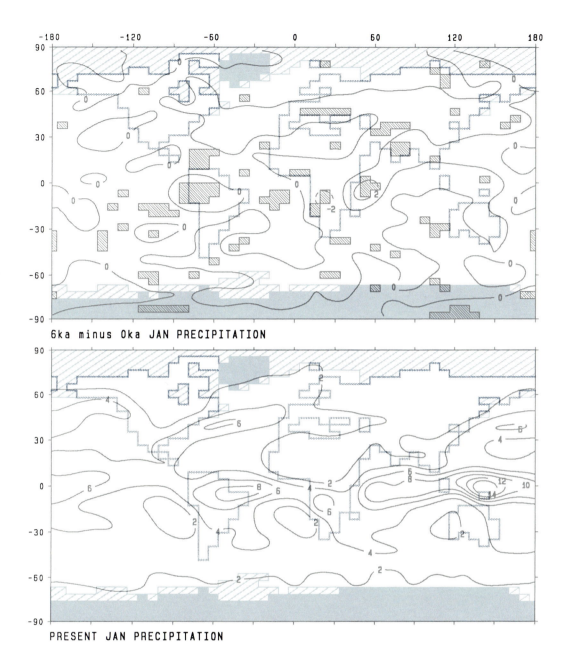

6ka minus 0ka JAN PRECIPITATION

PRESENT JAN PRECIPITATION

Fig. 4.10. Differences (mm/day) between simulated past (18, 12, 9, and 6 ka) and present (0-ka control) January precipitation. Precipitation is generally lower than at present over the regions covered by glacial ice and sea ice and south of the Eurasian ice sheets. Precipitation is greater than at present across parts of North America south of the ice sheet and across the North Atlantic south of the sea-ice border at 18 ka, but this difference is statistically significant only in southwestern North America. In western North America, a narrow strip of reduced precipitation occurs south of the ice sheet, but this feature is not statistically significant. Precipitation differences are generally small at 12 ka and thereafter. From 12 to 6 ka precipitation is lower than at present in the southern summer-monsoon regions of South America and southern Africa because of solar-radiation changes associated with orbital changes.

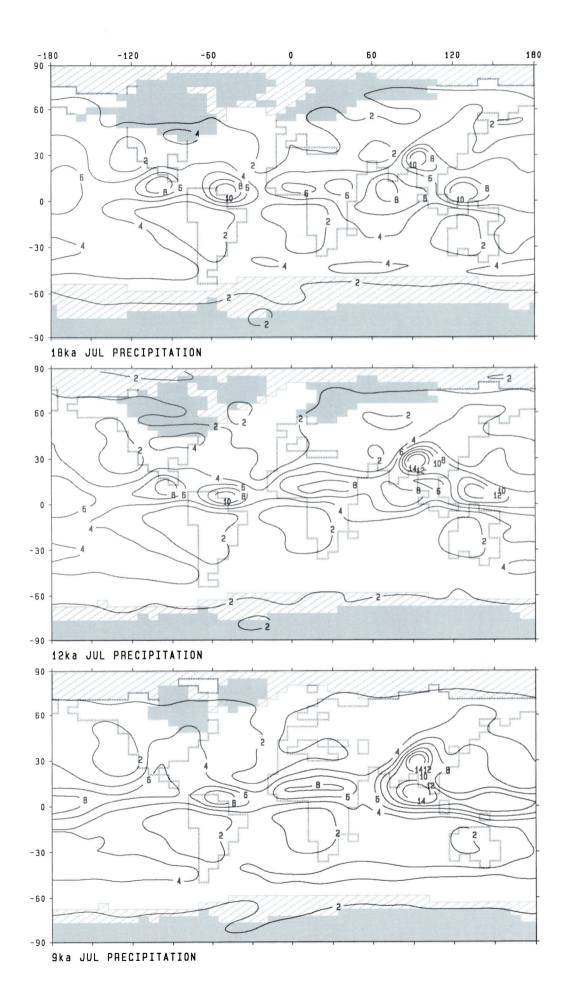

18ka JUL PRECIPITATION

12ka JUL PRECIPITATION

9ka JUL PRECIPITATION

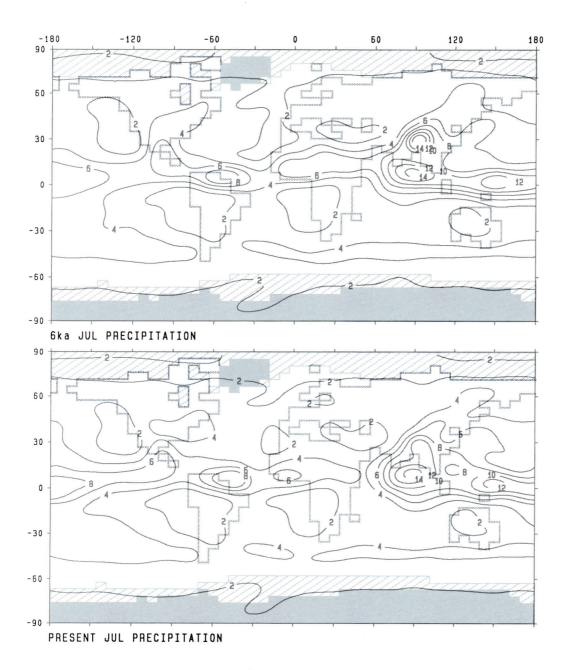

6ka JUL PRECIPITATION

PRESENT JUL PRECIPITATION

Fig. 4.11. Simulated July precipitation (mm/day) for 18, 12, 9, 6, and 0 ka. Precipitation is at a maximum along the southern edge of the North American ice sheet at 18-12 ka. A band of precipitation associated with the intertropical convergence zone occurs along and north of the equator on each map. Summer-monsoon precipitation is increased in both magnitude and areal extent at 12-6 ka across parts of northern Africa and southern Asia because of solar-radiation changes associated with orbital changes.

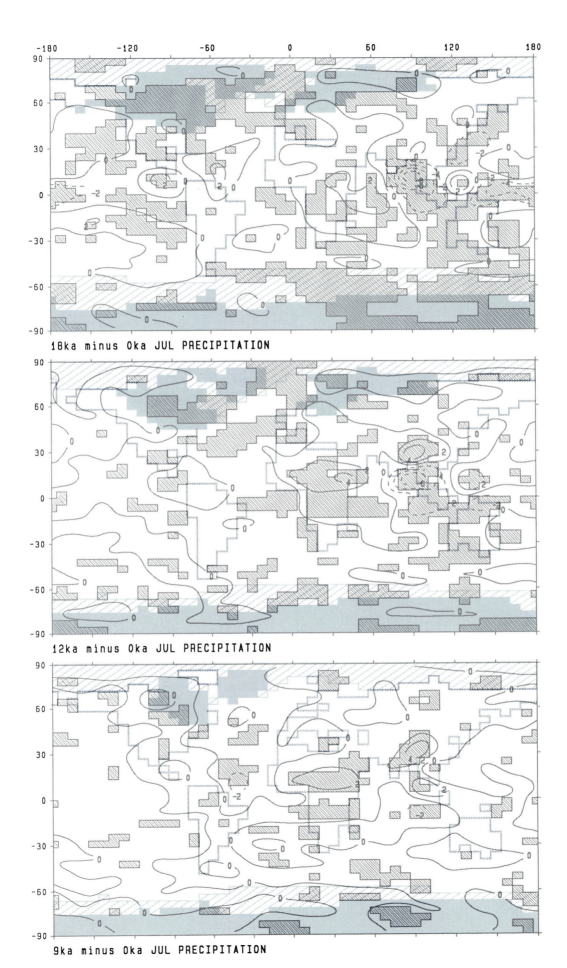

18ka minus 0ka JUL PRECIPITATION

12ka minus 0ka JUL PRECIPITATION

9ka minus 0ka JUL PRECIPITATION

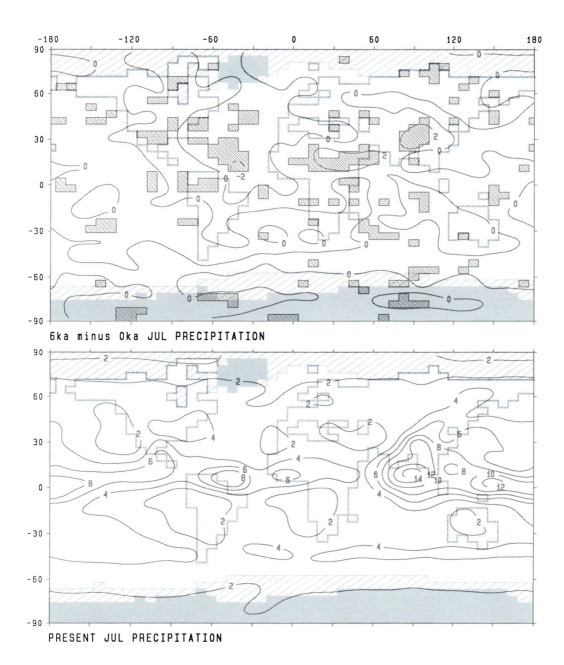

6ka minus 0ka JUL PRECIPITATION

PRESENT JUL PRECIPITATION

Fig. 4.12. Differences (mm/day) between simulated past (18, 12, 9, and 6 ka) and present (0-ka control) July precipitation. Precipitation is generally lower than at present over the northern continents and ice sheets at 18 ka. Precipitation is lower than at present in southern Asia and greater than at present in eastern Africa at 18 ka in part because of specified glacial-age sea-surface temperatures in the Indian Ocean (see COHMAP Members, 1988). Summer monsoon precipitation is greater than at present at 12-6 ka across parts of northern Africa and southern Asia and, to a much lesser extent, across parts of southern North America because of solar-radiation changes associated with orbital changes.

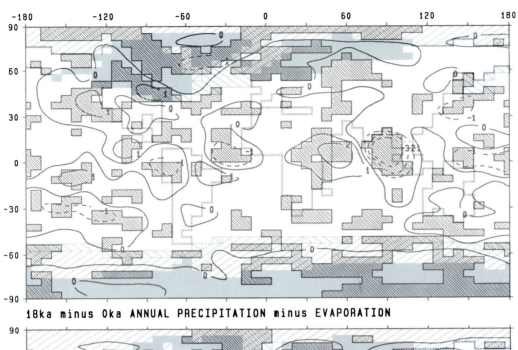

18ka minus 0ka ANNUAL PRECIPITATION minus EVAPORATION

12ka minus 0ka ANNUAL PRECIPITATION minus EVAPORATION

9ka minus 0ka ANNUAL PRECIPITATION minus EVAPORATION

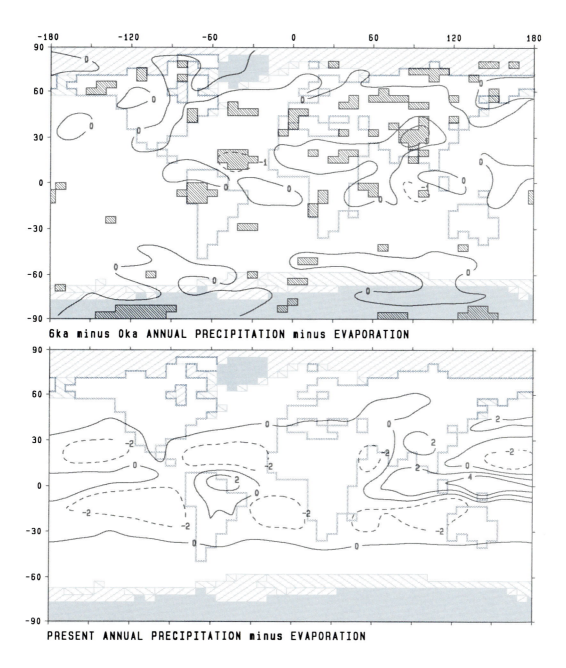

6ka minus 0ka ANNUAL PRECIPITATION minus EVAPORATION

PRESENT ANNUAL PRECIPITATION minus EVAPORATION

Fig. 4.13. Differences (mm/day) between simulated past (18, 12, 9, and 6 ka) and present (0-ka control) estimated annual average precipitation-minus-evaporation (P-E). P-E is generally greater than at present over the ice sheets (except over the northern part of the North American ice sheet) and less than at present over much of the Northern Hemisphere land south of the ice sheets at 18 ka, but the statistical significance of the drier conditions is high only in parts of Asia. The most noticeable exception to these drier conditions is over eastern Africa (see discussion in COHMAP Members, 1988). At 12 and 9 ka, P-E remains greater than at present over the North American ice sheets. From 12 to 6 ka, P-E is greater than at present over much of northern Africa and southern Asia and less than at present in the continental interior of Eurasia (and to a much lesser extent in interior North America). At these same times, P-E is less than at present in parts of the monsoon lands of tropical South America and southern Africa. The conditions of strengthened northern (summer) monsoons and weakened southern (summer) monsoons from 12 to 6 ka that dominate the annual-average P-E differences are caused by solar-radiation changes associated with orbital changes.

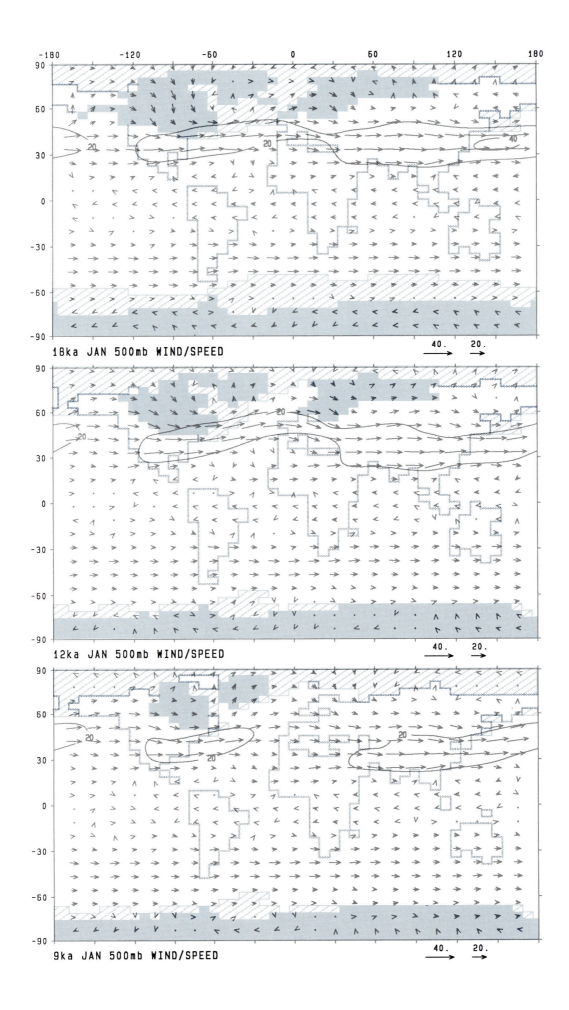

18ka JAN 500mb WIND/SPEED

12ka JAN 500mb WIND/SPEED

9ka JAN 500mb WIND/SPEED

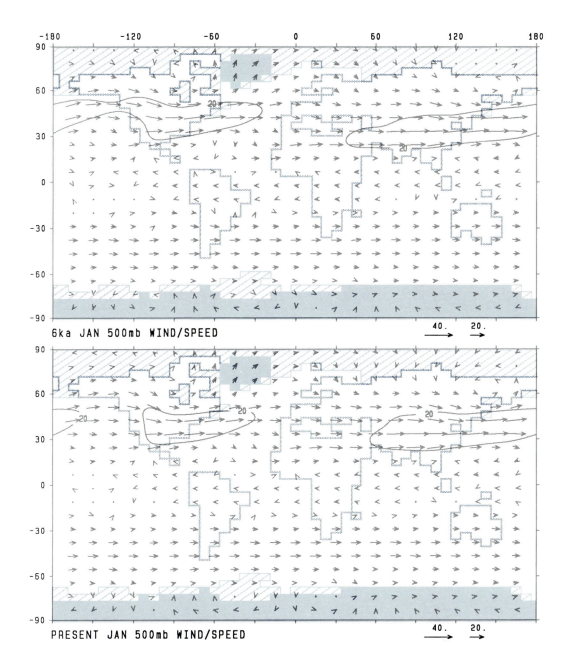

Fig. 4.14. Simulated January winds (m/s) at 500 mb (about 5.5 km) for 18, 12, 9, 6, and 0 ka. Arrows show direction and speed, and speed contours of 20 and 40 m/s are indicated. The normal northern winter jet stream is strengthened at 18 ka, with winds reaching 40 m/s along the southern edge of the North American ice sheet and North Atlantic sea-ice margin. Winds exceed 20 m/s in almost a continuous band around the Northern Hemisphere. Wind arrows show that the jet splits into two branches over North America; the weaker branch is located along the northern margins of the ice sheet (see Kutzbach and Wright, 1985). At 12 ka, with the reduced height and extent of the northern ice sheets, the winds have weakened (although they remain stronger than at present), and the split jet has largely disappeared. Strong westerlies have returned to the Gulf of Alaska and the Pacific Northwest of North America. At 9 and 6 ka the wind patterns are similar to the present.

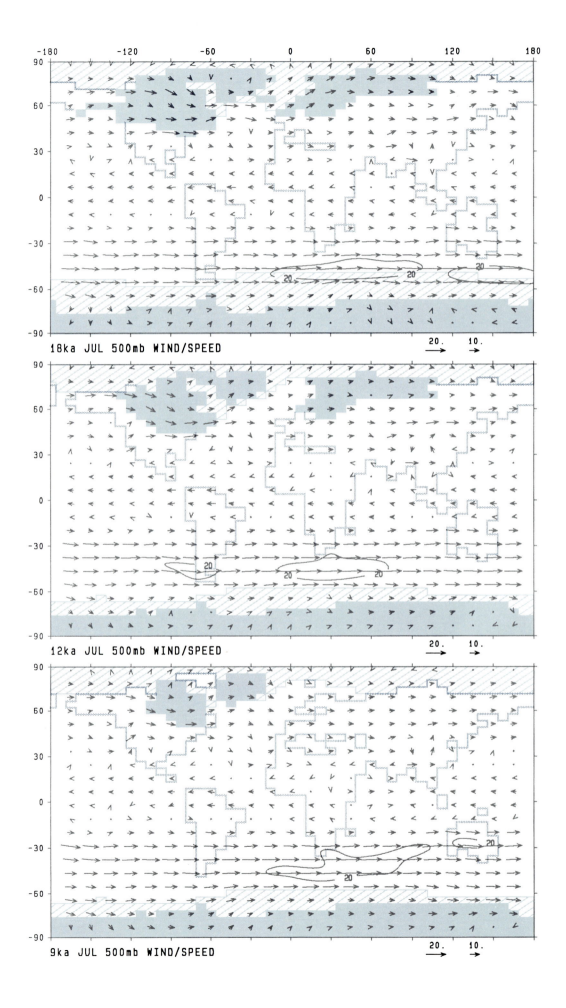

18ka JUL 500mb WIND/SPEED   20. →   10. →

12ka JUL 500mb WIND/SPEED   20. →   10. →

9ka JUL 500mb WIND/SPEED   20. →   10. →

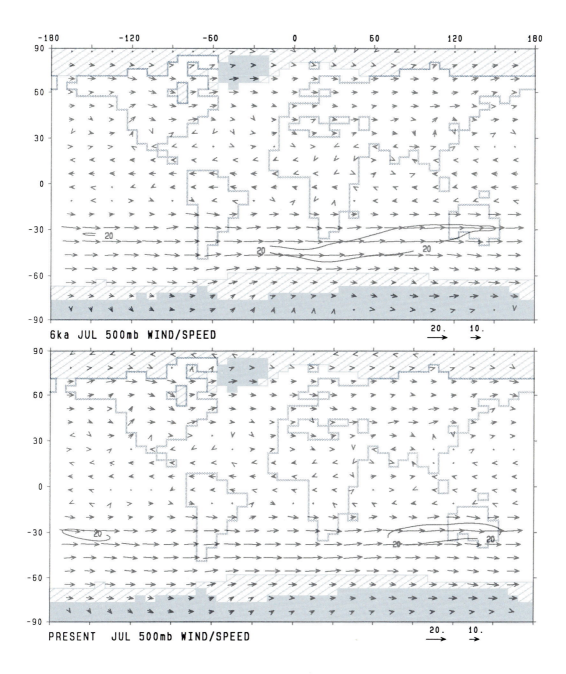

Fig. 4.15. Simulated July winds (m/s) at 500 mb (about 5.5 km) for 18, 12, 9, 6, and 0 ka. Arrows show direction and speed, and speed contours of 20 and 40 m/s are indicated. At 18 and 12 ka the core of maximum winds of the southern winter jet stream is shifted south of its present location and lies along or near the extended southern sea-ice margin. The northern westerlies are stronger at 18 and 12 ka than at present. At 9 and 6 ka the wind patterns are similar to the present.

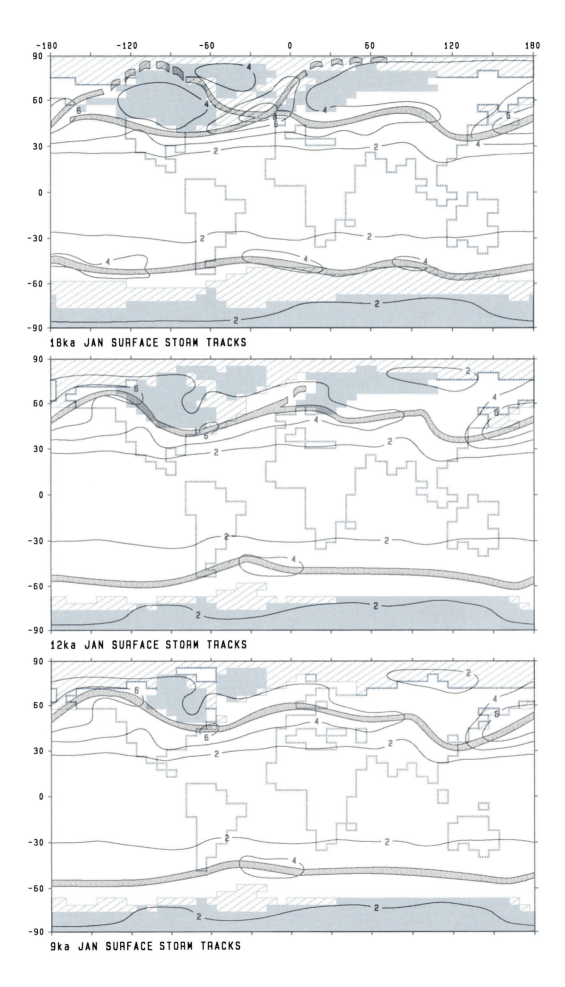

18ka JAN SURFACE STORM TRACKS

12ka JAN SURFACE STORM TRACKS

9ka JAN SURFACE STORM TRACKS

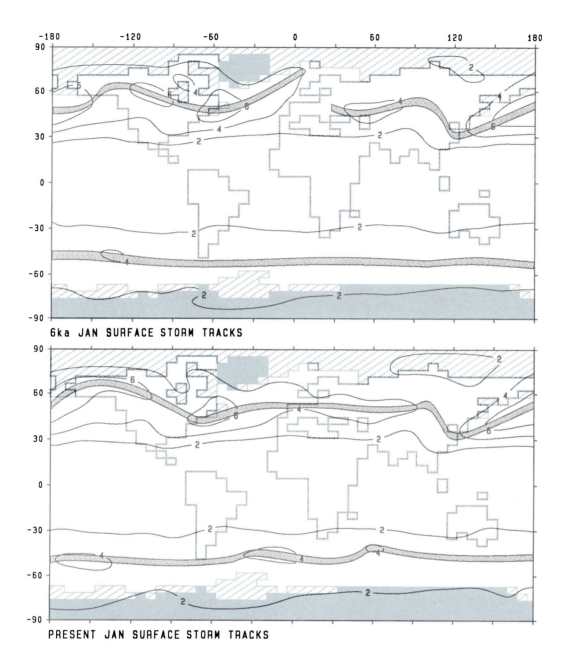

6ka JAN SURFACE STORM TRACKS

PRESENT JAN SURFACE STORM TRACKS

Fig. 4.16. Simulated January storm tracks (standard deviation of surface pressure [mb]) for 18, 12, 9, 6, and 0 ka. The standard deviation is computed from the variance of surface pressure contained between periods of 2.5 and 6 days. The band-pass filter described by Blackmon and Lau (1980) was used to filter the time series of surface pressure. At 18 ka there is a major storm track along the southern margins of the North American and Eurasian ice sheets and the North Atlantic sea ice. A secondary track follows the northern branch of the split jet around the northern margins of the ice sheets. A second major track runs along the Southern Ocean sea-ice margin. At 12 ka, and at 9 ka over North America, the northern major storm track has shifted north following the retreating ice. At 6 ka the track is similar to the present.

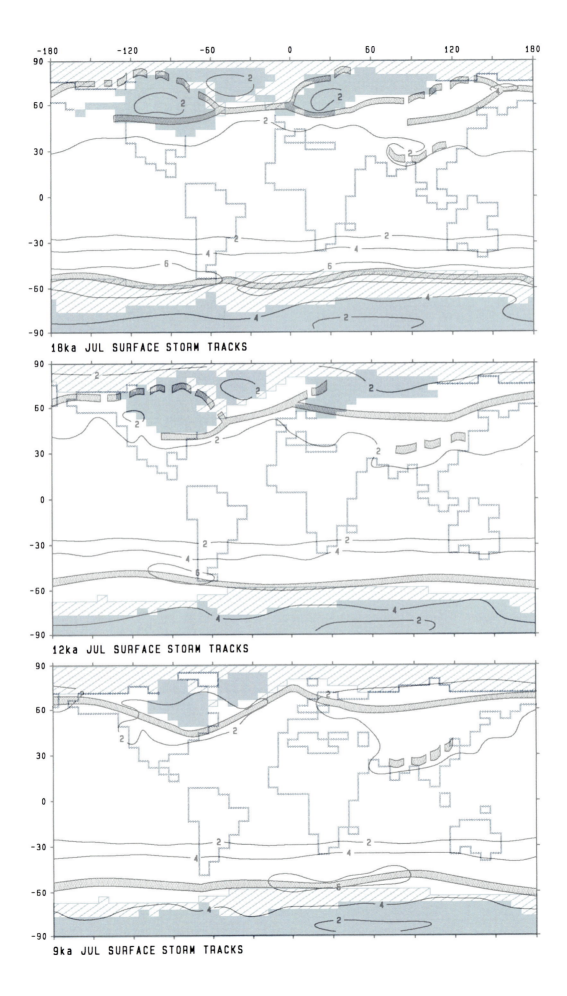

18ka JUL SURFACE STORM TRACKS

12ka JUL SURFACE STORM TRACKS

9ka JUL SURFACE STORM TRACKS

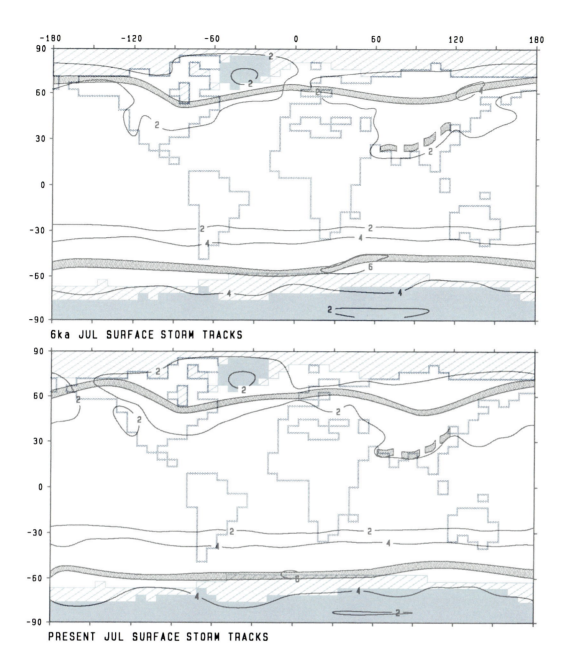

6ka JUL SURFACE STORM TRACKS

PRESENT JUL SURFACE STORM TRACKS

Fig. 4.17. Simulated July storm tracks (standard deviation of surface pressure [mb]) for 18, 12, 9, 6, and 0 ka. The standard deviation is computed from the variance of surface pressure contained between periods of 2.5 and 6 days. The band-pass filter described by Blackmon and Lau (1980) was used to filter the time series of surface pressure. At 18 and 12 ka there is a major storm track along the southern margins of the North American and Eurasian ice sheets and the North Atlantic sea ice. A secondary track follows the northern margins of the two ice sheets. A second major track runs along the sea-ice margin in the Southern Ocean. At 9 and 6 ka the storm tracks are similar to the present. The one exception is at 9 ka over North America, where increased storminess (compared to the present) persists along the southern margin of the retreating ice sheet.

precipitation (compared to the present) for 18, 12, 9, 6, and 0 ka for January (Figs. 4.18–4.22) and July (Figs. 4.23-4.27).

Finally, we present maps from several additional simulations for 18 ka that used alternative specifications of glacial-age surface boundary conditions. Figures 4.28 and 4.29 show the same information as Figures 4.18 (January, 18 ka) and 4.23 (July, 18 ka), respectively, but derived from a model with ice sheets in North America 20% lower than specified in the original model. Figure 4.30 shows the same information as Figure 4.18 but derived from a model with less North Atlantic sea ice, and Figures 4.31 and 4.32 show the same information as Figures 4.18 and 4.23, respectively, but derived from a model with a lower atmospheric concentration of CO_2 than the original model.

Several features of the maps and simulations should be kept in mind when this atlas is used. First, the continents are shown exactly as they are coded in the climate model; that is, as boxes with dimensions of 4.4° latitude by 7.5° longitude. Local coastal features and even major islands, such as New Zealand, are not represented. Second, the heights of mountains and ice sheets have been smoothed with the same spectral smoothing applied to the horizontal motions simulated by the climate model. Thus rainshadow effects and local climate modifications associated with lakes (such as the Great Lakes) are excluded from these simulations. The results should therefore be examined from a regional (subcontinental) perspective rather than from the perspective of individual model grids.

The results of the COHMAP experiments have been summarized in two additional forms. The temperature and precipitation values have been used to construct maps of the Köppen climate classification (Guetter and Kutzbach, 1990) and maps of estimated net primary productivity (Meyer, 1988).

Limitations in the Model and in the Experimental Design

The simulation of past climates is an evolutionary process in the sense that models are constantly being improved and experimental designs are constantly being revised. The set of COHMAP experiments described in this book used a version of the NCAR CCM with relatively coarse spatial resolution, perpetual January and July insolation, and prescribed (noninteractive) soil moisture, snow cover, SST, and sea ice. Although these limitations of resolution and parameterizations are very significant, this first version of the NCAR CCM simulates some general features of the present climate rather accurately (Pitcher et al., 1983; Ramanathan et al., 1983). The patterns of monsoon circulations and middle-latitude winds are approximately correct, as are the patterns of warm or cold and wet or dry climates. However, when we examine the model results in greater detail, we see that the simulated temperatures are higher than observed temperatures in summer and lower than observed temperatures in winter (Webb et al., 1987) and that simulated precipitation is greater than observed precipitation in many regions.

Given that the simulation of the present-day climate is not exact when examined in detail, we should also expect inaccuracies in the simulated response of climate variables to changes in insolation or surface boundary conditions. From this perspective, we might ascribe differences between simulated and observed paleoclimates to inaccurate or incomplete models, but they might also arise from incorrect paleoclimatic observations, inappropriate experimental design (discussed below), or combinations of all of these sources.

Models are continually being improved. For example, models with higher resolution, capable of full seasonal-cycle simulations, and with interactive parameterizations for soil moisture, snow cover, sea-surface (ocean mixed-layer) temperature, and sea ice are now available (see Kutzbach, Bartlein, et al., this vol.). We have taken initial steps to repeat the original COHMAP experiments with improved models.

We repeated the simulation of the climate of 9 ka with a low-resolution atmospheric model with interactive soil moisture coupled to a 50-m mixed-layer ocean model with interactive sea ice (Kutzbach and Gallimore, 1988; Gallimore and Kutzbach, 1989). This coupled model simulates the full seasonal cycle. The simulation with the improved model gave the same pattern of climatic change as the original COHMAP simulation with the NCAR CCM, but the simulated climatic changes over land were somewhat greater than those shown here because of soil-moisture feedbacks. We also repeated the simulation of the climate of 6 ka with a new version of the NCAR CCM that incorporates interactive snow cover and soil moisture, simulates the full seasonal cycle, and is coupled to a 50-m mixed-layer ocean model with interactive sea ice. These experiments with coupled atmosphere-ocean models confirmed the main points of the earlier experiments but also added new detail. The results showed that orbitally induced changes in insolation produced only small changes in ocean temperature, thereby indicating that the use of modern SSTs at 9, 6, and 3 ka was not grossly in error. On the other hand, Arctic sea-ice cover was reduced significantly.

Many other models differing somewhat from the NCAR CCM in spatial resolution and physical parameterizations have been used to simulate past climates. For example, Mitchell *et al.* (1988) and Street-Perrott *et al.* (1990) simulated the climate of 9 ka, Royer *et al.* (1984) simulated the climate of the previous interglacial, and Williams *et al.* (1974), Gates (1976a,b), Manabe and Hahn (1977), Hansen *et al.* (1984), Manabe and Broccoli (1985a,b), Broccoli and Manabe (1987), Rind (1987), Joussaume *et al.* (1989), and Lautenschlager and Herterich (1989) simulated glacial conditions. However, comparing the results of these many experiments has proved difficult because the different models specify somewhat different boundary conditions. S. Joussaume (personal communication, 1990) summarized many of the simulations of glacial-maximum climate and showed that they differ in important details, such as the degree of splitting of the jet stream around the north and south slopes of the North American ice sheet, the location and intensity of the North Atlantic jet stream, and the associated temperature and precipitation patterns. In the future it will be important for different modeling groups to coordinate experimentation. If different models use identical paleoclimatic boundary conditions, then differences in the paleoclimatic simulations can be attributed to differences in model parameterizations. Comparisons of the model results with paleoclimatic observations can then lead to assessments of the relative accuracy of climate models.

None of the above-mentioned models employs ocean dynamics, fully interactive vegetation dynamics, or interactive biogeochemical cycles. Because these processes may also be important factors in climatic change, we can anticipate that paleoclimatic simulations made with these more detailed models will differ from simulations made with the models available now.

Limitations in the accuracy of the boundary conditions are a final important consideration in interpreting the COHMAP results. For example, all of the COHMAP experiments shown in this atlas use the CLIMAP (1981) "maximum" ice sheet, but there is increasing evidence that the "minimum" ice sheet is a more accurate representation of the height of the 18-ka ice sheet. Kutzbach and Ruddiman (this vol.) describe the results of a sensitivity test that reduced North American ice-sheet heights by 20% compared to the standard experiment. The January 18-ka simulations with the reduced (Fig. 4.28) and "maximum" (Fig. 4.18) North American ice sheet show significantly different patterns of winds, sea-level pressure, and surface temperature. For example, the experiment with the smaller ice sheet simulates a smaller glacial

anticyclone, colder conditions west of the ice sheet (Alaska), and warmer conditions south of the ice sheet than the experiment with the large ice sheet. Shinn and Barron (1989) also explored the climate's sensitivity to ice-sheet height and configuration.

Another uncertainty concerns the amount of sea ice in the North Atlantic at the glacial maximum. Kutzbach and Ruddiman (this vol.) conducted a sensitivity test with less extensive North Atlantic sea ice. The experiment with reduced sea-ice cover (Fig. 4.30) simulates less severe glacial-age cooling downstream from the North Atlantic over western Europe than the standard experiment (Fig. 4.18).

The model does not yet include the possible effects of glacial-age aerosols (see Kutzbach and Ruddiman, this vol.), nor does the prescribed atmospheric carbon dioxide concentration systematically change to match the observed concentration through the 18,000-yr period. A sensitivity experiment with lowered CO_2 concentration for 18 ka (Figs. 4.31 and 4.32) simulated slightly lower surface temperatures and precipitation rates over land than in the standard experiment (Fig. 4.18) (see Kutzbach and Ruddiman [this vol.] and Kutzbach and Guetter [1986]). In general, the use of modern CO_2 levels produces a small bias toward warmer conditions in the paleoclimatic simulations. However, because SSTs and sea ice are specified at glacial-age values (or fractions thereof) from 18 to 12 ka, this bias should be generally limited to land areas beyond the influence of temperature advection from oceanic regions. If the model had included an interactive ocean, then it would have been essential to use the correct carbon dioxide concentration.

Possible errors in the CLIMAP-based estimates of SST at 18, 15, and 12 ka, or in our use of these estimates (see Kutzbach and Ruddiman, this vol.), could also influence the results. At 9, 6, and 3 ka, the model prescribes SSTs at modern values; recent experiments (mentioned above) confirm that this is a reasonable first approximation.

A final limitation of these experiments is the intentional focus on simulating the low-frequency, long-period "envelope" of climatic change that could be produced by slowly changed orbital conditions and surface boundary conditions. The experiments were not designed to study short-term perturbations of climate such as occurred around 11-10 ka (the Younger Dryas). Other investigators have designed experiments to test whether these abrupt changes could have been associated with the presence of glacial meltwater in the North Atlantic (Rind *et al.,* 1986; Schneider *et al.,* 1987; Overpeck *et al.,* 1989) or the Gulf of Mexico (Oglesby *et al.,* 1989). Our goal has been to help establish the framework of general

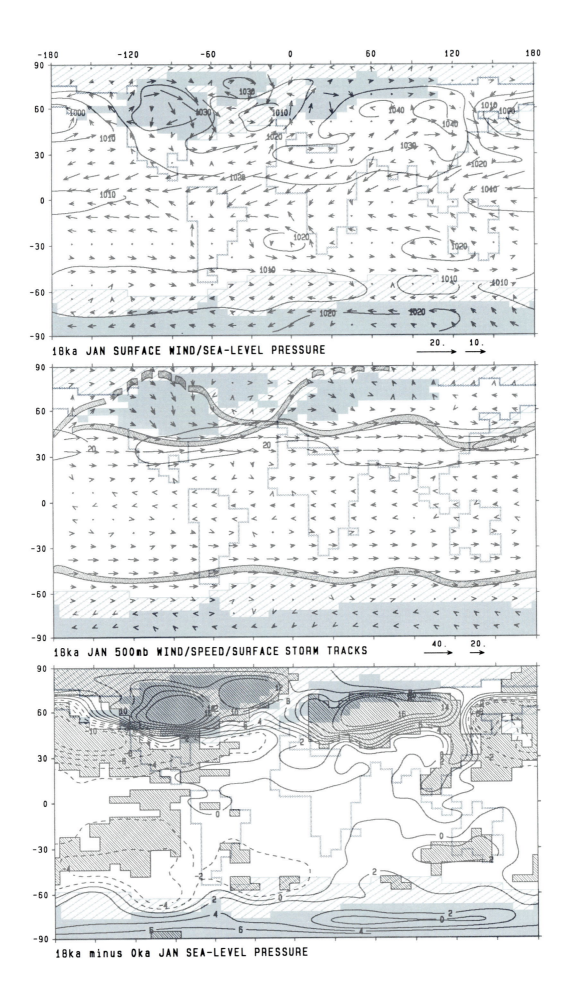

18ka JAN SURFACE WIND/SEA-LEVEL PRESSURE

18ka JAN 500mb WIND/SPEED/SURFACE STORM TRACKS

18ka minus 0ka JAN SEA-LEVEL PRESSURE

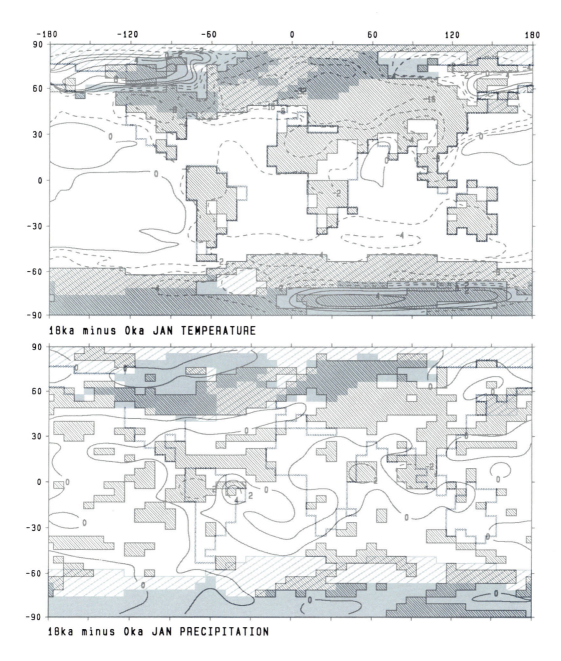

18ka minus 0ka JAN TEMPERATURE

18ka minus 0ka JAN PRECIPITATION

Fig. 4.18. Simulated climate variables for January, 18 ka: sea-level pressure and surface winds; 500-mb (about 5.5-km) winds, wind speed contours, and surface storm tracks; and differences between simulated and present (0-ka control) sea-level pressure, surface temperature, and precipitation. Large glacial anticyclones over the North American and Greenland ice sheets and over and along the southern margin of the Eurasian ice sheet are accompanied by generally cold, dry conditions. Notable exceptions are the higher-than-present temperatures over Alaska (associated with increased southerly flow) and the wetter conditions in parts of southwestern North America and along the North American/North Atlantic ice-sheet/sea-ice border (associated with the southward-shifted storm track and jet stream—see the region of lowered sea-level pressure). The 500-mb flow pattern shows the jet stream intensified (compared to the present) and split into two branches around the northern and southern margins of the northern ice sheets. In high southern latitudes, temperatures are lower than at present over the extended sea ice and higher than at present over parts of the Antarctic continent.

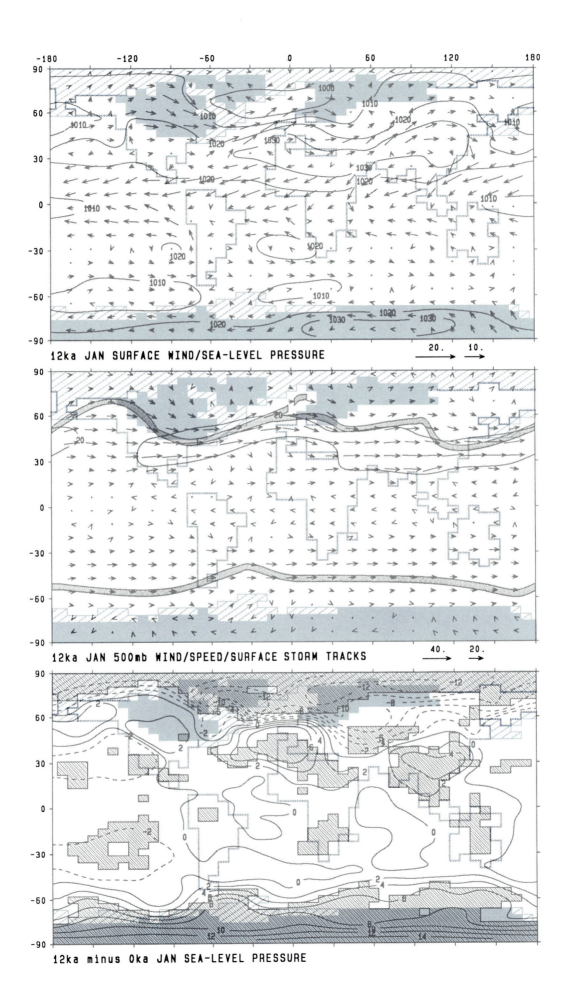

12ka JAN SURFACE WIND/SEA-LEVEL PRESSURE   20. →   10. →

12ka JAN 500mb WIND/SPEED/SURFACE STORM TRACKS   40. →   20. →

12ka minus 0ka JAN SEA-LEVEL PRESSURE

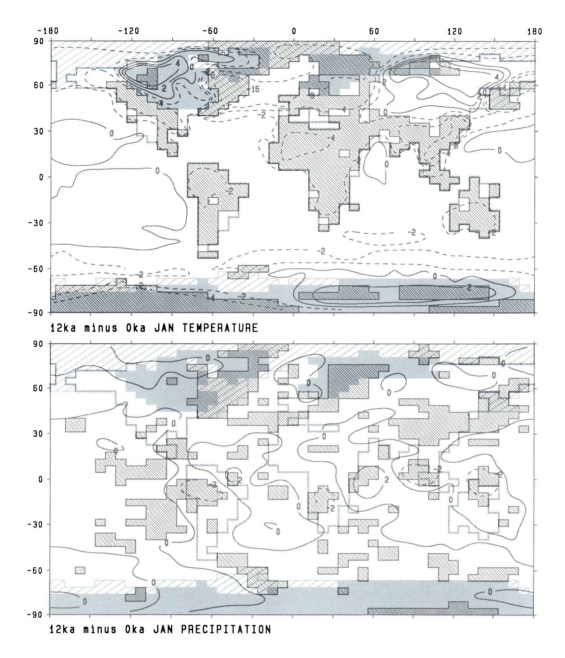

Fig. 4.19. Simulated climate variables for January, 12 ka: sea-level pressure and surface winds; 500-mb (about 5.5-km) winds, wind speed contours, and surface storm tracks; and differences between simulated and present (0-ka control) sea-level pressure, surface temperature, and precipitation. The glacial anticyclones are much weaker than at 18 ka (Fig. 4.18), and the reduced sea-ice cover in the North Atlantic has produced stronger middle-latitude westerlies and a considerable warming in the eastern North Atlantic and across northern Eurasia compared to 18 ka. However, most northern lands remain colder and generally drier than at present. With the lowering of the ice sheets, the split jet flow has largely disappeared, and westerlies have returned to the Gulf of Alaska and the Pacific Northwest of North America. In tropical lands, temperatures are lower and sea-level pressure is higher than at present. In the southern tropics of South America and southern Africa, summer monsoons are weaker and precipitation is lower than at present; these tropical effects are caused by solar-radiation changes associated with orbital changes.

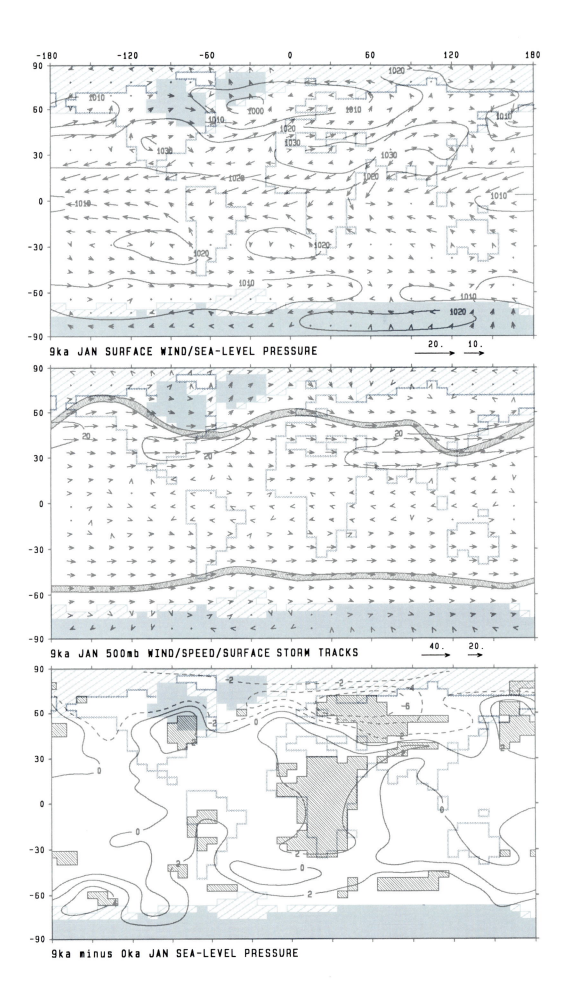

9ka JAN SURFACE WIND/SEA-LEVEL PRESSURE   20. 10.

9ka JAN 500mb WIND/SPEED/SURFACE STORM TRACKS   40. 20.

9ka minus 0ka JAN SEA-LEVEL PRESSURE

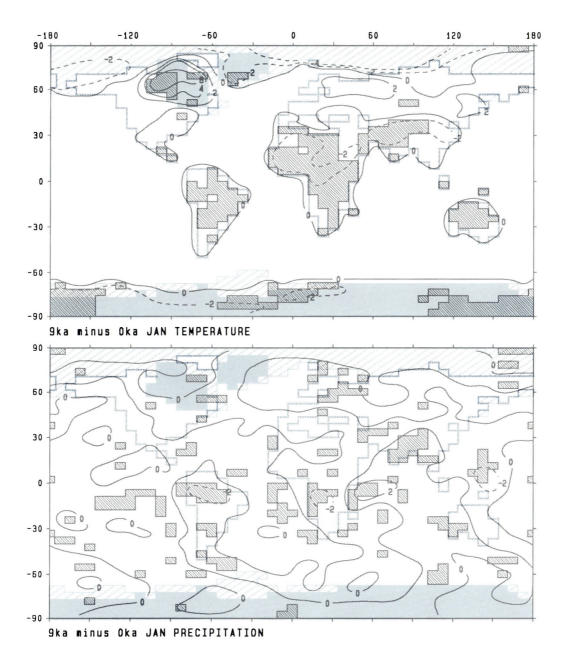

Fig. 4.20. Simulated climate variables for January, 9 ka: sea-level pressure and surface winds; 500-mb (about 5.5-km) winds, wind speed contours, and surface storm tracks; and differences between simulated and present (0-ka control) sea-level pressure, surface temperature, and precipitation. The small residual North American ice sheet only slightly intensifies the wintertime high-pressure center over North America, compared to the present (see sea-level pressure difference). In the northern and southern tropics, temperatures are slightly lower and sea-level pressure slightly higher than at present because of solar-radiation changes associated with orbital changes. In northern midlatitudes, the increased north-south pressure gradient along the northern margin of the region of increased sea-level pressure helps to strengthen the westerlies and bring warmer conditions to northern Eurasia. In the southern tropics, the weaker summer monsoons of South America and southern Africa are associated with reduced precipitation.

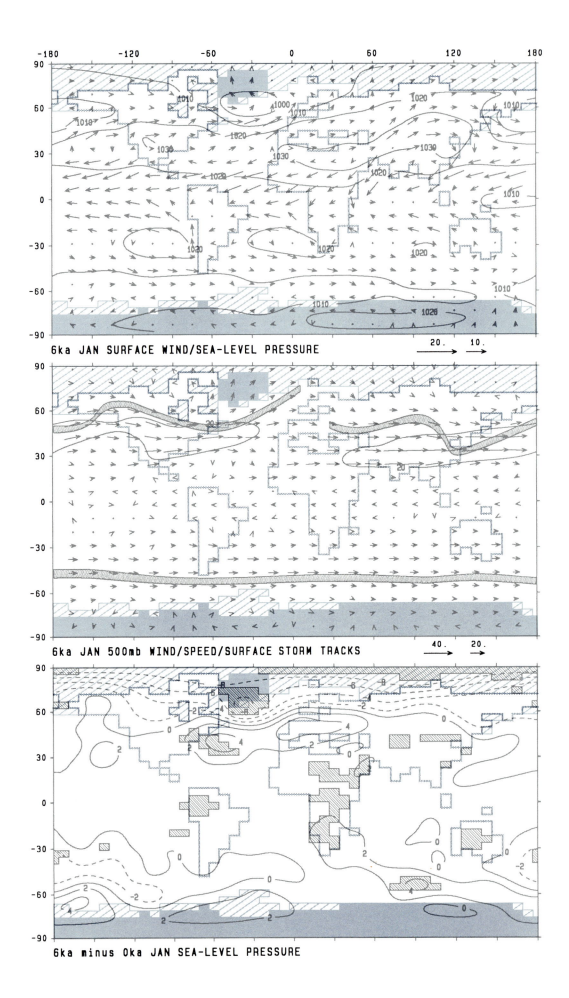

6ka JAN SURFACE WIND/SEA-LEVEL PRESSURE   20. → 10. →

6ka JAN 500mb WIND/SPEED/SURFACE STORM TRACKS   40. → 20. →

6ka minus 0ka JAN SEA-LEVEL PRESSURE

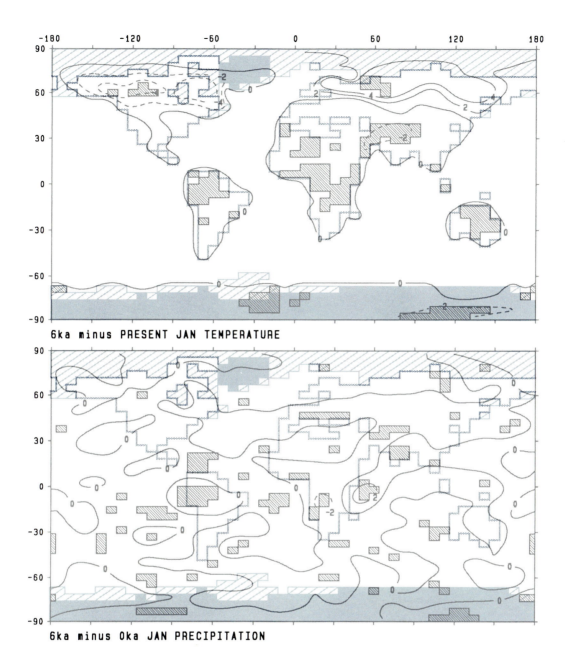

6ka minus PRESENT JAN TEMPERATURE

6ka minus 0ka JAN PRECIPITATION

Fig. 4.21. Simulated climate variables for January, 6 ka: sea-level pressure and surface winds; 500-mb (about 5.5-km) winds, wind speed contours, and surface storm tracks; and differences between simulated and present (0-ka control) sea-level pressure, surface temperature, and precipitation. In the northern and southern tropics and parts of the northern midlatitudes, temperatures are slightly lower and sea-level pressure slightly higher than at present because of solar-radiation changes associated with orbital changes. In the northern midlatitudes, the higher sea-level pressure helps to strengthen the westerlies and bring warmer conditions to northern Eurasia. In the southern tropics, the weaker summer monsoons of South America and southern Africa are associated with reduced precipitation.

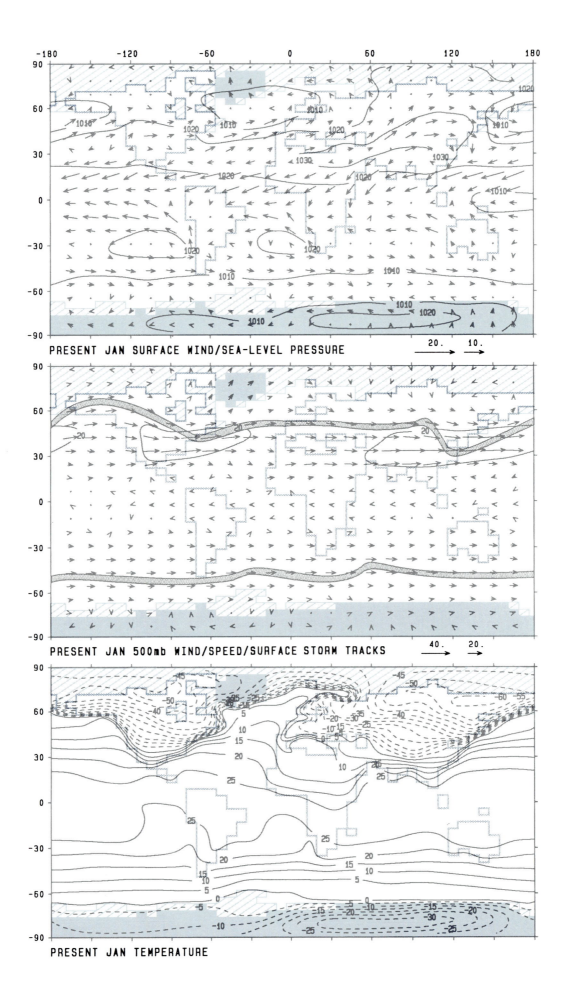

PRESENT JAN SURFACE WIND/SEA-LEVEL PRESSURE

PRESENT JAN 500mb WIND/SPEED/SURFACE STORM TRACKS

PRESENT JAN TEMPERATURE

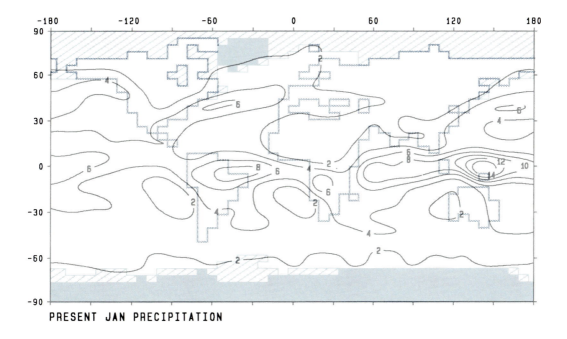

PRESENT JAN PRECIPITATION

Fig. 4.22. Simulated modern (0-ka control) climate variables for January: sea-level pressure and surface winds; 500-mb (about 5.5-km) winds, wind speed contours, and surface storm tracks; surface
temperature; and precipitation. The northern oceanic lows (Aleutian and Icelandic), continental
highs (Siberian and North American), and southern westerlies are simulated in generally realistic
fashion. The center of the Siberian anticyclone is located south of the observed position. Northern
continental interiors are 5–10°C colder than observed. Precipitation maxima occur over southern
tropical lands, along the oceanic intertropical convergence zone, and along the east coasts of North
America and Asia.

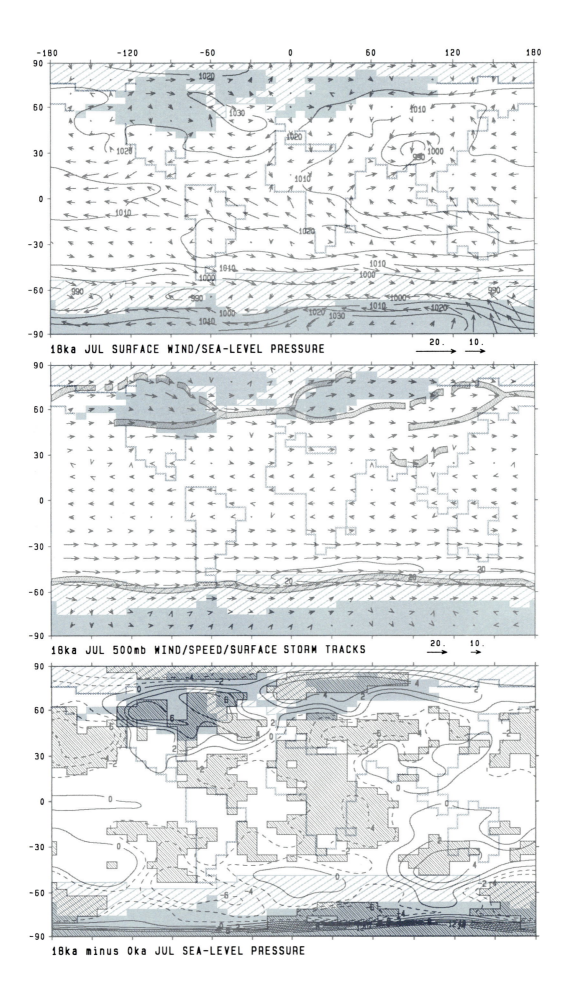

18ka JUL SURFACE WIND/SEA-LEVEL PRESSURE   20. → 10. →

18ka JUL 500mb WIND/SPEED/SURFACE STORM TRACKS   20. → 10. →

18ka minus 0ka JUL SEA-LEVEL PRESSURE

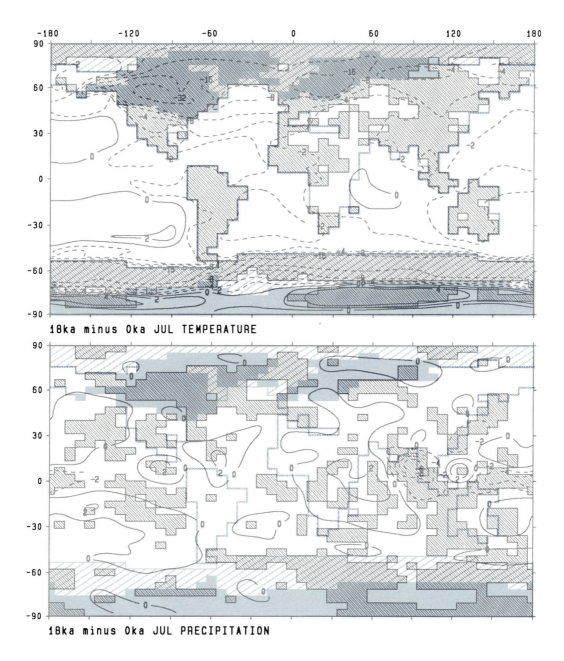

Fig. 4.23. Simulated climate variables for July, 18 ka: sea-level pressure and surface winds; 500-mb (about 5.5-km) winds, wind speed contours, and surface storm tracks; and differences between simulated and present (0-ka control) sea-level pressure, surface temperature, and precipitation. A very large anticyclone spans the North American, Greenland, and Eurasian ice sheets and the sea ice-covered North Atlantic. Temperatures in these regions and over the continents south of the ice sheets are much lower than at present. The 500-mb winds are stronger and are shifted south of their modern positions. Temperatures are also much lower than at present over the extended Southern Hemisphere sea ice. Precipitation is generally lower than at present over the ice sheets and the land south of the ice sheets. Precipitation is greater than at present in parts of eastern Africa. Higher precipitation in the eastern Mediterranean and the Near East, while not statistically significant, is associated with the southward-displaced storm track over the Eurasian continent, as reflected in the sea-level pressure differences. Lower precipitation over southern Asia and higher precipitation over eastern Africa are associated with the specified glacial-age sea-surface temperature pattern in the Indian Ocean (see COHMAP Members, 1988).

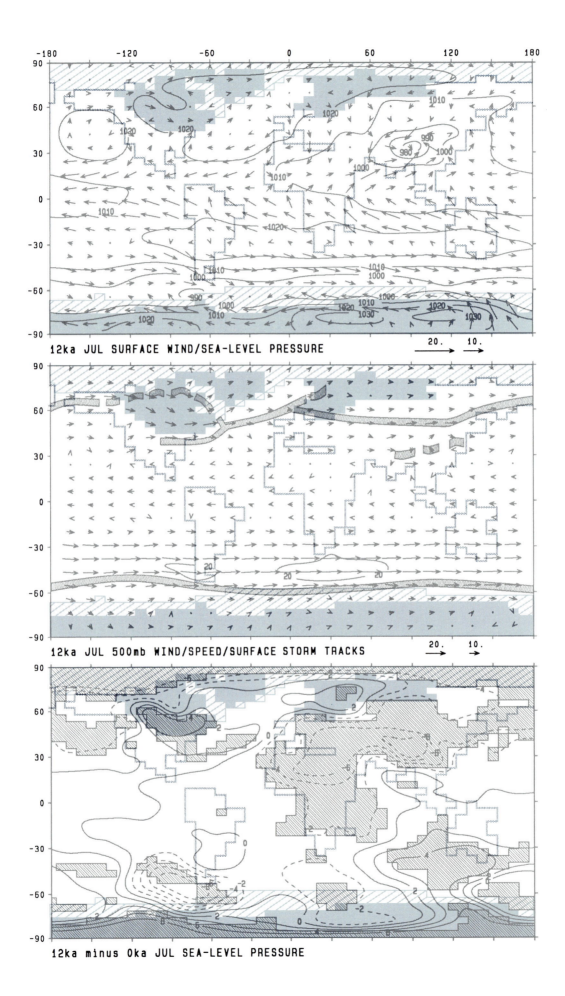

12ka JUL SURFACE WIND/SEA-LEVEL PRESSURE   20. ⟶   10. ⟶

12ka JUL 500mb WIND/SPEED/SURFACE STORM TRACKS   20. ⟶   10. →

12ka minus 0ka JUL SEA-LEVEL PRESSURE

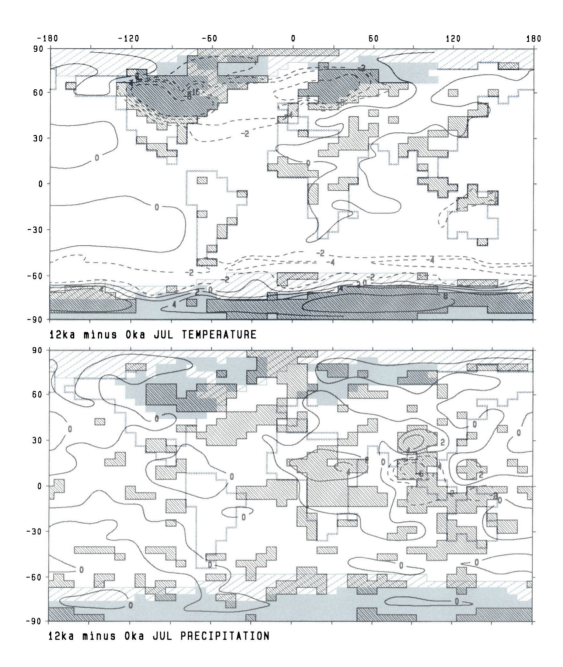

Fig. 4.24. Simulated climate variables for July, 12 ka: sea-level pressure and surface winds; 500-mb (about 5.5-km) winds, wind speed contours, and surface storm tracks; and differences between simulated and present (0-ka control) sea-level pressure, surface temperature, and precipitation. A small glacial anticyclone is associated with the North American and European ice sheets, where temperatures remain much colder than at present. The 500-mb westerlies are weaker than at 18 ka but still stronger than at present. Away from the ice sheet, the effects of solar-radiation changes associated with orbital changes are apparent. Surface temperatures are higher than at present, and sea-level pressure is lower than at present across northern Africa and Asia. This lower pressure is associated with a strengthened African-Asian summer monsoon and greater precipitation in parts of northern Africa and southern Asia.

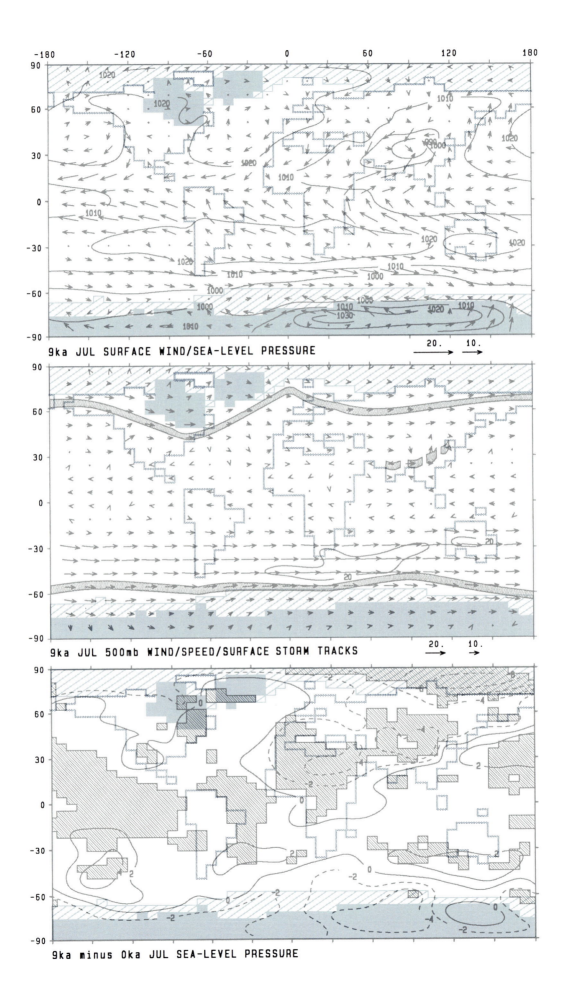

9ka JUL SURFACE WIND/SEA-LEVEL PRESSURE

20. → 10. →

9ka JUL 500mb WIND/SPEED/SURFACE STORM TRACKS

20. → 10. →

9ka minus 0ka JUL SEA-LEVEL PRESSURE

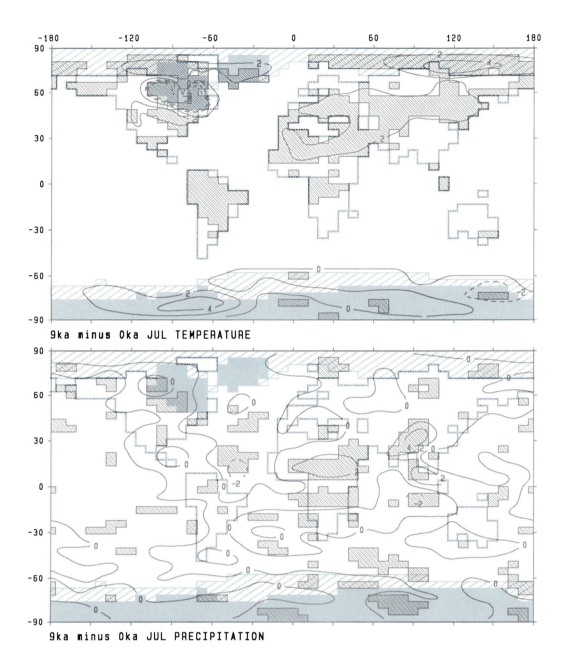

Fig. 4.25. Simulated climate variables for July, 9 ka: sea-level pressure and surface winds; 500-mb (about 5.5-km) winds, wind speed contours, and surface storm tracks; and differences between simulated and present (0-ka control) sea-level pressure, surface temperature, and precipitation. The small North American ice sheet is associated with lower temperatures, but there is no longer a large area of higher sea-level pressure. Elsewhere, the effects of solar-radiation changes associated with orbital changes are apparent. Surface temperatures are higher than at present in North America away from the ice sheet and over all other continents. Sea-level pressure is lower than at present in western North America and across northern Africa and southern Asia in association with strengthened summer-monsoon circulations. Precipitation is higher than at present in North America near the ice-sheet margin and in northern Africa and southern Asia, associated with the strengthened summer monsoon. The oceanic subtropical highs are stronger than at present.

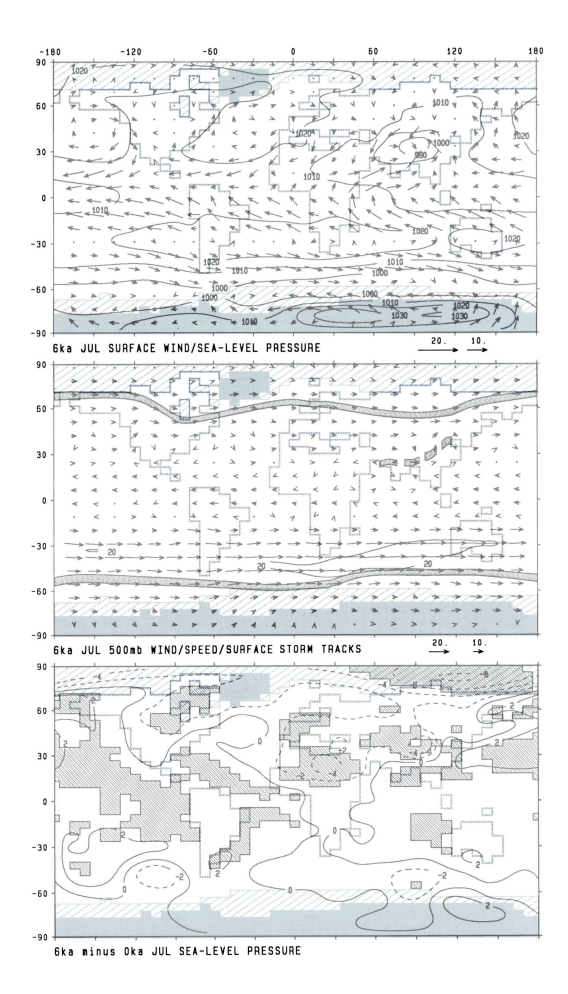

6ka JUL SURFACE WIND/SEA-LEVEL PRESSURE   20. → 10. →

6ka JUL 500mb WIND/SPEED/SURFACE STORM TRACKS   20. → 10. →

6ka minus 0ka JUL SEA-LEVEL PRESSURE

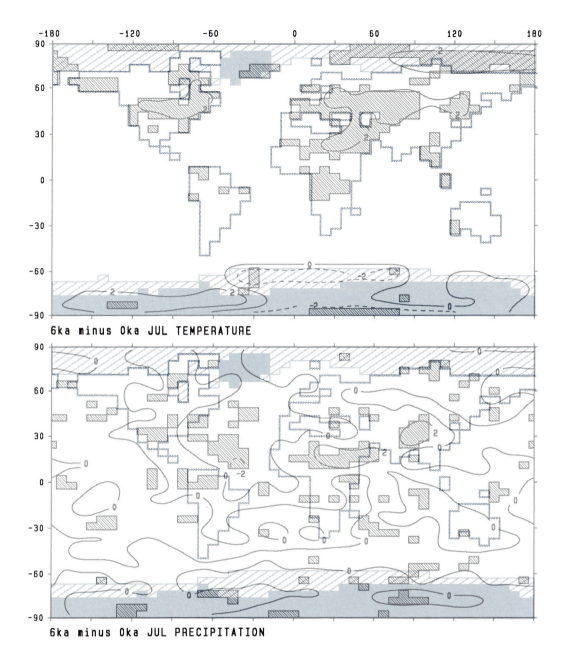

Fig. 4.26. Simulated climate variables for July, 6 ka: sea-level pressure and surface winds; 500-mb (about 5.5-km) winds, wind speed contours, and surface storm tracks; and differences between simulated and present (0-ka control) sea-level pressure, surface temperature, and precipitation. With the North American ice sheet absent, the effects of solar-radiation changes associated with orbital changes are apparent in North America in terms of raised temperature, lowered sea-level pressure, and slightly increased precipitation. On the larger North African-Eurasian landmass, the region of increased temperature and lowered sea-level pressure spans the continent. Associated with the stronger summer monsoon, precipitation is higher than at present across North Africa and southern Asia and parts of southern North America. The oceanic subtropical highs are stronger than at present.

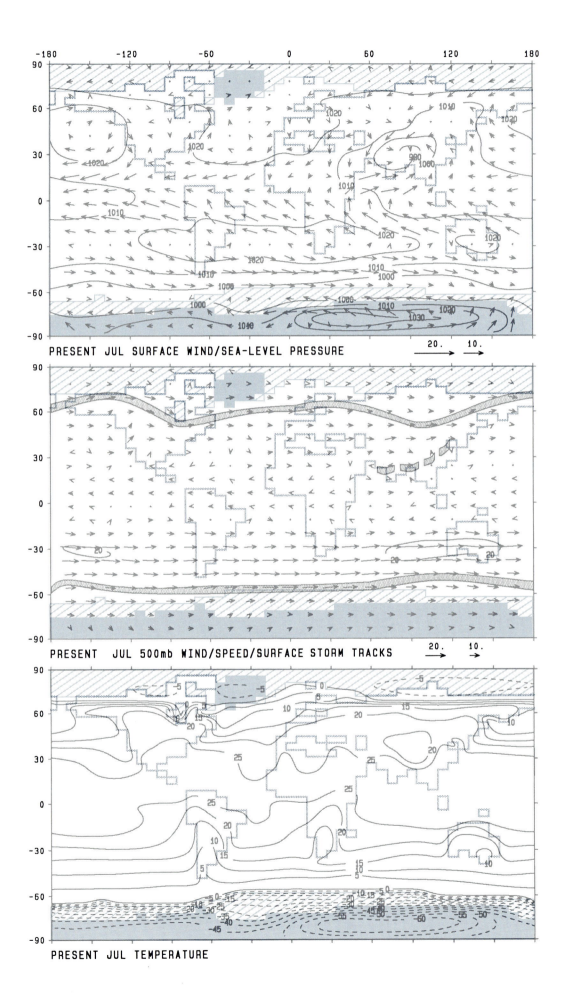

PRESENT JUL SURFACE WIND/SEA-LEVEL PRESSURE   20. 10.

PRESENT   JUL 500mb WIND/SPEED/SURFACE STORM TRACKS   20. 10.

PRESENT JUL TEMPERATURE

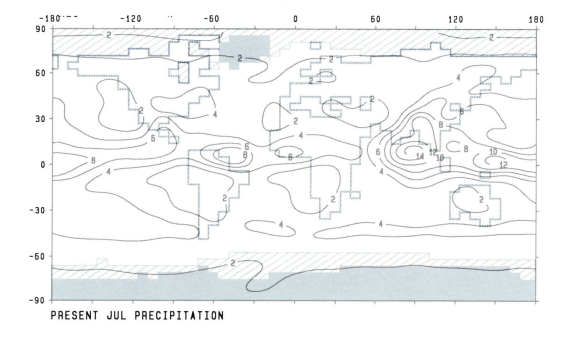

PRESENT JUL PRECIPITATION

Fig. 4.27. Simulated modern (0-ka control) climate variables for July: sea-level pressure and surface winds; 500-mb (about 5.5-km) winds, wind speed contours, and surface storm tracks; surface temperature; and precipitation. The northern African-Asian monsoon circulation, the northern oceanic subtropical highs, and the southern westerlies are simulated. Temperatures are lowest in the polar regions and over the southern parts of the southern continents. Precipitation maxima occur in Central America, northern Africa, southern Asia, and along the oceanic intertropical convergence zone.

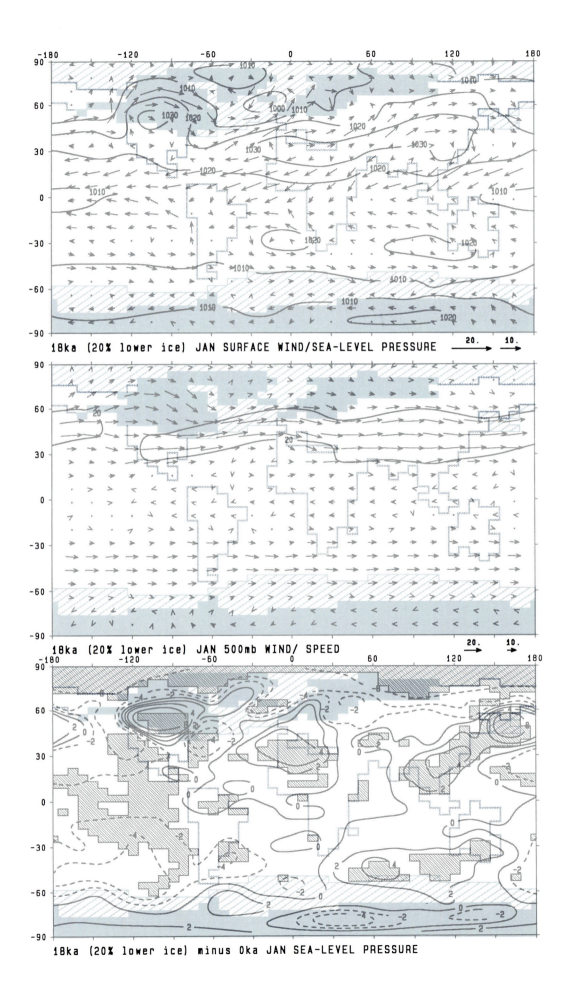

18ka (20% lower ice) JAN SURFACE WIND/SEA-LEVEL PRESSURE   20. → 10. →

18ka (20% lower ice) JAN 500mb WIND/ SPEED   20. → 10. →

18ka (20% lower ice) minus 0ka JAN SEA-LEVEL PRESSURE

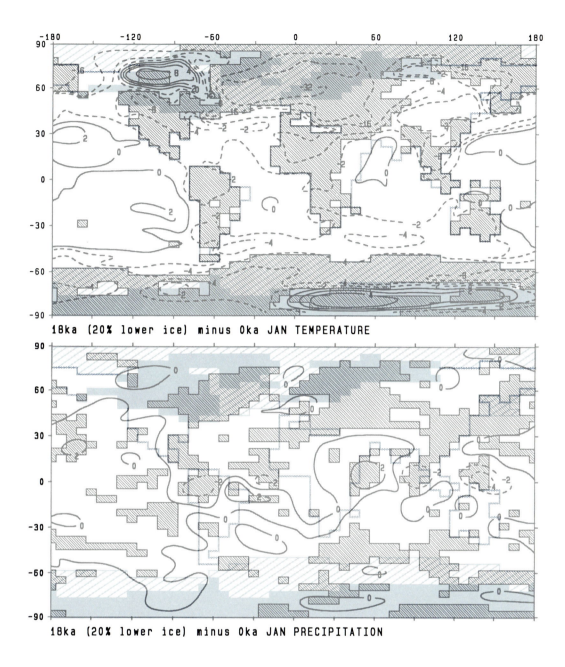

Fig. 4.28. Simulated climate variables for January, 18 ka, with reduced height of ice sheets in North America: sea-level pressure and surface winds; 500-mb (about 5.5-km) winds and wind speed contours; and differences between simulated and present (0-ka control) sea-level pressure, surface temperature, and precipitation. In this simulation, ice sheets over North America are 20% lower than the "maximum" heights used in the main COHMAP experiment (Fig. 4.18). Compared to the simulation with the maximum ice sheet (Fig. 4.18), this simulation has a smaller glacial anticyclone, less extreme splitting of the 500-mb flow pattern into two branches, colder conditions in Alaska, and less cold conditions south of the ice sheet in the central United States.

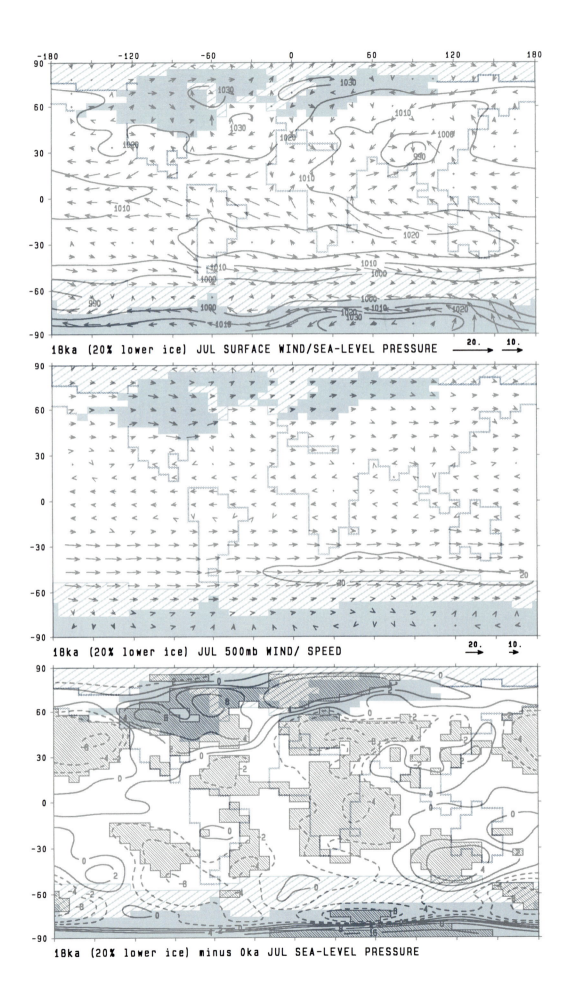

18ka (20% lower ice) JUL SURFACE WIND/SEA-LEVEL PRESSURE   20. →   10. →

18ka (20% lower ice) JUL 500mb WIND/ SPEED   20. →   10. →

18ka (20% lower ice) minus 0ka JUL SEA-LEVEL PRESSURE

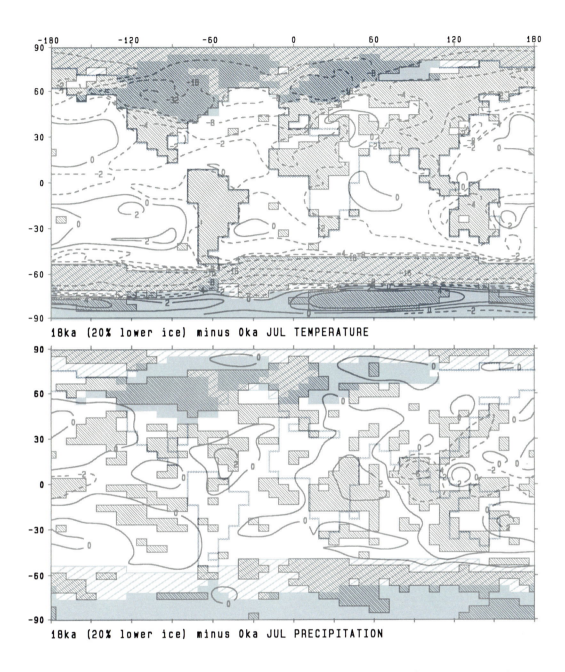

18ka (20% lower ice) minus 0ka JUL TEMPERATURE

18ka (20% lower ice) minus 0ka JUL PRECIPITATION

Fig. 4.29. Simulated climate variables for July, 18 ka, with reduced height of ice sheets in North America: sea-level pressure and surface winds; 500-mb (about 5.5-km) winds and wind speed contours; and differences between simulated and present (0-ka control) sea-level pressure, surface temperature, and precipitation. In this simulation, ice sheets over North America are 20% lower than the "maximum" heights used in the main COHMAP experiment (Fig. 4.23). Compared to the simulation with the maximum ice sheet (Fig. 4.23), this simulation has somewhat less extreme cold in the vicinity of the ice sheets. However, the climate simulation is less sensitive to ice-sheet height in July (compare Figs. 4.29 and 4.23) than in January (compare Figs. 4.28 and 4.18).

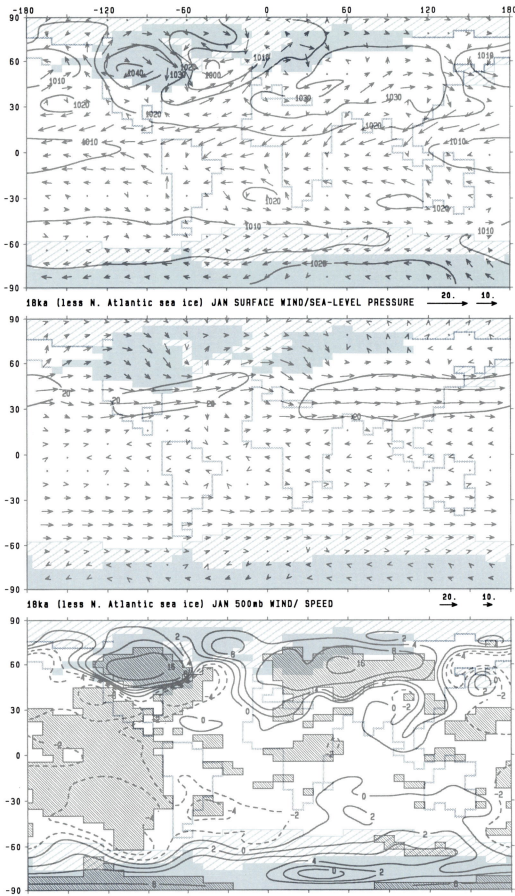

18ka (less N. Atlantic sea ice) JAN SURFACE WIND/SEA-LEVEL PRESSURE   20. → 10. →

18ka (less N. Atlantic sea ice) JAN 500mb WIND/ SPEED   20. → 10. →

18ka (less N. Atlantic sea ice) minus 0ka JAN SEA-LEVEL PRESSURE

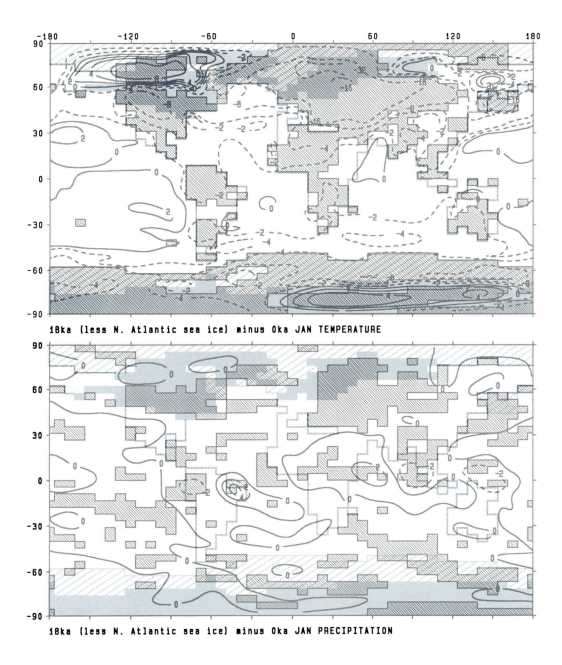

Fig. 4.30. Simulated climate variables for January, 18 ka, with less North Atlantic sea ice: sea-level pressure and surface winds; 500-mb (about 5.5-km) winds and wind speed contours; and differences between simulated and present (0-ka control) sea-level pressure, surface temperature, and precipitation. In this simulation, the southern limit of North Atlantic sea ice is approximately 15° latitude farther north than in the main COHMAP experiment (Fig. 4.18). Compared to the simulation with more extensive sea ice (Fig. 4.18), this simulation has lower sea-level pressure, northward-shifted 500-mb winds over the North Atlantic, and higher temperatures over western Europe, downstream from the region of reduced sea-ice cover.

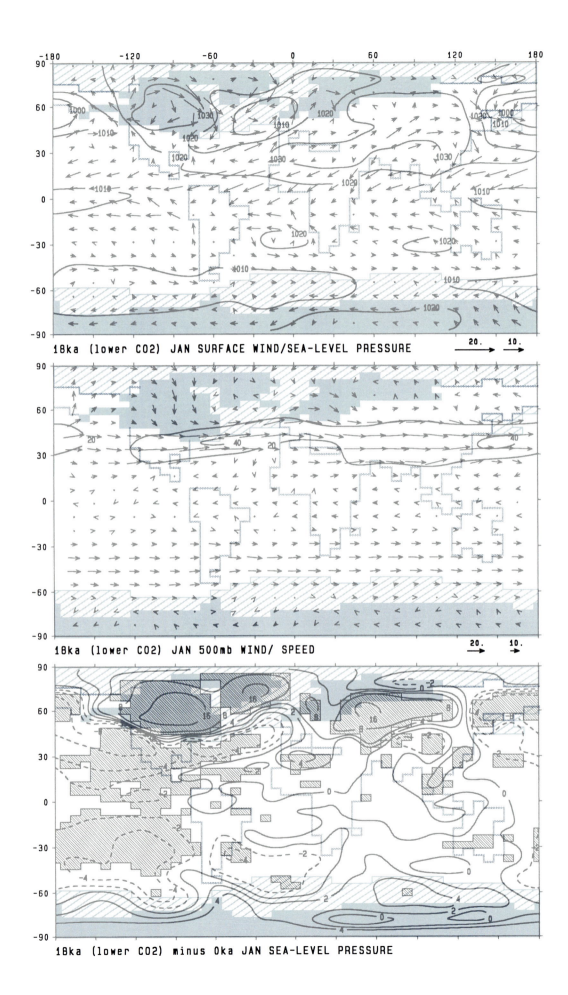

18ka (lower CO2) JAN SURFACE WIND/SEA-LEVEL PRESSURE

18ka (lower CO2) JAN 500mb WIND/ SPEED

18ka (lower CO2) minus 0ka JAN SEA-LEVEL PRESSURE

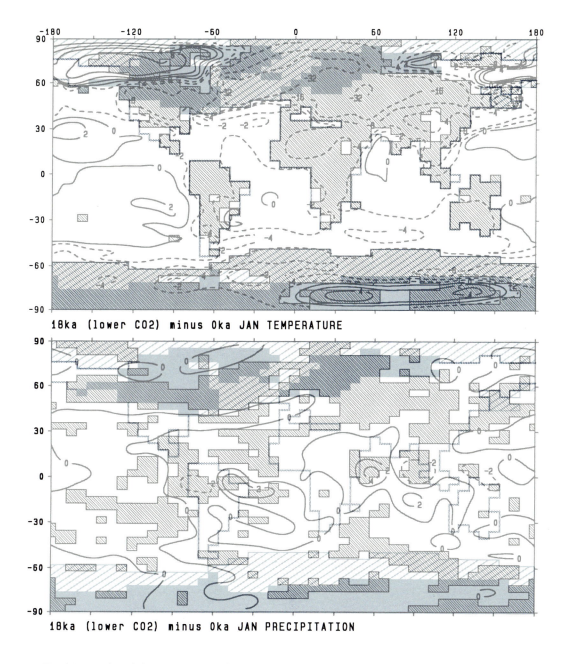

18ka (lower CO2) minus Oka JAN TEMPERATURE

18ka (lower CO2) minus Oka JAN PRECIPITATION

Fig. 4.31. Simulated climate variables for January, 18 ka, with lower carbon dioxide concentration: sea-level pressure and surface winds; 500-mb (about 5.5-km) winds and wind speed contours; and differences between simulated and present (0-ka control) sea-level pressure, surface temperature, and precipitation. In this simulation, atmospheric concentration of CO_2 was set at 200 ppmv, compared to 330 ppmv in the main COHMAP experiment (Fig. 4.18). Both simulations used the same values for sea-surface temperature, land and sea ice, and land and ice albedo. Compared to the simulation with higher CO_2 levels (Fig. 4.18), this simulation has slightly lower surface temperature and slightly lower precipitation rates over land.

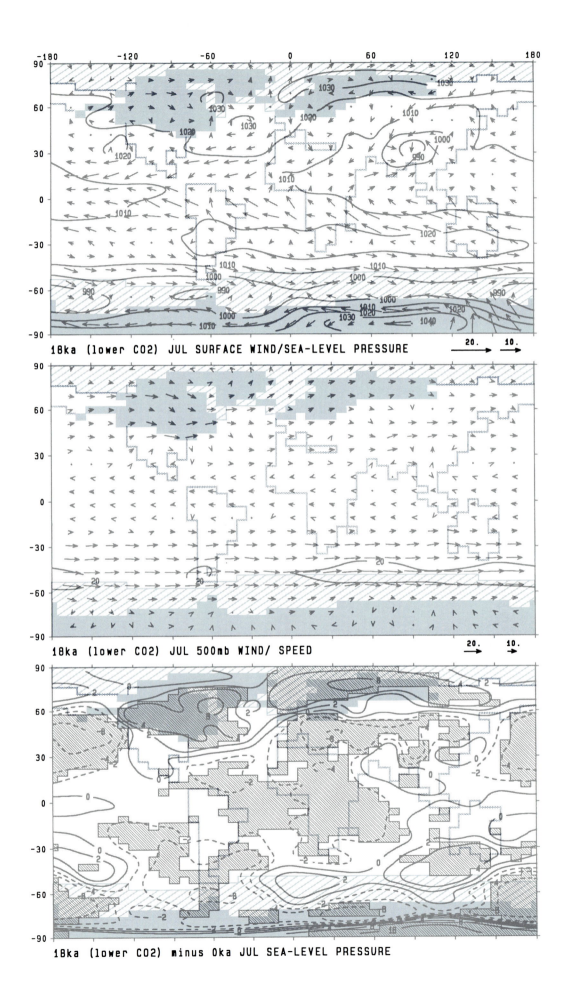

18ka (lower CO2) JUL SURFACE WIND/SEA-LEVEL PRESSURE

18ka (lower CO2) JUL 500mb WIND/ SPEED

18ka (lower CO2) minus 0ka JUL SEA-LEVEL PRESSURE

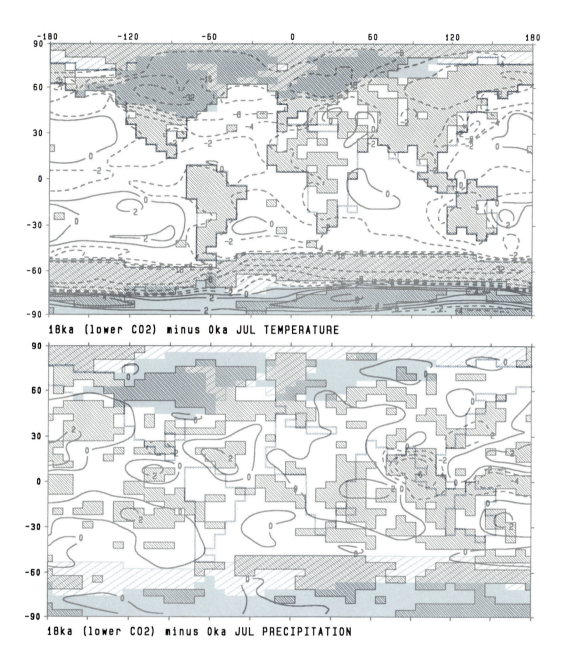

18ka (lower CO2) minus 0ka JUL TEMPERATURE

18ka (lower CO2) minus 0ka JUL PRECIPITATION

Fig. 4.32. Simulated climate variables for July, 18 ka, with lower carbon dioxide concentration: sea-level pressure and surface winds; 500-mb (about 5.5-km) winds and wind speed contours; and differences between simulated and present (0-ka control) sea-level pressure, surface temperature, and precipitation. In this simulation, atmospheric concentration of CO_2 was set at 200 ppmv, compared to 330 ppmv in the main COHMAP experiment (Fig. 4.18). Both simulations used the same values for sea-surface temperature, land and sea ice, and land and ice albedo. Compared to the simulation with higher CO_2 levels (Fig. 4.18), this simulation has slightly lower surface temperature and slightly lower precipitation rates over land.

climatic conditions within which abrupt, short-term events are embedded.

Conclusion

The COHMAP experiments were the first attempt to simulate a sequence of paleoclimates by the "snapshot" technique. The approach has proved useful in examining the causes, mechanisms, and patterns of regional and continent-scale climatic change since the last glacial maximum. We look forward to designing new experiments with improved models to provide further insights into the mechanisms of climatic change.

Appendix
Other Publications That
Have Used Results from
the COHMAP Simulations

We have provided information from the history tapes of the COHMAP simulation experiments for a number of projects that are not described in this book. The history tapes of the COHMAP experiments are archived at NCAR for use by other investigators. In most cases some special programming was needed to transform the information on the history tapes into the format required for other studies.

- Arctic ice sheet and ice-shelf cover: Lindstrom and MacAyeal (1986, 1989)
- Ocean-atmosphere exchanges: Birchfield (1987), Erickson (1990)
- Detailed comparison of CCM and energy-balance model simulations of climatic changes over the past 18,000 yr: Hyde *et al.* (1989)
- Estimates of environmental conditions and possible relations to archaeology: Price (1987)
- Forcing of dynamical ocean models with wind stress from CCM paleoclimatic experiments: Luther *et al.* (1990), Prell *et al.* (1990)

References

Birchfield, G. E. (1987). Changes in deep-ocean water $\delta^{18}O$ and temperature from the last glacial maximum to the present. *Paleoceanography* 2, 431–442.

Blackmon, M. L., and Lau, N. C. (1980). Regional characteristics of the Northern Hemisphere wintertime circulation: A comparison of the simulation of a GFDL general circulation model with observations. *Journal of the Atmospheric Sciences* 37, 497-513.

Broccoli, A. J., and Manabe, S. (1987). The influence of continental ice, atmospheric CO_2, and land albedo on the climate of the last glacial maximum. *Climate Dynamics* 1, 87-99.

CLIMAP Project Members (1976). The surface of the ice-age earth. *Science* 191, 1138-1144.

———. (1981). Seasonal reconstructions of the earth's surface at the last glacial maximum. *Geological Society of America Map and Chart Series* MC-36.

COHMAP Members (1988). Climatic changes of the last 18,000 years: Observations and model simulations. *Science* 241, 1043-1052.

Erickson, D. III (1990). Ocean-atmosphere CO_2 exchange during the last glacial maximum. *In* "Science of Gaia," pp. 256-260. MIT Press, Cambridge, Mass.

Gallimore, R. G., and Kutzbach, J. E. (1989). Effects of soil moisture on the sensitivity of a climate model to earth orbital forcing at 9000 yr B.P. *Climatic Change* 14, 175-205.

Gates, W. L. (1976a). Modeling the ice-age climate. *Science* 191, 1138-1144.

———. (1976b). The numerical simulation of ice-age climate with a global general circulation model. *Journal of the Atmospheric Sciences* 33, 1844-1873.

Guetter, P. J., and Kutzbach, J. E. (1990). A modified Köppen classification applied to model simulations of glacial and interglacial climates. *Climatic Change* 16, 193-215.

Hansen, J., Lacis, A., Rind, D., Russell, G., Stone, P., Fung, I., Ruedy, R., and Lerner, J. (1984). Climate sensitivity; analysis of feedback mechanisms. *In* "Climate Processes and Climate Sensitivity" (J. E. Hansen and T. Takahashi, Eds.), pp. 130-163. Maurice Ewing Series No. 5. American Geophysical Union, Boulder, Colo.

Hyde, W. T., Crowley, T. J., Kim, K.-Y., and North, G. R. (1989). A comparison of GCM and energy balance model simulations of seasonal temperature changes over the past 18,000 years. *Journal of Climate* 2, 864-887.

Joussaume, S., Jouzel, J., and Sadourny, R. (1989). Simulations of the last glacial maximum with an atmospheric general circulation model including paleoclimatic tracer cycles. *In* "Understanding Climate Change" (A. Berger, R. E. Dickinson, and J. W. Kidson, Eds.), pp. 159-162. Geophysical Monograph 52, Vol. 7. International Union of Geodesy and Geophysics (IUGG) and American Geophysical Union (AGU), Boulder, Colo.

Kutzbach, J. E. (1981). Monsoon climate of the early Holocene: Climate experiment with the earth's orbital parameters for 9000 years ago. *Science* 214, 59-61.

Kutzbach, J. E., and Gallimore, R. G. (1988). Sensitivity of a coupled atmosphere/mixed-layer ocean model to changes in orbital forcing at 9000 years B.P. *Journal of Geophysical Research* 93(D1), 803-821.

Kutzbach, J. E., and Guetter, P. J. (1986). The influence of changing orbital parameters and surface boundary conditions on climate simulations for the past 18,000 years. *Journal of the Atmospheric Sciences* 43, 1726-1759.

Kutzbach, J. E., and Otto-Bliesner, B. L. (1982). The sensitivity of the African-Asian monsoonal climate to orbital parameter changes for 9000 yr B.P. in a low-resolution general circulation model. *Journal of the Atmospheric Sciences* 39, 1177-1188.

Kutzbach, J. E., and Wright, H. E., Jr. (1985). Simulation of the climate of 18,000 yr B.P.: Results for the North American/North Atlantic/European sector and comparison with the geologic record of North America. *Quaternary Science Reviews* 4, 147-187.

Lautenschlager, M., and Herterich, K. (1989). "Climatic Response to Ice-Age Conditions. Part 1: The Atmospheric Circulation." Report 42. Max-Planck Institut für Meteorologie, Hamburg.

Lindstrom, D. R., and MacAyeal, D. R. (1986). Paleoclimatic constraints on the maintenance of possible ice-shelf cover in the Norwegian and Greenland seas. *Paleoceanography* 1, 313-337.

———. (1989). Scandinavian, Siberian, and Arctic Ocean glaciation: Effect of Holocene atmospheric CO_2 variations. *Science* 245, 628-631.

Luther, M. E., O'Brien, J. J., and Prell, W. L. (1990). Variability in up-welling fields in the northwestern Indian Ocean. 1. Model experiments for the past 18,000 years. *Paleoceanography* 5, 433-446.

Manabe, S., and Broccoli, A. J. (1985a). The influence of continental ice sheets on the climate of an ice age. *Journal of Geophysical Research* 90, 2167-2190.

———. (1985b). A comparison of climate model sensitivity with data from the last glacial maximum. *Journal of the Atmospheric Sciences* 42, 2643-2651.

Manabe, S., and Hahn, D. G. (1977). Simulation of the tropical climate of an ice age. *Journal of Geophysical Research* 82, 3889-3911.

Meyer, M. K. (1988). "Net Primary Productivity Estimates for the Last 18,000 Years Evaluated from Simulations by a Global Climate Model." Unpublished M.S. thesis, Department of Meteorology, University of Wisconsin-Madison.

Mitchell, J. F. B., Grahame, N. S., and Needham, K. J. (1988). Climate simulations for 9000 years before present: Seasonal variations and effect of the Laurentide ice sheet. *Journal of Geophysical Research* 93, 8283-8303.

Oglesby, R. J., Maasch, K. A., and Saltzman, B. (1989). Glacial meltwater cooling of the Gulf of Mexico: GCM implications for Holocene and present-day climates. *Climate Dynamics* 3, 115-133.

Overpeck, J. T., Peterson, L. C., Kipp, N., Imbrie, J., and Rind, D. (1989). Climate change in the circum-North Atlantic region during the last deglaciation. *Nature* 338, 553-557.

Pitcher, E. J., Malone, R. C., Ramanathan, V., Blackmon, M. L., Puri, K., and Bourke, W. (1983). January and July simulations with a spectral general circulation model. *Journal of the Atmospheric Sciences* 40, 580-604.

Prell, W. L., Marvil, R. E., and Luther, M. E. (1990). Variability in up-welling fields in the northwestern Indian Ocean. 2. Data-model comparison at 9000 years B.P. *Paleoceanography* 5, 447-457.

Price, T. D. (1987). The Mesolithic of Western Europe. *Journal of World Prehistory* 1, 225-305.

Ramanathan, V., Pitcher, E. J., Malone, R. C., and Blackmon, M. L. (1983). The response of a spectral general circulation model to refinements in radiative processes. *Journal of the Atmospheric Sciences* 40, 605-630.

Rind, D. (1987). Components of the ice age circulation. *Journal of Geophysical Research* 92, 4241-4281.

Rind, D., Peteet, D., Broecker, W., McIntyre, A., and Ruddiman, W. F. (1986). The impact of cold North Atlantic sea-surface temperature on climate: Implications for the Younger Dryas cooling (11-10K). *Climate Dynamics* 1, 3-33.

Royer, J. F., Deque, M., and Pestiaux, P. (1984). A sensitivity experiment to astronomical forcing with a spectral GCM: Simulation of the annual cycle at 125,000 B.P. and 115,000 B.P. *In* "Milankovitch and Climate," Part 2 (A. Berger, J. Imbrie, J. Hays, G. Kukla, and B. Saltzman, Eds.), pp. 733-763. D. Reidel, Dordrecht, The Netherlands.

Schneider, S. H., Peteet, D. M., and North, G. R. (1987). A climate model intercomparison for the Younger Dryas and its implications for paleoclimatic data collection. *In* "Abrupt Climatic Change" (W. H. Berger and L. D. Labeyrie, Eds.), pp. 399-417. D. Reidel, Dordrecht, The Netherlands.

Shinn, R. A., and Barron, E. J. (1989). Climate sensitivity to continental ice sheet size and configuration. *Journal of Climate* 2, 1517-1537.

Street-Perrott, F. A., Mitchell, J. F. B., Marchand, D. S., and Brunner, J. S. (1990). Milankovitch and albedo forcing of the tropical monsoons: A comparison of geologic evidence and numerical simulations for 9,000 yr B.P. *Transactions of the Royal Society of Edinburgh, Earth Sciences* 81, 407-427.

Webb, T. III, Bartlein, P. J., and Kutzbach, J. E. (1987). Climatic change in eastern North America during the past 18,000 years: Comparisons of pollen data with model results. *In* "North America and Adjacent Oceans during the Last Deglaciation" (W. F. Ruddiman and H. E. Wright, Jr., Eds.), pp. 447-462. The Geology of North America, Vol. K-3. The Geological Society of America, Boulder, Colo.

Williams, J. R. G., Barry, R. G., and Washington, W. M. (1974). Simulation of the atmospheric circulation using the NCAR global circulation model with ice age boundary conditions. *Journal of Applied Meteorology* 13, 305-317.

CHAPTER 5

The North and Equatorial Atlantic at 9000 and 6000 yr B.P.

W. F. Ruddiman and A. C. Mix

In this chapter we examine North Atlantic and equatorial Atlantic cores for which estimates of sea-surface temperature (SST) are available for 9000 and 6000 yr B.P. and compare those estimates to modern (atlas or core-top) temperatures. Modern SST trends for summer and winter in the world ocean are shown in Figure 5.1. Over long time scales (more than 10,000 yr), the most thermally reactive areas of the North Atlantic Ocean are the high latitudes (40-60°N), the coastal upwelling regions, and the equatorial divergences, as shown by maps of differences between glacial and interglacial SSTs (Fig. 5.2). Most of the Atlantic cores we selected for study (Fig. 5.3 and Table 5.1) came from these reactive areas, partly because these regions naturally yielded many cores that met our quality standards as climatic monitors. In addition, because we knew at the outset that SST differences between today and 9000 or 6000 yr B.P. would generally be small, we chose to concentrate on areas that would maximize those differences.

Core Selection and Stratigraphic Ranking

Our approach was mainly to select cores with sedimentation rates of 2-6 cm per 1000 yr (Table 5.2), somewhat above the mean rate of 2-3 cm per 1000 yr that is characteristic of carbonate-rich sectors of the North Atlantic. Cores with rates below 2 cm per 1000 yr generally give signals that are highly attenuated by mixing at the sea floor. Rates above 6 cm per 1000 yr are generally indicative of large-scale lateral transport along the sea floor under high-energy conditions that may invalidate ^{14}C dating of bulk or fine-fraction components and otherwise disrupt orderly pelagic deposition. However, we did examine cores with higher deposition rates (Table 5.2) to test for more detailed signals.

Table 5.3 lists all ^{14}C dates that are new to this study. All dates previously reported in the literature are referenced in Table 5.2. For many of the equatorial Atlantic cores, oxygen-isotope records are also available as a cross-check on the ^{14}C dates. Isotopic curves are widely recognized as globally correlative signals, except for minor differences due to local factors. With most of the equatorial cores, it was thus possible to check the chronology derived from radiocarbon dating against the between-core alignment of the oxygen-isotope curves.

For cores lacking radiocarbon control above the 9000- or 6000-yr B.P. levels, we assigned an age of 1500 yr B.P. to the top of the core. This choice was governed by ^{14}C ages previously measured in Atlantic cores with relatively high deposition rates (Ruddiman *et al.*, 1980; Ruddiman and McIntyre, 1981), although whether any core top actually represents the sediment-water interface is still uncertain.

Our basic strategy for choosing the 9000- and 6000-yr B.P. levels was to interpolate their depths in each core from surrounding ^{14}C dates (if available) or between other stratigraphic controls with ages known from other sources. We then selected the faunal count (and associated SST estimates) nearest the interpolated levels (Table 5.2). With counts at time intervals averaging about 1000 yr, the chosen count generally was offset by no more than a few hundred years from the targeted age. In a few instances the two counts closest to the interpolated 9000- or 6000-yr B.P. levels were offset by roughly equal ages; when the average age of the two samples fell within 100 yr of the desired level, we used the average of the two SST values.

We ranked each sample used for the 9000- or 6000-yr B.P. level for stratigraphic quality according to the degree of chronologic control provided by the adjacent ^{14}C dates (Table 5.2). The ranking scheme is identical to that used for land-based pollen records in other chapters in this volume.

This strategy for establishing chronologies worked with varying degrees of success in different areas, depending on local problems. It worked best in the equatorial Atlantic because of the excellent control provided by ^{14}C dating and oxygen-isotope curves.

Cores from the higher latitudes of the North Atlantic vary more in stratigraphic quality. One major complication is the invalidation of many late-glacial ^{14}C dates because of suppression of biogenic carbonate productivity and the accompanying increase in deposition of ice-rafted limestone from the continents. In addition, high-energy near-bottom conditions in many regions of the high-latitude oceans cause much lateral redistribution of sediment. As a result, even some of the highly ranked cores north of 30°N may not be useful if the ^{14}C dates are invalid or if the sediment is redistributed. On the other hand, some of the cores from high latitudes appear very suitable for this kind of study because of excellent carbonate preservation and moderately high deposition rates. These regions are discussed in more detail later.

Faunal Analysis and SST Estimates

Planktonic foraminifera in the North and equatorial Atlantic are as clearly correlated to water-mass conditions (including atlas SST values) as any planktonic fossil group in the world ocean (Imbrie and Kipp, 1971; Kipp, 1976). Five transfer functions were used to generate the SST estimates reported here (Table 5.4). One is the basic North Atlantic equation F13 published by Kipp (1976). Two others are closely related variants: equation F13B-4CE (referred to in Table 5.4 as F13′), which involves only minor taxonomic changes (Ruddiman and Glover, 1975), and unpublished equation F20 of N. G. Kipp (personal communication), which incorporates a more comprehensive and slightly edited core-top data set as well as newer atlas SST values. Also cited for cores along the northwest coast of Africa is an unnamed equation from Molina-Cruz and Thiede (1978). For core V23–82 in the North Atlantic, we report SST estimates from Sancetta et al. (1973) based on equation F3 (Imbrie et al., 1973).

For all cores in this study, standard practice required counting of more than 300 individuals where available in order to obtain adequate precision (Imbrie et al., 1973). The standard error of the SST estimate from all these equations is in the range of ±1.0–1.6°C, which is acceptable for studies of glacial-interglacial changes of 5–10°C or more but marginal for the smaller Holocene SST changes, which are generally less

than 2°C. The standard error reflects the ability of the equations to reproduce the winter and summer atlas SST values from all core tops in the Atlantic.

The near-equatorial regions present a special problem for seasonal SST estimates. Because it is difficult to define past seasonal positions of the thermal equator, it is hard to know which caloric hemisphere the "winter" and "summer" SST estimates should be assigned to in this region. Thus, "winter" and "summer" could as easily be February as August in these cores.

For the purposes of this study, we used the CLIMAP atlas thermal equator (shown in Fig. 1b of Mix et al., 1986) for the modern thermal equator. Based on evidence in Mix et al. (1986), we assumed that the thermal equator at 6000 yr B.P. was in the same position as today. For 9000 yr B.P. the situation is more complex. Mix et al. (1986) showed that the mean annual thermal equator based on transfer function estimates could be placed either in the Northern Hemisphere at essentially the same place as today, or about 5° south of the geographic equator. We examine the consequences of both alternatives for the anomaly patterns.

Finally, we had to decide what basis of comparison to use for calculating SST anomalies at 9000 and 6000 yr B.P. Following CLIMAP Project Members (1984), we selected atlas SST values (CLIMAP Project Members, 1981) as the initial standard of comparison; however, we also calculated differences relative to core-top SST estimates. This closer examination is particularly valuable for cores where the core-top estimates are offset from the atlas values by more than 1.0°C, because the difficulty in calibrating the transfer function to the atlas SST values indicates a need for caution in accepting SST anomalies at older levels.

Atlas SST values are shown in Table 5.1. SST estimates for 9000 and 6000 yr B.P. and the calculated SST anomalies relative to modern values are given in Table 5.5. The data in Table 5.5 were calculated assuming that the thermal equator at 9000 yr B.P. was in the Northern Hemisphere.

Regional Assessment: Credibility of Results

As noted above, each region for which we have numerous cores may be subject to problems and complications. Each also tends to have its own patterns of response. The discussion in this section is designed to provide (1) a background understanding of the longer-term changes in which the 9000- and 6000-yr B.P. estimates are embedded, and (2) a basis for assessing the credibility of the SST values and anomalies reported in the succeeding sections.

(a)

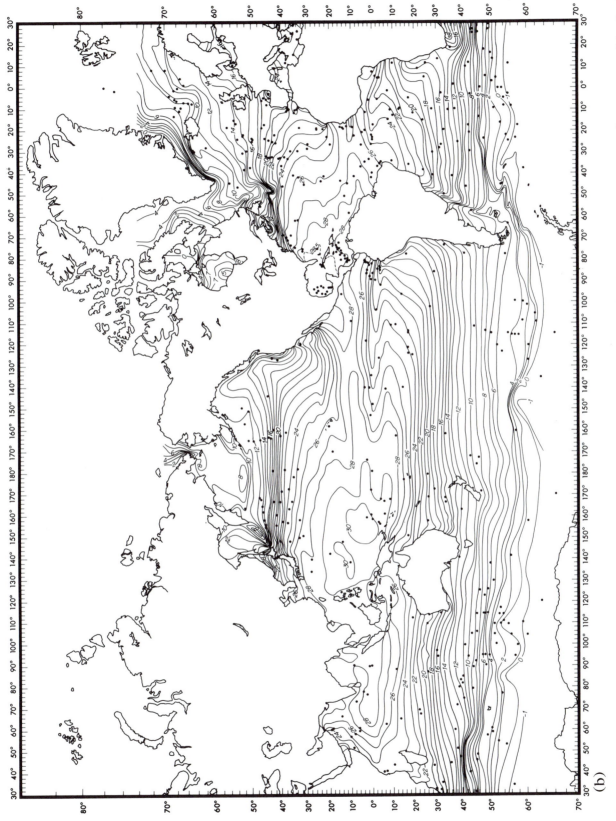

(b)

Fig 5.1. Modern sea-surface temperatures in the world ocean in (a) February and (b) August. From CLIMAP Project Members (1981); with permission of the Geological Society of America, Boulder, Colo.

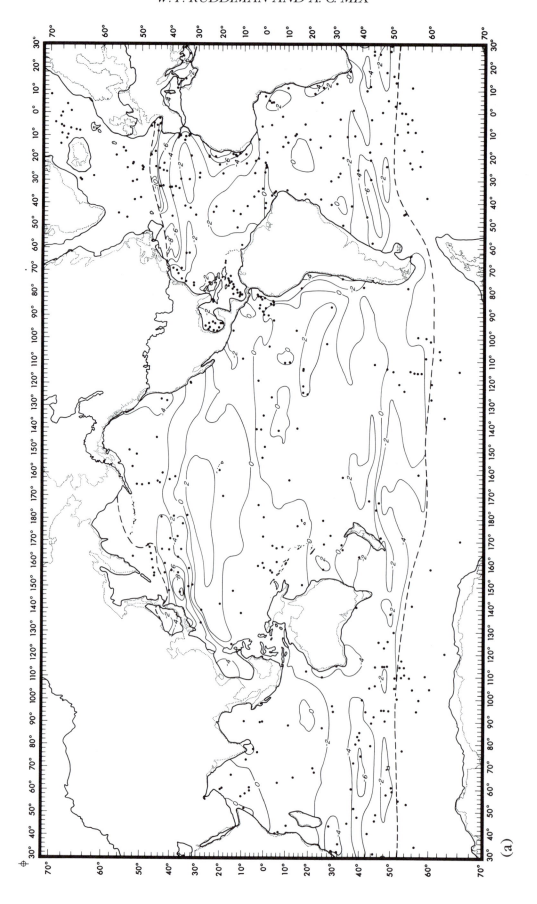

(a)

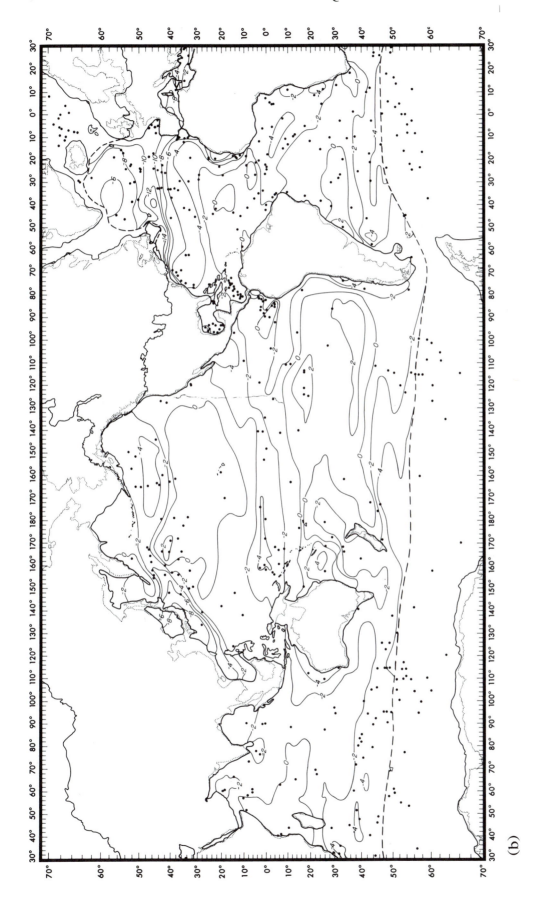

Fig 5.2 Differences between glacial and interglacial sea-surface temperatures in (a) February and (b) August. From CLIMAP Project Members (1981); with permission of the Geological Society of America, Boulder, Colo.

(b)

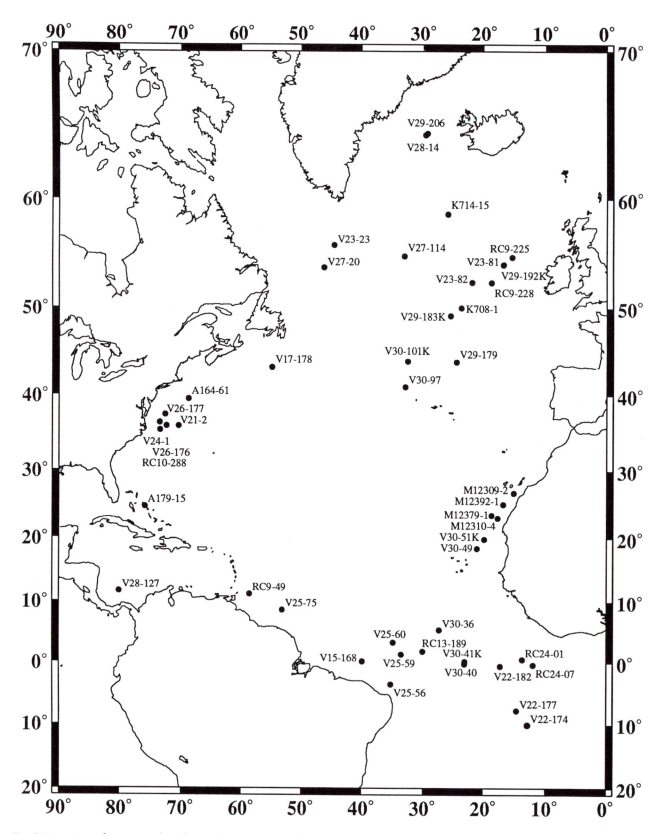

Fig. 5.3. Location of cores used in this study (see also Table 5.1).

Table 5.1. Core location, water depth, and modern atlas sea-surface temperature (SST)

Core	Location Latitude	Longitude	Depth (m)	Modern atlas SST (°C) February	August
A164-61	39°32'N	68°47'W	2722	6.8	—
A179-15	24°48'N	75°56'W	3109	23.7	28.6
K708-1	50°00'N	23°45'W	4053	10.5	16.3
K714-15	58°46'N	25°57'W	2598	8.3	11.9
M12309-2	26°50'N	15°07'W	2820	17.5	21.3
M12310-4	23°30'N	18°43'W	3080	18.3	21.9
M12379-1	23°08'N	17°45'W	2066	17.8	21.0
M12392-1	25°10'N	16°51'W	2575	18.0	21.3
RC9-49	11°11'N	58°35'W	1851	26.1	28.1
RC9-225	54°59'N	15°24'W	2334	10.0	14.6
RC9-228	52°33'N	18°45'W	3981	10.2	15.2
RC10-288	35°32'N	73°25'W	3678	19.1	—
RC13-189	1°52'N	30°00'W	3233	26.6	26.1
RC24-01	0°34'N	13°39'W	3837	26.9	23.5
RC24-07	0°21'S	11°55'W	3899	26.5	23.2
V15-168	0°12'N	39°54'W	4219	27.2	26.7
V17-178	43°23'N	54°52'W	4006	6.2	—
V21-2	36°05'N	70°24'W	4455	18.7	—
V22-174	10°04'S	12°49'W	2630	25.4	23.2
V22-177	7°45'S	14°37'W	3290	26.1	24.0
V22-182	0°33'S	17°16'W	3614-3937	26.9	23.7
V23-23	56°05'N	44°33'W	3292	3.8	9.4
V23-81	54°15'N	16°50'W	2393	10.1	14.8
V23-82	52°35'N	21°56'W	3974	10.1	15.3
V24-1	36°30'N	73°30'W	3012	17.5	—
V25-56	3°33'S	35°14'W	3512	27.3	26.0
V25-59	1°22'N	33°29'W	3824	26.7	26.4
V25-60	3°17'N	34°50'W	3749	26.5	26.7
V25-75	8°35'N	53°10'W	2743	26.3	28.1
V26-176	36°03'N	72°23'W	3942	19.0	—
V26-177	37°33'N	72°34'W	2979	14.0	—
V27-20	54°00'N	46°12'W	3510	3.9	11.0
V27-114	55°03'N	33°04'W	2532	7.0	12.0
V28-14	64°47'N	29°34'W	1855	3.2	9.5
V28-127	11°39'N	80°08'W	3227	26.1	28.0
V29-179	44°01'N	24°32'W	3331	13.2	19.8
V29-183K	49°08'N	25°30'W	3643	11.1	16.7
V29-192K	54°16'N	16°47'W	2365	10.0	14.6
V29-206	64°54'N	29°17'W	1624	3.5	10.0
V30-36	5°21'N	27°19'W	4245	26.2	26.3
V30-40	0°12'S	23°09'W	3706	27.5	24.4
V30-41K	0°13'N	23°04'W	3874	27.5	24.5
V30-49	18°26'N	21°05'W	3093	20.0	24.8
V30-51K	19°52'N	19°55'W	3409	19.0	24.3
V30-97	41°00'N	32°56'W	3371	14.6	22.6
V30-101K	44°06'N	32°30'W	3504	13.2	20.5

Equatorial Atlantic

Mix *et al.* (1986) applied empirical orthogonal function analysis to the equatorial Atlantic group of cores (Fig. 5.3) and found two basic patterns of variation, examples of which are displayed in Figure 5.4. Most of the equatorial Atlantic SST variation over the last 20,000 yr is a combination of these two patterns.

The dominant signal is the cold-glacial/warm-interglacial pattern shown by the winter SST trend in core V22-177 from the South Atlantic (Fig. 5.4). This record shows the typical response of cores in this region. Be-

cause the basic change from cold to warm SST occurs between 14,000 and 8000 yr B.P., this pattern tends not to affect the 6000-yr B.P. SST estimates but is manifested as negative (cool) SST anomalies for many of the 9000-yr B.P. estimates. A few cores along the coast of Brazil show some suggestion that this response pattern reverses sign (at least for the warm season), with glacial SST estimates slightly warmer than modern values in core V25-75 (Fig. 5.4).

In the second pattern defined by Mix *et al.* (1986), temperature extremes (either cold or warm) are centered on the early deglaciation (12,000–11,000 yr B.P.). This secondary signal is superimposed on the more dominant first pattern. In cores north of the equator, this deglacial SST extreme is a negative (cold) anomaly (e.g., core V30–49 in Fig. 5.4); to the south of the thermal equator, it is a positive (warm) anomaly. The changes associated with this pattern occurred close enough to 9000 yr B.P. to influence the 9000-yr B.P. SST estimates but do not affect the 6000-yr B.P. values.

Because of the high quality of the equatorial Atlantic records, the largest constraint on the validity of the SST anomalies for these regions is the impact of mixing on the smoothing and translation of SST signals. This problem is discussed in detail in CLIMAP (1984). The broader question of the validity of low-latitude SST estimates in general is discussed by Kutzbach and Ruddiman (this vol.).

NORTHWEST AFRICAN MARGIN

All four METEOR (M) cores off the northwest coast of Africa show basically similar Holocene patterns, with temperatures rising from 10,000 yr B.P. to a maximum at 5000 or 4000 yr B.P. and then in several cores declining toward the present. This interval of high SST values in the middle to late Holocene appears time-correlative with the weakening or cessation of upwelling inferred from marine mollusks along the adjacent African coast (Petit-Maire, 1980).

Table 5.2. Number of ^{14}C dates, source of published dates (^{14}C and stratigraphy), mean Holocene sedimentation rate, and map level rank of cores used in this study

Core	[14]C dates[a] (no.)	Source of published dates	Sedimentation rate (cm/1000 yr)	Depth (cm) of map level pick[b] 6000 yr B.P.	9000 yr B.P.	Map level rank[c] 6000 yr B.P.	9000 yr B.P.
A164–61	1	Balsam (1981)	6.8	NA	NA	2II	7II
A179–15	5	Broecker *et al.* (1956); Broecker and Kulp (1957); Mix and Ruddiman (1985)	11.2	72.0	102.0	1I	1I
K708–1	0	Ruddiman and McIntyre (1981)	5.8	30.0	45.0	5I	6I,II
K714–15	3	Ruddiman and McIntyre (1981); this chapter	25.9	80.0	180.0	4I	1I
M12309–2	1	M. Sarnthein (unpublished)	5.5	27.5	47.5	4I	2II
M12310–4	2	Koopmann (1979); M. Sarnthein (unpublished)	13.2	2.5	34.5	4II	1III
M12379–1	1	Koopmann (1979)	8.0	37.5	67.5	5I	5II
M12392–1	3	M. Sarnthein (unpublished); Koopmann (1979) (Koopmann date of 9430 yr B.P. not used)	4.2	20.0	32.5	5I	5II
RC9–49	6	Be *et al.* (1976); Mix and Ruddiman (1985); this chapter	3.7	13.0	30.0	3II	1I
RC9–225	1	Ruddiman and McIntyre (1981); this chapter	6.9	25.0	50.0	4I	1I
RC9–228	3	Ruddiman and McIntyre (1981); this chapter (date of 9445 yr B.P. not used)	9.3	47.5*	72.0	2I	1I
RC10–288	1	Balsam (1981)	10.3	NA	NA	3II	7II
RC13–189	5	Mix and Ruddiman (1985) (date of 16,160 yr B.P. not used)	2.8	15.0	23.75*	4I	2I,II
RC24–01	6	Mix and Ruddiman (1985)	2.8	12.0	20.0	2I	1I
RC24–07	2	Mix and Ruddiman (1985)	4.7	25.0	40.0	4I	5II
V15–168	4	Be *et al.* (1976); Mix and Ruddiman (1985) (date of 15,700 yr B.P. not used)	6.8	20.0	47.5*	4I	2I
V17–178	2	Balsam (1981)	24.5	NA	NA	5I	6II
V21–2	1	Balsam (1981)	13.8	NA	NA	6II	7II
V22–174	5	Prell and Damuth (1978)	2.7	15.0	20.0	4I	1I
V22–177	5	Mix and Ruddiman (1985)	2.9	17.5	25.0	1I	2I
V22–182	4	Mix and Ruddiman (1985)	3.8	20.0	32.0	4I	3II
V23–23	0	Ruddiman and McIntyre (1981)	3.4	17.5*	30.0	5I	6I,II
V23–81	3	Ruddiman and McIntyre (1981); this chapter (date of 14,825 yr B.P. not used)	15.5	59.0	102.0*	2I	1I
V23–82	2	Ruddiman and McIntyre (1981) (date of 10,220 yr B.P. not used)	5.9	30.0	50.0	5I	6I,II
V24–1	2	Balsam (1981)	10.1	NA	NA	4I	1I
V25–56	4	Mix and Ruddiman (1985)	4.6	20.0	30.0	2I	2I
V25–59	6	Be *et al.* (1976); Mix and Ruddiman (1985)	2.9	15.0	25.0	2II	3I

The SST signal in core M12392-1 (Fig. 5.5) typifies the pattern observed off northwest Africa, and it agrees very closely with the SST trends in nearby core V30-49 (Fig. 5.4). This pattern appears to reflect a superposition of the cold-deglacial signal of Mix *et al.* (1986) on the more dominant cold-glacial/warm-interglacial temperature response. The major difference in Holocene responses among these cores is the highly variable late-Holocene SST decrease, which is negligible in M12392-1 but as large as 4-5°C in M12379-1.

There are substantial mismatches between the core-top SST estimates and the atlas values overlying these sites (Thiede, 1977; Molina-Cruz and Thiede, 1978). These residuals are less than 1°C in core M12309-2 to the north but are as large as 2°C in core M12392-1 to the south and 5°C in core M12379-1 still farther south. This is also an area of significant residuals between core-top and atlas SST estimates for transfer function F13 of Kipp (1976), although a transfer function developed by Pflaumann (1985) does not give such large SST residuals. Because of the inability of the available transfer functions to give reliable core-top SST estimates in this area, we compared the 9000- and 6000-yr B.P. values to both atlas SST values and core-top estimates (see Figs. 5.9 and 5.12 and Table 5.5). By either comparison, SST values for this region generally indicate cooler conditions at 9000 and 6000 yr B.P. than today.

Table 5.2. Number of ^{14}C dates, source of published dates (^{14}C and stratigraphy), mean Holocene sedimentation rate, and map level rank of cores used in this study (continued)

Core	^{14}C dates[a] (no.)	Source of published dates	Sedimentation rate (cm/1000 yr)	Depth (cm) of map level pick[b] 6000 yr B.P.	9000 yr B.P.	Map level rank[c] 6000 yr B.P.	9000 yr B.P.
V25-60	4	Mix and Ruddiman (1985)	2.9	15.0	22.5	2I	1II
V25-75	2	Mix and Ruddiman (1985)	7.4	35.0	60.0	4I	3II
V26-176	3	Balsam (1981)	26.6	NA	NA	3I	1II
V26-177	1	Balsam (1981)	11.2	NA	NA	4I	3II
V27-20	0	Ruddiman and McIntyre (1981)	4.5	22.5*	35.0	5I	6I,II
V27-114	3	Ruddiman and McIntyre (1981); this chapter	23.9	70.0	150.0	1I	1II
V28-14	6	Kellogg (1984); Ruddiman and McIntyre (1981)	11.8	66.25*	102.5	1I	1II
V28-127	3	Prell (1978); CLIMAP Project Members (1981)	4.3	18.0*	36.0	3II	3I
V29-179	0	Ruddiman and McIntyre (1981)	3.8	20.0	32.0*	5I	6I,II
V29-183K	12	Ruddiman and McIntyre (1981)	2.6	15.0	21.0	2I	1II
V29-192K	10	Ruddiman and McIntyre (1981) (date of 13,890 yr B.P. not used)	3.7	24.0	33.0	1I	1I,II
V29-206	8	Kellogg (1984); Ruddiman and McIntyre (1981)	13.0	70.0	105.0	1I	1II
V30-36	3	Mix and Ruddiman (1985)	1.8	10.0	17.0*	4I	1II
V30-40	5	Mix and Ruddiman (1985) (date of 12,805 yr B.P. not used)	3.1	18.0	27.0	3I	1II
V30-41K	9	Jones and Ruddiman (1982); Mix and Ruddiman (1985)	2.4	17.5*	22.5	1I	1II
V30-49	3	Mix and Ruddiman (1985)	3.7	18.0*	32.0	1II	3I
V30-51K	8	Mix and Ruddiman (1985)	3.6	20.0	30.0	4I	1II
V30-97	1	Ruddiman and McIntyre (1981)	4.3	20.0	35.0	7I,II	7I,II
V30-101K	17	Ruddiman and McIntyre (1981) (dates of 3960, 18,300, 18,590, 27,200, 29,500, and 31,400 yr B.P. not used)	2.7	16.0	22.5	1I	2I

[a]^{14}C dates from Mix and Ruddiman (1985) and this chapter (except RC9-49) are ^{13}C-corrected (we subtracted 400 yr before using them in this study).

[b]NA indicates that depths are not available. An asterisk indicates that the map level pick is an average of samples above and below. The depth listed is the average for these samples and is within 100 yr of the calculated map level pick.

[c]Ranking scheme was as follows: For case I where dates bracket map level, if bracketing dates are within ±2000 yr of map level, rank is 1; between +2000 and –4000 yr or conversely, rank 2; within ±4000 yr, rank 3; between +4000 and –6000 yr or conversely, rank 4; within ±6000 yr, rank 5; between +6000 and –8000 yr or conversely, rank 6; and poorly dated, rank 7. For case II where date is close to map level, if date is within ±250 yr of map level, rank is 1; ±500 yr, rank 2; ±750 yr, rank 3; ±1000 yr, rank 4; ±1250 yr, rank 5; ±2000 yr, rank 6; and more than 2000 yr, rank 7. Whichever case gave the better rank was used. If both cases gave the same rank, then both are noted.

Our choice of the 6000-yr B.P. levels disagrees in some cases with those published by Pflaumann (1980) and Sarnthein *et al.* (1982). In some cases this reflects new ¹⁴C dates made available to us by M. Sarnthein (personal communication, 1987). Our procedure was to reject all ¹⁴C dates based on organic carbon (Geyh, 1979) and to rely solely on dates made on carbonate fractions (both bulk sediment and coarse-fraction carbonate). We then used linear interpolation between the ¹⁴C dates and selected the nearest actual sample value. Core-top ages were set at 1500 yr B.P. except for core M12310-4, which has a mid-Holocene age near the top.

The SST estimates for 6000 yr B.P. that we selected are generally slightly cooler (1–3°C) than both the atlas values and the core-top values (Table 5.5). In contrast, Pflaumann (1980) and Sarnthein *et al.* (1982) found positive (warm) SST anomalies of 1–3°C. Their 6000-yr B.P. levels were chosen closer to the peak of the warm SST anomaly that we date by interpolation to 5000 or 4000 yr B.P., whereas our 6000- yr B.P. choices fall well below this Holocene SST maximum at deeper (and cooler) points on the warming trend toward this SST maximum.

No ¹⁴C dates on CaCO₃ are available in these METE-OR cores from the middle to late Holocene (6000–4000 yr B.P.) to provide a choice between our 6000-yr B.P. selections and those of Pflaumann (1980) and Sarnthein *et al.* (1982). Core V30-49, located just to the south of the METEOR cores (Fig. 5.3), has a ¹⁴C date of 6000 yr B.P. at a level well below the winter SST maximum but within the lower part of the summer SST maximum (Fig. 5.4).

WESTERN SUBTROPICAL NORTH ATLANTIC

The basic pattern of SST response in the western subtropical North Atlantic region (see Fig. 6 of Balsam,

Table 5.3. Previously unpublished ¹⁴C dates of cores used in this study[a]

Core	Depth (cm)	Date (yr B.P.)	Error (yr)	Sample
K714-15	84–110	7560	±270	GX10146
	110–120	7950	±240	GX10147
RC9-49	23–27	8260	±600	GX2861
	32–37	9900	±525	GX2863
RC9-225	33–35	8155	±215	GX8262
RC9-228	40–44	5650	±180	GX8624
	53–55	9445	±300	GX8626
	67–69	9100	±260	GX8625
V23-81	84–87	8960	±195	GX8263
V27-114	40–46	4805	±185	GX8623
	81–84	7070	±215	GX8264
	153–157	9610	±225	GX8265

[a]All ¹⁴C analyses by Geochron on total sample. All dates except those for core RC9-49 are ¹³C-corrected. We subtracted 400 yr from these dates before using them in this study.

Table 5.4. Source of transfer functions and sea-surface temperature (SST) estimates

Core	Transfer function	Source of transfer function	Source of SST estimates
A164-61	F13	Kipp (1976)	Balsam (1981)
A179-15	F13	Kipp (1976)	Kipp (1976)
K708-1	F13'	Ruddiman and Glover (1975)	Ruddiman *et al.* (1977)
K714-15	F13'	Ruddiman and Glover (1975)	This chapter
M12309-2		Molina-Cruz and Thiede (1978)	Thiede (1977)
M12310-4		Molina-Cruz and Thiede (1978)	Thiede (1977)
M12379-1		Molina-Cruz and Thiede (1978)	Thiede (1977)
M12392-1	F20	N. G. Kipp (unpublished)	Thiede (1977); Mix *et al.* (1986)[a]
RC9-49	F20	N. G. Kipp (unpublished)	Be *et al.* (1976); Mix *et al.* (1986)[b]
RC9-225	F13'	Ruddiman and Glover (1975)	Ruddiman *et al.* (1977); this chapter[c]
RC9-228	F13'	Ruddiman and Glover (1975)	Ruddiman *et al.* (1977)[d]
RC10-288	F13	Kipp (1976)	Balsam (1981)
RC13-189	F20	N. G. Kipp (unpublished)	Mix *et al.* (1986)
RC24-01	F20	N. G. Kipp (unpublished)	Mix *et al.* (1986)
RC24-07	F20	N. G. Kipp (unpublished)	Mix *et al.* (1986)
V15-168	F20	N. G. Kipp (unpublished)	Be *et al.* (1976); Mix *et al.* (1986)[b]
V17-178	F13	Kipp (1976)	Balsam (1981)
V21-2	F13	Kipp (1976)	Balsam (1981)
V22-174	F20	N. G. Kipp (unpublished)	Mix *et al.* (1986)
V22-177	F20	N. G. Kipp (unpublished)	Mix *et al.* (1986)
V22-182	F20	N. G. Kipp (unpublished)	Mix *et al.* (1986)
V23-23	F13'	Ruddiman and Glover (1975)	This chapter
V23-81	F13'	Ruddiman and Glover (1975)	Ruddiman *et al.* (1977)[d]
V23-82	F3	Imbrie *et al.* (1973)	Sancetta *et al.* (1973)
V24-1	F13	Kipp (1976)	Balsam (1981)
V25-56	F20	N. G. Kipp (unpublished)	Be *et al.* (1976); Mix *et al.* (1986)[a]
V25-59	F20	N. G. Kipp (unpublished)	Mix *et al.* (1986)
V25-60	F20	N. G. Kipp (unpublished)	Mix *et al.* (1986)

1981) is colder SST values at the maximum of the last glaciation, a warming between 14,000 and 10,000 yr B.P., an SST maximum warmer than today between 10,000 and 6000 yr B.P., and a late-Holocene cooling to modern values. Because Balsam (1981) reported only winter SST estimates, there are no summer SST values for any of these cores.

Because of the large changes in SST near the end of the last glaciation, all of the SST anomalies at 9000 yr B.P. are highly sensitive to the accuracy of Balsam's chronology. Selecting levels just a few hundred (interpolated) years higher or lower in these cores alters the calculated anomalies significantly. The accuracy of the chosen 9000-yr B.P. level is limited by (1) the lack of ¹⁴C dates in the older, deeper sections of many cores, (2) the likelihood of contamination of the older ¹⁴C dates by nonbiogenic carbon, and (3) the possibility that dissolution has altered the 9000-yr B.P. SST estimates, as it has in immediately underlying sediment layers (Balsam and Heusser, 1976).

These problems are much diminished at the 6000-yr B.P. level. There are more ¹⁴C dates, including several within 1000 yr of 6000 yr B.P., and contamination is less likely to have been a problem. Furthermore, small errors in selecting the 6000-yr B.P. level would generally have a small impact on the SST anomalies because of the slowly declining trends of the SST curves during this period.

The 6000-yr B.P. levels thus appear stratigraphically valid, and most cores show small positive SST anomalies. One core (A164-61) giving a very large SST anomaly (almost 10°C) at 6000 yr B.P. also has large offsets between core-top and atlas values. The anomaly relative to the core-top value is considerably smaller (Table 5.5).

The North Atlantic transfer function is particularly difficult to apply in this region (Balsam, 1981), because the seasonal SST contrast is larger than in most of the Atlantic. For this reason, the amplitude of the winter SST anomalies is open to question.

High-Latitude North Atlantic

The high-latitude North Atlantic cores show two major patterns of SST response over the last 20,000 yr. The first is centered north of 45°N in a region dominated by periodic advances and retreats of cold polar water that occur roughly in phase with advances and retreats of the ice sheets (Ruddiman and McIntyre, 1984). Here the basic SST response pattern over the last 20,000 yr has been cold-glacial/warm-interglacial, with a brief return toward glacial conditions centered at 10,500 yr B.P. (Ruddiman and McIntyre, 1981). The transition from glacial to interglacial SST values, however, is both abrupt and time-transgressive, occurring earlier (13,000 yr B.P.) in cores to the southeast near Portugal, later (10,000 yr B.P.) in the western North Atlantic, and still later (9000-7000 yr B.P.) in the Labrador

Table 5.4. Source of transfer functions and sea-surface temperature (SST) estimates (continued)

Core	Transfer function	Source of transfer function	Source of SST estimates
V25-75	F20	N. G. Kipp (unpublished)	Mix *et al.* (1986)
V26-176	F13	Kipp (1976)	Balsam (1981)
V26-177	F13	Kipp (1976)	Balsam (1981)
V27-20	F13'	Ruddiman and Glover (1975)	Ruddiman and McIntyre (1984)
V27-114	F13'	Ruddiman and Glover (1975)	This chapter
V28-14	F13'	Ruddiman and Glover (1975)	Kellogg (1984); this chapter[a]
V28-127	F20	N. G. Kipp (unpublished)	This chapter
V29-179	F13'	Ruddiman and Glover (1975)	Ruddiman and McIntyre (1984); this chapter[c]
V29-183K	F13'	Ruddiman and Glover (1975)	This chapter
V29-192K	F13'	Ruddiman and Glover (1975)	This chapter
V29-206	F13'	Ruddiman and Glover (1975)	Kellogg (1984); this chapter[a]
V30-36	F20	N. G. Kipp (unpublished)	Mix *et al.* (1986)
V30-40	F20	N. G. Kipp (unpublished)	Mix *et al.* (1986)
V30-41K	F20	N. G. Kipp (unpublished)	Mix *et al.* (1986)
V30-49	F20	N. G. Kipp (unpublished)	Mix *et al.* (1986)
V30-51K	F20	N. G. Kipp (unpublished)	Mix *et al.* (1986)
V30-97	F13'	Ruddiman and Glover (1975)	Ruddiman and McIntyre (1984)
V30-101K	F13'	Ruddiman and Glover (1975)	This chapter

[a]Faunal counts are from the earlier publication. Temperature estimates were rerun in the later publication on the same faunal date but with a different transfer function.

[b]Faunal counts are from both sources (with additional counts done for the later publication); a different transfer function was used on all faunal data in the later publication.

[c]Additional faunal counts were done for this study, and the same transfer function was used as in the previous publication.

[d]Some temperatures in Table 5.5 and Figure 5.6 are slightly different from those previously published. The previous temperatures were calculated without adding *pachyderma* (dextral) intergrade to *Neogloboquadrina pachyderma* (dextral). Transfer function F13' calls for adding them together.

Table 5.5. August and February sea-surface temperature (SST) estimates for 6000 and 9000 yr B.P. and SST anomalies relative to modern atlas temperatures and core-top SST estimates assuming that the thermal equator at 9000 yr B.P. was in the Northern Hemisphere

Core	6000 yr B.P.[a] SST (°C) Feb.	Aug.	6000 yr B.P. SST anomaly[b] (°C) Atlas Feb.	Aug.	Core top Feb.	Aug.	9000 yr B.P.[a] SST (°C) Feb.	Aug.	9000 yr B.P. SST anomaly[b] (°C) Atlas Feb.	Aug.	Core top Feb.	Aug.
A164–61[c]	16.7 *	—	9.9	—	3.9	—	17.4	—	10.6	—	4.6	—
A179–15[c]	24.0	27.2	0.3	-1.4	1.0	0.0	22.9	27.1	-0.8	-1.5	-0.1	-0.1
K708-1	11.6	17.6	1.1	1.3	0.6	0.6	11.1	17.3	0.6	1.0	0.1	0.3
K714-15[c]	8.1	12.3	-0.2	0.4	-1.0	-1.2	8.2	12.7	-0.1	0.8	-0.9	-0.8
M12309-2[c]	16.6	20.0	-0.9	-1.3	-0.4	-1.0	13.9	18.1	-3.6	-3.2	-3.1	-2.9
M12310-4[d]	16.4	21.4	-1.9	-0.5	—	—	11.9	17.0	-6.4	-4.9	—	—
M12379-1[c]	13.4	18.6	-4.4	-2.4	1.3	2.0	12.4	17.3	-5.4	-3.7	0.3	0.7
M12392-1[c]	15.8	22.8	-2.2	1.5	-0.8	-0.6	13.0	19.5	-5.0	-1.8	-3.6	-3.9
RC9-49[c]	24.6	27.6	-1.5	-0.5	-0.4	0.5	23.5	27.8	-2.6	-0.3	-1.5	0.7
RC9-225[c]	10.7	16.3	0.7	1.7	1.4	2.7	9.4 x	14.1	-0.6	-0.5	0.1	0.5
RC9-228[c]	11.1 *	16.6	0.9	1.4	1.5	1.6	9.0	13.8	-1.2	-1.4	-0.6	-1.2
RC10-288	19.0	—	-0.1	—	0.2	—	17.3 *	—	-1.8	—	-1.5	—
RC13-189[f]	27.2	25.4	0.6	-0.7	0.1	0.1	26.3 *	23.8	-0.3	-2.3	-0.8	-1.5
RC24-01	25.2	23.0	-1.7	-0.5	-0.7	-0.2	23.1	22.1	-3.8	-1.4	-2.8	-1.1
RC24-07	25.5	24.0	-1.0	0.8	-0.9	0.6	23.8	22.7	-2.7	-0.5	-2.6	-0.7
V15-168	27.2	24.9	0.0	-1.8	0.1	-0.3	27.0 *	24.8	-0.2	-1.9	-0.1	-0.4
V17-178[c]	9.2	—	3.0	—	2.2	—	6.4	—	0.2	—	-0.6	—
V21-2	19.6	—	0.9	—	2.0	—	20.2	—	1.5	—	2.6	—
V22-174	26.8	24.0	1.4	0.8	-0.5	-0.7	26.7	24.0	1.3	0.8	-0.6	-0.7
V22-177[f]	25.7	23.8	-0.4	-0.2	0.2	0.7	25.5	23.7	-0.6	-0.3	0.0	0.6
V22-182[f]	25.6	23.1	-1.3	-0.6	-0.7	-0.9	24.5	22.6	-2.4	-1.1	-1.8	-1.4
V23-23[c]	1.6 *	7.2	-2.2	-2.2	-1.3	-0.7	0.4	6.4	-3.4	-3.0	-2.5	-1.5
V23-81[e]	9.5	14.3	-0.6	-0.5	-0.7	-1.2	9.1 *	13.8	-1.0	-1.0	-1.1	-1.7
V23-82[c]	9.8	14.2	-0.3	-1.1	0.5	0.2	9.5	14.0	-0.6	-1.3	0.2	0.0
V24-1	15.9	—	-1.6	—	2.4	—	14.4	—	-3.1	—	0.9	—
V25-56[c]	26.9	24.7	-0.4	-1.3	-0.2	0.1	27.2	25.2	-0.1	-0.8	0.1	0.6
V25-59	27.4	25.1	0.7	-1.3	-0.1	-0.2	26.9 x	24.4	0.2	-2.0	-0.6	-0.9
V25-60	27.4	24.7	0.9	-2.0	-0.3	-0.6	26.8	24.8	0.3	-1.9	-0.9	-0.5
V25-75	25.1	27.5	-1.2	-0.6	1.1	-0.9	24.7	27.7	-1.6	-0.4	0.7	-0.7
V26-176[c]	19.6 *	—	0.6	—	0.6	—	15.7 *	—	-3.3	—	-3.3	—
V26-177[c]	16.6	—	2.6	—	1.4	—	16.7	—	2.7	—	1.5	—
V27-20[e]	4.5 *	9.9	0.6	-1.1	-0.1	-0.9	3.6	10.9	-0.3	-0.1	-1.0	0.1
V27-114[c,e]	9.1	14.3	2.1	2.3	0.7	2.3	9.5	15.0	2.5	3.0	1.1	3.0
V28-14	6.8 *	10.9	3.6	1.4	-1.6	-1.9	1.9	7.8	-1.3	-1.7	-6.5	-5.0
V28-127[c]	24.9 *	27.2	-1.2	-0.8	-0.3	-0.4	22.9	27.0	-3.2	-1.0	-2.3	-0.6
V29-179[e]	14.2	20.7	1.0	0.9	2.1	2.7	11.8 *	17.8	-1.4	-2.0	-0.3	-0.2
V29-183K[e]	12.5	19.3	1.4	2.6	1.6	3.0	11.0	18.0	-0.1	1.3	0.1	1.7
V29-192K[c]	9.9	14.8	-0.1	0.2	0.1	-0.2	9.4	14.1	-0.6	-0.5	-0.4	-0.9
V29-206	7.1	13.5	3.6	3.5	-0.7	-0.4	5.9	10.8	2.4	0.8	-1.9	-3.1
V30-36	25.8	28.3	-0.4	2.0	-0.5	-0.3	25.4 *	27.7	-0.8	1.4	-0.9	-0.9
V30-40	26.0	24.0	-1.5	-0.4	-1.0	0.1	24.7	22.4	-2.8	-2.0	-2.3	-1.5
V30-41K[c]	26.7 *	24.4	-0.8	-0.1	1.3	0.5	25.2	23.8	-2.3	-0.7	-0.2	-0.1
V30-49[f]	17.7 *	26.2	-2.3	1.4	-0.8	2.3	15.9	23.7	-4.1	-1.1	-2.6	-0.2
V30-51K	14.3	21.4	-4.7	-2.9	-1.0	-1.8	13.7	21.2	-5.3	-3.1	-1.6	-2.0
V30- 97[c]	14.5	21.6	-0.1	-1.0	0.0	0.0	17.4	22.0	2.8	-0.6	2.9	0.4
V30-101K[c]	12.6	18.0	-0.6	-2.5	-1.3	-1.5	12.0	16.8	-1.2	-3.7	-1.9	-2.7

[a]An asterisk indicates that the SST estimate is an average of two samples above and below map level pick; the offset from map level for the samples was within 100 yr. An *x* indicates that the SST estimate is an average of two samples at the same depth.

[b]A positive (negative) anomaly indicates temperatures were warmer (cooler) in the past than today.

[c]Core top not at 0 cm but within 5 cm of top.

[d]No core top.

[e]Core-top SST estimate from F13' core-top data set (Ruddiman and Glover, 1975).

[f]Core-top SST estimate from F20 core-top data set (N. G. Kipp, unpublished).

Winter (•) and Summer (○) SST (°C)

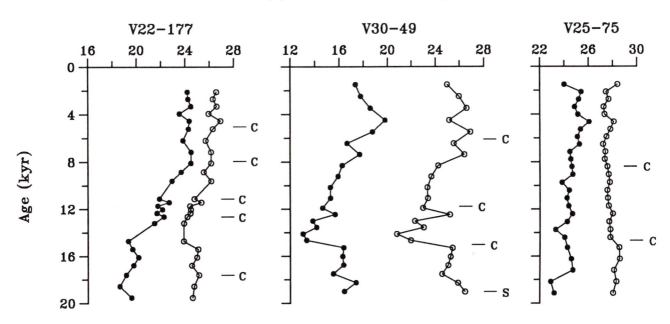

Fig. 5.4. Selected time series of the major patterns of variation in winter (•) and summer (○) sea-surface temperature (°C) in equatorial Atlantic cores (data from Mix *et al.*, 1986). Lines indicate dated levels: C = level dated by ^{14}C; S = stratigraphic pick from δ^{18}O records.

Winter (•) and Summer (○) SST (°C)

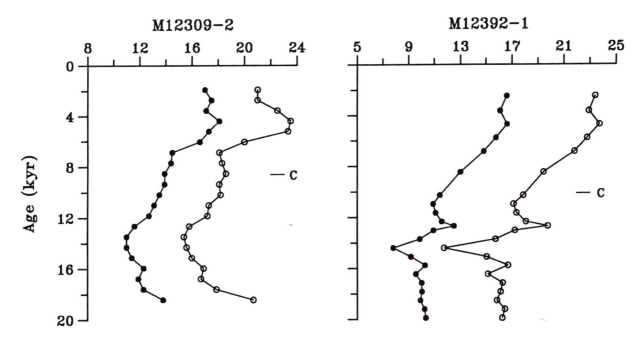

Fig. 5.5. Late-glacial and Holocene winter (•) and summer (○) sea-surface temperature (°C) in cores M12309-2 and M12392-1 off northwest Africa (data from Thiede, 1977). Age scales were established by radiocarbon dates (C) from Geyh (1979) and M. Sarnthein (personal communication, 1987).

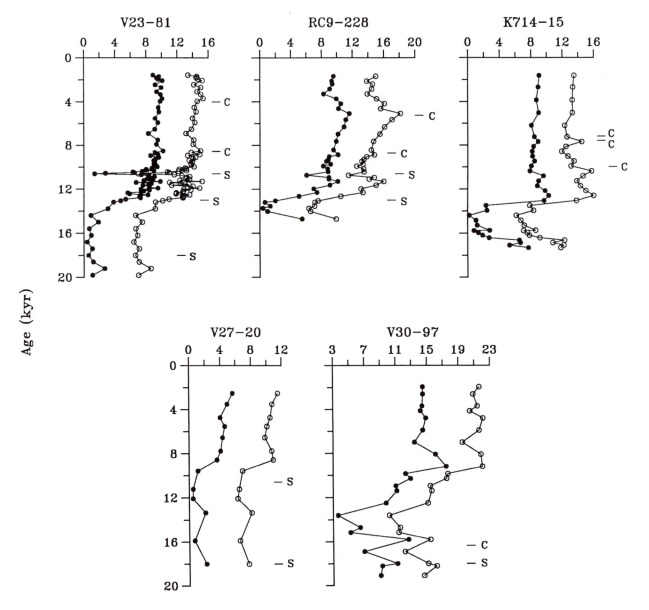

Fig. 5.6. Selected time series of winter (•) and summer (○) sea-surface temperature (SST) (°C) from North Atlantic cores. The three cores at the top and core V27-20 at bottom left show the differential timing of the SST warming associated with the time-transgressive retreat of the North Atlantic polar front. The three top cores also show the uncertain timing of warming during the early, middle, and late Holocene. Core V30-97 shows the cold-deglacial pattern of the northwestern subtropical gyre. C = radiocarbon dates; S = stratigraphic pick based on first polar-front retreat at 13,000 yr B.P.

Sea (Duplessy *et al.,* 1981; Ruddiman and McIntyre, 1981). Kellogg (1984) also found evidence that the East Greenland-Irminger Current region warmed after 10,000 yr B.P.

Examples of the effects of this fundamental glacial-interglacial polar-front response on estimated SST in several cores from the high-latitude North Atlantic are shown in Figure 5.6. The three cores at the top of Figure 5.6 show the 13,000-yr B.P. warming, and two of the

three cores show the 10,000-yr B.P. warming. Labrador Sea core V27-20 (Fig. 5.6, bottom left) shows only the latest warming (at or after 9000 yr B.P.). This pattern of gradual polar-front retreat causes negative (cold) SST anomalies at 9000 yr B.P. in the Labrador Sea (core V27-20) and in the East Greenland and Irminger currents but no definite anomalies at 6000 yr B.P.

Recent ^{14}C dating using accelerator mass spectrometers (Duplessy *et al.,* 1986; Bard *et al.,* 1987; Broecker *et al.,* 1988) has confirmed a Younger Dryas age of 10,500 yr B.P. for the prominent cold SST event evident in Figure 5.6 in cores V23-81 and RC9-228. Estimates of the apparent age of the first warming of the eastern North Atlantic vary widely, however, from as old as 14,500 yr B.P. (Broecker *et al.,* 1988) to as young as 11,600 yr B.P. (Duplessy *et al.,* 1986). The estimate of Ruddiman and McIntyre (1981) of 13,000 yr B.P. falls roughly in the middle of the range.

All SST signals show much less variation once the polar front has passed over (retreated from) a given core site. Nonetheless, several records indicate considerable warmth in the early to middle Holocene, with estimated SST values at various times as much as 1-3°C greater than today. These cores also show other intervals cooler than today. However, the timing of these anomalies is highly variable from core to core, and the anomalies show no coherent spatial pattern (top three cores in Fig. 5.6).

These uncorrelated or poorly correlated trends are open to several interpretations. They could mean that SSTs differed significantly at the indicated time intervals on a small spatial scale. However, cores RC9-228 and V23-81 (Fig. 5.6) are separated by less than 100 km, and it is implausible that the apparent mid-Holocene warmth in core RC9-228 and the apparent late-Holocene warmth in core V23-81 were caused by different responses over such a small distance. This example does not disprove the possibility of real differences in response between more widely separated cores in the high-latitude Atlantic, but a literal acceptance of all interpolated SST estimates from this area would result in a jumbled pattern of positive SST anomalies juxtaposed with negative anomalies nearby (as shown in Figs. 5.9 and 5.12). Whether these values are real or artifactual, averaging them regionally would give a rather small SST anomaly for the area as a whole.

Two other interpretations are possible. First, warmer SSTs may actually have occurred across the entire high-latitude North Atlantic at some time during the Holocene, but the signal may have been dated at very different ages because of contaminated ^{14}C dates or other effects of sediment redistribution, such as highly nonlinear sedimentation rates (Ruddiman and Bowles, 1976). Second, the apparent Holocene warmth may be an artifact created by local complications such as sediment redistribution or dissolution.

After unsuccessful attempts to resolve this problem definitively by numerous strategies, we can only conclude that the available evidence at this time neither rules out an episode of regional high-latitude North Atlantic warmth in the mid-Holocene nor convincingly documents such an episode. We suggest that any 9000- and 6000-yr B.P. SST anomalies north of 50°N be regarded for now as artifacts caused by local factors, except for the negative (cold) anomalies in the Labrador Sea at 9000 yr B.P.

To select 9000- and 6000-yr B.P. levels in North Atlantic cores north of 50° for the maps of SSTs and SST anomalies, we used direct interpolation between ^{14}C dates. Even though many of these dates are probably too old because of contamination, we could find no means (biostratigraphic or other) to improve on these age estimates in an objective way. Future work dating the coarser (and less mobile) $CaCO_3$ fractions with accelerator mass spectrometers may yet help to resolve these uncertainties.

The second pattern of surface-ocean response in the North Atlantic is observed in cores from the northwestern part of the subtropical gyre between latitudes 35 and 50°N. In this area a cold SST pulse is centered on the early part of the deglaciation between 15,000 and 10,000 yr B.P., with warmer values in the preceding glacial maximum and the succeeding interglaciation. This pattern is shown by core V30-97 in Figure 5.6. The cold SSTs associated with this pulse have largely dissipated by 9000 yr B.P., although sediment mixing may in this case smooth the transition to Holocene warmth.

Results

Estimated SSTs for February and August at 9000 yr B.P. are shown in Figures 5.7 and 5.8 for the two possible positions of the thermal equator indicated by Mix *et al.* (1986). In Figure 5.7, the mean annual thermal equator is assumed to be in the Northern Hemisphere near its present location; in Figure 5.8, it is placed in the Southern Hemisphere. SST anomalies for these two 9000-yr B.P. configurations relative to both atlas and selected core-top values are shown in Figures 5.9 and 5.10. Estimated SSTs for February and August at 6000 yr B.P. are shown in Figure 5.11, with the thermal equator in a position equivalent to that of today (Mix *et al.,* 1986). SST anomalies for 6000 yr B.P. are shown in Figure 5.12.

All SST estimates and anomalies are listed in Table 5.5 (9000-yr B.P. thermal equator in the Northern Hemisphere) and Table 5.6 (9000-yr B.P. thermal equator in the Southern Hemisphere). Relative to atlas SST values, roughly half of the anomalies at 9000 yr B.P. and one-third at 6000 yr B.P. exceed the standard errors of the equations (1-1.5°C one-way error at the 80% confidence interval). As summarized in the previous sections, some of the larger anomalies shown are credible, but many others must be regarded as suspect. The following discussion is directed toward anomalies that (1) exceed the standard error of estimate when calculated relative to atlas values, or exceed the 0.5°C count reproducibility when calculated relative to core-top values, and (2) form coherent clusters in specific regions.

W. F. RUDDIMAN AND A. C. MIX

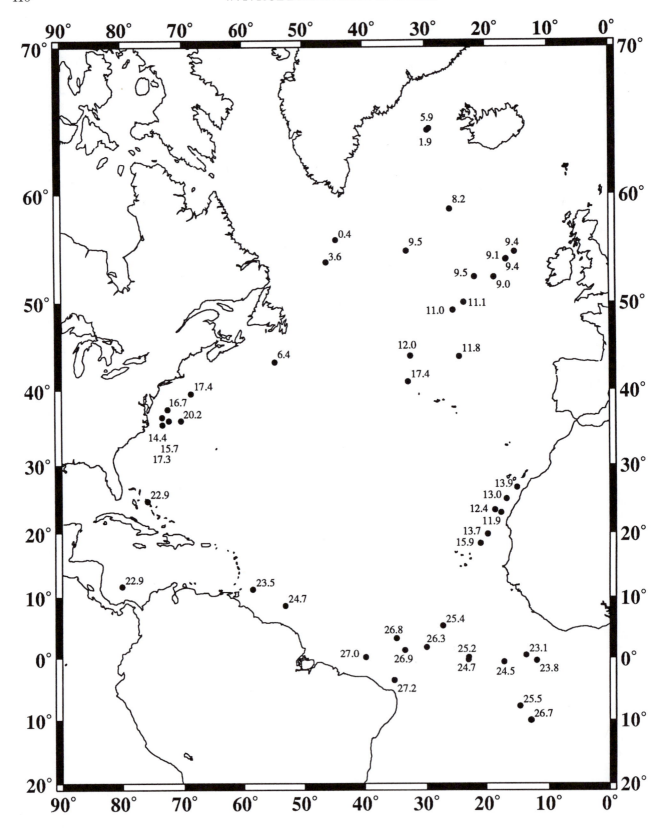

Fig. 5.7. Estimated sea-surface temperatures at 9000 yr B.P. in (a) February and (b) August, based on the assumption that the ther-

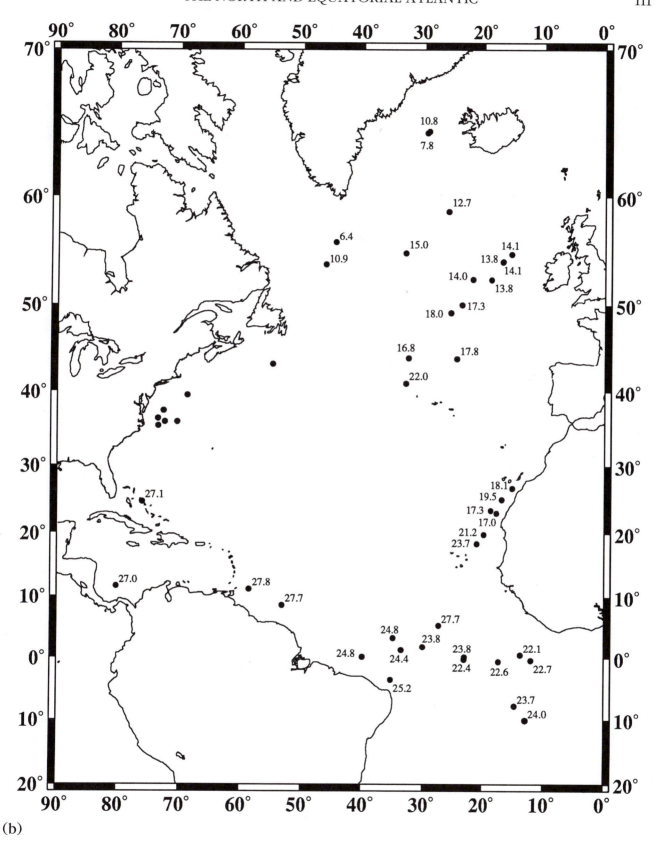

(b)

mal equator was in the modern position (Mix *et al.*, 1986).

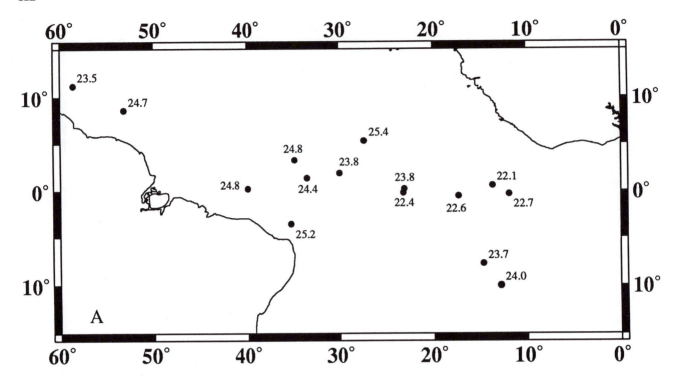

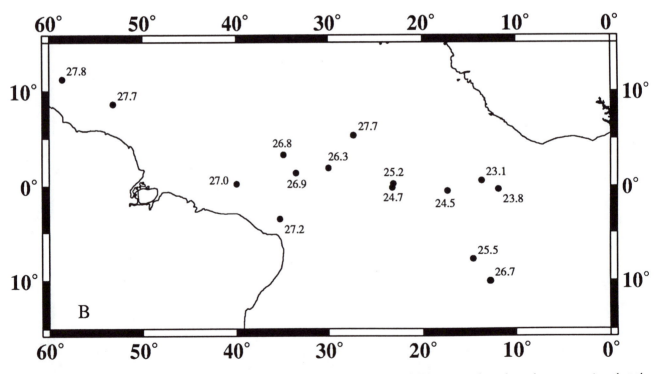

Fig. 5.8. Estimated sea-surface temperatures at 9000 yr B.P. in (A) February and (B) August, based on the assumption that the thermal equator was in the Southern Hemisphere (Mix *et al.*, 1986).

Table 5.6. August and February sea-surface temperature (SST) estimates for 9000 yr B.P. and SST anomalies relative to modern atlas temperatures and core-top SST estimates based on the assumption that the thermal equator was in the Southern Hemisphere

| | SST (°C) | | SST anomaly[a] (°C) | | | |
| | | | Atlas | | Core top | |
Core	Feb.	Aug.	Feb.	Aug.	Feb.	Aug.
RC9–49[b]	23.5	27.8	-2.6	-0.3	-1.5	0.7
RC13–189[c]	23.8	26.3	-2.8	0.2	-3.3	1.0
RC24–01	22.1	23.1	-4.8	-0.4	-3.8	-0.1
RC24–07	22.7	23.8	-3.8	0.6	-3.7	0.4
V15–168	24.8	27.0	-2.4	0.3	-2.3	1.8
V22–174	24.0	26.7	-1.4	3.5	-3.3	2.0
V22–177[c]	23.7	25.5	-2.4	1.5	-1.8	2.4
V22–182[c]	22.6	24.5	-4.3	0.8	-3.7	0.5
V25–56[b]	25.2	27.2	-2.1	1.2	-1.9	2.6
V25–59	24.4	26.9	-2.3	0.5	-3.1	1.6
V25–60	24.8	26.8	-1.7	0.1	-2.9	1.5
V25–75	24.7	27.7	-1.6	-0.4	0.7	-0.7
V30–36	25.4	27.7	-0.8	1.4	-0.9	-0.9
V30–40	22.4	24.7	-5.1	0.3	-4.6	0.8
V30–41K[b]	23.8	25.2	-3.7	0.7	-1.7	1.3

[a]A postive (negative) anomaly indicates temperatures were warmer (cooler) than today.
[b]Core top not at 0 cm but within 5 cm of top.
[c]Core-top SST estimate from F20 core-top data set (N. G. Kipp, unpublished).

EQUATORIAL ATLANTIC

Three clusters of cores have significant anomalies in the 9000-yr B.P. comparison based on the Northern Hemisphere thermal equator position (Fig. 5.9). Negative February SST anomalies in the Caribbean-Brazil Current area and along the equator in the eastern Atlantic (Fig. 5.9a) persist relative to both atlas and core-top values. And seven cores along the equator have negative August anomalies at 9000 yr B.P. that exceed 1°C relative to atlas values; these anomalies are smaller relative to core-top values but still exceed the count reproducibility (Fig. 5.9b).

The negative February SST anomalies in the equatorial Atlantic are much larger for the 9000-yr B.P. comparison based on the Southern Hemisphere thermal equator position, essentially encompassing all equatorial waters between 5°N and 10°S (Fig. 5.10a). These anomalies are almost as large relative to core-top estimates as they are relative to atlas values. In August positive SST anomalies relative to atlas values are observed in the southeastern equatorial Atlantic (Fig. 5.10b); in the core-top comparison, these positive anomalies stretch northward beyond the equator in the western Atlantic.

At 6000 yr B.P., with the thermal equator unambiguously in the Northern Hemisphere position, the negative February SST anomalies are clustered in the Caribbean-Brazil Current area and in the eastern equa-

torial Atlantic (Fig. 5.12a). In the Caribbean-Brazil Current region, the anomalies apparent in the atlas data disappear in the core-top comparison. In the equatorial Atlantic, the negative SST anomalies at 6000 yr B.P. are smaller than those at 9000 yr B.P. but are still significant in both the atlas and the core-top comparisons. In August at 6000 yr B.P., negative SST anomalies relative to atlas values show up only in the western equatorial Atlantic (Fig. 5.12b); they disappear in the core-top comparisons, however, leaving no region of significant August SST anomalies at 6000 yr B.P.

In general, the 6000-yr B.P. anomalies are most similar to the 9000-yr B.P. anomalies derived when the thermal equator is assumed to be in the Northern Hemisphere, and this is true for both August and February. This observation suggests that the thermal equator remained in the Northern Hemisphere during the deglaciation, despite the obvious decrease in thermal imbalance between the hemispheres noted by Mix et al. (1986).

In summary, negative February anomalies in the Caribbean-Brazil Current area at 9000 yr B.P. have weakened by 6000 yr B.P. Large negative February SST anomalies mapped at 9000 yr B.P. in the eastern equatorial Atlantic for both positions of the thermal equator persist at a reduced magnitude to 6000 yr B.P. The August SST anomaly patterns at 9000 yr B.P. differ greatly for the two thermal equator positions; choosing the northern position results in a negative anomaly along the modern thermal equator, which persists in weakened form at 6000 yr B.P.

NORTHWEST AFRICAN MARGIN

Most of the cores in the northwest African margin have negative SST anomalies for both seasons at both 9000 and 6000 yr B.P. These anomalies are large relative to the atlas SST values but smaller relative to core-top estimates.

The negative anomalies for both seasons at 9000 yr B.P. range from 3 to 6°C compared to atlas values, all well above the standard error of estimate of the transfer function. In some cases, however, the anomalies were smaller relative to core-top values. Core M12379-1 differs from the other METEOR cores in having smaller SST anomalies versus core-top values because it has a stronger late-Holocene SST cooling. We conclude that SSTs in this region of coastal upwelling and offshore Canaries Current flow were several degrees (1-5°C) cooler than today at 9000 yr B.P. in both seasons.

SST anomalies at 6000 yr B.P. are generally within the range of the standard error (1-2°C) for both atlas and core-top comparisons. Negative (cool) SST anomalies were slightly more common than positive (warm) anomalies, and no cores had positive anomalies for both core-top and atlas comparisons in either season. We conclude that SSTs in this region could have continued to be cooler at 6000 yr B.P. than at pres-

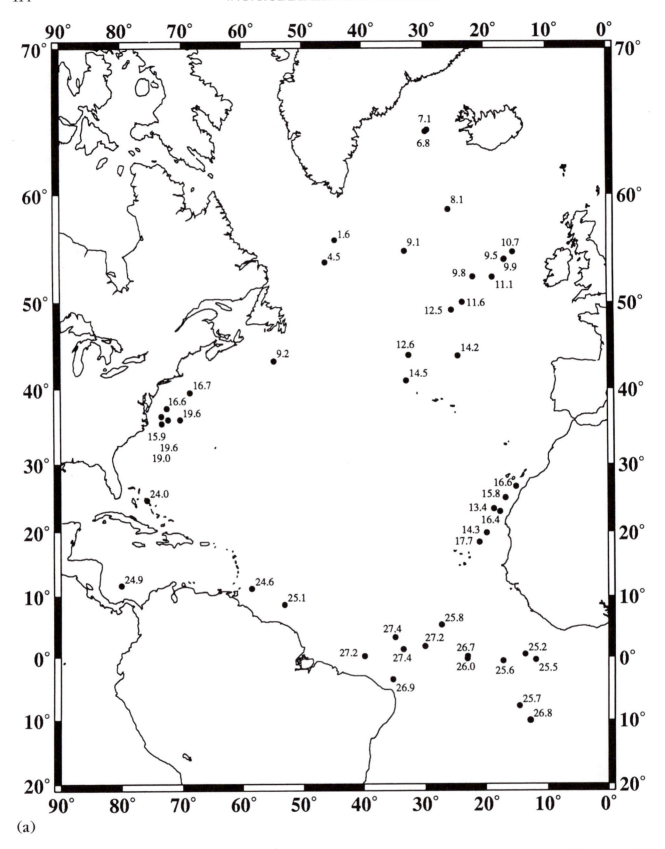

(a)

Fig. 5.9. Differences between estimated sea-surface temperatures (SSTs) at 9000 yr B.P. and at present in (a) February and (b) August, based on the assumption that the thermal equator at 9000 yr B.P. was in the modern position (Mix *et al.,* 1986). Anom-

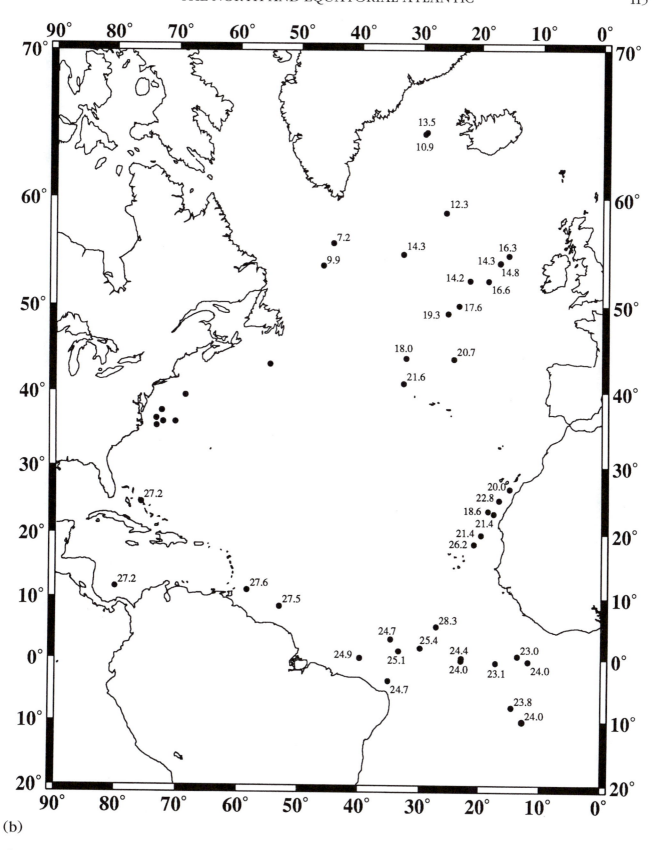

(b)

alies were calculated relative to atlas SST values, but anomalies relative to core-top SST estimates are shown in parentheses for cores with atlas anomalies larger than 1°C.

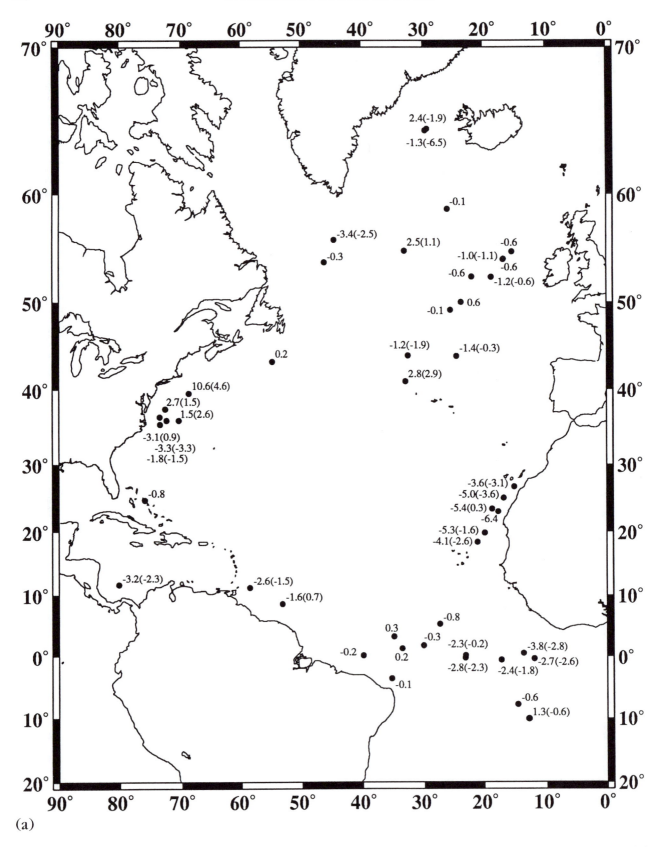

(a)

Fig. 5.10. Differences between estimated sea-surface temperatures (SSTs) at 9000 yr B.P. and at present in (a) February and (b) August, based on the assumption that the thermal equator at 9000 yr B.P. was in the Southern Hemisphere (Mix *et al.,* 1986).

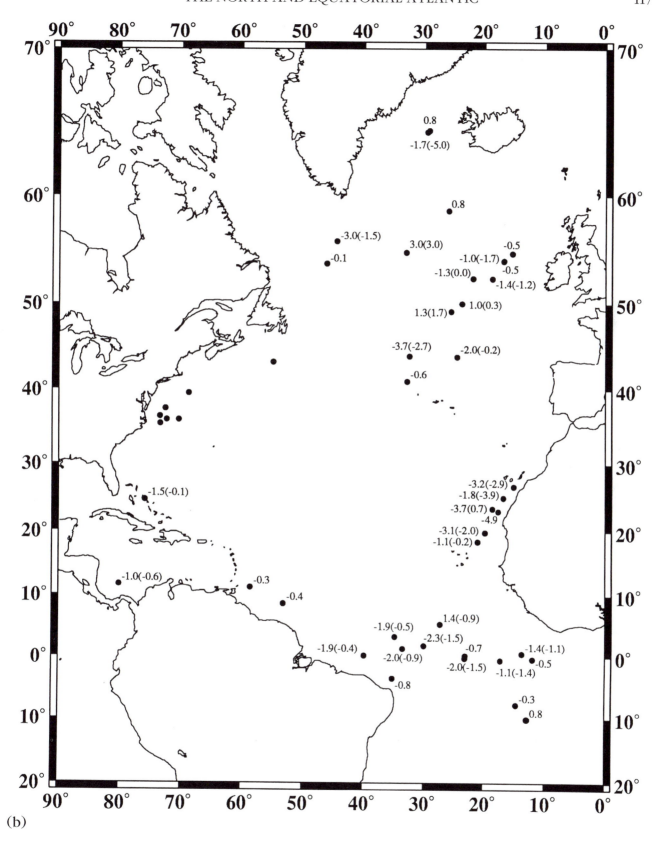

(b)

Anomalies were calculated relative to atlas SST values, but anomalies relative to core-top SST estimates are shown in parentheses for cores with atlas anomalies larger than 1°C.

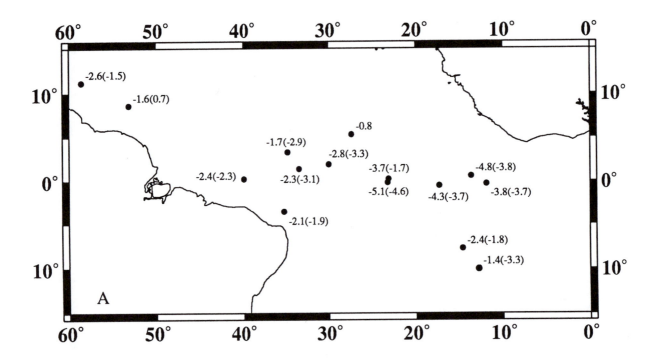

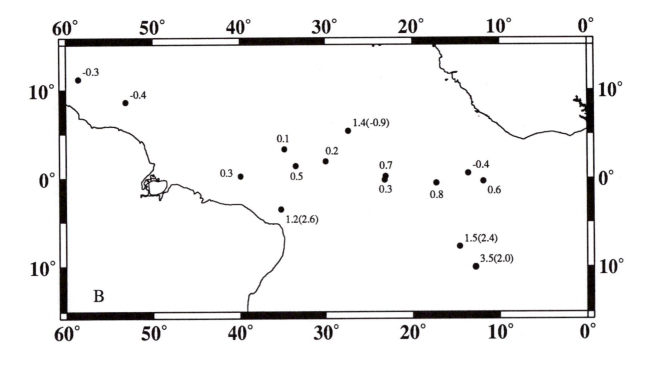

Fig. 5.11. Estimated sea-surface temperatures at 6000 yr B.P. in (a) February and (b) August.

ent by as much as 1-2°C, but the conflicting evidence on both anomaly strength and sign among these cores precludes any firm conclusions.

WESTERN SUBTROPICAL ATLANTIC

The winter (February) SST anomalies at 9000 yr B.P. in the western subtropical Atlantic are generally negative near Cape Hatteras and positive to the north. However, we reject all 9000-yr B.P. anomalies in this area because of inadequate stratigraphic control.

At 6000 yr B.P. the SST anomalies are generally positive. If we ignore the very large anomaly calculated relative to atlas values in core A164-61 off New Jersey and use instead the smaller anomaly calculated relative to the core-top estimates, then the winter SST anomalies at 6000 yr B.P. range from 0 to +4°C and average about +1.5°C. Although the amplitude of the SST anomalies is open to question, the consistent spatial and temporal trends argue for a significant positive SST anomaly in winter for this region at 6000 yr B.P. In contrast, Bartlein and Webb (1985) found that July temperatures in the lower atmosphere at 6000 yr B.P. were at least 1°C warmer than today in a latitudinal band across the Great Lakes, southern Canada, and New England somewhat north of the latitude of these western subtropical Atlantic cores.

HIGH-LATITUDE NORTH ATLANTIC

Most of the large SST anomalies at 9000 and 6000 yr B.P. in the high-latitude North Atlantic are regarded as artifacts of local complications, with positive anomalies in some cores canceling negative anomalies in others (Figs. 5.9 and 5.12). The cold SST anomalies at 9000 yr B.P. at the mouth of the Labrador Sea and along the East Greenland-Irminger Current region are regarded as real; they are also consistent with evidence from the coasts of Greenland, Baffin Island, and Labrador that the first warm subpolar waters arrived at or after 8000 yr B.P. (Andrews, 1972). None of the scattered positive and negative SST anomalies anywhere in the high-latitude North Atlantic at 6000 yr B.P. are accepted as valid. In contrast, Huntley and Prentice (this vol.) found summer temperatures several degrees warmer than at present in northern Europe at 6000 yr B.P.

Other Circum-Atlantic Oceanic Evidence

Seas and basins adjacent to the North Atlantic also hold evidence of sea-surface conditions at 9000 and 6000 yr B.P. Because of their smaller size and more restricted circulation, these basins tend to be subject to larger precipitation-evaporation imbalances than oceanic bodies, and their bottom waters tended to stagnate periodically during the Pleistocene.

GULF OF MEXICO

Meltwater runoff from North American ice entered the Gulf of Mexico during the last deglaciation and left an oxygen-isotope record in the shells of planktonic foraminifera. The timing of maximum runoff is still uncertain because of contaminated ^{14}C dates, but all published marine chronologies indicate that detectable runoff had ceased by 10,000 yr B.P. (see Fig. 3 in Leventer et al., 1982). Drainage patterns from proglacial lakes also indicate that all meltwater flowed eastward to the Atlantic after 9500 yr B.P. (Teller and Last, 1981). The available evidence thus argues against meltwater influence in the Gulf of Mexico by 9000 yr B.P.

Brunner (1982) published several records of estimated late-Quaternary SST in cores from the Gulf of Mexico. Although the Holocene layers are 30-90 cm thick, none of the cores had ^{14}C dating, and the wide sampling spacing (20 cm) equates to time intervals of 2000-6000 yr. Most of the records show core-top summer temperatures warmer than those earlier in the Holocene (including the presumed 9000- and 6000-yr B.P. levels), but winter SST values in cores from the Florida Straits fell slightly in the late Holocene. Most differences between modern values and estimated temperatures at the presumed 6000- and 9000-yr B.P. levels are less than 1°C.

MEDITERRANEAN

Oxygen-isotope evidence from planktonic foraminifera indicates that the surface waters of the eastern Mediterranean were stratified by an influx of low-salinity water beginning very late in the Pleistocene or early in the Holocene and continuing until about 7000 yr B.P. (Williams and Thunell, 1979). Although this influx has been widely attributed to deglacial meltwater, the Scandinavian ice sheet had largely disintegrated by 9000 yr B.P.

Rossignol-Strick et al. (1982) attributed the influx of low-salinity water to Nile runoff from a monsoon-driven increase in precipitation over east-central Africa. Because of the increased density stratification, overturning of surface waters diminished, ventilation of the deeper basin ceased, and bottom waters became stagnant from lack of oxygen, thus allowing the formation of organic muds called sapropels. The evidence thus indicates that a low-salinity lid existed in the eastern Mediterranean at 9000 yr B.P. but had disappeared by 6000 yr B.P.

Thunell et al. (1977) showed that no-analogue planktonic foraminiferal faunas were present during many of the low-salinity sapropel episodes in the eastern Mediterranean during the Quaternary. Although data in Thunell and Lohmann (1979) do not specify whether a no-analogue fauna existed during the early-Holocene sapropel episode, SST estimates are not available.

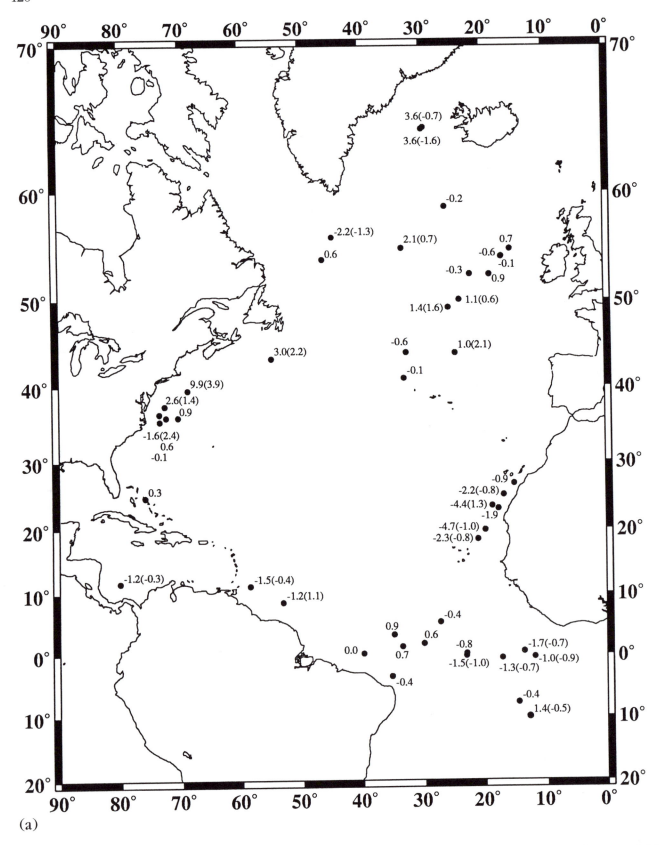

Fig. 5.12. Differences between estimated sea-surface temperatures (SSTs) at 6000 yr B.P. and at present in (a) February and (b) August. Anomalies were calculated relative to atlas SST values, but anomalies relative to core-top SST estimates are shown in

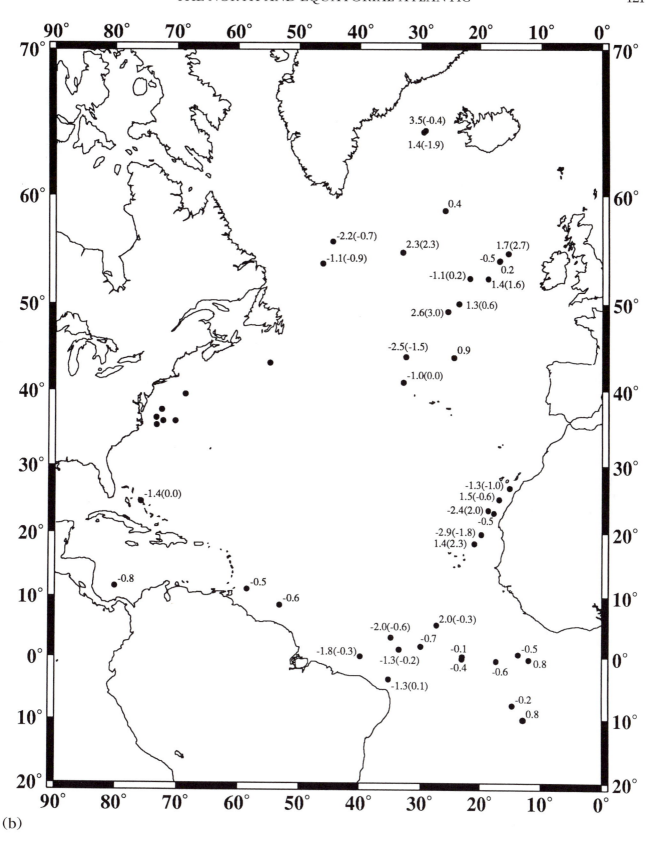

(b)

parentheses for cores with atlas anomalies larger than 1°C.

Discussion and Conclusions

For many of the regions summarized in this chapter, estimates show little or no significant difference between SSTs at 9000 or 6000 yr B.P. and at present. This basic similarity is in keeping with climatic theory, for several reasons.

First, ice is widely regarded as one of the major means by which insolation forcing is converted into a climatic response on earth. By this logic, the rapid areal retreat (and volumetric shrinkage) of Northern Hemisphere ice from 14,000 until 7000 yr B.P. should have reduced the large late-glacial cooling effect of the ice on the rest of the earth (and especially the Northern Hemisphere oceans) by the early Holocene (9000 yr B.P.). This kind of impact is borne out by the first-order SST signal in the north and equatorial oceans over the last 18,000 yr; most regions show the major change from cooler glacial to warmer Holocene SSTs within 2000 yr of 11,000 yr B.P. Thus by 9000 yr B.P. the thermal forcing effect of the ice on the northern and equatorial Atlantic oceans was much diminished, and by 6000 yr B.P. it was no different from today.

This interpretation is also consistent with the localized delay in warming of the Labrador Sea, which remained near the colder glacial SST values until the early Holocene because of the persistence of extensive Laurentide ice over eastern North America (Prest, 1969; Denton and Hughes, 1981). In a general circulation model (GCM) simulation, Mitchell *et al.* (1988) found that even a high-standing North American ice sheet at 9000 yr B.P. would have had a negligible effect on the North Atlantic except for the Labrador Sea.

Atmospheric CO_2 is a second important forcing function. Although the 70-ppm deglacial increase in CO_2 is not precisely dated, it appears to have occurred between 13,000 and 9000 yr B.P. (Oeschger *et al.*, 1984). If these age estimates are correct, CO_2 concentrations must have been close to their preindustrial levels by 9000 and 6000 yr B.P. Again, the lack of difference in CO_2 forcing between the early Holocene and the (preindustrial) late Holocene is consistent with the generally small SST anomalies in the northern and equatorial Atlantic at these times.

In several regions, however, estimates of past SSTs differed significantly from modern SST values. Cold SST anomalies were noted in the Brazil Current, in the south equatorial Atlantic, and along the coast of northwest Africa at 9000 yr B.P. (and to a lesser extent also at 6000 yr B.P.). Warm SST anomalies were noted off Cape Hatteras at 6000 yr B.P. and possibly in the subpolar North Atlantic as well. These anomalies must be due to other factors.

One forcing function—insolation—clearly was different from today at 9000 and 6000 yr B.P. Insolation at all latitudes of the Northern Hemisphere was higher in summer and lower in winter than at present; at low and middle latitudes of the Southern Hemisphere, summer insolation was lower and winter values were

higher than at present (Berger, 1978). These insolation anomalies were particularly influential on thermal responses in regions remote from the ice sheets; closer to the ice sheets, they became important only after the pervasive ice-sheet effects had diminished during the early and middle Holocene.

Insolation variations probably do not have much direct impact on the low-latitude ocean. Calculations by Kutzbach and Otto-Bliesner (1982) based on modeling results from Schneider and Thompson (1979) suggest that the strong seasonal insolation signal is largely damped by the thermal inertia of the ocean and by physical mixing in the mixed layer, which extends to 50–100 m. Kutzbach and Otto-Bliesner (1982) calculated that the seasonal insolation changes of 7% translate into a change of only 0.5°C in the seasonal SST range, or just 0.25°C in any seasonal SST estimate. Recent sensitivity tests with GCMs (Kutzbach and Gallimore, 1988; Mitchell *et al.*, 1988) confirm this assessment. Such small changes are below the level of detection in our SST estimates. Thus direct insolation effects on the low-latitude ocean should not be an important factor in explaining SST anomalies.

Holocene insolation changes, however, may have had an impact on SST values via other factors, in particular sea ice and differential heating of land and sea. Sea ice responds to seasonal changes in insolation today, and it should have responded to the different insolation values of the early Holocene. Changes in sea-ice limits are climatologically important because of the marked contrast between ice and water albedos and because sea ice thermally isolates the lower atmosphere from the upper ocean. Because summer insolation dominates the annual-average insolation at higher latitudes, insolation trends predict less sea ice in the Northern Hemisphere and more in the Southern Hemisphere at 9000 and 6000 yr B.P. than at present.

Recent GCM studies (Kutzbach and Gallimore, 1988; Mitchell *et al.*, 1988) indicated that the insolation levels at 9000 yr B.P. greatly reduced year-round sea-ice thickness and autumn sea-ice extent in the Northern Hemisphere. Mitchell *et al.* (1988) showed that this warming would have been locally diminished by the downwind impact of a high-standing residual Laurentide ice sheet at 9000 yr B.P. but that the influence of stronger insolation generally prevailed, with reduced sea-ice cover and thickness and a warmer Arctic at 9000 yr B.P.

Although none of the cores in our study are from high-latitude areas that were covered by sea ice in the Holocene, the scattered suggestion of warm SST anomalies at 9000 and 6000 yr B.P. in the midlatitude North Atlantic could conceivably be the result of reduced sea ice; however, this conclusion is speculative at this point. Changes in Antarctic sea ice were negligible in the 9000-yr B.P. GCM experiments (Kutzbach and Gallimore, 1988; Mitchell *et al.*, 1988).

Insolation could also have significant local effects

on the ocean by directly heating adjacent land surfaces and altering the overlying atmospheric structure. Such effects would be especially evident in coastal upwelling areas (Kutzbach, 1981). This mechanism is potentially powerful because of the low heat capacity and large seasonal thermal response of land surfaces. We consider it likely that some of the seasonal SST anomalies observed at low latitudes and in coastal areas at 9000 and 6000 yr B.P. (Figs. 5.9, 5.10, and 5.12) reflect this kind of land-mediated insolation forcing.

Two GCM sensitivity tests in which insolation was set to 9000-yr B.P. values (Kutzbach and Gallimore, 1988; Mitchell *et al.,* 1988) produced generally small, and sometimes contradictory, changes in low-latitude SSTs. Our finding of cooler SSTs along the equator in February agrees with the winter values from Mitchell *et al.* (1988). It also agrees with the annual-average changes in Kutzbach and Gallimore (1988) but not with their (warmer) February values. Similarly, our estimate of cooler SSTs in both seasons off the coast of northwest Africa agrees with the results of Mitchell *et al.* (1988) but disagrees with the warmer August estimates of Kutzbach and Gallimore (1988).

The problem with these data-model comparisons is that both the equatorial Atlantic and the northwest African margin are sites of divergence and upwelling. The GCM models used in the sensitivity tests are coupled to mixed-layer ocean models that lack upwelling. As a result, more meaningful data-model comparisons for the reactive low-latitude regions, as well as a full understanding of the key dynamics, must await more sophisticated ocean models.

Acknowledgments

We thank Bill Balsam, Jean-Claude Duplessy, Tom Janecek, Joe Morley, Uwe Pflaumann, and Michael Sarnthein for reviews. Ann Esmay kept meticulous track of the ever-changing data tables, and Beatrice Rasmussen helped with the manuscript. This chapter was assembled with support from U.S. Department of Energy contract 79EV10097.000 subcontracted through Brown University.

References

Andrews, J. T. (1972). Recent and fossil growth rates of marine bivalves, Canadian Arctic, and late Quaternary Arctic marine environments. *Palaeogeography, Palaeoclimatology, Palaeoecology* 11, 157-176.

Balsam, W. (1981). Late Quaternary sedimentation in the western North Atlantic: Stratigraphy and paleoceanography. *Palaeogeography, Palaeoclimatology, Palaeoecology* 35, 215-240.

Balsam, W. L., and Heusser, L. K. (1976). Direct correlation of sea-surface paleotemperatures, deep circulation, and terrestrial paleoclimates: Foraminiferal and palynological evidence from two cores off Chesapeake Bay. *Marine Geology* 21, 121-147.

Bard, E., Arnold, M., Moyes, J., and Duplessy, J.-C. (1987). Reconstruction of the last deglaciation: Deconvolved records of δ18O profiles, micropaleontological variations and accelerator mass spectrometer 14C dating. *Climate Dynamics* 1, 101-112.

Bartlein, P. J., and Webb, T. III (1985). Mean July temperature at 6000 yr B.P. in eastern North America: Regression equations for estimates from fossil-pollen data. *Syllogeus* 55, 301-342.

Be, A. W. H., Damuth, J. E., Lott, L., and Free, R. (1976). Late Quaternary climatic record in western equatorial Atlantic sediment. *Geological Society of America Memoir* 145, 165-200.

Berger, A. L. (1978). Long-term variations of caloric insolation resulting from the earth's orbital elements. *Quaternary Research* 9, 139-167.

Broecker, W. S., and Kulp, J. L. (1957). Lamont natural radiocarbon measurements IV. *Science* 126, 1324-1334.

Broecker, W. S., Kulp, J. L., and Tucek, C. S. (1956). Lamont natural radiocarbon measurements III. *Science* 124, 154-165.

Broecker, W. S., Andree, M., Wolfli, W., Oeschger, H., Bonani, G., Peteet, D., and Kennett, J. (1988). The chronology of the last deglaciation: Implications to the cause of the Younger Dryas event. *Paleoceanography* 3, 659-669.

Brunner, C. A. (1982). Paleoceanography of surface waters in the Gulf of Mexico during the late Quaternary. *Quaternary Research* 17, 105-119.

CLIMAP Project Members (1981). Seasonal reconstructions of the earth's surface at the last glacial maximum. *Geological Society of America Map and Chart Series* MC-36.

——. (1984). The last interglacial ocean. *Quaternary Research* 21, 123-224.

Denton, G. H., and Hughes, T. J. (1981). "The Last Great Ice Sheets." Wiley Interscience, New York.

Duplessy, J. C., Delibrias, G., Turon, J. L., Pujol, C., and Duprat, J. (1981). Deglacial warming of the northeastern Atlantic Ocean: Correlation with the paleoclimatic evolution of the European continent. *Palaeogeography, Palaeoclimatology, Palaeoecology* 35, 121-144.

Duplessy, J. C., Arnold, M., Maurice, P., Bard, E., Duprat, J., and Moyes, J. (1986). Direct dating of the oxygen isotope record of the last deglaciation by 14C accelerator mass spectrometry. *Nature* 320, 350-352.

Geyh, M. A. (1979). 14C routine dating of marine sediments. *In* "Radiocarbon Dating" (R. Berger and H. E. Suess, Eds.), pp. 470-491. International Radiocarbon Conference, 9th Proceedings. University of California Press, Berkeley.

Imbrie, J., and Kipp, N. G. (1971). A new micropaleontological method for quantitative paleoclimatology: Application to a late Pleistocene Caribbean core. *In* "The Late Cenozoic Glacial Ages" (K. K. Turekian, Ed.), pp. 71-181. Yale University Press, New Haven.

Imbrie, J., van Donk, J., and Kipp, N. G. (1973). Paleoclimatic investigation of a late Pleistocene Caribbean deep-sea core: Comparison of isotopic and faunal methods. *Quaternary Research* 3, 10-38.

Jones, G. A., and Ruddiman, W. F. (1982). Assessing the global meltwater spike. *Quaternary Research* 17, 148-172.

Kellogg, T. B. (1984). Late-glacial-Holocene high-frequency climatic changes in deep-sea cores from the Denmark Strait. *In* "Climatic Changes on a Yearly to Millennial Basis" (N.-A. Mörner and W. Karlén, Eds.), pp. 123-133. D. Reidel, Hingham, Mass.

Kipp, N. G. (1976). New transfer function for estimating past sea-surface conditions from the seabed distribution of planktonic foraminiferal assemblages in the North Atlantic. *Geological Society of America Memoir* 145, 3–42.

Koopmann, B. (1979). "Saharastaub in Sedimenten des Subtropisch-tropischen Nordatlantik Wahrend der Letzten 20,000 Jahre." Unpublished Ph.D. thesis, University of Kiel.

Kutzbach, J. E. (1981). Monsoon climate of the early Holocene: Climate experiment with the earth's orbital parameters for 9000 years ago. *Science* 214, 59–61.

Kutzbach, J. E., and Gallimore, R. G. (1988). Sensitivity of a coupled atmosphere/mixed-layer ocean model to changes in orbital forcing at 9000 yr B.P. *Journal of Geophysical Research* 93, 803–821.

Kutzbach, J. E., and Otto-Bliesner, B. L. (1982). The sensitivity of the African-Asian monsoonal climate to orbital parameter changes for 9000 yr B.P. in a low-resolution general circulation model. *Journal of the Atmospheric Sciences* 39, 1177–1188.

Leventer, A., Williams, D. F., and Kennett, J. P. (1982). Dynamics of the Laurentide ice sheet during the last deglaciation: Evidence from the Gulf of Mexico. *Earth and Planetary Science Letters* 59, 11–17.

Mitchell, J. F. B., Grahame, N. S., and Needham, K. J. (1988). Climate simulations for 9000 years before present: Seasonal variations and effect of the Laurentide ice sheet. *Journal of Geophysical Research* 93, 8283–8303.

Mix, A. C., and Ruddiman, W. F. (1985). Structure and timing of the last deglaciation: Oxygen-isotopic evidence. *Quaternary Science Reviews* 4, 59–108.

Mix, A. C., Ruddiman, W. F., and McIntyre, A. (1986). Late Quaternary paleoceanography of the tropical Atlantic. 1: Spatial variability of annual mean sea-surface temperatures, 0–20,000 years B.P. *Paleoceanography* 1, 43–66.

Molina-Cruz, A., and Thiede, J. (1978). The glacial eastern boundary current along the Atlantic Eurafrican continental margin. *Deep-Sea Research* 25, 337–356.

Oeschger, H., Beer, J., Siegenthaler, U., Stauffer, B., Dansgaard, W., and Langway, C. C. (1984). Late glacial climate history from ice cores. *In* "Climate Processes and Climate Sensitivity" (J. E. Hansen and T. Takahashi, Eds.), pp. 299–306. Geophysical Monograph 29. American Geophysical Union, Washington, D.C.

Petit-Maire, N. (1980). Holocene biogeographical variations along the northwestern African coast (28°-19°): Paleoclimatic implications. *Palaeoecology of Africa* 12, 365–377.

Pflaumann, U. (1980). Variations of the surface-water temperatures at the eastern North Atlantic continental margin (sediment surface samples, Holocene climatic optimum, and last glacial maximum). *Palaeoecology of Africa* 12, 191–212.

——. (1985). Transfer-function '134/6'—A new approach to estimate sea-surface temperatures and salinities of the eastern North Atlantic from the planktonic foraminifera in the sediment. *"Meteor" Forschungs-Ergebnisse Reihe* C 39, 37–71.

Prell, W. L. (1978). Upper Quaternary sediments of the Colombia Basin: Spatial and stratigraphic variation. *Geological Society of America Bulletin* 89, 1241–1255.

Prell, W. L., and Damuth, J. E. (1978). The climate-related diachronous disappearance of *Pulleniatina obliquilocula-*

ta in late Quaternary sediments of the Atlantic and Caribbean. *Marine Micropaleontology* 3, 267–277.

Prest, V. K. (1969). "Retreat of Wisconsin and Recent Ice in North America." Geological Survey of Canada Map 1257A.

Rossignol-Strick, M., Nesteroff, V., Olive, P., and Vergnaud-Grazzini, C. (1982). After the deluge: Mediterranean stagnation and sapropel formation. *Nature* 295, 105–110.

Ruddiman, W. F., and Bowles, F. A. (1976). Early interglacial bottom-current sedimentation on the eastern Reykjanes Ridge. *Marine Geology* 21, 191–210.

Ruddiman, W. F., and Glover, L. K. (1975). Subpolar North Atlantic circulation at 9300 yr B.P.: Faunal evidence. *Quaternary Research* 5, 361–389.

Ruddiman, W. F., and McIntyre, A. (1981). The North Atlantic Ocean during the last deglaciation. *Palaeogeography, Palaeoclimatology, Palaeoecology* 35, 145–214.

——. (1984). Ice-age thermal response and climatic role of the surface North Atlantic Ocean, 40°N to 63°N. *Geological Society of America Bulletin* 95, 381–396.

Ruddiman, W. F., Sancetta, C. D., and McIntyre, A. (1977). Glacial/interglacial response rate of subpolar North Atlantic surface waters to climatic change: The record in oceanic sediments. *Philosophical Transactions of the Royal Society of London Series B* 280, 119–142.

Ruddiman, W. F., Jones, G. A., Peng, T.-H., Glover, L. K., Glass, B. P., and Liebertz, P. J. (1980). Tests for size and shape dependency in deep-sea mixing. *Sedimentary Geology* 25, 257–276.

Sancetta, C. D., Imbrie, J., and Kipp, N. G. (1973). Climatic record of the past 130,000 years in North Atlantic deep-sea core V23-82: Correlation with the terrestrial record. *Quaternary Research* 3, 110–116.

Sarnthein, M., Thiede, J., Pflaumann, U., Erlenkeuser, H., Futterer, D., Koopman, B., Lange, H., and Seibold, E. (1982). Atmospheric and oceanic circulation patterns off northwest Africa during the past 25 million years. *In* "Geology of the Northwest African Continental Margin" (U. von Rad, K. Hinz, M. Sarnthein, and E. Seibold, Eds.), pp. 545–604. Springer-Verlag, Berlin/New York.

Schneider, S. S., and Thompson, S. L. (1979). Ice ages and orbital variations: Some simple theory and modeling. *Quaternary Research* 12, 188–203.

Teller, J. T., and Last, W. M. (1981). Late Quaternary history of Lake Manitoba, Canada. *Quaternary Research* 16, 97–116.

Thiede, J. (1977). Aspects of the variability of the glacial and interglacial North Atlantic eastern boundary current (last 150,000 years). *"Meteor" Forschungs-Ergebnisse Reihe* C 28, 1–36.

Thunell, R. C., and Lohmann, G. P. (1979). Planktonic foraminiferal fauna associated with eastern Mediterranean Quaternary stagnations. *Nature* 281, 211–213.

Thunell, R. C., Williams, D. F., and Kennett, J. P. (1977). Late Quaternary paleoclimatology, stratigraphy, and sapropel history in eastern Mediterranean deep-sea sediments. *Marine Micropaleontology* 2, 371–388.

Williams, D. F., and Thunell, R. C. (1979). Faunal and oxygen isotopic evidence for surface water salinity changes during sapropel formation in the eastern Mediterranean. *Sedimentary Geology* 23, 81–93.

CHAPTER 6

Holocene Temperature Patterns in the South Atlantic, Southern, and Pacific Oceans

Joseph J. Morley and Beth A. Dworetzky

In this chapter we examine Holocene sea-surface temperature (SST) records generated from analyses of marine siliceous microfaunal assemblages. These studies were conducted on sediments from gravity and piston cores taken from the South Atlantic, Southern, and Pacific oceans. We compare temperature estimates for both 6000 and 9000 yr B.P. with modern (atlas) and core-top SSTs. These data, when combined with comparison studies of cores from the North Atlantic Ocean, provide a basis for assessing the degree of difference between SSTs in the middle and early Holocene and those observed today.

Unlike many nonmarine data bases, few well-dated marine sequences of Holocene climate change have sedimentation rates high enough to allow extraction of detailed climate data. Records with sedimentation rates higher than 2 cm per 1000 yr are essential, because in most parts of the ocean temperatures at 6000 and 9000 yr B.P. were not very different from those of today. Of the few well-dated records with high sedimentation rates, a small subset contains a suitable number of siliceous microfauna (radiolarians) from which paleotemperatures can be estimated with regression equations (transfer functions).

Various investigators, many associated with CLIMAP, have shown that assemblages produced from factor analyses of the abundance of specific siliceous flora and fauna can be regressed on modern SSTs to derive transfer functions for estimating paleotemperatures. These multivariate techniques have been successfully applied in the South Atlantic (Morley, 1979; Morley and Hays, 1979a), Southern Ocean

(Lozano and Hays, 1976; Hays *et al.*, 1976a), and Pacific Ocean (Sachs, 1973; Molina-Cruz, 1975; Robertson, 1975; Moore, 1978; Sancetta, 1979; Moore *et al.*, 1980).

Molfino *et al.* (1982) compared various transfer functions based on the abundance of foraminifera, coccolithophorids, and radiolarians in Atlantic Ocean core-top samples and showed that even though these three biotic groups did not have similar ecological controls or factor assemblage distribution patterns, the concordancy among temperature estimates derived from these various regression equations exceeded 87% on both modern and ice-age data sets. Thus abundance records of all three biotic groups appear equally suitable for estimating SSTs.

The 12 marine records discussed in this chapter (Fig. 6.1) are positioned within eastern or western boundary currents, eastern equatorial divergences, marginal seas, or the highly productive subantarctic circumpolar flow. In addition to the high sedimentation rates that are usually associated with these regions, results from CLIMAP Project Members (1981) show that temperature differences between today and the last glacial maximum (about 18,000 yr B.P.) were greatest in these areas (see Fig. 5.2 of Ruddiman and Mix, this vol.). Even though we examine a limited number of sites, the location of these sites in areas characterized by relatively large glacial-interglacial temperature changes ensures that any differences between mid- and early-Holocene SSTs and those of today will be more pronounced in these records than in records from most other ocean regions.

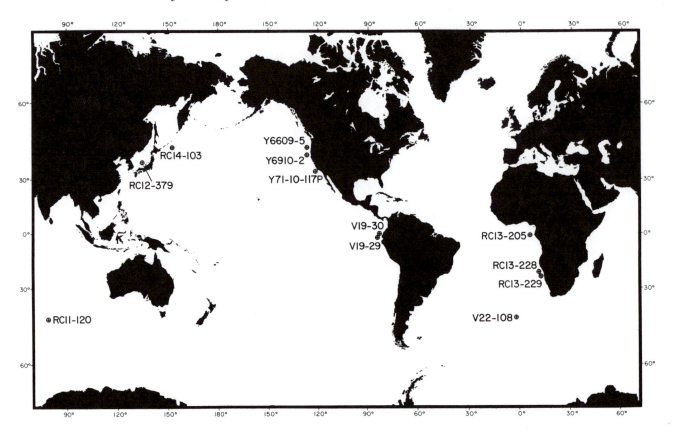

Fig. 6.1. Location and identification of the piston cores examined in this study (see also Table 6.1).

Data Sets

Table 6.1 gives the location, water depth, number of ^{14}C dates, ranking of dating control at 6000 and 9000 yr B.P. (assessment of stratigraphic quality based on type and precision of age indicators), type of data, and average sample interval for each Holocene section. Data from five of the deep-sea cores are original to this study; the other data sets were compiled from the literature. We describe the records in the context of five specific regions to facilitate the evaluation of the significance of any temperature anomalies and the detection of regional trends (note that data from one record [V22-108] are discussed under two regions).

In most of the cores, ^{14}C dates provide the chronostratigraphic control. Time control in the remaining sites is furnished by oxygen-isotope stratigraphy, annual laminations, or high-resolution radiolarian biostratigraphy. Table 6.2 lists the ^{14}C analyses conducted as part of this study. Oxygen-isotope stratigraphies in six of the eight cores with ^{14}C dates provide additional chronological control. Oxygen-isotope analyses have been published for two of the four remaining records, both of which are located in the vicinity of other

Holocene sequences cited in this study that have oxygen-isotope as well as ^{14}C control. In the Santa Barbara core (Y71-10-117-P), ages were determined from counts of annual layers preserved throughout much of the sediment record. Stratigraphy in RC14-103 from the northwest Pacific was provided by the continuous abundance variations of the radiolarian species *Cycladophora davisiana*, which have been correlated with oxygen-isotope stratigraphies (Hays *et al.,* 1976a,b; Morley and Hays, 1979b).

Because different investigators generated the faunal and paleotemperature records, the average interval between consecutive samples within the Holocene sequences in each of these cores varies widely from 25 to 2200 yr (Table 6.1). In all cases except the Santa Barbara core with its annual laminations, the depths of the 6000- and 9000-yr B.P. levels were determined by interpolation between the older stratigraphic marker (^{14}C date, oxygen-isotope maximum, or *C. davisiana* maximum) and a younger stratigraphic level (^{14}C date or core top). The sample depth in each data set nearest the interpolated (target) depth was selected, and its faunal counts and associated SSTs were used as the 6000- and 9000-yr B.P. values. In the six

Table 6.1. Description of the core sites and their data bases

Core	Ocean	Latitude	Longitude	Water depth (m)	[14]C dates (no.)	Rank[a] 6000 yr B.P.	Rank[a] 9000 yr B.P.	Data[b]	Sample interval (yr)
RC13-205	South Atlantic	2°17'S	5°11'E	3731	1	4I	4II	F,T	800
RC13-228	South Atlantic	22°20'S	11°12'E	3204	3	2I	2II	F,T	800
RC13-229	South Atlantic	25°30'S	11°18'E	4191	0	7I,II	7I,II	F,T	2200
V22-108	Southern	43°11'S	3°15'W	4171	1	5I	6II	F,T	2200
RC11-120	Southern	43°31'S	79°52'E	3193	1	4I	2II	F,T	2000
V19-29	Equatorial Pacific	3°35'S	83°56'W	3157	0	7I,II	7I,II	T	1400
V19-30	Equatorial Pacific	3°23'S	83°31'W	3091	2	1II	2I	T	900
RC14-103	Northwest Pacific	44°02'N	152°56'E	5365	0	7I,II	7I,II	F,T	500
RC12-379	Northwest Pacific	36°54'N	134°33'E	1010	3	3I	3II	F	800
Y6609-5	Northeast Pacific	43°34'N	126°28'W	2978	1	7I,II	7I,II	T	1500
Y6910-2	Northeast Pacific	41°16'N	127°01'W	2615	2	7I,II	6I	T	1400
Y71-10-117-P	Northeast Pacific	34°00'N	120°00'W	627	0	2I		T	25

[a]Stratigraphic quality of chronological control (for explanation, see Table 5.2 of Ruddiman and Mix [this vol.]).
[b]F = faunal, T = temperature.

Table 6.2. Previously unpublished 14C dates

Core	Sample depth (cm) Top	Sample depth (cm) Bottom	[14]C age (yr B.P.)
V19-30	30.5	35.0	6090 ± 180
V19-30	63.5	70.0	9935 ± 270
RC13-228	19.0	21.0	4725 ± 170
RC13-228	49.0	51.0	8740 ± 185
RC13-228	99.9	101.0	13,375 ± 435
V22-108	36.0	38.0	10,310 ± 215
RC12-379	40.0	60.0	3160 ± 80
RC12-379	251.0	279.0	9540 ± 105

Holocene records sampled at intervals of less than 1000 yr, the difference in age between the target depth and the depth of faunal counts was less than several hundred years. In only two cores (RC13-229 and V22-108) did the estimated age of the depth sampled for faunal analysis differ from the target age by more than 400 yr.

SOUTH ATLANTIC REGION

The four South Atlantic cores (RC13-205, RC13-228, RC13-229, and V22-108) form a transect extending from the eastern equatorial Atlantic southward through the eastern subtropical region, across the subtropical convergence, and into subantarctic waters (Fig. 6.1). Three of the four cores contain one or more ^{14}C dates in addition to oxygen-isotope stratigraphies (Tables 6.2 and 6.3). Based on these chronologies, the average interval between consecutive Holocene samples was 2200 yr in RC13-229 and V22-108 and 800 yr in RC13-228 and RC13-205.

Winter and summer SST estimates extending back through the last glacial maximum for each of the four cores are shown in Figure 6.2. The paleotemperature estimates in three of the four cores are the product of transfer functions derived from analyses of the abundance of 24 radiolarian species in 57 surface-sediment samples from the South Atlantic (Morley, 1979). These transfer functions have standard errors of estimate of 1.4°C for both winter and summer. The transfer functions that yielded the temperature estimates in V22-108 are based on the abundance of 19 radiolarian species in 72 core-top samples from the Southern Ocean (J. D. Hays and J. J. Morley, unpublished). The standard errors of estimate for these Southern Ocean transfer functions are 1.1 and 1.4°C for winter and summer, respectively.

SOUTHERN OCEAN REGION

Two Holocene records from the Southern Ocean are included in this study: RC11-120 from the southeast Indian sector and V22-108 from the South Atlantic sector (Fig. 6.1). The ^{14}C ages in both are substantiated by oxygen-isotope stratigraphies (Tables 6.2 and 6.3). Neither data set was a direct outgrowth of this study, and the low-density sampling of the Holocene interval in these cores yielded relatively long time intervals between consecutive samples (2000 and 2200 yr in RC11-120 and V22-108, respectively). The same transfer functions developed from the abundance of 19 radiolarian species in 72 Southern Ocean core tops (J. D. Hays and J. J. Morley, unpublished) were used to produce the temperature estimates in these two records (Fig. 6.2b).

Table 6.3. Stratigraphic control and basis for paleoestimates

Region and core	^{14}C dates (no.)	Other stratigraphic control[a]	Basis for paleoestimates[a]
South Atlantic			
RC13-205	1	Oxygen-isotope data (13)	Faunal counts (7) and transfer functions (7, 18)
RC13-228	3	Oxygen-isotope data (10)	Faunal counts (7, 10) and transfer functions (7, 18)
RC13-229	0	Oxygen-isotope data (1)	Faunal counts (1) and transfer functions (7, 18)
V22-108	1	Oxygen-isotope data (13)	Faunal counts (3) and transfer functions (17)
Southern Ocean			
V22-108	1	Oxygen-isotope data (13)	Faunal counts (3) and transfer functions (17)
RC11-120	1	Oxygen-isotope data (2, 13)	Faunal counts (2, 3) and transfer functions (17)
Eastern equatorial Pacific			
V19-29	0	Oxygen-isotope data (12)	Faunal counts (5) and transfer functions (5)
V19-30	2	Oxygen-isotope data (14)	Faunal counts (6) and transfer functions (6)
Northwest Pacific			
RC14-103	0	*Cycladophora davisiana* correlation (9)	Faunal counts (4) and transfer functions (19)
RC12-379	3	*C. davisiana* correlation (8)	Faunal data (8)
Northeast Pacific			
Y6609-5	1		Faunal counts (15) and transfer functions (15)
Y6910-2	2	Oxygen-isotope data (16)	Faunal counts (15) and transfer functions (15)
Y71-10-117-P	0	Varve counts (11)	Faunal counts (11) and transfer functions (11)

[a]Numbers in parentheses refer to the following sources: 1 = Embley and Morley (1980); 2 = Hays *et al.* (1976b); 3 = J. D. Hays (unpublished data); 4 = Heusser and Morley (1985); 5 = Molina-Cruz (1975); 6 = T. C. Moore, Jr. (unpublished data); 7 = Morley (1977); 8 = Morley *et al.* (1986); 9 = Morley *et al.* (1982); 10 = Morley and Shackleton (1984); 11 = Pisias (1978); 12 = Shackleton (1977); 13 = N. J. Shackleton (unpublished data); 14 = Shackleton *et al.* (1983); 15 = Moore (1973); 16 = Moore *et al.* (1977); 17 = J. D. Hays and J. J. Morley (unpublished data); 18 = Morley (1979); 19 = Robertson (1975).

EASTERN EQUATORIAL PACIFIC REGION

Two Panama Basin cores make up the Holocene data set from the eastern equatorial Pacific (Fig. 6.1). Oxygen-isotope stratigraphies have been published for both cores (Shackleton, 1977; Shackleton *et al.,* 1983), and two ^{14}C dates in V19-30 give Holocene ages (Tables 6.2 and 6.3). Based on these stratigraphic controls, the Holocene sequences in V19-29 and V19-30 were sampled at intervals of 1400 and 900 yr, respectively.

Different transfer functions were used to produce the SST curves for these two cores (Fig. 6.3). The transfer functions that were applied to the Holocene record in V19-29 were derived from analyses of 58 radiolarian species in 47 core tops from the eastern equatorial and southeastern Pacific (Molina-Cruz, 1975, 1977) and have standard errors of estimate of 1.2 and 0.5°C for winter and summer, respectively. The temperature estimates for V19-30 were generated from a set of transfer functions specifically developed to interpret the unusual radiolarian faunal associations sometimes found in Panama Basin sediments. These equations were derived from analysis of the abundance of 42 radiolarian species in core tops specifically from the Panama Basin (T. C. Moore, Jr., unpublished). A limited comparison of the two sets of transfer functions (those based solely on Panama

Basin core tops and those derived by Molina-Cruz for the eastern equatorial Pacific) shows that the Panama Basin equations produce slightly warmer temperature maxima (1-2°C) during the Holocene and slightly cooler temperatures (≤0.5°C) during the late stages of the last glacial and the early deglacial in Panama Basin records.

NORTHWEST PACIFIC REGION

We examined the Holocene sequence in two cores from the northwest Pacific Ocean. The northern core, RC14-103, sampled sediments underlying the southwestern portion of the subpolar gyre, north of today's subarctic front and east of Hokkaido (Fig. 6.1). The other core, RC12-379, came from the southeastern Sea of Japan (Fig. 6.1). The relative abundance curve for the radiolarian species *C. davisiana* provided the stratigraphy for RC14-103 (Table 6.3). Comparison of this species' abundance pattern in late-Pleistocene sediments with oxygen-isotope records in the North Atlantic (Morley and Hays, 1979b) and the Southern Ocean (Hays *et al.,* 1976a,b) and with tephrastratigraphic and biostratigraphic datum levels in the northwest Pacific (Morley *et al.,* 1982) shows that its major features are similar and changes in its relative abundance are synchronous in all of these regions. Therefore, *C. davisiana* abundance records provide a high-

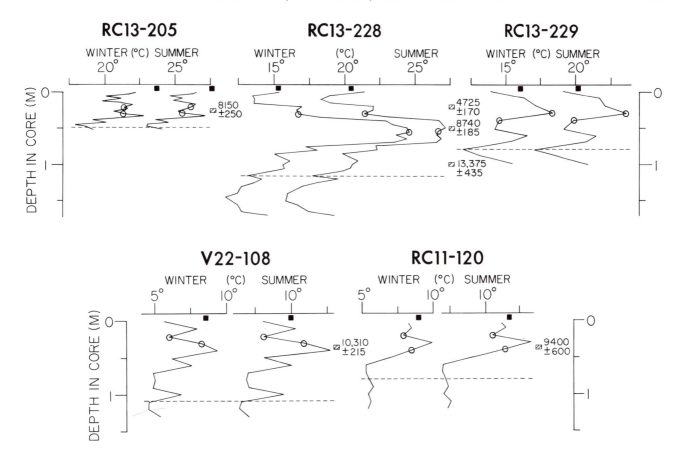

Fig. 6.2. Estimated winter and summer sea-surface temperatures (SSTs) since the last glacial maximum (about 18,000 yr B.P.) in three South Atlantic cores (above) and two Southern Ocean cores (below). The dashed lines represent the 18,000-yr B.P. level in each core. Open circles designate the 6000- and 9000-yr B.P. temperature estimates. Modern SSTs (averages for February and August) are indicated by the solid squares above each curve. The depth intervals sampled for ^{14}C dates are indicated by the hatched boxes to the right of the temperature curves.

resolution stratigraphy in high-latitude sediments devoid of calcium carbonate. ^{14}C ages were determined at three depths in RC12-379, two of which give Holocene ages (Table 6.2). The average age intervals between consecutive Holocene samples from RC14-103 and RC12-379 are 500 and 800 yr, respectively.

Although radiolarian abundances were calculated for samples from both cores, Holocene temperature estimates were generated from faunal counts only for samples from RC14-103 (Fig. 6.4) because of the low concentration of radiolarians in many of the glacial and early-Holocene samples from RC12-379. Radiolarian abundance data from 66 surface-sediment samples from the northwest Pacific provided the data base for development of the transfer functions that produced these paleotemperature estimates (Robertson, 1975). The standard errors of estimate for these transfer functions are 1.3°C for winter and 0.9°C for summer.

NORTHEAST PACIFIC REGION

We examined three Holocene records from the northeast Pacific Ocean region. The two cores located off the Washington-Oregon coast (Fig. 6.1) both contain ^{14}C dates; however, all three ages from these analyses are pre-Holocene. Additional stratigraphic control is available in Y6910-2 from oxygen-isotope analyses (Moore *et al.*, 1977). Varve stratigraphy provides the chronological control in the nearly 8000-yr record from the Santa Barbara Basin (Table 6.3). Sediment-trap studies off California (Anderson *et al.*, 1987) have confirmed that the alternating light- and dark-colored varves in sediments of late-Pleistocene through Holocene age from this region are deposited on an annual cycle. Based on these chronologies, the age intervals between consecutive samples in Y6609-5, Y6910-2, and Y71-10-117-P averaged 1500, 1400, and 25 yr, respectively.

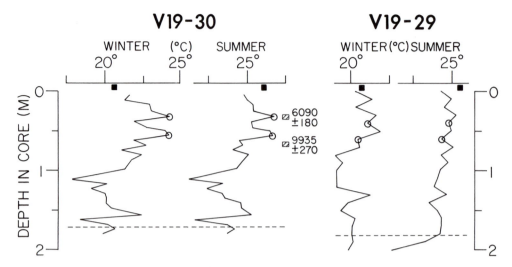

Fig. 63. Estimated winter and summer sea-surface temperatures (SSTs) since the last glacial maximum (about 18,000 yr B.P.) in two eastern equatorial Pacific cores. The dashed lines represent the 18,000-yr B.P. level in each core. Open circles designate the 6000- and 9000-yr B.P. temperature estimates. Modern SSTs (averages for February and August) are indicated by the solid squares above each curve. The depth intervals sampled for ^{14}C dates are indicated by the hatched boxes to the right of the temperature curves.

The transfer function applied to faunal counts from Y6609-5 and Y6910-2 was developed from abundance data for 44 radiolarian species in 183 surface-sediment samples from the Pacific and has a standard error of estimate of 1.5°C for summer (Moore, 1978). The paleotemperature estimates for winter in the Santa Barbara record were produced from a transfer function based on abundance data for 32 radiolarian species in 38 core-top samples from the Pacific Ocean off the California and Mexican coasts (Pisias, 1978). The standard error of estimate for this transfer function is 1.4°C.

Results

Table 6.4 shows the differences between SST estimates for winter and summer at 6000 and 9000 yr B.P. and core-top estimates as well as modern (atlas) temperatures. A large discrepancy between the temperature difference (anomaly) calculated relative to the core-top temperature estimate and that calculated relative to the atlas temperature value may indicate that the core-top sediments do not record the latest Holocene conditions or that the transfer function is unable to estimate the SSTs accurately based on the faunal assemblages in the sediments at a particular site. If it cannot be proved that core-top sediments record the latest Holocene conditions, then temperature estimates generated for the core in question should be viewed with caution.

Several other variables must also be considered in assessing the significance of a specific temperature anomaly. First, the quality of the chronological control should be reviewed. Because most of the sites do not have multiple ^{14}C dates within their Holocene sequence, it is likely that the depths of the projected 6000- and 9000-yr B.P. levels would change as additional dates became available. Second, the difference between the depth at which faunal data were available and the interpolated depth of the 6000- and 9000-yr B.P. target levels is as large as 400 yr in some records (RC13-229 and V22-108). Therefore, improvements in either the chronological control or the sampling density could alter the temperature estimates for 6000 and 9000 yr B.P. Changes in both factors could make a major difference in the calculation of the temperature anomalies in several of the Holocene records described in this chapter.

To be statistically significant, a temperature anomaly should exceed the standard error of estimate of the transfer function that produced the temperature estimate. Finally, more than one Holocene record should be examined within each region to determine whether the direction of the temperature anomalies (warmer or colder than modern) is the same. Regardless of the magnitude of a temperature anomaly, its value is suspect if its direction differs from that of anomalies in other records from the same region.

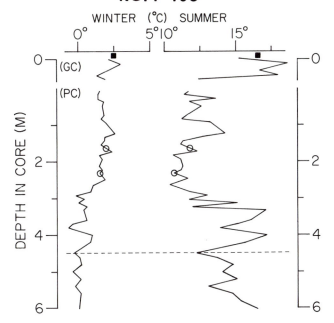

RC14-103

Fig 6.4. Estimated winter and summer sea-surface temperatures (SSTs) in gravity core (GC) and the upper portion of piston core (PC) RC14-103 from the northwest Pacific. The dashed line represents the 18,000-yr B.P. level in the core. Open circles designate the 6000- and 9000-yr B.P. temperature estimates. Modern SSTs (averages for February and August) are indicated by the solid squares above the curves.

SOUTH ATLANTIC REGION

Holocene SSTs in the four cores that make up our South Atlantic transect are 2-5°C above glacial-maximum values (CLIMAP, 1981). The temperature curves of the four cores from this region (Fig. 6.2) all show this warming trend across the glacial-interglacial transition.

Estimated temperatures throughout the Holocene record from the eastern equatorial Atlantic (RC13-205) never exceed modern (atlas) values. Although the differences between estimated winter and summer temperatures at both 6000 and 9000 yr B.P. and modern temperatures for the corresponding seasons (Table 6.4) exceed the standard error of estimate of the transfer functions that produced the temperature estimates, these cooler temperature values may simply reflect the inability of the transfer functions to predict temperatures as warm as those of today at the core site. Other data, however, suggest that these estimates may indeed be accurate. The winter and summer temperature curves covering the last glacial maximum through the present from RC13-205 are similar to temperature records derived from foraminiferal analyses in cores south of the thermal equator in the eastern equatorial Atlantic (Mix *et al.*, 1986; Ruddiman and Mix, this vol.). Also, the communalities of all siliceous faunal samples analyzed throughout the mid- and early-Holocene sequence from RC13-205 are more

Table 6.4. Differences (°C) between temperature estimates for 9000 and 6000 yr b.p. and core-top estimates and modern (atlas) sea-surface temperatures (SSTs) at core sites

| Region and core | 9000 yr B.P. | | | | 6000 yr B.P. | | | |
| | Core top | | Modern SSTs | | Core top | | Modern SSTs | |
	Winter	Summer	Winter	Summer	Winter	Summer	Winter	Summer
South Atlantic								
RC13-205	-0.8	-1.0	-2.4	-2.1	-0.8	-0.4	-2.4	-1.5
RC13-228	9.2	5.2	9.3	6.2	1.4	0.0	1.4	1.0
RC13-229	0.6	0.8	-1.6	-0.3	4.5	4.5	2.3	3.4
V22-108	2.7	2.9	-0.2	0.9	0.2	0.1	-2.7	-1.9
Southern Ocean								
V22-108	2.7	2.9	-0.2	0.9	0.2	0.1	-2.7	-1.9
RC11-120	0.2	0.2	-0.5	-0.4	-0.4	-0.7	-1.1	-1.3
Eastern equatorial Pacific								
V19-29	0.2	0.0	-0.2	-1.1	0.8	0.4	0.4	-0.7
V19-30	2.7	1.9	3.7	0.6	2.6	2.0	3.6	0.7
Northwest Pacific								
RC14-103[a]	-0.6	-4.5	-1.0	-5.9	-0.2	-3.4	-0.6	-4.8
Northeast Pacific								
Y6609-5	—	-0.2	—	0.0	—	-0.2	—	0.0
Y6910-2	—	-0.2	—	-0.6	—	0.4	—	0.0
Y71-10-117-P	—	—	—	—	-0.7	—	1.5	—

[a]Core-top estimates are from the top of a gravity core taken at the site.

than 0.8 (a value of 1.0 indicates that the factor model describes all the original faunal assemblage information in the sample).

Because RC13-228 and RC13-229 are from the same region of the South Atlantic, a degree of similarity might be expected in their Holocene temperature patterns, and indeed the Holocene records in these cores both show a period within the Holocene when SSTs were much warmer than modern values (Fig. 6.2a). In RC13-228, this temperature maximum is well dated to 9000 yr B.P. Temperature estimates for 6000 yr B.P. in this core are also higher than both modern and core-top values (Table 6.4). Charles and Morley (1988) showed that the transfer functions may have over-estimated temperatures during the late-deglacial interval in RC13-228 by 3-4°C relative to those estimated for RC13-229. However, a significant early-Holocene temperature anomaly remains even after adjustments are made to correct for this overestimation.

The Holocene temperature maximum in RC13-229 occurs at an age estimated from isotopic stratigraphy as 6000 yr B.P. Estimated temperatures for 9000 yr B.P. in this core are similar to core-top estimates and are slightly cooler than modern (atlas) temperatures. If, however, 20 cm of the latest Holocene was missing from the top of this record, the interpolated age of the Holocene maximum would be revised to 9000 yr B.P. Comparison of the temperature curves for these two records (Fig. 6.2a) gives some indication that the core top of RC13-229 may indeed not record the latest Holocene conditions, because temperature estimates fail to increase toward modern values from the latest Holocene temperature minimum recorded at both sites.

The Holocene temperature pattern in V22-108 (Fig. 6.2b) also shows a Holocene temperature maximum that is warmer than modern and core-top temperatures. Neither of the depths selected to represent the 6000- and 9000-yr B.P. levels, however, coincides with the depth of this temperature maximum. Be-

cause of the relatively low sample density in the Holocene sequence and the absence of ^{14}C dates younger than 10,300 yr B.P., the depths calculated for 6000 and 9000 yr B.P. and their associated temperature estimates would most likely change with improved chronological control and additional faunal data.

The Holocene temperature pattern in each of the four cores from this South Atlantic transect shows an SST maximum in the early and/or middle Holocene. In three of the four records (RC13-228, RC13-229, and V22-108), this temperature maximum is warmer than core-top estimates and modern SSTs. In the three cores with ^{14}C control (RC13-205, RC13-228, and V22-108), the estimated age of this temperature maximum ranges from 9200 to 10,600 yr B.P. (Table 6.5).

SOUTHERN OCEAN REGION

Temperature anomaly maps published by CLIMAP (1981) show that temperatures at the last glacial maximum in the vicinity of V22-108 and RC11-120 were 2-4°C cooler than modern values. The interpolated depths of the 6000- and 9000-yr B.P. levels in these cores occur at temperature minima or between temperature minima and maxima. Increased sample density of the Holocene sediments in these cores (the average age interval between consecutive samples currently is 2000 yr or more [Table 6.1]) and additional ^{14}C dating of the Holocene section might cause the depths of the 6000- and 9000-yr B.P. levels in these cores to be revised and thus might significantly modify the temperature anomalies calculated for these specific ages in these records (Table 6.4).

The temperature patterns for both Southern Ocean cores described in this study (Fig. 6.2b) show a distinct early-Holocene maximum in winter and a summer temperature that is warmer than both modern and core-top temperatures for that season. Based on the single early-Holocene ^{14}C date in each of these cores, the estimated age of this Holocene temperature maxi-

Table 6.5. Age of Holocene sea-surface temperature (SST) maximum in records with at least one Holocene 14C date

Core	SST maximum (cm)	[14]C depth (cm)	[14]C age (yr B.P.)	SST maximum (yr B.P.)	Temperature anomaly at SST maximum			
					Core top		Modern SST	
					Winter	Summer	Winter	Summer
RC13-205	32.5	25-30	8150 ± 250	10,339 ± 306	0.7	0.6	-0.9	-0.5
RC13-228	55	49-51	8740 ± 185	9200 ± 247	9.2	5.2	9.3	6.2
		99.9-101.0	13,375 ± 435					
V22-108	40	36-38	10,310 ± 215	10,626 ± 223	3.7	4.7	0.8	2.7
RC11-120	30	36-39	9400 ± 600	7520 ± 480	1.5	2.0	0.8	1.5
V19-30	56	30.5-35	6090 ± 180	8719 ± 487	2.6	1.9	3.7	0.6
		63.5-70	9935 ± 270					

mum is 7500 yr B.P. in RC11-120 and 10,600 yr B.P. in V22-108 (Table 6.5).

EASTERN EQUATORIAL PACIFIC REGION

Glacial SSTs in the eastern equatorial Pacific were similar to modern values, with temperature anomalies ranging between 0 and 2°C (CLIMAP, 1981). Winter temperature estimates for both 6000 and 9000 yr B.P. in V19-30 are warmer than modern and core-top temperatures (Table 6.4 and Fig. 6.3). The summer temperature anomalies at these two levels are also slightly positive, although they do not exceed the standard error of estimate of the transfer function in all cases. In V19-29, however, SST estimates for 6000 and 9000 yr B.P. are, with one minor exception, within 1°C of modern and core-top temperatures.

As mentioned in the previous section, temperature estimates in these two records were produced by different transfer functions. Assuming that both sets of transfer functions produce similar temperature estimates (within their standard errors of estimate) on identical faunal assemblages, the results in Table 6.4 indicate that at 6000 and 9000 yr B.P. SSTs at these two sites in the eastern equatorial Pacific were similar to today's (atlas) values (six of the eight anomalies relative to modern SSTs are less than or equal to 1.1°C).

NORTHWEST PACIFIC REGION

SSTs during the last glacial maximum are estimated to have been 4-8°C cooler than today in the vicinity of RC14-103 (CLIMAP, 1981). Both winter and summer temperature estimates for this site at 6000 and 9000 yr B.P. are colder than core-top and modern temperatures (Table 6.4 and Fig. 6.4). Only the summer anomalies, however, exceed the standard error of estimate for the northwest Pacific transfer functions. Future refinements in the Holocene chronology will not alter these negative summer SST anomalies, because summer temperature estimates throughout much of the Holocene record in this core are 2-5°C cooler than today. Winter SSTs gradually increase throughout the early Holocene, reaching mid-Holocene values similar to modern values. The seasonality of SSTs (difference between summer and winter temperatures) was much lower during the early and middle Holocene than during the latest Holocene and deglacial intervals at this site. A zonal increase in the seasonal variability of the subarctic front could account for the heightened seasonality in this region during the latest Pleistocene and latest Holocene.

Chinzei *et al.* (1987) presented results of a detailed floral and faunal analysis of sediments from three short cores located farther south than RC14-103, off eastern Honshu. As part of their study they devised a semiquantitative method for estimating SSTs based on the mean of normalized ratios of warm-water (Kuroshio) flora-fauna to total flora-fauna. SSTs derived by this method for 6000 and 9000 yr B.P. in the two southern sites along their north-south core transect were within 1°C of core-top values. In their northernmost core at 36°N, SSTs at 6000 yr B.P. were approximately 3°C warmer than today, whereas temperatures at 9000 yr B.P. were estimated to be only slightly warmer than at present.

NORTHEAST PACIFIC REGION

Temperatures off the Washington-Oregon coast during the last glacial maximum were 1-3°C cooler than today (CLIMAP, 1981). None of the summer SST anomalies calculated from the Holocene records in cores from the northeast Pacific exceeds the standard error of estimate of its associated transfer function (Table 6.4). The absence of Holocene ^{14}C dates, in addition to the relatively low sample density (1400-1500 yr) in Y6609-5 and Y6910-2, however, makes it difficult to assess the significance of the anomalies relative to modern summer temperatures at these two sites.

In the Holocene record from the Santa Barbara Basin, the winter temperature estimate for 6000 yr B.P. is 0.7°C cooler than the core-top temperature estimate and 1.5°C warmer than today's (atlas) value. Only the latter anomaly exceeds the standard error of estimate of the associated transfer function.

Discussion and Conclusions

The Holocene temperature pattern in all the Southern Hemisphere middle- and high-latitude sites contains an early-Holocene temperature maximum (Fig. 6.2). Estimated temperatures at these maxima are all warmer than modern SSTs from these regions. The age of this temperature peak, estimated in three of these records with Holocene ^{14}C dates (Table 6.5), ranges from 7500 to 10,600 yr B.P. This spread in age estimates would probably narrow with additional ^{14}C dating and increased resolution of the faunal records, especially since RC11-120 and V22-108 are currently sampled at a relatively low density (≥2000 yr).

Except for winter temperature estimates from V19-30, the three low-latitude records (Figs. 6.2a and 6.3) all show mid- and early-Holocene temperatures similar to or cooler than modern values. Estimated SSTs for 6000 and 9000 yr B.P. from the three northeast Pacific sites are also similar to modern SSTs. Summer SST estimates at the northwest Pacific site are cooler at both 6000 and 9000 yr B.P. than at present, and winter SST estimates are similar to modern values.

Although this global data base is extremely limited, we briefly examine climatic forcing functions that may have differed from today during the middle and early Holocene and that might account for the temperature anomalies identified in this study. General circulation model studies indicate that the thermal forcing effect of the ice sheets on the oceans was significantly reduced from glacial-maximum levels by 9000 yr B.P., except in the high-latitude regions of the western and central North Atlantic (Mitchell *et al.,* 1988). Presumably by 6000 yr B.P. this effect was no different from today. Therefore one cannot invoke the effect of glacial ice sheets to explain the difference between mid- or early-Holocene SSTs and modern values.

Cooler temperatures relative to today recorded during the middle and early Holocene in the eastern tropical Atlantic and northwest Pacific sites might be ascribed to different levels of CO_2 during this time interval. Oeschger *et al.* (1984), however, showed that CO_2 levels had nearly recovered from their glacial minimum to preindustrial values by 9000 yr B.P.

Another proposed explanation for the mid- and early-Holocene temperature anomalies is the effect of differences in solar radiation between 6000 and 9000 yr B.P. and today. At 6000 and 9000 yr B.P. summer insolation was lower and winter insolation higher in the Southern Hemisphere, whereas in the Northern Hemisphere insolation was higher in summer and lower in winter (Berger, 1978). Solar-radiation values were also affected by the greater axial tilt at 9000 yr B.P. compared with today (24.24 vs. 23.44°).

Recent analyses using coupled atmosphere/mixed-layer ocean models (Kutzbach and Gallimore, 1988; Mitchell *et al.,* 1988) show modest responses in SST to variations in the distribution of solar radiation. The model results indicate that seasonal (June-July-August and December-January-February) surface-ocean temperatures were similar to or slightly cooler than modern values in the eastern equatorial Pacific and Atlantic at 9000 yr B.P. This finding is in agreement with results from our analysis of marine Holocene records from these low-latitude regions. Other areas of agreement between model experiments and marine records are the northeast Pacific, where estimated mid- and early-Holocene SSTs were similar to modern temperatures, and the high-latitude Southern Hemisphere, where estimated paleotemperatures were similar to or slightly warmer than today's values.

Results from the coupled atmosphere-ocean models differ significantly from marine records in the northwest Pacific. The model experiments of Kutzbach and Gallimore (1988) indicate cooler winter SSTs in this region, while those of Mitchell *et al.* (1988) show warmer winter temperatures compared with modern readings. The marine faunal assemblages, however, yield similar (within the standard error of estimate) winter temperatures and colder summer temperatures relative to modern values. Also, in the southeastern subtropical South Atlantic both models predict similar to slightly cooler SSTs than at present, whereas the marine records indicate warmer mid- and early-Holocene temperatures.

This comparison of results from model experiments and marine records suggests that some of the apparent differences between mid- and early-Holocene SSTs and those observed today may result from differences in the seasonal and latitudinal distribution of solar radiation. However, in some ocean regions atmosphere-ocean model results differ significantly from the marine record. These differences may be resolved with the development of more sophisticated coupled atmosphere-ocean models incorporating advection and upwelling effects. Advances in model sophistication, along with additional detailed Holocene records, should verify whether changes in solar radiation can account for all the differences in the global SST pattern between the middle or early Holocene and today or whether other forcing mechanisms must be invoked to explain specific temperature anomalies.

Acknowledgments

We thank W. F. Ruddiman, J. E. Kutzbach, T. C. Moore, Jr., A. C. Mix, W. L. Prell, and T. Webb III for reviewing earlier versions of this manuscript. J. D. Hays, T. C. Moore, Jr., N. G. Pisias, M. Leinen, C. Schramm, and N. J. Shackleton lent valuable raw data and unpublished material. A. Pesanell provided laboratory assistance. Illustrations were prepared by N. Katz. This research was directly supported by a U.S. Department of Energy grant to Brown University under the Carbon Dioxide and Climate Research Program. Lamont-Doherty Geological Observatory of Columbia University Contribution 4909.

References

Anderson, R. Y., Hemphill-Haley, E., and Gardner, J. V. (1987). Persistent late-Pleistocene-Holocene seasonal upwelling and varves off the coast of California. *Quaternary Research* 28, 307-313.

Berger, A. L. (1978). Long-term variations of caloric insolation resulting from the earth's orbital elements. *Quaternary Research* 9, 139-167.

Charles, C. D., and Morley, J. J. (1988). The paleoceanographic significance of the radiolarian *Didymocyrtis tetrathalamus* in eastern Cape Basin sediments. *Palaeogeography, Palaeoclimatology, Palaeoecology* 66, 113-126.

Chinzei, K., Fujioka, K., Kitazato, H., Koizumi, I., Oba, T., Oda, M., Okada, H., Sakai, T., and Tanimura, Y. (1987). Postglacial environmental change of the Pacific Ocean off the coasts of central Japan. *Marine Micropaleontology* 11, 273-291.

CLIMAP Project Members (1981). Seasonal reconstructions of the earth's surface at the last glacial maximum. *Geological Society of America Map and Chart Series* MC-36.

Embley, R. W., and Morley, J. J. (1980). Sedimentation and paleoenvironmental studies off Nambia (south-west Africa). *Marine Geology* 36, 183-204.

Hays, J. D., Lozano, J. A., Shackleton, N. J., and Irving, G. (1976a). Reconstruction of the Atlantic and western Indian Ocean sectors of the 18,000 B.P. Antarctic Ocean. *In* "Investigation of Late Quaternary Paleoceanography and Paleoclimatology" (R. M. Cline and J. D. Hays, Eds.), pp. 337-372. Memoir 145, Geological Society of America, Boulder, Colo.

Hays, J. D., Imbrie, J., and Shackleton, N. J. (1976b). Variations in the earth's orbit: Pacemaker of the ice ages. *Science* 194, 1121-1132.

Heusser, L. E., and Morley, J. J. (1985). Pollen and radiolarian records from deep-sea core RC14-103: Climatic reconstructions of northeast Japan and northwest Pacific for the last 90,000 years. *Quaternary Research* 24, 60-72.

Kutzbach, J. E., and Gallimore, R. G. (1988). Sensitivity of a coupled atmosphere/mixed-layer ocean model to changes in orbital forcing at 9000 yr B.P. *Journal of Geophysical Research* 93, 803-821.

Lozano, J. A., and Hays, J. D. (1976). Relationship of radiolarian assemblages to sediment types and physical oceanography in the Atlantic and western Indian Ocean sectors of the Antarctic Ocean. *In* "Investigation of Late Quaternary Paleoceanography and Paleoclimatology" (R. M. Cline and J. D. Hays, Eds.), pp. 303-336. Memoir 145, Geological Society of America, Boulder, Colo.

Mitchell, J. F. B., Grahame, N. S., and Needham, K. J. (1988). Climate simulations for 9000 years before present: Seasonal variations and effect of the Laurentide ice sheet. *Journal of Geophysical Research* 93, 8283-8303.

Mix, A. C., Ruddiman, W. F., and McIntyre, A. (1986). Late Quaternary paleoceanography of the tropical Atlantic. 1: Spatial variability of annual mean sea-surface temperatures, 0-20,000 years B.P. *Paleoceanography* 1, 43-66.

Molfino, B., Kipp, N. G., and Morley, J. J. (1982). Comparison of foraminiferal, coccolithophorid, and radiolarian paleotemperature equations: Assemblage coherency and estimate concordancy. *Quaternary Research* 17, 279-313.

Molina-Cruz, A. (1975). "Paleoceanography of the Subtropical Southeastern Pacific during Late Quaternary: A Study of Radiolaria, Opal and Quartz Contents of Deep-Sea Sediments." Unpublished M.S. thesis, Oregon State University, Corvallis.

———. (1977). Radiolarian assemblages and their relationship to the oceanography of the subtropical Pacific. *Marine Micropaleontology* 2, 315-352.

Moore, T. C., Jr. (1973). Late Pleistocene-Holocene oceanographic changes in the northeast Pacific. *Quaternary Research* 3, 99-109.

———. (1978). The distribution of radiolarian assemblages in the modern and ice-age Pacific. *Marine Micropaleontology* 3, 229-266.

Moore, T. C., Jr., Pisias, N. G., and Heath, G. R. (1977). Climate changes and lags in Pacific carbonate preservation, sea-surface temperature and global ice volume. *In* "The Fate of Fossil Fuel CO_2 in the Oceans" (N. R. Andersen and A. Malahoff, Eds.), pp. 145-165. Plenum Publishing Corp., New York.

Moore, T. C., Jr., Burckle, L. H., Geitzenauer, K., Luz, B., Molina-Cruz, A., Robertson, J. H., Sachs, H., Sancetta, C., Thiede, J., Thompson, P., and Wenkam, C. (1980). The reconstruction of sea surface temperatures in the Pacific Ocean of 18,000 B.P. *Marine Micropaleontology* 5, 215-247.

Morley, J. J. (1977). "Upper Pleistocene Climatic Variations in the South Atlantic Derived from a Quantitative Radiolarian Analysis: Accent on the Last 18,000 Years." Unpublished Ph.D. dissertation, Columbia University, New York.

———. (1979). A transfer function for estimating paleoceanographic conditions based on deep-sea surface sediment distribution of radiolarian assemblages in the South Atlantic. *Quaternary Research* 12, 381-395.

Morley, J. J., and Hays, J. D. (1979a). Comparison of glacial and interglacial oceanographic conditions in the South Atlantic from variations in calcium carbonate and radiolarian distributions. *Quaternary Research* 12, 396-408.

———. (1979b). *Cycladophora davisiana:* A stratigraphic tool for Pleistocene North Atlantic and interhemispheric correlation. *Earth and Planetary Science Letters* 44, 383-389.

Morley, J. J., and Shackleton, N. J. (1984). The effect of accumulation rate on the spectrum of geologic time series: Evidence from two South Atlantic sediment cores. *In* "Milankovitch and Climate" (A. Berger, J. Imbrie, J. Hays, G. Kukla, and B. Saltzman, Eds.), Part 1, pp. 467-480. D. Reidel, Boston.

Morley, J. J., Hays, J. D., and Robertson, J. H. (1982). Stratigraphic framework for the late Pleistocene in the northwest Pacific Ocean. *Deep-Sea Research* 29, 1485-1499.

Morley, J. J., Heusser, L. E., and Sarro, T. (1986). Latest Pleistocene and Holocene palaeoenvironment of Japan and its marginal sea. *Palaeogeography, Palaeoclimatology, Palaeoecology* 53, 349-358.

Oeschger, H., Beer, J., Siegenthaler, U., Stauffer, B., Dansgaard, W., and Langway, C. C. (1984). Late glacial climate history from ice cores. *In* "Climate Processes and Climate Sensitivity" (J. E. Hansen and T. Takahashi, Eds.), pp. 299-306. Geophysical Monograph 5. American Geophysical Union, Washington, D.C.

Pisias, N. G. (1978). Paleoceanography of the Santa Barbara Basin during the last 8000 years. *Quaternary Research* 10, 366-384.

Robertson, J. H. (1975). "Glacial to Interglacial Oceanographic Changes in the Northwest Pacific, Including a Continuous Record of the Last 450,000 Years." Unpublished Ph.D. dissertation, Columbia University, New York.

Sachs, H. M. (1973). North Pacific radiolarian assemblages and their relationship to oceanographic parameters. *Quaternary Research* 3, 73-98.

Sancetta, C. (1979). Oceanography of the North Pacific during the last 18,000 years: Evidence from fossil diatoms. *Marine Micropaleontology* 4, 103-123.

Shackleton, N. J. (1977). The oxygen isotope stratigraphic record of the late Pleistocene. *Philosophical Transactions of the Royal Society of London Series B* 280, 169-182.

Shackleton, N. J., Imbrie, J., and Hall, M. A. (1983). Oxygen and carbon isotope record of east Pacific core V19-30: Implications for the formation of deep water in the late Pleistocene North Atlantic. *Earth and Planetary Science Letters* 65, 233-244.

CHAPTER 7

Holocene Vegetation and Climates of Europe

Brian Huntley and I. Colin Prentice

Plant distributions, and the competitive balance among plant species, respond sensitively to changes in summer warmth, winter cold, and moisture balance. The responses are individualistic in the sense that no two taxa have identical responses to these climate variables. Climatic changes during the Quaternary have continually reshuffled the species mix, resulting in large and incongruent changes in the areas occupied by different species and changes in the composition of the vegetation types that dominate the landscape (Huntley and Webb, 1988). These changes are recorded in pollen diagrams, which provide a starting point for the reconstruction and explanation of past climates.

Europe is rich in late-Quaternary pollen data, yet the climatic changes in Europe during the past 20,000 yr—especially those of the Holocene—are still poorly understood. Holocene vegetational changes, as summarized by Huntley (1988), cannot all be explained by the classical model of early-Holocene warming, mid-Holocene "climatic optimum" (i.e., thermal maximum), and subsequent cooling. Changes in the broad patterns of vegetation across Europe that seemed to be inconsistent with this simplistic model of climate change have been attributed to differential lags in dispersal and ecesis (e.g., Iversen, 1960) or to human impact. Holocene palynology in Europe during the last two decades has concentrated largely on the evidence in the pollen record for changes in agriculture rather than on the evidence for regional changes in climate. Apart from pioneer studies by Iversen (1944) and Grichuk (1969), European palynologists have not until recently (e.g., Klimanov, 1984; Guiot, 1987; Huntley and Prentice, 1988) taken on the challenge of interpreting the broad patterns of vegetation change explicitly in terms of changing regional and seasonal climates.

Complex climatic changes must nevertheless have taken place during the Holocene in Europe as in other continents. Changes in seasonal temperatures and moisture balance in different regions reflect a part of the earth system's response to changes in insolation through direct effects on heating and evapotranspiration and indirect effects on the atmospheric circulation (Webb, 1986; Kutzbach, 1987; Bartlein, 1988; COHMAP Members, 1988). Changes in the seasonal and latitudinal distribution of insolation during the Holocene have been substantial and have altered climate and vegetation worldwide.

The effects of climatic changes are transmitted to vegetation through effects on the viability, regeneration, and growth rates of physiologically different types of plants, which in turn affect succession patterns and indirectly influence the frequency of different types of natural disturbance (Grimm, 1984; Prentice, 1986, 1992). Modern ecology envisages natural landscapes not as composed of homeostatic climax vegetation but rather as subject to disturbance by agencies such as wind and fire that allow succession to start afresh on different patches of land at different times (Pickett and Thompson, 1978; White, 1979; Heinselman, 1981; Shugart, 1984; Pickett and White, 1985). Landscapes approach an equilibrium species composition determined by the succession patterns and statistical properties of the disturbance regime. With disturbance frequencies on the order of one every 50–300 yr (e.g., Zackrisson, 1977; Heinselman, 1981; Runkle, 1985), boreal and temperate forest landscapes offer little resistance to invasion by new taxa on time scales of thousands of years. Natural disturbance allows the species composition to remain close to dynamic equilibrium with a changing climate (Davis and Botkin, 1985; Huntley and Webb, 1989; Prentice et al., 1991).

These considerations permit the use of equilibrium models to interpret large-scale vegetation changes during the Holocene. "Snapshot" climate simulations generated from atmospheric general circulation models (GCMs) (e.g., Kutzbach and Guetter, 1986) can be used to translate past global boundary conditions into estimates of paleoclimatic variables, and response surface methods (e.g., Bartlein *et al.*, 1986) can be used to relate past vegetation and climate. Using these techniques, Webb *et al.* (1987), Webb, Bartlein, *et al.* (this vol.), and Prentice *et al.* (1991) showed that complex but predictable climatic changes, combined with the rapid and individualistic responses of plant taxa, can account for a large part of the mapped space-time variation in pollen assemblages during the late Quaternary in eastern North America. The phenomena explained in this way include differences in Holocene species-migration patterns previously attributed to migrational lags.

This chapter represents a first attempt at a comparable reconstruction of Holocene climatic change and vegetation response in the geographically more heterogeneous setting of Europe. We present maps summarizing the broadest-scale palynologic changes of the European Holocene and qualitative reconstructions of changes in the distribution of vegetation formations. The changes in pollen and vegetation are interpreted climatically with the help of a set of response surfaces illustrating how the abundance of taxa in surface pollen samples relates to present-day spatial variation in summer and winter temperatures and annual precipitation. The inferred climatic anomaly patterns are compared with snapshot simulations for 9000, 6000, and 3000 yr B.P. made with the Community Climate Model (CCM) of the National Center for Atmospheric Research (Kutzbach and Guetter, 1986; COHMAP Members, 1988).

Topography and Climate

Europe for the purposes of this chapter extends from about 30°W to 30°E and from 35 to 75°N, including Iceland and islands of the western and central Mediterranean but excluding Greenland, Svalbard, and all but the western fringe of the European portion of the former Soviet Union. Europe thus defined (Fig. 7.1) is a topographically complex area with three semienclosed seas (the Mediterranean, the Baltic, and the North Sea), two major mountain ranges (the Alps and the Pyrenees), two elevated plateau areas (the Balkan and Iberian peninsulas), and numerous lesser ranges.

Unlike in North America, there is no high mountain range running north-south for the length of the continent. The Scandinavian mountain chain runs from about 56 to 70°N but is not high enough to consistently block the penetration of westerly storms. Much of northern and central Europe is a plain at low elevation that offers free passage to airstreams from the At-

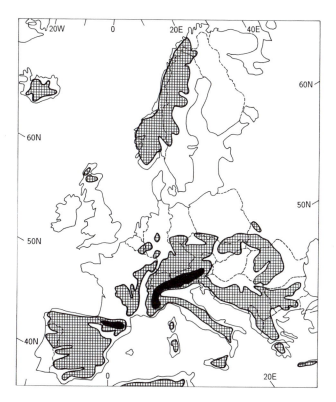

Fig. 7.1. Land in Europe above 400 m (hatched) and 2000 m (shaded).

lantic, and the entire continent is strongly influenced by Atlantic weather patterns.

The zonal gradient of annual insolation and mean temperature is modified by the proximity of the Atlantic (Fig. 7.2). In winter the subtropical anticyclone over the midlatitude Atlantic (the Azores high) is confined to lower latitudes, while the subpolar low-pressure cell (the Icelandic low) is deep and expanded southward over the warm North Atlantic. Southwesterly to westerly winds around the south of the Icelandic low bring winter warmth and moisture to western Europe. During "blocking" episodes an anticyclone forms over northern Europe, bringing much colder and drier conditions; on average, however, warm southwesterly winds and mild, wet winters prevail from the Mediterranean to the Arctic. The winter temperature gradient runs approximately east-west over north and central Europe, with progressively colder winters inland; only in southern Europe does the gradient turn north-south, partly because of the buffering effect of the Mediterranean. The mild winters are accompanied by high precipitation, especially along the west coast, at higher elevations in mountain areas, and in areas of local cyclogenesis around the Mediterranean.

In summer the subtropical anticyclone belt in the Northern Hemisphere shifts northward and becomes stronger, but westerly to northwesterly flow around

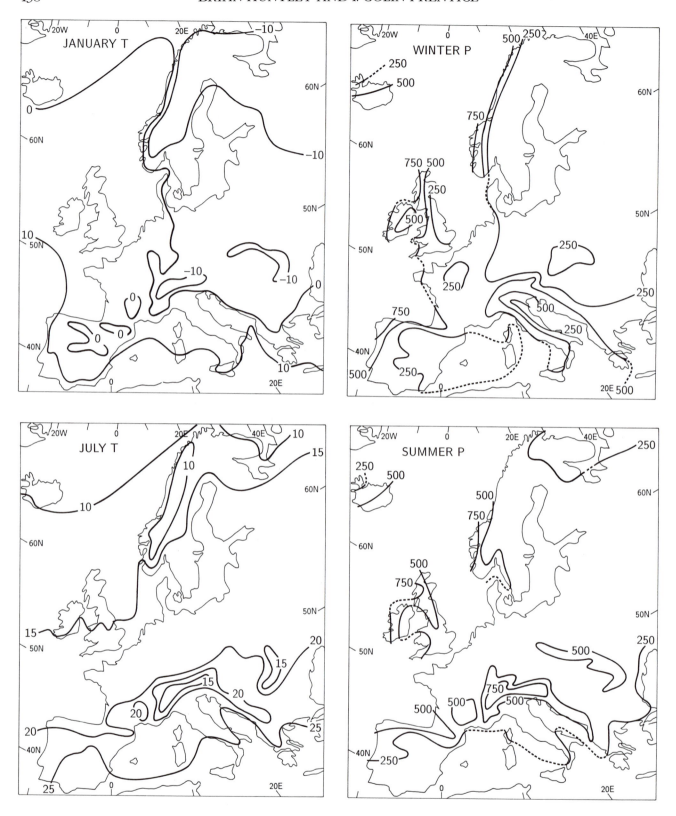

Fig. 7.2. Modern European climate: mean temperature (T) for January and July, precipitation (P) in winter (November-April) and summer (May-October), and annual precipitation (P) (data from Fullard and Darby, 1973), and annual moisture balance

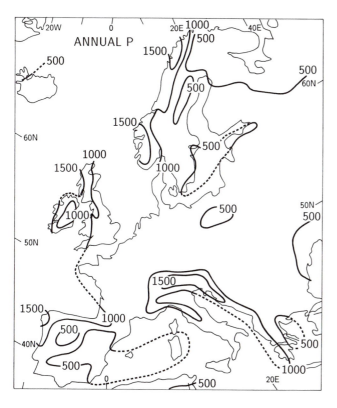

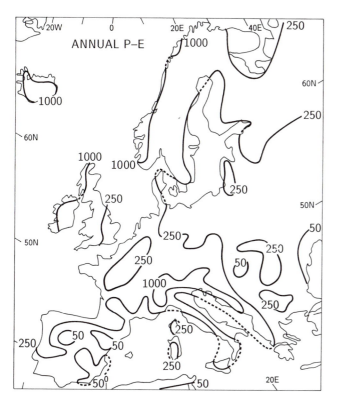

(precipitation minus evapotranspiration [P-E]) (data from Baumgartner and Reichel, 1975).

the north of the expanded Azores high keeps north-western Europe cool. The summer temperature gradient runs from northwest to southeast across northern and central Europe, with highest precipitation in the northwest and in the mountains. Farther south the predominant winds are offshore, and the summers in southern Europe are sunny, dry, and hot. Accumulated temperatures for the growing season are influenced by spring and autumn as well as summer temperatures, and the accumulated temperature gradient runs approximately north-south throughout the continent.

Total annual precipitation is highest along the west coasts of Scandinavia and the British Isles and in the high Alps and lowest in southeastern Spain. Annual moisture balance (precipitation minus actual evapotranspiration) shows a similar pattern but declines more steeply southward and eastward to less than 50 mm over the central Hungarian plain, the western coast of the Black Sea, and parts of interior and southeastern Spain.

Vegetation

The above features of Europe's climate are reflected in the present distribution of natural vegetation types (Fig. 7.3) (see, e.g., Sjörs, 1967; Dahl, 1980; Ellenberg, 1982; Noirfalise, 1987). In the far north, where the growing season is too cold to support tree growth, the natural vegetation is tundra, with dwarf birch (*Betula nana*), *Salix* shrubs, Ericales, Cyperaceae, and grasses. In much of Fennoscandia, and at higher elevations in the mountains of central Europe, trees can grow but the cold winters and short growing seasons favor the northern evergreen conifers Scots pine (*Pinus sylvestris*) and Norway spruce (*Picea abies*), along with cold-tolerant deciduous trees such as birch (*Betula* spp.), aspen (*Populus tremula*), and grey alder (*Alnus incana*); the natural vegetation is classified as boreal or montane conifer forest. Subarctic birch and birch-pine woods characterize the transition from the tundra to the boreal forest.

Several temperate deciduous trees, including oak (*Quercus robur*), ash (*Fraxinus excelsior*), elm (*Ulmus* spp.), and linden (*Tilia cordata*)—the Quercetum mixtum group of classical palynology—tolerate relatively cold winters but require higher growing-season temperatures than the boreal trees. They also thrive in areas with mild winters. Thus, temperate deciduous forest occurs south and west of the boreal forest, and a broad intervening belt of mixed forest occurs on the eastern (continental) side and at intermediate elevations in central Europe.

The composition of the temperate deciduous forest changes along the winter and summer temperature gradients. Along the winter temperature gradient, taxa that tolerate cold winters, such as *T. cordata*, are

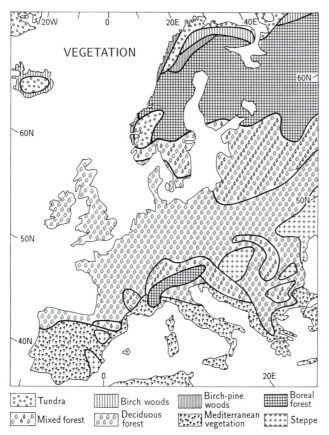

Fig. 7.3. European vegetation regions (data from Fullard and Darby, 1973; Noirfalise, 1987).

abundant in the northeast, and less cold-tolerant taxa such as beech (*Fagus sylvatica*) are abundant in the southwest. Along the summer temperature gradient, taxa tolerant of cool summers, such as hazel (*Corylus avellana*) and common alder (*A. glutinosa*), are abundant in the northwest but give way to taxa requiring warm summers, such as hop hornbeam (*Ostrya carpinifolia*), oriental hornbeam (*Carpinus orientalis*), and the more thermophilic deciduous oaks such as *Q. pubescens* and *Q. frainetto*, which are abundant in areas with adequate rainfall in the southeast. Taxa characteristic of the mixed forest in central Europe include beech, hornbeam (*Carpinus betulus*), and fir (*Abies alba*), as well as other taxa of the temperate deciduous and montane conifer forests.

At low elevations around the Mediterranean, summers are too dry for winter-deciduous trees, whereas winters are warm enough for broad-leaved evergreens. These conditions favor evergreen vegetation with evergreen oaks such as *Q. ilex*, southern pines such as *Pinus halepensis*, and Mediterranean sclerophyll shrubs such as olive (*Olea europaea*) and species of *Pistacia* and *Phillyrea*. In the Pannonian region (the Hungarian plain and the Black Sea coasts of Romania and western Ukraine) and in parts of interior and southeastern Spain, the combination of low

annual moisture balance with cold winters excludes forests and sclerophyll vegetation but allows the development of natural steppe-woodlands (with deciduous oaks) or steppe, with abundant *Artemisia*, Chenopodiaceae, and grasses.

Data and Methods

DATA

The maps we present are based on a compilation of pollen data from literature and unpublished sources. The data consist of relative pollen counts of major taxa in samples of lake or mire sediment. Data were extracted for 423 sites, listed in Huntley and Birks (1983). The sites were selected as much as possible to give an even geographic coverage, although this could not be achieved everywhere. Special site types, such as mor humus or very large lakes, were excluded.

Sixty-five percent of the sites are ^{14}C-dated; the rest were dated by pollen correlation with nearby ^{14}C-dated sites or by comparison with standard ^{14}C-dated regional pollen stratigraphies. The average number of dates per site was 5.5 (range one to thirty-eight). Figure 7.4 shows site locations for 9000, 6000, 3000, and 500 yr B.P. The symbols in the figure indicate the reliability of dating control on a four-level scale: (1) sites marked with a square were dated by linear interpolation between two ^{14}C dates or between one ^{14}C date and the sediment surface; (2) sites marked with a circle were also dated by linear interpolation between two dates, but one bracketing date was doubtful, a major lithostratigraphic change occurred between the bracketing dates, the bracketing dates were more than 4000 yr apart, or the bracketing dates were dated by linear interpolation between one ^{14}C date and a pollen-stratigraphic event that was dated in a nearby site; (3) sites marked with a triangle were dated by linear interpolation between two pollen-stratigraphic events or close to one such event; and (4) sites marked with a plus sign were poorly dated.

An expanded data base of surface pollen samples was used to derive the response surfaces (Huntley *et al.,* 1989). The expanded pollen data base included samples from the former Soviet Union as far east as 60°E in order to broaden the range of modern analogues for past climates. Climate data were derived from Meteorological Office (1972).

POLLEN MAPPING

We present isopoll maps for 9000, 6000, 3000, and 500 yr B.P. for a set of abundant taxa that show clear, spatially coherent patterns of change in abundance among these times. A standard set of percentage isopolls (5, 20, 40, 60, and 80%) was used, plus 2% for certain less abundant taxa. Percentages of tree and shrub pollen types were based on a pollen sum of all

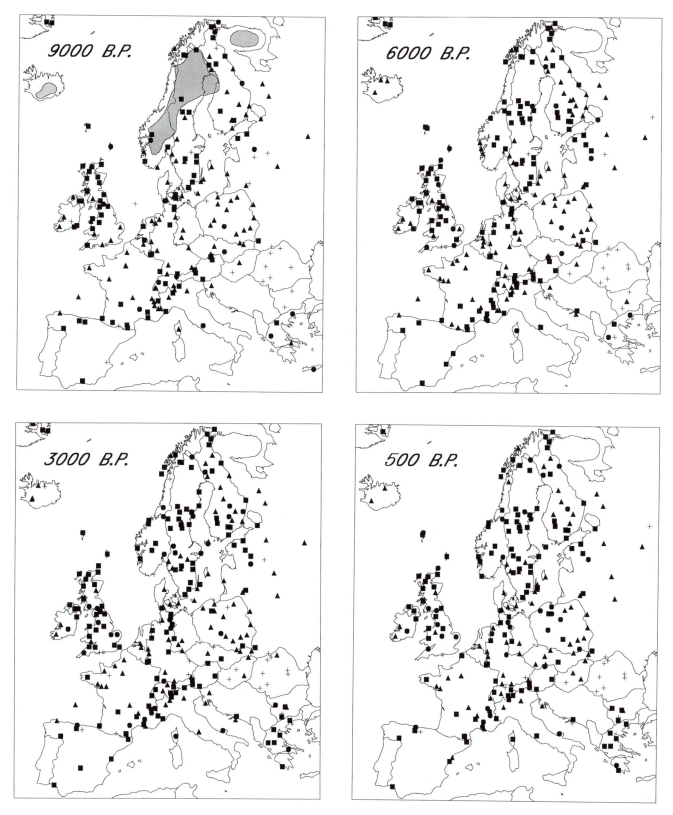

Fig. 7.4. Sites with fossil pollen data. Dating control: ■ = class 1 (best), ● = class 2, ▲ = class 3, + = class 4 (worst). (See text for additional explanation of codes.)

trees and shrubs; percentages of *Artemisia* and Chenopodiaceae were based on a pollen sum of all terrestrial pollen types.

Taxonomic details of the pollen types can be found in Huntley and Birks (1983). Note that *Betula* and *Alnus* include shrubs (*B. nana, B. humilis, A. viridis*) as well as trees; *Pinus* refers to subgenus Diploxylon, which includes both northern and southern pines; and *Ostrya* type includes *Carpinus orientalis*. Three Mediterranean shrub taxa (*Olea, Pistacia,* and *Phillyrea*) and two steppe forb taxa (*Artemisia* and Chenopodiaceae) were combined into composite types for mapping purposes.

Mapped patterns based on surface samples (0 yr B.P.) do not differ greatly from those based on data for 500 yr B.P., but the geographic distribution of available sites is more even at 500 yr B.P., and some taxa (notably *Corylus*) show more coherent patterns at 500 yr B.P., probably because of increased forest clearance since that time. We used 500 yr B.P. to indicate how the pollen record senses the broad patterns of vegetation, allowing direct comparison with earlier Holocene times.

VEGETATION RECONSTRUCTION

Palynologic criteria for the reconstruction of vegetation regions (Table 7.1) were deduced from a comparison of the expanded surface-sample data set with the natural vegetation map (Fig. 7.3). Qualitative reconstructions of vegetation regions were made for 9000, 6000, 3000, and 0 yr B.P. The reconstruction for 0 yr B.P. was based on surface-sample data.

Table 7.1. Pollen percentage criteria for vegetation reconstructions

Formation	Taxon group[a]					
	MD	M	MB	S	Be	TT
Tundra	<5	—	<10	—	<50	<50
Birch forest	<5	—	<10	—	>50	>50
Birch-pine forest	<5	—	<10	—	<50	>50
Boreal and montane forest	<5	—	>10	—	—	>50
Mixed forest	>5	<5	>10	—	—	>50
Temperate deciduous forest	>5	<5	<10	—	—	>50
Mediterranean vegetation	—	>5	<10	—	—	>50
Steppe	—	<5	<10	>10	—	—

[a]Pollen sum is total tree pollen unless otherwise stated. MD = *Acer, Carpinus, Corylus, Fagus, Fraxinus excelsior*-type, *Ilex, Quercus* (deciduous), *Taxus, Tilia,* and *Ulmus*. M = *Buxus, Fraxinus ornus, Olea, Ostrya, Pistacia, Phillyrea,* and *Quercus* (evergreen). MB = *Abies, Larix, Picea,* and *Pinus* (Haploxylon). S = *Artemisia* and Chenopodiaceae (both as a percentage of total land pollen). Be = *Betula*. TT = total tree pollen (as a percentage of total land pollen).

RESPONSE SURFACES AND CLIMATIC INFERENCE

The use of pollen-climate response surfaces (Bartlein *et al.,* 1986) to infer past climates is an extension of the "climatic envelope" method introduced by Iversen (1944) and applied by Hintikka (1963) and Grichuk (1969). Using methods described by Huntley *et al.* (1989), we derived response surfaces for all of the mapped taxa. The surface pollen data were interpolated to the locations of the weather stations by inverse weighting functions for horizontal distance and elevation difference, with a maximum horizontal radius of 100 km and a maximum elevation difference of 100 m. Response surfaces were fitted to the interpolated data by a locally weighted regression method with mean temperatures for January and July and annual mean precipitation as the predictor variables. The response surfaces express the general pattern of abundance of each taxon as a function of the three predictor variables. We used these response surfaces to help interpret in climatic terms the broad-scale patterns of change in abundance shown in the isopoll maps.

The method of response surface construction smooths over the local and nonclimatic components of variation in abundance, leaving the climatic signal (Bartlein *et al.,* 1986; Webb *et al.,* 1987). With a set of response surfaces, fossil pollen spectra that have approximate modern analogues can be "located" at the combination of climatic values corresponding to the most similar modern assemblage type as assessed, for example, by the chord-distance criterion (Prentice *et al.,* 1991). This procedure leads directly to quantitative estimates of paleoclimate. Interpretation is more uncertain for pollen spectra without modern analogues; however, response surfaces can still be used in an informal way, together with data on the climatic limits of species (e.g., Prentice and Helmisaari, 1991), to infer the probable direction of climatic changes.

CLIMATE-MODEL RESULTS

We compared climatic anomaly patterns inferred from the pollen data for 9000, 6000, and 3000 yr B.P. with anomaly patterns simulated in the COHMAP series of GCM experiments (Kutzbach and Guetter, 1986; COHMAP Members, 1988; Kutzbach, Guetter, *et al.,* this vol.).

Results obtained from GCM experiments are subject to two kinds of uncertainty. By simulating weather patterns, GCMs generate "natural variability" comparable to the day-to-day and year-to-year variability of the climate system. Climate anomalies simulated at individual grid points are therefore often not statistically significant, and robust deductions can be made only from broad-scale spatial patterns based on the results at many grid points. Numerical models also cannot be expected to reproduce the full complexity of the climate system; their spatial resolution is necessarily limited, and many processes, such as the generation of precipitation, evaporation, and clouds, are unavoidably simplified. In general, GCMs are best at simulating the broad features of the atmospheric circulation and temperature fields and are less good at simulating detailed regional features of climate, including the spatial distribution of the components of moisture balance (Schlesinger and Mitchell, 1987).

This study focuses on the broadest-scale patterns of vegetational and climatic change across Europe. Grid-point results from the CCM experiments have therefore been mapped at a suitably coarse scale for comparison with these patterns. Maps are shown for simulated July and January temperatures and annual moisture balance. Annual moisture balance was approximated by averaging simulated precipitation-minus-evaporation (P-E) for January and July, with a weighting to reflect orbital changes in the duration of the winter and summer half-years (Berger, 1979; Kutzbach and Guetter, 1986). To minimize the impact of uncertainties in the reconstructed and simulated climate variables, data-model comparisons are restricted to qualitative, continental-scale spatial patterns observed in both data sets. The anomaly patterns are examined in the light of a mechanistic analysis of simulated changes in atmospheric circulation over the North Atlantic (Harrison *et al.*, 1991a, 1992).

Results

Modern Pollen Distribution and Response Surfaces

Isopoll maps for 500 yr B.P. (Fig. 7.5) demonstrate that subrecent pollen spectra reflect broad-scale vegetation patterns (Fig. 7.3). Despite the extensive human impact on Europe's vegetation, the natural distribution of vegetation formations is reflected in the maps of individual taxon abundance and in the vegetation maps reconstructed from surface samples (Fig. 7.6). Post-Neolithic land use may have shifted some of the boundaries, for example by converting open woodlands to seminatural tundra or steppe (Behre, 1988), but probably not to an extent that would be significant on a synoptic map.

Response surfaces (Fig. 7.7) illustrate in a more analytical way the dependence of broad-scale vegetation patterns on climate. Each response surface is unique; the individual surfaces relate the distinct geographic distribution of each taxon to spatial variation in climate. *Picea*, for example, is seen to require cold winters and to be absent from regions with hot summers. *Betula* has a greater tolerance of mild winters and cool summers and is most abundant in climates wetter than those inhabited by *Picea*. Both taxa have effective mechanisms for resisting damage by temperatures lower than -40°C (Woodward, 1987). *Pinus* (Diploxylon) has a wide climatic range and a complex

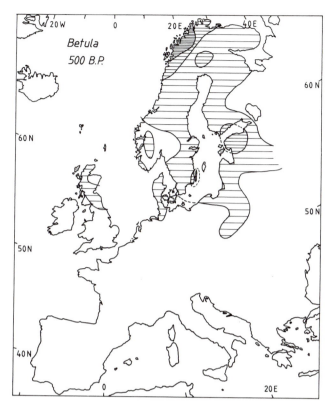

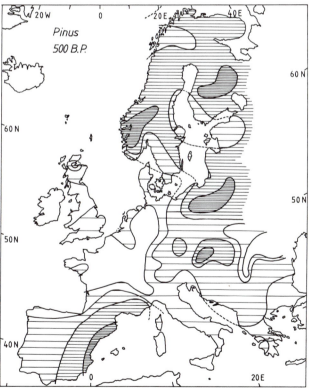

Fig. 7.5. Isopoll maps for main pollen taxa at 500 yr B.P. Stippling (where used) denotes pollen percentages in the 2-5% range. Increasingly dense hatching denotes pollen percentages of 5-20, 20-40, 40-60, and 60-80%. (Figure continues on pp. 144-45.)

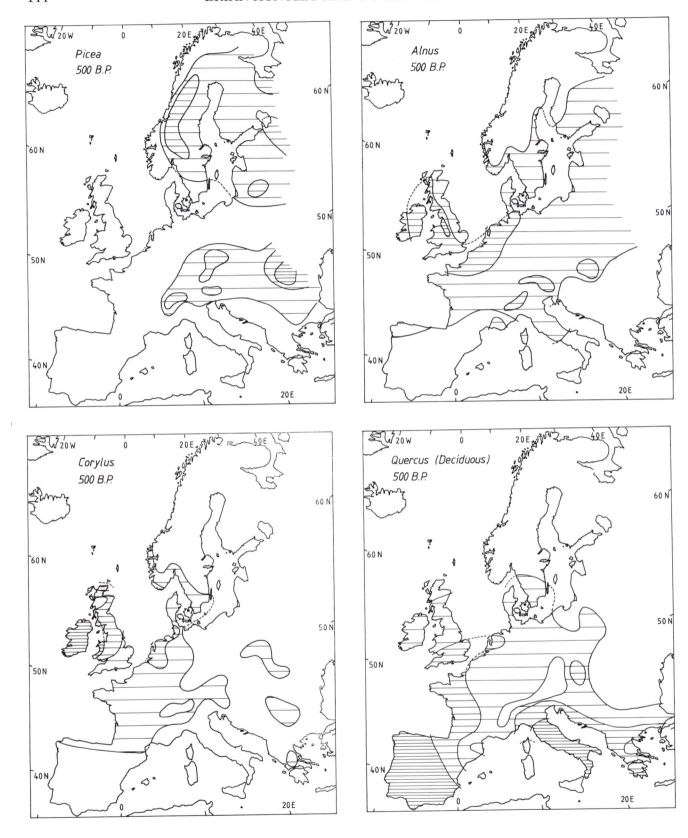

Fig. 7.5, continued from p. 143.

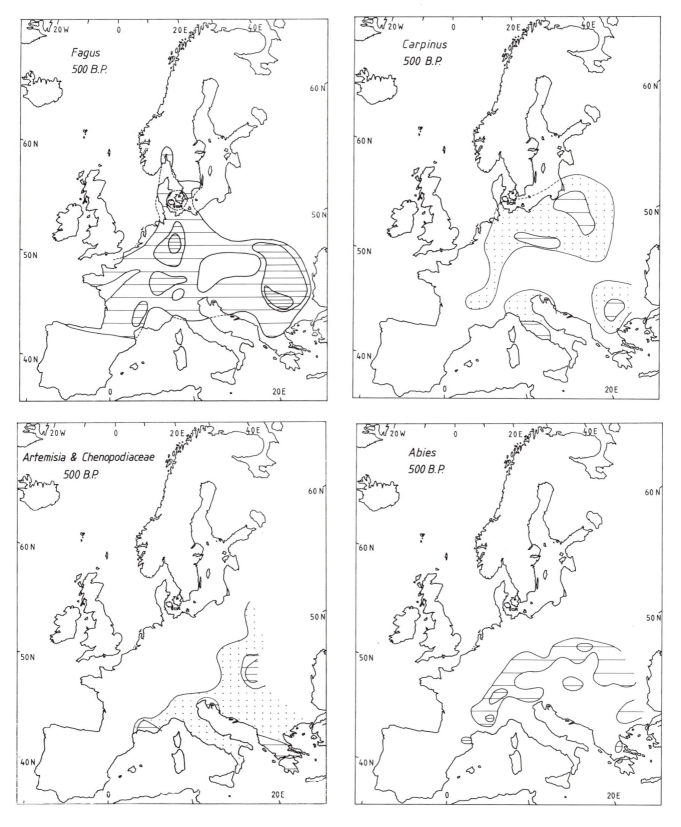

Fig. 7.5, continued from p. 144. For explanation, see p. 143.

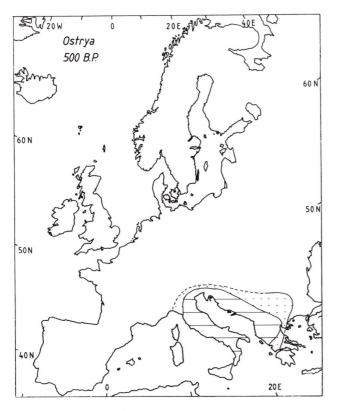

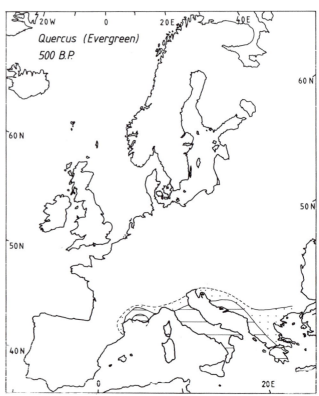

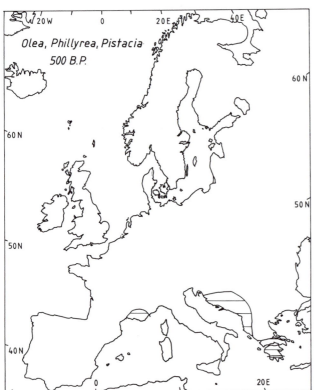

Fig. 7.5, continued from p. 145. For explanation, see p. 143.

surface because it includes both northern and southern pines. Its pollen percentage maximum reflects the abundance of *P. sylvestris* in the northern mixed forest.

Temperate deciduous trees are shown to require greater overall warmth than the boreal taxa. The trend of the isopolls reflecting their poleward limits (from upper left to lower right in Fig. 7.7) indicates a threshold requirement for accumulated temperature during the growing season (Hintikka, 1963; Dahl, 1980). Among the temperate deciduous trees, *Corylus* is shown to be abundant at the lowest summer temperatures; its pollen percentage maximum occurs at a lower July temperature than that for *Picea*, even though *Corylus* requires more growing degree-days (Prentice and Helmisaari, 1991). This emphasizes the maritime nature of the climate in which *Corylus* is most abundant today, but *Corylus* can also grow in climates with much colder winters (Prentice and Helmisaari, 1991). *Carpinus betulus, Fagus,* and *Abies* require warmer summers and warmer winters than does *Corylus,* and their highest pollen percentages occur at much higher summer temperatures. The pollen percentage maximum for *Ostrya* type occurs at high summer temperatures combined with warm winters and high precipitation. Deciduous *Quercus* spp. collectively span a wide range of summer temperatures but reach maximum pollen percentages in the drier, more continental climates of the steppe-forest transition with high summer temperatures, low winter temperatures, and low precipitation.

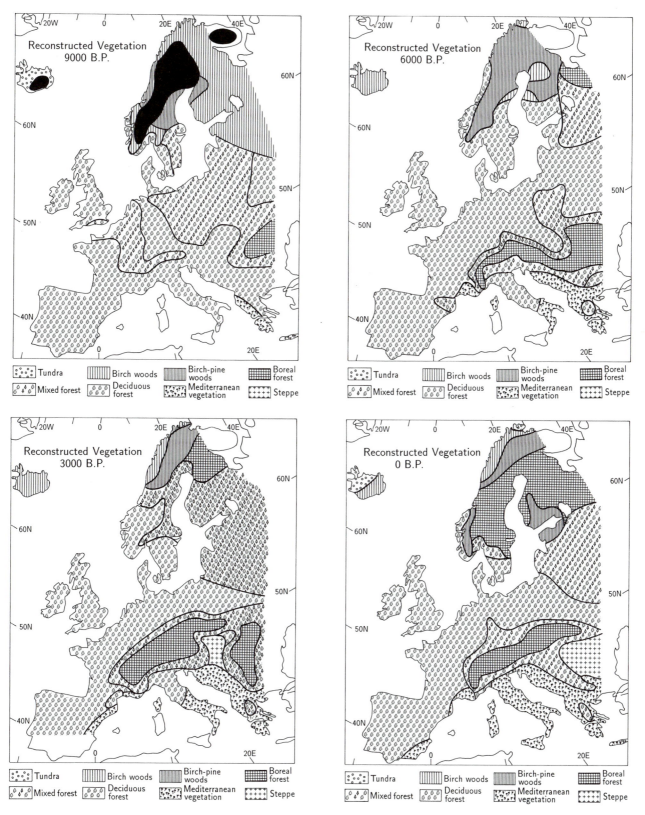

Fig. 7.6. Vegetation regions reconstructed from pollen data for 9000, 6000, 3000, and 0 yr B.P. (See Table 7.1 for reconstruction criteria.)

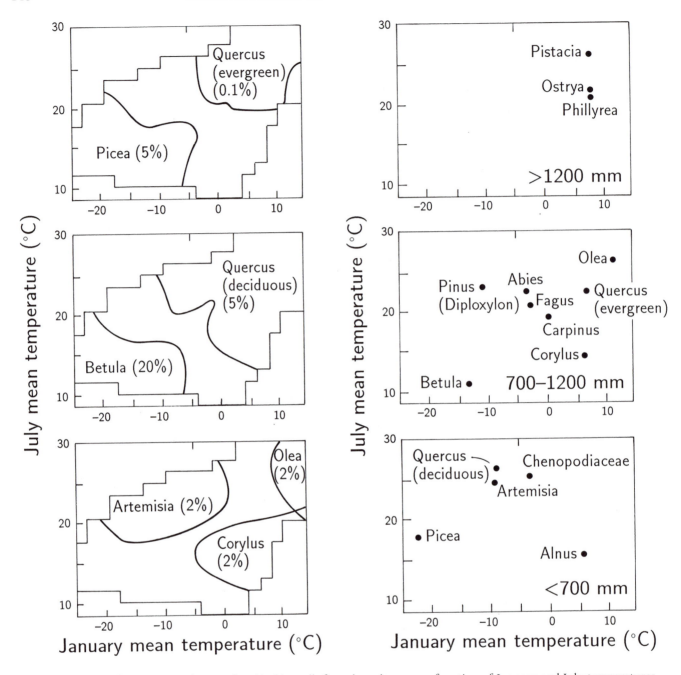

Fig. 7.7. Summary of response surfaces. Left: critical isopolls for selected taxa as a function of January and July temperatures. The diagram indicates regions of climate space where pollen percentages exceed the values given in parentheses after the taxon names. Right: locations of pollen percentage maxima in terms of January and July temperatures and annual precipitation.

Broad-leaved evergreens are shown to require warm winters, consistent with the relatively high temperatures at which they are damaged by frost (Woodward, 1987). Evergreen *Quercus* spp. are somewhat more frost-tolerant than the sclerophyll shrubs (*Olea, Pistacia,* and *Phillyrea*). The optimum environment for these shrubs has relatively high annual precipitation but is characterized by high evaporation and

summer drought. Steppe forbs (*Artemisia* and Chenopodiaceae) are seen to be most abundant in climates with warm summers, cold winters, and low precipitation that do not support closed forest. The low precipitation optimum for *Alnus* spp. reflects their occurrence in riparian environments within dry climates.

HOLOCENE CHANGES IN POLLEN DISTRIBUTION AND THEIR CLIMATIC INTERPRETATION

Isopoll maps corresponding to "snapshots" of the vegetation pattern at 3000-yr intervals during the Holocene (Fig. 7.8) show individualistic patterns whose broad spatial coherence implies overall control by changes in macroclimate (Huntley and Prentice, 1988). *Betula* was abundant over most of unglaciated northern Europe by 9000 yr B.P. Its southern limit shifted northward between 9000 and 6000 yr B.P., readvancing slightly toward the present. Birch-pine woodland extended almost to the Arctic coast by 6000 yr B.P.; the northern limit of pine retreated gradually after 6000 yr B.P. (Hyvärinen, 1976). Deciduous *Quercus* spp. expanded northward from 9000 to 6000 yr B.P. and retreated from their maximum northward position after 6000 yr B.P.

The maps also show that the upper elevational limits of *Quercus* and the other temperate deciduous trees in central Europe were highest around 6000 yr B.P. *Corylus,* which requires somewhat less summer warmth, was already abundant in northwestern Europe and the mountains of southern and central Europe by 9000 yr B.P. It was partially replaced by other deciduous trees by 6000 yr B.P., and its northern and elevational limits retreated after 6000 yr B.P. *Alnus* was restricted at 9000 yr B.P. but was widespread by 6000 yr B.P., and its northern limit also retreated southward after 6000 yr B.P.

These patterns all suggest a general increase in growing-season warmth over northern and central Europe from 9000 to 6000 yr B.P., followed by a decline. That is, maximum summer warmth came later than the insolation maximum but was approximately contemporaneous with the time of warmest summers in eastern North America (Webb, Bartlein, *et al.,* this vol.). This interpretation is consistent with the classical climatic optimum for northern and central Europe.

Other patterns of vegetation change cannot be explained by the classical model. Evergreen *Quercus* spp. and Mediterranean sclerophyll shrubs were restricted at 9000 yr B.P. and were abundant only in the eastern Mediterranean. These taxa gradually expanded westward, with sclerophyll vegetation mainly replacing deciduous forests. Forest clearance is thought to have accelerated this change (Behre, 1988), and human impact is believed to have extended the area of Mediterranean shrublands at the expense of broadleaved evergreen forests. Nevertheless, the large-scale, spatially coherent nature of the shift in dominance from deciduous trees to broad-leaved evergreens implies a climatic driving force. Huntley and Prentice (1988) suggested that summers in southern Europe were cooler and/or moister than at present at 9000 yr B.P. and became gradually warmer and/or drier toward the present, causing Mediterranean taxa to expand toward the west. The temperate deciduous trees are more cold-tolerant than the broad-leaved ever-

green trees and shrubs that replaced them, so early- and mid-Holocene winters in southern Europe may also have been cooler than at present.

Picea is another taxon showing essentially unidirectional, east-west movement during the Holocene. *Picea* was apparently absent from Fennoscandia and the western Alps at 9000 yr B.P. It subsequently spread westward. Changes in the location of its abundance maximum in northern Europe paralleled the changes in its range limit, and a retreat on the eastern front accompanied its advance on the western front. Both observations imply a response to a continuing climatic change (Huntley, 1988). In this case the east-west movement suggests the involvement of winter temperature changes. *Picea* in Europe is confined to areas with coldest-month temperatures below -15°C, and high pollen abundances indicate winters colder than this. The early- and mid-Holocene prevalence of forests with deciduous trees and pine across northern Europe indicates a mild, maritime winter climate. As in western Norway and Great Britain today, winter temperatures even at the elevational treeline (determined by growing-season temperatures) were apparently too mild for the regeneration of *Picea.* Progressive winter cooling during the Holocene then explains the spread of *Picea* into Fennoscandia.

Abies, Fagus, Carpinus, and *Ostrya* were all very restricted at 9000 yr B.P., but by 6000 yr B.P. *Abies* was present in abundance in the southern Alps, the Pyrenees, and mountains in Greece, and the other taxa had become established in the mountains of south and east-central Europe. Presumably the combination of warm summers and rainfall that these taxa require was restricted in the early Holocene. Their restricted distributions at 9000 yr B.P. suggest that the early Holocene in central Europe was drier than at present.

Isopoll maps for *Artemisia* and Chenopodiaceae show the development of steppe vegetation in southeastern Europe after 6000 yr B.P., partly at the expense of deciduous trees such as *Quercus* and *Fagus.* This development was paralleled by a westward shift across Europe of the isopolls for *Alnus* and *Corylus.* These patterns suggest that precipitation was at its maximum in central Europe around 6000 B.P. and declined thereafter, although conditions never again became as dry as they were at 9000 yr B.P.

Abies, Carpinus, and *Fagus* expanded northwestward after 6000 yr B.P.; the latter two taxa reached southern Great Britain and Scandinavia. At the same time *Carpinus* declined in southeastern Europe. These changes can be partially explained by cold winters in the continental interior at 6000 yr B.P., which subsequently warmed gradually toward the present; that is, the reverse of the trend hypothesized for northern and western Europe (Huntley *et al.,* 1989).

The main features described above are summarized in the reconstructed vegetation maps (Fig. 7.6). At 9000 yr B.P. vegetation zone boundaries were gen-

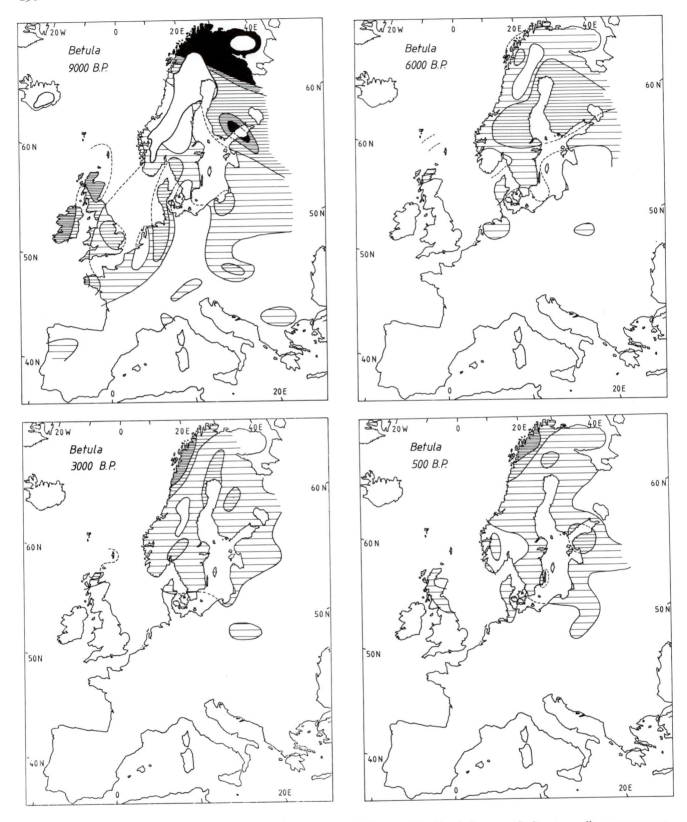

Fig. 7.8. Isopoll maps for main pollen taxa at 9000, 6000, 3000, and 500 yr B.P. Stippling (where used) denotes pollen percentages in the 2–5% range. Increasingly dense hatching denotes pollen percentages of 5–20, 20–40, 40–60, and 60–80%. (Figure continues on pp. 151–62.)

Fig. 7.8, continued from p. 150.

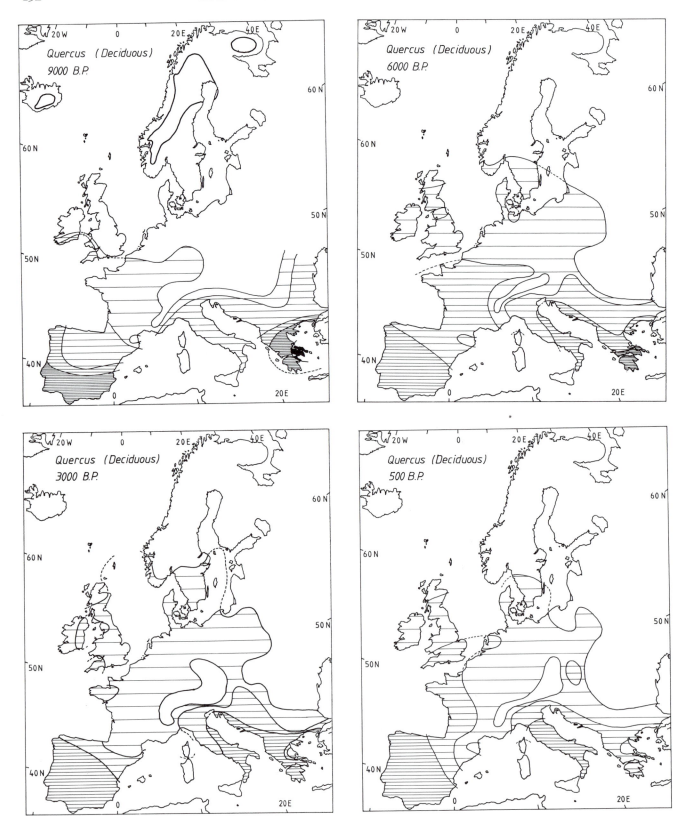

Fig. 7.8, continued from p. 151. For explanation, see p. 150.

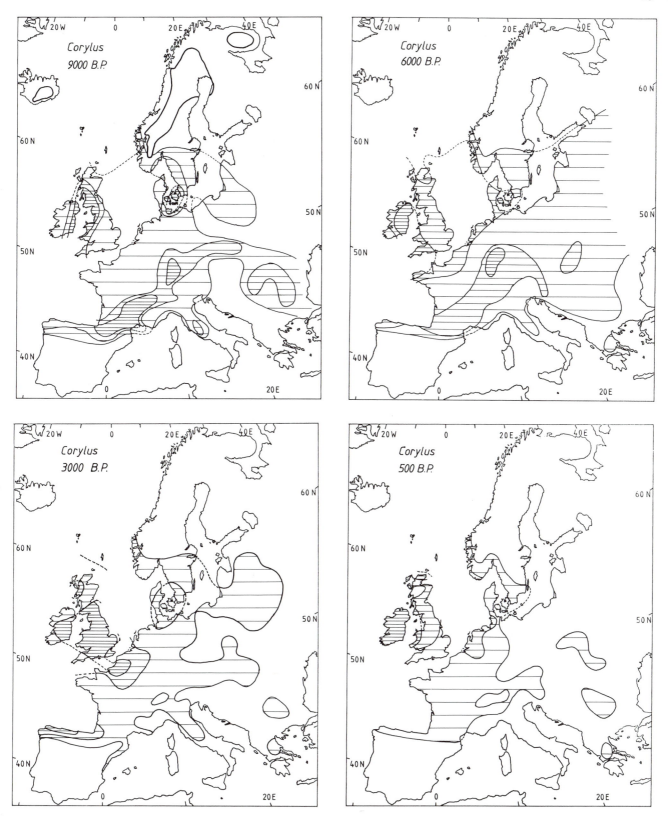

Fig. 7.8, continued from p. 152. For explanation, see p. 150.

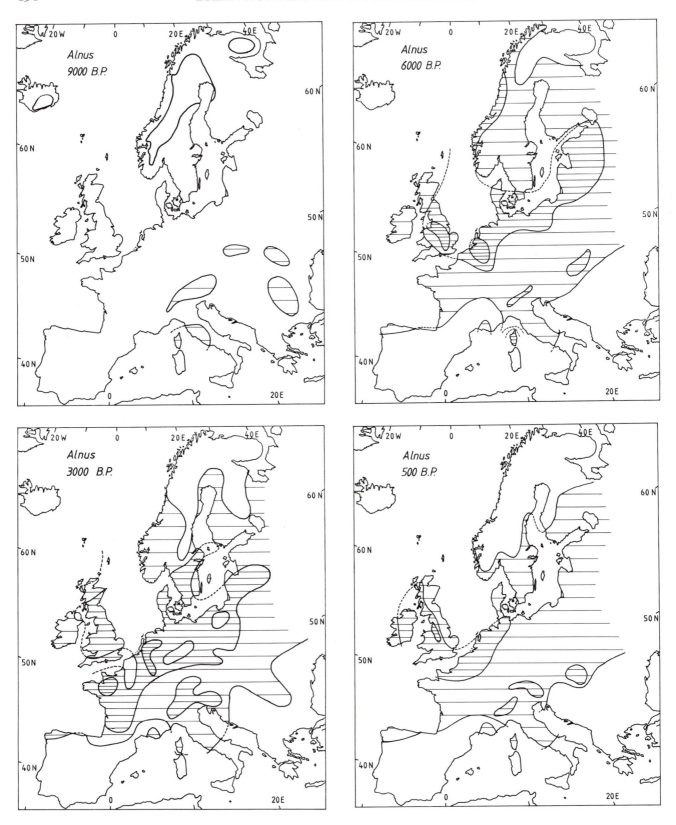

Fig. 7.8, continued from p. 153. For explanation, see p. 150.

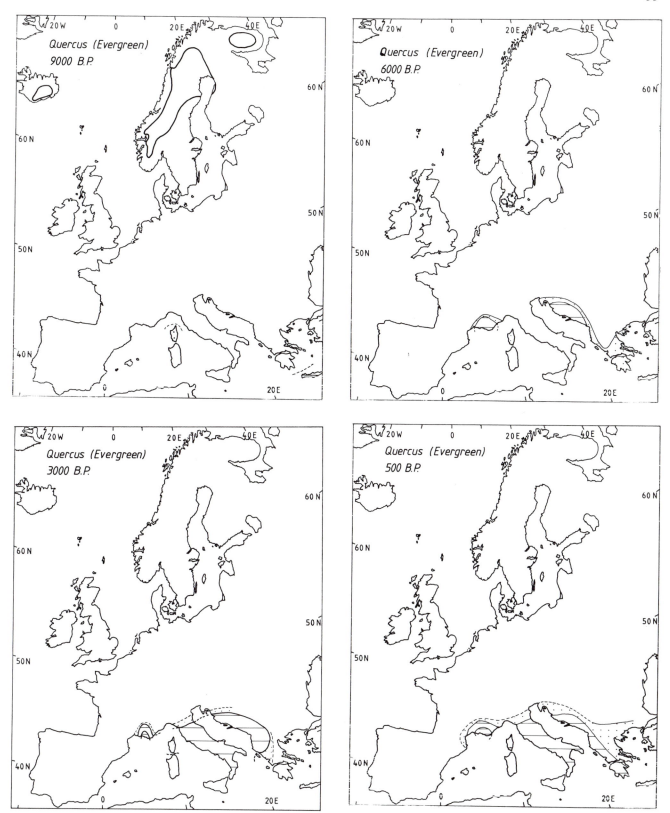

Fig. 7.8, continued from p. 154. For explanation, see p. 150.

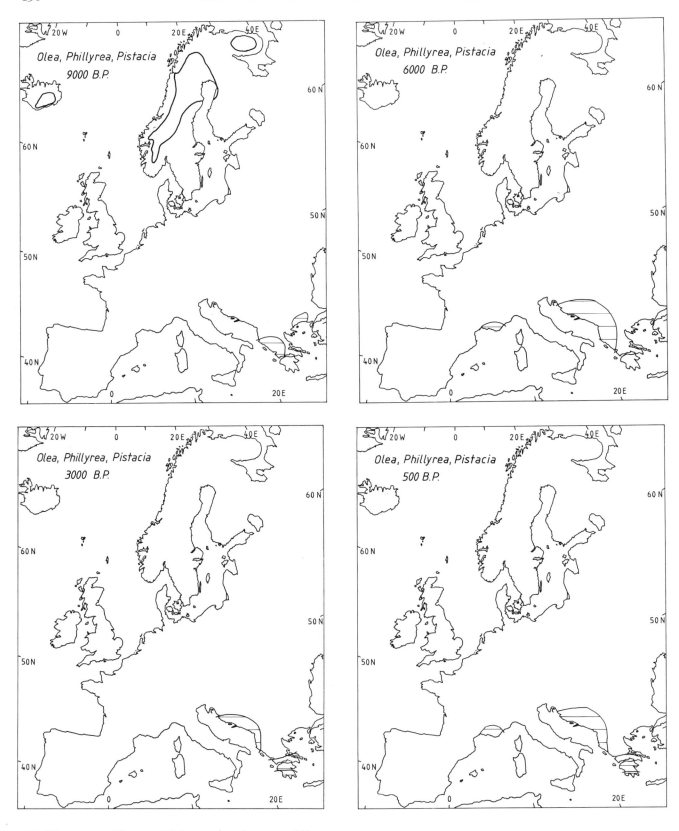

Fig. 7.8, continued from p. 155. For explanation, see p. 150.

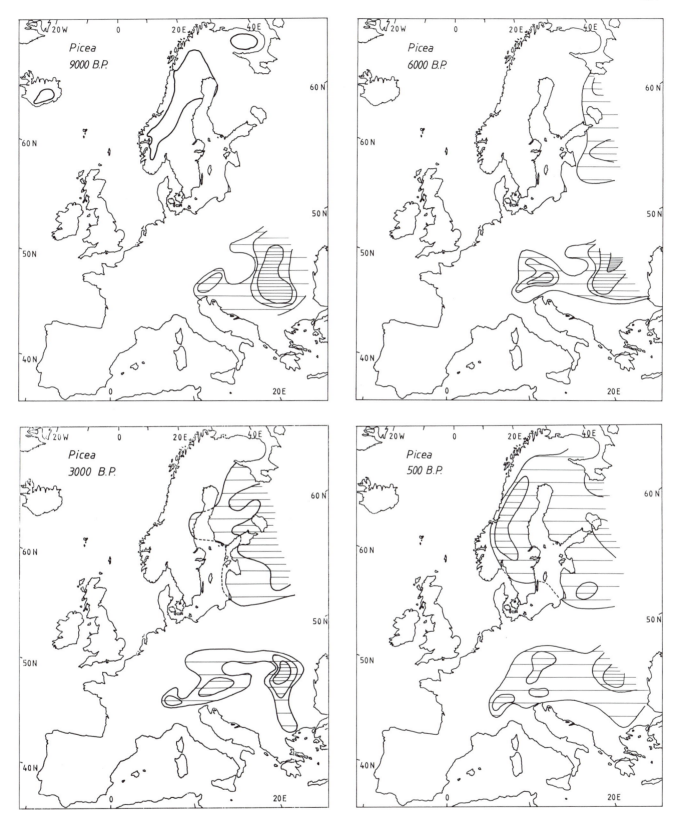

Fig. 7.8, continued from p. 156. For explanation, see p. 150.

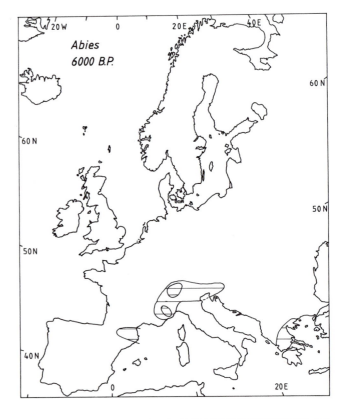

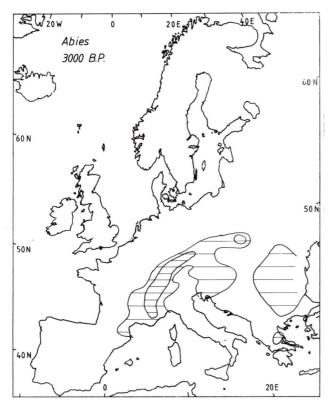

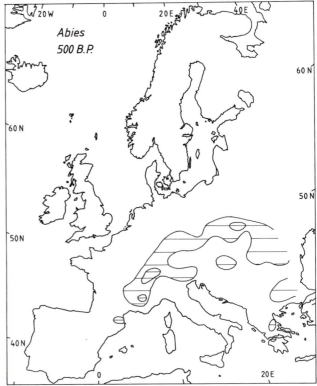

Fig. 7.8, continued from p. 157. For explanation, see p. 150.

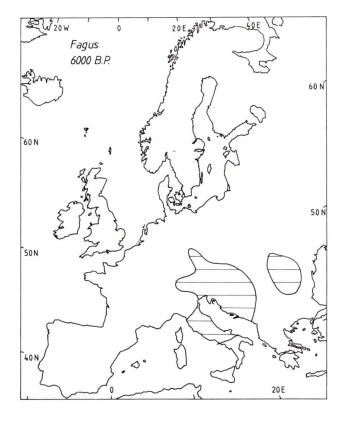

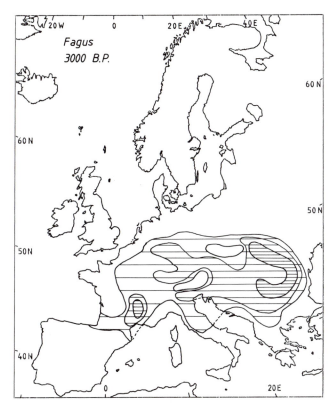

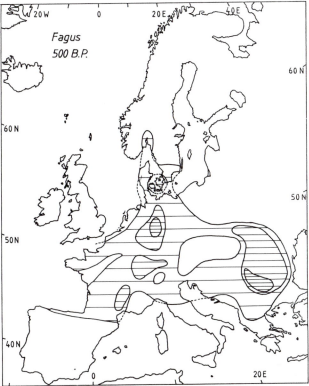

Fig. 7.8, continued from p. 158. For explanation, see p. 150.

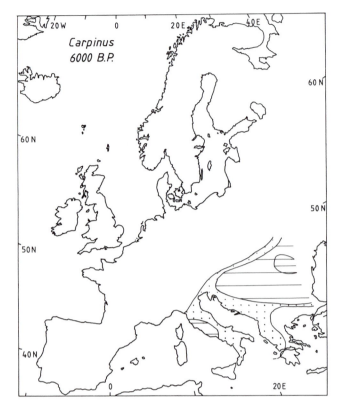

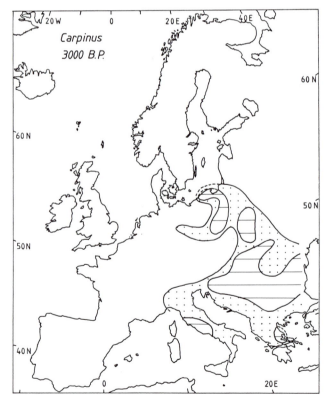

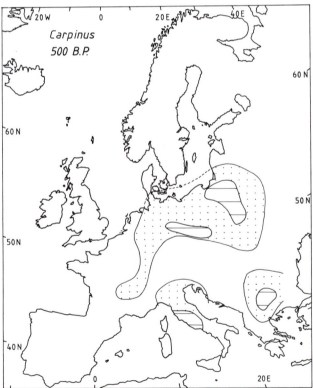

Fig. 7.8, continued from p. 159. For explanation, see p. 150.

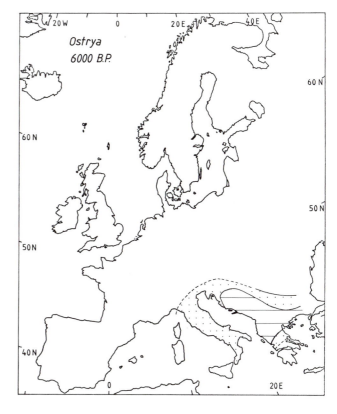

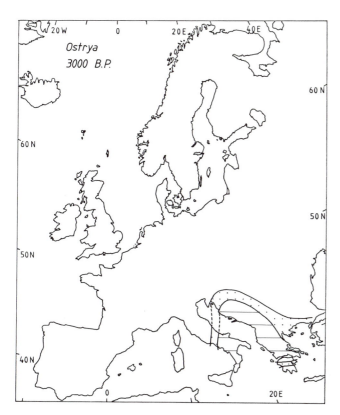

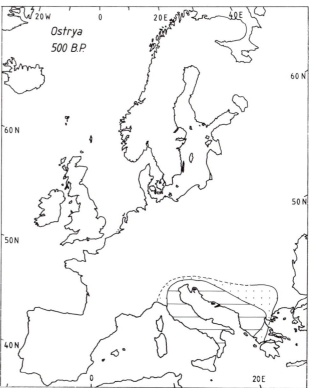

Fig. 7.8, continued from p. 160. For explanation, see p. 150.

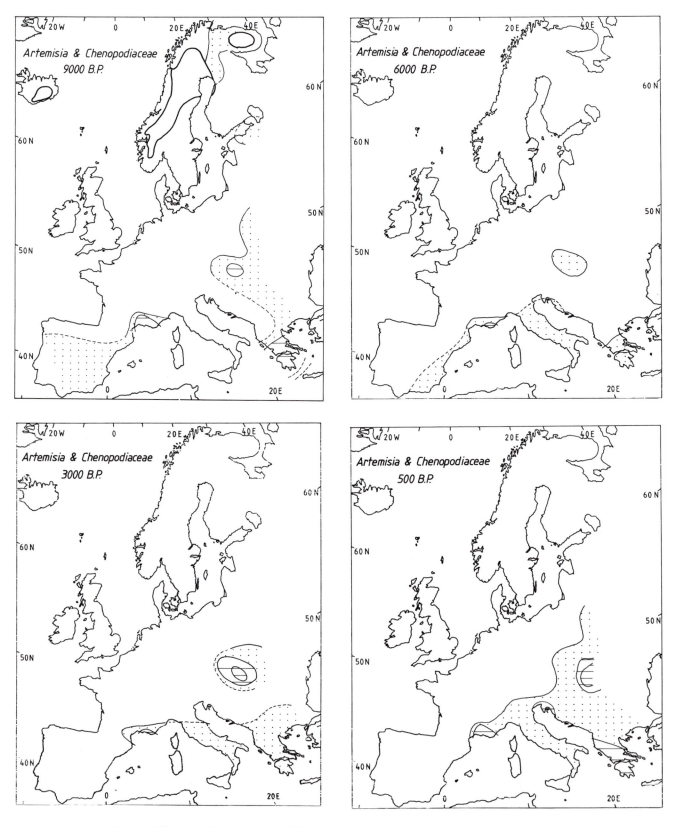

Fig. 7.8, continued from p. 161. For explanation, see p. 150.

erally much farther south than at present, the boreal forest was absent from most of its present area, and Mediterranean vegetation was restricted to a small area. By 6000 yr B.P. temperate deciduous forest extended higher and farther north than at present, and boreal forest and Mediterranean vegetation extended farther west than before but were still less extensive than today. Steppe vegetation was not in evidence in southeastern Europe until after 6000 yr B.P. The map for 3000 yr B.P. shows a vegetation pattern intermediate between those of 6000 yr B.P. and the present, suggesting basically unidirectional long-term trends in climate since 6000 yr B.P.

What these vegetation maps do not show is detailed variation in composition of the mapped assemblages, particularly the temperate deciduous and mixed forests. These variations can be inferred directly from the isopoll maps. Huntley (1990) showed that the pollen assemblages from much of southern, central, and western Europe at 9000 and 6000 yr B.P. have no known modern analogue at the formation level and that no-analogue vegetation was still extensive at 3000 yr B.P. In this respect the climatic interpretation of vegetation changes during the European Holocene is more uncertain than that for eastern North America, where no-analogue vegetation was restricted in space and time (Overpeck *et al.*, 1985). The no-analogue problem means that many pollen assemblages cannot be located precisely on Figure 7.7, which implies that the climate was probably somewhat outside the range of climates represented in the surface-sample data set. For example, early- to mid-Holocene pollen spectra from the Mediterranean region may reflect higher rainfall than occurs in warm-summer climates in present-day Europe; contemporaneous pollen spectra from Fennoscandia may reflect lower precipitation than occurs in modern maritime climates.

INFERRED CLIMATIC ANOMALY PATTERNS

Figure 7.9 summarizes inferred directions of climatic anomalies for 9000 and 6000 yr B.P. relative to the present in the form of spatial patterns across the continent. Results for 3000 yr B.P. are not considered in detail because both inferred and simulated anomalies for 3000 yr B.P. were in most respects similar to those for 6000 yr B.P., although they were reduced in magnitude.

Climatic interpretation of the vegetation pattern in Fennoscandia at 9000 yr B.P. is complicated by the recency of deglaciation and the proximity of some areas to the residual ice sheet. The warm summers shown in Figure 7.9 for northern Europe at 9000 yr B.P. are supported by other evidence (Iversen, 1973; Berglund, 1966; Berglund *et al.*, 1984) that taxa requiring warm summers, particularly aquatics, were already present north of their present limits in Scandinavia shortly after 10,000 yr B.P. In addition, some components of the temperate deciduous forest (notably *Tilia corda-*

ta) were already north of their present limits in the Baltic region by 9000 yr B.P. (Huntley and Birks, 1983), suggesting that this region too was experiencing summers warmer than at present as early as 9000 yr B.P.

For 6000 yr B.P. the inferred summer-temperature anomaly pattern is supported by the reconstruction of Huntley and Prentice (1988), who used a regionalized transfer function method to derive quantitative estimates of mean July temperatures across Europe. This reconstruction showed warmer summers over most of Europe by an average of about 2°C, summers more than 2°C warmer than at present in midcontinent and in the far north and cooler than at present in the Mediterranean region, and a reduced environmental lapse rate, with high-elevation sites recording the highest positive anomalies. Guiot (1987) also reconstructed large positive anomalies for high elevations in southern France.

Our 6000-yr B.P. reconstruction is also consistent with independent estimates by Klimanov (1984) for the European portion of the former Soviet Union. Klimanov made quantitative reconstructions of July and January temperatures and annual precipitation for 6000–5000 yr B.P. in the area north of about 45°N. Inferred January temperatures were 1–1.5°C warmer than at present in the Baltic region; the warming increased toward the north and east. July temperatures were 3–4°C warmer than at present north of 65°N; the warming decreased southward, nearly reaching zero by 45°N. Annual precipitation was greater than at present, especially in the far north and near the present-day steppe-forest boundary in the southeast.

COMPARISON WITH LAKE-LEVEL DATA

Our reconstructions of past moisture conditions from the pollen record are called "effective moisture" in Figure 7.9 because high summer insolation during the early to middle Holocene is likely to have produced evapotranspiration rates higher than at present and thus to have reduced the proportion of precipitation available to plants. These reconstructions show broad-scale patterns similar to those shown in preliminary reconstructions of Holocene changes in effective moisture based on lake-level records (Harrison *et al.*, 1991b; S. P. Harrison and G. Digerfeldt, unpublished). The lake-level data also show dry conditions over much of Europe at 9000 yr B.P., wet conditions in northeastern Europe at 6000 yr B.P., and conditions wetter than at present in the early and middle Holocene at sites near the Mediterranean, especially in the west.

COMPARISON WITH CLIMATE-MODEL RESULTS

Figure 7.9 shows both similarities and differences between the inferred climatic anomalies and those simulated in the CCM experiments. For winter temperatures there is qualitative agreement. The January

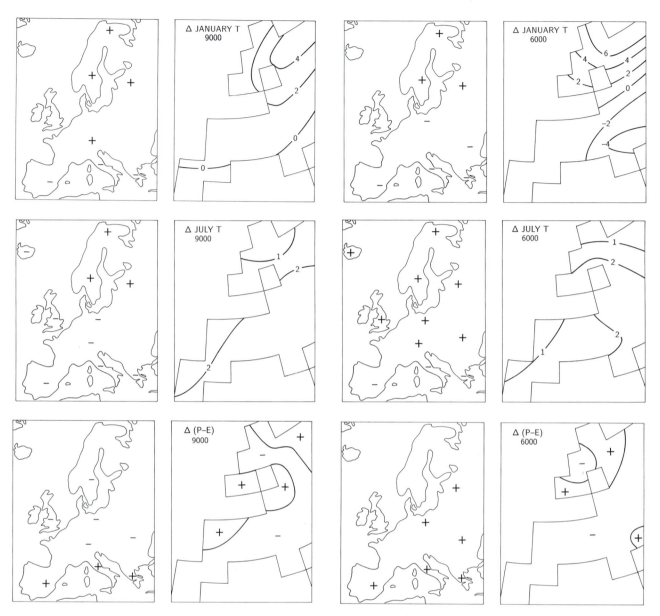

Fig. 7.9. Climatic anomaly patterns for January and July temperatures (T) and effective moisture (precipitation minus evapotranspiration [P-E]) for 9000 and 6000 yr B.P. inferred from pollen data (left-hand maps) and simulated by the Community Climate Model (right-hand maps).

simulation for 9000 yr B.P. shows positive temperature anomalies (more than 2°C) in Fennoscandia, especially toward the east, and negative anomalies in southern Europe. The January simulation for 6000 yr B.P. shows still larger positive anomalies (more than 4°C) in Fennoscandia and negative anomalies in southern and eastern Europe. These anomalies form part of a large-scale pattern in which Holocene winters significantly colder than at present are simulated for the low- and middle-latitude continental interiors of Africa and Asia, while warm winters are simulated for northern Eurasia. Maximum simulated warm-winter

anomalies in western Siberia at 6000 yr B.P. exceed 8°C and are statistically highly significant.

There is less agreement about summer temperatures. The July simulation for 9000 yr B.P. shows summers significantly warmer than at present across the whole continent, with the highest anomalies in the south. The reconstruction in contrast shows summers cooler than at present except in the north. The July simulation for 6000 yr B.P. is consistent with the reconstruction in Figure 7.9 and with the quantitative estimates of Huntley and Prentice (1988), in that it produces summers warmer by about 2°C in midcontinent and smaller positive anomalies elsewhere. However, the simulation does not show the large (greater than 2°C) anomalies reconstructed by Huntley and Prentice (1988) for the far north and, as in the 9000-yr B.P. case, it produces summers significantly warmer (rather than cooler) than at present in southern Europe.

Reconstructed and simulated moisture patterns also differ. Simulated January precipitation at 9000 yr B.P. was greater than at present in the area of mild-winter anomalies in northern and central Europe but significantly less than at present in southern Europe. Simulated July precipitation was also high in the north and low in the south. Simulated precipitation patterns for 6000 and 3000 yr B.P. were qualitatively similar to those for 9000 yr B.P. Simulated annual moisture-balance anomalies were positive in limited areas of western and northeastern Europe and negative elsewhere. The simulations thus correctly predict the dryness of much of Europe at 9000 yr B.P. but not the wetter-than-present conditions inferred for central Europe at 6000 yr B.P., and they show conditions drier (rather than wetter) than at present in southern Europe at 9000 and 6000 yr B.P.

Discussion: Mechanisms of Climatic Change in Europe

At 9000 yr B.P. the combination of July perihelion and high obliquity produced high Northern Hemisphere summer insolation, especially at high latitudes, low Northern Hemisphere winter insolation, especially at low latitudes, and consequently high land-sea temperature and pressure contrasts and a reduced latitudinal temperature gradient in both seasons (Kutzbach and Guetter, 1986; Mitchell *et al.*, 1988; Harrison *et al.*, 1991a, 1992). The insolation anomaly at 6000 yr B.P. was similar but somewhat reduced in magnitude relative to 9000 yr B.P.

The CCM simulations show that low insolation in winter at 9000 and 6000 yr B.P. caused significant cooling of the low- to middle-latitude continental areas of the Northern Hemisphere. This cooling enhanced the general hemispheric land-sea temperature and pressure contrast, producing among other things a stronger-than-present Icelandic low (Harrison *et al.*, 1991a, 1992). The subpolar low-pressure belt was also shifted northward, in response to the reduced latitudinal insolation gradient. These changes in the circulation over the North Atlantic brought strongly increased westerly flow into northern Eurasia and greatly *increased* winter temperatures in northern Europe.

In contrast to this "dynamic" control of winter temperatures, the simulated July temperature anomalies for 9000 and 6000 yr B.P. are primarily a direct response to insolation forcing. These anomalies were greatest in midcontinent because advection of cool air

from the ocean affected maritime areas. The simulated summer warming of northern Europe at 9000 yr B.P. was less than that at 6000 yr B.P. despite the greater insolation anomaly at 9000 yr B.P. because the "downstream" cooling effect of the Labrador ice sheet was superimposed on the insolation effect. Kutzbach and Guetter (1984) and Mitchell *et al.* (1988) showed this cooling effect in model experiments with and without a Laurentide ice sheet and estimated its magnitude as about 2°C for western Europe.

The simulations thus correctly predict the delayed "climatic optimum" in northwestern Europe, warm summers in northernmost Europe at 9000 yr B.P., and warm summers over the midcontinent at 6000 yr B.P. However, the reconstructions suggest that the cooling effect of the ice sheet at 9000 yr B.P. was larger than the simulations indicate (larger than the opposing effect of high summer insolation over much of Europe) and that the summer temperature anomaly at 6000 yr B.P. was at least as large in the far north as in the midcontinent (Huntley and Prentice, 1988).

An explanation for these discrepancies may lie in the North Atlantic and Arctic sea-surface temperatures (SSTs). The CCM simulations for the Holocene assumed SSTs similar to those of the present, but meltwater input from the wasting ice sheet may have produced lower-than-present North Atlantic SSTs (Ruddiman and Mix, this vol.). On the other hand, Arctic waters were warmer than at present (Koc Karpuz and Schrader, 1990) because of the high total annual insolation at high latitudes. Kutzbach and Gallimore (1988) and Mitchell *et al.* (1988) simulated this warming in experiments with GCMs that included a full seasonal cycle and a mixed-layer ocean model.

These experiments (Kutzbach and Gallimore, 1988; Mitchell *et al.*, 1988) further showed that the effect of high summer insolation at the high latitudes would be carried over into winter and would reduce the extent and duration of Arctic sea ice, thus strongly enhancing the *winter* warming of high-latitude land. This sea-ice effect thus provides a mechanism generating still greater winter temperature anomalies in the early to middle Holocene in the north.

The CCM simulations correctly predict the general dryness of Europe south of about 55°N in the early Holocene. The stronger westerly flow in winter produced simulated increases in precipitation over much of Europe, but these were counteracted by increased evaporation caused by high insolation, leaving positive moisture-balance anomalies only in the areas most strongly affected by the circulation change.

However, essentially the same pattern persists in the 6000-yr B.P. simulations, whereas the reconstructions show wetter-than-present conditions in central Europe. S. P. Harrison and G. Digerfeldt (unpublished) link changes in moisture balance from 9000 to 6000 yr B.P. to the summer subtropical anticyclone, which the simulations show as stronger than at present and shifted about 10° north at 9000 yr B.P. and as still strong but closer to its present position at 6000 yr B.P. These changes produced predominantly offshore flow at 50°N in the 9000-yr B.P. simulation but slightly stronger-than-present onshore flow in the 6000-yr B.P. experiment. The circulation changes may therefore have reinforced the direct insolation effect on the moisture balance at 9000 yr B.P. and opposed it at 6000 yr B.P. This mechanism would explain the wetter conditions at 6000 yr B.P. if we postulate that the effects of these shifts on summer precipitation are slightly larger than the simulations indicate.

The most consistent discrepancies between the data and model results relate to southern Europe. The model results for this region indicate warm and dry summers in the early to middle Holocene as a result of the high insolation and the strong subtropical anticyclone, whereas the pollen data show cool and/or moist summers. The model results also indicate dry winters, in contrast to evidence in the lake-level data of a more positive annual water balance (S. P. Harrison and G. Digerfeldt, unpublished).

The reasons for these discrepancies are unknown. Huntley and Prentice (1988) suggested that the increased land-sea contrast may have had effects on a scale too fine to be resolved by the GCM. The July simulations do show an increase in the onshore component of the surface winds in the Mediterranean region, which could have increased orographic precipitation and summer cloudiness. This mechanism might have been particularly effective in the western Mediterranean, where the Iberian plateau is large enough to produce a summer "heat low" and an associated local circulation that even today brings summer rains to the east coast of Spain.

Conclusions

The results presented in this chapter are based on the assumption that the major continental-scale patterns of change during the Holocene as shown in small-scale isopoll maps can be interpreted directly in terms of climatic forcing. This assumption is supported by the spatially coherent nature of the changes, by the agreement between reconstructed climatic changes and lake-level evidence for changes in moisture balance, and by the extent to which the reconstructed climatic changes can be explained qualitatively by forcings and mechanisms included in the climate-model experiments. The results therefore encourage us to attempt a more quantitative analysis of the record and to continue to seek mechanistic explanations for the remaining features of the reconstructions that are not reflected in the available model experiments.

The reliability of pollen-based climatic reconstructions depends on the amount and the quality of available data, and in these respects the data set could be improved. Ideally, the data set should allow mapping with adequate spatial resolution across the whole continent, unambiguous separation of horizontal and elevational gradients in areas of high relief, reliable pollen identification, and confident cross-correlation to within 500 yr based on adequate numbers of radiocarbon dates from every site. This ideal has not yet been realized, but advances since the compilation of the data base used in this chapter have produced a considerable volume of published and unpublished data, including sites in key areas of deficiency such as southeastern Europe and the Mediterranean, whose incorporation would considerably upgrade the data set. The current development of a European data base of high-quality palynologic data at Arles, France, is therefore extremely important for the future of paleoclimatic research in Europe.

Equally important is the application of a robust, quantitative methodology to infer past climates from the available data. We are attempting a quantitative reconstruction of late-Quaternary paleoclimates in Europe based on an expanded pollen data set analyzed with the help of response surface methods. Guiot (1990, 1991) developed a bootstrap procedure for quantitative paleoclimatic inference, in which fossil pollen samples are directly compared with a large set of surface samples, and a continent-wide reconstruction using this method is also being attempted. The reliability of response surface and analogue methods, however, ultimately depends on the availability of the correct modern analogues. We need (1) to continue to enlarge the surface-sample data set to include samples from the widest possible range of climates and (2) to develop diagnostic procedures based on the intrinsic limits of different plant types in order to allow unambiguous reconstruction of climates that lack modern analogues.

Because of the topographic complexity of Europe, spatial resolution is likely to remain for the foreseeable future a limiting factor in the reliability of GCM simulations of regional European climates. Nevertheless, improvements in GCM capability will strengthen the basis for testing hypotheses regarding the mechanisms of climatic change. The most obviously applicable improvements in recently developed GCMs are in spatial resolution and in the inclusion of a full seasonal cycle incorporating the response of the ocean. It is also important to try to link the broad-scale circulation features simulated by GCMs to more local weather patterns in order to bridge the scale gap between the coarse GCM resolution and the much finer resolution of the pollen record.

This preliminary analysis has emphasized the importance of changes in atmospheric circulation over the North Atlantic for European climate, as well as the possible role of changes in North Atlantic SSTs in explaining Holocene climates. Changes in SSTs caused by processes not included in atmospheric GCMs, such as changing meltwater inputs and upwelling rates, need to be better known. Our analysis suggests that we already have some ability to "predict" past climates but that the predictions can be misleading, possibly because of incomplete specification of boundary conditions such as SSTs in addition to the limited spatial resolution of current GCMs. We have drawn attention to problems that will be resolved only through a coordinated effort among scientists concerned with atmospheric, oceanic, and ecological modeling and the paleoclimatic record of Europe and its surrounding seas.

Acknowledgments

We thank John Birks for his part in the development of the data base and COHMAP project members—especially Pat Bartlein, Sandy Harrison, John Kutzbach, and Tom Webb—for scientific input. I. C. Prentice acknowledges research support from the Swedish Natural Science Research Council (NFR) to the project "Simulation Modelling of Natural Forest Dynamics." The final revision was undertaken while I. C. Prentice was a research associate of the biosphere dynamics project at the International Institute of Applied Systems Analysis.

References

Bartlein, P. J. (1988). Late-Tertiary and Quaternary palaeoenvironments. *In* "Vegetation History" (B. Huntley and T. Webb III, Eds.), pp. 113-152. Kluwer, Dordrecht.

Bartlein, P. J., Prentice, I. C., and Webb, T. III (1986). Climatic response surfaces from pollen data for some eastern North American taxa. *Journal of Biogeography* 13, 35-57.

Baumgartner, A., and Reichel, E. (1975). "The World Water Balance." Elsevier, Amsterdam.

Behre, K.-E. (1988). The role of man in European vegetation history. *In* "Vegetation History" (B. Huntley and T. Webb III, Eds.), pp. 633-672. Kluwer, Dordrecht.

Berger, A. L. (1979). Long-term variations of daily insolation and Quaternary climatic changes. *Journal of the Atmospheric Sciences* 35, 2362-2367.

Berglund, B. E. (1966). Late-Quaternary vegetation in eastern Blekinge, south-eastern Sweden: A pollen-analytical study. II. Post-glacial time. *Opera Botanica* 12(2), 1-190.

Berglund, B. E., Lemdahl, G., Liedberg-Jonsson, B., and Persson, T. (1984). Biotic response to climatic changes during the time span 13,000-10,000 B.P.: A case study from SW Sweden. *In* "Climatic Changes on a Yearly to Millennial Basis" (N. A. Mörner and W. Karlén, Eds.), pp. 25-36. D. Reidel, Dordrecht.

COHMAP Members (1988). Climatic changes of the last 18,000 years: Observations and model simulations. *Science* 241, 1043-1052.

Dahl, E. (1980). "Relations entre la repartition des plantes dans la nature et les facteurs climatiques." Université scientifique et medicale de Grenoble, Grenoble, France.

Davis, M. B., and Botkin, D. B. (1985). Sensitivity of cool-temperate forests and their fossil pollen record to rapid temperature change. *Quaternary Research* 23, 327-340.

Ellenberg, H. (1982). "Vegetation Mitteleuropas mit den Alpen," 3rd ed. Ulmer, Stuttgart.

Fullard, H., and Darby, H. C., Eds. (1973). "Atlas General Larousse." George Philip and Son Ltd., Paris.

Grichuk, V. P. (1969). Experience in reconstructing certain climatic elements of the Northern Hemisphere in the Atlantic period of the Holocene (in Russian). *In* "The Holocene" (M. I. Neustadt, Ed.), pp. 41-57. Nauka Press, Moscow.

Grimm, E. C. (1984). Fire and other factors controlling the Big Woods vegetation of Minnesota in the mid-nineteenth century. *Ecological Monographs* 54, 291-311.

Guiot, J. (1987). Late Quaternary climatic change in France estimated from multivariate pollen time series. *Quaternary Research* 28, 100-118.

——. (1990). Methodology of the last climatic cycle reconstruction in France from pollen data. *Palaeogeography, Palaeoclimatology, Palaeoecology* 80, 49-69.

——. (1991). Structural characteristics of proxy data and methods for quantitative climate reconstruction. *Paläoklimaforschung* 6, 271-285.

Harrison, S. P., Prentice, I. C., and Bartlein, P. J. (1991a). What climate models can tell us about the Holocene palaeoclimates of Europe. *Paläoklimaforschung* 6, 285-300.

Harrison, S. P., Saarse, L., and Digerfeldt, G. (1991b). Holocene changes in lake levels as climate proxy data in Europe. *Paläoklimaforschung* 6, 159-179.

Harrison, S. P., Prentice, I. C., and Bartlein, P. J. (1992). Influence of insolation and glaciation on atmospheric circulation in the North Atlantic sector: Implications of general circulation model experiments for the late Quaternary climatology of Europe. *Quaternary Science Reviews* 11, 283-300.

Heinselman, M. L. (1981). Fire and succession in the conifer forests of northern North America. *In* "Forest Succession: Concepts and Application" (D. C. West, H. H. Shugart, and D. B. Botkin, Eds.), pp. 374-405. Springer-Verlag, New York.

Hintikka, V. (1963). Über das Grossklima einiger Pflanzenareale in zwei Klimakoordinaten systemen dargestellt. *Annales Botanici Societatis "Vanamo"* 34(5), 1-64.

Huntley, B. (1988). Glacial and Holocene vegetation history: Europe. *In* "Vegetation History" (B. Huntley and T. Webb III, Eds.), pp. 341-383. Kluwer, Dordrecht.

——. (1990). Dissimilarity mapping between fossil and contemporary pollen spectra in Europe for the past 13,000 years. *Quaternary Research* 33, 360-376.

Huntley, B., and Birks, H. J. B. (1983). "An Atlas of Past and Present Pollen Maps for Europe 0-13,000 Years Ago." Cambridge University Press, Cambridge.

Huntley, B., and Prentice, I. C. (1988). July temperatures in Europe from pollen data, 6000 years before present. *Science* 241, 687-690.

Huntley, B., and Webb, T. III, Eds. (1988). "Vegetation History." Kluwer, Dordrecht.

——. (1989). Migration: Species' response to climatic variations caused by changes in the earth's orbit. *Journal of Biogeography* 16, 5-19.

Huntley, B., Bartlein, P. J., and Prentice, I. C. (1989). Climatic control of the distribution and abundance of beech (*Fagus* L.) in Europe and North America. *Journal of Biogeography* 16, 551-560.

Hyvärinen, H. (1976). Flandrian pollen deposition rates and tree-line history in northern Fennoscandia. *Boreas* 5, 163-175.

Iversen, J. (1944). *Viscum, Hedera* and *Ilex* as climatic indicators. *Geologiska Föreningens Stockholm Förhandlingar* 66, 463-483.

—. (1960). Problems of the early post-glacial forest development in Denmark. *Danmarks Geologiske Undersögelse IV* 4(3), 1-32.

—. (1973). The development of Denmark's nature since the last glacial. *Danmarks Geologiske Undersögelse V* 7-C, 1-126.

Klimanov, V. A. (1984). Paleoclimatic reconstructions based on the information statistical method. *In* "Late Quaternary Environments of the Soviet Union" (A. A. Velichko, Ed.; H. E. Wright, Jr., and C. W. Barnosky, Eds., English-language ed.), pp. 297-303. University of Minnesota Press, Minneapolis.

Koc Karpuz, N., and Schrader, H. (1990). Surface sediment distribution and Holocene paleotemperature variations in the Greenland, Iceland and Norwegian Sea. *Paleoceanography* 5, 557-580.

Kutzbach, J. E. (1987). Model simulations of the climatic patterns during the deglaciation of North America. *In* "North America and Adjacent Oceans during the Last Deglaciation" (W. F. Ruddiman and H. E. Wright, Jr., Eds.), pp. 425-446. The Geology of North America, Vol. K-3. The Geological Society of America, Boulder, Colo.

Kutzbach, J. E., and Gallimore, R. G. (1988). Sensitivity of a coupled atmosphere/mixed-layer ocean model to changes in orbital forcing at 9000 years B.P. *Journal of Geophysical Research* 93, 803-821.

Kutzbach, J. E., and Guetter, P. J. (1984). The sensitivity of monsoon climates to orbital parameter changes for 9000 years B.P.: Experiments with the NCAR general circulation model. *In* "Milankovitch and Climate" (A. Berger, J. Imbrie, J. Hays, G. Kukla, and B. Saltzman, Eds.), Part 2, pp. 801-820. Reidel, Dordrecht.

—. (1986). The influence of changing orbital parameters and surface boundary conditions on climate simulations for the past 18,000 years. *Journal of the Atmospheric Sciences* 43, 1726-1759.

Meteorological Office (1972). "Tables of Temperature, Relative Humidity, Precipitation and Sunshine for the World. Part III. Europe and the Azores." Met. O. 856C. Her Majesty's Stationery Office, London.

Mitchell, J. F. B., Grahame, N. S., and Needham, K. J. (1988). Climate simulations for 9000 years before present: Seasonal variations and effect of the Laurentide ice sheet. *Journal of Geophysical Research* 93, 8283-8303.

Noirfalise, A., Ed. (1987). "Map of the Natural Vegetation of the Member Countries of the European Community and the Council of Europe," 2nd ed. Council of Europe and Commission of the European Community, Strasbourg, France.

Overpeck, J., Webb, T. III, and Prentice, I. C. (1985). Quantitative interpretation of fossil pollen spectra: Dissimilarity coefficients and the method of modern analogs for pollen data. *Quaternary Research* 23, 87-108.

Pickett, S. T. A., and Thompson, J. N. (1978). Patch dynamics and the design of nature reserves. *Biological Conservation* 13, 27-37.

Pickett, S. T. A., and White, P. S., Eds. (1985). "The Ecology of Natural Disturbance and Patch Dynamics." Academic Press, Orlando.

Prentice, I. C. (1986). Vegetation responses to past climatic variation. *Vegetatio* 67, 131-141.

—. (1992). Climate change and long-term vegetation dynamics. *In* "Plant Succession: Theory and Prediction" (D. C. Glenn-Lewin, R. A. Peet, and T. Veblen, Eds.), pp. 293-339. Chapman and Hall, New York.

Prentice, I. C., and Helmisaari, H. (1991). Silvics of north European trees: Compilation, comparisons and implications for forest succession modelling. *Forest Ecology and Management,* 42, 79-93.

Prentice, I. C., Bartlein, P. J., and Webb, T. III (1991). Vegetation and climate changes in eastern North America since the last glacial maximum: A response to continuous climatic forcing. *Ecology* 72, 2038-2056.

Runkle, J. R. (1985). Disturbance regimes in temperate forests. *In* "The Ecology of Natural Disturbance and Patch Dynamics" (S. T. A. Pickett and P. S. White, Eds.), pp. 17-33. Academic Press, Orlando.

Schlesinger, M. E., and Mitchell, J. F. B. (1987). Climate model simulations of the equilibrium climatic response to increased carbon dioxide. *Reviews of Geophysics* 25, 760-798.

Shugart, H. H. (1984). "A Theory of Forest Dynamics." Springer-Verlag, New York.

Sjörs, H. (1967). "Nordisk Växtgeografi," 2nd ed. Svenska Bokförlaget, Stockholm.

Webb, T. III (1986). Is vegetation in equilibrium with climate? How to interpret late-Quaternary pollen data. *Vegetatio* 67, 75-91.

Webb, T. III, Bartlein, P. J., and Kutzbach, J. E. (1987). Climatic change in eastern North America during the past 18,000 years: Comparisons of pollen data with model results. *In* "North America and Adjacent Oceans during the Last Deglaciation" (W. F. Ruddiman and H. E. Wright, Jr., Eds.), pp. 447-462. The Geology of North America, Vol. K-3. The Geological Society of America, Boulder, Colo.

White, P. S. (1979). Pattern, process and natural disturbance in vegetation. *Botanical Review* 45, 229-299.

Woodward, F. I. (1987). "Climate and Plant Distribution." Cambridge University Press, Cambridge.

Zackrisson, O. (1977). Influence of forest fires on the North Swedish boreal forest. *Oikos* 29, 22-32.

CHAPTER 8

Vegetational and Climatic History of the Western Former Soviet Union

G. M. Peterson

The area occupied by the former Soviet Union represents about one-sixth of the earth's land surface and includes both a wide variety of climatic zones and the geographic range limits of many tree genera. West of 100°E the topography is relatively flat, and the Russian Plain is separated from the West Siberian Lowland by the Ural Mountains at 60°E (Fig. 8.1). These vast plains in the western former Soviet Union have yielded a substantial record of Holocene movements in major forest tree genera (Peterson, 1983a). The eastern half of the region is mountainous and has provided few well-dated Holocene pollen diagrams (Fig. 8.2 and Table 8.1).

In the last few decades Soviet scientists published many articles describing the late-Quaternary environments in northern Eurasia (Peterson, 1979). They were

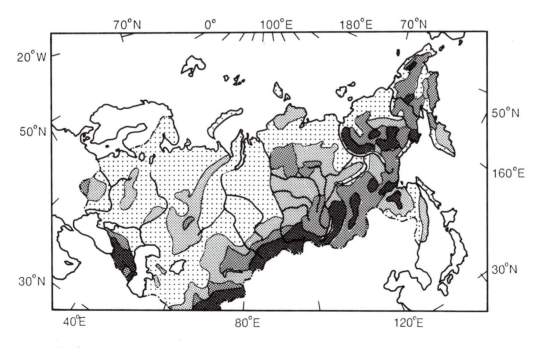

Fig. 8.1. The former Soviet Union. Note the relatively flat Russian Plain in the west (European part of the former Soviet Union), the Ural Mountains at 60°E, the West Siberian Lowland between 60 and 100°E, and the more mountainous regions of eastern Siberia and the Far East. Dotted areas are below 200 m in elevation; lightly shaded areas are 200–500 m; moderately shaded areas are 500–1000 m; and heavily shaded areas are above 1000 m.

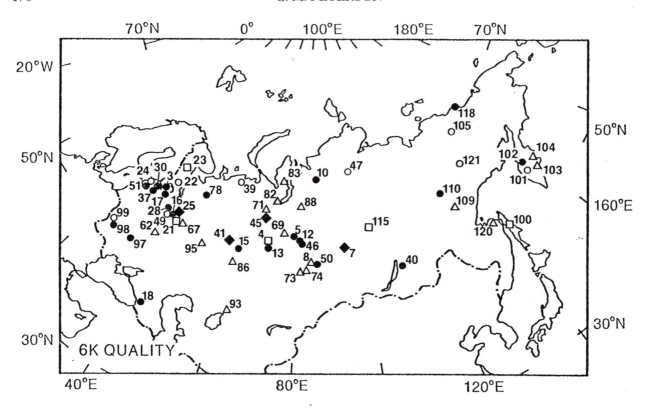

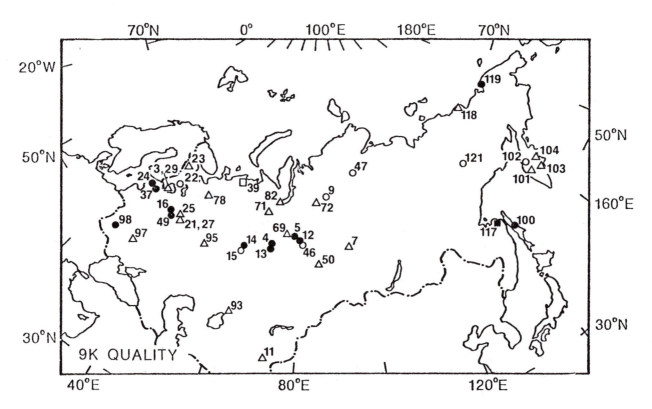

Fig. 8.2 Locations of Holocene cores mapped for 6000 and 9000 yr B.P. Numbers refer to sites listed in Table 8.1. Symbols indicate the quality of radiocarbon control: ● = 1 (highest), ○ = 2, ■ = 3, □ = 4, ◆ = 5, ◇ = 6, and △ = 7 (lowest).

Table 8.1. Location of sites, time levels mapped, and bibliographic sources

Site no.	Site name[a]	North Latitude	East Longitude	^{14}C dates (no.)	Quality[b] (0, 6, 9000 yr B.P.)	Elevation[c] (m)	Reference
3	SHuvalovskoe	60°03'	30°20'	7	217	77	Neishtadt (1965)
4	Bol. Pershino	59°21'	69°00'	2	741	77	Neishtadt (1967)
5	Nizhne-Vartovsk	60°56'	76°38'	13	-11	77	Neishtadt (1976a)
7	Pit-Gorodok	59°18'	93°50'	1	-57	458	KHotinskii (1977)
8	Vasiugan'e I	56°52'	83°05'	1	47-	77	KHotinskii (1977)
9	Igarka	67°28'	86°35'	3	4-2	77	Kind (1969)
10	Karginskii	70°00'	84°00'	13	21-	77	Firsov et al. (1974)
11	Kerkidon	40°29'	72°04'	3	--7	458	Serebriannyi et al. (1976)
12	Lukashin Iar	60°20'	78°24'	13	211	77	Glebov et al. (1974)
13	Gorno-Slinkino	58°45'	68°49'	10	111	77	Volkov et al. (1973)
14	Gorbunovskoe	57°55'	60°30'	2	7-1	229	KHotinskii (1977)
15	Aiatskoe	57°00'	60°05'	7	212	229	KHotinskii (1977)
16	Osechenskoe	57°31'	34°50'	6	111	229	Neishtadt (1965)
17	Tesovo-Netyl.	58°57'	31°04'	6	11-	77	Neishtadt (1965)
18	Imnatskoe	42°05'	41°43'	4	11-	458	Neishtadt (1965)
21	Somino	56°51'	38°39'	6	-47	77	KHotinskii (1977)
22	Bezdonnoe	61°45'	32°12'	3	422	77	P'iavchenko et al. (1976)
23	No-Suo	64°35'	31°05'	1	647	229	P'iavchenko et al. (1976)
24	Vakharu	58°51'	24°47'	10	-21	77	Sarv and Il'ves (1971)
25	Polovetsko-Kup.	57°34'	37°54'	2	757	77	KHotinskii (1977)
27	Ivanovskoe 3	56°50'	39°00'	2	4-7	77	Krainov and KHotinskii (1977)
28	IAzykovo 1	57°25'	37°09'	4	-1-	77	Krainov and KHotinskii (1977)
29	Lakhtinskoe	60°00'	30°10'	3	-47	77	Kleimenova (1975)
37	Kuiksilla	57°47'	26°02'	9	-11	77	Valk et al. (1966)
39	Markhida	67°10'	52°33'	2	424	77	Nikiforova et al. (1975)
40	Davshe	54°20'	110°02'	8	11-	458	Kol'tsova et al. (1979)
41	Mulianka	57°47'	56°19'	1	55-	229	Nemkova (1976)
45	Tugiian-IUgan	63°33'	65°43'	1	-5-	77	Arkhipov et al. (1980)
46	R. Entarnoe	60°02'	79°01'	4	112	77	Arkhipov et al. (1980)
47	R. B. Romanikha	70°45'	98°36'	3	-22	77	Nikol'skaia (1980)
49	Orshinskii Mokh	56°57'	35°57'	3	121	77	Peterson (1983)
50	R. Tom'	56°50'	84°27'	6	117	77	Arkhipov and Votakh (1980)
51	Myksi	58°09'	24°58'	6	11-	77	Sarv and Il'ves (1976)
53	Vasiugan'e 2	56°52'	83°05'	1	2--	77	KHotinskii et al. (1970)
54	Laplandiia	67°30'	37°00'	3	1--	77	P'iavchenko et al. (1976)
57	Retiazhi 8	52°36'	35°56'	2	2--	229	Serebriannaia (1976)
60	Pesochnia	52°19'	35°21'	4	1--	229	Serebriannaia and Il'ves (1972)
61	Bezengi 1	43°12'	43°15'	2	2--	77	Serebriannyi et al. (1980)
62	Balkashkinskii	53°02'	35°22'	2	27-	77	Serebriannaia (1980)
67	Sakhtysh 1	56°48'	40°25'	0	-7-	77	Krainov and KHotinskii (1977)
69	Surgut	61°14'	73°20'	0	777	77	Neishtadt (1976b)
71	Sartynia	64°10'	65°28'	0	777	77	KHotinskii (1977)
72	Momchik	66°34'	82°46'	0	--7	77	KHotinskii (1977)
73	Ubinskii Riam	55°19'	80°00'	0	77-	77	KHotinskii (1977)
74	Beglianskii Riam	55°30'	81°34'	0	77-	77	KHotinskii (1977)
77	Poluia	66°31'	66°33'	0	-7-	77	Kats and Kats (1958)
78	Paden'g	62°48'	42°56'	3	-17	77	Nikiforova (1979)
82	Glukharinoe	66°00'	69°00'	0	777	77	Levkovskaia (1971)
83	IUribei	69°00'	70°00'	0	77-	77	Levkovskaia (1971)
86	Tiuliukskoe	54°40'	59°10'	0	77-	458	Makovskii and Panova (1977)
88	Iamsovei	65°40'	78°15'	0	77-	77	Levkovskaia (1965)
92	R. B. Kebezh	53°10'	92°57'	3	1--	458	Savina (1976)
93	Aral Sea	46°40'	61°30'	0	-77	77	Veinbergs and Stelle (1975)
94	Kichi-Kaindy	42°20'	78°00'	0	7--	1068	Neishtadt (1970)
95	Ulanovo	55°33'	48°43'	0	777	77	SHalandina (1977)
96	SHur	54°13'	47°32'	0	7--	229	SHalandina (1977)
97	Helmiazevskoe	49°49'	31°21'	7	117	77	Artiushenko et al. (1982)
98	Zalozhtsy II	49°40'	25°30'	15	111	229	Artiushenko et al. (1982)

Table 8.1 continued on p. 172.

Table 8.1. Location of sites, time levels mapped, and bibliographic sources (continued)

Site no.	Site name[a]	North Latitude	East Longitude	[14]C dates (no.)	Quality[b] (0, 6, 9000 yr B.P.)	Elevation[c] (m)	Reference
99	Stoianov II	50°22'	24°39'	5	12-	229	Artiushenko et al. (1982)
100	Uanda	51°24'	142°05'	4	141	229	KHotinskii (1977)
101	Ichi	55°34'	155°59'	3	-27	77	KHotinskii (1977)
102	Ust-KHairiuzovo	57°08'	156°47'	4	-12	77	KHotinskii (1977)
103	Kirganskaia	54°48'	158°48'	0	177	150	KHotinskii (1977)
104	Ushkovskii	56°13'	159°58'	1	177	150	Kuprina (1970)
105	Sort	68°50'	148°00'	2	-2-	0	Boiarskaia and Kaplina (1979)
109	Belkachi	59°09'	131°59'	0	-7-	458	Popova (1969)
110	Kradenoe	62°00'	129°35'	4	11-	229	KHotinskii (1977)
115	CHunia	61°45'	102°48'	1	-4-	229	Kutaf'eva (1973)
117	Sakhalinskii Za	53°25'	140°50'	2	--4	305	Boiarskaia et al. (1975)
118	B Kurapotochia	71°04'	156°30'	4	-17	77	Lozhtin et al. (1975)
119	B Routan	69°45'	170°10'	5	--1	77	Veinbergs et al. (1976)
120	CHernyi IAr	52°20'	140°27'	2	17-	77	Korotkii et al. (1976)
121	Selerikan	64°18'	141°52'	2	222	458	Belorusova et al. (1977)

[a]See U.S. Board on Geographic Names (1970) for information on names.
[b] There is one entry for each time interval: 1 indicates two dates within 2000 yr of sample; 2 indicates one date within 2000 yr and second date within 4000 yr of sample; 3 indicates two dates within 4000 yr; 4, one date within 4000 yr and second date within 6000 yr; 5, two dates within 6000 yr; 6, one date within 6000 yr and second date within 8000 yr; 7, two dates more than 6000 yr from sample or no dates; a dash indicates no data.
[c]Mean of class intervals estimated from Goode (1978).

among the first to recognize the importance of geographic patterns in paleoclimatic proxy data and to produce continent-wide maps of Holocene pollen data (Neishtadt, 1957; Neustadt, 1959, 1966; KHotinskii, 1977). Soviet scientists also made impressive contributions to the quantitative climatic calibration of pollen data (Grichuk, 1969; Geleta and Spiridonova, 1981; Klimanov, 1982; Savina and KHotinskii, 1982). Soviet-American cooperation made some of this research available to English-speaking scientists (Velichko, 1984), but the pollen data still need to be mapped in a form that will permit comparison with the results of similar studies in North America (Bernabo and Webb, 1977; Webb, 1987; Webb, Bartlein, et al., this vol.) and western Europe (Huntley and Birks, 1983). In this chapter I present mapped summaries of the radiocarbon-dated Holocene pollen data for 9000, 6000, and 0 yr B.P. from the former Soviet Union and a climatic calibration of the Holocene pollen data from four sites in European Russia.

Methods

The data were compiled from published pollen diagrams, and the pollen data were recorded at 1000-yr intervals (Peterson, 1983b). For each time level I selected the closest pollen spectrum in the profile rather than interpolating between pollen spectra. Each pollen diagram was assigned geographic coordinates with the help of a gazetteer of the U.S.S.R. (U.S. Board on Geographic Names, 1970). Soviet palynologists published their pollen diagrams in a variety of formats and used various pollen sums. In some cases the pollen data had to be recalculated to percentages of the standard pollen sum (see below).

I compiled data from 95 pollen diagrams from Holocene sites with radiocarbon dating. Table 8.1 lists those sites with data for 0, 6000, and/or 9000 yr B.P. Well-dated diagrams are heavily concentrated in the European part of the former Soviet Union between 20 and 40°E, and several sites in the Baltic Republics have more than 12 radiocarbon dates. In contrast, the entire eastern half of the area is represented by fewer than 10 sites with radiocarbon-dated sediments (Fig. 8.2).

The maps present pollen percentages based on a sum of tree pollen (arboreal pollen [AP]) and herb pollen (nonarboreal pollen [NAP]), including Pinus, Picea, Larix, Carpinus, Abies, Betula, B. nana, Salix, Alnus, Castanea, Corylus, Populus, Fraxinus, Quercus, Tilia, Ulmus, and Acer as well as herbs such as Artemisia, Chenopodiaceae, Gramineae, Ericaceae, Cyperaceae, Compositae excluding Artemisia, and other NAP. The standardized pollen percentages were plotted on maps. Percentage values were rounded to the nearest integer. Values of less than 0.5% were plotted as zeros. The maps were then contoured by hand.

Results

TOTAL ARBOREAL POLLEN

Maps of modern pollen (Peterson, 1983b) suggest that the northern and southern forest limits parallel the 80 and 60% isopolls, respectively, of total AP. At

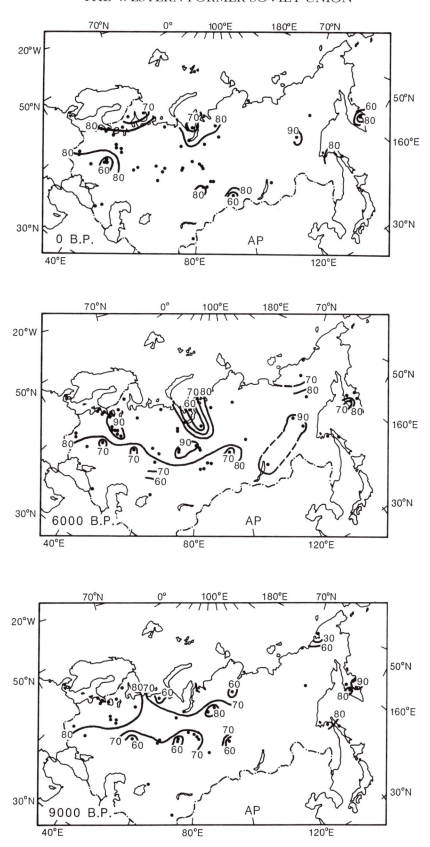

Fig. 83. Isopoll maps for total arboreal pollen (AP) at 0, 6000, and 9000 yr B.P.

9000 yr B.P. (Fig. 8.3) AP values were less than at present, and at 6000 yr B.P. they were greater than at present, especially in the European part of the former Soviet Union, where deciduous AP reached maximum values in the mid-Holocene.

Betula

A strong pattern of northward increase in *Betula* pollen was established by 9000 yr B.P. and continued at 6000 and 0 yr B.P. (Fig. 8.4). The slightly higher percentages of *Betula* pollen at 6000 yr B.P. than at 9000 or 0 yr B.P. parallel the generally higher AP percentages at 6000 yr B.P.

B. nana pollen percentages (Fig. 8.5) at 9000 yr B.P. also increased northward from 0.5% to more than 20%. At 6000 yr B.P. the lower pollen percentages in the west signaled the replacement of *B. nana* by more thermophilic tree types.

Pinus

The gradients in the percentages of *Pinus* pollen contrast with gradients in *Betula* pollen in that they increase to the south (Fig. 8.6). The pine pollen values were higher in the northwest at 9000 yr B.P. than at 6000 yr B.P. After 9000 yr B.P. pine trees there were replaced by thermophilic deciduous trees. *Pinus sibirica* pollen was centered in western Siberia, and the population of this species expanded northward throughout the Holocene (Fig. 8.7).

Picea

At 9000 yr B.P. the distribution center for *Picea* pollen was located in western Siberia (Fig. 8.8). By 6000 yr B.P. two populations appear on the map, one centered in central and eastern Siberia (10% contour) and one that moved continuously to the west from 9000 yr B.P. to the present (Peterson, 1983a). At present (0 yr B.P.) the main center of *Picea* pollen is entirely west of the Ural Mountains (60°E). This westward movement of the eastern limit of *Picea* parallels the westward movement of the western limit of *Picea* pollen in western Europe (Huntley and Birks, 1983).

Larix

Although *Larix* spp. are the dominant trees in much of the eastern part of the former Soviet Union, their pollen is greatly underrepresented in the pollen record (Fig. 8.9), and the 1% isopoll may represent regions of *Larix* dominance. The widespread area of *Larix* pollen at 9000 yr B.P. had contracted and moved to the east by 6000 and 0 yr B.P. In western Siberia *Larix* trees were partly replaced by *Pinus sibirica* as it expanded northward during the Holocene (Fig. 8.7).

Abies

Abies is a minor pollen type with slightly increasing percentages between 9000 and 6000 yr B.P. (Fig. 8.10). After 6000 yr B.P. and especially after 3000 yr B.P., *Abies* pollen percentages decreased in the north and increased in the south (Peterson, 1983a). This pattern is consistent with decreasing temperatures in this region after the mid-Holocene.

Alnus

Large values of *Alnus* pollen of 10% or more (Fig. 8.11) may reflect early-Holocene warming in the far east at 9000 yr B.P. (KHotinskii, 1977). By 6000 yr B.P. *Alnus* pollen percentages had increased the most in the European part of the former Soviet Union, where the pollen from alder trees and other thermophilic deciduous trees reached maximum percentages in the mid-Holocene (Figs. 8.12 and 8.13). After 6000 yr B.P. percentages of *Alnus* pollen decreased in the west, along with a slight southward retreat in the far east.

Quercetum mixtum and *Corylus*

Between 9000 and 6000 yr B.P. percentages of Quercetum mixtum (*Quercus + Tilia + Ulmus*) pollen increased, and the northern 0.5% contour moved north in the European part of the region (Fig. 8.12). These changes were reversed after 6000 yr B.P., and Quercetum mixtum and other thermophilic deciduous trees were replaced by *Picea, Pinus*, and *Betula*. These patterns are consistent with mid-Holocene warming, especially in July temperature (Peterson, 1983a), and with subsequent cooling in the late Holocene in the European part of the former Soviet Union. Changes in *Corylus* pollen percentages were similar to those for Quercetum mixtum, with maximum values at 6000 yr B.P. in European Russia (Fig. 8.13).

Artemisia and Chenopodiaceae

Artemisia pollen percentages were highest in the southwest near the southern forest boundary (Fig. 8.14). Modern pollen maps suggest that the 5% *Artemisia* contour parallels the forest-steppe boundary (Peterson, 1983b). The general pattern of *Artemisia* pollen is one of decreasing values throughout the Holocene, with a slight increase at 0 yr B.P. that may be related to human activity. Peterson (1983a) concluded that the generally decreasing values of *Artemisia*, Chenopodiaceae, and other NAP were consistent with a unidirectional southward movement of the southern forest-steppe boundary in the Holocene. That is, the forest-steppe boundary in the former Soviet Union did not advance and retreat as did the prairie-forest boundary in the central United States (Bernabo and Webb, 1977; Webb *et al.*, 1983).

The 2% isopoll of Chenopodiaceae parallels the forest-steppe boundary today (Peterson, 1983b). Between 9000 and 6000 yr B.P. the 2% isopoll moved southward (Fig. 8.15). The slight northward movement by 0 yr B.P. may be related to human disturbance.

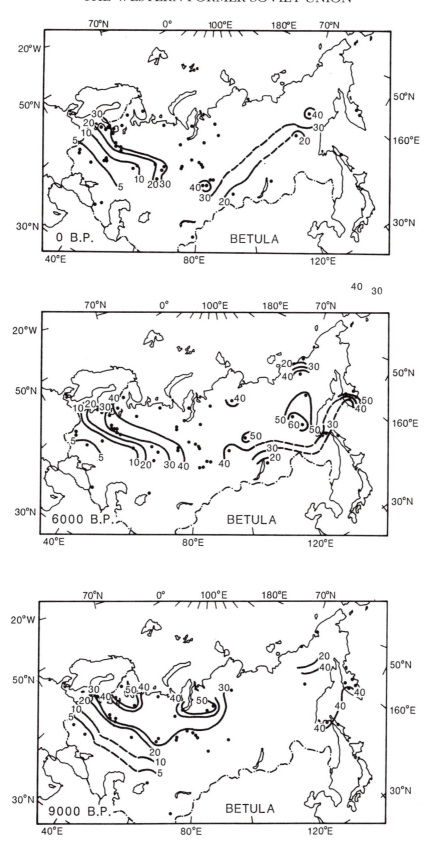

Fig. 8.4. Isopoll maps for total *Betula* pollen at 0, 6000, and 9000 yr B.P.

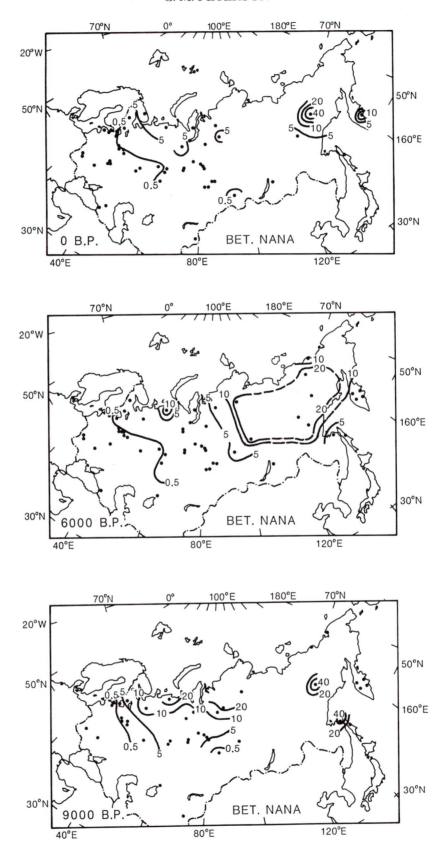

Fig. 8.5. Isopoll maps for *Betula nana* type at 0, 6000, and 9000 yr B.P.

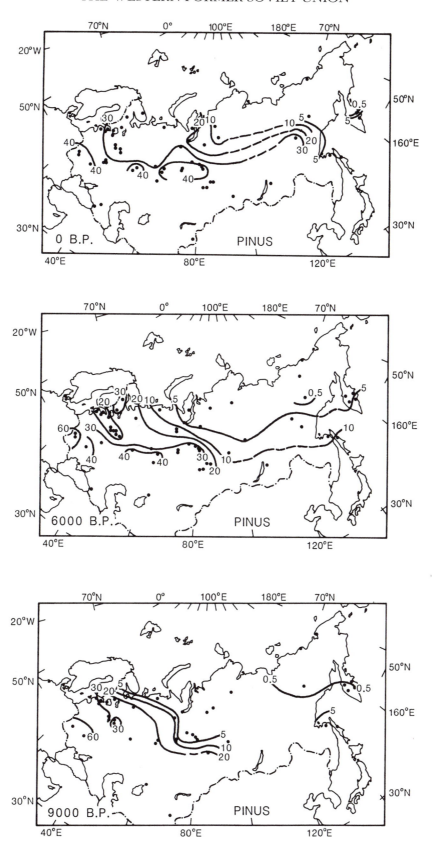

Fig. 8.6. Isopoll maps for *Pinus* at 0, 6000, and 9000 yr B.P.

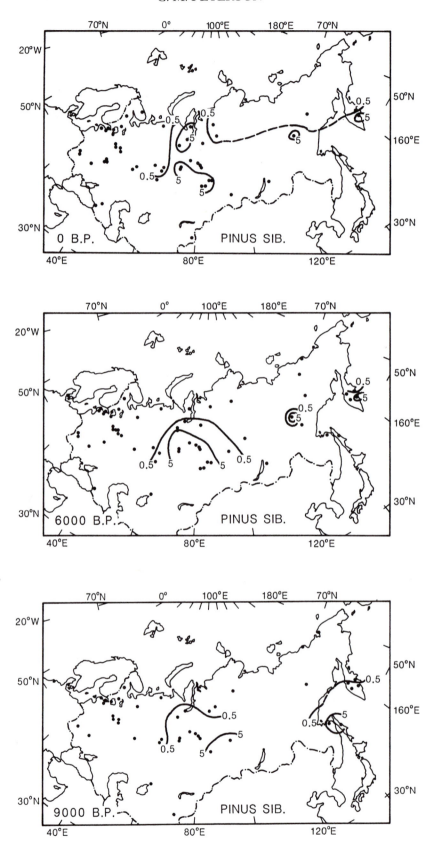

Fig. 8.7. Isopoll maps for *Pinus sibirica* at 0, 6000, and 9000 yr B.P.

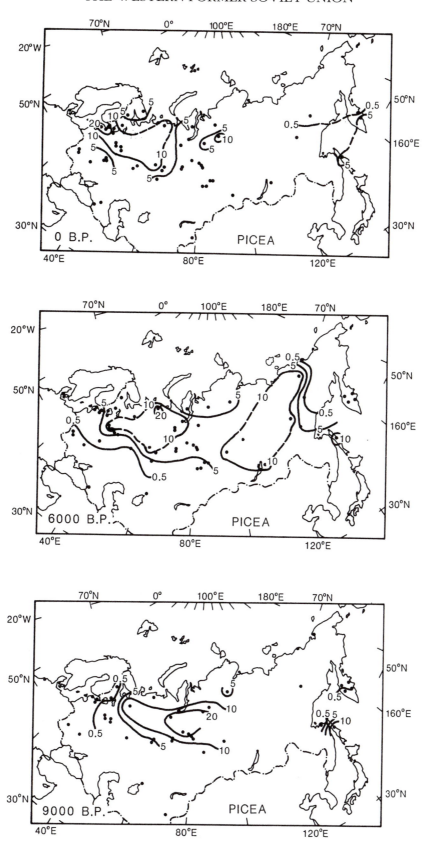

Fig. 8.8. Isopoll maps for *Picea* at 0, 6000, and 9000 yr B.P.

G. M. PETERSON

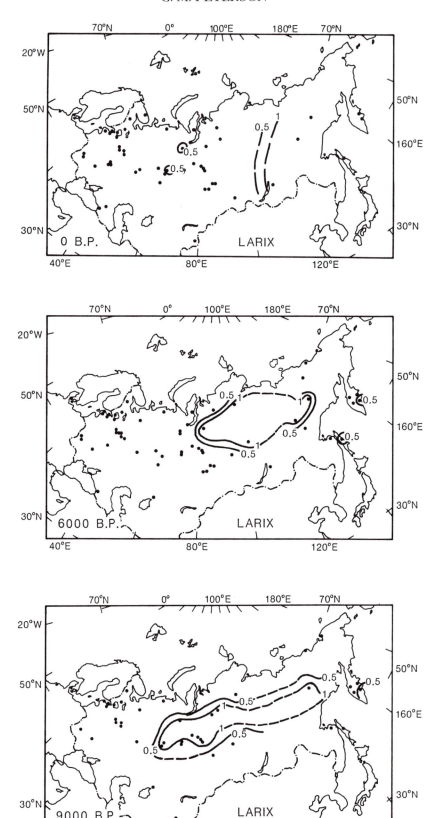

Fig. 8.9. Isopoll maps for *Larix* at 0, 6000, and 9000 yr B.P.

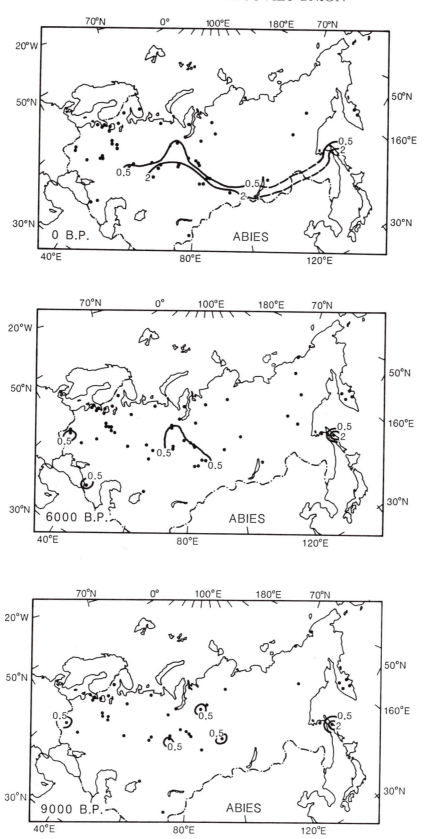

Fig. 8.10. Isopoll maps for *Abies* at 0, 6000, and 9000 yr B.P.

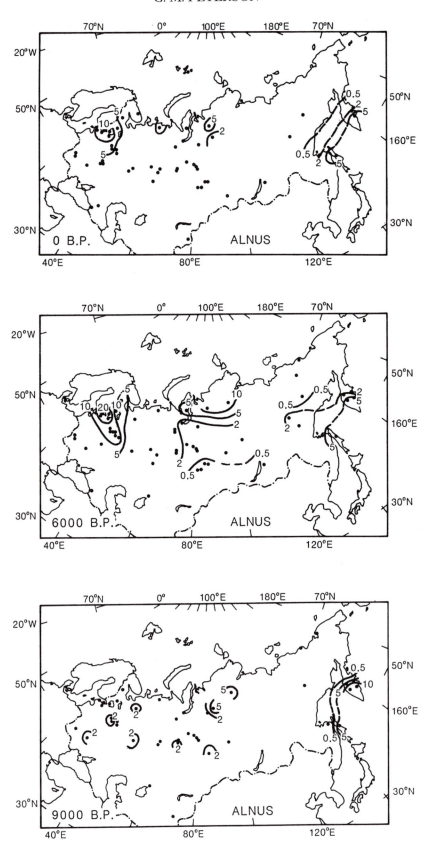

Fig. 8.11. Isopoll maps for *Alnus* at 0, 6000, and 9000 yr B.P.

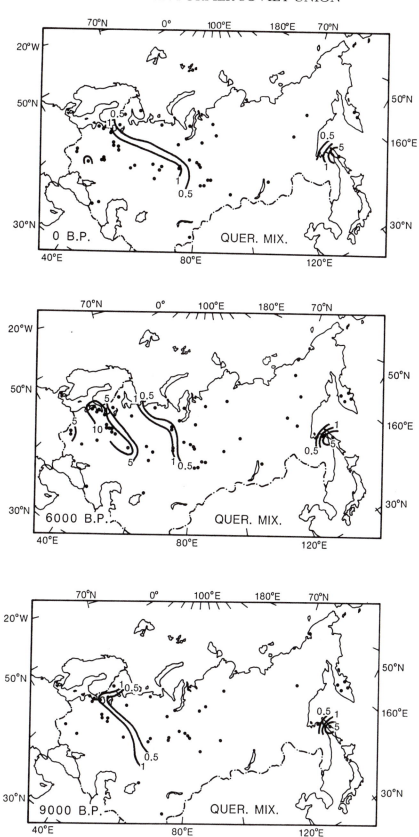

Fig. 8.12. Isopoll maps for Quercetum mixtum (*Quercus* + *Tilia* + *Ulmus*) at 0, 6000, and 9000 yr B.P.

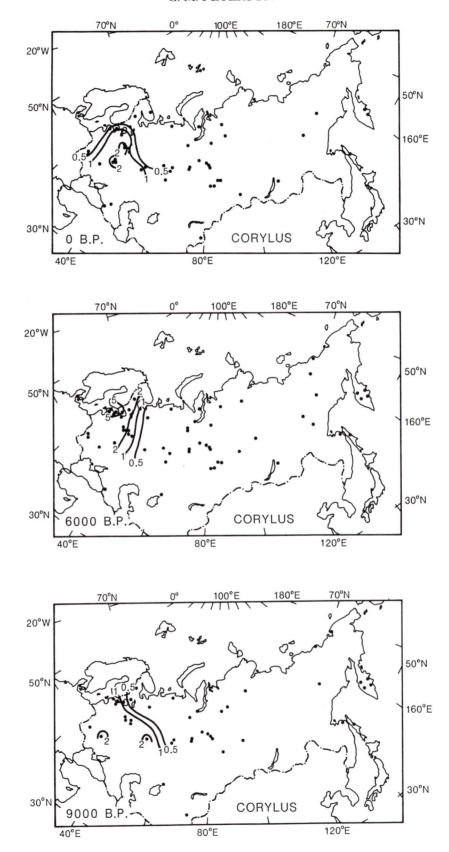

Fig. 8.13. Isopoll maps for *Corylus* at 0, 6000, and 9000 yr B.P.

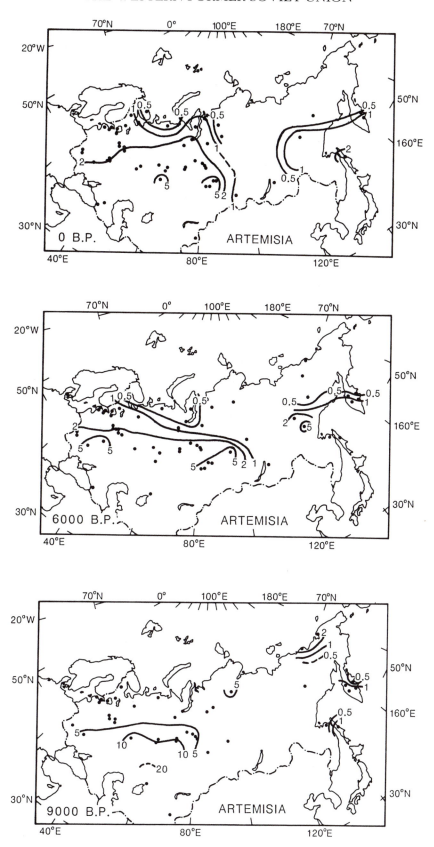

Fig. 8.14. Isopoll maps for *Artemisia* at 0, 6000, and 9000 yr B.P.

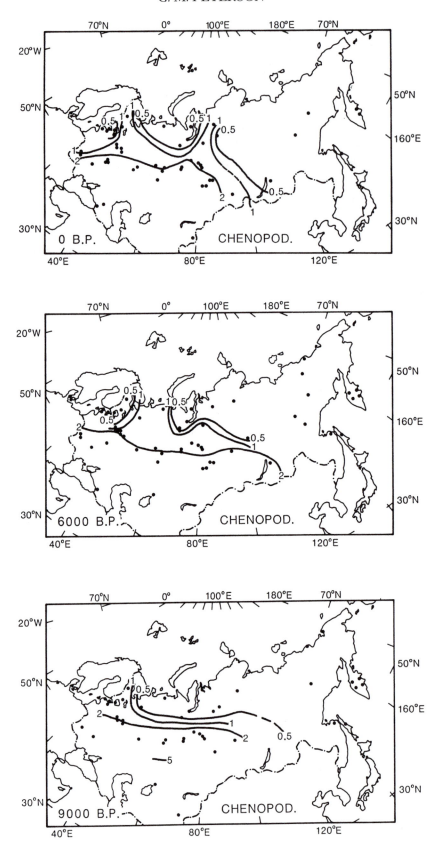

Fig. 8.15. Isopoll maps for Chenopodiaceae at 0, 6000, and 9000 yr B.P.

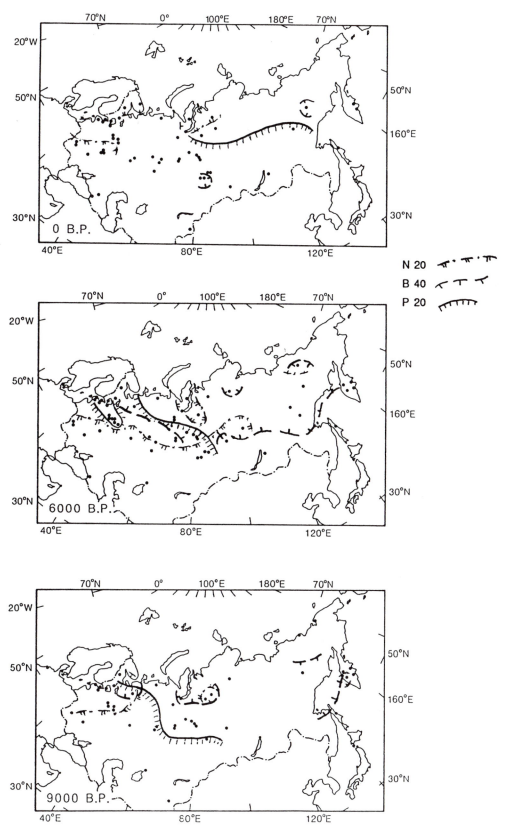

Fig. 8.16. Selected isopolls for widespread pollen types at 0, 6000, and 9000 yr B.P.: P20 represents 20% *Pinus* pollen; B40, 40% *Betula* pollen; and N20, 20% nonarboreal pollen..

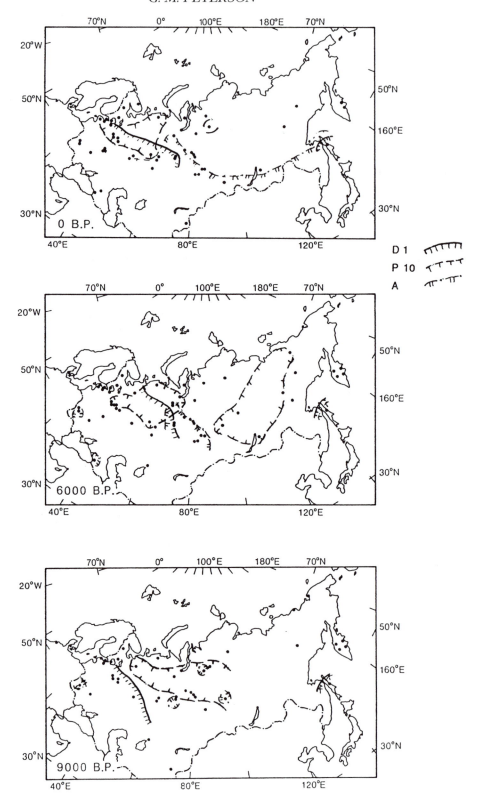

Fig. 8.17. Selected isopolls for important regional pollen types at 0, 6000, and 9000 yr B.P.: D1 represents 1% Quercetum mixtum (*Quercus + Tilia + Ulmus*) pollen and parallels the northern limit of the mixed conifer-hardwood forest in the modern data; P10 represents the 10% contour of *Picea*, which indicates a substantial presence of *Picea* trees in the modern data; and A represents the 0.5% contour of *Abies*, which may indicate areas of dominant *Abies* trees.

Holocene Vegetation Changes

Changes in vegetation composition are suggested by combining selected isopolls for (1) pollen types that cover a large geographic area in the former Soviet Union (*Pinus* 20%, *Betula* 40%, and NAP 20%) (Fig. 8.16) and (2) pollen types that are regionally important (Quercetum mixtum 1%, *Picea* 10%, and *Abies* 0.5%) (Fig. 8.17).

By 9000 yr B.P. forest had extended over much of its present-day area, although NAP percentages still exceeded 20% except in the more oceanic parts of eastern Europe. Birch species were established in the region of the present-day northern boreal forest and remained throughout the Holocene, although *B. nana* type, which today grows in the northern boreal forest and forest-tundra, played a greater role at 9000 yr B.P. than at present, especially in eastern Europe and western Siberia (Fig. 8.5). *Pinus* spp. were similarly established in the southern forest zones of eastern Europe and western Siberia. In eastern Europe small quantities of Quercetum mixtum and other thermophilic trees were present in the southwest. In western Siberia *Picea* was an important part of the forest, and *Pinus sibirica* was present in the southeast (Fig. 8.7). Although probably greatly underrepresented in the fossil pollen spectra, *Larix* trees were probably a major component of the forests from western Siberia to the Far East. In more continental parts in the east, larch forests remained throughout the Holocene. The few sites in the Far East indicate low NAP values in the oceanic regions (Sakhalin and Kamchatka), where birch and alder were important components of the vegetation. A mixture of tree taxa grew on Sakhalin Island, including spruce, haploxylon pines, fir, and Quercetum mixtum.

By 6000 yr B.P. NAP had declined overall, as forests continued to expand. *Betula* spp. occupied large areas in the north. *B. nana* (Fig. 8.5) decreased in the northern part of eastern Europe and western Siberia, but it remained in the continental parts of eastern Siberia, together with other birch species, larch, and spruce. In eastern Europe the more thermophilic trees, such as Quercetum mixtum, European tree alder (Fig. 8.11), and hazel (Fig. 8.13), attained their maximum values during the mid-Holocene. In western Siberia *Picea* populations decreased, possibly because of increased climatic continentality at 9000 yr B.P. (Kutzbach and Otto-Bliesner, 1982; Kutzbach and Guetter, 1986) (see Huntley and Prentice [this vol.] for discussion).

Slight increases in NAP at 0 yr B.P. probably reflect human disturbance of the vegetation, such as forest cutting. Except for the last 1000 yr, the overall trend in the Holocene at this map scale is toward expansion of forests. In eastern Europe an increase in spruce, pine, and birch forests at the expense of the more thermophilic Quercetum mixtum, hazel (Fig. 8.13), and European alder (Fig. 8.11) was consistent with late-Holocene cooling of summer temperatures. In western Siberia the less temperature-sensitive *Pinus sibirica* continued to increase (Fig. 8.7), while the somewhat more temperature-sensitive *Abies* decreased in the north and increased in the south (Fig. 8.10).

Climatic Calibration

I used multiple regression equations to estimate mean July temperature from the pollen data from four Holocene cores located in the Russian Plain (Peterson, 1983a). I chose a set of modern pollen data from between 35 and 60°E, in which percentages of *Picea* pollen decrease from 20% at 60°N to near zero at 50°N, whereas Quercetum mixtum pollen increases sharply south of 60°N (Figs. 8.8 and 8.12). These trends represent the ecotone between boreal forest and mixed forest. They reflect the southward increase in mean July temperature from 16°C to more than 22°C.

Pollen diagrams from the Russian Plain show an increase in deciduous forest pollen between 8000 and about 4500 yr B.P. (Peterson, 1983a) and subsequent replacement of deciduous forest types by spruce. I examined the climatic implications of these changes in the ecotone between the mixed and boreal forests by analyzing pollen diagrams from four sites on the Russian Plain: Osechenskoe (no. 16 in Fig. 8.2 and Table 8.1), Polovetsko-Kupanskoe (no. 25), Ivanovskoe (no. 27), and Orshinskii Mokh (no. 49).

The methods used to calculate the multiple regression equations are described in Arigo *et al.* (1986). A number of pollen types were included in the regression model, but the strongest ecologically interpretable relationships were found for spruce and for deciduous forest pollen types. Although other pollen types, such as pine and birch, showed some correlation with climatic variables, they represent geographically widespread tree species with a wide range of ecological tolerances and did not seem to yield a climatically interpretable signal.

Mean July temperature was selected as the dependent variable from a number of similarly distributed variables that reflect growing-season temperature. The calibration region yielded the following regression equation (Peterson, 1983a):

$$T_{July} = 16.80 + 1.40 \times (Qm)^{0.25} - 0.20 \times (P)^{0.5},$$

where Qm = Quercetum mixtum pollen percentage and P = *Picea* pollen percentage (Fig. 8.18). The equation has an R^2 of 0.64 and a standard error of 0.76°C.

The reconstructed July temperatures (Fig. 8.19) suggest that summer temperatures may have been somewhat warmer at 9000 yr B.P. than at present, that they increased thereafter and remained at generally higher levels until about 4500 yr B.P., and that they then decreased. The magnitude of change was about 2°C.

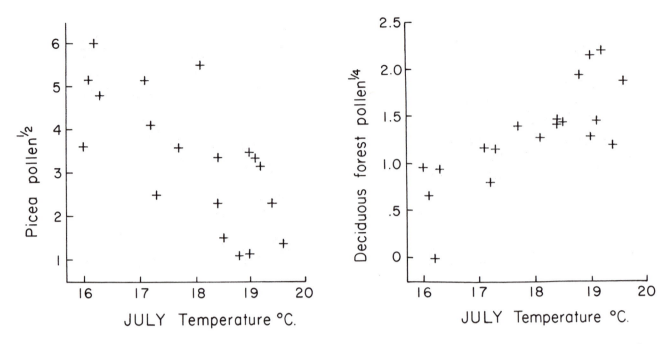

Fig. 8.18. Scattergrams of the square root of *Picea* pollen percentage and the fourth root of Quercetum mixtum (deciduous forest) pollen percentage versus mean July temperature in modern pollen samples in the former Soviet Union between 35 and 60°E and 48 and 65°N.

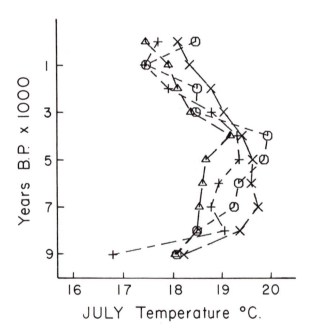

Fig. 8.19. Reconstructed mean July temperatures for four sites on the Russian Plain: Orshinskii Mokh (○), Osechenskoe (△), Ivanovskoe (+), and Polovetsko-Kupanskoe (X). These sites are numbers 49, 16, 27, and 25, respectively, in Figure 8.2 and Table 8.1.

Discussion and Conclusions

The reconstructed mean July temperatures (Fig. 8.19) can be compared to results obtained by Soviet scientists. KHotinskii (1977) presented a qualitative interpretation of the Holocene changes in temperature and moisture in the Russian Plain. His summary diagram for this region indicates increasing temperatures since at least the Boreal (about 9000 yr B.P.). The rising temperatures were accompanied by gradually increasing humidity until the late Holocene, when temperatures fell and humidity continued to increase. This general pattern for temperature is in agreement with the quantitative reconstructions presented in this chapter. Burashnikova *et al.* (1982) used a statistical model to reconstruct mean July temperature and annual precipitation for the mid-Holocene (5000–6000 yr B.P.). Their results at Polovetsko-Kupanskoe and Ivanovskoe are similar both in magnitude and sign to those reported here. Klimanov (1982) also inferred higher mean July temperatures on the Russian Plain at about 5500 yr B.P.

The present results from the Russian Plain may also be compared to those of Bartlein *et al.* (1984) from central North America. While Bartlein *et al.* (1984) found that mean annual temperatures were 2.5°C cooler than at present at many sites at 9000 yr B.P., results from the Russian Plain imply that July temperatures at 9000 yr B.P. were the same as or slightly warmer than at present. This temperature pattern is consistent with the modeling results of Kutzbach (1981) and Kutzbach and Otto-Bliesner (1982) and of

Kutzbach and Guetter (1986) for eastern Europe. As in North America, temperatures in the Russian Plain rose between 9000 and 6000 yr B.P.

Bryson *et al.* (1970) and Bartlein *et al.* (1984) attributed the warm and dry mid-Holocene conditions in North America to a northward retreat of Arctic airmasses and increased westerly flow, which increased the influence of dry Pacific airmasses. A similar change may have occurred in eastern Europe in the mid-Holocene. Wendland and Bryson (1981) and R. A. Bryson (personal communication) showed that the spruce and mixed deciduous forests in eastern Europe lie in a region dominated by Atlantic airmasses. The spruce forest is further separated from the mixed deciduous forest to the south by the two-month contour of Arctic airmass duration. Thus the expansion of mixed deciduous forests in the Russian Plain from 9000 to 6000 yr B.P. may be related to increased zonal flow and dominance of Atlantic airmasses and a simultaneous northward retreat of Arctic airmasses.

My study shows that the Russian Plain, like central North America, is an important region for studies of the Holocene climates of the Northern Hemisphere. The present reconstructions generally agree with the results of Soviet researchers.

The most striking feature of the Holocene pollen data from the former Soviet Union is the uneven distribution of dated cores. Although cores are generally sufficiently abundant west of 100°E to permit mapping (Peterson, 1983a), the number of cores in the eastern half of the region is far too small for any detailed analysis of Holocene vegetation patterns. Even in the west, more dated cores are needed from the northern and southern forest limits in order to map the changes in these important ecotones.

The Russian Plain appears to be a favorable region for climatic calibration of the Holocene pollen record, which showed sharp changes in the relative abundance of coniferous and deciduous trees. The eastern former Soviet Union, with its more mountainous terrain and smaller number of dated cores, may be a substantially more difficult region for quantitative paleoclimatic reconstruction. The maps show that there are abundant opportunities for improving the quantity of Holocene pollen data from the eastern part of the former Soviet Union.

Acknowledgments

This chapter was prepared with support from grants ATM 81-11455, ATM 79-26039, and ATM 82-19079 from the Climate Dynamics Program of the National Science Foundation, and I thank John E. Kutzbach for making available the excellent facilities of the Center for Climatic Research. While revising this paper I was a research associate at the Institute of Polar Studies, Ohio State University. Rachael Purington drafted the maps.

References

Arigo, R., Howe, S. E., and Webb, T. III (1986). Climatic calibration of pollen data: An example and annotated computing instructions. *In* "Handbook of Holocene Palaeoecology and Palaeohydrology" (B. E. Berglund, Ed.), pp. 817-849. John Wiley and Sons, Chichester.

Arkhipov, S. A., and Votakh, M. R. (1980). Palinologicheskaia kharakteristika i absoliutnyi vozrast torfianika v ust'e r. Tomi. *In* "Paleopalinologiia Sibiri," pp. 118-122. Nauka, Moscow.

Arkhipov, S. A., Levina, T. P., and Panychev, V. A. (1980). Palinologicheskaia kharakteristika dvukh golotsenovykh torfianikov iz doliny Srednei i Nizhnei Obi. *In* "Paleopalinologiia Sibiri," pp. 123-127. Nauka, Moscow.

Artiushenko, A. T., Arap, R. I. A., Bezus'ko, L. G., Il'ves, E. O., Kaiutkina, T. M., and Kovaliukh, N. N. (1982). Novye dannye o rastitel'nosti Ukrainy v golotsene. *In* "Razvitie prirody territorii SSSR v pozdnem Pleistotsene i Golotsene," pp. 173-179. Nauka, Moscow.

Bartlein, P. J., Webb, T. III, and Fleri, E. (1984). Holocene climatic change in the northern Midwest: Pollen derived estimates. *Quaternary Research* 22, 361-374.

Belorusova, ZH. M., Lovelius, N. V., and Ukraintseva, V. V. (1977). Paleogeografiia pozdnego pleistotsena i golotsena v raione nakhodki selerikanskoi loshadi. *In* "Fauna i flora Antropogena severo-vostoka Sibiri," pp. 265-276. Leningrad.

Bernabo, J. C., and Webb, T. III (1977). Changing patterns in the Holocene pollen record of northeastern North America: A mapped summary. *Quaternary Research* 8, 64-96.

Boiarskaia, T. D., and Kaplina, T. N. (1979). Novye dannye o razvitii rastitel'nosti severnoi IAkutii v Golotsene. *Vestnik Moskovskogo Universiteta Ser. 5, Geografiia* 5, 70-75.

Boiarskaia, T. D., Savchenko, I. F., Sokhina, E. N., and CHerniuk, A. V. (1975). Sopriazhennoe izuchenie razreza noveishikh otlozhenii iuzhnogo poberezh'ia Sakhalinskogo zaliva. *Voprosy geografii dalnego vostoka* Sbornik 16, 158-165.

Bryson, R. A., Baerreis, D., and Wendland, W. (1970). The character of late glacial and post-glacial climatic changes. *In* "Pleistocene and Recent Environments of the Central Great Plains," pp. 53-74. University of Kansas, Lawrence.

Burashnikova, T. A., Muratova, M. V., and Suetova, I. A. (1982). Klimaticheskii model' territorii sovetskogo soiuza vo vremia golotsenovogo optimuma. *In* "Razvitie prirody territorii SSSR v Pozdnem Pleistotsene i Golotsene," pp. 245-251. Nauka, Moscow.

Firsov, L. V., Troitskii, S. L., Levina, T. P., Nikitin, V. P., and Panychev, V. A. (1974). Absoliutnyi vozrast i pervaia dlia severa sibiri standartnaia pyl'tsevaia diagramma golotsenovogo torfianika. *Biulleten' Komissii po Izucheniiu CHetvertichnogo Perioda* 41, 121-127.

Geleta, I. F., and Spiridonova, E. A. (1981). Opyt vosstanovleniia klimata golotsena po dannym palinologii metodami mnogomernogo statisticheskogo analiza. *In* "Palinologiia Pleistotsena i Golotsena," pp. 52-70. Leningradskii Gosvdarstvenuyi Universitet, Leningrad.

Glebov, F. Z., Toleiko, L. S., Starikov, E. V., and ZHidovlenko, V. A. (1974). Palinologicheskaia kharakteristika i datirovanie po C14 torfianika v aleksandrovskom raione tomskoi oblasti (srednetaezhnaia zona). *In* "Tipy Bolot SSSR i Printsipy ikh Klassifikatsii," pp. 194-199. Leningrad.

Goode, J. P. (1978). Goode's World Atlas, 15th ed. (E. B. Espenshade, Jr., Ed.). Rand McNally, Chicago.

Grichuk, V. P. (1969). Opyt rekonstruktsii nekotorykh elementov klimata severnogo polushariia v atlanticheskii period golotsena (An attempt to reconstruct certain elements of the climate of the Northern Hemisphere in the Atlantic period of the Holocene). *In* "Golotsen," pp. 41-57. Nauka, Moscow. (English translation by G. M. Peterson, Center for Climatic Research, University of Wisconsin-Madison)

Huntley, B., and Birks, H. J. B. (1983). "An Atlas of Past and Present Pollen Maps for Europe 0-13,000 Years Ago." Cambridge University Press, Cambridge.

Kats, N. IA., and Kats, S. V. (1958). K istorii flory i rastitel'nosti severa zapadnoi sibiri v poslelednikovoe i pozdnelednikovoe vremia. *Botanicheskii ZHurnal* 43, 998-1014.

KHotinskii, N. A. (1977). "Golotsen Severnoi Evrazii." Nauka, Moscow.

KHotinskii, N. A., Devirts, A. L., and Markova, N. G. (1970). Vozrast i istoriia formirovaniia bolot vostochnoi okrainy vasiugan'ia. *Moskovskoe Obshchestvo Ispytatelei Prirody Biulleten', Otdelenie Biologicheskoe* 75(5), 82-91.

Kind, N. V. (1969). Pozdne- i poslelednikov'e sibiri (novye materialy po absoliutnoi khronologii). *In* "Golotsen," pp. 195-201. Nauka, Moscow.

Kleimenova, G. I. (1975). Palinologicheskie issledovaniia poslelednikovykh otlozhenii po razrezam lakhtinskogo i shuvalovakogo bolot. *Leningrad Universitet Vestnik Geologiia-Geografiia* 12(2), 94-103.

Klimanov, V. A. (1982). Klimat vostochnnoi evropy v klimaticheskim optimum golotsena (po dannym palinologii). *In* "Razvitie prirody territorii SSSR v Pozdnem Pleistotsene i Golotsene," pp. 251-258. Nauka, Moscow.

Kol'tsova, V. G., Starikov, E. V., and ZHidovlenko, V. A. (1979). Razvitie rastitel'nosti i vozrast torfianika v doline r. Davshe (Barguzinskii Zapovednik). *Biulleten' Komissii po Izucheniiu CHetvertichnogo Perioda* 49, 121-124.

Korotkii, A. M., Karaulova, L. P., and Pushkar', V. S. (1976). Klimat i kolebaniia vertikal'nykh landshaftnykh zon sikhote-alinia v golotsene. *In* "Geomorfologiia i chetvertichnaia geologiia Dal'nego Vostoka," pp. 112-129. Vladivostok.

Krainov, D. A., and KHotinskii, N. A. (1977). Verkhnevolzhskaia ranneneoliticheskaia kul'tura. *Sovetskaia Arkheologiia* 3, 42-67.

Kuprina, N. P. (1970). Stratigrafiia i istoriia osadkonakopleniia pleistotsenovykh otlozhenii tsentral'noi kamchatki. AN SSSR Geologicheskii Institut, Moskva; *Trudy* vyp. 216.

Kutaf'eva, T. K. (1973). Istoriia lesov iuzhnoi evenki po dannym sporovo-pyl'tsevogo analiza torfianykh zalezhei. *In* "Palinologiia Golotsena i Marinopalinologiia," pp. 71-75. Nauka, Moscow.

Kutzbach, J. E. (1981). Monsoon climate of the early Holocene: Climate experiment with the earth's orbital parameters for 9000 years ago. *Science* 214, 59-61.

Kutzbach, J. E., and Guetter, P. (1986). The influence of changing orbital parameters and surface boundary conditions on climate simulations for the past 18,000 years. *Journal of the Atmospheric Sciences* 43, 1726-1759.

Kutzbach, J. E., and Otto-Bliesner, B. L. (1982). The sensitivity of the African-Asian monsoonal climate to orbital parameter changes for 9000 years B.P. in a low-resolution general circulation model. *Journal of the Atmospheric Sciences* 39, 1177-1188.

Levkovskaia, G. M. (1965). Stratigrafiia golotsenovykh otlozhenii severa zapadnoi sibiri po dannym sporovo-

pyl'tsevogo analiza. *In* "Problemy Paleogeografii," pp. 214-227. Nauka, Leningrad.

——. (1971). O granitsakh razlichnykh gorizontov golotsena na severe zapadnoi sibiri. *In* "Kainozoiskie Flory Sibiri po Palinologicheskim Dannym," pp. 111-123. Nauka, Moscow.

Lozhkin, A. V., Prokhorova, T. P., and Parii, V. P. (1975). Radiouglerodnye datirovki i palinologicheskaia kharakteristika otlozhenii alasnogo kompleksa kolymskoi nizmennosti. *AN SSSR Doklady* 224, 1395-1398.

Makovskii, V. I., and Panova, N. K. (1977). Formirovanie rastitel'nosti verkhnego gornogo poiasa iuzhnogo urala v golotsene. *In* "Razvitie Lesoobrazovatel'nogo Protsessa na Urale," pp. 3-17. Sverdlovsk.

Neishtadt, M. I. (1957). "Istoriia lesov i Paleogeografiia SSSR v Golotsene." Akademiia Nauk, Moscow.

——. (1965). "Paleogeografiia i KHronologiia Verkhnego Pleistotsena i Golotsena po Dannym Radiouglerodnogo Metoda." Nauka, Moscow.

——. (1967). Ob absoliutnom vozraste torfianykh bolot zapadnoi sibiri. *Revue Roumaine de biologie, série de botanique* 12, 181-187.

——. (1970). K paleogeografii predgorii khrebta terskei ala-too v pozdnem golotsene. *In* "Fizicheskaia Geografiia Priissykkul'ia," pp. 58-63. Frunze.

——. (1976a). Golotsenovye protsessy v zapadnoi sibiri i voznikaiushchie v sviazi s etim problemy. *In* "Izuchenie i Osvoenie Prirodnoi Sredy," pp. 90-99. Moscow.

——. (1976b). Regional'nye zakonomernosti istorii fitotsenozov SSSR v Golotsene po palinologicheskim dannym. *In* "Istoriia Biogeotsenozov SSSR v Golotsene," pp. 79-91. Nauka, Moscow.

Nemkova, V. K. (1976). Istoriia rastitel'nosti predural'ia za pozdne- i poslelednikovoe vremia. *In* "Aktual'nye Voprosy Sovremennoi Geokhronologii," pp. 259-275. Moscow.

Neustadt, M. I. (1959). Geschichte der UDSSR im Holozan. *Grana Palynologica* 2, 69-77.

——. (1966). Present achievements in a palynological study of the Quaternary period in the USSR. *The Palaeobotanist* 15, 213-227.

Nikiforova, L. D. (1979). Personal communication. Geological Institute, USSR Academy of Sciences, Moscow.

Nikiforova, L. D., Parunin, O. B., and Zaitseva, G. IA. (1975). Palinologicheskie i radiouglerodnye issledovaniia golotsenovykh otlozhenii bol'shezemel'skoi tundry. Unpublished manuscript, Vsesoiuzuyi Institut Nauchnoi i Tekhnicuescoi Infermatsii (VINITI), Moscow.

Nikol'skaia, M. V. (1980). Paleobotanicheskaia kharakteristika verkhne-pleistotsenovykh i golotsenovykh otlozhenii Taimyra. *In* "Paleopalinologiia Sibiri," pp. 97-111. Nauka, Moscow.

Peterson, G. M. (1979). Keeping up with Soviet Quaternary science. *Quaternary Research* 12, 150-151.

——. (1983a). "Holocene Vegetation and Climate in the Western USSR." Unpublished Ph.D. dissertation, University of Wisconsin-Madison.

——. (1983b). Recent pollen spectra and zonal vegetation in the western USSR. *Quaternary Science Review* 2, 281-321.

P'iavchenko, N. I., Elina, G. A., and CHachkhiani, V. N. (1976). Osnovnye etapy istorii rastitel'nosti i torfianonakopleniia na vostoke baltiiskogo shchita v golotsene. *Biulleten' Komissii po Izucheniiu CHetvertichnogo Perioda* 45, 3-24.

Popova, A. I. (1969). Sporovo-pyl'tsevye spektry otlozhenii levogo berega reki aldan u poselka bel'kachi. *In* "Mnogosloinaia stoianka Bel'kachi I i periodizatsiia Kamennogo Veka IAkutii" (IU. A. Mochanov, Ed.). Moscow.

Sarv, A. A., and Il'ves, E. O. (1971). Datirovanie po radiouglerodu ozerno-bolotnykh otlozhenii bol. Vakharu (S-Z Estoniia). *In* "Palinologicheskie Issledovaniia v Pribaltike," pp. 143-149. Riga.

——. (1976). Stratigrafiia i geokhronologiia golotsenovykh ozernykh i bolotnykh otlozhenii iugo-zapadnoi chasti Estonii. *In* "Palinologiia v kontinental'nykh i morskikh geologicheskikh issledovaniiakh," pp. 47-59. Riga.

Savina, L. N. (1976). "Noveishaia Istoriia Lesov Zapadnogo Saiana." Novosibirsk.

Savina, S. S., and KHotinskii, N. A. (1982). Zonal'nyi metod rekonstruktsii paleoklimatov golotsena. *In* "Razvitie prirody territorii SSSR v Pozdnem Pleistotsene i Golotsene," pp. 321-244. Nauka, Moscow.

Serebriannaia, T. A. (1976). Vzaimootnosheniia lesa i stepi na srednerusskoi vozvyshennosti v golotsene (po palinologicheskim i radiouglerodnym dannym). *In* "Istoriia Biogeotsenozov SSSR v Golotsene," pp. 159-166. Nauka, Moscow.

——. (1980). K golotsenovoi istorii lesov zapada srednerusskoi vozvyshennosti. *Biulleten' Komissii po Izucheniiu CHetvertichnogo Perioda* 50, 178-185.

Serebriannaia, T. A., and Il'ves, E. O. (1972). Pervye dannye po palinologii i vozrastu vodorazdel'nogo torfianika v tsentral'noi chasti sredne-russkoi vozvyshennosti bliz g. ZHeleznogorska. *Akademiia Nauk Estonskoi SSR Izvestiia KHimiia-Geologiia* 21(2), 161-170.

Serebriannyi, L. R., Pshenin, G. N., and KHalmukhamedova, R. A. (1976). Evoliutsiia aridnykh landshaftov iuzhnoi fergany v golotsene. *In* "Istoriia Biogeotsenozov SSSR v Golotsene," pp. 221-229. Nauka, Moscow.

Serebriannyi, L. R., Gei, N. A., Dzhinoridze, R. N., Il'ves, E. O., Maliasova, E. S., and Skobeeva, E. I. (1980). Rastitel'nost' tsentral'noi chasti vysokogornogo kavkaza v golotsene. *Biulleten' Komissii po Izucheniiu CHetvertichnogo Perioda* 50, 123-137.

SHalandina, V. T. (1977). O razvitii rastitel'nogo pokrova severo-vostoka privolzhskoi vozvyshennosti v golotsene.

Nauchnye Doklady Vysshei SHkoly, Seriia Biologicheskikh Nauk 12, 87-96.

U.S. Board on Geographic Names (1970). "U.S.S.R." Official Standard Names Gazetteer No. 42, 2nd ed., 7 vols. Geographic Names Division, U.S. Army Topographic Command, Washington, D.C.

Valk, U. A., Il'ves, E. O., and Miannil', R. P. (1966). Datirovanie faz razvitiia lesov po C14 po materialam bolota Kuiksilla iuzhnoi Estonii. *In* "Palinologiia v Geologicheskikh Issledovaniiakh Pribaltiki," pp. 120-127. Riga.

Veinbergs, I. G., and Stelle, V. IA. (1975). Istoriia razvitiia arala v pozdnem pleistotsene i golotsene. *In* "Istoriia Ozer i Vnutrennykh Morei Aridnoi Zony. IV Vsesoiuznoe Simposium po Istorii Ozer," Tezisy Dokladov, Vol. 4, pp. 53-64.

Veinbergs, I. G., Voshchiklo, M. I., and Stelle, V. IA. (1976). Sporovo- pyl'tsevye kompleksy pozdnechetvertichnykh otlozhenii i izmenenie klimata i rastitel'nosti raiona chaunskoi guby. *In* "Palinologiia v kontinental'nykh i morskikh geologicheskikh issledovaniiakh," pp. 119-132. Riga.

Velichko, A. A., Ed. (1984). "Late Quaternary Environments of the Soviet Union." English-language edition edited by H. E. Wright, Jr., and C. W. Barnosky. University of Minnesota Press, Minneapolis.

Volkov, I. A., Gurtovaia, E. E., Firsov, L. V., Panychev, V. A., and Orlova, L. A. (1973). Stroenie, vozrast i istoriia formirovaniia golotsenovogo torfianika u s. Gorno-Slinkina na Irtyshe. *In* "Pleistotsen Sibiri i Smezhnykh Oblastei," pp. 34-39. Nauka, Moscow.

Webb, T. III (1987). The appearance and disappearance of major vegetational assemblages: Long-term vegetational dynamics in eastern North America. *Vegetatio* 69, 177-187.

Webb, T. III, Cushing, E. J., and Wright, H. E., Jr. (1983). Holocene changes in the vegetation of the Midwest. *In* "Late Quaternary Environments of the United States, Vol. 2: The Holocene" (H. E. Wright, Jr., Ed.), pp. 142-165. University of Minnesota Press, Minneapolis.

Wendland, W. M., and Bryson, R. A. (1981). Northern Hemisphere airstream regions. *Monthly Weather Review* 109, 255-270.

CHAPTER 9

Vegetational, Lake-Level, and Climatic History of the Near East and Southwest Asia

Neil Roberts and H. E. Wright, Jr.

Late-Quaternary paleoclimatic data available for southwest Asia and the Near East are outstanding neither in quality nor in quantity. Nonetheless this region, which for the present purpose includes the Arabian peninsula, the Levant, Anatolia, Iraq, Iran, Greece, Caucasia, and parts of south and central Asia, is potentially of great significance in any global reconstruction of Holocene climates by virtue of its strategic position at the junction of three continents. The records of Holocene climate and environment from Europe and Africa in particular are relatively rich and internally consistent (see Huntley and Prentice [this vol.] and Street-Perrott and Perrott [this vol.]) but differ from each other, and the transition zone must occur around 30–40° north latitude. The paleoclimatic record of southwest Asia and the Near East is therefore likely to reflect changes in adjacent regions rather than any distinctive pattern of its own (Roberts, 1984). Lake-level as well as palynological data are available for this region, and integrated paleoecological investigations of some sites (e.g., Lake Zeribar) have provided records in both categories.

Present-Day Environments

CLIMATE

Lying between 13 and 42°N, southwest Asia and the Near East straddle the subtropical high-pressure belt, so that much of the area is today climatically arid or semiarid (Köppen classes BWh, BSh, and BSk). The northern sector of the region receives most of its precipitation during the winter half of the year, in association with middle- to high-latitude westerly depressions whose tracks are steered by the subtropical jet stream (Wigley and Farmer, 1982). The absence of topographic barriers south of the Alps allows Atlantic low-pressure cells to progress as far east as Afghanistan, and the warm waters of the Mediterranean Sea provide a secondary moisture source so that storms maintain or renew their strength as they move eastward.

In the summer, on the other hand, the belt of moist westerly winds shifts northward, and the subtropical high-pressure area with its subsiding air brings warm and dry conditions to the entire Mediterranean region. In contrast, the southern part of the Arabian peninsula, as well as the Indian subcontinent, receives monsoonal precipitation in summer. This precipitation is controlled by the South Asian monsoon and is related to summer heating of the Asian landmass.

Note that the seasonal Mediterranean-type (Csa) climate prevails on the western side of all the continents around latitude 30–35°, including California, Chile, Australia, and South Africa, but because of the absence of mountain barriers and the supplementary moisture provided by the Mediterranean Sea, only in southern Europe does this climatic type extend so far eastward (Wright, 1976). Away from the Mediterranean coast, it is modified by the effects of elevation and continentality, so that much of Anatolia, Iran, and Afghanistan experiences mean January temperatures below freezing.

These climatic controls play a dominant role in determining the pattern of natural vegetation and lake water balances in the Near East.

VEGETATION RELATIONS

In the Mediterranean sector of Greece and southwest Asia, many plants are closely adjusted to a climatic regime of winter rain and summer drought. Because these plants include some major pollen pro-

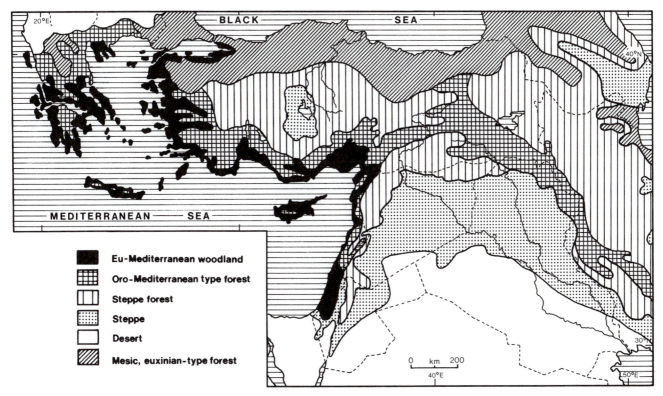

Fig. 9.1. Modern vegetation map of eastern Mediterranean region (data from Zohary [1973] and other sources).

ducers, we can attempt to reconstruct the history of the Mediterranean-type climatic regime from palynologic data. At the same time, the absolute amount of annual moisture is important to many plants. In particular, trees almost always require more rainfall than the shrubs and herbs that dominate the steppe, as can be seen today in the restriction of forest cover to the mountains, where precipitation is greater. Most of the northern shores of the Mediterranean Sea were forested before historical times, and tree growth was restricted directly by temperature only at elevations above 2000 m. The basic distinction between woodland associations on the one hand and steppe-desert on the other therefore reflects variations in precipitation across the region.

Six major vegetation types are identified here for the Near East (Zohary, 1973) (Fig. 9.1):

Desert: low-density xerophytic shrubs, herbs, and sedges. Mean annual precipitation less than 100 mm (winter maximum) or less than 250 mm (summer maximum). Arboreal pollen (AP) less than 10%; low levels of pollen concentration; *Calligonum, Cyperus,* and *Aerva* dominant; chenopods also often important.

Steppe: treeless except along watercourses, but dwarf shrubs important as well as grasses. Mean annual precipitation 100–300 mm (winter maximum). AP less than 15%; chenopods and *Artemisia* usually dominant, often associated with Gramineae, Compositae, and *Plantago.*

Steppe-forest: transitional zone, ranging from open xeric woodland to steppe with scattered trees and shrubs; deciduous pine, oak, and pistachio plus juniper; *Amygdalus scoparia* important as well as Gramineae. Interior areas receive mean annual precipitation of 300–600 mm (winter maximum). AP 15–50%; less than 20% eu-Mediterranean elements.

Eu-Mediterranean woodland (including xeric variety): includes *Quercus ilex, Q. coccifera, Pinus halepensis, Pistacia lentiscus, Olea europaea* var. *oleaster,* and *Ceratonia siliqua,* plus Gramineae and herbs. Summer-dry climate; near Mediterranean coasts up to about 400-m elevation; mean annual precipitation 300–1000 mm (winter maximum). AP more than 30%; more than 20% evergreen oak plus *Olea* plus *Pistacia.*

Oro-Mediterranean forest: true forest, but varied in composition across region; deciduous oak dominant in Greece and the Zagros Mountains; pine and cedar in southwestern Turkey and Lebanon. Elevations between 400 and more than 2000 m (lower in northern Greece); mean annual precipitation more than 600 mm (winter maximum). AP more than 50% (deciduous oak and conifers dominant); eu-Mediterranean elements less than 20%.

Mesic Euxinian forest: summer-green vegetation; includes *Fagus, Carpinus,* and *Castanea* as well as deciduous oak and pine; rhododendron near coasts. Year-round precipitation averaging more than 600 mm per year. More than 50% AP; detailed palynologic composition currently being investigated.

These six major types can be subdivided further on floristic grounds (Zohary, 1973). Alpine, halophytic, marshland, and riverine associations are not included in this group. The term *steppe* is used here for all climatically caused treeless vegetation other than desert, savanna, and tundra, but it is not used for temperate grassland with summer rain and deep soils (compare Freitag, 1977).

Most pollen sites in the Near East come from Mediterranean forest and steppe-forest associations. There are very few pollen diagrams from either desert or steppe zones, nor are there more than a handful for the summer-green mesic forests of the Caspian and Black Sea lowlands (Roberts, 1982b; Bottema, 1986). For this reason, the pollen maps presented here are restricted to the eastern Mediterranean part of the region (see Fig. 9.2 and Table 9.1 for locations). The pollen and lake-level maps incorporate data pub-

lished up to 1985, although we refer in the text to some later publications.

LAKE LEVELS

Virtually no permanent lakes are found south of 30°N in southwest Asia at present, because potential annual evaporation loss exceeds inputs of precipitation and runoff (Fig. 9.3). All but two of the lake basins included in the lake-level data set from this southern area (Fig. 9.4) are interdune depressions, namely, Rub' al Khali (McClure, 1978, 1984) and Rajasthan (Singh *et al.*, 1972; Wasson *et al.*, 1983). Even north of 30°N many lakes are either playas (e.g., on the Iranian plateau [Krinsley, 1970]) or are hypersaline (e.g., the Dead Sea), or both. Most extant freshwater or brackish-water lakes in the Near East and adjacent regions are found in Anatolia, the Zagros Mountains, Caucasia, the Levant, and northern Greece, where winter precipitation is not largely lost through evaporation from the lake or its catchment. Midwinter is also a time when low temper-

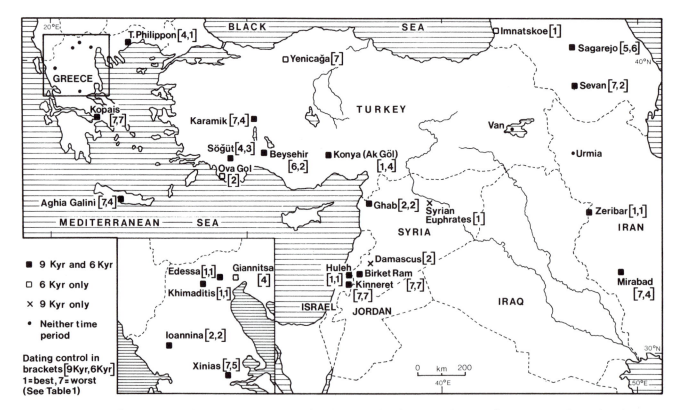

Fig. 9.2. Eastern Mediterranean pollen sites. Dating control is shown in brackets ([9000 yr B.P., 6000 yr B.P.]). The coding for continuous sequences was based on the following criteria: 1, bracketing dates within 2000 yr of the selected date and core top can be assigned a date of 0 yr B.P., or a single date within 200 or 250 yr of the selected level; 2, one bracketing date within 2000 yr and the other within 4000 yr of the selected date; 3, bracketing dates within 4000 yr; 4, one bracketing date within 4000 yr and the other within 6000 yr; 5, bracketing dates within 6000 yr; 6, one bracketing date within 6000 yr and the other within 8000 yr, or one within 4000 yr and the other within 10,000 yr; 7, undated at selected date (no dates in the core, no bracketing dates, no top to core and no date within 8000 yr of the selected date, or bracketing dates more than 14,000 yr apart); 0, status uncodable for selected date regardless of the availability of radiocarbon dates. For discontinuous sequences, ranks 1–6 were assigned when at least one radiocarbon date was within 250 (rank 1), 500 (rank 2), 750 (rank 3), 1000 (rank 4), 1250 (rank 5), or 2000 (rank 6) years of the selected level; rank 7 was assigned if no radiocarbon date was within 2000 yr of the selected level; rank 0 was assigned when the status was uncodable for the selected level regardless of the availability of radiocarbon dates.

Table 9.1. Sites used in the preparation of the paleoclimatic maps

Site name	Country or region	Latitude (°N)	Longitude (°E)	Elevation (m)	References
LAKE LEVELS					
Al-Nafud	Saudi Arabia	28°2'	40°56'	700	Garrard *et al.*, 1981; Whitney, 1983; Whitney *et al.*, 1983
Chatyrkel, Kirgiz	Former Soviet Union	40°36'	75°18'	530	Shnitnikov, 1976; Shnitnikov *et al.*, 1978
Damascus	Syria	33°37'	35°31'	600	Kaiser *et al.*, 1973
Gebel Maghara	Egypt	30°45'	33°24'	400	Bar-Josef and Phillips, 1977
Konya	Turkey	37°30'	33°0'	990	Erol, 1978; Roberts *et al.*, 1979; Roberts, 1980, 1983; Bottema and Woldring, 1984
Mundafan, Rub' al Khali	Saudi Arabia	18°32'	45°25'	870	McClure, 1976, 1984; Whitney, 1983
Nefud As-sirr	Saudi Arabia	20°10'	43°19'	700	Whitney *et al.*, 1983
Nefud-Urayq	Saudi Arabia	25°36'	42°39'	900	Whitney *et al.*, 1983
Rub' al Khali, 16	Saudi Arabia	20°29'	46°37'	500	McClure, 1984
Rub' al Khali, 18	Saudi Arabia	21°53'	47°43'	210	McClure, 1976, 1984
Sambhar	India	27°0'	75°0'	360	Agrawal *et al.*, 1971; Singh *et al.*, 1972, 1973
Seistan	Afghanistan	30°51'	61°55'	500	Pias, 1972
Van	Turkey	38°30'	43°0'	1646	Schweizer, 1975; Degens and Kurtman, 1978
Wady al-Luhy	Saudi Arabia	24°26'	46°36'	500	Al-Sayari and Zotl, 1978
Wadi Dawasir	Saudi Arabia	20°21'	45°0'	600	Zarins *et al.*, 1979; McClure, 1984
Zeribar	Iran	35°32'	46°7'	1300	Hutchinson and Cowgill, 1963; van Zeist and Wright, 1963; Megard, 1967; van Zeist, 1967; Wasylikowa, 1967; van Zeist and Bottema, 1977
POLLEN					
Aghia Galini	Crete	35°8'	24°35'	1	Bottema, 1980
Beyşehir	Turkey	37°32'	31°30'	1220	van Zeist *et al.*, 1975
Birket Ram	Israel	33°13'	35°45'	940	Weinstein, 1976
Damascus	Syria	33°37'	35°31'	600	Bottema and Barkoudah, 1979
Didwana, Rajasthan	India	27°19'	74°34'	350	Singh *et al.*, 1973, 1974
Edessa	Greece	40°49'	21°57'	330	Bottema, 1974
Ghab	Syria	35°41'	36°18'	190	Niklewski and van Zeist, 1970
Giannitsa	Greece	40°40'	22°19'	5	Bottema, 1974
Hoyran	Turkey	38°17'	30°53'	920	van Zeist *et al.*, 1975
Huleh	Israel	33°4'	35°37'	70	M. Tsukada, unpublished; Cowgill, 1969
Ioannina	Greece	39°40'	20°51'	490	Bottema, 1974
Karamik	Turkey	38°25'	30°48'	1018	van Zeist *et al.*, 1975
Kinneret	Israel	32°48'	35°33'	-210	Horowitz, 1971
Khimaditis	Greece	40°37'	21°35'	573	Bottema, 1974
Konya (Akgöl)	Turkey	37°30'	33°51'	1002	Bottema and Woldring, 1984
Kopais	Greece	38°27'	23°01'	100	Greig and Turner, 1974; Turner and Greig, 1975
Lunkaransar, Rajasthan	India	28°30'	73°45'	200	Singh *et al.*, 1974
Mirabad	Iran	33°4'	47°43'	800	van Zeist and Bottema, 1977
Ova Gol	Turkey	36°19'	29°19'	5	Bottema and Woldring, 1984
Sagarejo	Georgia	41°31'	45°30'	500	Gogichaishvili, 1982
Sevan	Armenia	40°30'	45°30'	1900	Sajadjan, 1978
Sögüt	Turkey	37°3'	29°52'	1393	van Zeist *et al.*, 1975
Syrian Euphrates	Syria	35°30'	39°	275	Leroi-Gourhan, 1974
Tenaghi Philippon	Greece	41°10'	24°20'	40	Wijmstra, 1969
Yenicaga	Turkey	40°45'	32°3'	976	Beug, 1967
Xinias	Greece	39°4'	22°15'	500	Bottema, 1979
Zeribar	Iran	35°31'	46°7'	1300	van Zeist and Bottema, 1977

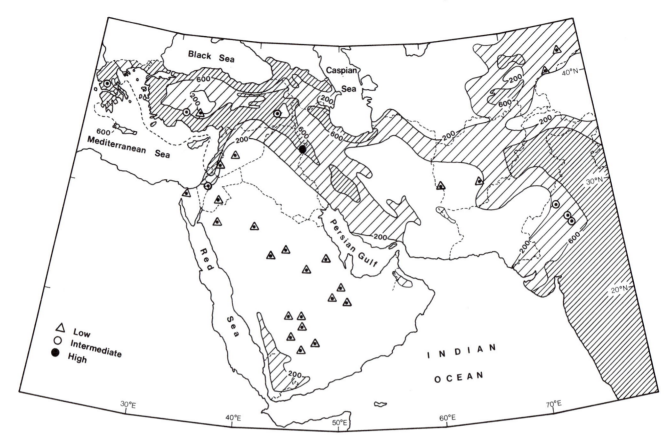

Fig 9.3. Annual rainfall (mm) and modern lake-level status in south and southwest Asia.

atures reduce respiration and hence transpiration by plants. All of the lakes included in the Near Eastern lake-level data set (Fig. 9.4) that are currently fresh or brackish lie at elevations at least 1000 m above sea level (Table 9.1). The contrasting modern intraregional pattern of lake water levels is shown in Figure 9.3, together with modern mean annual precipitation.

Late-Pleistocene Environments

Before 1960 it was believed that the Near East experienced a "pluvial" climate around the time of the last Northern Hemisphere ice-sheet maximum (e.g. Butzer, 1958). The knowledge that such lakes as the Dead Sea at one time were much more extensive than at present was used to support this glacial-pluvial correlation. Publication of the pollen diagram from Lake Zeribar (van Zeist and Wright, 1963) revealed that the present oak forest in the Zagros Mountains was steppe before about 11,000 yr B.P. Mediterranean-type woodland was apparently confined to the mountain slopes of the Levant and the Balkan Peninsula, where precipitation was a little greater than in the interior lowlands.

In the Zeribar area the late-Pleistocene lowland steppe apparently graded upslope to alpine vegeta-

tion, without intervening woodland (Wright, 1976). Such a vegetational pattern can be seen today only in the most arid mountain regions, such as at the very eastern end of the Zagros Mountains near Shiraz. The winter precipitation that occurs today in this region appears to have been greatly reduced during much of the glacial period, perhaps because of the increased strength and influence of the Siberian winter anticyclonic circulation, which made the entire Eurasian continent cold and dry. Freitag (1977) suggested that annual precipitation during the last pleniglacial was between 150 and 250 mm, or about a third of present-day values. Cold late-Pleistocene conditions in the Mediterranean are independently demonstrated by the extent of the glaciers, with a depression of the snowline of 1200–1500 m in the Zagros Mountains (Wright, 1961; Kuhle, 1976), although these glacial advances remain undated radiometrically.

At the same time the summers may have been less severely dry than they are today, according to two lines of evidence. First, the pollen diagrams from the northern part of the region indicate the absence or near absence during the last glacial period of three of the trees most indicative of a Mediterranean climate, namely olive, Mediterranean pistachio (*Pistacia lentiscus*), and evergreen oak. Of course these trees

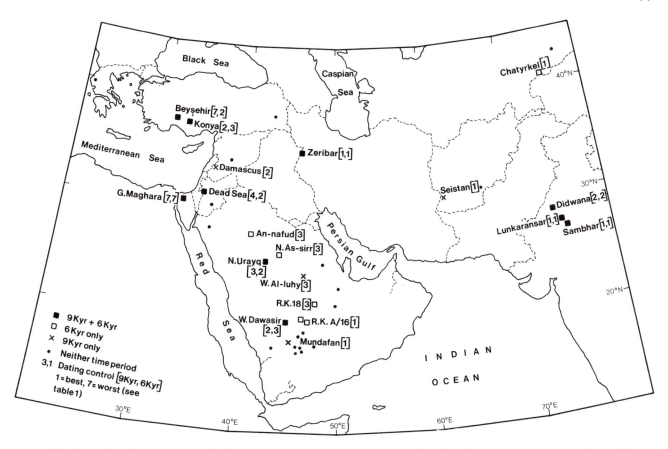

Fig. 9.4. South and southwest Asian lake-level sites. Dating control is shown in brackets ([9000 yr B.P., 6000 yr B.P.]), and dating control codes were assigned as in Figure 9.2.

may have been absent because of the low winter temperatures, for they are susceptible to heavy frost, but they are adapted to dry summers, unlike most other trees. Second, among the minor pollen types at one site is an herb of the chenopod family that is not found today in the Mediterranean region but rather occurs in the Pamir, which receives summer as well as winter rains (Smit and Wijmstra, 1970).

It may be that the glacial-age climate in the summer was characterized not by increased precipitation but simply by greater cloudiness and by greater effective moisture as a result of decreased temperature and evaporation. This should cause a rise in lake levels, and certainly there is evidence of high lake levels on the Iranian and Anatolian plateaus around 18,000–20,000 yr ago (see below). However, the intricate interactions of temperature and precipitation are difficult to decipher precisely (Farrand, 1981) and are a subject of considerable debate (El-Moslimany, 1984; Roberts, 1984). It is certain that the climate was too dry for trees at least at the lower elevations. Here steppe rather than tundra or alpine vegetation must be assumed, because of the positive evidence for trees at higher elevation in the Greek mountains at this time (see below).

Any climatic reconstruction of glacial-age conditions in the Mediterranean region must take into account the climatic conditions in adjacent areas, so that compatibility is assured. Europe north of the Alps and Carpathians was cold and dry, as indicated by the pollen record of tundra and steppe vegetation and the widespread occurrence of permafrost indicators. On the south side of the Mediterranean, aridity is indicated by dry lakes in basins that contained permanent freshwater during the early Holocene. The same picture applies to the southern sector of southwest Asia under consideration here: the period from 15,000 to 12,000 yr B.P. was more intensely arid than any other during the last 30,000 yr. In the Rubʿ al Khali sand sea of southern Arabia and in Rajasthan, the terminal Pleistocene was a phase of active eolian sand movement (McClure, 1978; Wasson *et al.*, 1983).

Holocene Vegetation in the Eastern Mediterranean

The early Holocene in the Mediterranean region was marked by the expansion of trees into areas previously dominated by steppe. This change was abrupt compared to the long interval of stable conditions that prevailed during the last cold period, but in detail it was marked by a transitional phase that reflected the adjustments of individual plant species to the changed climatic conditions. In some cases the transition from steppe to woodland lasted for thousands of years, up into the middle Holocene. Consequently the time level of 9000 yr ago is in the midst of this transition.

One of the most significant changes at the outset was the first substantial record of important indicators of Mediterranean-type climates—olive, Mediterranean pistachio, and evergreen oak. These indicators expanded westward during the course of the Holocene (Huntley and Birks, 1983). The initial change indicates that the first evidence of a Mediterranean-type climate, with winter rains and summer drought, appeared about 11,000 yr ago, when the climate of western Europe likewise was changing rapidly to its Holocene mode, and when Saharan lakes were rising to their early-Holocene maxima. On the other hand, Mediterranean bioclimatic indicators were very much less extensive at the beginning of the Holocene than they were later (see below).

In the following sections we describe the vegetational and climatic sequence at several important pollen sites. Recent summaries by principal researchers in this area (Bottema and van Zeist, 1981; van Zeist and Bottema, 1982, 1991; Bottema, 1987) have been helpful in providing an independent perspective.

ZAGROS MOUNTAINS OF IRAN AND EASTERN TURKEY

Modern research into the late-Quaternary vegetation history of the eastern Mediterranean region started in the Zagros Mountains of western Iran at Lake Zeribar. The lake, which is now only 4 m deep, is located at an elevation of 1300 m in an intermontane valley between fold ridges. The detailed pollen sequence shows a monotonous assemblage dominated by herbs such as *Artemisia*, chenopods, and umbellifers up until about 10,500 yr ago (van Zeist and Bottema, 1977) (Fig. 9.5). A modern analogue for this pollen assemblage may be the cold steppe of the Armenian or western Iranian interior plateaus (Wright *et al.*, 1967) or, perhaps more likely, the equally cold and dry steppe of northwestern Afghanistan, where umbellifers are an important component of the steppe (van Zeist and Bottema, 1977).

About 10,500 yr B.P. the pollen curves for oak and pistachio become continuous, and grasses abruptly replace *Artemisia* and chenopods as the dominant herbs. The equally abrupt appearance of *Plantago maritima* pollen indicates the inception of warmer conditions. By 9000 yr B.P. the vegetation could be described as a grass steppe with some trees. Tree pollen amounted to only 10% at this time. The vegetation may have resembled the oak-pistachio forest-steppe found today on south-facing slopes in the Zagros foothills near the lower treeline, where annual precip-

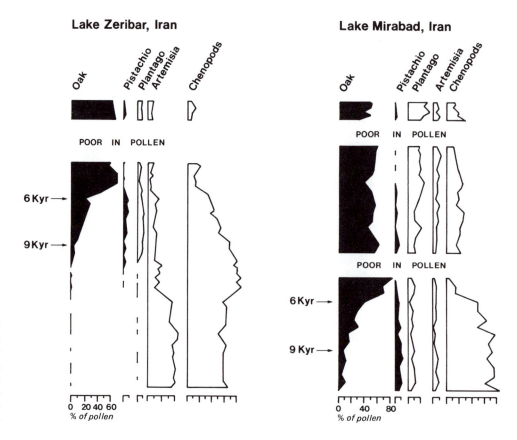

Fig. 9.5. Summary pollen diagrams from Zeribar and Mirabad, western Iran (redrawn from van Zeist, 1967). Radiocarbon dates in thousands of years B.P. (Kyr); tree pollen in black, herbs in white.

itation is about 500 mm, compared to about 700 mm in the area of Lake Zeribar. After 9000 yr B.P. the pollen percentages for oak and pistachio steadily rose, mostly at the expense of chenopods. Increasing moisture in the soils is implied. By 6000 yr B.P. the total value for oak, pistachio, and ash pollen was about 20%. Thereafter the percentage of oak pollen rose rapidly to levels of 40-55%, which have persisted to modern time.

The importance of the Zeribar pollen diagram is enhanced by a companion diagram from a lake 300 km to the southeast near Mirabad at the base of the Zagros woodland at an elevation of about 800 m, or about 500 m lower than Zeribar. It was originally presumed that during the time of glaciation the oak woodland missing from the Zeribar area had simply been depressed to lower elevations because of the reduced temperature and that the Mirabad pollen sequence would therefore show higher proportions of oak pollen before 10,500 yr B.P. But the Mirabad pollen diagram, although dated to only about 10,500 yr B.P. at the base, shows a sequence very similar to that for Lake Zeribar, with low oak values at the base and higher values above (van Zeist and Bottema, 1977) (Fig. 9.5). Thus the oak woodland was not simply depressed to lower elevations but was eliminated entirely from the area because of the aridity.

A third diagram comes from Lake Van in eastern Anatolia. Lake Van is a very large, deep lake in volcanic terrain. It lies at an elevation of about 1650 m, and the annual precipitation is about 400 mm. Because the sediment cores were obtained from deep water, the resulting pollen diagram represents entirely regional vegetation changes. The absence of local pollen is especially important in the case of some nonarboreal taxa, such as Chenopodiaceae, which can form part of either the dryland or the lake-margin flora. It is therefore doubly significant that the pollen diagram from Lake Van closely resembles that for Lake Zeribar in its general features. The lower part is dominated by herbs, and total tree pollen is less than 10%. This assemblage gradually changes upward as oak pollen increases at the expense of herbs (especially chenopods and *Ephedra*). The upper half of the sequence shows a steady 50% oak pollen to near the surface, where the increases in walnut and pine pollen are signs of cultural disturbance.

The Lake Van sequence implies initial steppe conditions that may have been even drier than those at Zeribar and Mirabad (as indicated by high values for the semidesert shrub *Ephedra*), followed by slowly increasing forest cover and then stability. The diagrams from the two areas are strikingly similar and would appear to represent contemporaneous changes in vegetation. At Van, however, the rise in oak pollen is dated as extending from 6400 to 3400 yr B.P. rather than from 10,500 to 5500 yr B.P. as at Zeribar. The dating at Van is based on varve counts (Kempe and Degens,

1978) rather than on radiocarbon analysis. Although the laminations are clearly rhythmic, the sequence includes some turbidites, tephra layers, and other interruptions that might reflect discontinuities. Furthermore, a diagram from a peat deposit at Söğütlü near Lake Van has a radiocarbon-dated pollen sequence that closely matches the Zeribar chronology during the early to middle Holocene (Bottema, 1987; van Zeist and Bottema, 1991). Because the Van varve-type chronology appears to be anomalous, the site is not included in the pollen or lake-level data sets used in this chapter, although its location is indicated in Table 9.1.

Farther east, a pollen diagram from Lake Urmia (Bottema, 1986) shows a pattern of early-Holocene vegetation change analogous to that at Zeribar and Mirabad, although AP totals never exceed 20% on account of the region's dry climate. *Artemisia* steppe was replaced by open forest-steppe between 9000 and 7000 yr B.P. (This site is not included in Table 9.1 because of its recent publication.)

THE LEVANT

The most important pollen site in the Levant is in the Ghab depression, which is a long, narrow marsh at the northern end of the Dead Sea-Jordan Valley rift. The initial core covered the last 50,000 yr in only 7 m of sediments (Niklewski and van Zeist, 1970). Supplemental cores at Ghab cover the Holocene in greater detail, but only a single radiocarbon date, at about 10,000 yr B.P., is available for the early Holocene (van Zeist and Woldring, 1980).

The glacial period at this site was dominated by herbaceous pollen throughout, but deciduous oak pollen reached 25% at some levels. By 9000 yr B.P. the area was well wooded, and the trees characteristic of a Mediterranean climatic regime (evergreen oak, pistachio, and olive) were well established (Fig. 9.6). By 6000 yr B.P. the proportion of deciduous oak pollen decreased to 20% as pine and the total tree cover diminished steadily from their early-Holocene maxima. If the extent of tree cover is a measure of precipitation levels (in the absence of significant human interference), then the climate was wetter 9000 yr ago than 6000 yr ago—the opposite conclusion to that drawn for the Zeribar area in Iran. On the other hand, an increase in summer drought severity as the Mediterranean climatic mode intensified could have led to a reduction in the cover of deciduous oak.

A well-dated pollen diagram from the Huleh marsh farther south in the same Jordan rift valley in Israel was compiled by M. Tsukada and summarized by Bottema and van Zeist (1981). The vegetation of the last glacial period included a significant amount of oak but was dominated by steppe vegetation between 25,000 and 14,000 yr B.P. Oak expanded greatly between 14,000 and 10,000 yr B.P. but by 9000 and 6000 yr B.P. was reduced about to its present level.

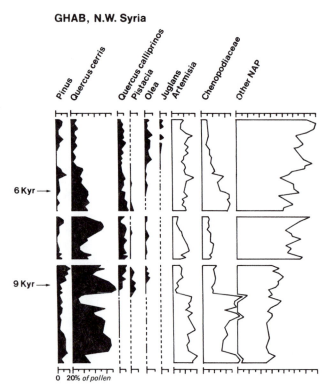

GHAB, N.W. Syria

Pinus
Quercus cerris
Quercus calliprinos
Pistacia
Olea
Juglans
Artemisia
Chenopodiaceae
Other NAP

6 Kyr →

9 Kyr →

0   20% of pollen

Fig. 9.6. Summary pollen diagram from Ghab, northwestern Syria (redrawn from van Zeist and Woldring [1980] and Niklewski and van Zeist [1974]). Radiocarbon dates in thousands of years B.P. (Kyr); tree pollen in black, herbs in white.

SOUTH-CENTRAL AND SOUTHWESTERN TURKEY

Pollen diagrams are available for several sites in the western Taurus Mountains of southwestern Turkey (van Zeist *et al.,* 1975; Bottema and Woldring, 1984). At least four of the detailed pollen diagrams completed for this region cover the times of 9000 and 6000 yr B.P.

Karamik Batakliği is located in the dry transition area between the forested mountains and the dry steppe of the Anatolian plateau. The diagram suggests steppe vegetation for the last glacial period and forest for the Holocene, but the chronology of the transformation cannot be detailed because only two radiocarbon dates are available (20,000 and 6500 yr B.P.). The proportions of herb pollen were still high enough at 9000 yr B.P. to imply drier conditions than prevailed later at 6000 yr B.P., when tree pollen totaled over 80%.

The sites of Söğüt and Beyşehir are located in intermontane valleys closer to the Mediterranean Sea and were in or near pine forest until modern clearance. Their pollen diagrams indicate that the last cold period was characterized by steppe assemblages (Fig. 9.7), although both record a temporary rise in pine during the terminal Pleistocene. The generally open vegetation conditions persisted until 9000 yr B.P., when pine, oak, cedar, and juniper began to expand. After about 6000 yr B.P. pine came to dominate at both sites, except

for an interval from about 3500 to 2000 yr B.P., when forest clearance and the cultivation of olive, walnut, manna ash, vines, hemp/hop, and probably grains prevailed. This is one of the most striking pollen records of a distinct episode of human disturbance in the Mediterranean region (Roberts, 1982b; Wright, 1984; Bottema and Woldring, 1990; van Zeist and Bottema, 1991).

A fourth southern Turkish site—Akgöl—lies at the eastern end of the Konya plain in an area that receives no more than 300 mm a year of precipitation on average. Although it is situated in the central Anatolian steppe, some of the highest peaks of the Taurus Mountains are only 40 km away to the south and east. The Akgöl pollen diagram (Bottema and Woldring, 1984) covers the late Pleistocene and early Holocene. In its lower part tree pollen accounts for 10% of the total. After 11,000 yr B.P. forest started to spread with a conspicuous expansion of birch, which was followed by rises in deciduous oak and pine. Pine became the dominant tree after 8000 yr B.P.

The transition from steppe to woodland in southern and southwestern Turkey indicates an appreciable rise in humidity during the early part of the Holocene. All four diagrams from the region show higher proportions of herb and grass pollen 9000 yr ago than subsequently, indicating drier conditions at the beginning of the Holocene than today.

GREECE

The Holocene vegetation history of Greece can be elucidated from several detailed pollen diagrams. In

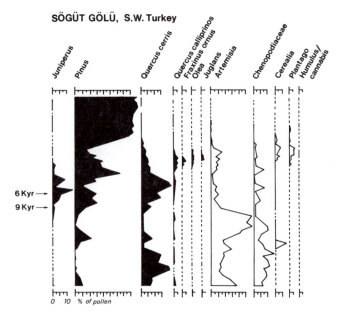

SÖGÜT GÖLÜ, S.W. Turkey

Juniperus
Pinus
Quercus cerris
Quercus calliprinos
Fraxinus ornus
Olea
Juglans
Artemisia
Chenopodiaceae
Cerealia
Plantago
Humulus/cannabis

6 Kyr →
9 Kyr →

0   10   % of pollen

Fig. 9.7. Summary pollen diagram from Söğüt, southwestern Turkey (redrawn from van Zeist *et al.,* 1975). Radiocarbon dates in thousands of years B.P. (Kyr); tree pollen in black, herbs in white.

northern Greece a series of pollen sites from east to west illustrates the contrast between the dry eastern side of the Pindus Mountains and the wet western side, as well as elevational and other geographic factors.

Farthest east in the series is Tenaghi Philippon on the Drama plain close to the Aegean coast, with annual precipitation of less than 500 mm and a vegetation marked by Mediterranean types, including evergreen oak (Wijmstra, 1969; Greig and Turner, 1974; Turner and Greig, 1975). The other sites are the work of Bottema (1974). On the Macedonian plain close to sea level is Giannitsa, with annual precipitation of about 500 mm. In a valley north of the Macedonian plain at an elevation of about 350 m and with annual precipitation of about 900 mm is Edessa, in deciduous forest with beech, fir, and an indication that the mountains of northern Greece provided refuge for temperate trees excluded from most of Europe by the low temperatures. At the same time, the substantial glaciation in the Greek mountains implies a deep depression of the snowline (to 1200 m below its assumed modern level of 3250 m, according to Messerli [1967]). The upper treeline today is estimated to be 1200 m below the snowline; if it were depressed as much as the snowline, as is usually assumed, and if Ioannina and Khimaditis (at elevations of about 500 m) were dominated by steppe (Fig. 9.8), then the calculations leave only a very few hundred meters for trees (Bottema, 1974). Furthermore, the trees included temperate trees as well as montane conifers, which may indicate the importance of local habitats of contrasting microclimates on the humid western side but not the drier eastern side of the mountains.

At the low-elevation site of Tenaghi Philippon, oak increased abruptly at about 13,500 yr B.P., as dated by straight interpolation between dates of 14,600 and 7850 yr B.P., and elm, linden, hazel, and pistachio started continuous curves. At the same time at the high-elevation site of Ioannina, pine and fir apparently moved downward from the mountain slope. By 9000 yr B.P. forest prevailed at all the northern sites. The forestation involved primarily oak, with elm as the most abundant secondary type. The high-elevation pine, fir, and beech did not participate in the forestation at any of the sites, so it appears that the lower treeline moved downward while the upper tree zones remained at the same elevation—that is, the effective climatic trend up until 9000 yr B.P. was toward increased moisture rather than increased temperature.

Central and southern Greece have yielded fewer diagrams than the northern part of the country; the most important are at Xinias in Thessaly and Copais in Boeotia (Greig and Turner, 1974; Turner and Greig, 1975; Bottema, 1979). Both sites lie near the Aegean coast, and some eu-Mediterranean taxa are found in their vicinities today. Their diagrams show a striking contrast between the steppe pollen assemblage (*Artemisia*, chenopods, and grasses, with some pine) of the last glacial period and the forest assemblage of the Holocene. At Xinias the change from steppe to forest is rather abrupt and is dated to around 10,680 yr B.P. (Fig. 9.9). The approximate level of 9000 yr B.P. is marked primarily by pollen of deciduous oak. The occurrence of *Quercus coccifera* here and at Copais indicates that an essentially Mediterranean climatic regime was established by this time. Also present at Xinias are minor percentages of such mesic types as

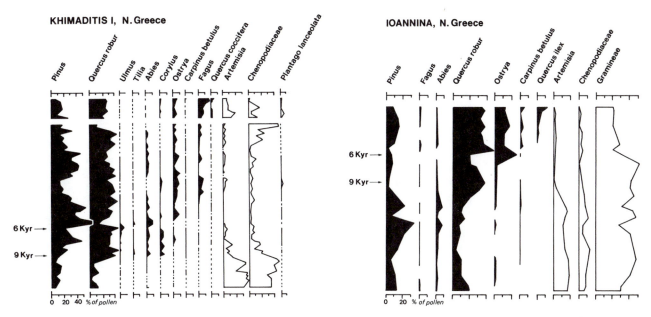

Fig. 9.8. Summary pollen diagrams from Khimaditis, northern Greece, and Ioannina, northwestern Greece (data from Bottema, 1974). Radiocarbon dates in thousands of years B.P. (Kyr); tree pollen in black, herbs in white.

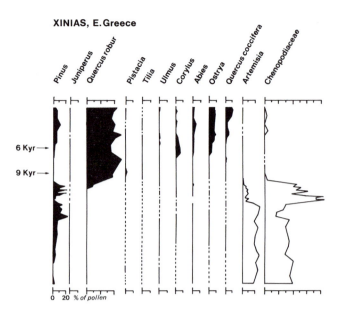

XINIAS, E. Greece

0   20 % of pollen

Fig. 9.9. Summary pollen diagram from Xinias, eastern Greece (data from Bottema, 1979). Radiocarbon dates in thousands of years B.P. (Kyr); tree pollen in black, herbs in white.

elm, linden, hazel, ash, hornbeam, and fir, whose importance increased during the early and middle Holocene. Significant human disturbance of the regional vegetation is recorded in the upper part of the Copais diagram.

In addition to these mainland sites there is one pollen diagram from the island of Crete, taken at sea level near Aghia Galini (Bottema, 1980). Although the diagram is incomplete and has only a single radiocarbon date, it nonetheless indicates a vegetation cover between 10,000 and 6000 yr B.P. very different from the *Q. coccifera* and *Ceratonia-Pistacia* maquis that exists today. The area was forested throughout the early Holocene, initially by pine and subsequently by deciduous oak, although *Asphodelus* pollen indicates that open vegetation was locally present. However, by about 5000 yr B.P. the area had been largely denuded of trees, and the landscape was comparable to that of the present day. Whether this deforestation was of natural or cultural origin is uncertain, although the latter must be viewed as a distinct possibility in view of the large Bronze Age population that Crete supported.

SUMMARY

The end of the Pleistocene and the beginning of the Holocene thus witnessed the advance of forest vegetation over large areas of the eastern Mediterranean region. At sites close to the sea this process was complete by 9000 yr B.P., and the vegetation in many cases was not substantially different from what would exist there naturally now. At sites in southwestern Turkey and parts of northern Greece, the ad-

vance of forest vegetation was not complete until after 9000 yr B.P., and the vegetation cover at 9000 yr B.P. was significantly more open than later in the Holocene (including both 6000 yr B.P. and the present). Most of these sites lie at elevations above 500 m and are somewhat removed from direct maritime influence. A third group of pollen sites, located in interior areas of Turkey and western Iran, responded even more slowly. Their pollen records indicate that the vegetation at 9000 yr B.P. was only marginally less open than during glacial times, and even by 6000 yr B.P. afforestation was incomplete. This delayed Holocene forest advance is best documented in western Iran at Zeribar and Mirabad but is also recorded at Van in eastern Turkey.

Holocene Lake Levels in the Eastern Mediterranean Area

Holocene lake-level sites in Greece, Turkey, Iran, and the Levant are less numerous and less informative than the pollen sites from the same region. The dominant feature of the late-Quaternary limnological record is a widespread phase of high water levels before and during the last pleniglacial (Roberts, 1982a).

At Lake Zeribar in the Zagros Mountains, the same sediment cores studied palynologically have also been analyzed for sediment chemistry, Cladocera, plant macrofossils, and diatoms (Hutchinson and Cowgill, 1963; Megard, 1967; Wasylikowa, 1967). The results indicate that from before 22,000 yr B.P. until around 14,000 yr B.P., the lake was at least 8-10 m deep. After this time diatoms and fruits of *Ruppia maritima* var. *rostellata* indicate that it shallowed and became periodically brackish. Water levels remained relatively low during the first part of the Holocene, and marked fluctuations allowed *Cyperus* and other annuals, and even *Salix*, to colonize the exposed shore zone. The presence of a number of tropical Cladocera and diatoms between 9000 and 6000 yr B.P. points to conditions of maximum summer warmth at this time. After 6000 yr B.P. the water level rose rapidly and became more stable, and a fringing sedge mat developed around the lake. Because of sedimentation the lake today is only 4 m deep and covers an area of about 10 km². In combination this evidence suggests that Zeribar must have been closed (i.e., without an outlet) at least episodically between 14,000 and 6000 yr B.P. but that it has had an outlet since then.

The Pisidian lake district of southwestern Turkey potentially contains some of the best lake-level sequences in the eastern Mediterranean region (Erol, 1978). So far the most detailed investigations have been made in the Konya basin on the southern margin of the Anatolian plateau (Roberts, 1980, 1983). This basin is now largely dry, with open water covering only about 100 km². During the late Pleistocene, how-

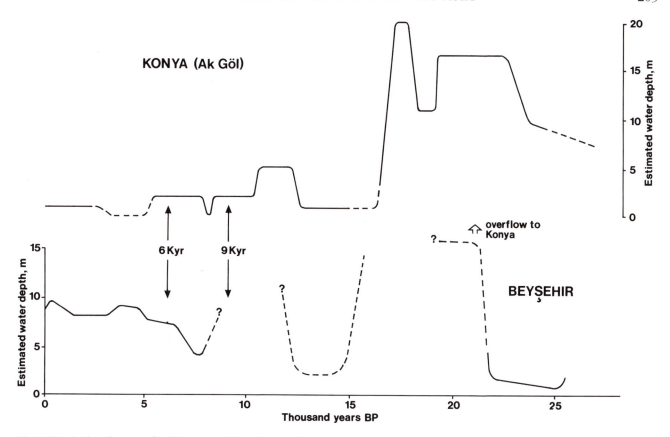

Fig. 9.10. Lake-level curves for Konya and Beyşehir, southern Turkey (data from Roberts, 1980, 1983).

ever, it was occupied by a lake covering more than 4000 km² (Fig. 9.10). Radiocarbon dates on mollusk shells from shoreline sediments show that high water levels endured between about 23,000 and 17,000 yr B.P. Paleolake Konya subsequently shrank to form several smaller residual lakes, one of which—Akgöl—has been cored and analyzed for pollen, diatoms, mollusks, and ostracods. Euryhaline diatoms and *Nitraria* pollen indicate saline conditions around 13,000 yr B.P. (Bottema and Woldring, 1984). Akgöl and other residual lakes of the Konya basin became more extensive and fresh about 12,000 yr B.P. During the early Holocene Akgöl was shallower but still fresh, except for a brief dry interval around 8000 yr B.P. During the second half of the Holocene, however, more long-lasting dry conditions were established, marked by a buried soil and subsequently by telmatic peat.

The climatic significance of Holocene changes in the level of Akgöl is uncertain, partly because the main river feeding Akgöl today is primarily spring-fed rather than rain-fed. In the Konya basin as a whole, lake-level fluctuations have been of only minor climatic significance since agricultural settlements first occupied the bed of the former lake 8500 yr ago.

Immediately west of Konya lies the intermontane Beyşehir-Sugla basin. Lake Beyşehir has a freshwater outlet at present, but sediment cores from its southern half revealed a series of buried peat layers. The lowest of these layers has a ¹⁴C date of 24,025 yr B.P., at which stage the lake was replaced by a marsh occupying less than one-tenth of the lake's present extent. The water level subsequently rose enough to cause overflow into the adjacent Konya basin, itself at a high level around this time (Roberts, 1980). The second organic layer from Lake Beyşehir is undated, but interpolated ages lie in the range of 15,000–12,000 yr B.P. During this time the lake was only about one-third of its present size, but since then it has expanded progressively to its present area of 650 km².

Although lake-level curves for Beyşehir and Konya (Fig. 9.10) show an essentially similar pattern of change, they are slightly out of phase. This may be a result of "hardwater" error in the ¹⁴C dates from Konya basin, which is surrounded by calcareous rocks. The different water-level status of the basins at present is the result of climatic differences. Beyşehir like Zeribar lies in a zone of relatively high annual precipitation (about 600 mm) and natural forest vegetation, whereas Konya receives only half this precipitation total and is located in steppe.

Probably the best-known lake site in the Near East is the Dead Sea, whose recent water-level oscillations have been documented since the 19th century. Lartet as early as 1865 demonstrated the existence of earlier high water levels from the so-called Lisan beds. These high levels have been dated by ¹⁴C and U-Th to between about 60,000 and 15,000 yr B.P. (Begin *et al.*, 1974; Vogel, 1980), after which time the lake shrank to approximately its modern dimensions. The Dead Sea today comprises a shallow southern basin and a deep northern basin. Boreholes from the former revealed a series of salt tongues formed when the southern basin reverted to a seasonal playa (Neev and Emery, 1967). These low stands are dated to 9850 to about 5500 yr B.P. and 4400 to about 1500 yr B.P. Upriver from the Dead Sea, Lake Huleh also has a well-dated limnological history, but because it does not appear to have been hydrologically closed (i.e., without an outlet) during the late Quaternary, we excluded it from detailed consideration here.

These and other proxy water-level records for the Near East and southwest Asia (Roberts, 1982a) have been incorporated into the Oxford lake-level data bank, which categorizes individual lake histories according to water-level status (low, intermediate, or high) at 1000-yr intervals (see Street-Perrott and Roberts [1983] for details). From this information differences in water-level status between the present and previous times can also be categorized as "no change in status" or as one or two steps lower or higher.

The water-level records of Zeribar, Beyşehir, and Konya lakes and the Dead Sea as classified in the Oxford data bank are summarized in Figure 9.11. Lake levels in these four basins were at or below present levels between 9000 and 6000 yr B.P., with Zeribar showing the most marked change in water level. This result is consistent with the Holocene pollen record, which shows a delayed rise in tree pollen in the Zagros Mountains during the early Holocene. Both low lake levels and more open vegetation are most easily explained—for the Holocene at least—in terms of a more arid climate than at present.

Other lakes in the northwestern sector of the Near East also record water levels at or below those of today during the early Holocene. Lake Burdur in southwestern Turkey was probably at a low or intermediate level at both 9000 and 6000 yr B.P. (N. Roberts, unpublished), and Lake Urmia in northwestern Iran was a shallow playa-type lake until around 9000 yr B.P. (Kelts and Shahrabi, 1986). The water level at Lake Van in eastern Turkey apparently fell sharply between 9500 and 6000 yr B.P., as dated by a varve-type chronology (Degens and Kurtman, 1978), although as

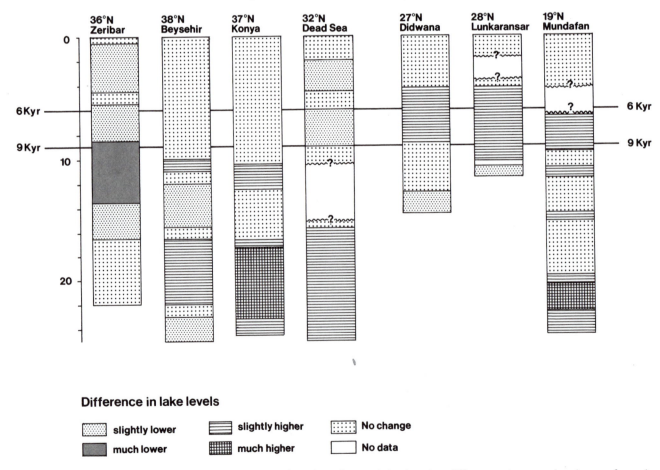

Difference in lake levels

slightly lower   slightly higher   No change

much lower   much higher   No data

Fig. 9.11. Lake-level sequences for selected sites in south and southwest Asia, showing differences in water-level status from the present.

noted above the dating of the Van cores is inconsistent with that for other pollen sequences in the same region.

Holocene Environments in Arabia and Northwest India

VEGETATION

The Holocene vegetational history of the arid southern sector of southwest Asia received almost no attention before the work of El-Moslimany (1983) in the Persian Gulf region and southern Arabia. Modern samples from the hyperarid Rub' al Khali sand sea not surprisingly indicate low pollen concentrations, with the sand sedge *Cyperus conglomeratus,* chenopods, and the xerophytic shrub *Calligonum* as the main contributors (Table 9.2). In contrast, pollen data from early-Holocene lake beds dated to between 9605 and 5890 yr B.P. show higher proportions of chenopods, Gramineae, and *Plantago,* plus *Tamarix* and *Pinus,* the latter presumably as a result of long-distance transport. A north-south increase in the proportion of chenopod pollen has been interpreted as possibly reflecting higher early-Holocene rainfall in the southern part of the Rub' al Khali. The near absence of *Cyperus* pollen further suggests that the sand dunes were stabilized at that time. Pollen analyses of early-Holocene peat at the head of the Persian Gulf also indicate a less markedly xerophytic-halophytic flora than at present (El-Moslimany, 1983).

The eastern part of southwest Asia also remains largely blank on the palynologic map, with the notable exception of the Rajasthan lake sites (Sambhar, Didwana, and Lunkaransar) investigated by Singh *et al.* (1974). Mean annual precipitation at these three sites ranges from 200 mm at Lunkaransar to 500 mm at Sambhar. During the first half of the Holocene, the pollen spectra were dominated by grasses, sedges, and herbs, with *Artemisia* notably more abundant than at present. Trees were sparse. Sedges and tree and shrub pollen increased suddenly after about 5000 yr B.P. Open steppe appears to have given way to a grassy savanna-type vegetation with some trees. In particular

Table 9.2. Modern and Holocene pollen percentages from the Rub' al Khali[a]

Pollen type	Modern	Holocene
Chenopodiaceae	24.0	70.0
Plantago	1.2	11.4
Calligonum	14.4	1.1
Gramineae	2.5	14.0
Cyperus	57.7	2.4
Pollen sum	1304.0	369.0

[a]Data from El-Moslimany (1983).

Table 9.3. Holocene precipitation estimates (mm) from Lunkaransar, Rajasthan[a]

Time level	Summer rainfall (June-October)	Winter rainfall (November-May)	Total
Present (actual)	269	35	304
Present (calculated)	262	43	305
6000 yr B.P. (calculated)[b]	490 (+82%)	38	528 (+74%)
9000 yr B.P. (calculated)[b]	495 (+84%)	39	534 (+76%)

[a]Data from Swain *et al.* (1983).
[b]Difference from present shown in parentheses.

the tree *Syzygium cuminii,* today found in areas where annual rainfall averages more than 850 mm, expanded into the semiarid zone.

This wetter phase did not persist, and after 4000–3000 yr B.P. conditions became unfavorable for pollen preservation. The uppermost spectra at Lunkaransar, dating to about the last 1000 yr, broadly reflect present conditions in this arid area (Fig. 9.11). "Sand-formation" taxa, including *Calligonum, Capparis, Aerva,* and Chenopodiaceae-Amaranthaceae predominate, while *Artemisia* and tree pollen are almost absent.

Swain *et al.* (1983) reanalyzed the pollen diagram from Lunkaransar to obtain quantitative estimates of Holocene rainfall changes in Rajasthan. They climatically calibrated pollen data from Singh *et al.* (1974) by means of stepwise multiple regression, using modern summer and winter precipitation values (means for 1930–1960) and pollen-rain samples from Singh *et al.* (1973). The calculations show that summer rainfall during the early and middle Holocene was nearly twice that of today, whereas winter rainfall was little changed (Table 9.3). Note, however, that the low values postulated by Swain *et al.* (1983) for the period 3000–1000 yr B.P. are based on an absence of pollen from the relevant layers—circumstances that make reliable numerical estimation of past precipitation difficult.

LAKE LEVELS

The Rajasthan lake cores have also been used to reconstruct limnological as well as vegetational histories (Singh *et al.,* 1972). Today Sambhar, Didwana, and Lunkaransar are all saline playa-type lakes located between stabilized sand dunes at the northeastern edge of the Thar dunefield. Their sediments include lacustrine silt, clay, and gypsum overlying sand. Singh *et al.* (1972) interpreted the basal sand as dune sand that was blown onto the dry lake bed, and it was dated by extrapolation as older than about 10,000 yr. After this phase the basins filled with water, which must have been fresh because pollen assemblages include the

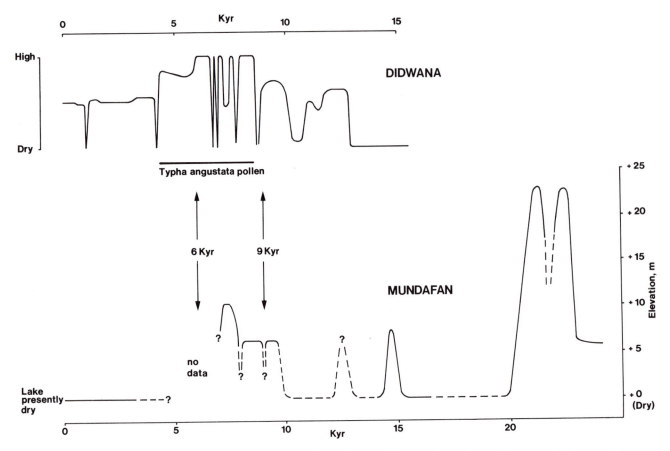

Fig. 9.12. Lake-level curves for Didwana, northwestern India, and Mundafan, southern Arabia (data from Wasson *et al.* [1983] and McClure [1984], respectively).

aquatic plant *Typha angustata.* After 4500–4000 yr B.P. *Typha* pollen disappears from the pollen record and halophytes increase, so conditions appear to have become saline, as at present.

Swain *et al.* (1983) also estimated precipitation from lake-level changes at Sambhar, using both energy- and water-balance approaches. They calculated that a 200-mm increase in annual precipitation would have led to a 21-m rise in water level, causing the Sambhar lake to overflow. Although such an increase is not inconsistent with field evidence, calibrated lake-level data suggest that 200–250 mm represents a maximal value for rainfall increase.

Wasson *et al.* (1983) published more detailed analyses of sediments from Didwana Lake, and a modified version of their lake-level curve is presented in Figure 9.12. They suggested that the basal sands of Singh *et al.* (1972) were not primary eolian deposits but rather were reworked nearshore lacustrine sediments derived from flanking dunes. Nonetheless, the data they present are in broad agreement with the changes proposed by Singh *et al.* (1972). Before 12,500 yr B.P. Didwana was hypersaline and shallow, and geochemical analyses suggest that the lake subsequently fluctuated between saline and freshwater conditions, with water

levels reaching a maximum around 6000 yr B.P. (Fig. 9.11). The onset of modern conditions is dated to about 4000 yr ago.

The most abundant evidence of Holocene paleoenvironments from the southeastern sector of southwest Asia comes from the fossil lake beds of the Arabian peninsula (al-Sayari and Zotl, 1978). Particularly important is the work of McClure (1976, 1978, 1984) on lacustrine deposits exposed around the flanks of interdune depressions in the Rub' al Khali sand desert. Although these lake beds do not form substantial lithostratigraphic sequences, this disadvantage has been largely overcome by extensive use of radiocarbon dating (some 56 dates are listed in McClure [1984]). By far the most important site is at Mundafan on the southwestern side of the erg adjacent to the Tuwaiq scarp. This basin is now completely dry, but it was occupied by a lake during two main periods within the range of radiocarbon dating (Fig. 9.11). The more recent period was in the early Holocene, with 13 ^{14}C dates on shell or marl between 9080 and 7040 yr B.P. It is possible that the lake did not in fact dry out until some time after 7000 yr B.P., for incipient soil development on the basin flanks is dated to around 6100 yr B.P.

The Holocene lake had a maximum water depth of just over 10 m and covered about 5-10 km².

Much larger and deeper than the Holocene lake was the late-Pleistocene lake, dated as older than 20,000 yr. In addition to freshwater gastropods, ostracods, and *Chara,* brackish-water foraminifera are found in some late-Pleistocene lake beds, indicating that water level and salinity fluctuated periodically. Between these two lacustrine intervals Mundafan was largely dry, and eolian activity may have been more intense than the limited sand movement observed at present. Figure 9.12 shows a water-level curve for Mundafan based on data in McClure (1984).

Similar although less detailed sequences are found elsewhere in Rub' al Khali. Many of these sites have been dated by McClure (1984) or Whitney (1983): six to the early Holocene, one to 12,315 yr B.P., and 10 as older than 19,000 yr B.P. All lakes were essentially of playa type and of relatively short duration. The early-Holocene dates cover a slightly wider time range than at Mundafan; the youngest date for this lacustrine phase is 5890 yr B.P.

In the northern deserts of the Arabian peninsula is another series of dated lake beds. The best-known lie in the al-Nafud dunefield, from whose center at Jubbah Garrard *et al.* (1981) described a sedimentary sequence comparable to that at Mundafan. The earlier lacustrine phase is represented by marl and diatomite dated to between 27,570 and 24,340 yr B.P. The Holocene phase has dates of 6685 and 5280 yr B.P.; the former dates an organic (marsh?) deposit between eolian sand layers. Although the humid climate responsible for creating these Holocene lakes clearly reached as far north as Jubbah (28°N), it does not appear to have created fully lacustrine conditions at this site. Thus the northern part of Arabia may have been somewhat less moist than the southern part during the early and middle Holocene. No mid-Holocene lake beds are recorded in the arid zone of southwest Asia anywhere north of Jubbah. No quantitative estimates of annual precipitation have yet been reported for Arabia, but Whitney *et al.* (1983) suggested a figure of about 250-300 mm, sufficiently higher than the 50-100 mm of today to turn the climate from arid to semiarid.

Maps for 9000 and 6000 Yr B.P.

POLLEN VALUES

Almost all eastern Mediterranean pollen cores come from shallow lakes or marshes. This means that pollen of local origin often forms a large part of the total pollen. Because the species involved are not easily distinguished palynologically from their counterparts in the dryland flora, nonarboreal pollen (NAP) can be difficult to disaggregate into its regional and local components. This applies especially to the Compositae, Gramineae, Chenopodiaceae, and Cyperaceae.

These taxa formed important parts of the dryland flora, as shown by studies of modern pollen rain (e.g., Wright *et al.,* 1967; Bottema and Barkoudah, 1979) and by pollen diagrams from deep-water sites, such as Lake Van (van Zeist and Woldring, 1978). In addition, Compositae are overrepresented in pollen spectra from archaeological sites (Bottema, 1975). In general, we followed designations by the original investigators in this analysis, but in some cases, such as the Rajasthan lakes (Singh *et al.,* 1974), where no distinction was made between local and regional pollen components, it was difficult to identify securely the regional vegetation component.

We also attempted to standardize the extent of present-day human impact on modern pollen spectra. We felt this was justified not least because most pollen cores were taken in recently drained lakes or swamps, where the uppermost sediment layers had been disturbed or even removed. Where there was clear palynologic evidence of 20th-century vegetation disturbance, a pollen spectrum below the topmost one was used as the modern reference. This does not mean, of course, that earlier human modification of natural vegetation is not recorded in pollen diagrams from the eastern Mediterranean; indeed, agricultural clearance is represented in Near Eastern pollen diagrams as early as 3500 yr B.P.

Figure 9.13 shows the proportions of AP versus NAP for 9000, 6000, and 0 yr B.P. and also indicates the importance of eu-Mediterranean elements in the flora. The trees included in this latter group are olive and evergreen oak (*Quercus calliprinos, Q. ilex, Q. coccifera*). Unfortunately, some investigators do not distinguish evergreen from deciduous oak in their pollen diagrams, although most do indicate qualitatively which form is more important. For this reason, the proportion of eu-Mediterranean trees at some sites is only an estimate.

The differences between AP values in the early and middle Holocene and those of recent times are shown in Figure 9.14. AP values at 6000 yr B.P. differed the most from modern values at Sagarejo in the Georgian Republic and Aghia Galini on Crete; at both sites AP percentages were higher than at present (Fig. 9.14, top). On the other hand, the human disturbance evident in both diagrams may have artificially exaggerated the magnitude of the climatic change. In contrast to these two localities, sites in the highland zone of Anatolia and western Iran show consistently lower AP values at 6000 yr B.P. than today.

The map for 9000 yr B.P. (Fig. 9.14, bottom) includes many of the same features as that for 6000 yr B.P. However, the dominant feature now is clearly the Anatolian-Iranian belt of lower AP, which also extended westward into northern Greece and northeastward into parts of Caucasia. Other Turkish sites not shown on this map (e.g., Lake Van) also show a very marked reduction in tree pollen at this time. Sites

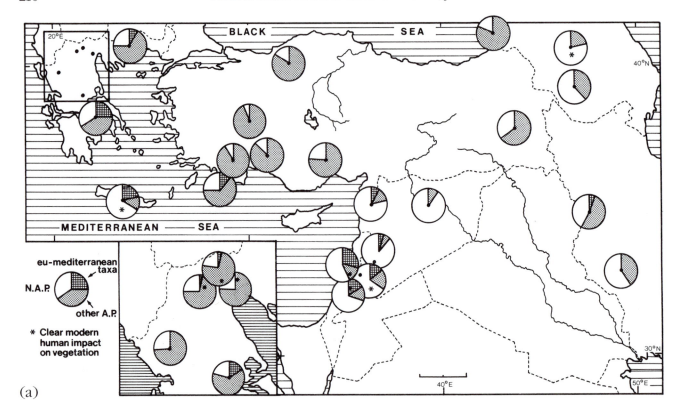

(a)

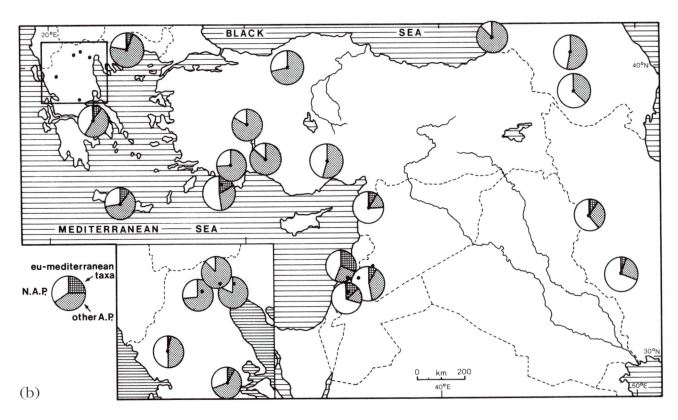

(b)

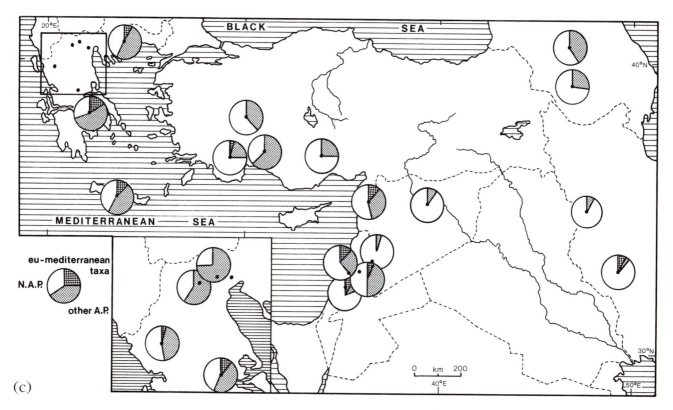

Fig. 9.13. Percentages of arboreal pollen (AP), nonarboreal pollen (NAP), and eu-Mediterranean pollen (olive and evergreen oak) at eastern Mediterranean sites at 0 yr B.P. (modern) (a), 6000 yr B.P. (b), and 9000 yr B.P. (c).

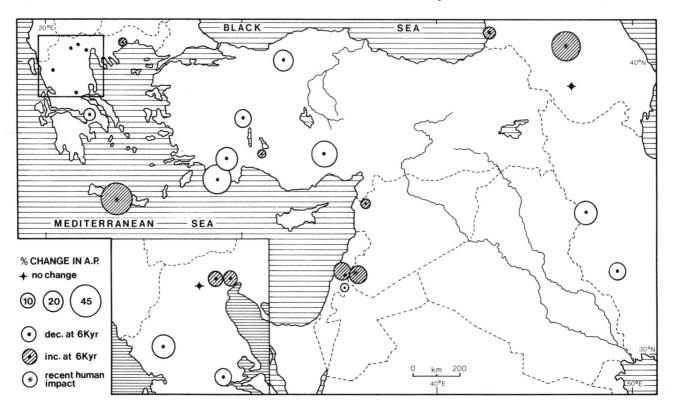

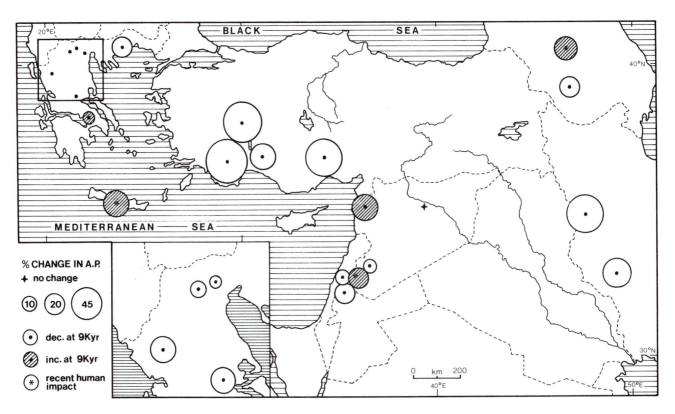

Fig. 9.14. Differences in the percentage of arboreal plus eu-Mediterranean pollen at eastern Mediterranean sites at 6000 yr B.P. (top) and 9000 yr B.P. (bottom) compared to the present.

in the Levant show no consistent regional pattern of pollen differences at either 9000 or 6000 yr B.P. compared to the present. AP values from Ghab (northwest Syria) were significantly higher 9000 yr ago than during the later Holocene, but a similar shift is not identifiable elsewhere. This inconsistency may partly reflect the rather poor quality of most of the paleoecological data currently available for most Levantine sites.

As might be anticipated from the foregoing discussion, the pollen maps for 6000 and 9000 yr B.P., although similar, are by no means identical. A map of the differences in AP percentages between 9000 and 6000 yr B.P. (Fig. 9.15) confirms that some significant changes in vegetation took place within the early Holocene. Most notable is the general increase in AP values between 9000 and 6000 yr B.P., a trend that affected all parts of the eastern Mediterranean region except perhaps the Levant and southern Greece.

The use of the AP-NAP ratio alone of course simplifies what was in reality a complex sequence of vegetation change. This oversimplification is especially significant for areas away from the forest-steppe boundary, such as the arid zone. The demarcation of modern steppe and desert vegetation zones itself presents considerable problems, and the proportion of AP is rarely a significant consideration in such an exercise. AP values can in fact be higher in the desert than in the steppe, because low levels of pollen pro-

duction in the former increase the relative importance of types (such as pine) preferentially transported over long distances. Pollen spectra from south and southwest Asian sites outside the area shown in Figures 9.13-9.15 nonetheless record some important vegetational changes.

In pollen diagrams from Rajasthan the modern spectra include several specifically desert forms, such as *Calligonum*, whereas those dated to 6000 and 9000 yr B.P. have a greater proportion of typically steppe elements such as *Artemisia* (Singh *et al.,* 1974). The sand-dune flora of the present Rajasthan desert seems to have been grass and *Artemisia*-dominated savanna during the early Holocene. Similarly, in the Rub' al Khali sand sea of southern Arabia, pollen from early-Holocene lake beds is dominated by Chenopodiaceae along with grasses and *Plantago* rather than by *Cyperus* and *Calligonum* as in modern pollen samples. This difference also suggests savanna rather than a hyperarid xerophytic flora.

LAKE LEVELS

The maps of lake levels for 6000 and 9000 yr B.P. (Fig. 9.16) show a clear contrast between the northwestern and southern-eastern sectors of southwest Asia. Lakes in the northwestern sector were at low or intermediate water levels at 9000 and 6000 yr B.P. Most of these lakes show no change in status compared to

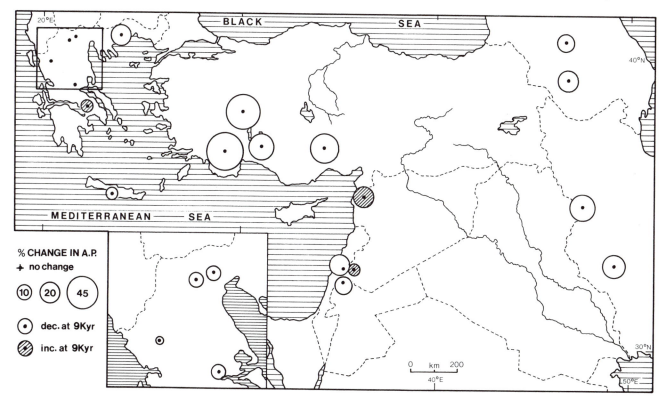

Fig. 9.15. Differences in the percentage of arboreal plus eu-Mediterranean pollen at eastern Mediterranean sites at 9000 yr B.P. compared to 6000 yr B.P.

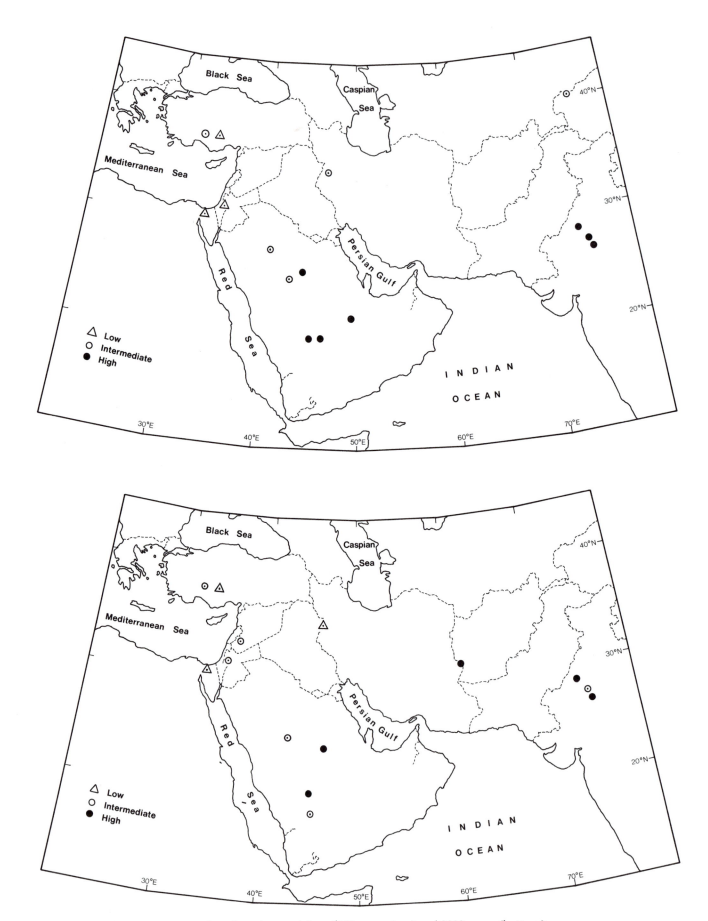

Fig. 9.16. Lake-level status in south and southwest Asia at 6000 yr B.P. (top) and 9000 yr B.P. (bottom).

the present, although what differences there are tend to indicate drier conditions during the early Holocene. This is most clearly marked at Lake Zeribar, which was low during the period up to 9000 yr ago and intermediate around 6000 yr B.P., whereas today its water level is relatively high. The southern basin of the Dead Sea also dried up during the early Holocene. Other lake sites such as Konya are dry at present, so a negative change in water balance would have no discernible limnological effect.

In contrast to the northwestern sector, lakes in the southern and eastern parts of southwest Asia were at high or intermediate water levels during the early Holocene. The northernmost area in Arabia to experience wetter conditions was al-Nafud desert at about 28°N (Garrard et al., 1981; Whitney et al., 1983). There is no evidence for higher Holocene lake levels in the southern Levant, so the boundary between the northern and southern zones may be placed at around 28–30°N. In the eastern part of the region, however, early-Holocene lakes are found at somewhat more northerly latitudes, probably extending into Afghanistan (31°N) and possibly into central Asia (Chatyrkel, 40°N).

No differences are clearly discernible between the patterns of lake levels at 9000 and 6000 yr B.P. in southwest Asia. Detailed examination of two of the best limnological records from the region, Didwana and Mundafan, reveals that whereas the water level at Didwana was higher 6000 than 9000 yr ago, the reverse was probably true for Mundafan. On the other hand, it is noteworthy that when viewed on a late-Quaternary time scale, the majority of lake sites show water levels and hence moisture levels only one and not two ranks above those of today. In northwest India this reflects the greater aridity during the late Pleistocene than at present, while in Arabia it is explained by higher lake levels before 20,000 yr B.P. than during the early Holocene.

Conclusions

Pollen diagrams from the eastern Mediterranean region show a diachronous pattern of early-Holocene vegetation change, with woodland returning later at interior than at coastal sites. Possible reasons for this pattern include the impact of Neolithic populations and the dispersal time required for trees migrating into new areas. Although Neolithic agriculture began before 9000 yr B.P. in the Near East, this explanation is unconvincing both in Anatolia (Roberts, 1982b) and in western Iran (Wright, 1980), the two areas that experienced the greatest time lag in forest readvance. Nor can dispersal times realistically account for the long delay in interior areas such as the Zagros Mountains. Early-Holocene climatic aridity in this area thus seems highly probable.

Holocene lake-level fluctuations in southwest Asia are, if anything, even more directly attributable to changes in water balance and hence climate. Records of water-level fluctuation are also rather widely distributed across the eastern Mediterranean, Arabia, and northwest India, and they indicate an early-Holocene belt of high water levels across the southern half of the region. This belt was not restricted to the Indo-Arabian sector but extended across the Red Sea into Egypt, Sudan, and Ethiopia, thence continuing westward across the full width of the Sahara (Street-Perrott and Perrott, this vol.). Its eastern limit at present is much less well defined. Lakes in southwest Asia north of 30°N and west of 50°E do not indicate high early-Holocene levels, but the evidence is ambiguous as to whether conditions were drier than or similar to those of today.

It is evident that in southwest Asia palynologic and lake-level records can and should be used in conjunction to reconstruct late-Quaternary paleoclimates. One advantage of such an integrated approach is that where one type of data is relatively sparse (e.g., pollen sites in the arid zone), the other may be more abundant. Furthermore, when both pollen and lake data are available and point to the same conclusion, the inferred climatic change is better founded than when it is based on only one line of evidence.

Integrated paleoclimatic maps based on both types of data are shown in Figure 9.17: they indicate differences in moisture between 6000 yr B.P. and today (Fig. 9.17, top) and between 9000 yr B.P. and today (Fig. 9.17, bottom). There is good agreement between pollen and lake data in Anatolia-western Iran and in the Indo-Arabian sector, although the direction of moisture change is opposite in these two areas. Wetter conditions than at present are clearly indicated for the Indo-Arabian sector for both 9000 and 6000 yr B.P. By contrast, in Anatolia and Iran lake levels and more importantly pollen records indicate less precipitation than at present, especially at 9000 yr B.P. (see Fig. 9.18), when drier conditions also affected northern Greece and possibly parts of Caucasia and central Asia.

Between these two sectors with rather clear patterns of moisture change is a zone where no clear trends are apparent (Fig. 9.17). This is most notable in the eastern Mediterranean area, including southern Greece and the Levant. The lack of convergent evidence here may be partly a product of inadequacies of the data, but it may also be related to the different responses of individual lake water balances and to an increase in summer rainfall to the south and a decrease in winter precipitation to the north.

In this regard it may be significant that xerophytic Mediterranean elements are less important in early-Holocene than in modern spectra of almost all pollen diagrams from Greece and the Levant. At Aghia Galini on Crete, for instance, the high proportion of deciduous oak during the early Holocene, on an island that

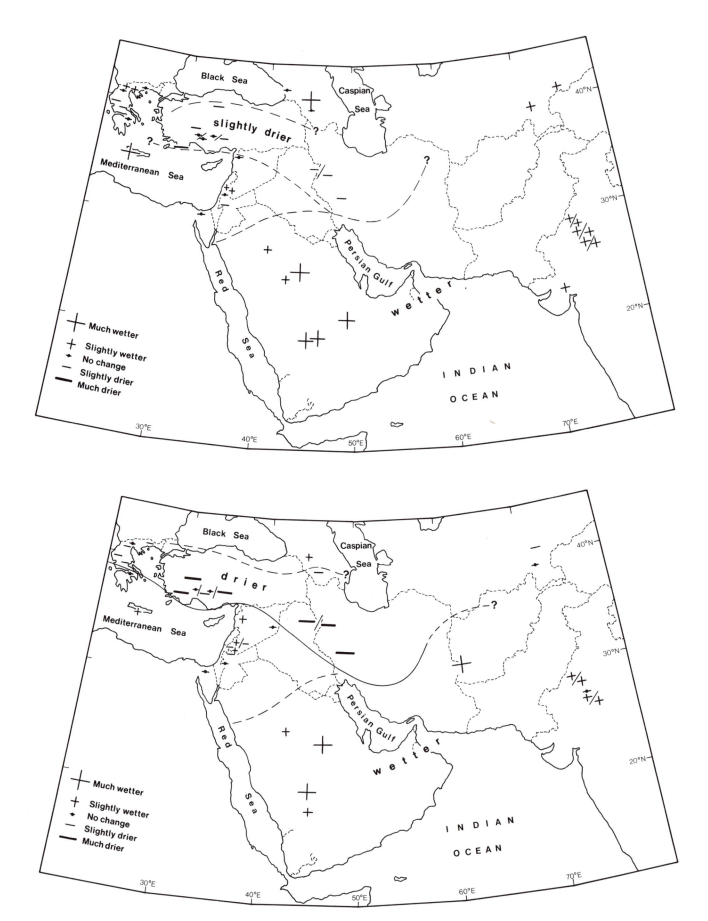

Fig. 9.17. Moisture differences at 6000 yr B.P. (top) and 9000 yr B.P. (bottom) relative to the present (all data).

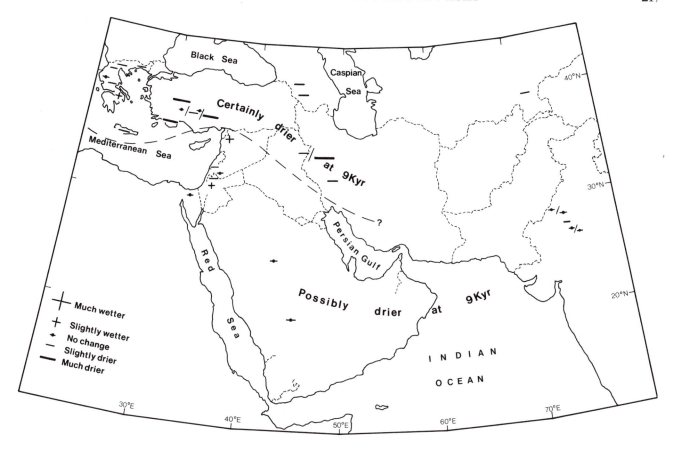

Fig. 9.18. Moisture differences between 9000 and 6000 yr B.P. (all data).

now contains an archetypically Mediterranean flora, points not only to greater precipitation (Rackham, 1982, p. 194) but perhaps more specifically to greater summer rainfall. The summer rains that appear to have reached Siwa in northwestern Egypt (Hassan, 1978; Street-Perrott and Perrott, this vol.) may therefore have extended partway across the Mediterranean Sea. Southern Crete, after all, is nearer to the North African coast than it is to Athens!

The paleoclimatic reconstructions presented in this chapter are supported by numerical modeling of the general circulation of the atmosphere with changed boundary conditions (COHMAP Members, 1988; Kutzbach, Guetter, *et al.,* this vol.). At 18,000 yr B.P. ice sheets were at their maxima, and sea levels and sea-surface temperatures were lower than at present (CLIMAP Project Members, 1976). Model experiments with these changed boundary conditions show for January in the Near East temperatures at least 10°C lower than at present, an enhanced Siberian high-pressure area, a strengthening and southward displacement of the westerly jet stream and associated storm tracks, and a decrease in precipitation (Kutzbach, Guetter, *et al.,* this vol.). For July, temperatures were 1-2°C lower than today, the subtropical high-pressure area was reduced, and storm tracks were displaced far

enough south so that the Mediterranean region experienced some precipitation. The associated cloudiness in the summer may have been especially important in reducing evaporation. Thus the apparent contradiction between aridity implied by the treeless vegetation and increased effective moisture implied by high lake levels may be explained primarily by the change in seasonality. For lake water balances the summer conditions of slightly lower temperatures, some precipitation, and increased cloudiness would have been critical. For vegetation the winter conditions of extreme cold and precipitation mainly in the form of snow would have been the dominant influences preventing tree growth. This climatic scenario would also reduce summer ablation of alpine glaciers and help to account for the deep depression of the snowline.

For the early Holocene too, numerical models confirm paleoenvironmental data in showing climatic conditions significantly different from those at present. According to the simulation experiments (Kutzbach, Guetter, *et al.,* this vol.), July temperatures over southwest Asia at 9000 yr B.P. were 2-3°C higher than at present as a result of the enhanced summer solar radiation in the Northern Hemisphere. The precessional cycle that caused this also caused winter temperatures over the Eurasian landmass to fall about 1.5°C below modern values.

The enhanced thermal contrast between seasons had a strong influence on early-Holocene moisture conditions across the region. Greater summer heating of the Asian landmass created an enhanced surface-pressure gradient, which pulled in moist maritime air from the Indian Ocean more powerfully than its modern monsoonal equivalent. It was consequently the southern-eastern sector of the Near East that benefited from the wider geographic extent of the early-Holocene summer monsoon. Paleoclimatic data from this sector, in the form of higher lake levels, stabilized sand dunes, and a shift from desert or semidesert scrub to savanna grassland vegetation, are in excellent agreement with model results. The calculated increase in annual precipitation of 250 mm for the monsoonal sector over land areas at 9000 yr B.P. is close to the paleoprecipitation estimates of Swain *et al.* (1983) for Rajasthan, northwest India (compare Table 9.3).

The general circulation model experiments (Kutzbach, Guetter, *et al.,* this vol.) show that an enhanced South Asian monsoon would also have strengthened northeasterly airflow across the northwestern sector of the Near East. This increased northeasterly airflow may have prevented monsoonal rains from penetrating as far north in Arabia and the Levant as they did, for instance, in South Asia and northwest Africa (Kutzbach, 1983; Street-Perrott and Roberts, 1983). It does not, however, explain why parts of the eastern Mediterranean region were drier than at present during the early Holocene, according to palynologic data. The eastern Mediterranean sector currently receives precipitation during the winter half of the year from depressions of westerly origin, and one explanation for the relative aridity would be a weakening of this moisture source during the early Holocene. The precise meteorological circumstances responsible for this weakening are uncertain, but one relevant component simulated by the model is a strengthened winter outflow of cold air from the Eurasian landmass. The pressure gradient created by lower winter temperatures may have acted to block the eastward movement of depressions from the Mediterranean Sea into interior areas such as Anatolia and the Zagros Mountains. This explanation is consistent with the paleoprecipitation gradient indicated in pollen data between coastal and interior areas at both 6000 and 9000 yr B.P. Alternatively, the early-Holocene dryness in the eastern Mediterranean sector may have been not absolute but relative. The higher summer temperatures led also to higher evapotranspiration, so that effective annual precipitation (P-E) was less than at present. Whatever the true cause, both model and paleoclimatic data agree that southwest Asia and the Near East straddled an important climatic hinge line during the early Holocene.

Acknowledgments

We thank Alayne Street-Perrott, Sandy Harrison, Harold McClure, and Sytze Bottema for information and comments on this chapter. Preparation of the manuscript was funded in part by the National Science Foundation. Diagrams were drafted by Anne Tarver and Rebne Karchefsky, and Val Pheby and Judith Jones assisted in typing the manuscript.

References

Agrawal, D. P., Gupta, S. K., and Kusumgar, S. (1971). Tata Institute radiocarbon date list VIII. *Radiocarbon* 13, 84-93.

Bar-Josef, O., and Phillips, J. L. (1977). "Prehistoric Investigations in Gebel Maghara, Northern Sinai." Monographs of the Institute of Archaeology, Hebrew University, Jerusalem.

Begin, Z. B., Ehrlich, A., and Nathan, Y. (1974). Lake Lisan, the Pleistocene precursor of the Dead Sea. *Geological Survey of Israel Bulletin* 63, 1-30.

Beug, H.-J. (1967). Contributions to the postglacial vegetational history of northern Turkey. *In* "Quaternary Paleoecology" (E. J. Cushing and H. E. Wright, Jr., Eds.), pp. 349-356. Yale University Press, New Haven, Conn.

Bottema, S. (1974). "Late Quaternary Vegetation History of Northwestern Greece." Unpublished Ph.D. dissertation, University of Groningen.

——. (1975). The interpretation of pollen spectra from prehistoric settlements (with special attention to Liguliflorae). *Palaeohistoria* 17, 17-35.

——. (1979). Pollen analytical investigations in Thessaly (Greece). *Palaeohistoria* 21, 19-40.

——. (1980). Palynological investigations on Crete. *Review of Palaeobotany and Palynology* 31, 193-217.

——. (1986). A late Quaternary pollen diagram from Lake Urmia (northwestern Iran). *Review of Palaeobotany and Palynology* 47, 241-261.

——. (1987). Chronology and climatic phases in the Near East from 16,000 to 10,000 B.P. *In* "Chronologies in the Near East" (O. Aurenche, J. Evin, and P. Hours, Eds.), pp. 295-310. B.A.R. International Series S379. British Archaeological Reports, Oxford.

Bottema, S., and Barkoudah, Y. (1979). Modern pollen precipitation in Syria and Lebanon and its relation to vegetation. *Pollen et spores* 21, 427-480.

Bottema, S., and van Zeist, W. (1981). Palynological evidence for the climatic history of the Near East, 50,000-6000 B.P. *In* "Préhistoire du Levant" (J. Cauvin and P. Sanlaville, Eds.), pp. 111-132. Actes du colloque international 598. Centre national de la recherche scientifique, Paris.

Bottema, S., and Woldring, H. (1984). Late Quaternary vegetation and climate of southwestern Turkey II. *Palaeohistoria* 26, 123-149.

Butzer, K. W. (1958). Quaternary stratigraphy and climate in the Near East. *Bonner Geographische Abhandlungen* 24, 1-157.

CLIMAP Project Members (1976). The surface of the ice-age earth. *Science* 191, 1138-1144.

COHMAP Members (1988). Climatic changes of the last 18,000 years: Observations and model simulations. *Science* 241, 1043-1052.

Cowgill, U. M. (1969). The waters of Merom: A study of Lake Huleh. *Archiv für Hydrobiologie* 66, 249-271.

Degens, E. T., and Kurtman, F., Eds. (1978). "The Geology of Lake Van." Institute Publication 169. Maden Tetkik ve Arama (M.T.A.), Ankara, Turkey.

Erol, O. (1978). The Quaternary history of the lake basins of central and southern Turkey. In "The Environmental History of the Near and Middle East since the Last Ice Age" (W. C. Brice, Ed.), pp. 111-139. Academic Press, London and New York.

Farrand, W. R. (1981). Pluvial climates and frost action during the last glacial cycle in the eastern Mediterranean—Evidence from archaeological sites. In "Quaternary Palaeoclimates" (W. C. Mahaney, Ed.), pp. 393-410. GeoAbstracts, Norwich, England.

Freitag, H. (1977). The pleniglacial, late glacial and early postglacial vegetations of Zeribar and their present-day counterparts. Palaeohistoria 19, 87-95.

Garrard, A. N., Harvey, C. P. D., and Switzur, V. R. (1981). Environment and settlement during the Upper Pleistocene and Holocene at Jubbah in the Great Nafud, northern Arabia. Atlal 5, 137-148.

Gogichaishvili, L. K. (1982). Vegetational and climatic history of the western part of the Kura river basin (Georgia). In "Palaeoclimates, Palaeoenvironments and Human Communities in the Eastern Mediterranean Region in Later Prehistory" (J. L. Bintliff and W. van Zeist, Eds.), pp. 325-341. British Archaeological Reports International Series 133.

Greig, J. R. A., and Turner, J. (1974). Some pollen diagrams from Greece and their archaeological significance. Journal of Archaeological Science 1, 177-194.

Hassan, F. A. (1978). Archaeological explorations of the Siwa oasis region, Egypt. Current Anthropology 19, 146-148.

Horowitz, A. (1971). Climatic and vegetational developments in northeastern Israel during Upper Pleistocene-Holocene times. Pollen et spores 13, 255-278.

Huntley, B., and Birks, H. J. B. (1983). "An Atlas of Past and Present Pollen Maps for Europe: 0-13,000 Years Ago." Cambridge University Press, Cambridge.

Hutchinson, G. E., and Cowgill, U. M. (1963). Chemical examination of a core from Lake Zeribar. Science 140, 67-69.

Kaiser, K., Kempf, E. K., Leroi-Gourhan, A., and Schutt, H. (1973). Quartärstratigraphische Untersuchungen aus dem Damaskusbecken und seiner Umgebung. Zeitschrift für Geomorphologie 17, 263-353.

Kelts, K., and Shahrabi, M. (1986). Holocene sedimentology of hypersaline Lake Urmia, northwest Iran. Palaeogeography, Palaeoclimatology, Palaeoecology 54, 105-130.

Kempe, S., and Degens, E. T. (1978). Lake Van varve record: The last 10,420 years. In "The Geology of Lake Van" (E. T. Degens and F. Kurtman, Eds.). Institute Publication 169. Maden Tetkik ve Arama (M.T.A.), Ankara, Turkey.

Krinsley, D. B. (1970). "A Geomorphological and Paleoclimatological Study of the Playas of Iran." U.S. Geological Survey, Washington, D.C.

Kuhle, M. (1976). Beitrage zur Quartärmorphologie SE-Iranische Hochgebirge. Die quartare vergletscherung des Kuh-i-Jupar. Göttinger Geographische Abhandlungen 67, 1-209.

Kutzbach, J. (1983). Monsoon rains of the late Pleistocene and early Holocene: Pattern, intensity and possible causes of changes. In "Variations in the Global Water Budget" (A. Street-Perrott, M. Beran, and R. Ratcliffe, Eds.), pp. 371-389. Reidel, Dordrecht.

Leroi-Gourhan, A. (1974). Etudes palynologiques des derniers 11,000 ans en Syrie semi-désertique. Paleorient 2, 443-451.

McClure, H. A. (1976). Radiocarbon chronology of late Quaternary lakes in the Arabian desert. Nature 263, 755-756.

——. (1978). Ar Rub' al Khali. In "Quaternary Period in Saudi Arabia," Vol. 1 (S. S. al-Sayari and J. G. Zotl, Eds.), pp. 252-263. Springer-Verlag, Vienna.

——. (1984). "Late Quaternary Palaeoenvironments of the Rub' al Khali." Unpublished Ph.D. thesis, London University.

Megard, R. O. (1967). Late Quaternary Cladocera of Lake Zeribar, western Iran. Ecology 48, 179-189.

Messerli, B. (1967). Die eiszeitliche und die gegenwartige Vergletscherung in Mittelmeerraum. Geographica Helvetica 22, 104-228.

El-Moslimany, A. P. (1983). "History of Vegetation and Climate of the Eastern Mediterranean and the Middle East from the Pleniglacial to the Mid-Holocene." Unpublished Ph.D. thesis, University of Washington, Seattle.

——. (1984). Comment on "Age, palaeoenvironments, and climatic significance of late Pleistocene Konya Lake, Turkey" by Neil Roberts. Quaternary Research 21, 115-116.

Neev, D., and Emery, K. O. (1967). The Dead Sea. Geological Survey of Israel Bulletin 41, 1-147.

Niklewski, J., and van Zeist, W. (1970). A late Quaternary pollen diagram from northwest Syria. Acta Botanica Neerlandica 19, 737-754.

Pias, J. (1972). Signification géologique, pedologique et palaeoclimatique des formations paleolacustres et deltaiques au Seistan (Afghanistan méridional). Comptes rendues de l'Académie des sciences, Paris, série D 279, 1143-1146.

Rackham, O. (1982). Land-use and the native vegetation of Greece. In "Archaeological Aspects of Woodland Ecology" (M. Bell and S. Limbrey, Eds.), pp. 177-198. B.A.R. International Series S146. British Archaeological Reports, Oxford.

Roberts, N. (1980). "Late Quaternary Geomorphology and Palaeoecology of the Konya Basin, Turkey." Unpublished Ph.D. thesis, London University.

——. (1982a). Lake levels as an indicator of Near Eastern palaeo-climates: A preliminary appraisal. In "Palaeoclimates, Palaeoenvironments and Human Communities in the Eastern Mediterranean Region in Later Prehistory" (J. L. Bintliff and W. van Zeist, Eds.), pp. 235-267. B.A.R. International Series S133. British Archaeological Reports, Oxford.

——. (1982b). Forest re-advance and the Anatolian Neolithic. In "Archaeological Aspects of Woodland Ecology" (M. Bell and S. Limbrey, Eds.), pp. 231-246. B.A.R. International Series S146. British Archaeological Reports, Oxford.

——. (1983). Age, palaeoenvironments, and climatic significance of late Pleistocene Konya Lake, Turkey. Quaternary Research 19, 154-171.

——. (1984). Reply to comments by Ann P. El-Moslimany. Quaternary Research 21, 117-120.

Roberts, N., Erol, O., de Meester, T., and Uerpmann, H.-P. (1979). Radiocarbon chronology of late Pleistocene Konya Lake, Turkey. Nature 281, 662-664.

Sajadjan, J. V. (1978). Armenien und die angrenzenden Gebiete in der Nacheiszeit. Zeitschrift für Archaologie 12, 15-37.

al-Sayari, S. S., and Zotl, J. G., Eds. (1978). "Quaternary Period in Saudi Arabia," Vol. 1. Springer-Verlag, Vienna.

Schweizer, G. (1975). Untersuchungen zur Physiogeographie von Ostanatolien und Nordwestiran. Tübingen Geographische Studien 60, 1-145.

Shnitnikov, A. V. (1976). Paleolimnology of Lake Chatyr-kel, Tien Shan. *In* "II International Symposium on Limnology (September 1976), Abstracts," p. 108.

Shnitnikov, A. V., Livja, A. A., Berdovskaya, G. N., and Sevastianov, D. V. (1978). Paleolimnology of Chatyrkel Lake (Tien Shan). *Polskie Archiwum Hydrobiologii* 25, 383-390.

Singh, G., Joshi, R. D., and Singh, A. B. (1972). Stratigraphic and radiocarbon evidence for the age and development of three salt lake deposits in Rajasthan, India. *Quaternary Research* 2, 496-505.

Singh, G., Chopra, S. K., and Singh, A. B. (1973). Pollen rain from the vegetation of north-west India. *New Phytologist* 72, 191-206.

Singh, G., Joshi, R. D., Chopra, S. K., and Singh, A. B. (1974). Late Quaternary history of vegetation and climate of the Rajasthan desert, India. *Philosophical Transactions of the Royal Society of London, Series B* 267, 467-501.

Smit, A., and Wijmstra, T. A. (1970). Application of transmission electron microscope analysis on the reconstruction of former vegetation. *Acta Botanica Neerlandica* 19, 867-876.

Street-Perrott, F. A., and Roberts, N. (1983). Fluctuations in closed basin lakes as an indicator of past atmospheric circulation patterns. *In* "Variations in the Global Water Budget" (A. Street-Perrott, M. Beran, and R. Ratcliffe, Eds.), pp. 331-345. Reidel, Dordrecht.

Swain, A. M., Kutzbach, J. E., and Hastenrath, S. (1983). Estimates of Holocene precipitation for Rajasthan, India, based on pollen and lake-level data. *Quaternary Research* 19, 1-17.

Turner, J., and Greig, J. R. A. (1975). Some Holocene pollen diagrams from Greece. *Review of Palaeobotany and Palynology* 20, 171-204.

van Zeist, W. (1967). Late Quaternary vegetation history of western Iran. *Reviews in Palaeobotany and Palynology* 2, 301-311.

van Zeist, W., and Bottema, S. (1977). Palynological investigations in western Iran. *Palaeohistoria* 19, 19-85.

——. (1982). Vegetational history of the eastern Mediterranean and the Near East during the last 20,000 years. *In* "Palaeoclimates, Palaeoenvironments and Human Communities in the Eastern Mediterranean Region in Later Prehistory" (J. L. Bintliff and W. van Zeist, Eds.), pp. 277-321. B.A.R. International Series S133. British Archaeological Reports, Oxford.

——. (1991). Late Quaternary vegetation of the Near East. Beihefte zum Tübinger Atlas des Vorderen Orients, Reihe A (Naturwissenschaften) Nr. 18. Reichert, Wiesbaden.

van Zeist, W., and Woldring, H. (1978). A postglacial pollen diagram from Lake Van in east Anatolia. *Review of Palaeobotany and Palynology* 26, 249-276.

——. (1980). Holocene vegetation and climate of northwestern Syria. *Palaeohistoria* 22, 111-125.

van Zeist, W., and Wright, H. E., Jr. (1963). Preliminary pollen studies at Lake Zeribar, Zagros Mountains, southwestern Iran. *Science* 140, 65-67.

van Zeist, W., Woldring, H., and Stapert, D. (1975). Late Quaternary vegetation and climate of southwestern Turkey. *Palaeohistoria* 17, 55-143.

Vogel, J. C. (1980). Accuracy of the radiocarbon time scale beyond 15,000 B.P. *Radiocarbon* 22, 210-218.

Wasson, R. J., Rajaguru, S. N., Misra, V. N., Agrawal, D. P., Dhir, R. P., Singhvi, A. K., and Kameswara, K. (1983). Geomorphology, late Quaternary stratigraphy and palaeoclimatology of the Thar dunefield. *Zeitschrift für Geomorphologie, Supplementband* 45, 117-151.

Wasylikowa, K. (1967). Late Quaternary plant macrofossils from Lake Zeribar, western Iran. *Review of Palaeobotany and Palynology* 2, 313-318.

Weinstein, M. (1976). The late Quaternary vegetation of the northern Golan. *Pollen et spores* 18, 553-562.

Whitney, J. W. (1983). "Erosional History and Surficial Geology of Western Saudi Arabia." Technical Record 04-1. U.S. Geological Survey, Washington, D.C.

Whitney, J. W., Faulkender, D. J., and Rubin, M. (1983). "The Environmental History and Present Conditions of the Northern Sand Seas of Saudi Arabia." Open File Report 83-749. U.S. Geological Survey, Washington, D.C.

Wigley, T. M. L., and Farmer, G. (1982). Climate of the eastern Mediterranean and Near East. *In* "Palaeoclimates, Palaeoenvironments and Human Communities in the Eastern Mediterranean Region in Later Prehistory" (J. L. Bintliff and W. van Zeist, Eds.), pp. 3-37. B.A.R. International Series S133. British Archaeological Reports, Oxford.

Wijmstra, T. A. (1969). Palynology of the first 30 m of a 120 m deep section in northern Greece. *Acta Botanica Neerlandica* 18, 511-527.

Wright, H. E., Jr. (1961). Pleistocene glaciations in Kurdi tan. *Eiszeitalter und Gegenwart* 12, 131-164.

——. (1976). Environmental setting for plant domestication in the Near East. *Science* 194, 385-389.

——. Climatic change and plant domestication in the Zagros Mountains. *Iran* 18, 145-148.

——. Palaeoecology, climatic change, and Aegean prehistory. *In* "Contributions to Aegean Prehistory: Studies in Honor of William A. McDonald" (N. C. Wilkie and W. D. E. Coulson, Eds.), pp. 183-195. Kendall/Hunt Publishing Co., Dubuque, Iowa.

Wright, H. E., Jr., McAndrews, J. H., and van Zeist, W. (1967). Modern pollen rain in western Iran, and its relation to plant geography and Quaternary vegetational history. *Journal of Ecology* 55, 415-443.

Zarins, J., Ibrahim, M., Potts, D., and Edens, C. (1979). Saudi Arabian archaeological reconnaissance 1978. *Atlal* 3, 9-42.

Zohary, M. (1973). "Geobotanical Foundations of the Middle East," 2 vols. Stuttgart and Amsterdam.

CHAPTER 10

The Late-Quaternary Vegetation and Climate of China

Marjorie G. Winkler and Pao K. Wang

Recent geologic, palynologic, and archaeologic studies in China provide much evidence for the paleoenvironmental changes that have taken place during the late Pleistocene and the Holocene. Many studies have been translated into English and published in synthesis volumes (Whyte, 1984a,b; Liu, 1985a,b; Walker, 1986) or have been summarized (Shih *et al.*, 1979; Verstappen, 1980; Wang, 1984a; Yang and Xie, 1984; Zhang Lansheng, 1984; Wang and Fan, 1987; Liu, 1988; Fang, 1991; Sun and Chen, 1991; and others). In this chapter we (1) review the biogeologic evidence from China in light of paleoecologic and paleoclimatologic findings from other regions of the world, (2) present maps of the paleoclimate and the paleovegetation of China at 3000-yr intervals between the last glaciation (the Dali glaciation) and early historic time in China, (3) compare the paleoclimate inferred from the data with that simulated by Kutzbach and Guetter (1984a,b; 1986) with the general circulation model of the National Center for Atmospheric Research (NCAR), and (4) discuss the relationship of the airstream climatology of China to the biotic regions at present and in the past.

Because the geography of China is not well known to many Western scientists, we first present a series of maps (Figs. 10.1–10.5) showing country boundaries, provinces and major cities, geographic divisions, mountain ranges and rivers, altitudinal zones, modern climate variables, and modern vegetation. These maps show the location of sites mentioned in the text and provide necessary background information for comparing paleoenvironments with present conditions.

The most abundant paleoenvironmental information comes from the eastern half of China, where research has been done on sea-level and vegetation changes in association with archaeologic findings (Fig. 10.1c). We also present results of recent research on loess and paleosol stratigraphy of central China (Liu,

1985a,b; Liu *et al.*, 1985), the glacial (Liu, 1985b) and paleolimnological (Li *et al.*, 1985; Wang and Fan, 1987; Fang, 1991) history of the Qinghai-Xizang Plateau and the Tian Shan and other regions of China, and vegetation and climate changes in Yunnan Province in southwestern China (Xu *et al.*, 1984; Walker, 1986), which is beginning to get wide circulation through translation. We also include studies from Japan in this chapter. Although climatic changes in China in historical times (since about 5000 yr B.P.) have been described in rich detail from the extensive observational records of weather (e.g., dust rain), phenological events (e.g., plum rains), and crop yields that were kept by successive dynasties (Chu, 1926, 1973; Hsieh, 1976; Wang, 1980; Wang *et al.*, 1980; Zhang, 1980; Zhang De'er, 1984; Sheng, 1987; Wang and Zhang, 1988, 1982; Zhang and Wang, 1989), the results from historical studies are not included here.

We focus on regional chronostratigraphic biologic and geomorphic data, including pollen, mollusks, lake levels, and paleosols, and use archaeologic data as a separate check. We compare the regional biogeologic and archaeologic evidence with the geography (Fig. 10.2), modern climate variables (Figs. 10.3 and 10.4), and modern vegetation of China (Fig. 10.5) to prepare maps representing the climate of China at 18,000, 12,000, 9000, 6000, and 3000 yr B.P. and the vegetation at 18,000, 12,000, 9000, and 6000 yr B.P.

Background

GEOGRAPHY, MODERN CLIMATE, AND MODERN VEGETATION

China, situated between latitudes 20 and 54°N and between longitudes 30 and 75°E (Fig. 10.1a), has climate

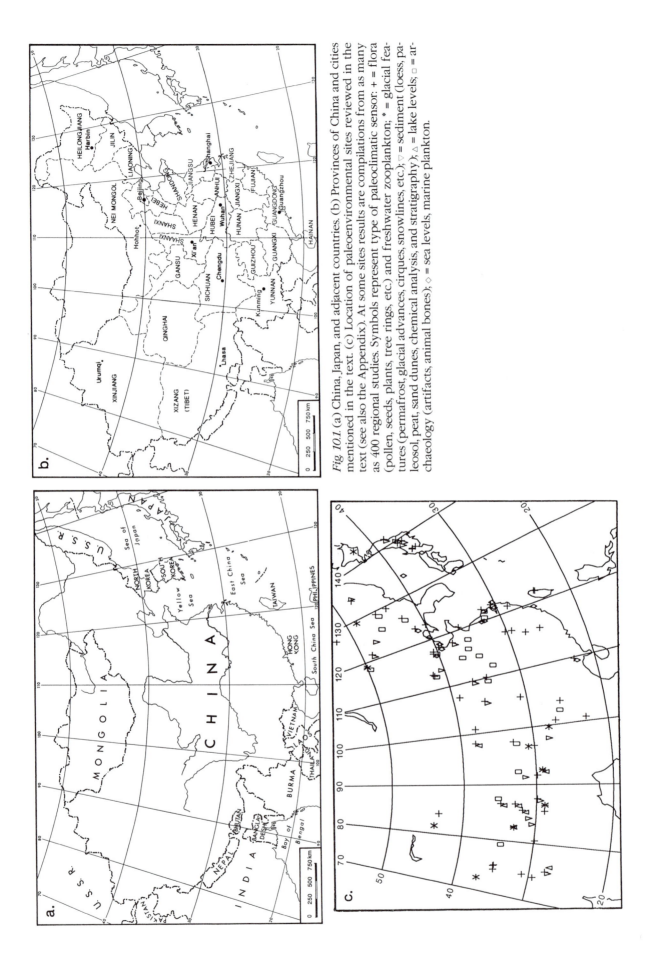

Fig 10.1. (a) China, Japan, and adjacent countries. (b) Provinces of China and cities mentioned in the text. (c) Location of paleoenvironmental sites reviewed in the text (see also the Appendix). At some sites results are compilations from as many as 400 regional studies. Symbols represent type of paleoclimatic sensor: + = flora (pollen, seeds, plants, tree rings, etc.) and freshwater zooplankton; * = glacial features (permafrost, glacial advances, cirques, snowlines, etc.); ▽ = sediment (loess, paleosol, peat, sand dunes, chemical analysis, and stratigraphy); △ = lake levels; □ = archaeology (artifacts, animal bones); ◇ = sea levels, marine plankton.

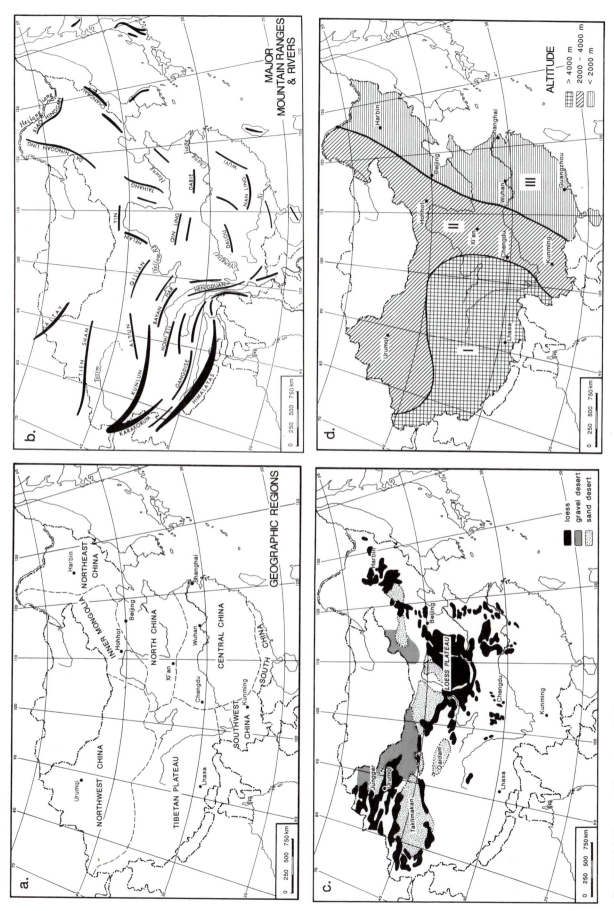

Fig 10.2 (a) Geographic regions of China. (b) Major mountain ranges and rivers. (c) Distribution of loess and desert (modified from Liu *et al.* [1985] and Zhang [1983]). (d) Altitudinal divisions: I = Tibetan (Qinghai-Xizang) Plateau and northwest Yunnan mountains; II = Xinjiang-Nei Mongol Plateau in the north, Loess Plateau in central China, and Yunnan-Guizhou Plateau in south-central China; III = coastal plain (modified from Liu and Ding [1984] and Teachers College of Northwest China [1984]). Except where otherwise noted, maps were modified from China Reconstructs Map (1981), National Geographic Society (1982), Teachers College of Northwest China (1984), and Ren *et al.* (1979, 1985).

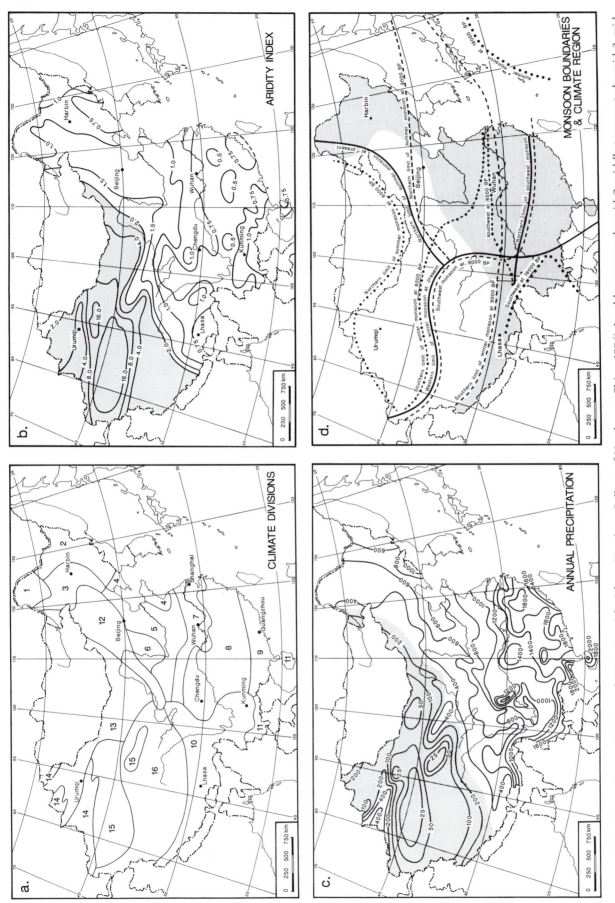

Fig 10.3 (a) Modern climate divisions of China (modified from Teachers College of Northwest China, 1984): 1, cold-temperate humid; 2, middle-temperate humid; 3, middle-temperate subhumid; 4, south-temperate subhumid; 5, south-temperate humid; 6, south-temperate subdry; 7, north-subtropical humid; 8, middle subtropical; 9, south-subtropical humid; 10, middle-subtropical subhumid and plateau (humid and subhumid); 11, tropical humid; 12, middle-temperate subdry; 13, middle-temperate dry; 14, middle-temperate alpine subdry; 15, south-temperate dry; 16, plateau (dry and subdry). (b) Aridity index (modified from Ren *et al.* [1985] and Teachers College of Northwest China [1984]). Areas with aridity values higher than 2 are stippled and include the desert zone (>4) and peripheral arid lands (between 2 and 4). (c) Modern mean annual precipitation (mm) (modified from Ren *et al.*, 1985). Areas that receive less than 300 mm are stippled and essentially coincide with areas where the aridity index is greater than 2. (d) Modern monsoon boundaries and modern monsoon climate region (stippled) (modified from Teachers College of Northwest China [1984] and Ren *et al.* [1985]) and paleomonsoon boundaries at 18,000, 9000, and 6000 yr B.P. as inferred from data.

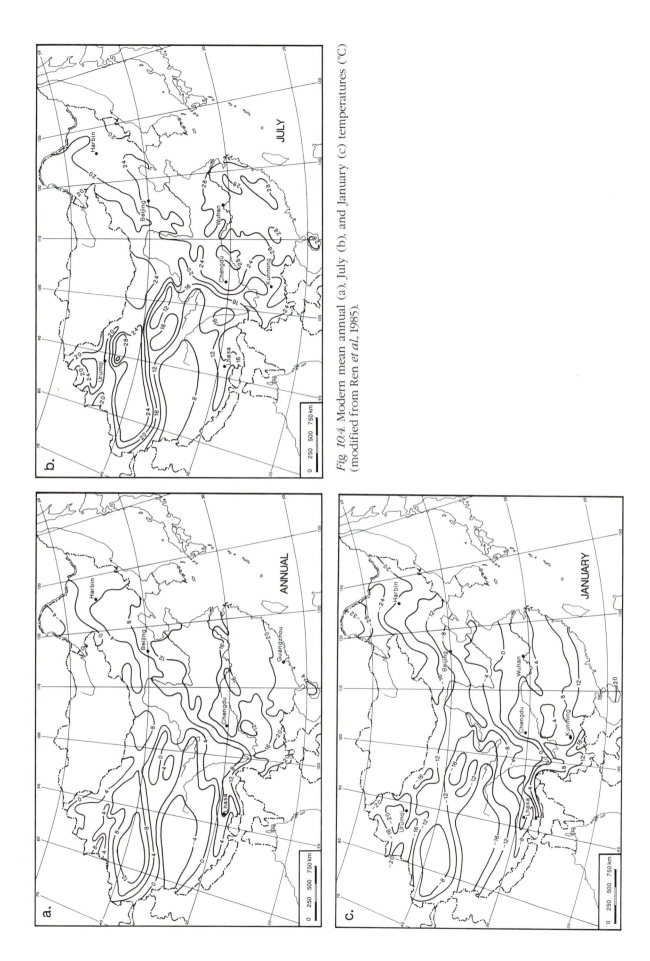

Fig. 10.4 Modern mean annual (a), July (b), and January (c) temperatures (°C) (modified from Ren et al, 1985).

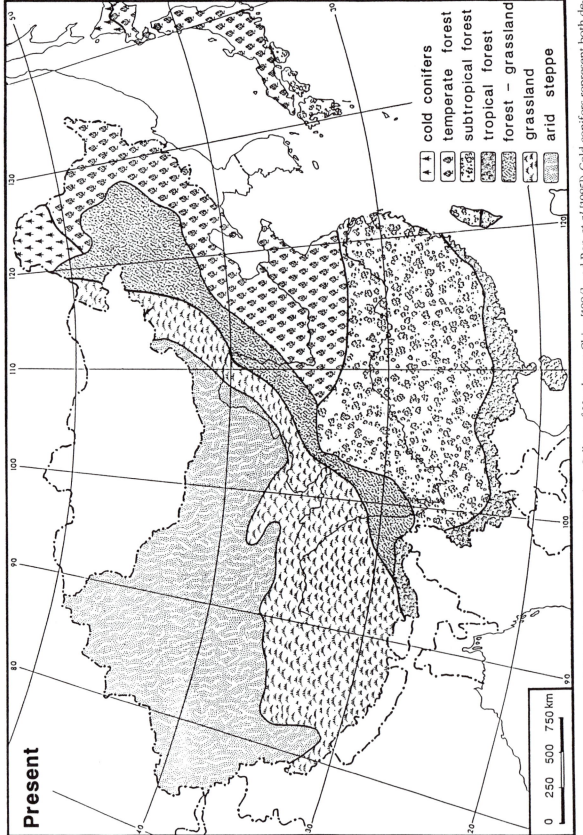

Present

Fig 105. Modern vegetation of China (data from Hou [1979], Teachers College of Northwest China [1984], and Ren et al. [1985]). Cold conifers represent both deciduous (larch) and evergreen (spruce, fir, and pine) coniferous forests. Temperate, subtropical, and tropical forests are general categories and include gradations caused by latitudinal and topographic temperature variations. Forest-grassland (savanna) represents the plant community structure and comprises open temperate woodlands in the north and open drier subtropical forests in south-central China as well as sparsely wooded slopes of the mountains on the Tibetan Plateau. Grassland and arid steppe also change character and species composition as latitudinal, altitudinal, and climatic boundaries are crossed.

regimes ranging from perennial snow on the high western mountains to deserts in the western lowlands, and from cold-temperate regions in the northeast to warm and humid tropics along the southern coast (Fig. 10.3a). The scattered islands that belong to China and extend south to 4°N are not considered in this chapter.

Although China had no late-Pleistocene continental ice sheet (Shi *et al.,* 1986), ice caps in the high western mountains advanced and retreated in response to climatic changes (Burbank and Kang, 1991), and contemporaneous edaphic and vegetational changes occurred throughout the country. The modern natural environment of China has been shaped by the fluctuations in climate during and since the Pleistocene and also by the tectonics of the western highlands (Lin and Wu, 1987; Ruddiman and Kutzbach, 1991). The South Asian continental plate is moving northward today in relation to the Tibetan landmass at an estimated rate of 5 cm/yr (Whyte, 1984a). Land movement had its greatest impact on the climate at the end of the Tertiary and during the Quaternary; the upward growth of the Himalayas on the southern rim of the Qinghai-Xizang (Tibetan) Plateau (Fig. 10.2a,b) increasingly blocked the southward flow of cold, dry continental air and the northward flow of warm, moist air from the South China Sea (Xu Ren, 1984). The rate of uplift of the Qinghai-Xizang Plateau has varied during the Quaternary but averaged about 1.9 mm/yr (Zheng and Shi, 1982). The uplift has accelerated since the late Pleistocene to about 10 mm/yr west of about 97°E (Xu Ren, 1981; Zheng and Shi, 1982) and to as much as 50 mm/yr at the eastern rim of the plateau in the northwest Yunnan mountains (Li, 1987). Today the northwestern and central part of the country north of 30°N latitude is arid (Fig. 10.3b,c), as exemplified by the wide distribution of sand and loess (Fig. 10.2c) (see Sheng, 1987), while east China (east of a diagonal line running from 50°N, 120°E to about 30°N, 100°E) has a monsoon climate (Fig. 10.3d) characterized by warm, rainy summers and cool, dry winters.

The topography of China can be described as a series of steps decreasing in altitude (Fig. 10.2d) from the Qinghai-Xizang Plateau in west-central China, with a mean altitude of 4000 m (highest elevation Qoomolangma Feng [Mount Everest], 8848 m), which makes up one-third of the country, to the Xinjiang-Nei Mongol, Loess, and Yunnan-Guizhou plateaus, descending to a mean altitude of 2000 m, and then a third descent eastward to a mean altitude of 200-500 m, extending down to the sea (Liu and Ding, 1984). These topographic variations divide China into three distinct climatic, edaphic, and vegetational regions (Fig. 10.5): the arid steppe of the northwest and north-central region, the grasslands of the east-central and southwestern part of the country, and the forests of the eastern coastal plain, southeast, and south (Hou, 1979; Xu Ren,

1984; Ren *et al.,* 1985). These regions are further divided by the topographic relief in the western portion of the country (Fig. 10.2b), where vegetation zonation is primarily altitudinal rather than latitudinal as in eastern China. The vegetation zones in eastern China include Arctic conifer forests in the extreme northeast and tropical rainforests in the southeast (Fig. 10.5). The boundaries between grassland, savanna (open forest), and forest in central China lie along a climatic tension zone. The movements of the Arctic, central Asian, Korean, and South Pacific airstreams (Bryson, 1986) all form a steep diagonal gradient in central China during some part of the year (discussed further in a later section).

INFERRING PALEOCLIMATE FROM FOSSIL EVIDENCE

Several researchers have calculated paleoclimatic variables such as mean annual temperature (MAT) and precipitation (MAP) based on fossil records. We cannot evaluate the calibration techniques for these calculations because they are not usually reported in the studies. However, we can compare the estimates of regional temperature and precipitation change inferred from the data with modern climate maps for China (Figs. 10.3 and 10.4).

A "heat zone" index (the number of degrees per year above 10°C) is currently used to differentiate climate zones (Fig. 10.3a) (Ren *et al.,* 1985). The natural regions of China can also be defined by an "aridity index" (Fig. 10.3b), an empirical measure of the ratio between potential evaporation and precipitation (Zhao and Xing, 1984). Today, the desert zone has an aridity index greater than 4, "peripheral arid lands" have aridity indices between 4 and 2 (this label may designate regions within a tension zone that with a small increase or decrease in effective moisture would become semiarid or desert, respectively), and semiarid lands have aridity values between 1.5 and 2 (Zhao and Xing, 1984). The "dryness scale" presented in some studies as part of a paleoclimatic interpretation from fossil data is probably comparable to the modern aridity index (Figs. 10.3b and 10.6a).

For Taiwan Tsukada (1967) proposed a lapse rate for calculating temperature change in altitudinal studies (0.5°C per 100 m adjusted by the "inferred range" of one or more plant species), estimated mean July and January paleotemperatures from "seasonal lapse rates" and modern weather data, and reported a thermal index of "total effective heat for plant growth" calculated by Kira (1949). The thermal index, $T = \sum_{i=0}^{12}(t\text{-}5)$, where t is the monthly mean temperature (°C) and i is the number of months when t exceeds 5°C, may have also been used by other investigators to infer past temperature from past vegetation. According to Tsukada (1967), this index can be used to define forest types by the number of "month-degrees C" received; that is, the timberline receives about 15 month-degrees C, boreal forest 15-55, cool-temperate forest 45–85, warm-temperate forest 85–180, subtropical forest 180–240, and tropical forest more than 240. Chu (1973) presented "climatic tolerance limits" for individual species of plants, while some investigators use the multispecies approach of Grichuk (1969).

DATING FOSSIL EVIDENCE

Accurate time control is essential for paleoclimatic reconstruction. In the work surveyed here, the dates of samples were determined by isotopic, thermoluminescence, paleomagnetic, lichenometric, and stratigraphic methods. Artifacts from archaeological sites were radiocarbon-dated and were also placed in a cultural context by artifact typology. Unfortunately, modern lake and bog sediments have not been systematically analyzed to provide pollen data from vegetation representative of modern climatic conditions. Modern vegetation-pollen-climate relations can be used to illuminate past conditions. However, intensive land use in China for the past 5000 yr may complicate the search for modern analogues of past vegetation associations.

Many of the pollen samples are individually dated, especially those associated with archaeological sites or loess-paleosol sequences. Although individually dated samples need no chronologic interpolation, the absence of contiguous and/or continuous information (e.g., core data) permits only "snapshot" interpretation in many cases. (Liu [1988] and Sun and Chen [1991] summarized pollen data from sediment cores from China with better dating control, and Fang [1991] summarized lake-level data.) Also some vegetation and climate estimates are based only on comparisons with data from other sites. However, when a large amount of data from many sources is synthesized, a rather coherent picture emerges of environmental change in China since the last glacial maximum.

Regional Paleoenvironmental Data

We discuss the paleoenvironmental data by geographic region (Ren *et al.*, 1979, 1985) (Fig. 10.2a): (1) east China, with subregions of northeast, north, central, south, and coastal China; (2) Inner Mongolia; (3) southwest China; (4) the Qinghai-Xizang (Tibetan) Plateau; (5) northwest China; (6) the Loess Plateau in north-central China (Fig. 10.2c); and (7) Japan (Fig. 10.1a). We give the province or region (Fig. 10.1b) and the latitude and longitude of each site at the first mention to help readers find the site on the maps (Figs. 10.1a,b and 10.2a). When an area rather than a point location is mentioned or when the information summarized represents more than a few locations, we give the latitude and longitude of the approximate center of the area. The latitude and longitude of the individual and composite sites (Fig. 10.1c) in each region are listed in the Appendix.

EAST CHINA

Northeast China

A cold climate with steppe vegetation in northeast China from about 21,000 to 11,500 yr B.P. is indicated by distribution of the bones of *Mammuthus primigenius* (woolly mammoth) and *Coelodonta antiquitatis* (woolly rhinoceros) at more than 200 sites in the region (Chow, 1978; Duan *et al.*, 1981; Qiu *et al.*, 1981). Bones found at Zhaoyuan in Heilongjiang Province (45°31'N, 125°09'E) were radiocarbon-dated at 21,200 ± 600 yr B.P. Permafrost features (buried involution layers and ice-wedge patterns) found at many sites in Heilongjiang Province and at sites even farther south provide evidence of continuous permafrost at latitudes as far south as 40°N during the late Pleistocene (Zhang Lansheng, 1984). Today the southern limit of continuous permafrost is at about latitude 51°N (Zhang Lansheng, 1984).

A human skull found in an involution layer near Jalainua (49°21'N, 117°35'E) was radiocarbon-dated at 11,460 ± 230 yr B.P. Pollen analysis from this site showed 13.4% arboreal pollen (AP), including 6% *Betula* (birch), 3% *Pinus sylvestris* var. *mongolica* (Mongolian Scotch pine), 2% *Abies* (fir), and a few grains each of several hardwoods (Kong and Du, 1981). *Artemisia* (wormwood) was the dominant taxon, contributing 56.7% of the total pollen, and 10.6% of the pollen came from nonarboreal plants other than *Artemisia*. Aquatic algae (primarily *Pediastrum*, *Mougeotia*, and *Zygnema* spp.) were also present. In contrast, an early-Holocene sample from the site had 10% AP, all *P. sylvestris* var. *mongolica*, 88% nonarboreal pollen composed solely of *Artemisia* (77%) and *Chenopodium* spp. (10%), and no hydrophytes. Because the AP represents a more mesophytic forest and because algae were present, we infer that the climate was more humid at about 12,000 yr B.P. than in the early Holocene. Kong and Du (1981) estimated that MAP was 500–600 mm around 11,000–12,000 yr B.P., compared to the present annual precipitation in this region of about 400 mm (Fig. 10.3c).

Cui and Xie (1985) also summarized the late-Pleistocene climate of northeast China as periglacial. They described two frigid intervals (35,000–26,000 yr B.P. and 23,000–12,000 yr B.P.) characterized by cold-adapted fauna dominated by woolly mammoth and woolly rhinoceros. The vegetation in this region at about 18,000 yr B.P. was dominated by tundra north of 45°N, open steppe to 42°N, and northern conifer forests to 39°N. Late-glacial temperatures inferred from the biogeologic evidence were 8–10°C lower than the modern MAT of 3°C at 45°N and about 11°C lower than the modern MAT of 6°C at 42°N (Fig. 10.4a). By 12,000 yr B.P. areas of mixed coniferous-deciduous forest had developed in the region as summer temperatures increased.

A group of researchers at the Guiyang Institute of Geochemistry (1977) analyzed pollen from sites in southern Liaoning Province (about 39°N, about 117°E). They interpreted the late-Pleistocene pollen assemblage as indicating that the climate was colder than today before 10,500 yr B.P., with MAT of 0°C and an aridity index value of less than 1. However, no late-Pleistocene pollen data are actually presented, and the reported paleotemperature and aridity index values for the late Pleistocene may be speculative or based on unpublished data (a problem in many of the older studies).

Other studies at Liaoning Province sites (Guiyang Institute of Geochemistry, 1977) provide a history of the vegetation and climate for three Holocene periods. The Pulandian period (Q_4^1) (10,300-8000 yr B.P.) was a *Betula* forest stage. *Betula* pollen was dominant (58-89%), along with 11-39% pollen from other trees, such as *Ulmus* (elm), *Quercus* (oak), and *Pinus*. However, nearer the coast *Artemisia* and Chenopodiaceae (chenopods) dominated the pollen profile. MAT was estimated to be about 6°C, with an aridity index of 1.50. Thus this period was colder and drier than the present: modern values for the region are MAT 8-10°C, annual precipitation 600-1000 mm, and aridity index 0.70-1.10.

The Dagushan period (Q_4^2) (8000-2500 yr B.P.) was a deciduous forest stage. *Quercus* and *Alnus* (alder) pollen were abundant, especially toward the northeast. Other deciduous species included *Tilia* (basswood), *Juglans* (walnut), *Carpinus* (hornbeam), and *Pterocarya* (wing nut). *Betula* pollen decreased abruptly at the beginning of this stage to less than 15%, while *Pinus* pollen steadily rose throughout the period from less than 20% to 52%. *Typha* (cattail), Cyperaceae (sedges), and Liliaceae (lilies) were present, indicating the development of wet meadows and bogs. *Artemisia,* Chenopodiaceae, Polygonaceae (smartweeds), Gramineae (grasses), and Compositae (sunflowers) grew along the seacoasts and slopes. The pollen assemblages indicate that early in this stage the climate was warm and humid, with MAT about 13°C (3-5°C warmer than at present) and an aridity index of less than 1. The climate remained warmer than at present (MAT about 12°C) in the later part of this period but became considerably drier after about 5000 yr B.P. (aridity index 1.50).

The Zhuanghe period (Q_4^3) (from about 2500 yr B.P. to the present) was characterized by a mixed coniferous-deciduous forest. AP fell as pollen from herbs rose. The climate was colder and drier, with MAT fluctuating within 1-2°C of the present value.

Figure 10.6a summarizes the AP changes, inferred temperature and dryness, and sea-level changes in southern Liaoning for the last 10,000 yr. The evidence suggests that the entire Dagushan period from 8000 to 2500 yr B.P. was warmer than at present and that the interval from about 8000 to 5000 yr B.P. was also wetter than at present.

Pollen studies in the Dunhua Basin area of Jilin Province (about 43°N, about 128°E) (Zhou *et al.,* 1984) showed changes during the Holocene similar to those described for Liaoning (Fig. 10.7). MAT was estimated to be 5°C higher than at present between 8000 and 2500 yr B.P. in Jilin Province.

North China

The geographic division of north China includes the Beijing lowlands as well as a large east-west extent of land in the middle of China (Fig. 10.2a). Archaeological activities in this region, known as "the cradle of Chinese culture," are extensive. Chen (1979) analyzed pollen, zooplankton, and animal fossil data at sites in the Beijing area of Hebei Province (39°55'N, 116°25'E) and divided the Holocene period into four time intervals: Q_4^1 (12,000-8000 yr B.P.), Q_4^2 (8000-6000 yr B.P.), Q_4^3 (6000-2000 yr B.P.), and Q_4^4 (2000 yr B.P.-present). All sites were ^{14}C-dated. A temperature 4-5°C lower than the present MAT of about 12°C was inferred for Q_4^1 because of the dominance of birch pollen. This abundance of birch pollen (see also Figs. 10.6a and 10.7) may have come from birch-shrub tundra similar to that growing concurrently in northeastern Russia, Beringia, and northern Alaska and the Yukon, according to pollen evidence cited in Colinvaux (1967), Hopkins *et al.* (1982), Anderson (1985), and Ritchie (1987).

As discussed in the preceding subsection on northeast China, Cui and Xie (1985) and Zhang Lansheng (1984) presented evidence for permafrost as far south as 40°N in the Beijing lowlands during the late Pleistocene (the earliest part of Q_4^1). Abundant ostracods in Q_4^1 sediments indicate the presence of numerous freshwater ponds in the Beijing lowlands, which may have developed as permafrost melted. The later Q_4^2 sediments contained *Abies* (fir) and *Picea* (spruce) pollen, indicative of a cold climate. The temperature for this interval was calculated as 7-8°C lower than at present. Pollen assemblages of Q_4^3 indicate conditions warmer than at present. The vegetation was mainly deciduous trees, but spores from ferns that now grow only south of the Changjiang (Yangtze River) (about 30°N) were also abundant. MAT was estimated to have been about 2-4°C higher than at present.

The combination of a cold period as late as 8000-6000 yr B.P. (Q_4^2) at Beijing (Chen, 1979) and peak warmth in southern Liaoning at about the same time (Guiyang Institute of Geochemistry, 1977) is climatologically unlikely. Although these areas are west and north, respectively, of Bo Hai (Bo Bay) and lie in different climatic regions today (divisions 5 and 4, respectively, in Fig. 10.3a), they were part of the same exposed coastal plain before about 6000 yr B.P., and broadly synchronous Holocene temperature changes would be expected. Moreover, if the early Q_4^1 birch pollen dominance is interpreted as shrub-tundra, the

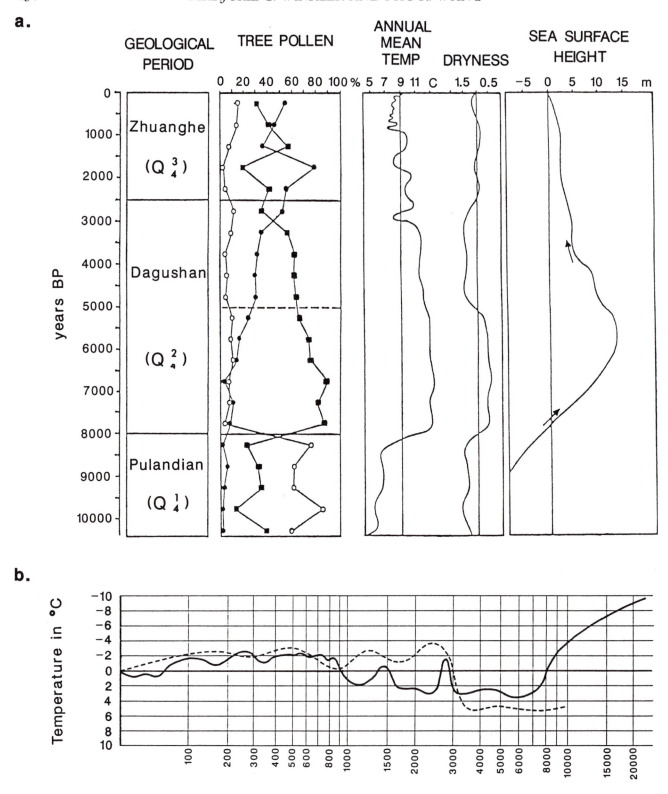

Fig. 10.6. (a) Summary of data for Liaoning Province (modified from Guiyang Institute of Geochemistry, 1977), showing major changes in arboreal pollen (○ = *Betula*, ■ = other deciduous, ● = *Pinus*), inferred mean annual temperature (°C), dryness scale (aridity index), and sea-level fluctuations. (b) Inferred mean annual temperature for China. The solid line is a composite temperature curve drawn from data in Chu (1973), Guiyang Institute of Geochemistry (1977), and Duan *et al.* (1980). The dashed line represents temperatures on the Tibetan Plateau (data from Wang and Fan, 1987).

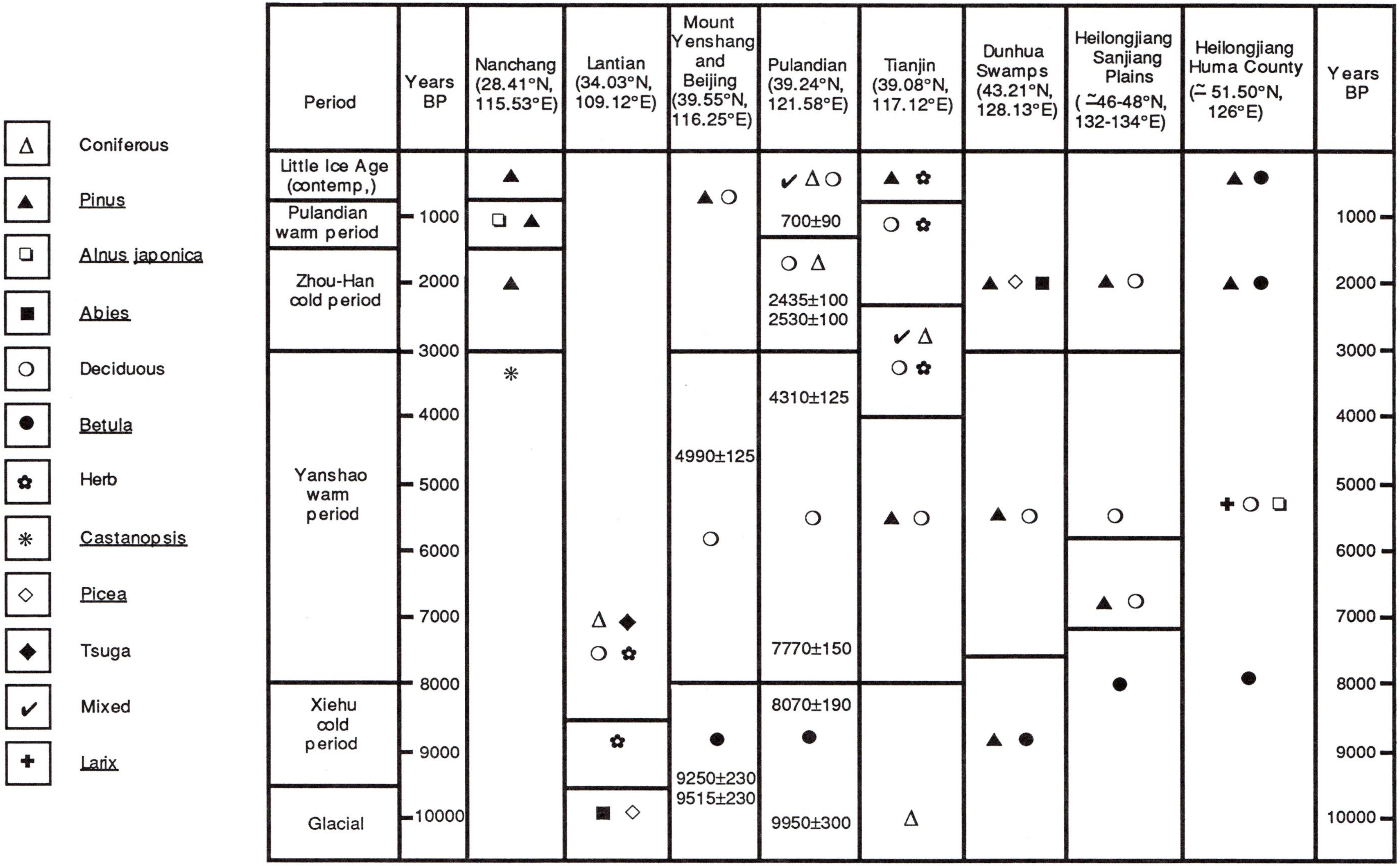

Fig. 10.7. Summary of pollen data from sites in eastern China. Translated and modified from Duan *et al.* (1981); stratigraphic data (columns from left to right) from Wang (1972); Institute of Botany (1966) and Lee and Huang (1966); Zhou (1965) and Liu *et al.* (1965); Chen *et al.* (1965); Hebei 7th Team (unpublished [1978]); Zhou *et al.* (1984); Changcun Institute of Geography (unpublished); and (Huabei) North China Institute of Geography (unpublished).

subsequent spruce-fir forest would be a natural transition from the cold late-glacial flora to the deciduous forest that grew during the warmer middle Holocene in north China. Thus it is likely that the radiocarbon chronology for this site is incorrect and that the spruce and fir forest dates from between 10,000 and 8000 yr B.P.—the Xiehu cold period (Fig. 10.7). In addition, evidence from pollen analyses from other sites in north China, including sites in Hebei Province (Liu *et al.*, 1965; Kong *et al.*, 1982; Wang *et al.*, 1982; Tang and Huang, 1985), and from sites summarized by Duan *et al.* (1981) (Fig. 10.7) and by Liu (1988) also suggests a cool early Holocene, warm middle Holocene, and cool late Holocene, as suggested by studies in southern Liaoning Province (Guiyang Institute of Geochemistry, 1977) and contrary to the results of Chen (1979) for the period 8000–6000 yr B.P. in Hebei Province. Because 6000 yr B.P. is the time of maximum transgression of the sea into the Beijing lowlands (Zhao, 1979; Chang and Wang, 1982; Zhao and Chin, 1982), the apparently anomalous pollen spectra found for this interval by Chen (1979) may be contaminated by pollen from sediment eroded by the water-level changes.

At sites in Lantian (34°03'N, 109°12'E) in Shaanxi Province, abundant spruce and fir pollen from regional forests and the presence of cold-water diatoms in lakes (Institute of Botany, 1966; Lee and Huang, 1966; Xu *et al.*, 1980) suggest a cool and wet late-Pleistocene climate in the western reaches of north China. However, dating control for the Lantian studies is poor.

Although the exact time of peak warmth has yet to be resolved, there is no doubt that the climate was warm and moist in the middle Holocene in north China. This interpretation is further supported by archaeological findings in the region. Ivory sculpture found at a Neolithic site occupied between 6000 and 7000 yr B.P. near Tengxian in Shandong Province (35°08'N, 117°10'E) (Chu, 1973; Shandong Group, 1980; Wang *et al.*, 1980) suggests that elephants inhabited more northerly latitudes in ancient China in the middle Holocene and that the climate was warmer and wetter than the modern climate. The discovery of elephant bones at Neolithic sites occupied during the middle Holocene in Henan Province (Jia and Zhang, 1977) provides further evidence for a warm and wet climate. A porcelain cup with a handle in the shape of a bamboo stem uncovered at another Neolithic site near Wangin (35°28'N, 115°46'E) (also radiocarbondated at about 6000 yr B.P.) suggests that bamboo grew in north China at 6000 yr B.P. (Shandong Group, 1979). In addition, carbonized bamboo, apparently used for fuel, was found in nearby Licheng (35°25'N, 119°25'E) (Chu, 1973). Today this area is too cold for bamboo (Chu, 1973). Bones of subtropical fauna, including *Hydropotes inermis* (water deer) and *Rhizomys sinensis* (bamboo rat), found at Neolithic sites from the Banpo culture (6800–5600 yr B.P.) in Banpo (34°15'N, 108°52'E) in north China also indicate a warm and wet climate.

In Henan Province, evidence of a mid-Holocene warming is even more abundant. Archaeological findings from the Yangshao cultural period (about 6500–4500 yr B.P.) have been summarized by Hu (1944), Chu (1973), and Huang (1984). The presence of rice and wild boar at this time indicates that the MAT was 2°C warmer than at present (Chu, 1973). Fossil fauna from a Neolithic site at Zichuan (33°02'N, 111°30'E) shows that giant pandas lived in the region (Jia and Zhang, 1977). Because pandas depend solely on bamboo for food, bamboo forests must have grown at least 3° north of their present latitudinal limit (Chu, 1973; Huang, 1984).

Several earlier Neolithic sites discovered in Hebei Province at Cishan (36°40'N, 114°12'E) and in Henan Province at Peiligang (34°25'N, 113°43'E) have ¹⁴C dates of 7500 and 9000–7000 yr B.P., respectively (An, 1979; Yen, 1979). The existence of well-developed cultures at these times is thought to indicate more favorable climatic conditions.

Wang (1984a) summarized radiocarbon-dated pollen data from northern and northeastern China and concluded that the Holocene climate in the region could be divided into three stages. The early Holocene (10,000 yr B.P. to 8000 or 7500 yr B.P.) was a period of increasing temperature, although the MAT remained at least 5–6°C lower than at present. The middle Holocene (8000–3000 yr B.P.) was warm, with MAT 2–4°C higher than at present. The late Holocene (3000 or 2500 yr B.P. to the present) was cooler than the middle Holocene, with frequent 2–4°C oscillations in MAT.

Liu (1986, 1988) also summarized late-Quaternary vegetational and climatic changes in north China and stated that five cold-warm cycles could be interpreted from pollen data for the Pleistocene from a site in the Beijing lowlands. Similar climatic fluctuations are recorded at many sites and are probably synchronous within the region. For example, Yang *et al.* (1979) found evidence of five cold and four warm intervals at sites in the Hebei plains (37°N, 117°E) from mineralogic changes in the sediments as well as vegetation changes inferred from pollen. The cold phases are represented in the pollen spectra by boreal conifer pollen from *Picea, Abies,* and *Pinus*. The warm phases had abundant pollen from temperate hardwoods such as *Ulmus* and *Quercus*. However, during the warm phase of the early Pleistocene, pollen from subtropical trees was evident, and this earlier interval was warmer than subsequent warm phases (Liu, 1988). Today *Picea* and *Abies* forests in northeast China (as well as those in the southwest) grow only at elevations above 1500 m, but during the cold periods of the Pleistocene they were widespread in the northern lowland regions at elevations of about 430 m and as far south as latitude 25°N (Xu *et al.*, 1980; Yang and Xu, 1981; Zhou *et al.*, 1982). From 18,000 to 13,000 yr B.P. a cold steppe existed in northern China, and the pollen profile was dominated by *Artemisia*. Between 13,000 and 11,600 yr B.P. pollen from boreal conifers such as *Larix* (larch), *Picea,* and *Abies* was again abundant. After 11,600 yr B.P. a *Tilia*-dominated hardwood forest covered parts of northern China (Kong and Du, 1980; Liu, 1986).

Liu (1988) reviewed in detail relatively well-dated palynologic data from 80 sites in northeast and north China, including many of the sites discussed above. He concluded that the vegetation near Beijing was cold and dry steppe-tundra before about 13,000 yr B.P. and was replaced by coniferous forest from 13,000 to about 11,400 yr B.P. as the climate became wetter but remained cold (about 6-9°C colder than at present). Peat began to accumulate at that time in the Beijing lowlands. A short dry period for the next 400 yr is indicated by increases in *Artemisia* pollen and in the spores of *Selaginella* in the Kong and Du (1980) Fenzhuang borehole site near Beijing (Liu, 1988). This dry period may be concurrent with the Younger Dryas period defined by investigators in Europe (Liu, 1988), although Liu considers such a placement of north China vegetation changes within a global context premature. After 11,000 yr B.P. a more mixed conifer-hardwood forest developed, and peatlands expanded as the climate became warmer and wetter (Liu, 1988). Pine increased in the early Holocene, and broadleaved hardwoods (such as *Quercus, Carpinus, Acer, Betula,* and *Juglans*) expanded in response to mid-Holocene warming from about 8000 to about 4000 (Liu, 1988) or 3000 yr B.P. (Duan *et al.,* 1981) (Fig. 10.7). The middle Holocene also was wetter in the first half and then became increasingly drier. Late-Holocene pine and oak forests in north China indicated by dominant pine and/or oak pollen percentages suggest cooling since about 4000 yr B.P. Duan *et al.* (1981) also summarized the Holocene pollen results from multiple sites on a transect from the northeast to the southeast (as far south as 28°N) and found changes similar to those presented by Liu (1988) (Fig. 10.7).

Fang (1991) summarized lake-level changes during the past 18,000 yr from east China, which he divided into north and central and central coastal plain regions. He concluded that lakes in the north and central region respond to precipitation and evaporation changes, while water-level changes in the coastal lakes were associated with Holocene sea levels and river water and channel changes. Low lake levels were evident at all sites from 18,000 to about 9000 yr B.P. The highest lake levels in the north and central regions were at about 6000-2000 yr B.P., with some basins showing high water a bit earlier, between 9000 and 7000 yr B.P. The coastal lakes in Fang's data base were all low until after 6000 yr B.P. Detailed analysis of pollen and lake-level changes at individual sites would provide a broader understanding of some of the asynchronous changes, such as the timing of the peak mid-Holocene wet period (earlier from pollen data and later from lake-level data), seen in the inter-pretation of paleoclimate from lake-level and pollen records from north and east China.

Central China

During the late Pleistocene the vegetation of central China was dominated by cold-climate coniferous forests. *Abies* and *Picea* forests expanded into the hills of the coastal lowlands of central China as far south as 28°N at that time (Xu *et al.,* 1980). Late-Pleistocene pollen samples from Mount Shefeng (25°N, 115°E) contained 16% *Abies* and 5% *Tsuga* (hemlock) pollen (Duan *et al.,* 1981). The evidence for glaciation in the eastern mountains (e.g., at Mount Lu [29°34'N, 115°58'E]) proposed by some researchers (Lee, 1975), however, is being seriously questioned (Huang, 1982; Shih, 1982; Shi *et al.,* 1986). Wang (1984b) documented the finding of a cold-water foraminifer, *Buccella frigida,* which now lives only above 38°N in the Yellow Sea, in cores dating from the late Pleistocene taken offshore at about latitude 30°N in the East China Sea.

Fossil evidence for a warm climate in central China during the Holocene is abundant. Wang *et al.* (1981) studied the Holocene evolution of the Changjiang (Yangtze River) Delta (an area roughly within 30.5-32.5°N and 120-122°E) using lithologic and geomorphologic evidence and the pollen and zooplankton stratigraphy of sediment cores taken from the delta. Evidence of coastal mangrove forests, which at present grow only along the coast of southern China, indicates a warm and humid climate during the middle Holocene. In the Changjiang Delta area the maximum sea transgression occurred at about 7500 yr B.P., when the coastline was about 225 km west (at about 32°N, 119°26'E) of its present position. In addition, Wang (1984b) showed that a warm-water foraminifer, *Asterorotalia subtrispinosa,* now found only south of 25°N in the East China Sea, grew at about the latitude of Shanghai (30°N) between about 7500 and 4500 yr B.P.

Fossil bones found at a Neolithic site (dated at 5000 yr B.P.) near Changzhou in Jiangsu Province (31°47'N, 119°57'E) include deer (*Cervus nippon, Hydropotes inermis,* and David's deer), wild boar, buffalo, turtle, tortoise, fish, and crab-eating mongoose (Huang, 1978), indicating a large forested region with many lakes and rivers and suggesting a warmer and wetter climate than at present.

Radiocarbon dates of objects from a site of the Hemudu Neolithic culture discovered near Yuyao in Zhejiang Province (30°04'N, 121°10'E) (Anonymous, 1976) place this culture between 6950 and 6570 yr B.P. Bones were found from 48 kinds of animals, all hunted by the ancient inhabitants, including elephant, rhinoceros, otter, crocodile, pelican, cormorant, and tortoise. Fossil plant remains included water chestnut, reeds, rice, camphor trees, and banyan. These faunal and floral assemblages are all found farther south today, and therefore a warmer (by about 2°C) and

more humid environment than at present was inferred.

At the Ziyang Neolithic site near Chengdu, Sichuan
Province (30°07'N, 104°39'E), pollen and spore assemblages from sediments radiocarbon-dated to between
7500 and 6740 yr B.P. contain subtropical species such
as *Keteleeria* (silver fir), *Castanea* (chestnut), *Pterocarya,* and palm trees as well as tropical fern spores
(Duan *et al.,* 1981). The MAT was estimated to be
about 3°C higher than today.

South China

Taiwan Island (Formosa), located off the southeast
coast of China in the South China Sea (Fig. 10.1a), has a
subtropical to tropical climate in the lowlands, with
annual precipitation between 1500 and 2500 mm. The
central mountains receive 4000 mm of rainfall per
year, and temperatures fall below freezing in only
one month (Tsukada, 1967).

Tsukada (1967) analyzed sediment cores taken
from lakes in central Taiwan to investigate changes in
vegetation during the last 60,000 yr. Two of the sites,
Jih Tan (23°52'N, 120°55'E; altitude 745.5 m) and
Tashuiku Tan (23°45'N, 120°46'E; altitude 640 m), are
summarized here. Both pollen diagrams show that the
late-glacial period (zone T3) was dominated by cool-
temperate forest trees such as *Quercus, Ulmus,* and
Zelkova. Near the end of zone T3, *Juglans cathayensis*
(Cathay walnut) appeared and then declined, indicating a rapidly warming climate from about 14,000 to
12,000 yr B.P. During the Holocene (zone R) the climate
was warmer than at present. Pollen from subtropical
plants such as *Liquidambar formosana* (sweet gum),
Mallotus paniculatus, and *Trema* cf. *orientalis* dominated, along with pollen from warm-temperate forest
trees, chiefly *Castanopsis* (chinquapin) and *Podocarpus* (plum yew). Peak warmth at about 8000 yr B.P. is
interpreted from the rapid rise in the percentage of
Castanopsis and *Trema* pollen, which started at 10,000
yr B.P. and reached a maximum at 8000 yr B.P. The sudden increase in chenopod and Cerealia pollen around
4000 yr B.P. indicates increased settlement of Taiwan
and land clearance for agriculture. The vegetation
changes caused by human activity after 4000 yr B.P.
make climate-induced vegetation changes hard to interpret from pollen evidence alone. This same problem also applies to late-Holocene pollen records from
mainland China.

Pollen studies by Wang *et al.* (1977) in tropical
south China in Beibu Bay (22°N, 109°E) showed wet-
dry cycles in vegetation during the Pleistocene paral-
leling the cold-warm cycles found in other regions.
During the wet intervals pollen from tropical and subtropical trees in the families Moraceae, Lauraceae (including camphor), Sapindaceae, and Palmae was
abundant. During the dry intervals pollen from savanna plants such as Gramineae and *Artemisia* increased.
South China was drier than at present during the last

glacial maximum, as was most of the eastern part of
the country (Zhang Lansheng, 1984). Wang (1984a) argued that the increased continentality of eastern and
southern China during the late glacial maximum contributed to the dry conditions at that time.

Coastal China

Global sea-level fluctuations during the Pleistocene
have been documented in many studies (see, e.g., Denton and Hughes, 1981). Liu and Ding (1984) summarized the results of some of the research on sea-level
changes for the east coast of China. Sea level was
about 120 m below the present level during the late
Pleistocene, similar to estimates for the coast of North
America (between about latitudes 25 and 45°N) (Denton and Hughes, 1981). In China the coastline at the last
glacial maximum was 600 km east of the present
coastline, Japan and Hainan Island (Fig. 10.1) were connected to the continent (Ren and Zong, 1980; Verstappen, 1980), and the South China Sea was almost
enclosed (Wang and Wang, 1990). As the glaciers receded, sea levels rose, but the land rebounded as well.
The rate of rebound in relation to the rate of sea-level
rise caused nonsynchronous local transgressions of
the sea, especially at more northerly latitudes. In north
China sea levels during the Holocene were highest
(5–10 m above present levels) at about 6000 yr B.P.,
with transgression inland as far as 100 km in the Beijing lowlands of the north China plain (Zhao, 1979;
Chang and Wang, 1982; Zhao and Chin, 1982).

Wang *et al.* (1981) and Wang and Wang (1980) studied sea transgressions in China at coastal sites between
latitudes 25 and 40°N. Paleobiological data and radiocarbon dates from a number of cores from onshore
and offshore sites suggest four transgressions (Fig.
10.8a), with the peak transgression at about 6000 yr B.P.
(during Q_4^2). In another study of beach morphometry,
coral reefs, and marine deposits at more than 60 sites,
Zhao and Zhang (1982, 1985) placed the peak transgression at about 5000–6000 yr B.P. at north China sites
(Fig. 10.8b), in agreement with the results of Guiyang
Institute of Geochemistry (1977) and Wang *et al.*
(1981) for north China and Matsushima (1976) for
Japan. However, because of the morphometry of the
Changjiang Delta region, the maximum transgression
of the sea in central China occurred at 7500 yr B.P.
(Zhao and Zhang, 1982, 1985).

In summary, based on the above evidence, Holocene sea-level changes on the east China coast can be
divided into three stages: (1) a rapid rise before 6000
yr B.P., (2) maximum sea level between 6000 and 5000
yr B.P. in the north latitudes (but at 7500 yr B.P. in the
Changjiang Delta region), and (3) relative stability in
the last 5000 yr (Fig. 10.8b). Wang *et al.* (1981) presented pollen evidence to suggest that sea-level transgressions in China since about 300,000 yr ago have coin-
cided with both melting of the glaciers in the western
mountains and paleotemperature maxima (Fig. 10.8c).

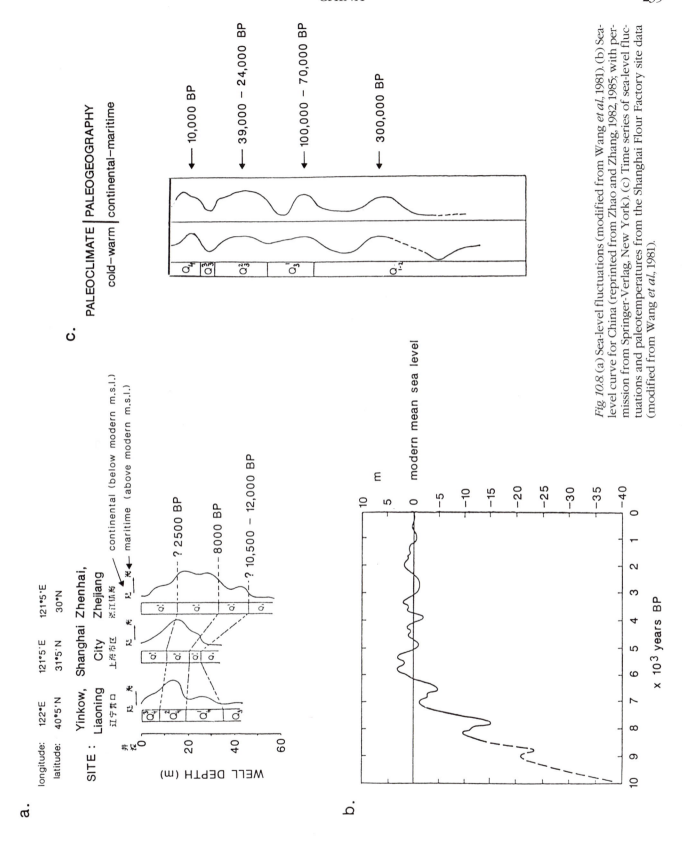

Fig. 10.8 (a) Sea-level fluctuations (modified from Wang *et al.*, 1981). (b) Sea-level curve for China (reprinted from Zhao and Zhang, 1982, 1985; with permission from Springer-Verlag, New York). (c) Time series of sea-level fluctuations and paleotemperatures from the Shanghai Flour Factory site data (modified from Wang *et al.*, 1981).

Yang and Xie (1984) also summarized sea-level changes in east China and correlated sea levels with climate. They found foraminiferal, sedimentary, and archaeological evidence for 10 marine advances and retreats by analyzing data from more than 400 cores and more than 450 cultural sites on the east China plain. Their results show that sea levels were lowest from 20,000 to 15,000 yr B.P. After 15,000 yr B.P. rapid sea-level fluctuations correlate with the three abrupt cold phases defined in Europe (the Older, Middle, and Younger Dryas). Yang and Xie (1984) also stated that conifer forest expansion and sea-level retreat were synchronous, but other researchers consider this speculation. The Younger Dryas event, dated from 11,050 to 10,900 yr B.P., was colder and drier than the more recent little ice age (Lamb, 1977) and was sensed by both conifer forest expansion and sea-level lowering in east China (Yang and Xie, 1984) and possibly by increases in herbaceous pollen in the north (Liu, 1988), decreased lake levels in western Tibet (Gasse *et al.,* 1991), and increased carbonate deposition and $\delta^{18}O$ values in Qinghai Lake on the Qinghai-Xizang Plateau (Lister *et al.,* 1989). The evidence for a cold Younger Dryas event at 11,050-10,900 yr B.P. in China as well as in the North Atlantic and in Europe (Ruddiman and Duplessy, 1985; Rind *et al.,* 1986) is important and may help to redefine boundary conditions for climate models.

Yang and Xie (1984) also inferred a mid-Holocene climate warmer than at present beginning at about 7500 yr B.P., but they found millennial-scale temperature oscillations since that time. They also compared sea-level fluctuations in China with Northern Hemisphere temperature curves for the past 2000 yr from historic, sediment, and tree-ring records and for the past 100 yr from observational data (the latter especially showing good agreement). Other sea-level researchers who have not found the Younger Dryas event or some of the other oscillations reported by Yang and Xie (1984) question these conclusions.

Broecker *et al.* (1988) studied sedimentation rate changes, foraminifera absolute abundances, and the oxygen-isotope record of a core from the South China Sea. They found that sedimentation rates were high in the late glacial and decreased by one-half between about 14,000 and 13,000 yr B.P. Also at about 13,000 yr B.P. two planktonic species of foraminifera abruptly increased. These changes may document a dry, windy glacial maximum in southeastern Asia and the transition from a cold-dry late-glacial climate to a warmer and wetter monsoon climate. A Younger Dryas oscillation was not identified in their South China Sea core record.

Sea-surface paleotemperatures have been reconstructed for the South China Sea from planktonic foraminifera analyzed from three marine cores (Wang and Wang, 1990). The results indicate a blocking of both Pacific and Indian Ocean tropical water into the South China Sea during glacial maxima. Wang and Wang (1990) estimated a decrease in sea-surface paleotemperatures of 6.8-9.3°C during the winter and 2-3°C during the summer at glacial maximum, greater seasonal changes than those estimated for the open western Pacific (Tsukada, 1983; Heusser and Morley, 1985).

INNER MONGOLIA

Inner Mongolia, a region of deserts and sparse grasslands (Fig. 10.5), currently has a semiarid continental climate, with MAT of about 4-6°C (Fig. 10.4a) and MAP of about 200-400 mm (Fig. 10.3c). Annual precipitation is critical in this region, because at least 400 mm is needed for farming without an irrigation system (Zhao and Xing, 1984). The weather has both seasonal and diurnal extremes. Desertification has increased in this region (Sheng, 1987; Wang and Li, 1990).

Few palynologic studies have been reported for this region. However, Chao (1977, 1981) studied a sequence of loess deposition and paleosol formation in the Hulun Buir sandy land (48°N, 120°E) (Fig. 10.2c) and found "three darkish buried layers" amid the sand dune systems in the semiarid steppe, each containing abundant *Quercus* pollen. This area lies in a transition belt between the semiarid steppe to the west and the more subhumid eastern part of Inner Mongolia, which is currently within the northern boundary of the summer monsoon (Figs. 10.3d and 10.5). Archaeologists estimated that the uppermost soil in this sequence (55-95 cm) formed between 5000 and 3000 yr ago, at the time of transition from the Neolithic to the Bronze Age in Inner Mongolia. The other paleosol strata, found 260-330 and 425-525 cm below the modern surface, represent earlier cycles of a more humid climate since the beginning of the Holocene.

Kong and Du (1981) analyzed sediments from a Paleolithic site near Hohhot (40°51'N, 111°40'E). Although the sediments were not radiocarbon-dated, the site is considered to date from the Dayao culture, estimated to be about 20,000 yr old. The pollen spectrum contained 25% AP, mainly *Pinus tabulaeformis* (Chinese pine), and about 74% pollen from a diverse assemblage of shrubs and herbs, including 40% *Artemisia,* 9% chenopods, and 8% each *Ephedra* and composites. The contemporary pollen sample at this site is 100% shrub and herb pollen, including 93% *Artemisia,* typical of a semiarid steppe. The "Paleolithic" pollen spectrum resembles mid-Holocene pollen from adjacent regions more than it does the pollen assemblage from frozen steppe plant communities found in late-glacial samples at other northern sites. Although Kong and Du (1981) concluded that the climate was warmer and more humid during the Paleolithic than at present because vegetation similar to that represented by their "Paleolithic" pollen sample is found today in parts of Inner Mongolia where the climate is warmer and wetter than at Hohhot (Figs. 10.2a and 10.5), the results from this study will remain in question until the older pollen samples from this site are radiocarbon-dated.

During the past 2000 yr the rise and collapse of successive dynasties that had settled in the region, as evidenced by dated cultural relics and historical records, may be linked to the progressive but episodic southeastward expansion of the desert in Inner Mongolia (see Sheng, 1987). Tree-ring analysis has identified several shorter cycles of dry and wet climates since 1700 A.D. (Zhao and Xing, 1984).

SOUTHWEST CHINA

Paleoclimatic information for southwest China (primarily Yunnan Province) comes from three sources: (1) studies of late-glacial permafrost features, snowlines, cirques, and paleomagnetism in the northwest mountains of Yunnan (Walker, 1986; Li, 1987; Li et al., 1987); (2) pollen analyses from a north-south transect of lakes in Yunnan (Lin et al., 1986; J. Liu et al., 1986; X. Sun et al., 1986; Walker, 1986); and (3) a study of Neolithic sites by Duan et al. (1981). The topographic diversity of this region (Fig. 10.2a,b) and its latitudinal extent dictate division of the region into three subregions. The northwest mountains of Yunnan are at the southeastern edge of the Tibetan Plateau, whereas central and eastern Yunnan (around Kunming [Fig. 10.2a]) are more like the central part of China to the east. The southwest corner of Yunnan is a third and unique division, where the present vegetation is the northernmost extension of the tropical evergreen rainforests that grow in the Southeast Asian lowlands to the south (Li and Walker, 1986).

Geomorphic studies of permafrost features and snowlines (Pu et al., 1982; Walker, 1986) and of fossil cirques (Li, 1987) dating from the late glacial maximum (about 18,000–20,000 yr B.P., according to paleomagnetic evidence [Li et al., 1987]) showed that glaciers had expanded in the northwest Yunnan mountains. These mountains, which reach elevations of 6000 m today, are at the southeast edge of the Xizang Plateau, which contains similar evidence of increased snow and more extensive glaciation at the late glacial maximum. Permafrost features were found at 1800 m below the present snowline, and the permanent snowline at the late glacial maximum was 1200 m lower than today. Late-Pleistocene temperatures were an estimated 4-5°C lower than at present in northwestern Yunnan. From the present relative positions of the cirques formed during the late glacial maximum, Li (1987) calculated the rate of crustal deformation in the northwest Yunnan mountains since 18,000 yr B.P. at about 40-50 mm/yr of vertical movement and 200-250 mm/yr of horizontal movement. This is a much higher estimate of uplift than that calculated for the larger Qinghai-Xizang Plateau (at most 10 mm/yr [Xu, 1981; Zheng and Shi, 1982]).

Pollen sites have been analyzed from all three geographic subregions of Yunnan. J. Liu et al. (1986) studied two small lakes in the extreme southwest of the province south of 22°N, 100°E, near Menghai. In central Yunnan near Er Yuan a large, shallow tectonic lake, Xi Hu (26°N, 100°E), was cored (Lin et al., 1986). X. Sun et al. (1986) studied a large plateau lake, Dian Chi (25°N, 103°E), in east-central Yunnan near Kunming. Several cores from each site were analyzed, and most were radiocarbon-dated. Modern pollen samples provide modern analogues for fossil pollen assemblages, and elevational and edaphic data were also examined for modern pollen assemblages (Li and Walker, 1986). The results from this work are summarized in Xu et al. (1984) and Walker (1986). The climatic changes interpreted for the late Quaternary from the pollen data differ for sites south of 22°N in the southwest region and sites from the central and northern parts of southwest China.

Unfortunately, only the record from 40,000 to 12,000 yr B.P. could be recovered from the Menghai lakes (elevation 1280 m) (J. Liu et al., 1986). At the late glacial maximum rainfall was greater in winter than today, but temperatures were only slightly lower. Drier winters since 18,000 yr B.P. were interpreted from pollen evidence of shifts to pine-oak scrubland, with abundant herbs from earlier pollen assemblages representing rainforest vegetation that included broad-leaved and seasonal species such as podocarps (Dacrydium and Dacrycarpus) and Fagaceae Lithocarpus, Cyclobalanopsis (evergreen oak), and Castanopsis.

At Dian Chi (elevation 1886 m) near Kunming, several taxa reach their northernmost and altitudinal limits today (X. Sun et al., 1986; Walker, 1986). High-altitude forests present before 9500 yr B.P. contained Tsuga dumosa, which now grows only above 2300 m east of Kunming. Precipitation may have been higher than at present before about 10,000 yr B.P., but the present monsoonal precipitation pattern was in place by about 9500 yr B.P. Many vegetation changes were taking place at about 10,000 yr B.P., and the composition of the mixed deciduous forest was more complex before 7000 yr B.P. A similar interpretation of plant dynamics from pollen data has been proposed for the same time period in northern and eastern North America (Ritchie, 1985; Jacobson and Grimm, 1986; Jacobson et al., 1987; Grimm and Jacobson, 1992), where species composition changed greatly before and during the early Holocene, whereas a more stable forest was evident after that time. This global similarity in forest dynamics in the Northern Hemisphere at about 9500 yr B.P. suggests that the increased seasonality (Kutzbach, 1981; Kutzbach and Otto-Bliesner, 1982) resulting from changes in the orbital parameters of the earth (the Milankovitch theory [see Imbrie and Imbrie, 1979; Xu Qinqi, 1981]) may have had a greater effect on the dynamics of vegetation change during deglaciation than did ice-sheet retreat. Vegetation instability in the middle latitudes of North America in the early Holocene has been attributed partly to the proximity of the waning ice sheet, but China had no ice sheet, and although the western mountain glaciers contracted after about 10,000 yr B.P., they are still features of the Asian landscape.

Brenner *et al.* (1989) studied the sediment chemistry as well as the pollen in a core from another large plateau lake, Qilu Hu (24°N, 102°45'E). The lake sediment core spans the Pleistocene-Holocene boundary (radiocarbon-dated at 11,790 ± 90 yr B.P.) and shows a change from a pollen assemblage dominated by *Quercus* to one with increased pollen from *Pinus* at the beginning of the Holocene.

Between about 7000 and 4000 yr B.P. broad-leaved evergreen forests expanded in the region around Dian Chi, and the climate was warmer and wetter than the modern climate. A Neolithic site at Tuanmou (25°38'N, 101°54'E) also near Kunming was occupied at about 3210 ± 90 yr B.P. and yielded fossil bones of *Rhizomys sinensis, Ursus thibetanus* (Tibetan bear), deer (both *Muntiacus* spp. and *Cervus elaphus*), and rabbit (Duan *et al.,* 1981). This fauna now inhabits the tropical forests of Zishuanbanna (about 22°N, 100°5'E) below the Tropic of Cancer, consistent with the pollen evidence that similar forests expanded in the middle Holocene at least as far north as the Kunming region of Yunnan.

At Xi Hu (elevation 1980 m) near Er Yuan in the foothills of the northwest mountains of Yunnan, the climate was drier than or similar to the present and colder than today until about 10,000 yr B.P. (Lin *et al.,* 1986; Walker, 1986). Alpine forests of *Picea, Abies,* and *Tsuga* occupied the slopes at altitudes at least 500 m lower than at present from 17,000 yr B.P., and the present altitudinal limit of 3000 m for these forests was gradually attained by 10,500 yr B.P. As at Dian Chi, the vegetation was more complex before 10,000 yr B.P. Dynamic shifts in species composition from about 14,000 to 10,000 yr B.P. were followed by a transition period to about 7500 yr B.P.; since then, the forests surrounding Xi Hu have been dominated by *Pinus,* although other plant communities persist because of the steep altitudinal gradients of the region.

QINGHAI-XIZANG (TIBETAN) PLATEAU

Paleoenvironmental research on the Qinghai-Xizang (Q-X) Plateau has increased in recent years, including studies of mountain glacier movements, lake-level changes, and peatland and sand dune development, chemical and sediment analyses of ice-cap cores, and archaeological investigations. Pollen analysis has also been used in most of these investigations.

Glacial Geology

Zheng and Shi (1982), Kuhle (1987), and Burbank and Kang (1991) studied the extent of the Pleistocene glaciation on the Q-X Plateau. Maximum glaciation occurred in the middle Pleistocene, but even then no large ice sheet was formed, although scattered piedmont glaciers covered the plateau. Glaciation on the Xizang Plateau during the last glacial maximum was only slightly more extensive than today (Zheng and Shi, 1982; Walker, 1986), indicating that precipitation was low at about 18,000 yr B.P., although evidence of expanded periglacial features suggests a colder climate than today. On the southeastern edge of the plateau adjacent to Yunnan, however, glaciers expanded during the last glacial maximum because of increased precipitation in that region.

The glacial fluctuations of the Holocene can be divided into three stages (Zheng and Shi, 1982): (1) a slow retreat in the early Holocene (10,000–8000 yr B.P.), (2) a rapid retreat in the middle Holocene (8000–3000 yr B.P.), and (3) alternating advances and retreats in the late Holocene (from 3000 yr B.P.). Li (1980) considered 2980 ± 150 yr B.P. as the time of the initiation of late-Holocene glacial activity in China, and glacial activity on the plateau since this time has been frequent.

Ice-Cap Core Analyses

Thompson et al. (1989) obtained three ice cores to bedrock from the Dunde ice cap in Qinghai (38°06'N, 96°24'E). $\delta^{18}O$, dust particle concentrations, and anion and ice-crystal-size stratigraphies suggest a change from a cold, dusty late-Pleistocene climate to a warmer Holocene climate. Although the authors interpret a lower concentration of dust particles and increased soluble anions as indications of increased dry surface in the Qaidam Basin throughout the Holocene, these changes may also be interpreted as suggesting decreased winds from the deserts of western Asia in the early Holocene and strengthened southwestern monsoon winds that deposited abundant sea-salt aerosols after about 10,750 yr ago. A mid-Holocene warming is indicated by increased $\delta^{18}O$ values in the cores, with further periods of higher temperatures alternating with colder periods during the past millennium and a more dramatic and systematic warming since the 1920s (Thompson *et al.,* 1989).

Lake Levels

Studies of the shorelines of and sediments from lakes on the Q-X Plateau provide evidence of large changes in lake levels during the Holocene. Xizang Province has more than 600 lakes (Wang and Fan, 1987). Those rimming the plateau are tectonic in origin and are fed by glacial meltwater. Closed-basin lakes in the interior of the plateau contracted in the Holocene (Li *et al.,* 1985; Holland *et al.,* 1991) and today occupy less than a third of their former area (Wang and Fan, 1987). After 9000 yr B.P. some lakes in the interior also became more saline (Chen *et al.,* 1981).

Between 7500 and 3000 yr B.P. the climate was generally warmer and wetter than at present in the plateau region of western China (as it appears to have been throughout the entire country), and glaciers melted on the high mountains. Some of the lakes that received this meltwater coalesced to form larger lakes. This high lake-level stage of the exterior lakes, especially in south Xizang, has been dated from peat and fossil snail deposits in high terraces surrounding the lakes. Lakes in the interior of the plateau that did not receive meltwater, however, had low water levels between 7500 and 3000 yr B.P., although evidence of layers of freshwater mud in the saline Lake Zhacancaka (32°30'N, 82°20'E), with bracketing dates of 8000–7000 yr B.P. and 6000–5000 yr B.P. (Wang and Fan, 1987) indicates that the desiccation was not a continuous process but rather was punctuated by millennial-scale wetter periods. The three Holocene lake terraces above the present border of saline Lake Zharinanmuco (31°N, 85°E) (Li et al., 1985) also provide evidence of wet periods alternating with dry periods on the Xizang Plateau.

After 3000 yr B.P. lake levels fell rapidly throughout the Xizang Plateau. As the climate cooled and glaciers expanded, meltwater was no longer abundant, and the water levels in some of the exterior lakes fell. Several of the larger lakes dropped as much as 10–20 m, remaining wet only in the deeper basins. The interior lakes continued to desiccate and fill with precipitates. Some of the lake-level changes on the Q-X Plateau during the Holocene were caused not only by changing climatic conditions, such as the intensification of the South Asian monsoon at about 10,000 yr B.P., but also by the continuing morphometric changes in the lake basins caused by uplift of the plateau.

In a study of the evolution of late-Pleistocene salt lakes in the Qaidam Basin (36–39°N, 91–99°E) in Qinghai, Chen and Bowler (1986) concluded from evidence of higher lakeshores and freshwater sediment deposition that water levels were higher before about 40,000 yr B.P. Smaller lake basins, halite deposition, and clastics identical in chemical composition to the Malan (late-Pleistocene) loess deposits indicate that lake levels fell between 25,000 and about 9000 yr B.P. during the cold, dry, and windy period of the last glacial maximum. During the Holocene the remaining lakes continued to desiccate, although Chen (Chen and Bowler, 1986) interpreted the sediment changes in some of the lake cores to show a more humid early Holocene followed by a drier middle and late Holocene.

Ostracod, gastropod, and pollen evidence was also used in these and other studies to define water-level changes in Qinghai. Lister et al. (1989) measured the $\delta^{18}O$ values of benthic ostracods in the closed-basin Qinghai Hu (37°N, 100°E) and concluded that the lake was low in the late Pleistocene, with a marked dry period during the Younger Dryas, although lake levels were rising by the end of the Pleistocene. An increased monsoon circulation is inferred after 10,500 yr B.P., and lake levels reached a maximum at about 8000 yr B.P. Between 8000 and 3500 yr B.P. lake levels decreased as evaporation increased. Higher lake levels

are again indicated after 3500 yr B.P. as Qinghai Hu approached the modern level.

Studies of the shorelines and a core for a closed-basin, regionally sensitive tectonic lake, Sumxi Co, in western Tibet by Gasse et al. (1991) document environmental changes from 13,000 yr B.P. From 13,000 to 10,000 yr B.P. cold and dry conditions were interpreted from desert Ephedra and chenopod pollen abundances, increased detritus from erosion, low water levels fed only by brief pulses of glacial meltwater (indicating brief warm periods), and reducing conditions due to long freezes. A Younger Dryas event is interpreted from decreased ^{18}O and ^{13}C content and the absence of organic matter and aquatic organisms between 11,000 and 10,000 yr B.P. From 10,000 to about 5000 yr B.P. warm and wet conditions prevailed. Artemisia steppe expanded, lake levels were high, and there is evidence from changes in diatom and ostracod species of mixing with more saline waters of a larger nearby lake, Longmu Co. After 3400 yr B.P. lower water levels, similar to present levels, were established. Event-scale changes were identified within all the longer periods of climatic change. Gasse et al. (1991) concluded that some of these events, like the Younger Dryas, were global in extent.

In an extensive study of lake-level fluctuations in China, Fang (1991) determined that lake levels in the western region of China including the Tibetan Plateau were highest between 9500 and 3000 yr B.P., but that some lakes had begun to rise as early as 12,000 yr B.P. Fang (1991) also concluded that lake levels were low on the Tibetan Plateau and throughout China during the late glacial maximum.

Development of Peatlands

Extensive peat deposits have formed since 10,000 yr B.P. in the wide plateau valleys of the Q-X Plateau above 4500 m and south of about 31°N. Some peatlands developed because of the abundant meltwater from the receding glaciers, and some developed because of increased moisture from the strengthened monsoon.

Three stages of peatland development can be distinguished (Li et al., 1985). The initial stage was in the early Holocene. Two peatlands in the Dangxiong District of Xizang had bottom radiocarbon dates of 9970 ± 135 and 8175 ± 200 yr B.P. as well as dates of 3575 ± 80 and 3050 ± 200 yr B.P. for the end of peat formation at those sites (Wang and Fan, 1987). In the middle Holocene peat developed at other sites on the plateau as well. Nine peatlands have basal ^{14}C dates from 7670 ± 250 to 3050 ± 120 yr B.P. (Li et al., 1985). Peat development declined in the late Holocene (from about 3000 yr B.P.) on the Q-X Plateau and ceased altogether at some sites. At other late-Holocene sites inorganic deposits (silts and gravels) interlayered with peat have been found, and a more detailed history of climatic and/or tectonic change during the past 3000 yr on the

Q-X Plateau has been inferred (Wang Peifang, 1981; Wang and Fan, 1987).

Landform Development

In a study of sand dune formation in the valley of the upper Yarlung Zangbo Jiang (Brahmaputra River) (31°N, 83°E), Yang (1984) indicated that active sand dunes are 10 times higher than stabilized dunes dating from the early Holocene. This evidence suggests that the westerlies are stronger and the climate is drier in the river valley today than in the early Holocene. Also, alluvial fans dating from the early Holocene are at higher elevations in the river valley, suggesting higher water levels from the receding glaciers at that time. Sediments from alluvial fans formed in the middle Holocene, however, contain both loess and organic material, indicating that paleosols formed during this time.

Vegetation Changes

Pollen data (Huang and Liang, 1981; Wang Peifang, 1981; Wang and Fan, 1987) indicate that the beginning of the Holocene was marked by expansion of trees such as *Picea* and *Pinus* on the Q-X Plateau, although nonarboreal plants (primarily *Artemisia* and other Compositae) were still dominant. In the middle Holocene (7500-3000 yr B.P.) pollen was more abundant, the number of species contributing pollen to a site increased, and AP was at times 50% of the total pollen sum. The AP consisted mainly of *Pinus* and Betulaceae (e.g., *Betula, Alnus, Carpinus, Corylus*), but with increasing percentages of other trees (*Quercus, Tsuga*) and shrubs (Rosaceae and Ericaceae). After 3000 yr B.P. the number of species and the percentage of AP dropped markedly.

In the western Q-X Plateau Gasse *et al.* (1991) found pollen evidence of a dry chenopod and *Ephedra* desert before 10,000 yr B.P., an expansion of *Artemisia* steppe during a warm and wet interval between 10,000 and 5000 yr B.P., and subsequent expansion of desert up to the present.

Archaeological Evidence

Paleolithic and Neolithic artifacts found at sites throughout the plateau date from the early to middle Holocene. Especially notable are sites in northern Xizang at about 35°N and at altitudes as high as 5200 m, where no humans live today (An *et al.*, 1980; Liu and Wang, 1981). The evidence of human activity on the plateau before and during the middle Holocene suggests that environmental conditions were more favorable at those times than at present.

Fossil bones found at a Neolithic site at Karub (near Qamdo; 32°N, 97°E) and dated at 4690 ± 135 yr B.P. include deer and goat species (*Hydropotes inermis* and *Capricornis* spp.) that now live only in south China or at lower elevations and *Capreolus capreolus* (roe deer) and *Cervus elaphus* (red deer), which are now limited to habitats south of the Changjiang (Yangtze River) (Li *et al.*, 1985). This faunal assemblage supports the interpretation of warmer and wetter conditions in the early and middle Holocene, as concluded from other lines of evidence.

In summary, even with the uncertainties of the effects of tectonics, a warmer-than-present climate during the middle Holocene in the Tibetan Plateau region is well established. The temperature was an estimated 3-5°C higher than at present (Duan *et al.*, 1981; Wang and Fan, 1987) from about 8000 to 3000 yr B.P. (Fig. 10.6b). After 3000 yr B.P. the temperature on the Q-X Plateau was estimated to have been as much as 2°C lower than the present MAT and even colder during the intervals of glacial advance and expansion of permafrost boundaries (Wang and Fan, 1987). The climate change after 3000 yr B.P. is the major difference between the temperature curves inferred for the Holocene on the Q-X Plateau and for all of China (Fig. 10.6b). In most of China the inferred temperature remains close to or slightly above the modern value from 3000 yr B.P. until after 1000 yr B.P. except for short cold periods at 3000 and 1500 yr B.P. Interestingly, the temperature fluctuations of short duration in the late Holocene are synchronous throughout China even though the direction of change varies (Fig. 10.6b), which suggests that these changes were caused primarily by climate change and not by anthropogenic factors.

The pattern of climatic fluctuations on the Q-X Plateau in the early and middle Holocene is consistent with the pattern in some of the other regions of China we have examined, as determined from pollen, faunal, sea-level, or sedimentary data (Fig. 10.9). Although the continued uplift and extreme orography of the plateau affect local vegetational, hydrologic, and climatic patterns, the evidence of regional climatic change summarized by Li *et al.* (1985) and Wang and Fan (1987) attests to the fact that the Holocene warming on the plateau was synchronous, within the limits of inference from paleodata, with the warm period in the rest of the country except the extreme northeast (Figs. 10.6b, 10.9, and 10.10).

NORTHWEST CHINA

Relatively little information is available from northwest China. Zhang (1981) and Wang Jingtai (1981) investigated snowline fluctuations in the Tian Shan range (Xinjiang Province, 43°N, 81°E) at the upper reaches of the Urumqi River. The late-Pleistocene snowline was 400 m lower than the present snowline (4000 m), whereas during the early and middle Holocene the snowline was 300-400 m higher than at present. The MAT was estimated to be 2°C higher than at present in the middle Holocene.

Zhou *et al.* (1981) analyzed pollen samples collected from sites near the Red May Bridge (elevation 2516 m). Three samples were radiocarbon-dated. The first sam-

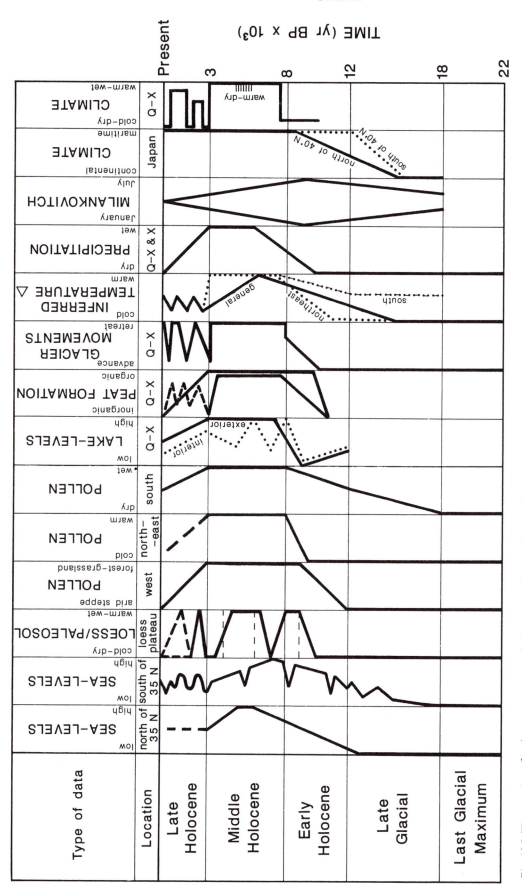

Fig. 10.9. Time series of paleoenvironmental changes and paleoclimate from biogeologic evidence compared to Milankovitch solar-radiation changes since the last glacial maximum.

ple (7320 ± 200 yr B.P.) shows a pollen assemblage dominated by 98.5% *Picea schrenkiana*. At present there are no *Picea* forests in this area, and only a few spruce trees grow in isolation at lower elevations in the Urumqi River valley (elevation about 2300 m). From this information Zhou *et al.* (1981) estimated that the temperature at 7000 yr B.P. was about 1°C higher than at present. The treeline was higher in the Tian Shan during the middle Holocene (Liu, 1985b) at the same time that the east China forests expanded to the north and west. The second sample (3950 ± 150 yr B.P.) contained 92.6% *Ephedra* and *Artemisia* pollen and documents a cooler and drier climate than the earlier sample. A third, more recent sediment sample had pollen assemblages similar to the second sample.

THE LOESS PLATEAU OF NORTH-CENTRAL CHINA

Loess (*huangtu*, "yellow earth"), an eolian sediment composed of calcareous silt, forms during arid, steppe-like conditions. Loess deposits are widely distributed in China (Fig. 10.2c) north and east of the Qin Ling Mountains (elevation 3800 m) (Fig. 10.2b), which form the southern boundary at 34°N, 106–110°E of the area referred to as the Loess Plateau (Fig. 10.2c). The Qin Ling range also marks the present-day boundary between warm-temperate and subtropical climatic zones in central China (Fig. 10.3a), for it provides a barrier to monsoon winds from the south and to cold airmasses from the north (Figs. 10.2b and 10.3d). The Loess Plateau of north-central China lies primarily within the middle reaches of the Huanghe (Yellow River) (Fig. 10.2b) and includes portions of several of the geographic regions that have already been discussed (i.e., north China, Inner Mongolia, and northwest China [Fig. 10.2a]). The greatest concentration of loess is found within this area of deeply dissected terrain, most typically on the downwind side of the great deserts (sandy lands) of China (Fig. 10.2c).

Loess sequences from this area have been intensively studied (Liu Tungsheng *et al.*, 1964, 1985, 1986, 1987; Liu and Yuan, 1982; Liu and Ding, 1984; Liu, 1985a,b; An *et al.*, 1991; Forman, 1991) by sedimentary, chemical, and palynological techniques (Liu, 1985a). The stratigraphy of alternating loess and paleosols dating from 2.4 million years ago to the present has been correlated by paleomagnetic analyses and thermoluminescence dating with oxygen-isotope records from ocean cores (Kukla, 1977; Lu and An, 1979; Liu and Yuan, 1982; Liu *et al.*, 1985; Kukla *et al.*, 1987; An *et al.*, 1991; Forman, 1991). The Brunhes-Matuyama boundary was identified at 53.05 m from the top of the Luochuan loess sequence, and the eight loess-paleosol cycles during the Pleistocene above this boundary are believed to represent global climatic cycles, with the loess layers representing cold, dry episodes and the paleosols representing warmer and wetter climatic intervals (Lu and An, 1979). The Quaternary loess deposits are (1) early Pleistocene (Wucheng), (2)

middle Pleistocene (Lishi), (3) late Pleistocene (Malan), and (4) Holocene. Similar Quaternary fluctuations have been identified in the loess stratigraphies in eastern Europe, primarily in Hungary (Pecsi, 1987), the former Czechoslovakia, Austria, and the former Soviet Union (An Zhisheng, personal communication, 1987).

In China loess deposits can be found as far northeast as 49°N in Heilongjiang Province and as far southeast of the Loess Plateau as Shanghai (about 30°N). In northwestern China loess is found in the foothills of the mountains surrounding the Tarim and Junggar basins and Taklimakan Desert in Xinjiang and in the saline Qaidam Basin in Qinghai (Fig. 10.2c). The loess deposits in northwest and east China are discontinuous, whereas the sequences in the Loess Plateau (in the middle reaches of the Yellow River) are thick and continuous. Complete stratigraphies exist at Dingxi, Luochuan, Jixian, and Longxi in both Gansu and Shaanxi provinces (Liu, 1985a). As desertification progresses, the area of loess deposits expands as well.

Holocene loess-paleosol deposits have been less well studied, but they have been identified in the central Loess Plateau and also in the eastern plains (Liu, 1985a). The Holocene deposits are less than 1 m thick and include three early- and mid-Holocene paleosols (black loam) formed at about 9000, 6000, and 4000 yr B.P. (An Zhisheng, personal communication, 1987). The formation of several paleosols during the early and middle Holocene suggests that desiccation in the north-central region of China was discontinuous, interrupted by long warm and wet intervals possibly synchronous with similar intervals interpreted for the Q-X Plateau and the northwest. The transport and deposition of loess, on the other hand, is a function of the strength and the direction of the winds in the central Asian and Arctic airmasses that dominate northwest and north-central China in the winter and spring. These winds are modified by the blocking and tunneling effects of the mountains north of the Kunlun and the Qin Ling ranges (Fig. 10.2b,c).

Loess is the aerosol in the "dust rains" that affect the eastern cities of China (Chu, 1973; Zhang De'er, 1984). Studies of dust rain and wind direction from historical documents indicate that the source areas of the dust are the deserts of northwest China (Sheng, 1987). This result is also supported by chemical and size analyses of loess (Zhao and Xing, 1984; Liu *et al.*, 1985). The dust deposition is most frequent during the cold and/or dry seasons of the year, and dustfall in the last millennium has occurred in the same regions as the ancient loess (Zhang De'er, 1983, 1984, 1985). An increase in dustfall from about 1500 to 1900 A.D. (Zhang De'er, 1984) coincides with the cold climatic event called the little ice age in North America and Europe (Lamb, 1977, 1984).

Further interpretation of the loess stratigraphies is presented in the preceding discussion of the regional data. Wang and Li (1990) discussed recent increased desertification of the semiarid part of the Loess Plateau in relation to El Niño/Southern Oscillation changes.

JAPAN

The paleoenvironment of Japan has been interpreted from palynologic data (Horie, 1957, 1974; Tsukada, 1972, 1983, 1986; Fuji, 1974; Yasuda, 1984; Heusser and Morley, 1985), glacial features (Kerschner, 1987; Korotkiy, 1987), zooplankton stratigraphies from both lake and ocean cores (Yasuda, 1984; Heusser and Morley, 1985), sediment changes and peatland development (Sakaguchi, 1979), and spectral analysis of grain size in lake sediment cores (Kashiwaya *et al.*, 1987). At the last glacial maximum, when sea level was 120 m below the present level, Japan was connected to the Asian continent and the climate was cold and dry (Yasuda, 1984; Tsukada, 1986; Kerschner, 1987). The forests of Japan were dominated by *Picea, Abies, Tsuga,* and haploxylon *Pinus* (Tsukada, 1972; Yasuda, 1984), MAP was at least one-third lower than at present, and MAT was about 8°C lower than at present (Yasuda, 1983). Some of this work has been summarized by Horie (1974), Yasuda (1984), and Tsukada (1983, 1986).

Between 15,000 and 10,000 yr B.P. *Betula, Alnus, Salix,* and *Quercus* forests increased, indicating that the climate was warming. *Fagus crenata,* a species of beech tree that now grows under moist maritime conditions with high snowfall, was abundant by 11,500 yr B.P. but only at sites south of latitude 40°N. Even by 10,000 yr B.P. warm-water diatoms were found only in ocean cores taken south of 40°N, indicating that the warm Tsushima current could not yet penetrate very far north in the Sea of Japan. This evidence suggests that sea level remained low near the entrance to the Sea of Japan (Figs. 10.1a and 10.9–10.11), affecting the climate of northeast China and western Japan north of 40°N as late as 9000 yr B.P.

From 10,000 to 6500 yr B.P. pollen from conifers decreased. *Picea* disappeared, and pollen from a mixed mesophytic forest comprising *Cryptomeria, Fagus, Quercus, Carpinus, Juglans, Ulmus,* and *Zelkova* increased, indicating a warm and wet climate. By 8500 yr B.P. *F. crenata* was present north of 40°N, and warm-water foraminifera and radiolaria were found in ocean cores taken throughout the Sea of Japan. These results indicate that the Tsushima current expanded to latitudes north of 40°N at about 8500 yr B.P., and the climate in Japan changed from continental to maritime (Fig. 10.9).

After 6500 yr B.P. rapid expansion of subtropical *Cyclobalanopsis* and *Castanopsis* forests suggests an even warmer and possibly drier climate. Sakaguchi (1979) studied the development of peatlands in the mountains of Hokkaido and Honshu as well as along the coast. The upland peat deposits date from the early to middle Holocene, whereas the coastal peat basins formed after the mid-Holocene transgression

of the sea, which was at its maximum at about 6000 yr B.P. in Japan (Matsushima, 1976). A wetter climate in Japan after about 4000 yr B.P. is suggested by the expansion of *Cryptomeria,* which requires the moisture from heavy winter snowfall, into sites as far north as 39°N (Tsukada, 1986).

Vegetational and Climatic Reconstructions from the Biogeologic Evidence

From the regional biogeologic evidence presented in the previous section, we prepared maps representing the climate of China at 18,000, 12,000, 9000, 6000, and 3000 yr B.P. (Fig. 10.10) and the vegetation at 18,000, 12,000, 9000, and 6000 yr B.P. (Fig. 10.11). Because of limitations in the data, the chronology of change at the sites from which the data used for the maps were gathered may deviate as much as ±1500 yr from the nominal dates on the maps. A much narrower range (say ±500 yr) would be preferable but would have excluded most of the evidence.

In this synthesis we use the more direct evidence of radiocarbon-dated botanical and geomorphological changes as well as evidence from archaeological sites dated by cross-correlation with other sites. In the maps we generally followed the original authors' interpretation of climate and vegetation. The paleoprecipitation maps (Fig. 10.10) depict the sign (direction) of the interpreted condition (wetter or drier) relative to the modern climate of China (Figs. 10.3 and 10.4). The paleovegetation maps (Fig. 10.11) can be compared to the modern vegetation distribution (Figs. 10.5 and 10.11). Although questions remain concerning the mechanics of gathering, processing, analyzing, and dating the sediments, when all the data from the many sources is synthesized (Fig. 10.1c and the Appendix), a rather coherent picture of environmental change in China since 18,000 yr B.P. emerges. However, we expect this picture to be modified as new radiocarbon-dated evidence is collected.

18,000 YR B.P.

At 18,000 yr B.P., the last glacial maximum, most of China and all of Japan were cold and dry (Fig. 10.10). Although China and Japan were not covered by an ice sheet in the last glaciation, mountain glaciers expanded in western and northwestern China and in the northern mountains of Japan. Sea level was 120 m below the present level, and Japan, Taiwan, and Hainan Island were connected to the continent. *Buccella frigida,* a cold-water foraminifer, shifted 5° southward in latitude and grew in the East China Sea. Spruce and fir forests covered most of Japan and expanded in China. Spruce and fir grew at least 1200 m lower than at present in both the eastern and western mountain ranges of China, while frozen steppe cov-

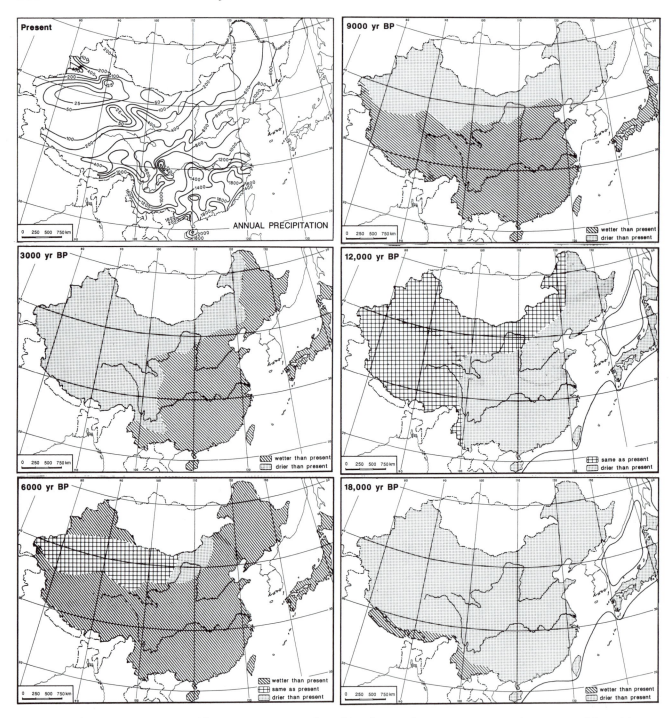

Fig. 10.10. Paleoprecipitation maps (showing areas wetter than, drier than, or the same as at present) constructed from biogeo-logic evidence for 18,000, 12,000, 9000, 6000, and 3000 yr B.P., and present mean annual precipitation (mm) (modified from Ren *et al.*, 1985).

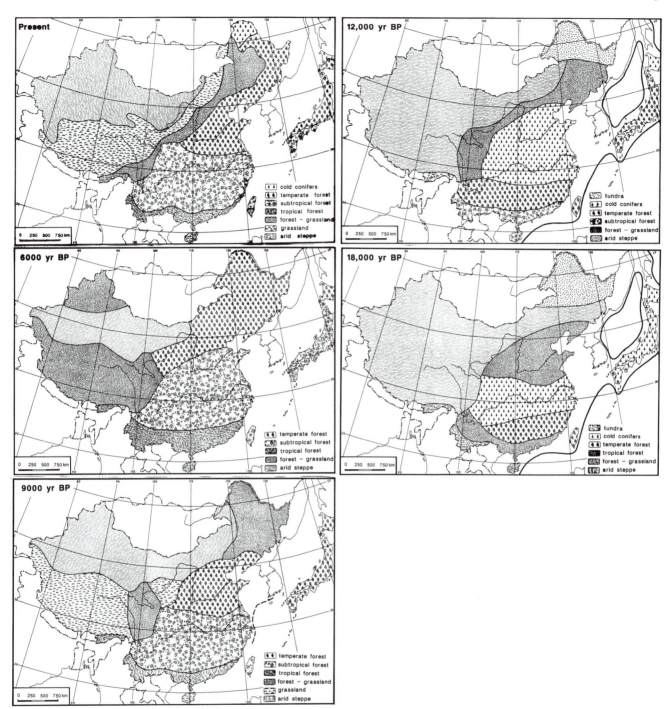

Fig. 10.11. Paleovegetation maps constructed from pollen and macrofossil evidence for 18,000, 12,000, 9000, and 6000 yr B.P., and present vegetation zones (data from Hou [1979], Teachers College of Northwest China [1984], and Ren *et al.* [1985]). As in the key to Figure 10.5, cold conifers represent both larch and spruce forests of today and spruce, fir, larch, and pine forests that were more extensive at 18,000 and 12,000 yr B.P. In the maps for 9000 yr B.P. to the present, the temperate forest community includes the northern conifer-hardwood forests, which also include spruce and fir. Tundra plants are shown only in the maps for 18,000 and 12,000 yr B.P. Arid steppe and forest-grassland communities represent frozen steppe and the sparse and/or dry vegetation of northern lands in 18,000–9000 yr B.P., the open dry subtropical forests in south China at 18,000 yr B.P., and the semiwooded grasslands of the Tibetan Plateau as they became more moist (9000 and 6000 yr B.P.).

ered large regions in the northern half of the country (Fig. 10.11). Vegetation zones in eastern China were displaced steeply southward at 18,000 yr B.P., partly because the central plateau formed a "cold trough" (Yang and Xu, 1981) where cold continental air (central east Asian, Arctic, and Korean airstreams) could reach farther south. Dated permafrost features indicate that the southern limit of continuous permafrost in the east moved about 11° south, while the permanent snowline in the western mountains was at least 1200 m lower than at present. There is pollen evidence of wetter forests at 18,000 yr B.P. in Nepal, the lower valleys along the southern face of the Himalayas, the southeastern corner of the Tibetan Plateau, and Yunnan Province in southwest China, indicating that moist air from the Bay of Bengal may have reached inland in summer as far as Kunming (25°N, 103°E) in Yunnan. Moist air would also cause increased snowfall at high altitudes and help account for the expansion of glaciers in the southeastern corner of the Tibetan Plateau and in northwest Yunnan. Lakes throughout China, however, were dry or had low water levels.

Although all of China was colder than at present at 18,000 yr B.P., the largest departures from modern temperatures were in northern China (Fig. 10.6b) and Japan and were probably the result of increased continentality caused by lowered sea levels at that time. Zhang Lansheng's (1984) reconstructions of mean annual, July, and January temperatures and annual precipitation in east China during the late glacial maximum similarly indicate that east China was cold and dry at 18,000 yr B.P.

12,000 YR B.P.

China remained colder than at present (though warmer than at 18,000 yr B.P.) at 12,000 yr B.P. (Figs. 10.6b and 10.9). The north and northeastern parts of the country were the coldest (annual temperatures were 6–10°C below modern values), and permafrost was still extensive in that region. Moisture increased somewhat in the north, the northeast, and parts of central China, but much of the rest of the country and most of Japan continued drier than at present (Fig. 10.10). The increased wetness in the north may have been caused by summer melting of permafrost as summer solar radiation increased as a result of changes in the earth's orbital parameters. The presence of extensive meltwater pools in permafrost could explain the increased moisture at northerly sites in China at 12,000 yr B.P. without a large increase in annual precipitation in the region.

Warming was more rapid in eastern Japan (at about 36°N), where forests dominated by *Fagus crenata* grew in highlands by 12,000 yr B.P. Warm-temperate forests also grew in Taiwan and southern China (Fig. 10.11). The vegetation changes also indicate that moisture increased in these areas. Lake levels rose

in Yunnan Province and in western Tibet. This evidence suggests that the continental airmasses may have weakened somewhat at 12,000 yr B.P. and that the monsoon from the Pacific may have been stronger than at 18,000 yr B.P. (Fig. 10.3d).

9000 YR B.P.

Glacial conditions were over in most parts of China by 9000 yr B.P. (Fig. 10.10). Temperatures were generally warmer than at present, perhaps by 1–3°C, and the regions that were warmer also had more moisture (Fig. 10.9). North China was still colder than at present according to the interpretations of the evidence, but even there temperatures were considerably higher than at 12,000 yr B.P. (Fig. 10.6a,b).

The monsoonal climate was strengthened by 9000 yr B.P. (Fig. 10.10) in the southern half of China, although drier conditions above 40°N in China and Japan suggest that the summer monsoon boundary was still displaced southward compared to the present boundary (Fig. 10.3d) to about 40°N latitude in the east. Evidence of vegetation similar to the present in Taiwan, the expansion of mesic and subtropical forests (Fig. 10.11) evident in pollen diagrams from the southern half of China, and the development of peatlands on the Tibetan Plateau (Fig. 10.9) all indicate that effective moisture throughout this region was greater than at present (Fig. 10.10) and that the northern and western limits of the southwestern monsoon had expanded. Pollen sites south of 40°N in Japan provide evidence for a decrease in conifers and an increase in forests dominated by *Cryptomeria, Fagus crenata, Quercus, Ulmus,* and *Zelkova,* similarly indicating warmer and wetter conditions. By 11,500 yr B.P. the climate had changed from continental to marine south of 40°N in Japan (Fig. 10.9), and by 8500 yr B.P. the presence of warm-water foraminifera and radiolaria in ocean cores and pollen from mesophytic forests in upland cores north of 40°N indicate that all of Japan was dominated by a maritime climate and that the warm Tsushima current reached farther north in the Sea of Japan (Fig. 10.10).

Further evidence for intensified monsoons at about 9000 yr B.P. comes from dated high lake terraces on the Qinghai-Xizang Plateau, indicating higher lake levels (Fig. 10.9). Western China warmed, and lakes that received meltwater from the receding glaciers were at high levels, as were the closed-basin lakes on the plateau. These findings suggest warmer and wetter summers (melting of ice and filling of lakes) and also drier winters (no renewal of the snowpack), indicating increased seasonality. Between about 10,000 and 8000 yr B.P. peatlands developed in southern and central Xizang, and spruce and pine forests expanded on the plateau and in north-central China (Fig. 10.11). The development of paleosols in the north-central regions of the Loess Plateau (Fig. 10.9), with *Quercus* and other AP dominating, also indicates a warmer and wetter early Holocene in these regions, although the middle Holocene was the warmest and most humid period in China since the last glacial maximum. Extreme northwest and northeast China and Japan (north of about 40°N), on the other hand, were still experiencing a continental climate and, although warmer than at 18,000 yr B.P., were still cool and dry, with dominant steppe vegetation (*Artemisia* and chenopods) in the northeast and birch and pine forests in the Beijing lowlands (Fig. 10.11).

6000 YR B.P.

Evidence from all regions of China and Japan shows that at 6000 yr B.P. both temperature (Figs. 10.6a,b and 10.9) and moisture (Fig. 10.10) were higher than at present. The data indicate that the warming may have exceeded 3°C in some regions (Fig. 10.6b). The interval of peak warmth and moisture occurred between about 8000 and 3000 yr B.P. throughout the region (Fig. 10.9), although the maps depict conditions at 6000 yr B.P. (Figs. 10.10 and 10.11).

By 7500 yr B.P. until about 3000 yr B.P. eastern China was warmer and wetter than at present, and forest belts shifted northwest (Fig. 10.11). Lake levels were high throughout most of China. Warm and dry conditions prevailed only in parts of north-central and northwestern China (Fig. 10.10), as evidenced by lower lake levels in Qinghai, continued desiccation of the western inland lakes on the Loess Plateau, and expansion of loess deposits in north-central China. Japan was warmer but may have experienced fluctuations of slightly drier-than-present conditions at times within this interval. Paleosols in loess in the middle reaches of the Yellow River dating from between 7000 and 4000 yr B.P. (Fig. 10.9) contain *Quercus, Pinus,* Betulaceae, and Ulmaceae pollen (Liu, 1985a), indicating warmer and wetter periods during the middle Holocene in north-central China. The snowline in northwest China was higher than at present, and spruce forests replaced steppe at lower altitudes in the Tian Shan, indicating a warmer and somewhat wetter climate there during the middle Holocene (Fig. 10.11). On the east coast of China between 7000 and 5000 yr B.P., the northern limit of the warm-water foraminifer *Asterorotalia subtrispinosa* was displaced northward about 4° latitude. Sea levels approached modern coastline configurations by about 8000 yr B.P. but were even higher than at present in the coastal lowlands of south China at 7500 yr B.P. and in north China and Japan at 6000 yr B.P. Neolithic cultures expanded in central, northern, and western China during this time. Evidence of human settlement was found on the Tibetan Plateau as far north as the Kunlun Mountains and at altitudes as high as 5200 m, where no people live today, suggesting a warmer, more habitable climate. Northward expansion of the subtropical and tropical forests of southeastern China (Fig. 10.11) is indicated by bamboo and mangrove fos-

sils and by bones of warm-climate fauna (bamboo rat, giant panda, elephant, and others). There is also evidence of coral reefs marking the boundaries of the mid-Holocene transgression of the sea in the central and south China lowlands. Some of these ancient corals are found north of the present growth limit for coral in China and Taiwan, indicating higher sea-surface temperatures in parts of the East and South China seas during the middle Holocene.

3000 YR B.P.

From about 3000 yr B.P. to the present, the climate cooled throughout China (Figs. 10.6b and 10.9). Glaciers expanded in the Tian Shan and the Tibetan Plateau at 3000 yr B.P., from 1920 ± 110 to 1540 ± 85 yr B.P., and from 1600 to 1900 A.D. (the little ice age). Permafrost also expanded during these cold periods. Spruce and fir forests increased in the northeast. The northern and western lands, however, continued arid (Fig. 10.10), lake levels on the Tibetan Plateau decreased, and loess deposits and deserts expanded. The eastern regions, under the influence of a monsoonal climate, remained wet.

Land use has been intensive in China for the last several thousand years. Vegetation changes associated with agriculture have been documented in pollen records from Taiwan dating from 4000 yr B.P. and from Japan from about 3000 yr B.P.

Comparison of Inferred Climate with Model Simulations

To help elucidate the dynamics of climate change, computer models of climate have been developed based on observed boundary conditions and on conditions defined by theoretical constraints. Comparison of model-simulated climates with "observed" climates leads to validation and refinement of the models. In this section we examine the results for China derived from a global climate model—the NCAR Community Climate Model (CCM) (Kutzbach and Guetter, 1986)—in light of the biogeologic data from China.

The CCM is based on changes in solar radiation associated with the orbital changes calculated by Milankovitch (Milankovitch, 1941; Imbrie and Imbrie, 1979; Kutzbach, 1981; Xu Qinqi, 1981, 1984; Kutzbach and Otto-Bliesner, 1982; Kutzbach and Guetter, 1984a,b, 1986). Milankovitch (1941) argued that variations in the earth's orbital parameters could have caused major glacial-interglacial changes in climate. This theory is supported by the isotope stratigraphy of ocean sediment cores (Hays *et al.*, 1976; Lyle *et al.*, 1992), the loess-paleosol sequences in China (Liu *et al.*, 1985; Kukla *et al.*, 1987) and Hungary (Pecsi, 1987), and the chronostratigraphy of the size of sediment grains in cores taken from Lake Biwa, Japan (Kashiwaya *et al.*, 1987). Kutzbach (1981) found that average solar radiation in July in the Northern Hemisphere 9000 yr ago was at least 7% greater than at present because of the differences in the orbital parameters and that this higher solar radiation drove a monsoonal flow more intense than today. Kutzbach and Guetter (1984a,b, 1986) used the CCM to simulate the response of climate to the combined effects of changes in the orbital parameters and in glacial-age boundary conditions, including the altered South Asian geography associated with sea-level lowering. Figure 10.12 shows the results calculated from the model that pertain to China. We compare these model simulations with the climate and vegetation maps (Figs. 10.10 and 10.11) generated from the biogeologic data.

The graphs in Figure 10.12 chart the variations in January, July, and annual temperature, precipitation, and precipitation-minus-evaporation (P-E, or effective moisture) simulated by the model at 3000-yr intervals from 18,000 yr B.P. to the present averaged over the whole of China (Fig. 10.12a) and region by region (Fig. 10.12b-h). For China as a whole, the simulated temperature rises from glacial conditions at 18,000 yr B.P. (about 7°C below present in January and 3°C below present in July) to a peak at 9000 yr B.P. (about 1.8°C warmer than at present in July but still about 0.5°C cooler than at present in January). After 9000 yr B.P. July temperature decreases gradually to the present level, while January temperature rises slightly toward the present level. This general pattern is similar to the climatic trends inferred from the biogeologic evidence (Figs. 10.6b and 10.9). For example, pollen samples from sites in west, northeast, and south China all show vegetation shifting from sparse, arid, or cold-climate plant associations to more thermophilic and mesophytic plant communities at 9000–8000 yr B.P. (Figs. 10.9 and 10.11), and Holocene paleosol development is evident at sites on the Loess Plateau at about 9000 yr B.P. (Fig. 10.9). The model predicts peak warmth at 9000 yr B.P. throughout China (Fig. 10.12). The data, however, suggest that peak warmth occurred on average throughout China at about 7500 yr B.P. (Figs. 10.6b and 10.9) and that the climate remained warmer than at present until about 3000 yr B.P. (Fig. 10.6b).

The model simulates precipitation as low as 0.7 mm/day below modern values at 18,000 yr B.P. throughout most of China for both July and January (Fig. 10.12). This represents a decrease in MAP of about 19% compared to the simulated modern values averaged for all of China (Fig. 10.12a); the decreased precipitation ranges from 3% in July in south China (Fig. 10.12c) to 32% in January in east China (Fig. 10.12d). The paleoprecipitation map constructed from the biogeologic evidence similarly shows that most of China was drier at 18,000 yr B.P. than at present (Fig. 10.10).

Peak mean July precipitation (close to 2 mm/day greater than at present) was simulated for most of China by 9000 yr B.P. and represents an 18% increase in

precipitation compared to the simulated modern mean values for July for all of China (Fig. 10.12a). However, in the southwest (Tibetan Plateau) (Fig. 10.12h) and indeed in the entire south (Fig. 10.12c) and west (Fig. 10.12e), simulated July precipitation had already increased by 15,000 yr B.P. and was between 2 and 5 mm/day greater than at present by 12,000 yr B.P. (46% higher on the Tibetan Plateau and 18% higher in south China than the simulated modern values for these regions).

Simulated January precipitation remained as much as 1 mm/day below present values throughout China until about 9000 yr B.P., and January P-E for the same regions also remained below present levels until 9000 yr B.P. Coastal east China (Fig. 10.12d) and northwest China (Fig. 10.12f) were exceptions to higher-than-present January precipitation after 9000 yr B.P.; in these regions simulated January precipitation and P-E remained below present values. In east China even July P-E did not exceed modern values until after 6000 yr B.P. (Fig. 10.12d).

The paleoclimatic maps drawn from the data (Fig. 10.10) similarly show that most of China and Japan south of latitude 40°N was wetter than at present at 9000 yr B.P., but they also indicate that by 6000 yr B.P. more of China and all of Japan north of 40°N were also wetter than at present. The data also show that at 6000 yr B.P. the Taklimakan Desert and parts of the Loess Plateau were the only regions in China that were as dry as at present or drier. These regions were also warmer than at present, as was the rest of China at that time (Fig. 10.9).

It is interesting to note that the simulated temperature and precipitation patterns for west China (Fig. 10.12e) and the Tibetan Plateau (Fig. 10.12h) are the most similar to those for China as a whole (Fig. 10.12a) and that the western and central regions are very different from the coastal regions in the simulated evolution of climate for the past 18,000 yr. Another interesting result is the decreased seasonality since about 6000 yr B.P. shown in both the temperature and precipitation curves for all of China; the July curves show the most dramatic change in this respect except in south China (Fig. 10.12c) and in the southwest on the Tibetan Plateau (Fig. 10.12h), where the January temperature curves change the most.

In summary the NCAR CCM appears to simulate the long-term trends of climatic change in China from the late Pleistocene to the present. The dynamical changes simulated in the model experiments may therefore suggest ways that the real climate changed since 18,000 yr B.P. as a function of orbital changes and glacial-age boundary condition changes. The most pronounced differences between the data and the model results are the timing of peak moisture throughout China (although both the model and the data indicate increased precipitation at both 9000 and 6000 yr B.P.) and the delayed warming in the northeast that is indicated by the data but not reflected in the model. Four factors may account for these differences:

(1) Late-Pleistocene and Holocene sea-level/coastline changes. The climate simulation for east China (Fig. 10.12d), which includes all of coastal China, differs the most from that of the western, northern, and central inland regions. The sea level was still lower than at present at 9000 yr B.P. (Fig. 10.10), which may have caused continental conditions to linger in the northeast, especially in January, when according to Milankovitch theory, solar radiation in the Northern Hemisphere was lower than at present. By 6000 yr B.P., however, sea level was at its present level or higher, and northeast China became warmer and wetter than at present. The lowered sea level along the east China coast at 9000 yr B.P. was not considered in the model.

(2) The response of biologic systems to orbitally induced seasonality changes. Orbital changes enhanced Northern Hemisphere seasonality before and during the early Holocene and lessened the seasonal contrast markedly after 6000 yr B.P. Seasonality, therefore, is a separate climatic factor that may affect biologic systems more directly than geologic ones. Even though temperature and moisture had increased by 9000 yr B.P., the more extreme cold in the winter half-year would inhibit the expansion of plants and animals into colder regions (higher latitudes and higher altitudes). At 6000 yr B.P. summer warmth and wetness remained within growth-permitting levels, while winter warmth increased as the seasonal contrast decreased. Therefore it is not surprising to find greatest expansion of plants and animals and human settlement at 6000 yr B.P. in China despite the increased warmth and wetness evident by 9000 yr B.P. in the geologic record. Seasonality may be the most important factor limiting biotic systems with broad temperature and moisture requirements (as opposed to biotic systems adapted to extremes) from being in equilibrium with climatic change. Geologic and hydrologic processes such as glacier movements, lake-level changes, and loess-paleosol development may therefore react sooner than most recorded biologic processes (phytoplankton, which have a short generation time, may be an exception) in Arctic and temperate regions.

(3) The response of geologic systems to orbitally induced climatic changes. Fluctuations in snowfall and changes in glacial dynamics on the Tibetan Plateau and the central mountains since the late Pleistocene may have produced feedback mechanisms that intensified or limited the climatic changes. Models of the dynamics of climate change may need to incorporate a stronger atmosphere-ocean-land process linkage.

(4) Possible dating problems. As mentioned above, some of the mapped sites may vary as much as 1500 yr from the nominal date. Dating errors may be additive, making the chronology of climate inferences from data even less precise.

Monsoons, Airstreams, and Biotic Regions

MODERN AIRSTREAM CLIMATOLOGY

The climate of China is affected on a global scale by boundary constraints, orbital dynamics, and latitudinal differences in solar radiation. Regional climatic constraints include the ocean currents off the east and south coasts and the extreme topography of the Tibetan Plateau in the western part of the country. The Tibetan Plateau not only blocks airstreams coming from the Arctic, Pacific, Atlantic, and Indian oceans, but also its elevation intensifies the land-sea temperature contrasts during winter and summer and contributes to seasonal precipitation extremes that result in a monsoon circulation. The Pacific Ocean currents east of China contribute to climatic conditions inland. However, eastern China, unlike the Atlantic coast of North America, where the continental climate is improved by the proximity of the warm Gulf Stream, does not benefit from the warm Kuroshio current, which lies far offshore in the western Pacific. One branch of the Kuroshio current touches the southeastern coast of Japan, and another branch, the Tsushima current, enters the Sea of Japan. The winter monsoon from the interior of China is strong enough to displace the warm flow even farther to the east, leading to an intensification of the cold climate in winter. Also, the cold littoral drift flowing south from the Bohai Sea (at 40°N) to the Taiwan Straits (about 25°N) is stronger in winter (Ren et al., 1985), making the coastal climates even colder (Fig. 10.4c).

In the west of China, the Tibetan Plateau blocks the midlatitude westerlies and results in a split jet that moves to the north and the south of the plateau. Cooling of the airmasses over the plateau gives rise to the central Asian airstream, which dominates the northwest and west-central parts of China with varying strength throughout the year and moves out of the plateau in winter to become the winter monsoon (cold, dry air moving from the Tibetan Plateau to the north, east, and south) throughout China (Bryson, 1986).

According to Ren et al. (1985) and Bryson (1986), weather patterns do a complete reversal from summer to winter. Low pressure over the plateau is centered at about 70-80°E in the summer from about June through August. High pressure north of this low brings drought to northwest China (Xinjiang and the western part of the Loess Plateau [Fig. 10.2a,c]) in summer as well as winter—hence the arid steppe vegetation (Fig. 10.5) and the deserts (Fig. 10.2c) in those regions. In winter, from about October through April, a cold high-pressure ridge is centered at about 80-90°E over the Tibetan Plateau, which brings rain to the southern foot of the Himalayas, an area that also receives rain from the southwest monsoon in summer.

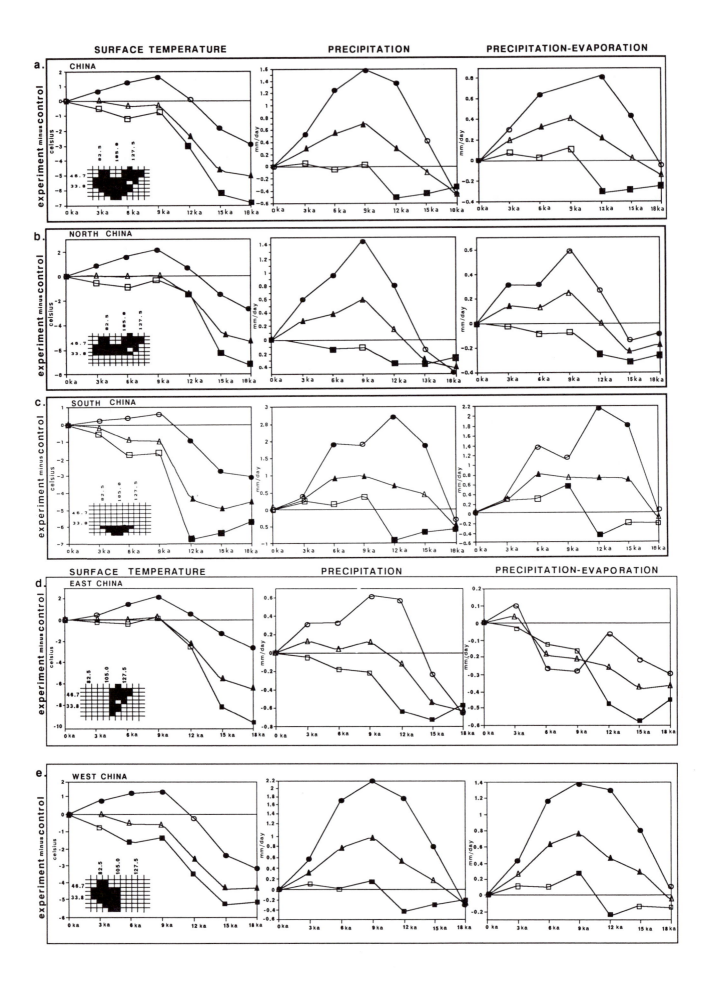

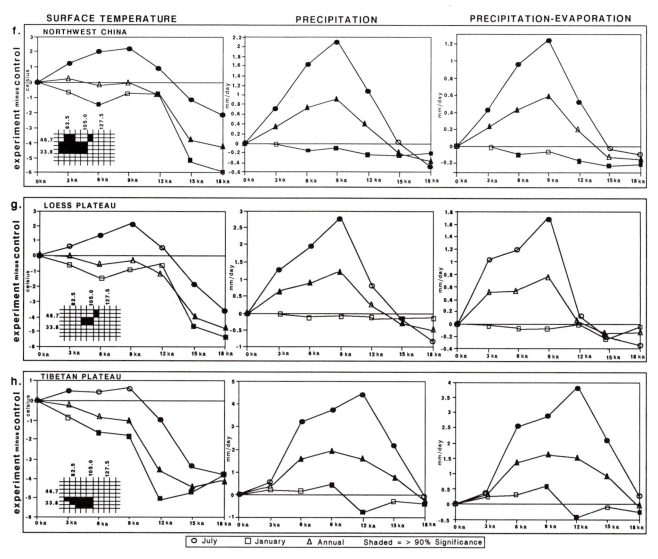

Fig. 10.12. Paleoclimatic variables simulated by the Community Climate Model (CCM): surface temperature (°C), precipitation (mm/day), and precipitation-minus-evaporation (P-E) (mm/day) for (a) all of China, (b) north China, (c) south China, (d) east China, (e) west China, (f) northwest China, (g) Loess Plateau, and (h) Tibetan Plateau. Mean July (○), January (□), and annual (△) values are plotted for each variable. Shaded symbols indicate values with statistical significance greater than 90%. Inset maps of the CCM map grid with each region show latitude and longitude and number of grid squares averaged to obtain the data for each entry.

A coastal low-pressure trough exists in winter and a coastal pressure ridge in summer. The transition in atmospheric circulation takes place in May and September.

In May the northwest jet disappears and the southwest jet pushes northward, strengthening the southwest monsoon (the Indian Ocean airstream) (Sun Shuqing *et al.,* 1986). The northward movement of the Indian Ocean airstream and the southwest jet (narrowed and concentrated by the north-south trend in the boundary of the eastern part of the Tibetan Plateau [Sun Shuqing *et al.,* 1986]) for a short time in May or early June causes rainfall on the eastern portion of the Tibetan Plateau, watering the grass steppe

(the plateau is only semiarid, in contrast to northwest China [Fig. 10.5]) and permitting forest growth on the windward slopes of the mountains in central China. In effect, the eastern rim of the Tibetan Plateau creates the southwest boundary of a tension zone (Fig. 10.13) and funnels moisture northeast to the coastal plain (Fig. 10.2d) and to Japan (Luo and Yanai, 1986). The South Pacific high-pressure system reaches the southeastern coast of China by May and moves slowly northward on the east coast (Fig. 10.13). Referred to as the *mei-yu* ("plum rain") front (Luo and Yanai, 1986), it brings the "plum rains" to north China in mid-June and July, when the plum trees ripen (Ren *et al.,* 1985). If there is a strong high-pressure system in

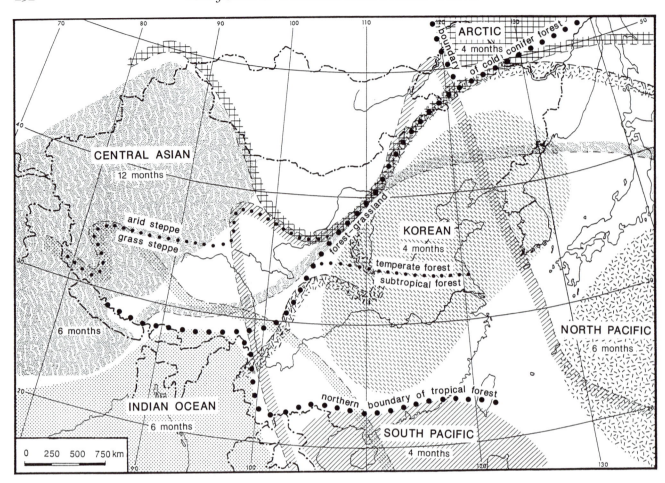

Fig. 10.13. Southeast Asian airstream positions (patterned areas labeled with capital letters) and durations (data from Bryson, 1986) compared with vegetation regimes in China (see Fig. 10.5). Five of the six dominant airstreams (those lasting longer than four months) are associated with a similarly patterned line that represents the position of that airstream for at least one month. The large dots demarcate major vegetation boundaries strongly influenced by the position of one or more airstreams today. The central diagonal large-dotted line marks the forest-grassland boundary, which lies along a tension zone formed by boundaries of five of the six dominant airstreams. The dotted line in the northeast corner of China marks the cold conifer forest boundary. The tropical forest boundary in south China coincides with the position of the Indian Ocean and South Pacific airstreams for more than four months. Lines of smaller dots represent less well-defined relationships between airstream position and biotic region, such as the boundary between grass steppe and arid steppe in the west (which may be orographic) and that between subtropical and temperate forests in the east.

northeast China—the Korean airstream (Bryson, 1986) or eastern Arctic airstream (Ren *et al.,* 1985)—the South Pacific air is stalled in the southeast, causing floods south of the Changjiang (32°N) and drought north of that latitude in July (Ren *et al.,* 1985).

In early March the southwest monsoon associated with the Indian Ocean airstream begins to affect south China. The Indian Ocean airstream (Fig. 10.13) and the accompanying warm summer rains dominate this part of China for about six months (Bryson, 1986). The airstream arrives at its most northerly position at about Kunming in August. Therefore, two airstreams bring summer rain to China, one from the Indian Ocean in the southwest and one from the South Pacific in the south and southeast. The intensity, timing,

and duration of these airstreams determine the northern border of the tropical forest vegetation in south China (Figs. 10.5 and 10.13).

In the northeast, Arctic air dominates in the summer (above about 50°N) and determines the southern boundary of the northern conifer forest (Figs. 10.5 and 10.13) composed primarily of the deciduous conifer *Larix* (larch), which is well adapted for growth in a cold and windy climate with seasonal extremes (mean July temperature 20°C, mean January temperature –30°C [Fig. 10.4b,c]). The Arctic airstream reaches its southernmost extent in May and forms a steep diagonal cold front stretching from 50°N, 140°E to 30°N, 80°E in north-central China (Bryson, 1986). This diagonal parallels the grassland and forest-grassland (savanna) region and forms a tension zone (Fig. 10.13) that divides the vegetation of China (Fig. 10.5) into wet (east) and dry (west) zones. The extent of grassland, savanna, and forest in central China along this tension zone (Figs. 10.5 and 10.13) is determined by the movements of the Arctic, central Asian, Korean, and South Pacific airstreams, which all form a steep diagonal gradient in central China but at different times of the year (Fig. 10.13). Until September the Qin Ling and other central mountain ranges block the cold, dry, continental central Asian airstream from extending into the northeastern part of the county, but the continental airstreams strengthen in late fall, and by November the flow of central Asian air (the winter monsoon) again moves strongly south-north in the middle of China.

PALEO-AIRSTREAM CLIMATOLOGY

The paleoclimate and paleovegetation maps (Figs. 10.10 and 10.11) indicate a warm climate and a strengthened monsoonal circulation by at least 9000 yr B.P. By 6000 yr B.P., however, the maps indicate that China was wetter than at present and also wetter than at 9000 yr B.P. Although summers were warmer at 9000 yr B.P., winters were also colder (Kutzbach, 1981), and severe winters would limit the expansion of human settlements, faunal species, and most temperate plants. Because of the orbital changes, Northern Hemisphere seasonality decreased after 9000 yr B.P. and decreased even more after 6000 yr B.P., and the more equitable seasons may have promoted the spread of Neolithic cultures in all of China and the expansion of forests in the northwest and east (Fig. 10.11) that is evident from the biogeologic data. After 6000 yr B.P. the moist summer monsoon circulation weakened. By 3000 yr B.P. most of China was dominated by the cold winter monsoon, and forests again declined while steppe vegetation expanded.

The monsoonal circulation in China clearly intensified at 9000 yr B.P. (Fig. 10.10), when seasonality was enhanced in the Northern Hemisphere and solar radiation (Fig. 10.9) in the summer half-year increased (Kutzbach, 1981, 1987). However, the presence of taiga (cold scrub forest-grassland) in northeast China (Fig. 10.11) suggests that the summer monsoon boundary was at about latitude 40°N at 9000 yr B.P. (Fig. 10.3d). Probably a concurrent strong east Arctic airstream prevented more northward movement of the wet *mei-yu* front during early-Holocene summers. The vegetation changed dramatically between 7500 and 3000 yr B.P. (Figs. 10.10 and 10.11), when winters became milder, P-E increased in most of Southeast Asia, and the coastline configuration approached the modern land-sea distribution. After 6000 yr B.P. the monsoon circulation gradually decreased to its present pattern (Fig. 10.10). The summer monsoon and the southwestern monsoons are less strong today (Fig. 10.3d), and the tropical forest belt determined by these boundaries has contracted since 6000 yr B.P. (Fig. 10.11). Because of the orbital changes, winters are warmer today than at 9000 yr B.P. (the winter monsoon [Fig. 10.3d] is less intense), and summers are cooler.

The cooler summers and decreased moisture (a less intense southwest monsoon [Fig. 10.3d]) both predicted in the model and inferred from the biogeologic data (Fig. 10.9) accentuate the contrast between the wet (east) and dry (west) parts of the country observed in the modern vegetation (Figs. 10.5 and 10.11) and climate (Fig. 10.3b). The tension zone (Fig. 10.13), which appears to be both a climatic and a physical (orographic) boundary today coincident with the present monsoon boundary (Figs. 10.3d, 10.11, and 10.13), was displaced in the past 18,000 yr as a result of orbital and surface boundary condition changes (Figs. 10.10, 10.11, and 10.13). The changing monsoon boundaries diagramed in Figures 10.3d and 10.13 were inferred from the biogeologic evidence of climate change since 18,000 yr B.P. A comparison of these inferred boundaries with those suggested by the past circulation patterns simulated by the CCM is underway.

OTHER MONSOON REGIONS

Prell and Kutzbach (1987) compared the CCM results with the paleoclimate records from southwest Asia and equatorial Africa and suggested that both Milankovitch solar-radiation changes (Fig. 10.9) and glacial-age boundary conditions are necessary to explain features of monsoonal climates on a longer time scale and that the effects of these variables are additive. Clemens *et al.* (1991) studied the forcing mechanisms of the summer monsoon from biogeochemical changes in sediment cores from the Indian Ocean. They concluded that although external forcing from cyclic changes in solar radiation is important, a major additional forcing component is internal and is associated with phases of latent heat transport from the southern Indian Ocean to the Tibetan Plateau.

The good fit of our "observed" paleoclimatic changes with the modeled climates for China and Japan (Southeast Asia) suggests that glacial-interglacial boundary conditions (such as changes in landmass due to sea-level changes), orbitally induced solar-radiation changes (increased seasonality and increased solar radiation in July), and the thermodynamic changes caused by the orography of the Tibetan Plateau (which may amplify the effects of the other variables) are all necessary to explain the changes in the monsoonal climate in Southeast Asia since 18,000 yr B.P.

Increased moisture from about 10,000 to 5000 yr B.P. has been inferred from mesic pollen and high lake levels in India, Kashmir, Nepal, and Uttar Pradesh, all southwest of the Tibetan Plateau (Singh *et al.*, 1974; Gupta, 1975; Singh and Agrawal, 1976; Swain *et al.*, 1983; Dodia *et al.*, 1984; Vishnu-Mittre, 1984; Wasson *et al.*,

1984; Kutzbach, 1987), high tropical lake levels in Africa (Street-Perrott and Harrison, 1984), an increase in "monsoon" pollen (windborne from the land) in ocean cores taken in the Gulf of Aden and near the southwestern coast of India in the Arabian Sea (van Campo et al., 1982; van Campo, 1986), an aridity index showing a moisture peak at that time in west equatorial Africa (Pokras and Mix, 1985), and the high salinity in the Bay of Bengal (because high-salinity water is transported from the western Indian Ocean when warm southwestern monsoonal winds are prevalent) (Prell and Kutzbach, 1987). Extreme aridity is also inferred at some of the same sites during the last glacial maximum (about 22,000–18,000 yr B.P.) (Street-Perrott and Harrison, 1984; van Campo, 1986; Kutzbach, 1987), and increasing aridity since about 5000 yr B.P. in northwest India and Kashmir (Singh et al., 1974; Bryson and Swain, 1981; Swain et al., 1983; Dodia et al., 1984; van Campo, 1986) and North Africa (Street-Perrott and Harrison, 1984; Kutzbach and Street-Perrott, 1985) is also evident from the paleoclimatic record. The chronology and direction of climate change in these areas are the same as those inferred in this chapter for Southeast Asia. The spatial and temporal similarity between rainfall patterns in Southeast Asia and those in southwest Asia and north to central Africa from the last glacial maximum to the present suggests that solar-radiation changes associated with orbital changes play the most important role in the amplification or attenuation of the monsoonal circulation in this region.

OTHER CLIMATE MODELS

There is some evidence that the extent of glaciation and the amount of snow cover on the Tibetan Plateau may affect other important climatic events, such as the El Niño/Southern Oscillation phenomenon (Barnett et al., 1988), as well as the monsoonal circulation in Southeast Asia. Climate models have shown that regional climate is sensitive to changes in snow cover (Gallimore et al., 1986; Barnett et al., 1988). When spring snowfall is doubled—as may happen when the southern jet is strengthened over the Tibetan Plateau—the southwestern (Indian) monsoon may be delayed in onset and in intensity (Barnett et al., 1988). The weak summer monsoon at 18,000 yr B.P. may have been partly a result of the increased snowfall on the mountains in the southeastern corner of the plateau, although in general the geologic evidence for other areas of the plateau indicates that glaciers elsewhere were not much larger than today. The effects of amplified seasons and enhanced monsoons, along with decreased snowfall on the plateau at 9000 yr B.P. (evidence of receding glaciers at that time [Fig. 10.9]), support the argument of Barnett et al. (1988) that "processes over continental land masses" must be considered as well as atmosphere-ocean coupling when large-scale changes in the global climate system

are studied. This interaction also suggests that at 9000 yr B.P. the effects of changed orbital parameters, changed boundary conditions, and the subsequent changes in glacial dynamics on the Tibetan Plateau were additive.

Conclusions

During the last glacial maximum the climate was cold and dry in China and Japan. The region warmed gradually after that time, and the monsoon circulation intensified at about 9000 yr B.P. Pollen samples from sites in west, south, and northeast China show that the vegetation shifted from sparse, arid, or cold-climate associations to more thermophilic and mesophytic communities at 9000–8000 yr B.P. Paleosols developed on the Loess Plateau and peatlands formed in west and central China also at about 9000 yr B.P. According to the biologic evidence, the interval from 8000 to 3000 yr B.P. was the warmest and wettest in the last 18,000 yr in China and Japan. During this time seasonal contrasts waned, and plant and animal ranges and human settlements expanded. The climate has been cooling since 3000 yr B.P.

The orography of western and central China has an important influence on regional climate and vegetation change in China, accentuating the dryness in the west and the wetness in the east. The Tibetan Plateau blocks airstreams coming from the Arctic, Pacific, Atlantic, and Indian oceans. Consequently, the Arctic, central Asian, Korean, and South Pacific airmasses form a steep diagonal gradient—a climatic tension zone—in central China at different times of the year. The extent of grassland, savanna, and forest in central China is determined by the movements of these airstreams along the tension zone. The timing, duration, intensity, and location of airstreams is in turn determined largely by the dynamical response of monsoon circulations to changing orbital parameters. The tension zone was displaced northward in China between 9000 and 6000 yr B.P. and has moved southward again since that time.

Although large-scale climate changes in China for the past 18,000 yr can be predicted by the changing orbital parameters of the earth, millennial-scale changes in climate are indicated in all late-glacial and Holocene time intervals by the widely varying kinds of paleoclimatic data examined in the studies reviewed in this chapter (Fig. 10.9). Because of the generally poor dating control in many of these studies, it is impossible to evaluate whether some of these events were synchronous. However, climatic fluctuations on these shorter time scales are documented in the chronologies of loess-paleosol deposition, glacier movements, and sea-level fluctuations as well as in historical climate records, and they cannot be explained by orbital changes or by orography. A most important event-scale finding is the identification of an abrupt cooling from 11,050 to 10,900 yr B.P., sensed by conifer forest expansion and sea-level lowering in east China, increased herbaceous pollen in the northeast, and changes in $\delta^{18}O$ values in lake sediments from north-central China and western Tibet. A cold event at this time has also been identified in the North Atlantic and in Europe and is referred to as the Younger Dryas. The demonstration of a broader spatial extent of this short-term cooling may lead to redefinition of boundary conditions for climate models and reexamination of the data from other regions.

Similar interpretations of vegetation dynamics immediately before and during the early Holocene have been made from pollen records in southwestern China and in northern and eastern North America. This interval was characterized by enhanced Northern Hemisphere seasonality caused by orbital change and was a period of transition from glacial to interglacial climates. Pollen assemblages demonstrate dynamic expansion and replacement of species during this transition, in contrast to the more stable vegetation established after the early Holocene. This global similarity in forest dynamics in the Northern Hemisphere at about 9500 yr B.P. suggests that increased seasonality may have had a greater impact on vegetation dynamics during deglaciation than did ice-sheet retreat. Vegetation instability in North America in the early Holocene has been attributed partly to the proximity of the waning ice sheet, but China had no ice sheet, and although the western mountain glaciers contracted after about 10,000 yr B.P., they are still features of the Asian landscape.

Because of a combination of sea-level rise and crustal rebound, the coastline of northern China was at about the level of the Korean peninsula by 9000 yr B.P. This sea level admitted warm water into the Sea of Japan and initiated a maritime climate in most of Japan, but northeast China continued to have a more continental climate until close to 7500 yr B.P., when sea level approached or surpassed the modern coastline configuration. Sea-level changes probably modified climatic changes associated with orbital changes in northeast China and delayed warming in that region in the early Holocene.

Acknowledgments

Figures were drafted by W. Fraczek. We thank P. Behling, R. Selin, A. Anderle, M. Kennedy, M. Woodworth, Zhang De'er, and P. Sanford for their help with many aspects of this study. We also thank K.-b. Liu for critical review.

References

An Zhimin (1979). Peiligang, Cishan, and Yangshao. *Kaogu* 4, 335-346.

An Zhimin, Yin Zesheng, and Li Bingyan (1980). Paleolithic and microlithic artifacts at Shenzha and Shuanhu of northern Tibet. *Kaogu* 6, 481-494.

An Zhisheng, Kukla, G. J., Porter, S. C., and Xiao Jule (1991). Magnetic susceptibility evidence of monsoon variation on the Loess Plateau of central China during the last 130,000 years. *Quaternary Research* 36, 29-36.

Anderson, P. M. (1985). Late Quaternary vegetational change in the Kotzebue Sound area, northwestern Alaska. *Quaternary Research* 24, 307-321.

Anonymous (1976). The first stage excavation of Hemudu Neolithic site. *Wenwu* 8, 15-17.

Barnett, T. P., Dumenil, L., Schlese, U., and Roeckner, E. (1988). The effect of Eurasian snow cover on global climate. *Science* 239, 504-507.

Brenner, M., Song, X., Wang, Z., Long, R., Binford, M. W., and Moore, A. M. (1989). Palaeolimnology of Qilu Hu, Yunnan Province, China. *In* "Abstracts, Vth International Symposium on Palaeolimnology, Cumbria, U.K.," p. 15.

Broecker, W. S., Andree, M., Klas, M., Bonani, G., Wolfli, W., and Oeschger, H. (1988). New evidence from the South China Sea for an abrupt termination of the last glacial period. *Nature* 333, 156-158.

Bryson, R. A. (1986). Airstream climatology of Asia. *In* "Proceedings of International Symposium on the Qinghai-Xizang Plateau and Mountain Meteorology, March 20-24, 1984, Beijing, China," pp. 604-617. American Meteorological Society, Boston.

Bryson, R. A., and Swain, A. M. (1981). Holocene variations of monsoon rainfall in Rajasthan. *Quaternary Research* 16, 135-145.

Burbank, D. W., and Kang Jian Cheng (1991). Relative dating of Quaternary moraines, Rongbuk Valley, Mount Everest, Tibet: Implications for an ice sheet on the Tibetan Plateau. *Quaternary Research* 36, 1-18.

Chang Chengfa and Wang Jingtai (1982). Late Quaternary climate, glacier, and sea level changes in China. *In* "Proceedings of the Third Chinese National Quaternary Conference," pp. 111-121. Science Press, Beijing.

Chao Sung-chiao (1977). Eolian sands and dust bowl in the Hulunbuir sandy land. *In* "Physical Geography of the Chinese Arid Land."

——. (1981). The sandy deserts and the Gobi in China: A preliminary study of their origin and evolution. *In* "Desert Lands of China." International Center for Arid and Semi-Arid Land Studies (ICASALS), Texas Technical University, Lubbock.

Chen Chenghui, Chen Shuomin, and Zhou Kunsu (1965). Pollen analysis of the *Nelumbo nucifera* containing Holocene sediments near Pulandian of Liaodong Peninsula. *Quaternaria Sinica* 4, 167-173.

Chen Fangji (1979). The Holocene strata of Beijing area and the change of its natural environment. *Scientia Sinica* 9, 900-907.

Chen Kezao and Bowler, J. M. (1986). Late Pleistocene evolution of salt lakes in the Qaidam Basin, Qinghai Province, China. *Palaeogeography, Palaeoclimatology, Palaeoecology* 54, 87-104.

Chen Kezao, Yang Shaoxiu, and Zheng Xiyi (1981). The salt lakes on the Qinghai-Xizang Plateau (in Chinese). *Acta Geographica Sinica* 36, 13-21.

China Reconstructs Map (1981). "Map of the People's Republic of China." Cartographic Publishing House, Beijing.

Chow Benshun (1978). The distribution, paleoecology, and paleoclimatic significance of the *Mammuthus-Coelodonta* fauna (in Chinese). *Acta Paleovertebrata and Paleoanthropologia Sinica* 16, 47-57.

Chu K'o-chen (1926). Climate pulsations during historical times in China. *Geographical Review* 16, 274-282.

——. (1973). A preliminary study on the climatic fluctuations during the last 5000 years in China. *Scientia Sinica* 16, 225-256.

Clemens, S., Prell, W., Murray, D., Shimmield, G., and Weedon, G. (1991). Forcing mechanisms of the Indian Ocean monsoon. *Nature* 353, 720-725.

Colinvaux, P. A. (1967). Quaternary vegetational history of arctic Alaska. *In* "The Bering Land Bridge" (D. M. Hopkins, Ed.), pp. 207-231. Stanford University Press, Palo Alto, Calif.

Cui Zhi-jiu and Xie You-yu (1985). On late Pleistocene periglacial environments in the northern part of China. *In* "Quaternary Geology and Environment of China" (Liu T.s., Ed.), pp. 226-232. China Ocean Press, Beijing, and Springer-Verlag, New York. (Paper also in Chinese book of the same name, 1982, pp. 167-171.)

Denton, G. H., and Hughes, T. J., (1981). "The Last Great Ice Sheets." Wiley Interscience Press, New York.

Dodia, R., Agrawal, D. P., and Vora, A. B. (1984). New pollen data from the Kashmir bogs. *In* "The Evolution of the East Asian Environment" (R. O. Whyte, Ed.), pp. 569-573. Centre of Asian Studies, University of Hong Kong.

Duan Wanti, Pu Qingyu, and Wu Xihao (1980). Climatic variations in China during the Quaternary. *GeoJournal* 4, 515-524.

——. (1981). A preliminary study on the Quaternary climatic change in China. *In* "Proceedings of the National Climatic Change Conference, Beijing, China, 1979," pp. 7-17. Science Press, Beijing.

Fang Jin-Qi (1991). Lake evolution during the past 30,000 years in China, and its implications for environmental change. *Quaternary Research* 36, 37-60.

Forman, S. L. (1991). Late Pleistocene chronology of loess deposition near Luochuan, China. *Quaternary Research* 36, 19-28.

Fuji, N. (1974). Palynological investigations on 12-meter and 200-meter core samples of Lake Biwa in central Japan. *In* "Paleolimnology of Lake Biwa and the Japanese Pleistocene" (S. Horie, Ed.), pp. 227-235.

Gallimore, R. G., Otto-Bliesner, B. L., and Kutzbach, J. E. (1986). The effects of improved parameterizations for orography, snowcover, surface fluxes, and condensational processes on the climate of a low resolution GCM. *Journal of the Atmospheric Sciences* 43, 1961-1983.

Gasse, F., Arnold, M., Fontes, J. C., Fort, M., Gibert, E., Huc, A., Li Bingyan, Li Yuanfang, Li Qing, Melieres, F., Van Campo, E., Wang Fubao, and Zhang Qingsong (1991). A 13,000-year climate record from western Tibet. *Nature* 353, 742-745.

Grichuk, V. P. (1969). Opyt rekonstruktsii nekotorykh elementov klimata severnogo polushariia v Atlicheskii period golotsena (An attempt to reconstruct certain elements of the climate of the Northern Hemisphere in the Atlantic period of the Holocene). *In* "Golotsen: K VIII Kongressu INQUA, Paris" (M. I. Neishtadt, Ed.), pp. 41-57. Izd-vo Nauka, Moscow. (English translation by G. M. Peterson, Center for Climatic Research, University of Wisconsin-Madison.)

Grimm, E. C., and Jacobson, G. L., Jr. (1992). Fossil-pollen evidence for abrupt climate changes during the past 18,000 years in eastern North America. *Climate Dynamics* 6, 179-184.

Guiyang Institute of Geochemistry (1977). Environmental changes in southern Liaoning Province during the last 10,000 years (in Chinese). *Scientia Sinica* 6, 603-614.

Gupta, H. P. (1975). Pollen analytical reconnaissance of past glacial deposits from subtropical zone in Naini Tal district, Kumaon Himalaya. *The Palaeobotanist* 24, 215-244.

Hays, J. D., Imbrie, J., and Shackleton, N. J. (1976). Variations in the earth's orbit: Pacemaker of the ice ages. *Science* 194, 1121-1132.

Heusser, L. E., and Morley, J. J. (1985). Pollen and radiolarian records from deep-sea core RC14-103: Climatic reconstructions of northeast Japan and northwest Pacific for the last 90,000 years. *Quaternary Research* 24, 60-72.

Holland, H. D., Smith, G. I., Jannasch, H. W., Dickson, A. G., Zhang Mianping, and Ding Tiping (1991). Lake Zabuye and the climatic history of the Tibetan Plateau. *Die Geowissenschaften* 2, 37-44.

Hopkins, D. M., Matthews, J. V., Jr., Schweger, C. E., and Young, S. B., Eds. (1982). "Paleoecology of Beringia." Academic Press, New York.

Horie, S. (1957). Pollen analytical studies on bogs of central Japan, with references to the climatic changes in the alluvial age. *Japanese Journal of Botany* 16, 102-127.

——. (1974). Glacial features on Japanese high mountains (Part I). *In* "Paleolimnology of Lake Biwa and the Japanese Pleistocene" (S. Horie, Ed.), pp. 11-27.

Hou, H. Y., Ed. (1979). "Vegetation Map of China." Institute of Botany, Academia Sinica. Map Publisher of P.R.C., Beijing.

Hsieh Chiao-Min (1976). Chu K'o-Chen and China's climatic changes. *The Geographical Journal* 142, 248-256.

Hu, H. H. (1944). Climatic changes and a study of the climatic conditions of the Yin dynasty. *Bulletin of Chinese Studies* 4, 1-84.

Huang Ci-Xuan and Liang Yu-Lian (1981). Palynological study on the natural environment of the central and southern Qinghai-Xizang Plateau during the Holocene. *In* "Geological and Ecological Studies of Qinghai-Xizang Plateau," Vol. 1 (Liu Dongsheng, Ed.), pp. 215-224. Science Press of China, Beijing.

Huang Jinsen (1984). Changes in sea-level since the late Pleistocene in China. *In* "The Evolution of the East Asian Environment" (R. O. Whyte, Ed.), pp 309-319. Centre of Asian Studies, University of Hong Kong.

Huang Peihua (1982). Quaternary climatic changes in China and problems of Lushan glaciation remnants. *Journal of Glaciology and Cryopedology (Lanzhou)* 4, 1-14.

Huang Wenji (1978). The identification of fossil faunal remains of Yudun Neolithic site. *Kaogu* 4, 241-243.

Imbrie, J., and Imbrie, K. P. (1979). "Ice Ages: Solving the Mystery." Enslow Publishers, Short Hills, N.J.

Institute of Botany, Academia Sinica (1966). A study on the Cenozoic paleobotany of the Lantian area, Shaanxi Province. *In* "Proceedings of the Site Meeting on the Cenozoic, Lantian, Shaanxi," pp. 157-182. Science Press, Beijing.

Jacobson, G. L., Jr., and Grimm, E. C. (1986). Synchrony of rapid change in late-glacial vegetation south of the Laurentide ice sheet. *In* "Late Pleistocene and Early Holocene Paleoecology and Archaeology of the Eastern Great Lakes Region" (R. S. Laub, Ed.), pp. 31–38. Bulletin 33, Buffalo Society of Natural Sciences.

Jacobson, G. L., Jr., Webb, T. III, and Grimm, E. C. (1987). Patterns and rates of vegetation change during the deglaciation of eastern North America. *In* "North America and Adjacent Oceans during the Last Deglaciation" (W. F. Ruddiman and H. E. Wright, Jr., Eds.), pp. 277–288. The Geology of North America, Vol. K-3. Geological Society of America, Boulder, Colo.

Jia Lanpo and Zhang Zhengbiao (1977). The fossil fauna of Xiawanggang site at Cichuan, Henan. *Wenwu* 6, 41–49.

Kashiwaya, K., Yamamota, A., and Fukuyama, K. (1987). Time variations of erosional force and grain size in Pleistocene lake sediments. *Quaternary Research* 28, 61–68.

Kerschner, H. (1987). A paleoprecipitation model for full glacial Hokkaido as inferred from paleoglaciation data. *In* "Abstracts, XII International Congress of the International Union for Quaternary Research, Ottawa, Canada, July 1987," p. 200.

Kira, T. (1949). "Forest Zones of Japan." Technological Association, Tokyo and Sapporo.

Koizumi, I. (1981). Changes in the fossil diatom associations in the cores taken from the bottom of the Japan Sea since the last glacial age. *The Substance of Japan Association for Quaternary Research's Lecture* 11, 41–44.

Kong Zhaochen and Du Naiqiu (1980). Vegetational and climatic change in the past 30,000–10,000 years in Beijing. *Acta Botanica Sinica* 22, 330–339.

——. (1981). Sporo-pollen analysis of some archaeological sites and preliminary discussion on the past flora and climate of Neimonggol Zizhuqu. *Acta Phytoecologia et Geobotanica Sinica* 5, 193–202.

Kong Zhaochen, Du Naiqiu, and Zhang Zibin (1982). Vegetational development and climatic changes in the last 10,000 years in Beijing. *Acta Botanica Sinica* 24, 172–182.

Korotkiy, A. M. (1987). Dynamics of periglacial phenomena of the mountains of the eastern Asia. *In* "Abstracts, XII International Congress of the International Union for Quaternary Research, Ottawa, Canada, July 1987," p. 203.

Kuhle, M. (1987). The Pleistocene glaciation of Tibet and its impact on the global climate. *In* "Abstracts, XII International Congress of the International Union for Quaternary Research, Ottawa, Canada, July 1987," p. 204.

Kukla, G. J. (1977). Pleistocene land-sea correlations. *Earth Science Reviews* 13, 307–374.

Kukla, G., Heller, F., Liu, S. M., Liu, T. S., and Xu, T. C. (1987). Magnetic susceptibility of Chinese loess records climate and time. *In* "Abstracts, XII International Congress of the International Union for Quaternary Research, Ottawa, Canada, July 1987," p. 205.

Kutzbach, J. E. (1981). Monsoon climate of the early Holocene: Climate experiment using the earth's orbital parameters for 9000 years ago. *Science* 214, 59–61.

——. (1987). The changing pulse of the monsoon. *In* "Monsoons" (J. Fein and P. Stephens, Eds.), pp. 247–268. John Wiley & Sons, New York.

Kutzbach, J. E., and Guetter, P. J. (1984a). The sensitivity of monsoon climates to orbital parameter changes for 9000 years B.P.: Experiments with the NCAR general circulation model. *In* "Milankovitch and Climate," Part 2 (A. L. Berger, J. Imbrie, J. Hays, G. Kukla, and B. Saltzman, Eds.), pp. 801–820. D. Reidel Publishing Co., Dordrecht, The Netherlands.

——. (1984b). Sensitivity of late-glacial and Holocene climates to the combined effects of orbital parameter changes and lower boundary condition changes: "Snapshot" simulations with a general circulation model for 18-, 9-, and 6 ka B.P. *Annals of Glaciology* 5, 85–87.

——. (1986). The influence of changing orbital parameters and surface boundary conditions on climate simulations for the past 18,000 years. *Journal of the Atmospheric Sciences* 43, 1726–1759.

Kutzbach, J. E., and Otto-Bliesner, B. L. (1982). The sensitivity of the African-Asian monsoonal climate to orbital parameter changes for 9000 years B.P. in a low-resolution general circulation model. *Journal of the Atmospheric Sciences* 39, 1177–1188.

Kutzbach, J. E., and Street-Perrott, F. A. (1985). Milankovitch forcing of fluctuations in the level of tropical lakes from 18 to 0 kyr B.P. *Nature* 317, 130–134.

Lamb, H. H. (1977). "Climate: Present, Past, and Future; Vol. 2: Climatic History and the Future." Methuen, London.

——. (1984). Climate in the last thousand years. *In* "The Climate of Europe: Past, Present, and Future" (H. Flohn and R. Fantechi, Eds.), pp. 25–64. D. Reidel Publishing Co., Dordrecht, The Netherlands.

Lee, J. S. (1975). "Quaternary Glaciations of China." Science Press, Beijing.

Lee Jiaying and Huang Chengyen (1966). Holocene diatom fossils of Lantian, Shaanxi Province. *In* "Collected Papers of the Conference on the Cenozoic Site of Lantian, Shaanxi," pp. 197–223. Science Press, Beijing.

Li Bingyuan, Wang Fubao, Yang Yizhou, and Zhang Qingsong (1982). On the paleogeographical evolution of Xizang in the Holocene. *Geographical Research* 1, 27.

Li Bingyuan, Yang Yizhou, Zhang Qingsong, and Wang Fubao (1985). On the environmental evolution of Xizang (Tibet) in Holocene. *In* "Quaternary Geology and Environment of China" (Liu T.-s., Ed.), pp. 234–240. China Ocean Press, Beijing, and Springer-Verlag, New York. (Paper also in Chinese book of the same name, 1982, pp. 173–177.)

Li Dingrong (1987). Dali glaciation and rate of crustal deformation in post-glacial period in the northwest Yunnan. *In* "Abstracts, XII International Congress of the International Union for Quaternary Research, Ottawa, Canada, July 1987," p. 211.

Li Dingrong, Huang Xinggen, and Wang Ande (1987). The division of the Quaternary in the northwest of Yunnan. *In* "Abstracts, XII International Congress of the International Union for Quaternary Research, Ottawa, Canada, July 1987," p. 211.

Li Jijun (1980). Recent advances in the studies on the existing glaciers in Qinghai-Xizang Plateau. *Glaciology and Cryopedology* 2, 11–14.

Li, X., and Walker, D. (1986). The plant geography of Yunnan Province, southwest China. *Journal of Biogeography* 13, 367–397.

Lin, S., Qiao, Y., and Walker, D. (1986). Late Pleistocene and Holocene vegetation history at Xi Hu, Er Yuan, Yunnan Province, southwest China. *Journal of Biogeography* 13, 419–440.

Lin Zhenyao and Wu Xiangding (1987). Some features on climatic fluctuation over the Qinghai-Xizang Plateau. *In* "The Climate of China and Global Climate" (Ye Duzheng, Fu Congbin, Chao Jiping, and Yoshino, M., Eds.), pp. 116–123. Proceedings of the Beijing International Symposium on Climate in Beijing, China in 1984. China Ocean Press, Beijing.

Lister, G., Kelts, K., Chen, K.-z., and Yu, J. Q. (1989). Ostracoda in Lake Qinghai, central China: A $\delta^{18}O$ record for closed-basin lake level changes since the latest Pleistocene. *In* "Abstracts, Vth International Symposium on Palaeolimnology, Cumbria, U.K.," p. 68.

Liu Jingling, Li Wenyi, Sun Mengron, and Liu Mulin (1965). Palynological assemblages from peat-bogs of southern Mt. Yan Shan in Hebei Province. *Quaternaria Sinica* 4, 105-117.

Liu, J., Tang, L., Qiao, Y., Head, M. J., and Walker, D. (1986). Late Quaternary vegetation history at Menghai, Yunnan Province, southwest China. *Journal of Biogeography* 13, 399-418.

Liu Kam-biu (1986). The Pleistocene changes of vegetation and climate in China. *In* "Program and Abstracts of the 9th Biennial Meeting of the American Quaternary Association," p. 94. University of Illinois, Champaign-Urbana.

——. (1988). Quaternary history of the temperate forests of China. *Quaternary Science Reviews* 7, 1-20.

Liu Tungsheng, Ed. (1985a). "Loess and the Environment." China Ocean Press, Beijing.

——, Ed. (1985b). "Quaternary Geology and Environment of China." China Ocean Press, Beijing, and Springer-Verlag, New York.

Liu Dongsheng (Tungsheng) and Ding Menglin (1984). The characteristics and evolution of the palaeoenvironment of China since the late Tertiary. *In* "The Evolution of the East Asian Environment" (R. O. Whyte, Ed.), pp. 11-40. Centre of Asian Studies, University of Hong Kong.

Liu Tungsheng and Yuan Baoyin (1982). Quaternary climatic fluctuation—A correlation of records in loess with that of the deep sea core V28-238. *Research in Geology, Institute of Geology, Academia Sinica* 1, 113-121.

Liu Tungsheng, Yang Lihua, and Chen Chenghui (1964). Distributional characteristics of Quaternary sediments in China. *In* "Problems of Quaternary Geology," pp. 1-44. Science Press, Beijing.

Liu Tungsheng, An Zhisheng, Yuan Baoyin, and Han Jiaomao (1985). The loess-paleosol sequence in China and climatic history. *Episodes* 8, 21-28.

Liu Tungsheng, Zhang Shouxin, and Han Jiaomao (1986). Stratigraphy and paleoenvironmental changes in the loess of central China. *Quaternary Science Reviews* 5, 489-495.

Liu Tungsheng, Liu Jiaqi, and An Zhisheng (1987). Loess of China and global change. *In* "Abstracts, XII International Congress of the International Union for Quaternary Research, Ottawa, Canada, July 1987," p. 212.

Liu Ze-Chun and Wang Fu-Bao (1981). Microlithic tools unearthed in three locations on the northeastern bank of Lake Mafamu on Xizang Plateau. *Journal of Nanjing University Philosophy and Social Science* 4, 87-108.

Lu Yanchou and An Zhisheng (1979). The quest for series of nature environmental changes during the Brunhes Epoch. *Kexue Tongbao* 24, 221-224.

Luo Huibang and Yanai Michio (1986). The general circulation and heat sources over the Tibetan Plateau and surrounding areas during the onset of the 1979 summer monsoon. *In* "Proceedings of International Symposium on the Qinghai-Xizang Plateau and Mountain Meteorology, March 20-24, 1984, Beijing, China," pp. 731-767. American Meteorological Society, Boston.

Lyle, M. W., Prahl, F. G., and Sparrow, M. A. (1992). Upwelling and productivity changes inferred from a temperature record in the central equatorial Pacific. *Nature* 355, 812-815.

Matsushima, Y. (1976). The alluvial deposits in the southern part of the Miura Peninsula, Kanagawa Prefecture. *Bulletin of Natural Science of the Kanagawa Prefectural Museum* 9, 87-162.

Milankovitch, M. (1941). "Canon of the Earth's Insolation and its Application to the Ice-Age Problem." Royal Serbian Academy of Geography (Belgrade) Special Publication 132. (Translated by Israel Program for Scientific Translations, Jerusalem, 1969. U.S. Department of Commerce, Washington, D.C.)

National Geographic Society (1982). Map of the People's Republic of China.

Pecsi, M. (1987). The loess-paleosol and related subaerial sequence in Hungary. *In* "Abstracts, XII International Congress of the International Union for Quaternary Research, Ottawa, Canada, July 1987," p. 239.

Pokras, E. M., and Mix, A. C. (1985). Eolian evidence for spatial variability of late Quaternary climates in tropical Africa. *Quaternary Research* 24, 137-149.

Prell, W. L., and Kutzbach, J. E. (1987). Monsoon variability over the past 150,000 years. *Journal of Geophysical Research* 92, 397-415.

Pu Qingyu, Wu Xihao, and Qian Fang (1982). Problems of Quaternary geology along the Qinghai-Tibet highway in the Tonglha Mountain area (in Chinese). *In* "Contributions to Geology of Qinghai-Tibet Plateau, Vol. 4: Quaternary Geology and Glaciation," pp. 19-33. Geology Press, Beijing.

Qiu Shanwen, Jiang Peng, Li Fenghua, Xia Yumei, Wang Manghua, and Wang Peifang (1981). The study of environmental changes since late-glacial period in northeastern China. *Acta Geographica Sinica* 36, 315-327.

Ren Mei'e and Zong Chen-kai (1980). Late Quaternary continental shelf of east China. *Acta Oceanologica Sinica* 2, 94-111.

Ren Mei'e, Yang Reng-zhang, and Bao Hao-sheng (1979). "Elements of the Natural Geography of China" (in Chinese). Commercial Press, Beijing.

——. (1985). "An Outline of China's Physical Geography." China Knowledge Series. Foreign Languages Press, Beijing.

Rind, D., Peteet, D., Broecker, W., McIntyre, A., and Ruddiman, W. (1986). The impact of cold North Atlantic sea surface temperatures on climate: Implications for the Younger Dryas cooling (11-10k). *Climate Dynamics* 1, 3-33.

Ritchie, J. C. (1985). Late-Quaternary climatic and vegetational change in the Lower Mackenzie Basin, northwest Canada. *Ecology* 66, 612-621.

——. (1987). "Postglacial Vegetation of Canada." Cambridge University Press, Cambridge.

Ruddiman, W. F., and Duplessy, J.-C. (1985). Conference on the last deglaciation: Timing and mechanisms. *Quaternary Research* 23, 1-17.

Ruddiman, W. F., and Kutzbach, J. E. (1991). Plateau uplift and climatic change. *Scientific American* 264, 66-75.

Sakaguchi, Y. (1979). Distribution and genesis of Japanese peatlands. *Bulletin of the Department of Geography, University of Tokyo* 11, 17-42.

Shandong Group, Institute of Archaeology, and Jining Area Cultural Bureau (1979). A brief report on the excavation of Wangin Neolithic site at Yanzhou, Shandong. *Kaogu* 1, 5-14.

Shandong Group of the Institute of Archaeology of Chinese Academy of Social Science, and Museum of Tengxian (1980). A report on the ancient site in Tengxian, Shandong. *Kaogu* 1, 32-34.

Sheng Chenyu (1987). A historical survey of arid areas of China. *In* "The Climate of China and Global Climate" (Ye

Duzheng, Fu Congbin, Chao Jiping, and Yoshino, M., Eds.), pp. 45-56. Proceedings of the Beijing International Symposium on Climate in Beijing, China in 1984. China Ocean Press, Beijing.

Shih Yafeng (1982). Is there really Quaternary glaciation on Lushan? (in Chinese). *Journal of Glaciology and Cryopedology* 4, 64-69.

Shih Ya-feng, Li Chi-chun, and Cheng Hsin (1979). Quaternary China. *Geographic Magazine* 51:636-643.

Shi Yafeng, Ren Binghui, Wang Jingtai, and Derbyshire, E. (1986). Quaternary glaciation in China. *Quaternary Science Reviews* 5, 503-507.

Singh, G., and Agrawal, D. P. (1976). Radiocarbon evidence for deglaciation in north-western Himalaya, India. *Nature* 206, 232.

Singh, G., Joshi, R. D., Chopra, S. K., and Singh, A. B. (1974). Late Quaternary history of vegetation and climate of the Rajasthan desert, India. *Philosophical Transactions of the Royal Society of London, Series B* 267, 467-501.

Street-Perrott, F. A., and Harrison, S. P. (1984). Temporal variations in lake levels since 30,000 yr B.P.—An index of the global hydrological cycle. *In* "Climate Processes and Climate Sensitivity" (J. E. Hansen and T. Takahashi, Eds.), pp. 118-129. Geophysical Monograph Series 29. American Geophysical Union, Washington, D.C.

Sun Shuqing, Ji Liren, and Dell'Osso, L. (1986). The dynamic effect of the Qinghai-Xizang Plateau on the formation of the low level jet in east Asia. *In* "Proceedings of International Symposium on the Qinghai-Xizang Plateau and Mountain Meteorology, March 20-24, 1984, Beijing, China," pp. 702-728. American Meteorological Society, Boston.

Sun, Xiangjun and Chen Yinshuo (1991). Palynological records of the last 11,000 years in China. *Quaternary Science Reviews* 10, 537-544.

Sun, X., Wu, Y., Qiao, Y., and Walker, D. (1986). Late Pleistocene and Holocene vegetation history at Kunming, Yunnan Province, southwest China. *Journal of Biogeography* 13, 441-476.

Swain, A. M., Kutzbach, J. E., and Hastenrath, S. (1983). Estimates of Holocene precipitation for Rajasthan, India, based on pollen and lake-level data. *Quaternary Research* 19, 1-17.

Tang Lingyu and Huang Baoren (1985). Quaternary sporopollen and ostracod assemblages from the continental sediments in north China. *In* "Quaternary Geology and Environment of China" (Liu T.-s., Ed.), pp. 155-159. China Ocean Press, Beijing, and Springer-Verlag, New York.

Teachers College of Northwest China (1984). "Maps of the Natural Geography of China." Department of Geography, Teachers College of Northwest China. Special Publications Map Office, Beijing.

Thompson, L. G., Mosley-Thompson, E., Davis, M. E., Bolzan, J. F., Dai, J., Yao, T., Gundestrup, N., Wu, X., Klein, L., and Xie, Z. (1989). Holocene-late Pleistocene climatic ice core records from Qinghai-Tibetan Plateau. *Science* 246, 474-477.

Tsukada, M. (1967). Vegetation in subtropical Formosa during the Pleistocene glaciations and the Holocene. *Palaeogeography, Palaeoclimatology, Palaeoecology* 3, 49-64.

——. (1972). The history of Lake Nojiri, Japan. *Transactions of the Connecticut Academy of Arts and Sciences* 44, 338-365.

——. (1983). Vegetation and climate during the last glacial maximum in Japan. *Quaternary Research* 19, 212-235.

——. (1986). Altitudinal and latitudinal migration of *Cryptomeria japonica* for the past 20,000 years in Japan. *Quaternary Research* 26, 135-152.

Van Campo, E. (1986). Monsoon fluctuations in two 20,000-yr B.P. oxygen-isotope/pollen records off southwest India. *Quaternary Research* 26, 376-388.

Van Campo, E., Duplessy, J.-C., and Rossignol-Strick, M. (1982). Climatic conditions deduced from a 150,000 yr oxygen isotope-pollen record from the Arabian Sea. *Nature* 296, 56-59.

Verstappen, H. T. (1980). Quaternary climatic changes and natural environment in SE Asia. *GeoJournal* 4, 45-54.

Vishnu-Mittre (1984). Floristic change in the Himalaya (southern slopes) and Siwaliks from the mid-Tertiary to recent times. *In* "The Evolution of the East Asian Environment" (R. O. Whyte, Ed.), pp. 483-503. Centre of Asian Studies, University of Hong Kong.

Walker, D. (1986). Late Pleistocene-early Holocene vegetational and climatic changes in Yunnan Province, southwest China. *Journal of Biogeography* 13, 477-486.

Wang Fu-Bao and Fan, C. Y. (1987). Climatic changes in the Qinghai-Xizang (Tibetan) region of China during the Holocene. *Quaternary Research* 28, 50-60.

Wang Jingtai (1981). Ancient glaciers at the head of Urumqi River, Tian Shan. *Journal of Glaciology and Cryopedology (Lanzhou)* 31, 57-63.

Wang Jingtai and Wang Pinxian (1980). Relationship between sea-level changes and climatic fluctuations in east China since late Pleistocene. *Acta Geographica Sinica* 35, 299-312.

Wang Jingtai, Guo Zuming, Xu Shiyuan, Li Ping, and Li Congxian (1981). Evolution of the Holocene Changjiang Delta. *Acta Geologica Sinica* 1, 67-81.

Wang Kaifa (1972). Pollen analysis of peat in Lake Xiyao of Nanchang. *Acta Botanica* 16, No. 1. (Cited in Duan *et al.,* 1981.)

Wang Kaifa, Chang Yulan, Ye Zehua, and Sun Yuhua (1977). The occurrence of Quaternary sporo-pollen assemblage and its paleoclimatology along the coast of Bac Bo, China. *Kexue Tongbao* 22, 221-223.

Wang Luejiang and Wang Pinxian (1990). Late Quaternary paleoceanography of the South China Sea: Glacial-interglacial contrasts in an enclosed basin. *Paleoceanography* 5, 77-90.

Wang Manhua, Wang Peifang, and Xia Yumen (1982). Palynomorphs and algal assemblages from Holocene peat of the Three-River-Plain and their significance in paleovegetation (in Chinese). *In* "Selected Papers from the First Symposium of the Palynological Society of China," pp. 13-21. Science Press, Beijing.

Wang, P. K. (1980). On the relationship between winter thunder and the climatic change in China in the past 2200 years. *Climatic Change* 3, 37-46.

Wang, P. K., and Zhang De'er (1988). An introduction to some historical governmental weather records of China. *Bulletin of the American Meteorological Society* 69, 753-758.

——. (1992). Recent studies of the reconstruction of East Asian monsoon climate in the past using historical literature of China. *Journal of the Meteorological Society of Japan* 70, 423-446.

Wang Peifang (1981). The study of the sporo-pollen assemblages in peat and the evolution of the natural environment of south Xizang Plateau in the Holocene (in Chinese). *Scientia Geographica Sinica* 1, 144-152.

Wang Pinxian (1984a). Progress in late Cenozoic palaeoclimatology of China: A brief review. *In* "The Evolution of the East Asian Environment" (R. O. Whyte, Ed.), pp. 165-187. Centre of Asian Studies, University of Hong Kong.

——. (1984b). Neotectonic and palaeoenvironmental implications of Quaternary foraminiferal faunas from east China. *In* "The Evolution of the East Asian Environment" (R. O. Whyte, Ed.), pp. 349-362. Centre of Asian Studies, University of Hong Kong.

Wang Pinxian, Min Qiubao, Bian Yunhua, and Chen Xingrong (1981). Strata of Quaternary transgressions in the East China Sea: A preliminary study. *Acta Geologica Sinica* 55, 1-13.

Wang Shao-wu, Zhang Pei-yuan, and Zhang De'er (1980). Further studies on the climatic change during historical times in China. Lecture presented at the Seminar on Climate, World Meteorological Organization Meeting, Dec. 1980, Canton, China.

Wang Wei-Chyung and Li Kerang (1990). Precipitation fluctuation over semiarid region in northern China and the relationship with El Niño/Southern Oscillation. *Journal of Climate* 3, 769-783.

Wasson, R. J., Smith, G. I., and Agrawal, D. P. (1984). Late Quaternary sediments, minerals, and inferred geochemical history of Didwana Lake, Thar Desert, India. *Palaeogeography, Palaeoclimatology, Palaeoecology* 46, 345-372.

Whyte, R. O. (1984a). Scope and objectives. *In* "The Evolution of the East Asian Environment" (R. O. Whyte, Ed.), pp. 1-6. Centre of Asian Studies, University of Hong Kong.

——, Ed. (1984b). "The Evolution of the East Asian Environment: Vol. I, Geology and Palaeoclimatology; Vol. II, Palaeobotany, Palaeozoology, and Palaeoanthropology." Centre of Asian Studies, University of Hong Kong.

Xu Qinqi (1981). Climatic variations during the late Pleistocene and the earth's orbit. *Journal of Stratigraphy* 5, 226-230.

——. (1984). Relation between climatic variations and deviations of solar radiation over the past 300,000 years. *In* "The Evolution of the East Asian Environment" (R. O. Whyte, Ed.), pp. 220-229. Centre of Asian Studies, University of Hong Kong.

Xu Ren (Hsu Jen) (1981). Vegetational changes in the past and the uplift of Qinghai-Xizang Plateau. *In* "Proceedings of Symposium on Qinghai-Xizang Plateau; Geological and Ecological Studies of Qinghai-Xizang Plateau," Vol. I (Liu Dongsheng, Ed.), pp. 139-144. Science Press, Beijing.

Xu Ren (1984). Changes of the palaeoenvironment of southern east Asia since the late Tertiary. *In* "The Evolution of the East Asian Environment" (R. O. Whyte, Ed.), pp. 419-425. Centre of Asian Studies, University of Hong Kong.

Xu Ren (Hsu Jen), Kong Zhaochen, and Du Naiqui (1980). Pleistocene flora of *Picea* and *Abies* in China and its significance in Quaternary research (in Chinese). *Quaternaria Sinica* 5, 40-56.

Xu Ren, Sun Xiangjun, Wu Yushu, Liu Jinling, Tang Lingyu, Lin Shaomeng, and Qiau Yulou (1984). Environmental changes since 40,000 B.P. in China (based on palynological data). *In* "Collected Abstracts of the Beijing International Symposium on Climate, Oct. 30-Nov. 3, 1984," pp. R11-R12.

Yang Huaijen and Xie Zhiren (1984). Sea level changes in east China over the past 20,000 years. *In* "The Evolution of the East Asian Environment" (R. O. Whyte, Ed.), pp. 288-308. Centre of Asian Studies, University of Hong Kong.

Yang Huaijen and Xu (Hsu) Shin (1981). Quaternary environmental changes in eastern China. *Journal of Nanjing University (Natural Sciences Edition)* 1, 121-144.

Yang Yi-Chou (1984). Aeolian landforms on the banks of a river valley: Case study in Yaluzanbu River Valley (in Chinese). *Journal of Desert Research* 4, 9-15.

Yang Zigeng, Li Youjun, Ding Qiuling, and He Baocheng (1979). Some fundamental problems of Quaternary geology of eastern Hebei plain. *Acta Geologica Sinica* 4, 263-279.

Yasuda, Y. (1983). Climatic changes since 50,000 years in Japan. *Research Report, Climatology and Meteorology, Tsukuba University* 8, 2-12.

——. (1984). Oscillations of climatic and oceanographic conditions since the last glacial age in Japan. *In* "The Evolution of the East Asian Environment" (R. O. Whyte, Ed.), pp. 397-413. Centre of Asian Studies, University of Hong Kong.

Yen Wenming (1979). New discoveries of early Neolithic cultures in Yellow River drainages. *Kaogu* 1, 45-50.

Zhang De'er (1980). Winter temperature changes during the last 500 years in South China. *Kexue Tongbao* 25, 497-500.

——. (1983). Analysis of dust rain in the historic times of China. *Kexue Tongbao* 28, 361-366.

——. (1984). Synoptic-climatic studies of dust fall in China since historic times. *Scientia Sinica, Series B* 27, 825-836.

——. (1985). Meteorological characteristics of dust fall in China since the historic times. *In* "Quaternary Geology and Environment of China" (Liu T.-s., Ed.), pp. 101-106. China Ocean Press, Beijing, and Springer-Verlag, New York.

Zhang De'er and Wang, P. K. (1989). Reconstruction of the 18th century summer monthly precipitation series of Nanjing, Suzhou, and Hangzhou using the clear and rain records of Qing Dynasty. *Acta Meteorologica Sinica* 3, 261-278.

Zhang Lansheng (1984). Reconstruction of the climate of the late Pleistocene ice age in east China. *In* "The Evolution of the East Asian Environment" (R. O. Whyte, Ed.), pp. 252-270. Centre of Asian Studies, University of Hong Kong.

Zhang Zhenshuan (1981). Changes of snowline at the head of Urumqi River, Tian Shan. *Journal of Glaciology and Cryopedology (Lanzhou)* 31, 106-113.

Zhao Songling and Chin Yunshan (1982). Transgressions and sea-level changes in the eastern coastal region of China in the last 300,000 years. *In* "Quaternary Geology and Environment of China" (Liu Tung Sheng, Ed.), pp. 147-154. China Ocean Press, Beijing.

Zhao Songqiao and Xing Jiaming (1984). Origin and development of the Shamo (sandy deserts) and the Gobi (stony deserts) of China. *In* "The Evolution of the East Asian Environment" (R. O. Whyte, Ed.), pp. 230-251. Centre of Asian Studies, University of Hong Kong.

Zhao Xi-tao (1979). Ages of formation of the Luhuitou coral reefs, Hainan Island, and their effects on shoreline changes. *Kexue Tongbao* 24, 995-998.

Zhao Xitao and Zhang Jingwen (1982). Basic characteristics of the Holocene sea level changes along the coastal areas in China. *In* "Quaternary Geology and Environment of China" (Liu Tung Sheng, Ed.), pp. 155-160. China Ocean Press, Beijing. (English translation, 1985, pp. 210-217. Springer-Verlag, New York.)

——. (1985). Basic characteristics of the Holocene sea level changes along the coastal areas in China. *In* "Quaternary

Geology and Environment of China" (Liu T.-s., Ed.), pp. 210-217. China Ocean Press, Beijing, and Springer-Verlag, New York.

Zheng Benxing and Shi Yafeng (1982). Glacial variation since late Pleistocene on the Qinghai-Xizang (Tibet) Plateau of China. *In* "Quaternary Geology and Environment of China" (Liu Tung Sheng, Ed.), pp. 161-166. China Ocean Press, Beijing. (English translation, 1985, pp. 218-225. Springer-Verlag, New York.)

Zhou Kunshu (1965). Investigation and pollen analysis of two buried peat bogs near Beijing. *Quaternaria Sinica* 4, 118-134. (Cited in Duan *et al.*, 1981.)

Zhou Kun-shu and Hu Ji-lan (1985). The paleoenvironment of the Shuidonggou site of ancient cultural remains in Lingwu County, Ningxia. *In* "Quaternary Geology and Environment of China" (Liu T.-s., Ed.), p. 185. China Ocean Press, Beijing, and Springer-Verlag, New York.

Zhou Kunshu, Liang Xiulong, and Liu Ruiling (1981). Preliminary study of palynology of glacier ice and Quaternary deposits at the upper reaches of Urumqi River, Tian Shan. *Journal of Glaciology and Cryopedology (Lanzhou)* 3, 97-105.

Zhou Kunshu, Chen Shuomin, Ye Yongying, and Liang Xiulong (1984). Pollen analysis of peat in Dunhua, Jilin Province. *Scientia Geologica Sinica* 2, 130-139.

Zhou Tingru, Zhang Lansheng, and Li Huazhang (1982). The changes of climate in north China since the late glacial age of Pleistocene. *Journal of Natural Science of Beijing Normal University* 1, 77-88.

Appendix:
Location, Kinds of Analysis, and References for Sites Used in This Review

Region Latitude and longitude	Kinds of analysis	References
Northeast China		
51°50'N, 126°E	Pollen	
46-48°N, 132-134°E	Pollen	
45°31'N, 125°09'E	Fauna [14]C (more than 200 sites)	Duan *et al.*, 1981
49°21'N, 117°35'E	Permafrost features, fauna, pollen	Kong and Du, 1981
About 39°N, about 117°E	Pollen, sea levels	Guiyang Institute of Geochemistry, 1977
About 43°21'N, 128°13'E	Pollen	Zhou *et al.*, 1984
48°N, 128°E	Permafrost features	Cui and Xie, 1985; Zhang Lansheng, 1984
45°N, 128°E	Pollen, fauna	Cui and Xie, 1985
42°N, 125°E	Pollen, fauna	Cui and Xie, 1985
39°24'N, 121°58'E	Pollen	Chen *et al.*, 1965
Region review	Pollen	Liu, 1988
Japan		
36°49'N, 138°13'E (654 m)	Pollen, zooplankton	Tsukada, 1972
35°33'32"N, 135°53'40"E (sea level)	Pollen	Yasuda, 1983, 1984
35°N, 136°E	Pollen	Yasuda, 1984
35°30'N, 136°E	Pollen, diatoms, zooplankton, chemical analysis	Horie, 1974; Fuji, 1974
36°58'40"N, 139°18'10"E (1500 m)	Sediment stratigraphy	Sakaguchi, 1979
38°3'00"N, 140°11'00"E (220 m)	Sediment, seeds	Sakaguchi, 1979
36°43'20"N, 138°56'30"E (1000 m)	Sediment, pollen	Horie, 1957; Kashiwaya *et al.*, 1987; Sakaguchi, 1979
36°29'10"N, 137°53'20"E (945 m)	Sediment stratigraphy	Sakaguchi, 1979
36°7'00"N, 138°10'10"E (1630 m)	Sediment, pollen	Horie, 1957; Sakaguchi, 1979
39°N, 133°E (V28-267)	Diatoms, foraminifers, $\delta^{18}O$	Koizumi, 1981
42°N, 143°E	Glacial features	Kerschner, 1987
About 35°N, 140°E	Sea levels, marine fauna	Matsushima, 1976
Inner Mongolia		
48°N, 120°E	Paleosol formation, pollen	Chao, 1977
40°51'N, 111°40'E	Pollen	Kong and Du, 1981
40°N, 113°E	Dated soil formation	Cui and Xie, 1985
47°N, 123°E	Dated soil formation	Cui and Xie, 1985
About 41°N, 112°E	Tree rings, artifacts	Zhao and Xing, 1984
North China		
39°55'N, 116°25'E	Pollen, zooplankton, fauna [14]C	Chen, 1979; Zhou, 1965; Liu *et al.*, 1965; Tang and Huang, 1985
About 25°N, 119°E; 30°N, 121°E; 40°N, 120°E	Sea levels	Liu and Ding, 1984; Zhao, 1979; Chang and Wang, 1982 Wang *et al.*, 1981; Zhao and Zhang, 1982
40°N, 116°E	Pollen, sea levels	Chen, 1979; Kong and Du, 1980
41°N, 122°E	Pollen, sea levels	Chen, 1979
38°19'N, 106°29'E	Pollen	Zhou and Hu, 1985
35°08'N, 117°10'E	Fauna, dated artifacts	Chu, 1973
35°28'N, 115°46'E	Dated artifacts	Chu, 1973
35°25'N, 119°25'E	Plant macrofossils	Chu, 1973
34°15'N, 108°52'E	Fauna	Chu, 1973
34°03'N, 109°12'E	Diatoms, pollen	Xu *et al.*, 1980; Lee and Huang, 1966; Institute of Botany, 1966
33°02'N, 111°30'E	Fauna	Jia and Zhang, 1977
36°40'N, 114°12'E	Dated artifacts	An, 1979
34°25'N, 113°43'E	Dated artifacts	Yen, 1979
37°N, 117°E	Pollen, sediment chemistry	Yang *et al.*, 1979
39°N, 118°E (Bohai Gulf)	Marine fauna	Zhao and Chin, 1982
Region review	Pollen	Liu, 1988
Region review	Lake levels	Fang, 1991

Appendix (continued)

Region Latitude and longitude	Kinds of analysis	References
Central China		
25°N, 115°E	Pollen	Yang and Xu, 1981; Duan *et al.*, 1981
28°N, 120°E	Pollen	Xu *et al.*, 1980
30–32°5'N, 120–122°E	Pollen, fauna, geomorphology	Wang *et al.*, 1981; Wang, 1984a
32°N, 119°26'E	Sea levels	Wang *et al.*, 1981
32°24'N, 119°26'E	Sea levels	Wang *et al.*, 1981
31°47'N, 119°57'E	Fauna	Huang, 1978
30°04'N, 121°10'E	Fauna, plant macrofossils, dated artifacts	Anonymous, 1976
30°07'N, 104°39'E	Pollen	Duan *et al.*, 1981
27°N, 115°E	Pollen	Duan *et al.*, 1981
25°N, 119°E; 30°N, 121°E; 40°N, 120°E	Sea levels	Liu and Ding, 1984; Wang *et al.*, 1981
28°41'N, 115°53'E	Pollen	Wang, 1974
29°34'N, 115°58'E	Pollen	Huang, 1982
30°N, 122°E	Marine fauna	Wang, 1984b
Loess Plateau		
36°N, 109°E (Luochuan)	Paleosol, pollen	Liu *et al.*, 1985
35°N, 105°E (Longxi)	Paleosol, pollen	Liu *et al.*, 1985
25–30°N, 75–120°E	Loess	Liu and Ding, 1984
47°N, 134°E	Loess	Liu and Ding, 1984; Zhao and Xing, 1984
South China		
23°52'N, 120°55'E (Taiwan, 745.5 m)	Pollen	Tsukada, 1967
23°45'N, 120°46'E (Taiwan, 640 m)	Pollen	Tsukada, 1967
22°N, 109°E	Pollen	Wang *et al.*, 1977
22°N, 109°E	Pollen, sea levels	Liu, 1985b; Zhang Lansheng, 1984
Southwest China		
Yunnan lakes		
Menghai (Nonggong and Manyang), 22°N, 100°E (1280 m)	Pollen	Liu *et al.*, 1986; Walker, 1986
Er Yuan (Xi Hu), 26°N, 100°E (1980 m)	Pollen	Lin *et al.*, 1986; Walker, 1986
Kunming (Dianchi), 25°N, 103°E (1886 m)	Pollen	X. Sun *et al.*, 1986; Walker, 1986
Qilu Hu, 24°10'N, 102°45'E (1797 m)	Pollen, sediment chemistry	Brenner *et al.*, 1989
25°38'N, 101°54'E	Fauna	Duan *et al.*, 1981
27°N, 99°E	Glacial features	Pu *et al.*, 1982; Li, 1987
Qinghai-Xizang Plateau (Tibetan Plateau)		
36°N, 75°E–30°N, 96°E	Glacial fluctuations, geomorphology	Zheng and Shi, 1982; Kuhle, 1987; Li, 1980; Li, 1987
37°11'N, 97°15'E (Qaidam)	Fauna, lake levels	Li *et al.*, 1985
37°N, 100°E	Pollen, ostracods, $\delta^{18}O$	Tang and Huang, 1985; Lister *et al.*, 1989
38°06'N, 96°24'E	Ice-core dust concentration and chemistry, $\delta^{18}O$	Thompson *et al.*, 1989
32°N, 97°E (Qamdo)	Fauna, pollen	Li *et al.*, 1985
34°N, 86°E (interior lakes)	Pollen, lake levels	Li *et al.*, 1985
33°N (exterior lakes)	Pollen, lake levels	Li *et al.*, 1985
34°N, 82–90°E	Pollen, lake levels	Li *et al.*, 1985
32°N, 86°E (Coquen Lake)	Pollen, lake levels	Li *et al.*, 1985
31°N, 91°E (Wumaqu peat)	Peat development	Li *et al.*, 1985
31°N, 91°E (Quilongduo)	Peat development	Wang and Fan, 1987
29°N, 87°E (Ngalaly)	Pollen, peat development	Li *et al.*, 1985
29°N, 91°E	Pollen	Shih *et al.*, 1979
28°N, 86°E	Plants, glacial features, snowline	Zheng and Shi, 1982
34°N, 79°E–32°N, 92°E	Dated artifacts	Li *et al.*, 1985
30°30'N, 84°E (Dawelong)	Peat development	Wang Peifang, 1981
34°47'N, 87°09'E	Dated artifacts	Liu and Wang, 1981; Li *et al.*, 1985; An *et al.*, 1980

Appendix (continued)

Region Latitude and longitude	Kinds of analysis	References
28°20'N, 91°56'E (Noryongco, 4760 m)	Pollen, lake levels, permafrost	Wang and Fan, 1987
31°N, 85°E (Zharinanmuco, 4613 m)	Pollen, zooplankton, lake levels	Huang and Liang, 1981; Li *et al.*, 1982, 1985
Region review	Lake levels	Fang, 1991
31°20'N, 83°05'E (Lake Zabuye)	Lake levels	Holland *et al.*, 1991
32°30'N, 82°20'E (Zhacancaka)	Lake levels	Chen *et al.*, 1981
32°N, 82°E	Permafrost, glacial features	Yang, 1984; Wang and Fan, 1987
29°10'N, 96°04'E	Pollen, tree rings, lichens	Wang, 1984a
30°N, 83°E	Geomorphic features	Yang, 1984

34°30'N, 80°E (Sumxi Co)	Lake levels, diatoms, ostracods,
	    pollen, sediment chemistry,
	    ^{18}O, ^{13}C isotopes
		Gasse *et al.*, 1991

Northwest China		
43°N, 81°E	Snowlines	Zhang, 1981; Wang Jingtai, 1981
42°N, 43°N, 80-87°E	Pollen	Zhou *et al.*, 1981
Northwest India		
27°N, 75°E	Sediments, geochemistry	Wasson *et al.*, 1984
	Pollen	Singh *et al.*, 1974
30°N, 74°E	Pollen	Singh *et al.*, 1974
27°N, 76°E	Lake levels	Swain *et al.*, 1983
28°30'N, 73°45'E	Pollen	Singh *et al.*, 1974; Bryson and Swain, 1981
Kashmir		
34°N, 74°E (3120 m)	Pollen	Singh and Agrawal, 1976
34°N, 74°48'E	Pollen	Dodia *et al.*, 1984
34°06'N, 74°43'E	Pollen	Dodia *et al.*, 1984
Nepal and Uttar Pradesh		
29°N, 80°E (1330 m)	Pollen	Gupta, 1975
28°N, 85°E	Pollen	Vishnu-Mittre, 1984

CHAPTER 11

Climates of Australia and New Guinea since 18,000 yr B.P.

Sandy P. Harrison and John Dodson

In this chapter we reconstruct the hydrological and vegetational changes in Australia and Papua New Guinea during the last 18,000 yr from the evidence of changes in lake levels and pollen assemblages at a network of sites over the continent. We interpret the inferred hydrological and vegetational changes in terms of regional climatic changes. We then compare the climatic reconstructions with general circulation model (GCM) simulations to see how far the inferred changes in climatic variables can be explained by the effects of known changes in insolation and alterations in the size of the ice sheets as simulated in the COHMAP GCM experiments.

Modern Environments

Australia, situated at 10-39°S and 113-153°E, is the smallest of the continents. It extends 4100 km from west to east and about 3200 km from north to south. Tasmania, off the southeast corner, is the biggest island adjacent to the continent, with an area of 68,400 km². Australia is also the flattest and lowest continent, with a mean elevation of less than 215 m above sea level (UNESCO, 1971). Its only highland region lies along the east coast, where a discontinuous chain of mountains and tablelands forms the Great Dividing Range. The highest peak on the mainland is Mount Kosciusko (2230 m). Papua New Guinea (785,000 km²) is more mountainous; its highest peak is Mount Djaja (5029 m).

CLIMATE

The climate of Australia is dominated by the descending air of the subtropical belt of high pressure and divergence (Fig. 11.1). Eastward movement of anticyclones across the continent is not impeded by the formation of a locationally stable high-pressure cell, because the cool offshore current along the west coast (Western Australian current) is relatively weak and diffuse and is not associated with significant upwelling (Hamon and Godfrey, 1978). Furthermore, there are no high mountains along the west coast to block the passage of traveling anticyclones. The winds leaving the anticyclonic spirals on the equatorward side form the southeasterly trade winds (tropical easterlies).

The mean latitude of the belt of traveling anticyclones shifts seasonally, from about 29-32°S in winter to 37-38°S in summer (Russell, 1893; Kidson, 1925; Karelsky, 1965). In winter the high-pressure belt is sufficiently far north that the southern part of the continent comes under the intermittent influence of the westerlies. When the traveling anticyclones are farthest south, they displace the westerlies, so that they do not influence the continental landmass but allow tropical low-pressure systems (associated with the intertropical convergence zone) and monsoonal flow into the north and northeast of the continent and into Papua New Guinea. Monsoonal flow into the north is primarily a local phenomenon caused by heating of the Australian landmass, but the northern coast is occasionally penetrated by the Asian monsoon (Gentilli, 1971).

These seasonal changes in circulation result in distinct seasonal patterns in rainfall distribution (Fig. 11.2). Central Australia, dominated by traveling anticyclones throughout the year, is extremely dry. Troughs between the traveling anticyclones draw in moist-maritime northwesterly airflows and bring summer rains to the east coast and southeastern highlands (Gentilli, 1971). The southern part of the continent, dominated by traveling anticyclones in summer and the westerlies in winter, is characterized by winter

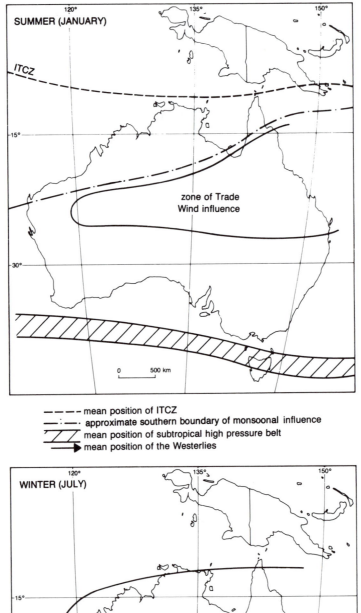

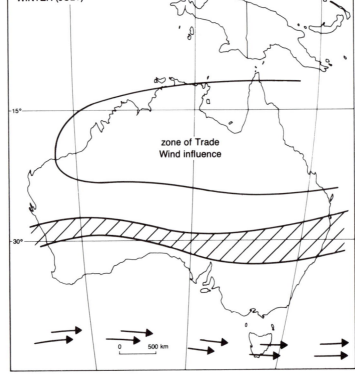

Fig. 11.1. Seasonal circulation patterns over Australia and New Guinea. ITCZ = intertropical convergence zone. Compiled from data in Flores and Balagot (1969), Gentilli (1971), and Karelsky (1954, 1965).

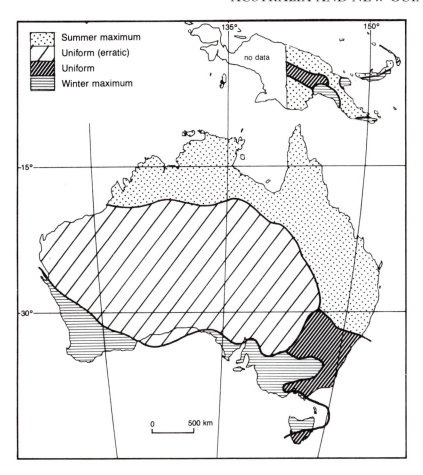

Fig. 11.2. Seasonal rainfall distribution in Australia and New Guinea.

rainfall and dry summers. Tasmania is sufficiently far south to receive precipitation from the westerlies all year. The far north, dominated by southeasterly trade winds in winter and monsoonal flow in summer, is characterized by dry winters and wet summers. Along the northeastern coast, however, onshore maritime airflow associated with the trade winds brings year-round precipitation. Tropical cyclones bring additional but erratic summer rainfall to most of the continent and are an important source of moisture in the arid interior.

Annual or longer-term shifts in the mean position of the subtropical high-pressure belt (Das, 1956; Lamb and Johnson, 1961, 1966) result in significant changes in regional rainfall (Deacon, 1953; Pittock, 1971, 1973, 1975, 1978). Pittock (1971) showed that interannual changes in the mean latitude of the subtropical high-pressure belt account for 30-50% of the variance in annual mean rainfall over much of coastal Australia. Rainfall anomalies are opposite on the southern and eastern coasts. Thus southward displacement of the subtropical high-pressure belt suppresses rainfall along the southern and southwestern coast (in the area north of Perth and in southern South Australia, coastal Victoria, and western Tasmania) but enhances it along the east coast from the Victorian border to about 25°S (Pittock, 1971, 1973, 1975, 1978).

Pittock (1975, 1978) also found significant correlations between rainfall distribution over the continent and the strength of the Walker Circulation. Interannual variations in the strength of the Walker Circulation are related to differences in sea-surface temperatures between the equatorial eastern Pacific and the Indonesian sector: relative cooling of the equatorial eastern Pacific favors a strong Walker Circulation, whereas relative cooling of the Indian Ocean leads to a weakened Walker Circulation. An enhanced east-west circulation anomaly, characteristic of a strong Walker Circulation, leads to increased surface flow toward the south over northern and eastern Australia. This increases rainfall, particularly in the eastern interior of the continent but also over much of the north and east. Because variations in the strength of the Walker Circulation are independent of changes in the mean latitudinal position of the subtropical high-pressure belt (Pittock, 1975, 1978), they provide an additional mechanism to explain changes in effective moisture regimes on a regional scale.

HYDROLOGY

The modern water balance over the continent is shown in Figure 11.3a. Permanent lakes are characteristic of Tasmania, the coastal regions of southeastern

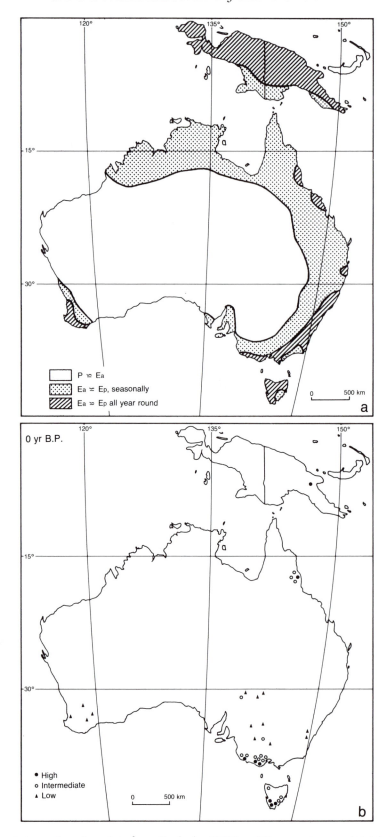

Fig. 11.3 (a) Regional water balance, based on data from Ceplecha (1971) and Baumgartner and Reichel (1975). In the arid zone, evapotranspiration (Ea) approximately equals precipitation (P), and runoff is approximately zero. In the intermediate zone, enough precipitation falls during some part of the year for Ea to equal potential evapotranspiration (Ep). In the humid zone, Ea equals Ep year-round. (b) Lake status at 0 yr B.P.

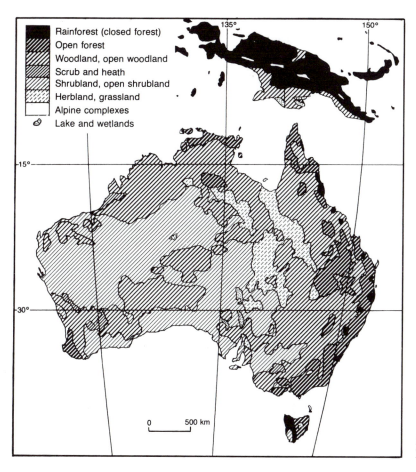

Fig. 11.4. Modern vegetation of Australia and New Guinea.

and southwestern Australia, northeastern Queensland, and Papua New Guinea. In contrast, lake basins in the interior of Western Australia, northern Australia, the interior regions of southeastern Australia, and the Southern Tablelands are either dry or contain only shallow or ephemeral lakes. These patterns are reflected in the modern-day status of the lakes used in this study (Fig. 11.3b). This distribution reflects the seasonal changes in atmospheric circulation over the continent. Permanent lakes in northeastern Queensland are maintained by year-round precipitation brought by the onshore trade winds. Permanent lakes in Tasmania are maintained by year-round precipitation brought by the westerlies. Permanent lakes are also found along the coast of Victoria in the region that receives abundant rains from the westerlies in winter. Lakes are dry or ephemeral where the subtropical high-pressure belt is dominant throughout the year and in regions characterized by cyclonic summer rainfall.

VEGETATION

The natural vegetation of Australia (Fig. 11.4) follows a structural continuum along the annual rainfall gradient (Specht, 1972) and thus also reflects the seasonal atmospheric circulation patterns. The vegetation is evergreen except for some northern monsoonal regions.

Closed forests, typically 10–20 m tall, are distributed throughout montane and lowland New Guinea and discontinuously along the east coast of Australia from northeastern Queensland to western Tasmania. The tropical and subtropical forests are species-rich and structurally complex, with lianas and epiphytes. The temperate forests are usually dominated by *Nothofagus cunninghamii* or *N. moorei* and contain few other tree species. All of these forests occupy humid sites with year-round rainfall. In monsoonal northern Australia low closed forests up to 10 m high occur as disjunct stands where subsoil seepage maintains moisture availability during the dry winter.

Tall but less dense forests, dominated by *Eucalyptus saligna, E. delegatensis, E. regnans,* and *E. diversicolor,* among others, occur throughout coastal and montane southeastern Australia and in southwestern Australia. They are usually over 30 m high and reach above 100 m in some areas. The lower stories contain trees, shrubs, and tree ferns, except that low heathy shrubs replace trees and tree ferns in mineral-deficient areas. These tall forests replace closed forest in areas where rainfall is more seasonal and available moisture is limiting during at least one month of the year and where the incidence of fire is higher.

Open forests and woodlands (typically 15–20 m tall) are much more widely distributed (Fig. 11.4). The dominant genera are *Eucalyptus, Angophora, Acacia,* and *Allocasuarina.* These formations occupy subhumid sites with summer or winter rainfall or both. Woodlands occupy the drier end of the spectrum and occur over large areas of the interior, wherever rainfall is sufficient regardless of source or season. The understory is usually grass-dominated, but sclerophyllous heathy taxa replace grasses on infertile soils. Fire is important in these systems.

Multistemmed eucalypts with well-developed lignotubers (mallee) form open scrublands up to 10 m high in areas of southern Australia where the dry season is long but where westerly flow brings significant rain in winter. Grasses or chenopod shrubs form an understory on the more fertile soils. Grasses are commonest in the higher-rainfall areas; chenopod shrubs, sometimes with xerophytic grasses, are important in drier areas.

Tall (4–8 m) shrubland dominated by *Acacia aneura* (mulga) occupies large areas of western and central Australia, where cyclonic summer rainfall dominates. The understory may contain low perennial trees and shrubs but always contains a variety of ephemeral herbs and grasses. Low chenopod shrublands, typically less than 2 m tall, occupy much of central and southern Australia from southeastern Western Australia to southwestern New South Wales. The major genera are *Atriplex, Maireana, Rhagodia,* and *Sclerolaena,* but several nonchenopodiaceous shrub taxa also occur. Much of the flora consists of ephemerals. These shrublands occur in arid areas on the extreme northern margin of winter westerly influence. The soils are generally calcareous or semisaline. The gray cracking clays of semiarid northern Australia support *Astrebla*-dominated tussock grasslands. These soils crack deeply during the dry winter, damaging root systems, so that few shrubs or small trees occur.

The driest areas of Australia are either treeless or have a scattered tree or shrub layer above hummock grasslands. These areas have species of *Triodia* and *Plectrachne* (prickly xerophilous grasses) that form spinifex rings. Many ephemeral species are present, and some heathland elements can also occur. Cane grass (*Zygochloa* spp.) dominates the tops of the extensive longitudinal sand ridges of arid central Australia.

The main patterns of vegetation cover are also affected by altitudinal variation, incidence of frost and fire, and soil properties unrelated to climate. High-altitude areas and frost hollows exclude tree growth and are occupied by grasslands and sometimes wetlands. These are restricted to New Guinea and the highest regions of southeastern Australia. Heathlands, sclerophyllous small-leaved plant communities dominated by shrubs generally 2–10 m high, abound in fire-prone regions of infertile soils in all the subhumid areas of Australia. Dry heaths occur in areas with seasonal droughts and wet heaths where waterlogging excludes trees. Heathy fellfields occur in alpine areas where cold or wind excludes all but scattered trees. Gray-brown cracking clays, solonized brown soils and low-nutrient sands, lithosols, and acid peaty soils all support important communities different from the main rainfall-dominated patterns of vegetation.

Data Sources and Methods

Lakes

The techniques used to reconstruct lake-level changes in closed-basin lakes in semiarid areas have been reviewed by Street-Perrott and Harrison (1985). Harrison (1988) and Harrison and Digerfeldt (in press) reviewed methods of reconstructing relative lake-level changes in humid areas. Bowler (1976), Bowler *et al.* (1976), Chappell and Grindrod (1983), Bowler and Wasson (1984), and Harrison *et al.* (1984) have discussed lake-level studies from Australia.

For the present study, we reconstructed changes in lake levels over the last 18,000 yr for 36 sites in Australia and Papua New Guinea (Table 11.1). The data were derived from published literature. Some sites were originally studied in order to reconstruct changes in regional vegetation, and the published information had to be reinterpreted in order to derive changes in water level. At other sites there were several published studies of water-level changes, each based on different types of data. The interpretation of lake-level changes at sites used in this study was based wherever possible on a consensus of the several lines of evidence. The chronology of lake-level changes was based solely on radiocarbon dates from individual sites. A discussion of the evidence and the detailed record of changing lake level at each site may be found in Harrison (1989).

The lakes lie between 6 and 43°S and between 115 and 150°E (Fig. 11.5). However, the spatial distribution of sites is extremely uneven; 75% lie in southeastern Australia, including Tasmania (south of 30°S, west of 139°E). Other sites are in the Atherton Plateau of northeastern Queensland (four sites), Papua New Guinea (one site), and southwestern Australia (four sites). With the exception of lakes in the highlands of Tasmania, the Southern Tablelands, and the Atherton Plateau, all of the lakes are at elevations less than 450 m above sea level. Eagle Tarn in Tasmania (1033 m) is the highest site. The lakes also range considerably in size and depth (Harrison, 1989), from small but deep crater lakes (e.g., Gnotuk, Keilambete, Bullenmerri, and Euramoo) to large structural lake basins that contain water only sporadically (e.g., George and Bancannia). Despite these differences, the major lake-level changes recorded over the last 18,000 yr show a geographic and temporal coherence that implies primary control by climatically induced changes in the hydro-

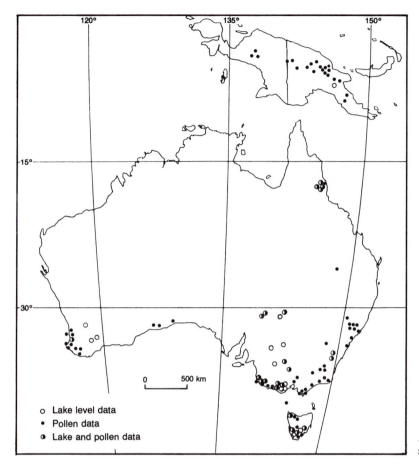

Fig. 11.5. The distribution of lake and pollen sites used in this study.

logic regime rather than by local factors in lake evolution.

The temporal distribution of lake-level information is also uneven. Only seven sites have continuous records over the last 18,000 yr. Coverage for the last 10,000 yr is significantly better; 20 sites have continuous Holocene records. Many of the pre-Holocene records are based on geomorphic evidence or discontinuous sedimentary data. Most of the Holocene data are derived from lacustrine cores, which provide quasi-continuous records.

To facilitate comparisons among sites, we categorized the records into "lake status" classes (Street and Grove, 1976). The maximum and minimum water depths registered within a particular basin were identified and used to define high and low status, respectively. Periods when the lake was in transition between these two states were assigned to an intermediate status category. Lake status was assessed at 1000-yr intervals from 18,000 yr B.P. to the present.

We present the lake-status data as histograms, to illustrate the temporal patterns of lake behavior within specified regions, and as a series of maps, to illustrate the spatial patterns in lake status through time. To facilitate comparison with the pollen data, we used the lake-level data to assess differences in effective moisture at

specified time periods. We then used the relationship between lake extent and circulation regimes over the continent at present (discussed above) to interpret the changes in lake status and effective moisture in terms of changes in climate over the past 18,000 yr.

POLLEN

Pollen data covering part or all of the last 18,000 yr have been compiled from 77 sites in Australia and Papua New Guinea (Table 11.2). These data were derived from published literature. Data from studies that concentrated on local geomorphic change or specific ecological problems were not included. Reliable data exist for coastal southeastern Australia, northeastern Queensland, and the mountainous areas of Tasmania and Papua New Guinea (Fig. 11.5). The arid zone, the lowland summer-rainfall areas, and a number of sensitive ecological boundaries either are underrepresented or have not been investigated.

Enough published modern pollen-rain data from Australia and New Guinea are available to establish the broad vegetation-pollen relationships (e.g., Flenley, 1967; Powell, 1970; Dodson, 1983; Dodson and Myers, 1986). Vegetation formations such as rainforest, sclerophyll forest, heath, and shrubland can be readily identified. Detailed variation within these formations

Table 11.1. The lake data base

Basin	Latitude (°S)	Longitude (°E)	Elevation (m)	14C dates (no.)	L	S	G	A	D	P	O	M	References[b]
New South Wales													
Lake Bancannia	30.82	141.93	107	3	*								52
Breadalbane	34.78	149.48	697	23	*	*		*		*			49, 62
Lake George	35.08	149.42	673	47	*			*				*	21, 31, 35–37, 41, 53, 63, 92, 93, 97, 98, 108
Salt Lake	30.05	142.18	78	1	*								52
Tysons Lake	33.85	142.83	350	2	*								15, 108
Lake Victoria	34.00	141.28	200	7	*								60
Willandra Lakes	30.50	140.00	200	70	*								1, 3, 4, 11–15, 19–21, 91, 108
Queensland													
Bromfield Swamp	17.38	145.55	755	5	*			*					67
Lake Euramoo	17.17	145.63	730	9	*			*					64, 68, 71
Lynch's Crater	17.37	145.70	760	17	*			*					66, 68–70, 72–74
Quincan Crater	17.30	145.58	790	4	*			*					65, 67
South Australia													
Lake Frome	30.75	139.83	-2	46	*	*	*						22, 25, 51, 105
Lake Leake	37.62	140.57	97	11	*				*	*			45, 46
Marshes Swamp	37.62	140.53	80	4	*	*			*	*			50
Valley Lake	37.85	140.77	100	6	*	*							7
Wyrie Swamp	37.65	140.30	100	15	*	*			*	*			47
Tasmania													
Beatties Tarn	42.67	146.63	990	2	*								77–79, 88
Crown Lagoon	42.28	147.63	375	2	*			*					75, 94–96
Eagle Tarn	42.67	146.58	1033	5	*			*	*				23, 77–79, 88
Pulbeena Swamp	40.89	145.12	30	23	*			*			*		5, 32–34, 40
Lake Tiberias	42.37	147.37	442	1	*			*					80
Lake Vera	42.27	145.87	560	3	*			*	*				23, 77–79, 87, 88
Victoria													
Lake Albacutya	35.75	141.97	90	1	*								11, 15
Lake Bullenmerri	38.25	143.12	146	53	*	*		*			*	*	6, 8, 24, 39, 42, 48, 58, 90, 103
Cobrico Swamp	38.30	143.03	80	1	*								24
Lake Colongulac	38.25	143.45	65	2	*								5, 39, 56–58, 61

is difficult to discern, however, because communities with different ecological relationships often have similar composition at the family and generic levels. In addition, most taxa, including many tree species, have poorly dispersed pollen, so most vegetation histories are local rather than regional.

Pollen analysis nevertheless provides the most detailed paleoclimatic information from the widest range of environments. The major genera *Eucalyptus* and *Acacia* are relatively short-lived and can be expected to respond to climatic change within a few decades. The most sensitive indicators of changes in available moisture appear to be the balance between rainforest and sclerophyll trees in moist environments and changes in the proportions of shrub, fern, and grassland species in humid to semiarid environments. In northern Australia and in montane areas it has been possible to detect temperature changes by pollen analysis; elsewhere the record is dominated by

changes in effective moisture. Quantitative reconstructions of temperature and/or precipitation changes have been made for some areas. At some sites close to the boundary between summer and winter rainfall—for example, at Lake Frome (South Australia) and Barrington Tops (New South Wales)—it has been possible to reconstruct changes in the seasonal distribution of precipitation. Some other vegetation types with high fire frequency (e.g., sclerophyll heath communities on poor soils) are relatively insensitive to climatic change.

Results

LAKES

The lake-level data for the continent as a whole (Fig. 11.6) show that conditions at 18,000 yr B.P. were on average drier than today. Most lakes (69%) were low.

Basin	Latitude (°S)	Longitude (°E)	Elevation (m)	[14]C dates (no.)	L	S	G	A	D	P	O	M	References[b]
Lake Corangamite	38.25	143.50	117	5	*								5, 21, 38, 39, 58, 59, 99
Lake Gnotuk	38.23	143.10	102	15	*		*	*	*		*	*	6, 8, 9, 21, 24, 31, 39, 42–44, 89, 90, 103, 104, 109
Lake Keilambete	38.20	143.87	150	40	*		*	*			*		6, 7, 9, 11, 16, 17, 21, 26–29, 42, 44, 59, 89, 91, 107
Kow Swamp	36.20	144.30	83	18	*								81, 82, 86, 102, 106
Lake Tyrrell	35.33	142.78	42	18	*	*							2, 18, 76, 83–85, 101
Western Australia													
Lake Grace	33.30	118.40	200	4	*								10, 15, 100, 108
Lake King	33.08	119.53	350	2	*								10, 15, 108
Myalup Swamp	33.12	115.72	6	1	*								30
Storeys Lake	31.52	118.03	300	2	*								10, 15
Papua New Guinea													
Lake Wanum	6.63	146.78	35	16	*	*		*					54, 55

[a]L = lithology/stratigraphy, S = sedimentation rates/hiatuses, G = geochemistry, A = aquatic pollen or macrofossils, D = diatoms, P = pollen preservation, O = ostracods, M = mollusks.

[b]1, Allen (1972); 2, An Zhisheng *et al.* (1986); 3, Barbetti and Allen (1972); 4, Barbetti and Polach (1973); 5, Barendsen *et al.* (1957); 6, Barton and Barbetti (1982); 7, Barton and McElhinny (1980); 8, Barton and McElhinny (1981); 9, Barton and Polach (1980); 10, Bettenay (1962); 11, Bowler (1970); 12, Bowler (1971); 13, Bowler (1973); 14, Bowler (1975); 15, Bowler (1976); 16, Bowler (1981); 17, Bowler and Hamada (1971); 18, Bowler and Teller (1986); 19, Bowler *et al.* (1970); 20, Bowler *et al.* (1972); 21, Bowler *et al.* (1976); 22, Bowler *et al.* (1986); 23, Bradbury (1986); 24, Buckley and Willis (1972); 25, Callen (1976); 26, Cann and De Deckker (1981); 27, Chivas *et al.* (1985); 28, Chivas *et al.* (1986a); 29, Chivas *et al.* (1986b); 30, Churchill (1968); 31, Churchill *et al.* (1978); 32, Colhoun (1978a); 33, Colhoun (1978b); 34, Colhoun *et al.* (1982); 35, Coventry (1973); 36, Coventry (1976); 37, Coventry and Walker (1977); 38, Currey (1964); 39, Currey (1970); 40, De Deckker (1982a); 41, De Deckker (1982b); 42, De Deckker (1982c); 43, Dodson (1971); 44, Dodson (1974a); 45, Dodson (1974b); 46, Dodson (1975a); 47, Dodson (1977); 48, Dodson (1979); 49, Dodson (1986); 50, Dodson and Wilson (1975); 51, Draper and Jensen (1976); 52, Dury (1973); 53, Galloway (1967); 54, Garrett-Jones (1979); 55, Garrett-Jones (1980); 56, Gill (1953); 57, Gill (1955); 58, Gill (1971); 59, Gill (1973a); 60, Gill (1973b); 61, Grayson and Mahoney (1910); 62, Harrison (1980); 63, Jacobson and Schuett (1979); 64, Kershaw (1970); 65, Kershaw (1971); 66, Kershaw (1974); 67, Kershaw (1975a); 68, Kershaw (1975b); 69, Kershaw (1976); 70, Kershaw (1978); 71, Kershaw (1979); 72, Kershaw (1981); 73, Kershaw (1983); 74, Kershaw (1985); 75, Lourandos (1970); 76, Luly *et al.* (1986); 77, Macphail (1975); 78, Macphail (1976); 79, Macphail (1979); 80, Macphail and Jackson (1978); 81, Macumber (1968); 82, Macumber (1977); 83, Macumber (1978); 84, Macumber (1980); 85, Macumber (1983); 86, Macumber and Thorne (1975); 87, V. Markgraf (personal communication, 1987); 88, Markgraf *et al.* (1986); 89, Ollier and Joyce (1964); 90, Polach and Barton (1983); 91, Polach *et al.* (1970); 92, Polach *et al.* (1978); 93, Sanson *et al.* (1980); 94, Sigleo (1979); 95, Sigleo and Colhoun (1981); 96, Sigleo and Colhoun (1982); 97, Singh *et al.* (1981a); 98, Singh *et al.* (1981b); 99, Skeats and James (1937); 100, Suzuki *et al.* (1982); 101, Teller *et al.* (1982); 102, Thorne and Macumber (1972); 103, Timms (1976); 104, Tudor (1973); 105, Ullman and McLeod (1986); 106, Wright (1975); 107, Yamasaki *et al.* (1970); 108, Yamasaki *et al.* (1977); 109, Yezdani (1970).

The number of lakes of intermediate status increased slightly at 16,000 yr B.P. (from 16 to 31%), but the number of high lakes fell (from 16 to 7%). A brief interval of wetter conditions occurred between 15,000 and 13,000 yr B.P., with a maximum number of high lakes (33%) at 14,000 yr B.P. Aridity reached a maximum at 12,000 yr B.P., when nearly all of the lakes (81%) were low. The number of high and intermediate lakes gradually increased between 11,000 and 7000 yr B.P. Wetness reached a maximum at 7000 yr B.P., when most lakes were high (58%) or intermediate (27%). After 7000 yr B.P. the number of high lakes declined gradually toward the present. The persistence of relatively moist conditions during the mid-Holocene is indicated by an increase in the number of intermediate lakes after 7000 yr B.P. Increasing aridity in the more recent past is indicated by an increase in the number of low lakes after 1000 yr B.P.

The spatial patterns in lake-level changes are illustrated by maps of lake status at 3000-yr intervals (Fig. 11.7) from 18,000 yr B.P. to the present. These intervals allow comparisons with the GCM simulations, although they do not always correspond with times of maximum expression of any given pattern.

At 18,000 yr B.P. (Fig. 11.7) the lakes in southwestern Western Australia were low, as were most lakes in coastal southeastern Australia. Lake George on the Southern Tablelands was also low. The lake-level records from the interior of southeastern Australia are more equivocal: Lake Frome, Lake Albacutya, and Kow Swamp were low, but Tysons Lake was high and the Willandra Lakes were intermediate. Pulbeena Swamp, the only record from Tasmania, was intermediate. At 17,000 yr B.P. the pattern of low lake levels in the extreme south of southeastern Australia and Tasmania was more pronounced, while high and intermediate lake levels were characteristic of the interior of southeastern Australia (Harrison, 1989).

At 15,000 yr B.P. (Fig. 11.7) low lake levels were characteristic of southwestern Western Australia, coastal

Table 11.2. The pollen data base

Site	Latitude (°S)	Longitude (°E)	Elevation (m)	[14]C dates (no.)	References
New South Wales					
Black Swamp	32.05	151.47	1450	3	Dodson *et al.* (1986)
Boggy Swamp	31.88	151.52	1160	11	Dodson *et al.* (1986)
Breadalbane Basin (northeast)	34.75	149.53	697	3	Dodson (1986)
Breadalbane Basin (northwest)	34.75	149.52	697	5	Dodson (1986)
Breadalbane Basin (southeast)	34.78	149.52	697	9	Dodson (1986)
Butchers Swamp	31.92	151.40	1230	14	Dodson *et al.* (1986)
Club Lake	36.41	148.28	1955	13	Martin (1986)
Fingal Bay Swamp	32.70	152.17	1	3	Macphail (1973)
Horse Swamp	31.93	151.38	1250	5	Dodson *et al.* (1986)
Killer Bog	32.70	151.55	1260	8	Dodson *et al.* (1986)
Lake George	35.08	149.35	673	5	Singh *et al.*, 1981a, b; Singh and Geissler, 1985
North Dee Why	33.46	151.30	2	1	Martin (1971)
Polblue Swamp	31.90	151.42	1430	4	Dodson *et al.* (1986)
Pound's Lake	36.38	148.32	1960	5	Martin (1986)
Sapphire Swamp	32.07	151.58	1260	1	Dodson *et al.* (1986)
Top Swamp	32.00	151.47	1530	4	Dodson *et al.* (1986)
Wet Lagoon	34.78	149.42	700	6	Dodson (1986)
Queensland					
Bromfield Swamp	17.38	145.55	755	5	Kershaw (1975a)
Lake Euramoo	17.17	145.63	730	7[a]	Kershaw (1970)
Lynch's Crater	17.37	145.70	760	6[a]	Kershaw (1974, 1976)
Quincan Crater	17.30	145.58	790	5[a]	Kershaw (1971)
South Australia					
Blue Tea Tree Swamp	37.62	140.52	150	1	Dodson and Wilson (1975)
Lake Frome	30.70	139.75	-2	3	Singh (1981)
Lake Leake	37.62	140.60	150	5	Dodson (1975a)
Lake Leake Lagoon	37.62	140.60	150	3	Dodson (1977)
Lashmar's Lagoon	35.82	138.05	2	5	Singh *et al.* (1981a)
Mount Burr Swamp	37.60	140.32	150	3	Dodson and Wilson (1975)
N145	31.63	129.17	100	5	Martin (1973)
Valley Lake	37.85	140.77	100	1	Dodson (1975b)
Wyrie Swamp	37.67	140.35	25	7	Dodson (1977)
Tasmania					
Beatties Tarn	42.67	146.50	990	2	Macphail (1979), Markgraf *et al.* (1986)
Brown Marsh	42.22	146.57	750	1	Macphail (1979)
Cave Bay	40.50	144.78	5	9	Hope (1978)
Eagle Tarn	42.67	146.58	1033	5	Macphail (1979), Markgraf *et al.* (1986)
Lake Tiberias	42.42	147.37	442	1	Macphail and Jackson (1978)
Lake Vera	42.27	145.87	560	3	Macphail (1979), Markgraf *et al.* (1986)
Ooze Lake	43.50	146.72	880	2	Macphail and Colhoun (1985)
Pulbeena Swamp	40.92	145.12	30	4	Colhoun *et al.* (1982)

southeastern Australia, and northeastern Queensland. High and intermediate lake levels were still predominant in the interior of southeastern Australia. By 12,000 yr B.P. (Fig. 11.7) lake levels were low at virtually all sites. According to the lake-level records, 12,000 yr B.P. was the most arid time in the last 18,000 yr.

After 12,000 yr B.P. there was a gradual transition to wetter conditions. This transition occurred first in Tasmania, where some lakes were already rising by 11,000 yr B.P. Rising lake levels were registered at some sites in coastal southeastern Australia and in northeastern

Queensland at 10,000 yr B.P. (Harrison, 1989). The onset of lacustrine sedimentation at Lake Wanum in Papua New Guinea at the beginning of the Holocene (Garrett-Jones, 1979) also indicates wetter conditions.

At 9000 yr B.P. (Fig. 11.7) most lakes in Tasmania were high or intermediate. Wetter conditions were also registered in northeastern Queensland and in coastal southeastern Australia. Lake levels were low in the Southern Tablelands and the interior of southeastern Australia. The trend to wetter conditions culminated at 7000 yr B.P., when Tasmania and coastal

Site	Latitude (°S)	Longitude (°E)	Elevation (m)	^{14}C dates (no.)	References
Tarraleah Bog	42.30	146.43	440	3	Macphail (1984)
Unnamed Cirque	43.35	146.82	960	1	Macphail (1979), Markgraf et al. (1986)
Unnamed Tarn	42.63	146.47	1158	3	Macphail (1979), Markgraf et al. (1986)
Wireless Hill	54.48	158.95	105	5	Selkirk et al. (1982)
Victoria					
Boomer Swamp	38.28	141.32	5	2	Head (1983, 1988)
Bridgewater Lagoon	38.32	141.42	5	2	Head (1983, 1988)
Bunyip Bog	36.77	146.75	1300	2	Binder (1978)
Cotters Lake	38.88	146.20	4	1	Hope (1974)
Darby Beach	38.97	146.22	0	1	Hope (1974)
Delegate River Swamp	37.08	148.82	800	3	Ladd (1979a)
Five Mile Beach	38.97	146.45	1	2	Ladd (1979c)
Hidden Swamp	38.03	147.65	10	2	Hooley et al. (1981)
Lake Bullenmerri	38.25	143.12	146	53	Dodson (1979)
Lake Curlip	37.77	148.57	5	2	Ladd (1978)
Lake Keilambete	38.20	142.87	150	4	Dodson (1974)
Mount Latrobe	39.00	146.37	600	1	Howard and Hope (1970)
Rooty Breaks Swamp	37.27	148.90	1100	1	Ladd (1979c)
Tidal River	39.03	146.30	5	2	Hope (1974)
Western Australia					
Boggy Lake	35.02	116.63	100	3	Churchill (1968)
Flinders Bay	34.32	115.15	15	2	Churchill (1968)
Madura Cave	32.03	127.08	100	3	Martin (1973)
Myalup Swamp	33.12	115.72	10	1	Churchill (1968)
Norina	32.07	127.93	80	3	Martin (1973)
Scott River Swamp	34.30	115.28	10	2	Churchill (1968)
Weld Swamp	34.68	116.52	100	2	Churchill (1968)
West Lake	34.47	116.60	100	1	Churchill (1968)
Papua New Guinea					
Birip	5.58	143.80	1900	3	Walker and Flenley (1979)
Brass Tarn	5.83	145.02	3910	5	Hope (1976)
Imbuka Bog	5.78	145.07	3550	3	Hope (1976)
Komanimambuno Mire	5.85	145.08	2740	9	Hope (1976)
Lake Inim	5.53	143.58	2550	6	Walker and Flenley (1979)
Lower Pindaunbe Lake	5.78	145.07	3510	3	Hope (1976)
Pegagl Mire	5.83	145.08	3680	1	Hope (1976)
Sirunki	5.48	143.52	2500	9	Walker and Flenley (1979)
Summit Bog	5.77	145.02	4420	3	Hope (1976)
West Irian					
Ertsberg	4.20	137.10	3620	3	Hope and Peterson (1976)
Ijomba Mire	4.10	137.13	3630	3	Hope and Peterson (1976)
Otomona Valley	4.35	137.40	1705	1	Hope and Peterson (1976)
Yellow Valley	4.25	137.38	4270	3	Hope and Peterson (1976)

[a]Includes only dates on core from which pollen counts were taken, not all dates from this site.

southeastern Australia were characterized by mostly high and some intermediate lake levels, lakes on the Southern Tablelands had also risen to high or intermediate levels, lakes in northeastern Queensland were all high, and Lake Wanum in Papua New Guinea had reached intermediate level (Harrison, 1989). Only lakes in the eastern part of the interior of southeastern Australia (e.g., the Willandra Lakes, Lake Bancannia, and Kow Swamp) remained low. More westerly sites from the interior (e.g., Lakes Frome, Victoria, and

Tyrrell) were intermediate. This pattern remained essentially unchanged at 6000 yr B.P. (Fig. 11.7).

After 6000 yr B.P. conditions became somewhat drier in southwestern Western Australia, northeastern Queensland, and coastal southeastern Australia and possibly slightly wetter in the interior of southeastern Australia. By 3000 yr B.P. (Fig. 11.7) lakes in southwestern Western Australia were low, lakes in northeastern Queensland were mostly intermediate, and several lakes in coastal southeastern Australia were interme-

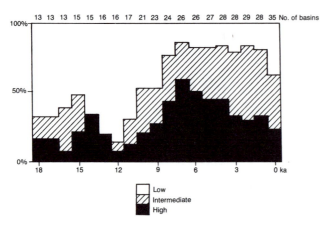

Fig. 11.6. Temporal variations in the frequency distribution of lake status for all sites in Australia and Papua New Guinea.

diate. Slightly wetter conditions in the interior of southeastern Australia were marked by high or intermediate lake levels at Lakes Frome, Bancannia, and Tyrrell. Lake levels in northeastern Queensland and Papua New Guinea were either high or intermediate.

Conditions in the interior of southeastern Australia and on the Southern Tablelands became gradually more arid after 3000 yr B.P. (Harrison, 1989). Wetter conditions in the far north are indicated by the continued rise of Lake Wanum in Papua New Guinea after 3000 yr B.P. The map of modern lake status (Fig. 11.3b) shows a contrast between high or intermediate lake levels in Tasmania and coastal southeastern Australia and low lake levels in southwestern Western Australia and the interior of southeastern Australia, with intermediate or high lake levels in Queensland and Papua New Guinea.

Thus, the magnitude and timing of hydrologic changes during the last 18,000 yr differed distinctly among regions (Fig. 11.8). Tasmania was relatively arid initially, but the lakes there had already begun to rise by 11,000 yr B.P., and the early Holocene was relatively wet. The lake records indicate maximum wetness by 7000 yr B.P. Conditions were slightly drier around 6000-5000 yr B.P., and conditions similar to present were established by 4000 yr B.P. Coastal southeastern Australia was also relatively arid initially and became wetter at the beginning of the Holocene. Wetness reached a maximum at 6000 yr B.P. A gradual shift back to drier conditions began at 5000 yr B.P. and culminated between 3000 and 2000 yr B.P.; by 1000 yr B.P. conditions were again wetter than today.

The data for the interior of southeastern Australia show a markedly different pattern. Although relatively dry at 18,000 yr B.P., the lakes were briefly higher at 15,000-14,000 yr B.P. The early Holocene (10,000-8000 yr B.P.) was dry, and the lakes did not begin to rise again until 7000 yr B.P. Maximum wetness occurred around 4000 yr B.P., after which conditions gradually

became drier again toward the present. The limited data from southwestern Western Australia suggest a somewhat similar pattern, with wet phases centered on 14,000 and 6000-5000 yr B.P.

Tropical northeastern Queensland was arid early and became wetter at the beginning of the Holocene. The lake records indicate maximum wetness at 7000 yr B.P., slightly earlier than in coastal southeastern Australia. A short dry phase is indicated around 5000 yr B.P., but conditions similar to the present were established by 3000 yr B.P. The one record from Papua New Guinea also shows an increase in wetness from 9000 yr B.P., but in contrast to sites in Queensland, no mid-Holocene maximum occurred, and the lake continued to rise after 3000 yr B.P.

POLLEN: VEGETATION HISTORY

The vegetation history of montane New Guinea has been described by Hope (1976), Hope and Peterson (1976), and Walker and Flenley (1979). From 18,000 to 14,000 yr B.P. snowline extended upward from 3200 m, and depauperate forests and warmth-demanding taxa were restricted to below 2200 m. By 12,000 yr B.P. lower montane areas (3200-2400 m) were occupied by subalpine grasslands and *Cyathea* woodlands, and tundra was found at higher altitudes. By 9000 yr B.P., as ice cover receded, grasslands were recorded at some alpine sites, and subalpine forests of *Phyllocladus, Podocarpus,* and *Castanopsis* were established in upper montane areas and *Nothofagus* forests in lower montane areas. In West Irian little change occurred after 9000 yr B.P. In Papua New Guinea at 6000 yr B.P. subalpine forests occurred in subalpine and upper montane areas, with mixed *Nothofagus* and some oak forests in lower montane areas. By 3000 yr B.P. tundra and subalpine grasslands were found at the highest altitudes, and subalpine forests occurred at lower montane sites. These data show that the greatest development of forests and shrublands occurred between 8600 and 5000 yr B.P. Glacial retreat accompanied these vegetation shifts. Forest cover has declined since 5000 yr B.P. as a result of cooling, human disturbance, and fire. The montane New Guinea sites show no sensitivity to precipitation changes, presumably because precipitation was never limiting.

Kershaw (1970, 1971, 1974, 1975a, 1976, 1978) described the history of four sites (Bromfield Swamp, Quincan Crater, Lake Euramoo, and Lynch's Crater) on the Atherton Plateau (700-800 m) in tropical northeastern Queensland. Sclerophyll vegetation dominated by Casuarinaceae and *Eucalyptus* existed from before 27,000 yr B.P. until around 12,000 yr B.P., when the vegetation shifted over 3500-4400 yr to warm-temperate rainforest dominated by Cunoniaceae but containing *Balanops* and *Sphenostemon.* Warm-temperate rainforest with Monimiaceae and *Quintinia* occurred between 7600 and 4000 yr B.P. at

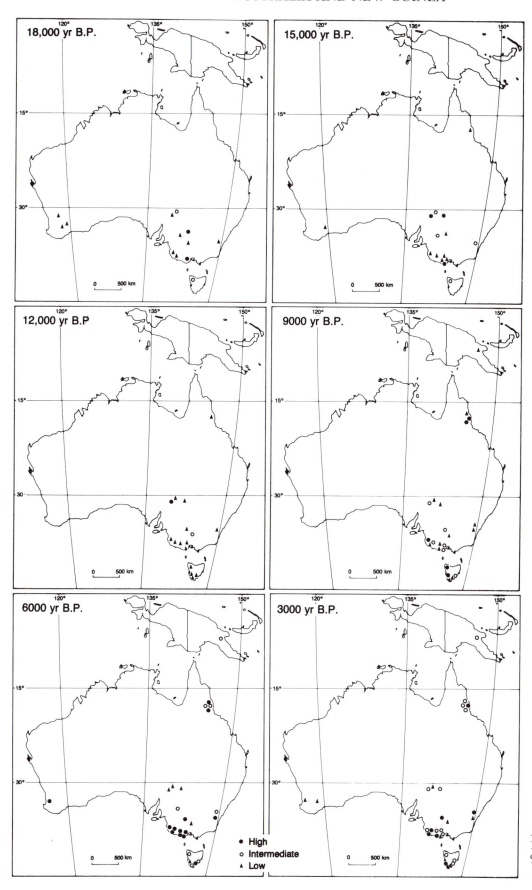

Fig. 11.7. Lake status at 18,000, 15,000, 12,000, 9000, 6000, and 3000 yr B.P.

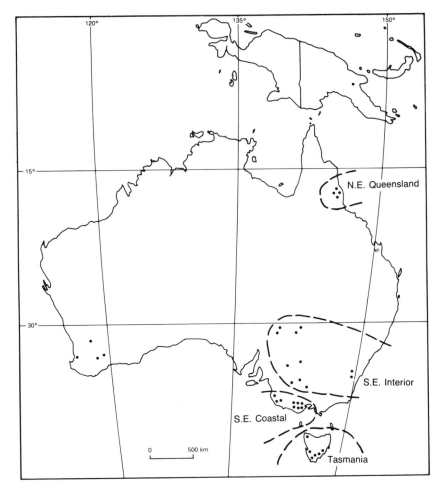

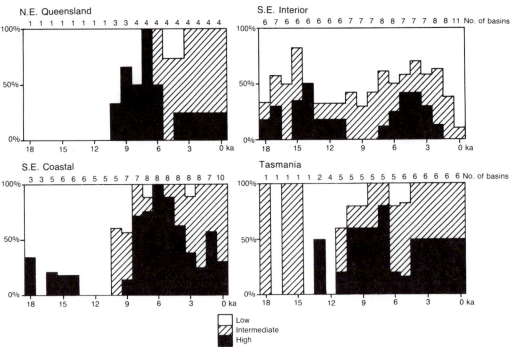

Fig. 11.8. Temporal variations in the frequency distribution of lake status for northeastern Queensland, the interior of southeastern Australia, coastal southeastern Australia, and Tasmania.

Lake Euramoo. At Bromfield Swamp warm-temperate rainforest was replaced between 6000 and 3800 yr B.P. by subtropical rainforest with *Elaeocarpus, Calamus,* and *Ilex.* The occurrence of dry subtropical forests with *Melia, Cedrela, Agathis, Macaranga-Mallotus,* and *Blepharocarya* at Bromfield Swamp and Lake Euramoo indicates drier conditions in the last 4000 yr. Drier conditions are also indicated by the presence of sclerophyll taxa at Quincan Crater after 2000 yr B.P. Thus the vegetation changes in northeastern Queensland indicate conditions warmer and moister than at present between 7000 and 4000 yr B.P. and drier and cooler after 4000 yr B.P.

Barrington Tops in New South Wales is a basalt and granite plateau (1100-1300 m) with a complex of vegetation types including subalpine woodlands, wet and dry sclerophyll forests, and cool and warm rainforest formations. The distribution of vegetation types on the plateau is related to precipitation and temperature gradients. Dodson *et al.* (1986) studied eight sites with vegetation histories covering part or all of the last 12,000 yr. Some eucalypt forest with steppe-like vegetation occurred around 12,000 yr B.P. but was replaced by eucalypt forests between 12,000 and 9000 yr B.P. on the western and drier parts of the plateau. The modern vegetation formations were all present by about 9000 yr B.P. Wetter formations, such as *Nothofagus moorei* rainforest and wet sclerophyll forests, attained their maximum extents between 6000 and 3500 yr B.P. Subalpine woodland increased after 3500 yr B.P., indicating drier conditions again. A small increase in *N. moorei* rainforest over the last one to two millennia probably indicates a return to wetter conditions.

Singh *et al.* (1981a,b) and Singh and Geissler (1985) described Pleistocene vegetation history at Lake George on the Southern Tablelands, and Dodson (1986) discussed Holocene changes at several sites around Breadalbane, about 20 km to the north. The data show that steppelike vegetation was replaced by eucalypt woodland between 9000 and 8000 yr B.P., indicating an increase in moisture.

Binder and Kershaw (1978) and Binder (1978) showed that a grassland-herbfield with only a sparse scatter of eucalypts occurred at Bunyip Bog in the Victorian highlands at 11,000 yr B.P. Wet sclerophyll forest gradually developed after 11,000 yr B.P. The presence of ferns, *Pomaderris,* and *Tasmannia* in the understory suggests that conditions were warmer and wetter. More xerophytic vegetation developed around 2000 yr B.P. Ladd (1979a) showed that similar patterns of vegetation change occurred in northeastern Victoria, except that the sclerophyll forest developed slightly earlier. Martin (1986) showed that an alpine desert in the Snowy Mountains at an elevation of about 2000 m was replaced by high alpine communities around 10,600 yr B.P. Between 7000 and 4400 yr B.P. mixed grassland-sedge and wet sclerophyll ele-

ments reached Holocene maxima. A small increase in some alpine taxa after 2000 yr B.P. may indicate a cooling trend.

Most of the vegetation history data from the continent are for lowland southeastern Australia, including coastal New South Wales (Martin, 1971; Macphail, 1973), coastal Victoria (Howard and Hope, 1970; Hope, 1974; Ladd, 1978, 1979a; Hooley *et al.,* 1981; Head, 1983, 1988), coastal South Australia (Dodson, 1977; Singh *et al.,* 1981a), coastal Tasmania (Hope, 1978; Colhoun *et al.,* 1982), and the basalt plains of Victoria and southeastern South Australia (Dodson, 1974a,b, 1975a,b, 1979; Dodson and Wilson, 1975). In the southernmost part of the region, cold steppe or open-wooded herbfield, with a semiarid component, was gradually replaced by sclerophyll forest between 15,000 and 10,000 yr B.P. At Wilson's Promontory (Hope, 1976) moister taxa such as *Nothofagus* expanded during 6000–4500 yr B.P. Many coastal sites reflect changes associated with dune building and swamp succession following the postglacial rise in sea level.

The vegetation history of montane Tasmania has been examined by Macphail (1979, 1984), Macphail and Colhoun (1985), and Markgraf *et al.* (1986). Macphail described the invasion of eucalypt forests into alpine herbfields and tundra over large areas around 11,500 yr B.P. Between 10,000 and 8600 yr B.P., *Nothofagus*-dominated rainforests expanded in the south, west, and central areas, and wet sclerophyll eucalypt forests indicated by *Pomaderris* expanded in the east. These wet taxa contracted around 6000 yr B.P., although mixed forests persisted, possibly maintained by fire. Data for some sites show conditions similar to the present over the last 3500 yr. Data for other sites, especially in western Tasmania, suggest a slight expansion of moist taxa in the last 2000 yr. Markgraf *et al.* (1986) reevaluated Macphail's fossil pollen data from southwestern Tasmania in light of additional modern pollen-rain samples. According to the reevaluation, the late-glacial environments are more likely to have been steppe and marshland, forest expansion had probably begun by 12,000 yr B.P., and the modern rainforest communities were established by 8000 yr B.P. These communities persisted until about 6000 yr B.P., when a short-term cooling episode reduced their extent to that of the modern rainforest cover. Reconciling the differences in these interpretations for Tasmania will require additional research.

The data from southwestern Western Australia cover the last 6000 yr continuously, with some spot data for earlier times. Churchill (1968) described changes in the relative abundance of some eucalypts that dominate the vegetation of the region. *Eucalyptus diversicolor,* which indicates greater effective precipitation, was more abundant than at present between 6000 and 4500 yr B.P. and around 2500 yr B.P.

Arid environments are underinvestigated. Martin (1973) published three pollen diagrams from cave deposits on the Nullabor Plain, and Singh (1981) described a pollen record from Lake Frome near the summer-winter rainfall boundary in South Australia. On the Nullabor, treeless chenopod shrubland was replaced by mallee around 9000 yr B.P. Eucalypt woodland with some shrubs and grasses occurred near Lake Frome between 9500 and 8000 yr B.P. The trees and shrubs declined until 7000 yr B.P. as a result of lower precipitation and fire but increased again between 7000 and 4200 yr B.P. before declining again. Singh (1981) attributed a small recovery after 2000 yr B.P. to increased winter rainfall.

Thus enough paleoecological evidence is available to reconstruct the broad patterns of climatic change during the past 18,000 yr (Fig. 11.9) independently of the lake-level evidence. Vegetation formations typical of generally cooler and drier conditions than at present were found throughout Australia before 12,000 yr B.P. New Guinea was also cooler. The period between 18,000 and 15,000 yr B.P. was cooler and drier than any subsequent time. Vegetation indicating moister and/or warmer conditions developed across the whole region after 12,000 yr B.P.

The data indicate distinct regional differences in the onset of wetter conditions and in the timing of peak effective moisture. These differences closely parallel the regional differences in lake-level history discussed above. In Tasmania, forest expansion marking the beginning of wetter conditions began around 12,000-11,500 yr B.P., consistent with the evidence for rising lake levels by 11,000 yr B.P. The vegetation indicates that the early Holocene was relatively wet, with maximum wetness around 8600-7000 yr B.P. This is also consistent with the lake-level record, which shows that relatively wet conditions during the early Holocene culminated at 7000 yr B.P. The contraction of rainforest and wet sclerophyll forest indicates slightly drier conditions around 6000-5000 yr B.P., as do the lake data. Conditions similar to present were established around 3500 yr B.P. Coastal southeastern Australia became gradually wetter during the early Holocene, reaching peak effective moisture between 6000 and 4500 yr B.P. This pattern is consistent with the lake-level records, except that the lake data suggest that effective moisture reached a peak at 6000 yr B.P. and that a gradual shift back to drier conditions began at 5000 yr B.P.

In tropical northeastern Queensland the shift from sclerophyll vegetation to warm-temperate rainforest, marking the beginning of wetter conditions, began around 12,000 yr B.P. The vegetation indicates peak wetness between 7600 and 4000 yr B.P.; the lake-level record shows peak effective moisture around 7000 yr B.P. Conditions similar to present were established by 4000 yr B.P. Conditions became wetter in Barrington Tops at the beginning of the Holocene, with peak effective moisture between 6000 and 3500 yr B.P. Slightly drier conditions, marked by an increase in the extent of subalpine woodland, occurred between 3500

and 2000 yr B.P., after which somewhat wetter conditions returned. The limited pollen evidence from southwestern Western Australia indicates wetter conditions than at present between 6000 and 4500 yr B.P., consistent with the lake-level records. The evidence from Lake Frome, the only pollen site in the arid interior of southeastern Australia, suggests moist conditions at 9500-8000 yr B.P., drier conditions at 8000-7000 yr B.P., and peak moisture at 7000-4200 yr B.P. Most of the lake-level records from the arid interior of southwestern Australia suggest that the early Holocene was dry, but they do show an increase in effective moisture starting around 7000 yr B.P. and culminating at 4000 yr B.P.

Quantitative Climatic Reconstructions from Pollen Data

Walker and Flenley (1979) used present relationships among vegetation, altitude, and temperature, calibrated by pollen analysis of surface samples, to infer temperature changes from fossil pollen data from montane New Guinea. They inferred a gradual increase in mean annual temperature at 2500 m from 1-6°C at 18,000-16,000 yr B.P. to 10-13°C at 12,000 yr B.P. Temperatures continued to rise, reaching 13-15°C between 6500 and 5000 yr B.P., before decreasing toward present values of 12-14°C.

Kershaw (1976) inferred mean annual precipitation and temperature at Lynch's Crater in northeastern Queensland by calibrating fossil pollen data with surface pollen spectra and modern vegetation patterns. He estimated that mean annual rainfall increased from 600 mm at 18,000-15,000 yr B.P. to 700 mm at 12,000 yr B.P. and 1600 mm at 9000 yr B.P. and reached a Holocene maximum of 3200 mm around 6000 yr B.P. Since 6000 yr B.P. rainfall has declined to the modern value of 2500 mm. The mean annual temperature was near 18°C at 9000 yr B.P., rose to a maximum of 20.5°C around 6000 yr B.P., then fell slightly to today's 20°C.

Dodson *et al.* (1986) suggested that the extension of *Nothofagus moorei* rainforest on Barrington Tops to near 1200 m around 6000 yr B.P. implies that the mean annual temperature was 0.5-1.5°C higher than the present mean of 10.2°C. Data from the Southern Tablelands of New South Wales, showing the establishment of *Eucalyptus* woodland around 9000 yr B.P. (Singh and Geissler, 1985; Dodson, 1986), have been interpreted as evidence that the mean temperature of the warmest month was below 10°C before 12,000 yr B.P., increased to near 16°C at 6000 yr B.P., and has changed little since. Vegetation histories from the Victorian highlands (Binder, 1978; Binder and Kershaw, 1978; Ladd, 1979a), showing that treeline was established around some sites at 1100 m, indicate that the mean temperature for the warmest month about 10,500 yr B.P. was less than 4°C below today's 14°C.

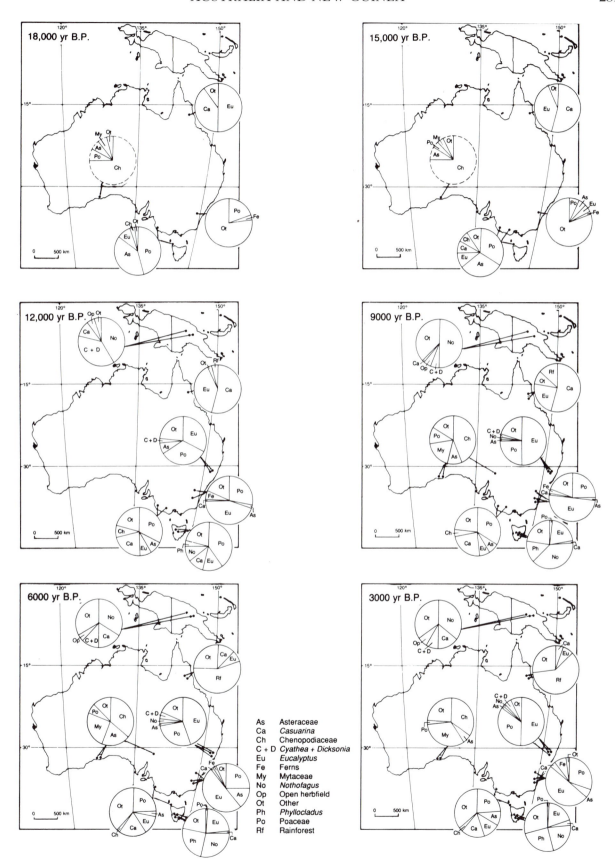

Fig. 11.9. Pollen data for 18,000, 15,000, 12,000, 9000, 6000, and 3000 yr B.P.

Macphail (1979) and Markgraf *et al.* (1986) inferred climatic variables from fossil pollen data from montane Tasmania by calibration with surface pollen spectra and modern vegetation. Macphail (1979) estimated that the summer temperature passed 10°C between 12,000 and 9000 yr B.P. Mean annual precipitation at 600 m was about 20% less than the present 1500 mm. The summer isotherm peaked 1°C above present around 6000 yr B.P. From 3000 yr B.P. to the present there was little change. Markgraf *et al.* (1986) disagreed with the earlier part of Macphail's interpretations and argued that the climate around 12,000 yr B.P. was marked by seasonal water stress, mean annual precipitation below 1000 mm, and mean annual temperature of about 6°C. By 9000 yr B.P. annual precipitation had increased to 1580 mm, seasonal water stress had ceased, and mean annual temperatures had risen to between 10 and 14.4°C. Precipitation further increased to more than 2500 mm, but mean temperatures declined by 0.7–1°C between 6000 and 4000 yr B.P.

Discussion

CHANGES IN EFFECTIVE MOISTURE AND ATMOSPHERIC CIRCULATION DURING THE LAST 18,000 YR

The regional estimates of qualitative changes in effective moisture derived from pollen and lake-level data are in good agreement. The reconstructed changes (Fig. 11.10) can be interpreted climatologically.

At 18,000–17,000 yr B.P. (Fig. 11.10) it was drier than at present in Tasmania, coastal southeastern Australia, southwestern Western Australia, the Southern Tablelands, and Queensland. Drier conditions in coastal southeastern Australia suggest that the subtropical high-pressure belt was displaced poleward of its modern position in winter, pushing the westerlies southward and depriving the coastal region of winter rain. Drier conditions in Tasmania suggest that the poleward displacement of the westerlies was sufficient to exclude even Tasmania from year-round westerly influence, and it seems likely that the westerlies lay south of Tasmania for most of the year.

In contrast, the southeastern interior was similar to or wetter than at present. The rain would have been brought by troughs between the traveling anticyclones, just as is observed today whenever the subtropical high-pressure belt is unusually far south (Pittock, 1971, 1973, 1975, 1978). Drier conditions on the Atherton Plateau at 18,000 yr B.P., however, suggest that the southward shift of the southern margin of the subtropical high-pressure belt was not accompanied by a southward displacement of the trade wind belt. Such a displacement would have placed northeastern Queensland closer to the intertropical conver-

gence zone and would have tended to increase rainfall. We infer that the subtropical high-pressure belt was expanded at 18,000 yr B.P., with its southern margin extended at least 8° south of its present position but its northern margin near or even north of its present position.

Aridity persisted over most of Australia until after 12,000 yr B.P. This suggests persistence of the circulation pattern just described. Slight variations in effective moisture in the interior of southeastern Australia and on the Southern Tablelands between 18,000 and 12,000 yr B.P. (Fig. 11.10) are consistent with slight variations in the position of the subtropical high-pressure belt or in the frequency of inter-anticyclonic troughs, but they do not imply major shifts in circulation regime.

The transition to wetter climates after 12,000 yr B.P., registered first at 11,000 yr B.P. in Tasmania and at 10,000–9000 yr B.P. in coastal southeastern Australia, indicates the equatorward advance of the southern margin of the subtropical high-pressure belt and the westerlies. Conditions on the southeastern coast were still drier than today at 9000 yr B.P., however, indicating that the southern margin of the high-pressure belt was still somewhat farther south than today. Conditions wetter than at present in coastal southeastern Australia between 8000 and 6000 yr B.P. (Fig. 11.10) may indicate that the southern margin of the subtropical high-pressure belt was by then north of its present mean position, so that coastal southeastern Australia lay within the zone of year-round westerly influence.

The steady increase in summer insolation through the Holocene, by increasing evaporation, may also have contributed to declining moisture levels after 7000–6000 yr B.P. in Queensland as well as farther south. However, in the monsoon region, including Papua New Guinea, low summer insolation in the early Holocene would be expected to have weakened the monsoon and produced conditions drier than today. Gradually increasing monsoon strength may explain the Lake Wanum record of increasing wetness from 9000 yr B.P. to the present.

The cause of the unusual spatial pattern at 6000–5000 yr B.P. (Fig. 11.10), with low lakes only in Tasmania and Queensland, is uncertain. The pattern may reflect a strong Walker Circulation, combined with a more contracted subtropical anticyclone belt. By 3000 yr B.P. conditions were similar to the present over most of the region, suggesting that the modern configuration of the circulation belts was already established.

PREVIOUS RECONSTRUCTIONS OF PALEOCIRCULATION PATTERNS AT THE GLACIAL MAXIMUM

Several earlier reconstructions of atmospheric circulation changes over Australia (e.g., Bowler, 1975; Bowler *et al.*, 1976; Webster and Streten, 1978; Harrison *et al.*, 1984) suggested that the subtropical high-pres-

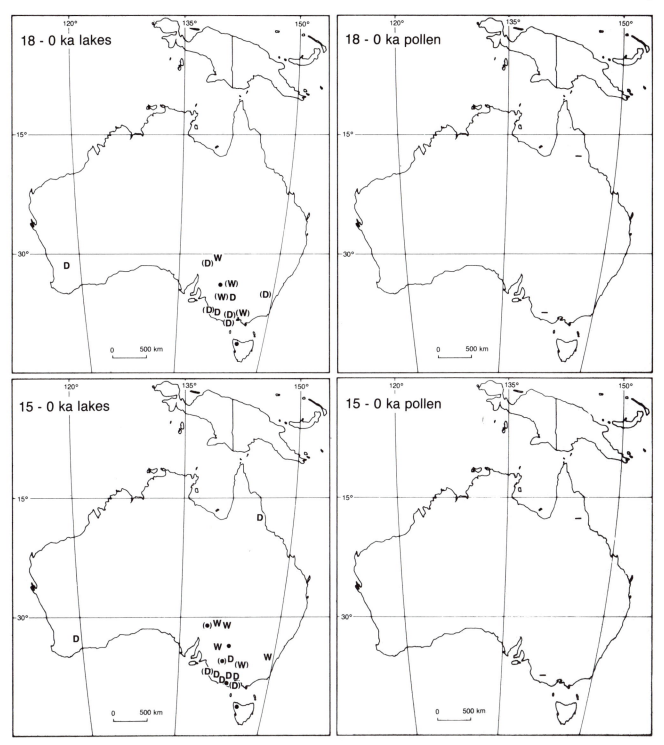

Fig. 11.10. Effective moisture anomalies at 18,000, 15,000, 12,000, 9000, 6000, and 3000 yr B.P. inferred from pollen and lake-level data. Differences in effective moisture were estimated from lake-level records at each lake site: W indicates more effective moisture, ● indicates similar effective moisture, and D indicates less effective moisture than today; parentheses around symbols indicate uncertainty. Pollen-based anomalies are not shown for individual pollen sites but were estimated from the regional signal: + indicates more effective moisture, ● indicates similar effective moisture, and - indicates less effective moisture than today; parentheses around symbols indicate uncertainty. (continued on pp. 284-85)

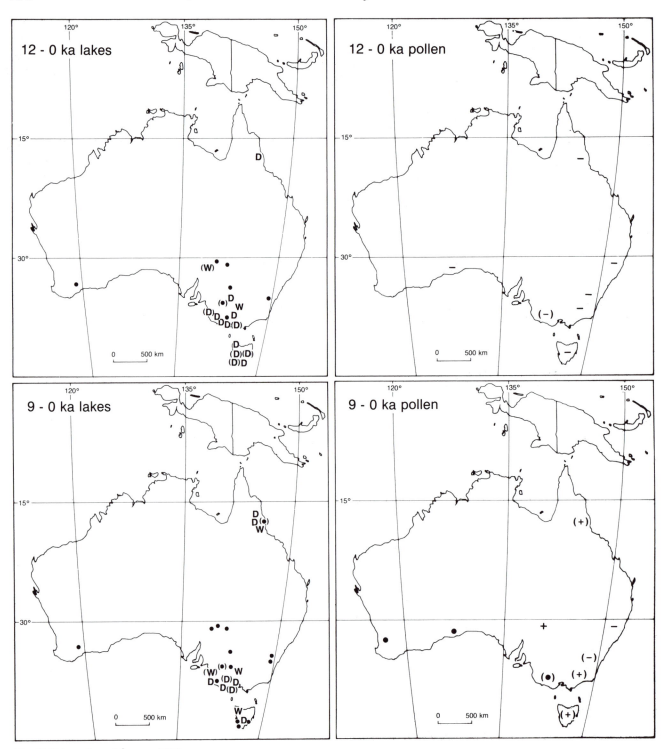

Fig. 11.10 (continued from p. 283)

sure belt and the westerlies were displaced *equator-ward* at the glacial maximum. Bowler (1975) and Bowler *et al.* (1976) argued that lunettes and linear dunes in the Murray Basin, an area dominated by southerly winds in summer today, were formed by summer winds from almost due west. The orientation of linear dunes and lunettes farther west on the Eyre

Peninsula reflects winds blowing from west-north-west, a trend maintained across the dunefields of the lower part of the Nullabor Plain. Bowler suggested that these dune orientations are consistent with summer anticyclonic systems positioned at least 5° farther north than today and argued that advective transfer of heat from the arid interior to more humid regions,

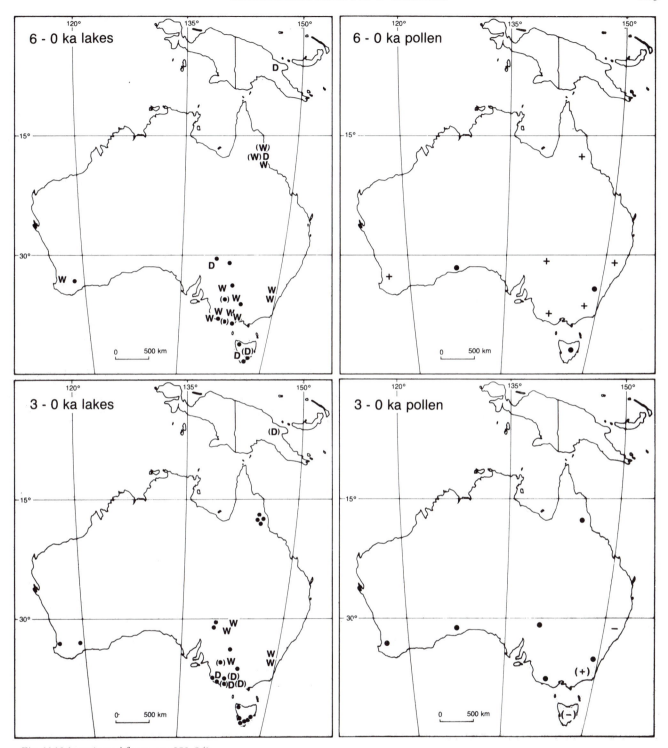

Fig. 11.10 (continued from pp. 283-84)

combined with greater continentality because of lowered sea level, would explain the drier conditions found along the southeastern continental margin.

Webster and Streten (1978) pointed out that the apparent aridity of the southern part of Australia at 18,000 yr B.P. was inconsistent with an equatorward shift in the westerlies. They postulated enhancement of the amplitude of the hemispheric longwave pattern, resulting in a semipermanent ridge pattern in the upper westerlies. They argued that this would result in predominantly northeasterly flow over southeastern Australia and the diversion of most rain-bearing disturbances over and south of Tasmania. This scenario is not consistent, however, with the evidence for drier conditions in Tasmania around the glacial maximum and during the late glacial.

Harrison *et al.* (1984) also inferred equatorward displacement of the subtropical high-pressure belt at 18,000 yr B.P. They invoked enhanced continentality as a result of sea-level lowering, more frequent winds blowing parallel to the coast rather than onshore because of strengthened westerly flow, and reduced convective precipitation from westerly disturbances because of the greater cooling of the land compared to the ocean to explain the "anomalous" dry conditions on the windward coasts of Victoria and Tasmania.

These various explanations for the aridity at the glacial maximum seem unnecessarily complicated and can be avoided by inferring that the circulation belts moved *poleward* rather than equatorward at the glacial maximum. Belief in the equatorward displacement of the subtropical high-pressure belt and the westerlies has been supported by the assumption that the equatorward extension of ice cover would cause stronger and more equatorward westerly regimes in both hemispheres. However, the Northern Hemisphere westerlies were displaced equatorward mainly because the ice sheet acted as an orographic barrier (Kutzbach and Wright, 1985). Antarctic sea ice would have no such physical blocking influence, and model simulations with the Community Climate Model (CCM) of the National Center for Atmospheric Research (NCAR) suggest that the Southern Hemisphere westerlies were indeed displaced poleward rather than equatorward in response to the steeper latitudinal temperature gradient caused by the extension of Antarctic sea ice at 18,000 yr B.P. (Barry and Williams, 1975; Kutzbach and Guetter, 1984, 1986).

Rognon and Williams (1977) pointed out that the orientation of dunes and lunettes in southwestern, south, and southeastern Australia could also be consistent with a southward displacement of the subtropical high-pressure belt, bringing dry summer winds to the southern borders of the mainland and displacing the westerlies southward. They therefore argued that the subtropical high-pressure belt expanded with a southward extension of its southern margin at the glacial maximum. This hypothesis agrees with our reconstruction of circulation patterns at 18,000–17,000 yr B.P.

A Reassessment of Dune Evidence for Paleocirculation Patterns at the Glacial Maximum

The desert dunes of Australia are arranged in a continent-wide counterclockwise whirl (Jennings, 1968) similar to the mean anticyclonic flow in winter. Dune-forming winds today occur mainly during the seasonal migration of the subtropical high-pressure belt (Brookfield, 1970). The last major phase of dune construction occurred between about 25,000 and 13,000 yr

B.P., peaking between 20,000 and 16,000 yr B.P. (Wasson, 1984, 1986). Ash and Wasson (1983) argued that because the present-day vegetative cover cannot prevent sand movement (given strong enough winds), conditions must have been even windier during the last glacial maximum.

Analyses of glacial-age dunes (Bowler and Wasson, 1984; Wasson, 1984) show that the orientation is remarkably similar to modern wind resultants in most of the Australian deserts, suggesting that the glacial circulation in the continental interior was substantially like today's. Glacial dune orientation was markedly different from modern-day wind resultants, however, in the Mallee, the southern Strzelecki Desert, and the coastal regions of the Great Sandy Desert (Bowler and Wasson, 1984; Wasson, 1984). In the Mallee, modern wind resultants show flow from the southwest, whereas glacial-age dunes are oriented west-east. In the southern Strzelecki Desert, modern wind resultants are southwesterly, whereas longitudinal dune trends are more nearly west-east. These differences are consistent with a southward expansion of the subtropical high-pressure belt, as inferred from the lake and pollen evidence. An expanded subtropical high-pressure belt would have reduced southerly flow across these deserts while leaving the circulation in the interior similar to (but stronger than) that of today.

Dunes are a unique source of evidence for circulation changes affecting the northern coastal region. In the Great Sandy Desert, modern wind resultants indicate northerly (onshore) flow, whereas the longitudinal dune trends are west-east (Bowler and Wasson, 1984; Wasson, 1984). This difference implies that the southeasterly trade winds were stronger around the glacial maximum and the monsoonal flow into northern Australia was weaker than at present. A weak summer monsoon is also indicated by the northward expansion of linear desert dunes in northern Western Australia (Bowler *et al.*, 1976), indicating a northward extension of the area affected by the southeasterly trade winds year-round.

Previous Reconstructions of Paleocirculation Patterns During the Late Glacial and Holocene

The persistence of aridity over most of Australia between 18,000 and 12,000 yr B.P. indicates that the southern margin of the subtropical high-pressure belt and the westerlies were displaced poleward of their modern positions until some time after 12,000 yr B.P. Markgraf *et al.* (1986) pointed out that drier conditions in Tasmania during the late glacial also suggest a poleward displacement of the westerlies.

Paleoenvironmental reconstructions from other Southern Hemisphere regions support the hypothesis that the westerlies were displaced poleward of their present position until after 12,000 yr B.P. (Markgraf,

1983, 1987, 1989a,b; Servant, 1984; Kessler, 1985). Markgraf (1987), for example, suggested that late-glacial aridity registered at a site now lying near the northern limit of *Nothofagus* forests in Argentina resulted from poleward displacement of the westerlies to south of 41°S. We postulate a shift of similar magnitude for Australia. Paleoenvironmental data from Tierra del Fuego (Markgraf, 1987) suggest that the zone of the westerlies was contracted compared to the present. This is consistent with the suggestion of Rognon and Williams (1977) that the Southern Hemisphere westerlies were more constricted during the glacial maximum and late glacial as a result of southward displacement of the southern margin of the subtropical high-pressure belt and the likely expansion of the Antarctic high.

Relatively few studies have attempted to reconstruct changes in Holocene circulation patterns. Our reconstructions of changes in effective moisture regimes based on lake-level and pollen data suggest that the subtropical high-pressure belt and the westerlies moved northward between 12,000 and 6000 yr B.P. Harrison *et al.* (1984) reached the same conclusion. Rognon and Williams (1977) also suggested that the southern boundary of the subtropical high-pressure belt shifted northward during the early Holocene. They implied that the subtropical high-pressure belt lay northward of its modern position at 6000 yr B.P. (consistent with the suggestion that coastal southeastern Australia lay within the zone of year-round westerly influence) and reexpanded during the last 6000 yr (consistent with a gradual poleward readvance of the southern margin of the subtropical high-pressure belt between 6000 yr B.P. and the present). Paleoenvironmental reconstructions from South America (Markgraf, 1987) support the hypothesis of equatorward migration of the southern westerlies during the early to middle Holocene, perhaps to north of their modern mean position, and a subsequent shift poleward to their present position after 4500 yr B.P.

Data-Model Comparisons

The changes in atmospheric circulation patterns over Australia and New Guinea from 18,000 to 9000 yr B.P. inferred from the lake-level, pollen, and dune evidence are broadly consistent with model simulations from experiments with the NCAR CCM (Kutzbach and Guetter, 1984, 1986; Kutzbach, Guetter, *et al.*, this vol.).

Simulations for 18,000 yr B.P. show an expanded subtropical high-pressure belt, with no change in the location of its northern margin but with the southern margin extended southward in winter (July) and the winter westerlies shifted poleward by about 8°. The southeasterly trade winds were stronger during summer (January), and southeasterly flow was stronger across northern Australia, implying a weaker Australian monsoon. These changes produced drier conditions over the whole continent, except on the east coast and southeastern interior, where wetter conditions are shown. The simulated circulation changes thus agree closely with the patterns we have inferred, and the magnitude of the simulated poleward shift of the westerlies agrees with the data. Simulated atmospheric circulation at 15,000 yr B.P. was similar to that at 18,000 yr B.P., as the data also suggest.

Simulations for 12,000 yr B.P. show the southern margin of the subtropical high-pressure belt in winter extended still farther southward, producing drier conditions over the southern part of the continent. The westerlies were shifted poleward by about 10°. The southeasterly trade winds, however, were weaker than before. The lake-level data from southern Australia also indicate maximum aridity at 12,000 yr B.P.

At 9000 yr B.P. the model shows the subtropical high-pressure belt and the westerlies still slightly south of their present positions, though much farther north than at 18,000 yr B.P., and conditions still somewhat drier than at present along the southern margin of the continent. This pattern is consistent with lake-level and pollen evidence for increasing effective moisture in Tasmania and coastal southeastern Australia after 12,000 yr B.P. and with the observation that coastal southeastern Australia was still drier than at present at 9000 yr B.P. The anomalous high rainfall in the eastern interior simulated for 18,000 yr B.P. disappears from the simulations by 9000 yr B.P., and indeed the lakes were dry by then.

The model simulations for 6000 and 3000 yr B.P. show a circulation pattern similar to that of the present. The lake-level and pollen data from coastal southeastern Australia, however, show conditions wetter than at present between 8000 and 6000 yr B.P. and a gradual decrease in effective moisture after 6000 yr B.P.

The model results do not agree with quantitative temperature estimates derived from pollen data. The model simulations for 18,000 yr B.P. show land-surface temperatures about 1–3°C cooler than at present in both seasons over the whole continent. Quantitative reconstructions from montane New Guinea do show cooler conditions, but the anomaly is much larger (6–13°C) than in the simulation and is inconsistent with the relatively warm prescribed sea-surface temperatures for this region. Both Webster and Streten (1978) and Rind and Peteet (1985) pointed out that the CLIMAP Project Members (1981) reconstructions of tropical sea-surface temperatures as warm as today are inconsistent with terrestrial evidence for much lower temperatures at the glacial maximum. This discrepancy remains unresolved.

Pollen-based reconstructions from Papua New Guinea, the Atherton Plateau, Barrington Tops, and Tasmania all indicate mean annual or summer temperatures about 0.5–1.5°C higher than at present at 6000 yr B.P. This is also inconsistent with model predictions, which show both summer temperature and mean annual temperature cooler than at present.

Conclusions

Compilations of lake-level and pollen data from Australia and New Guinea show concordant patterns of regional changes in effective moisture between 18,000 yr B.P. and the present that, in general, suggest simple explanations in terms of circulation changes. Paleodune evidence supports the changes hypothesized for the period around 18,000 yr B.P.

Between 18,000 and 12,000 yr B.P. conditions were drier than at present everywhere in the region except in the interior of southeastern Australia. This pattern is consistent with a southward expansion of the southern margin of the subtropical high-pressure belt, a southward displacement of the westerlies, and a weaker summer monsoon over northern Australia, all changes supported by dune evidence. Conditions became wetter around 11,000 yr B.P. in Tasmania and by 10,000 yr B.P. in coastal southeastern Australia, reflecting a gradual equatorward shift in the southern margin of the subtropical high-pressure belt and the westerlies after 12,000 yr B.P.

Conditions wetter than at present in coastal southeastern Australia between 8000 and 6000 yr B.P. suggest equatorward displacement of the subtropical high-pressure belt to north of its present position, bringing coastal southeastern Australia under the influence of the westerlies throughout the year. A gradual decrease in effective moisture in coastal southeastern Australia after 6000 yr B.P. and the establishment of modern conditions in Tasmania by 4000 yr B.P. may reflect the poleward retreat of the subtropical high-pressure belt between 6000 and 4000 yr B.P. The gradual increase in summer insolation may also have contributed to declining moisture levels during the Holocene. Increased effective moisture in the interior of southeastern Australia between 6000 and 4000 yr B.P. and a sudden temporary shift to conditions drier than at present in Tasmania and northeastern Queensland cannot be explained by latitudinal shifts in the circulation belts but might reflect an increase in the strength of the Walker Circulation.

The changes in the positions of the subtropical high-pressure belt and the westerlies inferred from lake-level and pollen data are qualitatively consistent with CCM simulations for 18,000-9000 yr B.P. The model also simulates cooler-than-present temperatures and a weak summer monsoon in northern Australia before 12,000 yr B.P., a relatively wet southeastern interior at 18,000-12,000 yr B.P., and maximum aridity in the south at 12,000 yr B.P., all features demonstrated in the data. However, the model produces little climatic change between 6000 yr B.P. and the present

and fails to explain conditions warmer and moister than at present in eastern Australia generally during the mid-Holocene. The model simulations probably could not reproduce Holocene circulation changes in the Southern Hemisphere midlatitudes because, although insolation is changed, sea-surface temperatures from 9000 yr B.P. are prescribed to be the same as today. The gradual rise in water level since 9000 yr B.P. at Lake Wanum in Papua New Guinea is consistent with the simulated Holocene increase in monsoon strength, but broader geographic coverage of records from northern Australia and Papua New Guinea is needed to evaluate simulated changes in the tropical circulation.

Acknowledgments

Partial financial support for this work (S. P. Harrison) was provided by the U.S. Department of Energy through contract no. DE-ACO2–79EV10097 to T. Webb III at Brown University and (J. R. Dodson) by the University of New South Wales. We would like to thank Patrick Bartlein, Brian Huntley, Vera Markgraf, Colin Prentice, Tom Webb III, and an anonymous reviewer for constructive criticism of the manuscript.

References

Allen, H. (1972). "Where the Crow Flies Backwards." Unpublished Ph.D. thesis, Australian National University, Canberra.

An Zhisheng, Bowler, J. M., Opdyke, N. D., Macumber, P. G., and Firman, J. B. (1986). Palaeomagnetic stratigraphy of Lake Bungunnia: Plio-Pleistocene precursor of aridity in the Murray Basin, southeastern Australia. *Palaeogeography, Palaeoclimatology, Palaeoecology* 54, 219–239.

Ash, J. E., and Wasson, R. J. (1983). Vegetation and sand mobility in the Australian desert dunefield. *Zeitschrift für Geomorphologie N.F. Supplementband* 45, 7–25.

Barbetti, M., and Allen, H. (1972). Prehistoric man at Lake Mungo, Australia, by 32,000 years B.P. *Nature* 240, 46–48.

Barbetti, M., and Polach, H. A. (1973). ANU radiocarbon date list V. *Radiocarbon* 15, 241–251.

Barendsen, G. W., Deevey, E. S., and Gralenski, L. J. (1957). Yale natural radiocarbon measurements III. *Science* 126, 908–919.

Barry, R. G., and Williams, J. (1975). Experiments with the NCAR global circulation model using glacial maximum boundary conditions: Southern Hemisphere results and interhemispheric comparison. *In* "Quaternary Studies" (R. P. Suggate and M. M. Cresswell, Eds.), pp. 57–66. The Royal Society of New Zealand, Wellington.

Barton, C. E., and Barbetti, M. (1982). Geomagnetic secular variation from recent lake sediments, ancient fireplaces and historical measurements in southeastern Australia. *Earth and Planetary Science Letters* 59, 375–387.

Barton, C. E., and McElhinny, M. W. (1980). Ages and ashes in lake floor sediment cores from Valley Lake, Mt. Gambier, South Australia. *Transactions of the Royal Society of South Australia* 104, 161–165.

——. (1981). A 10,000 year geomagnetic secular variation record from three Australian maars. *Geophysics Journal of the Royal Astronomical Society* 67, 465-485.

Barton, C. E., and Polach, H. A. (1980). [14]C ages and magnetic stratigraphy in three Australian maars. *Radiocarbon* 22, 728-739.

Baumgartner, A., and Reichel, E. (1975). "The World Water Balance: Mean Annual Global, Continental and Maritime Precipitation, Evaporation and Run-Off." Elsevier, Amsterdam.

Bettenay, E. (1962). The salt-lake systems and their associated aeolian features in the semi-arid regions of Western Australia. *Journal of Soil Science* 13, 10-17.

Binder, R. M. (1978). Stratigraphy and pollen analysis of a peat deposit, Bunyip Bog, Mt. Buffalo, Victoria. *Monash Publications in Geography* 19.

Binder, R. M., and Kershaw, A. P. (1978). A late-Quaternary pollen diagram from the south-eastern highlands of Australia. *Search* 9, 44-45.

Bowler, J. M. (1970). "Late Quaternary Environments: A Study of Lakes and Associated Sediments in South-Eastern Australia." Unpublished Ph.D. thesis, Australian National University, Canberra.

——. (1971). Pleistocene salinities and climatic change: Evidence from lakes and lunettes in south-eastern Australia. *In* "Aboriginal Man and Environment in Australia" (D. J. Mulvaney and J. Golson, Eds.), pp. 47-65. Australian National University Press, Canberra.

——. (1973). Clay dunes: Their occurrence, formation and environmental significance. *Earth-Science Reviews* 9, 315-338.

——. (1975). Deglacial events in southern Australia: Their age, nature, and palaeoclimatic significance. *In* "Quaternary Studies" (R. P. Suggate and M. M. Cresswell, Eds.), pp. 75-82. The Royal Society of New Zealand, Wellington.

——. (1976). Aridity in Australia: Age, origins and expression in aeolian landforms and sediments. *Earth-Science Reviews* 12, 279-310.

——. (1981). Australian salt lakes: A palaeohydrological approach. *Hydrobiologia* 82, 431-444.

Bowler, J. M., and Hamada, T. (1971). Late Quaternary stratigraphy and radiocarbon chronology of water level fluctuations in Lake Keilambete, Victoria. *Nature* 232, 331-334.

Bowler, J. M., and Teller, J. T. (1986). Quaternary evaporites and hydrological changes, Lake Tyrrell, north-west Victoria. *Australian Journal of Earth Sciences* 33, 43-63.

Bowler, J. M., and Wasson, R. J. (1984). Glacial age environments of inland Australia. *In* "Late Cainozoic Palaeoclimates of the Southern Hemisphere" (J. C. Vogel, Ed.), pp. 183-208. Balkema, Rotterdam.

Bowler, J. M., Jones, R., Allen, H., and Thorne, A. G. (1970). Pleistocene human remains from Australia: A living site and human cremation from Lake Mungo, western New South Wales. *World Archaeology* 1, 39-60.

Bowler, J. M., Thorne, A. G., and Polach, H. (1972). Pleistocene Man in Australia: Age and significance of the Mungo skeleton. *Nature* 240, 48-50.

Bowler, J. M., Hope, G. S., Jennings, J. N., Singh, G., and Walker, D. (1976). Late Quaternary climates of Australia and New Guinea. *Quaternary Research* 6, 359-394.

Bowler, J. M., Huang Qi, Chen Kezao, Head, M. J., and Yuan Baoyin (1986). Radiocarbon dating of playa-lake hydrologic changes: Examples from northwestern China and central Australia. *Palaeogeography, Palaeoclimatology, Palaeoecology* 54, 241-260.

Bradbury, J. P. (1986). Late Pleistocene and Holocene paleolimnology of two mountain lakes in western Tasmania. *Palaios* 1, 381-388.

Brookfield, M. (1970). Dune trends and wind regime in central Australia. *Zeitschrift für Geomorphologie N.F. Supplementband* 10, 121-153.

Buckley, J. D., and Willis, E. H. (1972). Isotopes' radiocarbon measurements IX. *Radiocarbon* 14, 114-139.

Callen, R. A. (1976). Tentative correlation of onshore and lacustrine stratigraphy, Lake Frome area. *BMR Journal of Australian Geology and Geophysics* 1, 248-250.

Cann, J. H., and De Deckker, P. (1981). Fossil Quaternary and living foraminifera from athalassic (non-marine) saline lakes, southern Australia. *Journal of Paleontology* 55, 660-670.

Ceplecha, V. J. (1971). The distribution of the main components of the water balance in Australia. *The Australian Geographer* 11, 455-462.

Chappell, J. M. A., and Grindrod, A., Eds. (1983). "CLIMANZ: A Symposium of Results and Discussions Concerned with Late Quaternary Climatic History of Australia, New Zealand and Surrounding Seas" (2 vols.). Department of Biogeography and Geomorphology, Research School of Pacific Studies, Australian National University, Canberra.

Chivas, A. R., De Deckker, P., and Shelley, J. M. G. (1985). Strontium content of ostracods indicates lacustrine palaeosalinity. *Nature* 316, 251-253.

——. (1986a). Magnesium content of non-marine ostracod shells: A new palaeosalinometer and palaeothermometer. *Palaeogeography, Palaeoclimatology, Palaeoecology* 54, 43-61.

——. (1986b). Magnesium and strontium in non-marine ostracod shells as indicators of palaeosalinity and palaeotemperature. *Hydrobiologia* 143, 135-142.

Churchill, D. M. (1968). The distribution and prehistory of *Eucalyptus diversicolor* F. Muell., *E. marginata* Donn ex Sm., and *E. calophylla* R. Br. in relation to rainfall. *Australian Journal of Botany* 16, 125-151.

Churchill, D. M., Galloway, R. W., and Singh, G. (1978). Closed lakes and the palaeoclimatic record. *In* "Climatic Change and Variability: A Southern Perspective" (A. B. Pittock, L. A. Frakes, J. A. Peterson, and J. W. Zillman, Eds.), pp. 97-108. Cambridge University Press, Cambridge.

CLIMAP Project Members (1981). Seasonal reconstruction of the earth's surface at the last glacial maximum. *Geological Society of America Map and Chart Series* MC-36.

Colhoun, E. A. (1978a). The Late Quaternary environment of Tasmania as a backdrop to Man's occupance. *Records of the Queen Victoria Museum Launceston* 61.

——. (1978b). Recent Quaternary studies in Tasmania. *Australian Quaternary Newsletter* 12, 2-15.

Colhoun, E. A., van de Geer, G., and Mook, W. G. (1982). Stratigraphy, pollen analysis and paleoclimatic interpretation of Pulbeena Swamp, northwestern Tasmania. *Quaternary Research* 18, 108-126.

Coventry, R. J. (1973). "Abandoned Shorelines and the Late Quaternary History of Lake George, N.S.W." Unpublished Ph.D. thesis, Australian National University, Canberra.

——. (1976). Abandoned shorelines and the late Quaternary history of Lake George, N.S.W. *Journal of the Geological Society of Australia* 23, 249-273.

Coventry, R. J., and Walker, P. H. (1977). Geomorphological significance of late Quaternary deposits of the Lake George area, N.S.W. *The Australian Geographer* 13, 369-376.

Currey, D. T. (1964). The former extent of Lake Corangamite. *Proceedings of the Royal Society of Victoria* 77, 377-387.

——. (1970). Lake systems, western Victoria. *Bulletin of the Australian Society for Limnology* 3, 1-13.

Das, S. C. (1956). Statistical analysis of Australian pressure data. *Australian Journal of Physics* 9, 394-399.

De Deckker, P. (1982a). Non-marine ostracods from two Quaternary profiles at Pulbeena and Mowbray Swamps, Tasmania. *Alcheringa* 6, 249-274.

——. (1982b). Late Quaternary ostracods from Lake George, New South Wales. *Alcheringa* 6, 305-318.

——. (1982c). Holocene ostracods, other invertebrates and fish remains from cores of four maar lakes in southeastern Australia. *Proceedings of the Royal Society of Victoria* 94, 183-220.

Deacon, E. L. (1953). Climatic change in Australia since 1880. *Australian Journal of Physics* 6, 209-218.

Dodson, J. R. (1971). "Holocene Vegetation History and Carbonate Sedimentation in Western Victoria." Unpublished M.Sc. thesis, Monash University, Melbourne.

——. (1974a). Vegetation and climatic history near Lake Keilambete, western Victoria. *Australian Journal of Botany* 22, 709-717.

——. (1974b). Vegetation history and water fluctuations at Lake Leake, southeastern South Australia. I. 10,000 B.P. to present. *Australian Journal of Botany* 22, 719-741.

——. (1975a). Vegetation history and water fluctuations at Lake Leake, southeastern South Australia. II. 50,000 B.P. to 10,000 B.P. *Australian Journal of Botany* 23, 815-831.

——. (1975b). The pre-settlement vegetation of the Mt. Gambier area, South Australia. *Transactions of the Royal Society of South Australia* 99, 89-92.

——. (1977). Late Quaternary palaeoecology of Wyrie Swamp, southeastern South Australia. *Quaternary Research* 8, 97-114.

——. (1979). Late Pleistocene vegetation and environments near Lake Bullenmerri, western Victoria. *Australian Journal of Ecology* 4, 419-427.

——. (1983). Modern pollen rain in southeastern New South Wales, Australia. *Review of Palaeobotany and Palynology* 38, 249-268.

——. (1986). Holocene vegetation and environments near Goulburn, New South Wales. *Australian Journal of Botany* 34, 231-249.

Dodson, J. R., and Myers, C. A. (1986). Vegetation and modern pollen rain from Barrington Tops and Upper Hunter River regions of New South Wales. *Australian Journal of Botany* 34, 293-304.

Dodson, J. R., and Wilson, I. B. (1975). Past and present vegetation of Marshes Swamp in south-eastern South Australia. *Australian Journal of Botany* 23, 123-150.

Dodson, J. R., Greenwood, P. W., and Jones, R. L. (1986). Holocene forest and wetland dynamics at Barrington Tops, New South Wales. *Journal of Biogeography* 13, 561-585.

Draper, J. J., and Jensen, A. R. (1976). The geochemistry of Lake Frome, a playa lake in South Australia. *BMR Journal of Australian Geology and Geophysics* 1, 83-104.

Dury, G. H. (1973). Paleohydrologic implications of some pluvial lakes in northwestern New South Wales, Australia. *Bulletin of the Geological Society of America* 84, 3663-3676.

Flenley, J. R. (1967). "The Present and Former Vegetation of the Wabag Region of New Guinea." Unpublished Ph.D. thesis, Australian National University, Canberra.

Flores, J. F., and Balagot, V. F. (1969). Climate of the Philippines. *In* "Climates of Northern and Eastern Asia" (H. Arakawa, Ed.), pp. 159-213. World Survey of Climatology, Vol. 8. Elsevier, Amsterdam.

Galloway, R. W. (1967). Dating of shore features at Lake George, New South Wales. *Australian Journal of Science* 29, 447.

Garrett-Jones, S. (1979). "Evidence for Changes in Holocene Vegetation and Lake Sedimentation in the Markham Valley, Papua New Guinea." Unpublished Ph.D. thesis, Australian National University, Canberra.

——. (1980). Holocene vegetation change in lowland Papua New Guinea. *In* "Abstracts, Fifth International Palynological Conference," p. 149. University of Cambridge, Cambridge.

Gentilli, J., Ed. (1971). "Climates of Australia and New Zealand." World Survey of Climatology, Vol. 13. Elsevier, Amsterdam.

Gill, E. D. (1953). Geological evidence in western Victoria relative to the antiquity of the Australian aborigines. *Memoirs of the Natural History Museum, Melbourne* 18, 25-92.

——. (1955). The Australian "arid period." *Australian Journal of Science* 17, 204-206.

——. (1971). Applications of radiocarbon dating in Victoria, Australia. *Proceedings of the Royal Society of Victoria* 84, 71-85.

——. (1973a). Second list of radiocarbon dates on samples from Victoria, Australia. *Proceedings of the Royal Society of Victoria* 86, 133-136.

——. (1973b). Geology and geomorphology of the Murray River region between Mildura and Renmark, Australia. *Memoirs of the Natural History Museum of Victoria* 34, 1-97.

Grayson, H. J., and Mahoney, D. J. (1910). The geology of the Camperdown and Mt. Elephant districts. *Memoirs of the Geological Survey of Victoria* 9.

Hamon, B. V., and Godfrey, J. S. (1978). The role of the oceans. *In* "Climatic Change and Variability: A Southern Perspective" (A. B. Pittock, L. A. Frakes, D. Jenssen, J. A. Peterson, and J. W. Zillman, Eds.), pp. 31-52. Cambridge University Press, Cambridge.

Harrison, S. P. (1980). "The Geomorphic History of the Breadalbane Basin, New South Wales." Unpublished M.Sc. thesis, Macquarie University, New South Wales.

——. (1988). Lake-level records from Canada and the eastern U.S.A. *Lundqua Report* 29. Department of Quaternary Geology, Lund University, Lund.

——. (1989). Lake-level records from Australia and New Guinea. *UNGI Rapport* 72. Department of Physical Geography, Uppsala University.

Harrison, S. P., and Digerfeldt, G. (In press). European lakes as palaeohydrological and palaeoclimatic indicators. *Quaternary Science Reviews*.

Harrison, S. P., Metcalfe, S. E., Pittock, A. B., Roberts, C. N., Salinger, N. J., and Street-Perrott, F. A. (1984). A climatic model of the last glacial/interglacial transition based on palaeotemperature and palaeohydrological evidence. *In* "Late Cainozoic Palaeoclimates of the Southern Hemisphere" (J. C. Vogel, Ed.), pp. 21-34. Balkema, Rotterdam.

Head, L. (1983). Environment as artefact: A geographic perspective on the Holocene occupation of southwestern Victoria. *Archaeology in Oceania* 18, 73-80.

——. (1988). Holocene vegetation, fire and environmental history of the Discovery Bay region, south-western Victoria. *Australian Journal of Ecology* 13, 21-49.

Hooley, A. D., Southern, W., and Kershaw, A. P. (1981). Holocene vegetation and environments of Sperm Whale Head, Victoria, Australia. *Journal of Biogeography* 7, 349-362.

Hope, G. S. 1974. The vegetation history from 6000 B.P. to present of Wilsons Promontory, Victoria, Australia. *New Phytologist* 73, 1035-1053.

——. (1976). The vegetational history of Mount Wilhelm, Papua New Guinea. *Journal of Ecology* 64, 627-663.

——. (1978). The late Pleistocene and Holocene vegetational history of Hunter Island, north-western Tasmania. *Australian Journal of Botany* 26, 493-514.

Hope, G. S., and Peterson, J. A. (1976). Palaeoenvironments. *In* "The Equatorial Glaciers of New Guinea" (G. S. Hope, J. A. Peterson, and U. Radok, Eds.), pp. 173-206. Balkema, Rotterdam.

Howard, T. M., and Hope, G. S. 1970. The past and present occurrence of beech (*Nothofagus cunninghamii oerst*) at Wilsons Promontory, Victoria, Australia. *Proceedings of the Royal Society of Victoria* 83, 199-210.

Jacobson, G., and Schuett, A. W. (1979). Water level, balance and chemistry of Lake George, New South Wales. *BMR Journal of Australian Geology and Geophysics* 4, 25-32.

Jennings, J. N. (1968). A revised map of the desert dunes of Australia. *The Australian Geographer* 10, 408-409.

Karelsky, S. (1954). Surface circulation in the Australian region. *Australian Bureau of Meteorology, Meteorological Studies* 3, 1-45.

——. (1965). "Monthly Geographical Distribution of Central Pressures in Surface Highs and Lows in the Australian Region, 1952-1963." Meteorological Summary, Australian Bureau of Meteorology, Melbourne.

Kershaw, A. P. (1970). A pollen diagram from Lake Euramoo, north-east Queensland, Australia. *New Phytologist* 69, 785-805.

——. (1971). A pollen diagram from Quincan Crater, north-east Queensland, Australia. *New Phytologist* 70, 669-681.

——. (1974). A long continuous pollen sequence from north-eastern Australia. *Nature* 251, 222-223.

——. (1975a). Stratigraphy and pollen analysis of Bromfield Swamp, north-eastern Queensland, Australia. *New Phytologist* 75, 173-191.

——. (1975b). Late Quaternary vegetation and climate in north-eastern Australia. *Bulletin of the Royal Society of New Zealand* 13, 181-187.

——. (1976). A late Pleistocene and Holocene pollen diagram from Lynch's Crater, north-eastern Queensland, Australia. *New Phytologist* 77, 469-498.

——. (1978). Record of last interglacial-glacial cycle from north-eastern Queensland. *Nature* 272, 159-161.

——. (1979). Local pollen deposition in aquatic sediments on the Atherton Tableland, north-eastern Australia. *Australian Journal of Ecology* 4, 253-263.

——. (1981). Quaternary vegetation and environments. *In* "Ecological Biogeography of Australia" (A. Keast, Ed.), pp. 81-101. Junk, The Hague, The Netherlands.

——. (1983). A Holocene pollen diagram from Lynch's Crater, north-eastern Queensland, Australia. *New Phytologist* 94, 669-682.

——. (1985). An extended late Quaternary vegetation record from north-eastern Queensland and its implications for the seasonal tropics of Australia. *In* "Ecology of the Wet-Dry Tropics" (M. G. Ridpath and L. K. Corbett, Eds.), pp. 179-189. Ecological Society of Australia.

Kessler, A. (1985). Zur Rekonstruktion von spätglazialen Klima und Wasserhaushalt auf dem peruanisch-bolivianischen Altiplano. *Zeitschrift für Gletscherkunde* 21, 107-114.

Kidson, E. (1925). Some periods in Australian weather. *Bulletin of the Commonwealth Bureau of Meteorology, Australia* 17.

Kutzbach, J. E., and Guetter, P. J. (1984). Sensitivity of late-glacial and Holocene climates to the combined effects of orbital parameter changes and lower boundary condition changes: "Snapshot" simulations with a general circulation model for 18, 9, and 6 ka B.P. *Annals of Glaciology* 5, 85-87.

——. (1986). The influence of changing orbital parameters and surface boundary conditions on climate simulations for the past 18,000 years. *Journal of the Atmospheric Sciences* 43, 1726-1759.

Kutzbach, J. E., and Wright, H. E., Jr. (1985). Simulation of the climate of 18,000 yr B.P.: Results for the North American/North Atlantic/European sector and comparison with the geologic record of North America. *Quaternary Science Reviews* 4, 147-187.

Ladd, P. G. (1978). Vegetation history at Lake Curlip in lowland eastern Victoria from 5200 B.P. to present. *Australian Journal of Botany* 26, 393-414.

——. (1979a). Past and present vegetation on the Delegate River in the highlands of eastern Victoria. II. Vegetation and climatic history from 12,000 B.P. to present. *Australian Journal of Botany* 27, 185-202.

——. (1979b). A short pollen diagram from rainforest in highland eastern Victoria. *Australian Journal of Ecology* 4, 229-237.

——. (1979c). A Holocene vegetation record from the eastern side of Wilsons Promontory, Victoria. *New Phytologist* 82, 265-276.

Lamb, H. H., and Johnson, A. I. (1961). Climatic variation and observed changes in the general circulation, III. *Geografiska Annaler* 43, 363-400.

——. (1966). Secular variations of the atmospheric circulation since 1750. *Geophysical Memoirs* 14(110). Meteorological Office, London.

Lourandos, H. (1970). "Coast and Hinterland: The Archaeological Sites of Eastern Tasmania." Unpublished Ph.D. thesis, Australian National University, Canberra.

Luly, J. G., Bowler, J. M., and Head, M. J. (1986). A radiocarbon chronology from the playa Lake Tyrrell, northwestern Victoria. *Palaeogeography, Palaeoclimatology, Palaeoecology* 54, 171-180.

Macphail, M. K. (1973). Pollen analysis of a buried organic deposit on the backshore at Fingal Bay, Port Stephens, New South Wales. *Proceedings of the Linnean Society of New South Wales* 98, 222-233.

——. (1975). Late Pleistocene environments in Tasmania. *Search* 6, 295-300.

——. (1976). "The History of the Vegetation and Climate in Southern Tasmania since the Late Pleistocene (ca. 13,000-0 B.P.)." Unpublished Ph.D. thesis, University of Tasmania.

——. (1979). Vegetation and climates in southern Tasmania since the last glaciation. *Quaternary Research* 11, 306-341.

——. (1984). Small-scale dynamics in an early Holocene wet sclerophyll forest in Tasmania. *New Phytologist* 96, 131-147.

Macphail, M. K., and Colhoun, E. A. (1985). Late last glacial vegetation, climates and fire activity in south west Tasmania. *Search* 16, 43-45.

Macphail, M. K., and Jackson, W. D. (1978). The late Pleistocene and Holocene history of the midlands of Tasmania, Australia: Pollen evidence from Lake Tiberias. *Proceedings of the Royal Society of Victoria* 90, 287-300.

Macumber, P. G. (1968). Interrelation between physiography, hydrology, sedimentation, and salinization of the Loddon River Plains, Australia. *Journal of Hydrology* 7, 39-67.

——. (1977). The geology and palaeohydrology of the Kow Swamp fossil hominid site, Victoria, Australia. *Journal of the Geological Society of Australia* 25, 307-320.

——. (1978). Evolution of the Murray River during the Tertiary period: Evidence from northern Victoria. *Proceedings of the Royal Society of Victoria* 90, 43-52.

——. (1980). The influence of groundwater discharge on the mallee landscapes. *In* "Aeolian Landscapes in the Semiarid Zone of Southeastern Australia" (R. R. Storrier, Ed.), pp. 67-84. Riverine Soil Science Society, Wagga Wagga, New South Wales.

——. (1983). "Interactions between Groundwater and Surface Systems in Northern Victoria." Unpublished Ph.D. thesis, University of Melbourne.

Macumber, P. G., and Thorne, A. G. (1975). The Cohuna cranium site—A re-appraisal. *Archaeology and Physical Anthropology in Oceania* 10, 67-72.

Markgraf, V. (1983). Late and postglacial vegetational and palaeoclimatic changes in subantarctic, temperate, and arid environments in Argentina. *Palynology* 7, 43-70.

——. (1987). Paleoenvironmental changes at the northern limit of the subantarctic *Nothofagus* forest, lat. 37°S, Argentina. *Quaternary Research* 28, 119-129.

——. (1989a). Palaeoclimates in Central and South America since 18,000 yr B.P. based on pollen and lake-level records. *Quaternary Science Reviews* 8, 1-24.

——. (1989b). Reply to C. J. Heusser's "Southern westerlies during the last glacial maximum." *Quaternary Research* 31, 426-432.

Markgraf, V., Bradbury, J. P., and Busby, J. R. (1986). Paleoclimates in southwestern Tasmania during the last 13,000 years. *Palaios* 1, 368-380.

Martin, A. R. H. (1971). The depositional environment of the organic deposits on the foreshore at North Dee Why, New South Wales. *Proceedings of the Linnean Society of New South Wales* 96, 278-281.

——. (1986). Late glacial and Holocene alpine pollen diagrams from the Kosciusko National Park, New South Wales, Australia. *Review of Palaeobotany and Palynology* 47, 367-409.

Martin, H. A. (1973). Palynology and historical ecology of some cave excavations in the Australian Nullabor. *Australian Journal of Botany* 21, 283-316.

Ollier, C. D., and Joyce, E. B. (1964). Volcanic physiography of the western plains of Victoria. *Proceedings of the Royal Society of Victoria* 77, 357-376.

Pittock, A. B. (1971). Rainfall and the general circulation. *In* "Proceedings of the International Conference on Weather Modification, Canberra, September 6-11, 1971," pp. 330-338. American Meteorological Society, Boston.

——. (1973). Global meridional interactions in stratosphere and troposphere. *Quarterly Journal of the Royal Meteorological Society* 99, 424-437.

——. (1975). Climatic change and the patterns of variation in Australian rainfall. *Search* 6, 498-504.

——. (1978). Patterns of variability in relation to the general circulation. *In* "Climatic Change and Variability: A Southern Perspective" (A. B. Pittock, L. A. Frakes, D. Jenssen, J. A.

Peterson, and J. W. Zillman, Eds.), pp. 167-179. Cambridge University Press, Cambridge.

Polach, H., and Barton, C. (1983). ANU radiocarbon date list X. *Radiocarbon* 25, 30-38.

Polach, H. A., Lovering, A. J., and Bowler, J. M. (1970). ANU radiocarbon date list IV. *Radiocarbon* 12, 1-18.

Polach, H. A., Head, M. J., and Gower, J. D. (1978). ANU radiocarbon date list VI. *Radiocarbon* 20, 360-385.

Powell, J. M. (1970). "The Impact of Man on the Vegetation of the Mt. Hagen Region, New Guinea." Unpublished Ph.D. thesis, Australian National University, Canberra.

Rind, D., and Peteet, D. (1985). Terrestrial conditions at the last glacial maximum and CLIMAP sea-surface temperature estimates: Are they consistent? *Quaternary Research* 24, 1-22.

Rognon, P., and Williams, M. A. J. (1977). Late Quaternary climatic changes in Australia and North Africa: A preliminary interpretation. *Palaeogeography, Palaeoclimatology, Palaeoecology* 21, 285-327.

Russell, H. C. (1893). Moving anticyclones in the Southern Hemisphere. *Quarterly Journal of the Royal Meteorological Society* 19, 23-34.

Sanson, G. D., Riley, S. J., and Williams, M. A. J. (1980). A late Quaternary Procoptodon fossil from Lake George, New South Wales. *Search* 11, 39-40.

Selkirk, D. R., Selkirk, P. M., and Griffin, K. (1982). Palynological evidence for Holocene environmental change and uplift on Wireless Hill, Macquarie Island. *Proceedings of the Linnean Society of New South Wales* 107, 1-17.

Servant, M. (1984). Climatic variations in the low continental latitudes during the last 30,000 years. *In* "Climatic Changes on a Yearly to Millennial Basis" (N.-A. Mörner and W. Karlén, Eds.), pp. 117-120. Reidel, Dordrecht.

Sigleo, W. R. (1979). "A Study of Late Quaternary Environment and Man from Four Sites in Southeastern Tasmania." Unpublished Ph.D. thesis, University of Tasmania.

Sigleo, W. R., and Colhoun, E. A. (1981). A short pollen diagram from Crown Lagoon in the midlands of Tasmania. *Papers and Proceedings of the Royal Society of Tasmania* 115, 181-188.

——. (1982). Terrestrial dunes, man and the late Quaternary environment in southern Tasmania. *Palaeogeography, Palaeoclimatology, Palaeoecology* 39, 87-121.

Singh, G. (1981). Late Quaternary pollen records and seasonal palaeoclimates of Lake Frome, South Australia. *Hydrobiologia* 82, 419-452.

Singh, G., and Geissler, E. A. (1985). Late Cainozoic history of vegetation, fire, lake levels and climate at Lake George, New South Wales, Australia. *Philosophical Transactions of the Royal Society (London) Series B* 311, 379-447.

Singh, G., Kershaw, A. P., and Clark, R. (1981a). Quaternary vegetation and fire history in Australia. *In* "Fire and the Australian Biota" (A. M. Gill, R. A. Groves, and I. R. Noble, Eds.), pp. 23-54. Australian Academy of Science, Canberra.

Singh, G., Opdyke, N. D., and Bowler, J. M. (1981b). Late Cainozoic stratigraphy, palaeomagnetic chronology and vegetational history from Lake George, N.S.W. *Journal of the Geological Society of Australia* 28, 435-452.

Skeats, E. W., and James, A. V. G. (1937). Basaltic barriers and other surface features of the new basalts of western Victoria. *Proceedings of the Royal Society of Victoria* 49, 245-292.

Specht, R. L. (1972). "The Vegetation of South Australia." Government Printer, Adelaide.

Street, F. A., and Grove, A. T. (1976). Environmental and climatic implications of late Quaternary lake-level fluctuations in Africa. *Nature* 261, 385-390.

Street-Perrott, F. A., and Harrison, S. P. (1985). Lake level and climate reconstruction. *In* "Paleoclimate Analysis and Modeling" (A. D. Hecht, Ed.), pp. 291-340. John Wiley and Sons, New York.

Suzuki, H., Uesugi, Y., Endo, K., Ohmori, H., Takeuchi, K., and Iwasaki, K. (1982). "Studies on the Holocene and Recent Climatic Fluctuations in Australia and New Zealand." Tokyo Metropolitan University, Tokyo.

Teller, J. T., Bowler, J. M., and Macumber, P. G. (1982). Modern sedimentation and hydrology in Lake Tyrrell, Victoria. *Journal of the Geological Society of Australia* 29, 159-175.

Thorne, A. G., and Macumber, P. G. (1972). Discoveries of late Pleistocene man at Kow Swamp, Australia. *Nature* 238, 316-319.

Timms, B. V. (1976). A comparative study of the limnology of three maar lakes in western Victoria. I. Physiography and physiochemical features. *Australian Journal of Marine and Freshwater Research* 27, 35-60.

Tudor, E. R. (1973). "Hydrological Interpretation of Diatom Assemblages in Two Victorian Western District Crater Lakes." Unpublished M.Sc. thesis, Monash University, Melbourne.

Ullman, W. J., and McLeod, L. C. (1986). The late-Quaternary salinity record of Lake Frome, South Australia: Evidence from Na$^+$ in stratigraphically-preserved gypsum. *Palaeogeography, Palaeoclimatology, Palaeoecology* 54, 153-169.

UNESCO (1971). "World Water Balance and Water Resources of the Earth." UNESCO Press, Paris.

Walker, D., and Flenley, J. R. (1979). Late Quaternary vegetational history of the Enga District of upland Papua New Guinea. *Philosophical Transactions of the Royal Society (London) Series B* 286, 265-344.

Wasson, R. J. (1984). Late Quaternary palaeoenvironments in the desert dunefields of Australia. *In* "Late Cainozoic Palaeoclimates of the Southern Hemisphere" (J. C. Vogel, Ed.), pp. 419-432. Balkema, Rotterdam.

——. (1986). Geomorphology and Quaternary history of the Australian continental dunefields. *Geographical Review of Japan* 59, 55-67.

Webster, P. J., and Streten, N. A. (1978). Late Quaternary ice age climates of tropical Australasia: Interpretations and reconstructions. *Quaternary Research* 10, 279-309.

Wright, R. V. S. (1975). Stone artifacts from Kow Swamp, with notes on their excavation and environmental context. *Archaeology and Physical Anthropology in Oceania* 10, 162-180.

Yamasaki, F., Hamada, T., and Hamada, C. (1970). Riken natural radiocarbon measurements VI. *Radiocarbon* 12, 559-576.

Yamasaki, F., Hamada, C., and Hamada, T. (1977). Riken natural radiocarbon measurements IX. *Radiocarbon* 19, 62-95.

Yezdani, G. H. (1970). "A Study of the Quaternary Vegetation History in the Volcanic Lakes Region of Western Victoria." Unpublished Ph.D. thesis, Monash University, Melbourne.

CHAPTER 12

Paleovegetation Studies of New Zealand's Climate since the Last Glacial Maximum

Matt S. McGlone, M. Jim Salinger, and Neville T. Moar

New Zealand is one of the few sizable landmasses in the Southern Ocean south of 35°S. Lying just poleward of the subtropical convergence, with its southern half embedded in the circumpolar westerly vortex, it is uniquely placed to record climatic changes in this vast expanse of ocean.

The pollen record of vegetation change in New Zealand provides good coverage of the Holocene and late Pleistocene, and several records extend back to 14,000–17,000 yr B.P. Despite the prevailing warm to cool-temperate oceanic climate of New Zealand, the pollen profiles record vegetation changes similar in magnitude to those from continental regions. Quantification of the pollen record in climatic terms has not yet progressed to a stage where the record can be used in reconstructions. Nevertheless, the emerging broad patterns of vegetation change are a good guide to the nature of past climates.

New Zealand's climatic record is of global importance for three main reasons. First, the high-resolution, well-dated record of change can be used to calibrate and corroborate evidence from deep-sea cores in the adjacent ocean. Because of New Zealand's small size and oceanic climate, the two sets of records should show a high degree of similarity. Second, New Zealand's remoteness from other landmasses means that it directly reflects global changes in the ocean-atmosphere system without the complexities associated with large continental areas. And third, New Zealand's small size and lack of large ice sheets at the glacial maximum ensured that the vegetation reacted quickly to climatic change, with a minimum of periglacial effects from decaying ice sheets or long migrational delays in vegetation recolonization. The potential for sensing and accurately dating the initiation of major global climatic change is therefore excellent.

The New Zealand Environment

New Zealand lies in the middle latitudes (34–47°S) of the Southern Hemisphere. It is long (about 1900 km), narrow (about 400 km at its widest point), and rugged. South Island is dominated by high axial ranges trending southwest-northeast (the Southern Alps) (Fig. 12.1). The southern half of North Island has somewhat lower ranges continuing the same trend. Thus the southern two-thirds of the country has a nearly continuous mountain barrier, reaching altitudes of more than 3000 m in central South Island. The northern half of North Island has no high mountains, but low, rugged ranges occur throughout.

Southern New Zealand is in the zone influenced by the circumpolar westerly vortex of atmospheric circulation, whereas the northern regions extend into the subtropical ridge of high pressures. A strong, persistent westerly circulation dominates the entire region. Eastward-moving anticyclones and intervening low-pressure troughs regularly progress over the country at intervals of six to ten days. At times, however, the weather patterns are complex, with occurrences of wave depressions and blocking anticyclones.

The generally mild and oceanic climate ranges from warm to cool-temperate. Only the inland basins of central Otago in the south of South Island approach a continental climate pattern of cold winters and warm summers. Alpine areas, although they have cold climates, are also generally oceanic, with damped seasonal cycles.

The high mountain barriers have a profound effect on local climates. Interaction of the eastward-migrating weather systems with the mountains creates complex patterns of contrasting temperature and precipitation regimes (Maunder, 1971). In broad outline the mean westerly airflow is uplifted orographically over

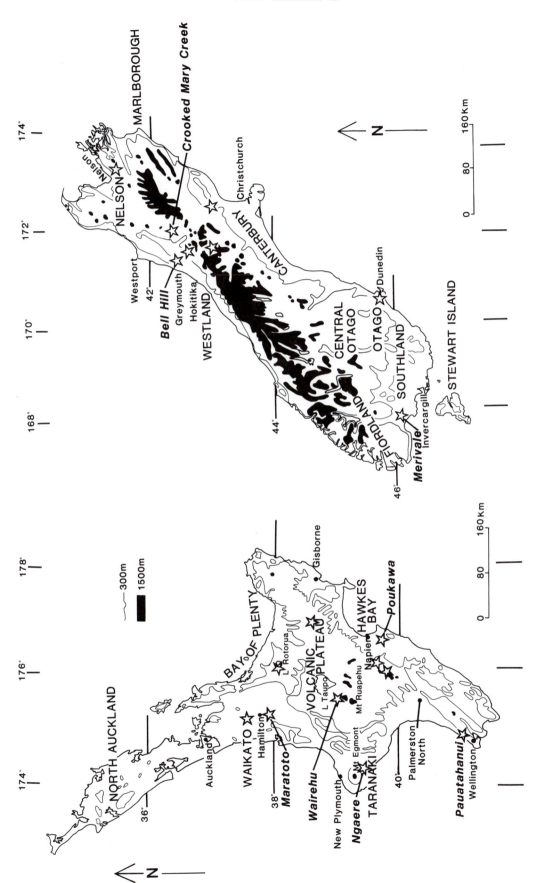

Fig 12.1. Site locations in North Island (left) and South Island (right) of New Zealand. Stars indicate locations of well-dated sites.

the axial ranges, bringing high annual rainfall to the west and drier conditions to the east (Fig. 12.2). These contrasts caused by orographic interaction are at their greatest in the center of South Island, where peak annual rainfall can reach 11,000 mm on the western side of the Southern Alps, while less than 30 km due east some inland valleys may have average annual precipitation of less than 800 mm (Griffiths and McSaveney, 1983). When westerly gradient airflow is strong, warm, dry foehn winds generate high evapotranspiration rates on the east coast, which may cause severe moisture stress in the vegetation.

New Zealand's weather and climate are highly sensitive to changes in the direction of the gradient airflow (Salinger, 1980a,b). Maps with isolines of equal degree of correlation show how the degree of zonality or meridionality of airflow is related to precipitation patterns (Fig. 12.3). When westerly flow is strong, precipitation increases in the west and along the south coast of South Island and in the central North Island mountains but decreases on the east coast of North Island. Conversely, when westerly flow is weak (i.e., enhanced easterly flow), the opposite pattern occurs. Enhanced southerly flow increases precipitation in the south of South Island and reduces precipitation in the north of both islands (Fig. 12.3b).

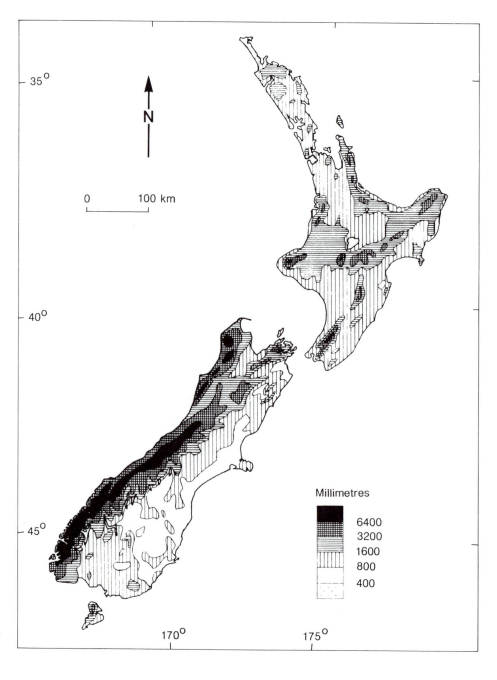

Fig. 12.2. Mean annual precipitation, 1951-1975.

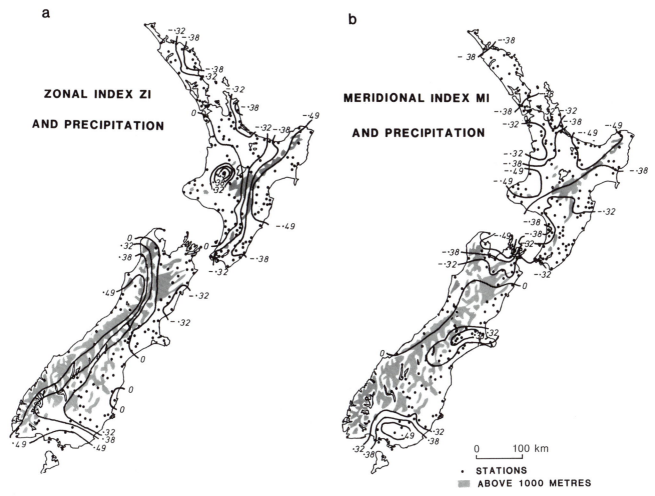

Fig. 12.3 (a) Zonal index ZI, a measure of zonal westerly airflow over New Zealand, correlated with annual precipitation for 1951–1975. (b) Meridional index MI, a measure of the meridional southerly airflow, correlated with annual precipitation.

Reduced southerly flow (i.e., enhanced northerly flow) has the reverse effect.

Vegetation

Before the deforestation that followed human settlement, New Zealand was largely covered with evergreen rainforest (Fig. 12.4). The only substantial area below treeline without complete forest cover was central Otago in southeastern South Island. Four broad floristic components of the forests (podocarps, hardwoods, *Nothofagus*, and *Agathis australis*) form a complex series of associations.

The coniferous podocarp group includes tall, often emergent trees that usually dominate the canopy of podocarp-hardwood forest, although occasionally they form nearly pure stands. Hardwoods are a broad category that includes all non-*Nothofagus* angiosperm trees. The many hardwood species range from tall trees rivaling the podocarps in size, such as *Laurelia novae-zelandiae*, to hardy, stunted small trees and

shrubs at treeline. Pure stands of hardwoods without *Nothofagus* or podocarps are restricted in area.

Four species of *Nothofagus* occur in New Zealand, but they belong to only two pollen taxa, *N. menziesii* type (one species) and *N. fusca* type (three species). All four species are tolerant of suboptimal environments and are often found as nearly pure, often single-species stands in montane and subalpine regions. *N. menziesii* and *N. solandri* are the dominant trees at treeline in most parts of the country. *Nothofagus* forest is therefore most abundant in the mountainous regions of South Island and the axial ranges of North Island. Nevertheless, stands are scattered throughout, even in the warm north of North Island.

Agathis australis is the largest of all forest trees in New Zealand and is confined to the north of North Island. It once formed extensive stands on the widespread nutrient-poor soils of that region and is still a conspicuous component of many northern forest associations.

Before deforestation some variant of conifer-hardwood forest commonly dominated the lowland and

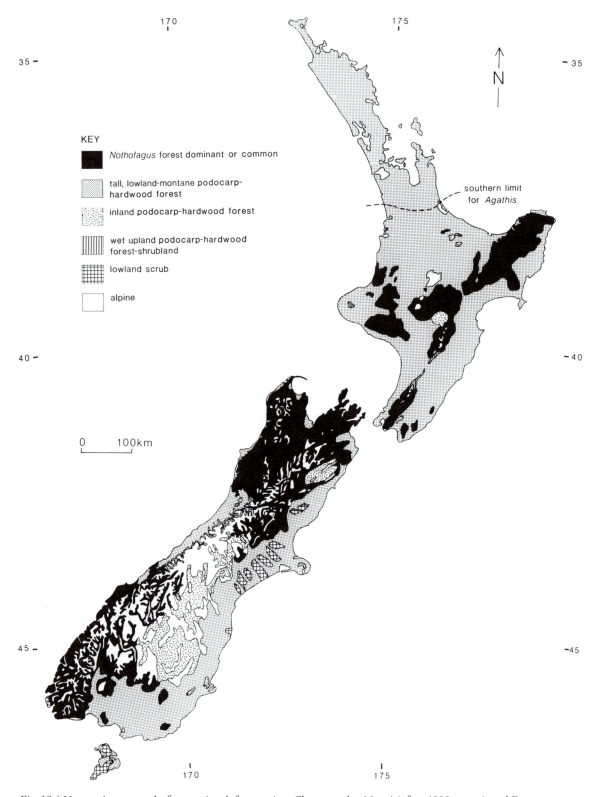

KEY

■ *Nothofagus* forest dominant or common

▨ tall, lowland-montane podocarp-hardwood forest

▨ inland podocarp-hardwood forest

▥ wet upland podocarp-hardwood forest-shrubland

▦ lowland scrub

☐ alpine

southern limit for *Agathis*

0   100km

Fig. 12.4. Vegetation cover before major deforestation. Clearance by Maori (after 1000 yr B.P.) and European settlers removed nearly all the lowland podocarp-hardwood forest.

montane regions of both islands, and *Nothofagus* was a minor component or was absent. Extensive clearance has since transformed the vegetation, and *Nothofagus*-dominant assemblages are now easily the most common indigenous forest type (Fig. 12.4).

Pollen Representation

Although wind-pollinated species are not rare and dominate some plant associations (e.g., *Nothofagus* forest), most New Zealand species are insect- or bird-pollinated. Thus most of the pollen recorded at a given site derives from a few wind-pollinated taxa that may represent but a small proportion of the source vegetation (Macphail and McQueen, 1983).

Natural nonseral grasslands and shrublands are almost all confined to subalpine and alpine locations. Pollen productivity is low in these areas, and only *Coprosma* (besides grasses and sedges) is both common and wind-pollinated. Lowland and montane forests are nearly always close to these upland sites, and pollen from these lower-elevation sources is frequently more abundant than that from the local vegetation (Moar, 1970; McGlone, 1982).

A typical podocarp-hardwood forest has an overstory or emergent layer of tall conifers, commonly including *Dacrydium cupressinum, Podocarpus totara, Prumnopitys taxifolia, Prumnopitys ferruginea,* and *Dacrycarpus dacrydioides,* over a dense main canopy of mixed hardwood species, commonly including *Weinmannia, Beilschmiedia, Metrosideros, Quintinia,* and tree ferns (*Cyathea* and *Dicksonia*). Pollen representation of the wind-pollinated emergent conifers tends to be very good, whereas the more numerous and diverse hardwoods are usually either poorly represented or absent.

An extreme example of the palynologic underrepresentation of main canopy trees is *Beilschmiedia.* Although it is one of the most abundant canopy trees in New Zealand, its pollen is rarely encountered, even at sites that it completely surrounds (Macphail, 1980). Modern pollen samples taken directly beneath a *Beilschmiedia* canopy record its pollen at less than 0.01% of the pollen count.

Nothofagus forests cover large areas of New Zealand, especially the uplands, but they also are a significant component of lowland forest in South Island (Wardle, 1984). Pollen production by *Nothofagus* is great, and the wind-transported pollen is dispersed efficiently. Thus *Nothofagus* pollen tends to be overrepresented in the pollen rain (Macphail and McQueen, 1983). Even high percentages of *Nothofagus* pollen do not necessarily imply local presence.

Pollen and Climate

The relationship between vegetation and rainfall regime in New Zealand is strong. Forests in the drier and drought-prone regions are dominated by *Podocarpus* and *Prumnopitys* in the lowlands and by *Nothofagus solandri* or *Phyllocladus-Podocarpus hallii* in the uplands. In wetter areas *Dacrydium cupressinum, Metrosideros, Weinmannia,* and tree ferns tend to be prominent. At higher altitudes in regions of high rainfall, *N. solandri* may be absent, and either *N. menziesii* or a diverse podocarp-hardwood low forest or shrubland may be present. Pollen diagrams clearly reflect changes in these dominants, and thus major changes in total precipitation or in the seasonal distribution of precipitation should be indicated. The ratio between *D. cupressinum* and *Podocarpus-Prumnopitys* has been used to demonstrate shifts in effective precipitation (Harris, 1963; McGlone and Topping, 1977; McGlone, 1988).

Temperature-vegetation relationships are difficult to establish with any confidence in New Zealand pollen diagrams. Many of the pollen taxa represented in pollen diagrams include palynologically indistinguishable species, which nevertheless have very different climatic preferences and geographic ranges. Moreover, it is not often possible to reconcile the mean annual temperature at the altitudinal limit of a species with that at its latitudinal limit. For example, *Agathis australis,* a widespread dominant in many northern North Island forests, has a rather abrupt southern limit (Fig. 12.4), which could be interpreted as temperature-related. Near the southern limit of its range, however, *A. australis* stands grow in upper montane as well as in lowland sites. No simple relationship to annual temperatures seems to account for this distribution pattern. Many other northern plants have similar distributions (Nicholls, 1983). For these reasons, changes in annual temperature cannot be directly inferred from the abundance of any one taxon or indeed any group of taxa.

Norton *et al.* (1986) used multiple regression analysis to calibrate transfer functions, which showed moderate relationships between modern pollen rain and mean temperature and precipitation (Fig. 12.5). The best results were obtained for annual precipitation and trees characteristic of low-rainfall areas (*Podocarpus totara, Podocarpus hallii,* and *Prumnopitys taxifolia*). Variations in annual precipitation also explained most of the variance in pollen from trees characteristic of high-rainfall areas (*Prumnopitys ferruginea, Weinmannia, Metrosideros,* and *Nothofagus menziesii*).

Some measure of the equability or seasonality of the climate can be inferred from the occurrence or predominance of some pollen types. One group of taxa (*Dacrydium cupressinum, Prumnopitys ferruginea, Weinmannia, Quintinia, Ascarina lucida,* and tree ferns) occurs over a wide range of annual temperatures but in regions where annual precipitation usually exceeds 1000 mm. This group is intolerant of water stress and severe frost and thus signals a mild, moist, equable climate. A second group (*N. fusca*

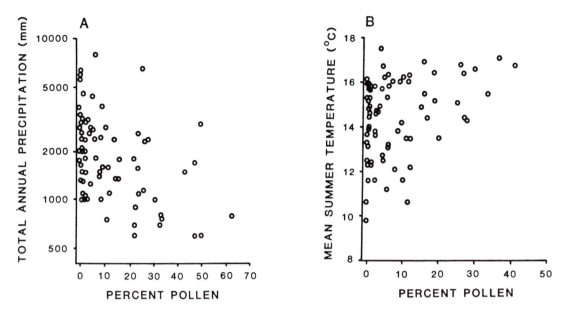

Fig. 12.5. Scatter diagrams of (a) total annual precipitation versus percentage of pollen of *Podocarpus-Prumnopitys* (*r* = -0.44) and (b) mean summer temperature versus spores of tree ferns (*r* = 0.33). From Norton *et al.* (1986).

type, *Libocedrus, Knightia excelsa, Phyllocladus,* and *Agathis australis*) is likewise found over a wide range of annual temperatures but tends to be favored at sites with climatic or edaphic extremes or with frequent disturbances of the vegetation. Although these groups are not absolute indicators of climatic variables, they give useful clues to the nature of the prevailing climate.

The Sites

New Zealand has a network of about 50 published pollen sites that span either all or a significant portion of the last 14,000 yr. Fewer than half the sites are sufficiently well dated to permit the selection of particular time horizons with any degree of certainty.

Most of the sites are in topogenous mires or raised peat bogs, although a few are located in lakes. Most peat bog sites have had a complex history. Many raised bogs began as basin peats and have had a dense cover of swamp forest at some point in their history. Many basin peats have also had lake phases, especially on the eastern side of both North and South islands.

The major peat-forming plants in raised bogs are restionaceous rushes and *Gleichenia* fern. Shrubland cover, often consisting of *Leptospermum,* is common on most bogs. *Sphagnum* moss is usually present but not often dominant. Basin peats have a wide variety of rushes and reeds. Trees growing on swamp peats often include *Dacrycarpus dacrydioides* (a tall podocarp tree) and *Syzygium maire.* A large number of trees, including *Dacrydium cupressinum,* can also exploit swampy margins.

Site Dating Control

Radiocarbon dates form the basis for late-Quaternary chronology. Dating control is poor at most sites; many have only one radiocarbon date. Widespread volcanic ash layers (tephra, derived from eruptions in central North Island) have been of great value in correlating and dating sites. Froggatt and Lowe (1990) compiled a radiocarbon-dated sequence of at least 23 widespread silicic tephras dating from the last glacial maximum and later.

Broad Trends in Vegetation and Climate since 18,000 yr B.P.

THE LAST GLACIAL MAXIMUM (22,000–14,000 YR B.P.)

During the last glacial maximum, glaciers covered much of the Southern Alps and extended to the coastal lowlands of the central west coast of South Island (Fig. 12.6) (Suggate, 1990). North Island mountains were free of permanent ice except for small valley glaciers on some of the highest peaks in central and southern districts. Erosion was active on steep slopes, and broad alluvial outwash plains and terraces formed, especially in the lowland east coast regions. Loess was deposited over most of South Island and in the southern half of North Island (McCraw, 1975).

Grass- and shrub-dominant communities were the norm at least from south of the latitude of Auckland (37°S); grassland was most abundant in the east, and shrubland was more common in the west (Moar, 1980;

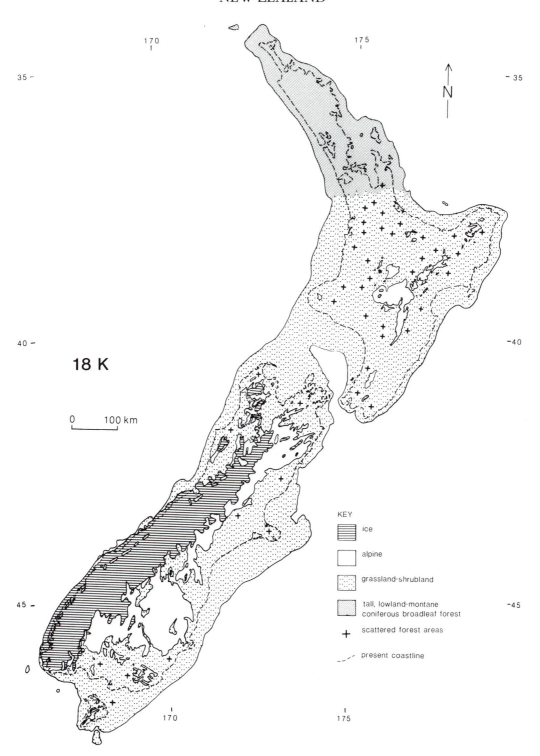

Fig. 12.6. New Zealand vegetation at the last glacial maximum.

McGlone, 1988) (Fig. 12.7). These communities varied greatly in composition and density even over short distances (McGlone and Topping, 1983). Forest pollen was usually in the range of 1-10% in lowland areas of western South Island and less than 2% on the eastern side (Moar, 1980). Nearly all recorded tree pollen is of *Nothofagus* or *Libocedrus.* Forest was therefore uncommon, and it must have consisted mainly of small patches of these trees. Low but consistently present percentages of podocarp pollen suggest that small scattered stands of podocarp trees were also present.

Forest was more common in most North Island dis-

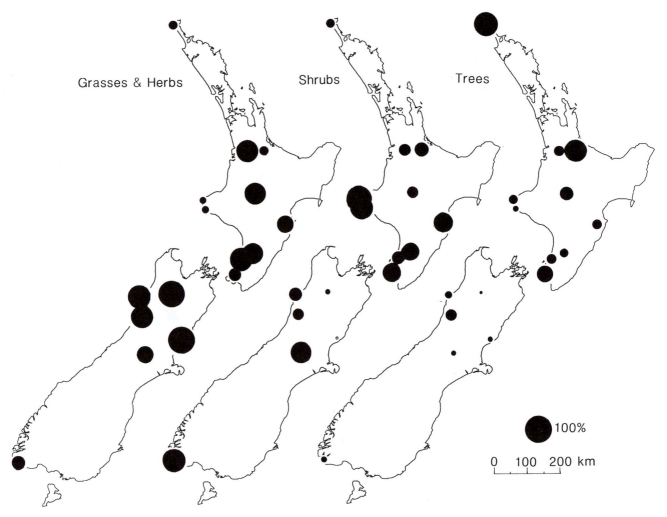

Fig. 12.7. Pollen spectra from selected sites for the last glacial maximum. The area of the circles is proportional to the percentage of the pollen sum.

tricts (Fig. 12.7). Tree pollen counts of 5–65% are recorded, although for most sites they lie between 10 and 30%. As in South Island, *Nothofagus* pollen is dominant, and *N. menziesii* has the highest pollen percentage of any tree taxon. *Libocedrus* and a range of other conifer taxa are also recorded, although generally at low levels. Extensive stands of forest probably persisted in North Island, especially in hilly areas, whereas grassland and shrublands were the most extensive vegetation cover on flat to rolling terrain. In the extreme north, continuous forest cover probably extended down to 36°S (R. Newman, personal communication, 1989).

Cool climates alone cannot explain the observed vegetation pattern. Annual temperatures about 4.5–5°C lower than at present (Soons, 1979) would imply that most of North Island and the northern half of South Island below 600–700 m in altitude should have had forest cover. Even in southern South Island, extensive lowland areas below 300–500 m in altitude should have been forested. Other climatic factors, such as

drought, invasion of cold maritime polar airmasses, and strong winds, may have acted to restrict forest to small patches of climatically favored environments (McGlone and Bathgate, 1983; McGlone, 1985, 1988).

LATE GLACIAL (14,000–10,000 YR B.P.)

The late-glacial period in New Zealand was characterized by progressive afforestation (Fig. 12.8). In broad terms, forest expanded in an apparently stepwise fashion from north to south. In nearly all cases the initial forest community was podocarp-hardwood. Reoccupation of a given region by forest was rapid. Upland sites apparently gained a forest cover at the same time as or even slightly earlier than lowland sites in the same region.

At about 14,000 yr B.P. podocarp-hardwood forest expanded throughout North Island except in the southwest (Figs. 12.8 and 12.9a,b). At about 12,500 yr B.P. forest occupied the northern half of the southwest sector. Not enough dates or diagrams are available to

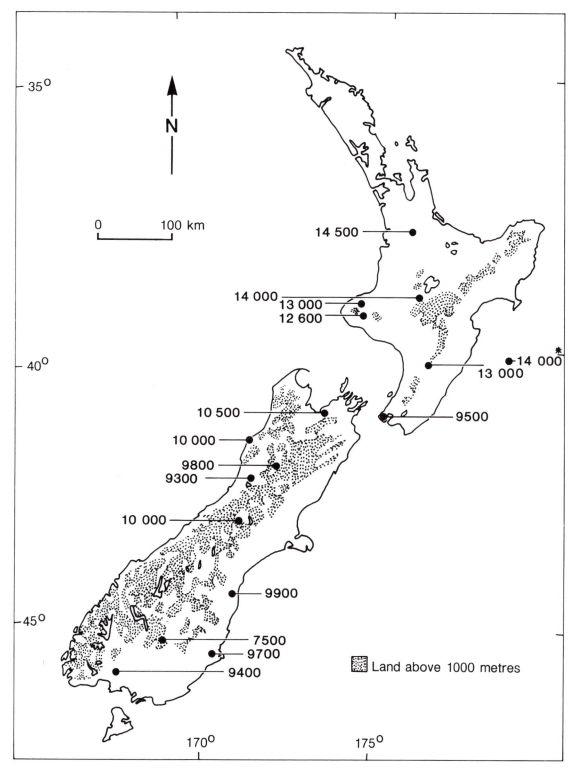

Fig. 12.8. Dates (yr B.P.) of initiation of major increase in arboreal pollen. The asterisk indicates a deep-sea core site. From McGlone (1988); reprinted by permission from Kluwer Academic Publishers, The Hague.

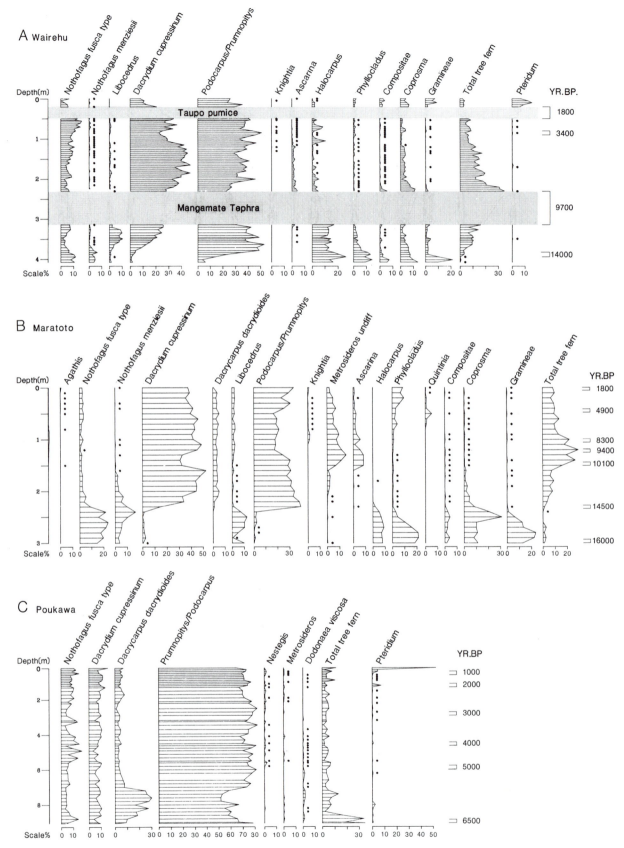

Fig. 12.9. North Island pollen diagrams from (a) Wairehu, central North Island; (b) Maratoto, Waikato; and (c) Poukawa, Hawke Bay. (For locations, see Fig. 12.1.)

place a precise limit to continuous forest at this time, but south of 40.5°S forest was probably confined to discontinuous patches. Nearly 90% of North Island below the contemporary treeline was therefore forested at 12,000 yr B.P. These late-glacial podocarp-hardwood forests, however, were not identical to Holocene forests. Montane and cool-climate species such as *Nothofagus menziesii* and *Libocedrus* were prominent in many late-glacial forests, and *Prumnopitys taxifolia* and *P. ferruginea* were nearly everywhere the dominant tall trees. *Dacrydium cupressinum* tended to increase with time (Figs. 12.9a, b and 12.11).

At about 12,000 yr B.P. parts of the wet central and southern portion of the west coast of South Island gained a low forest cover of *Weinmannia* and *Metrosideros,* similar to that found above the limit of tall podocarp forest in the region at present (N. T. Moar and M. S. McGlone, unpublished data). The transition from grassland to shrubland and low forest occurred between 12,000 and 10,000 yr B.P. at some eastern South Island sites (McGlone, 1988; Burrows and Russell, 1990).

Between 10,500 and 9300 yr B.P. the southern tip of North Island (Lewis and Mildenhall, 1985) and all of South Island except the dry southeastern interior basins became covered with tall podocarp-hardwood forest (Figs. 12.8 and 12.10). Forest may have become established nearly 1000 yr earlier in the north than in the south of South Island, but there are not enough dates to be conclusive on this point (Fig. 12.8).

The times centering on about 14,000 yr B.P. and about 10,000 yr B.P. stand out as periods of rapid vegetational and climatic change. These periods can be regarded as well-defined steps lasting perhaps 1000 yr. Climatic conditions appear to have improved steadily from 14,000 to 12,000 yr B.P., whereas the period from 12,000 to 10,000 yr B.P. seems to have had a slower rate of change.

The late-glacial forests in both North and South Island contained cool-climate elements, but where macrofossil evidence permits the determination of accurate altitudinal limits, annual temperatures seem to have been within 2°C of their present values by 12,000 yr B.P. or earlier (McGlone and Topping, 1977; Soons and Burrows, 1978). The continuing absence of forest from large areas of South Island seems therefore to have resulted from persistence of the extreme climatic conditions of the last glacial maximum rather than from low annual temperatures.

POSTGLACIAL (HOLOCENE)
(10,000 YR B.P. TO PRESENT)

The afforestation of New Zealand was complete by 10,000–9000 yr B.P. except for the central-southeastern districts of South Island (Fig. 12.8). Between 10,500 and 9300 yr B.P. *Dacrydium cupressinum* became the dominant podocarp tree in western and northern

North Island forests (Fig. 12.11), while forests in the east remained dominated by *Podocarpus-Prumnopitys* (Fig. 12.9c). A range of hardwood taxa, including *Metrosideros, Ascarina lucida,* and tree ferns, became common at the same time (Fig. 12.9a,b). The first postglacial forests on the west coast of South Island (excluding the earlier low forests of the central and southern region) were dominated by *D. cupressinum, Weinmannia,* and tree ferns (Fig. 12.10b). In the extreme southwest of South Island and Stewart Island, hardwoods (particularly *Metrosideros* and *Weinmannia*) and tree ferns continued to dominate until the mid-postglacial, although small numbers of podocarp trees were present (M. S. McGlone, unpublished data).

These strikingly uniform podocarp-hardwood forests, differentiated into a wet western (e.g., Fig. 12.10b) and a dry eastern (Fig. 12.10a) facies, extended throughout New Zealand. Even high-altitude subalpine sites generally had some type of podocarp-hardwood forest or shrubland.

By around 10,000 yr B.P. annual temperatures were at least equivalent to those of today. At a site near the limit of woody shrubs in the central Southern Alps of South Island, woody vegetation was dominant by this time (McGlone, 1988), and forest extended close to present treeline in southern South Island (McGlone and Bathgate, 1983) and in North Island (M. S. McGlone, unpublished data). Neither pollen nor macrofossil evidence points to treelines being significantly higher than at present during this period or later. However, speleothem evidence suggests that temperatures were 1–2°C warmer than at present (Hendy and Wilson, 1968), and this has been corroborated by microfossil analysis from a deep-sea core off the east coast of North Island (W. Prell, personal communication, 1987).

Pollen and macrofossil evidence strongly indicates milder, less frost-prone, and almost certainly less windy climates than at present between 10,000 and 7000 yr B.P. The abundance of the frost-sensitive *Ascarina* throughout western districts at that time (e.g., Fig. 12.9b) supports the conclusion that a milder climate prevailed. Tree ferns were abundant in the presently windy and exposed south to southwest coast of South Island (M. S. McGlone, unpublished data) and also appeared to have been more common at high altitudes (McGlone, 1988). The apparent extension of the ecological dominance of these trees, which are sensitive to exposure and cold, also argues for milder, less windy climates. Finally, *Nothofagus* trees, which are renowned for their tolerance of harsh, exposed conditions, were restricted, particularly in montane and subalpine sites where they dominate at present (Fig. 12.12).

In eastern South Island and southeastern North Island, the dominance of *Prumnopitys taxifolia* and *Podocarpus,* with only a small proportion of *Dacrydium cupressinum,* is evidence that effective precipi-

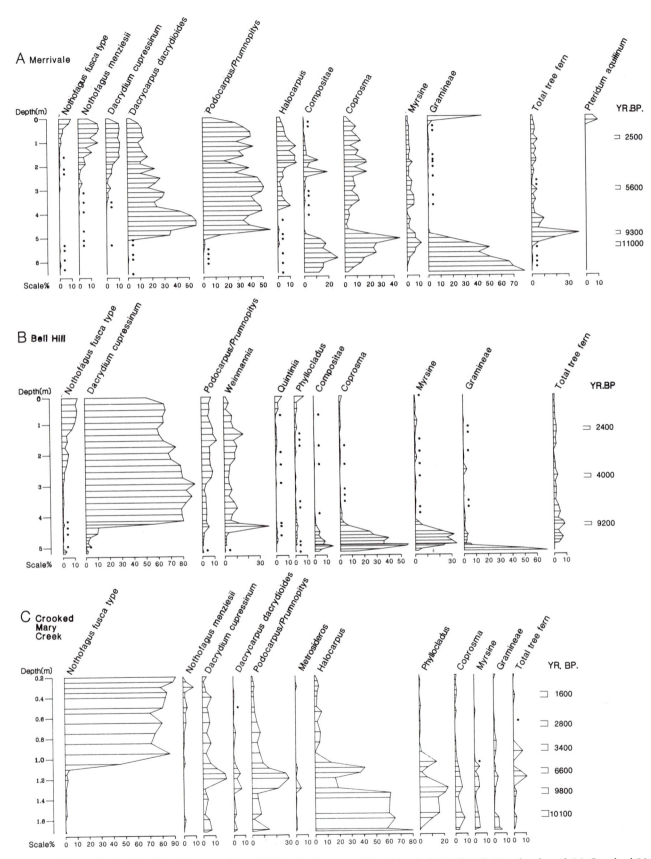

Fig. 12.10. South Island pollen diagrams from (a) Merrivale, coastal Southland; (b) Bell Hill, Westland; and (c) Crooked Mary Creek, Westland. (For locations, see Fig. 12.1.)

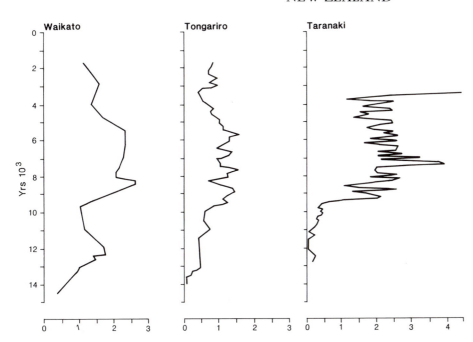

Fig. 12.11. Ratio of *Dacrydium cupressinum* to *Podocarpus-Prumnopitys* for Waikato Basin (Maratoto site), Tongariro (Wairehu site), and Taranaki (Ngaere site). (For locations, see Fig. 12.1.) From McGlone (1988); reprinted by permission from Kluwer Academic Publishers, The Hague.

tation was as low as or lower than at present. McGlone and Bathgate (1983) suggested that rainfall in Southland may have been as much as 30% lower. The absence of tall forest from central Otago in southeastern South Island makes a clear case for much lower precipitation in what is now the driest region of the country (Fig. 12.2). Pollen diagrams from Dunedin (McIntyre and McKellar, 1970) and Timaru (Moar, 1973) show that this early-postglacial dry phase was general in southeastern South Island. The widespread absence of organic sediments dating to this time throughout the east of both North and South Island is clearly related to much lower effective precipitation (M. S. McGlone, unpublished data).

The abundance of organic deposits and the prominence of both *Dacrydium cupressinum* (Figs. 12.11 and 12.13) and tree ferns (Fig. 12.14) in western and northern districts between 10,000 and 7000 yr B.P. would seem to support greater precipitation, but this may not have been the case, for three reasons. First, mountain bogs seem not to have become common until late in the postglacial (Rogers and McGlone, 1989; Froggatt and Rogers, 1990), suggesting that either overall rainfall was lower or its distribution was different. Second, although *D. cupressinum* and tree ferns need plentiful and well-distributed precipitation, there is certainly no simple linear relationship between increasing rainfall and the abundance of these taxa. And third, if eastern districts were drier than at present and western districts were wetter, we must infer that zonal westerly winds were stronger. As we have seen, what evidence there is points to weaker westerly airflow.

The fire history of eastern South Island also supports the interpretation of weaker westerly airflow. Despite lower eastern precipitation, there is no evi-

dence for large outbreaks of fire between 10,000 and 7500 yr B.P. in eastern districts. Deforestation by fire is a feature of the mid- to late-postglacial period (McGlone, 1988; Burrows and Russell, 1990). If southern New Zealand was under the influence of strong zonal westerlies in the early postglacial, the resultant regime of warm, dry foehn winds to the east of the axial mountain chains would have made devastating outbreaks of fire inevitable. We must assume that the westerly wind flow was reduced, particularly in late spring and summer.

Rogers and McGlone (1989) suggested that although overall precipitation was lower during the early postglacial, summers tended to be moist and cloudy and winters drier and clearer than now. We argue that early-postglacial summers were characteristically cloudy and moist because of the predominance of northerly and easterly winds. Winter, on the other hand, must have been often calm and anticyclonic, with few deep, vigorous fronts and a weaker overall southerly wind flow. Such a pattern fits both the vegetation pattern and the initiation of wetlands in the east and mountainous regions. For example, central Otago receives much of its rainfall from spillover rain from strong westerly wind flow and from deep frontal depressions moving from the southwest. A reduction of these features of the present precipitation pattern would dramatically reduce rainfall. In fact, rainfall in central Otago during the early postglacial was insufficient to support tall forest or lowland wetlands.

Most western districts receive so much rainfall from westerly flow that it is unlikely that a moderate reduction in total rainfall during the early postglacial would have had a noticeable effect on the vegetation. How-

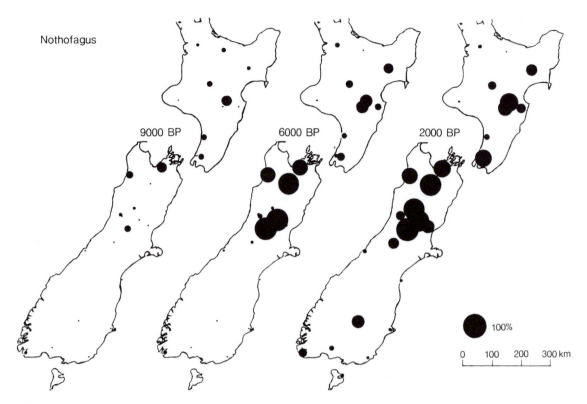

Fig. 12.12. Holocene *Nothofagus fusca*-type pollen spectra. The area of the circles is proportional to the percentage of pollen.

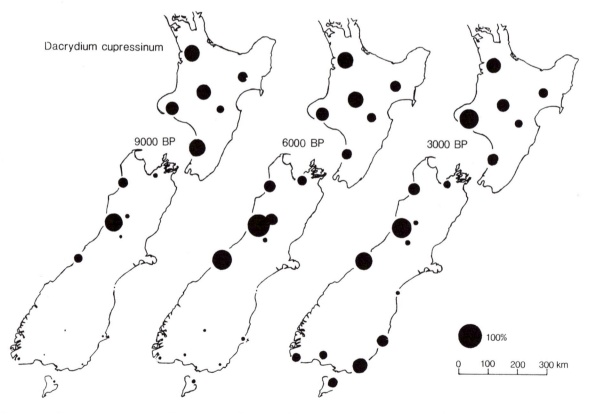

Fig. 12.13. Dacrydium cupressinum pollen spectra. The area of the circles is proportional to the percentage of pollen.

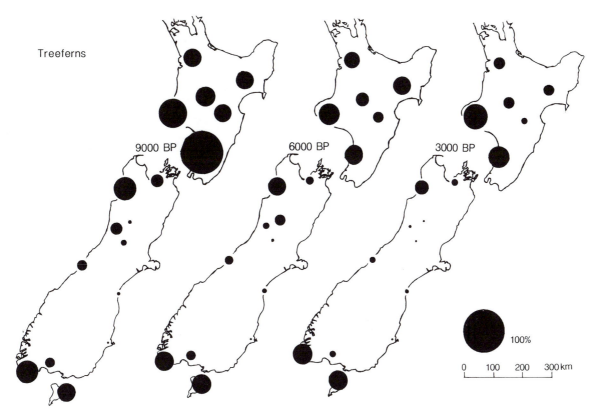

Fig. 12.14. Tree fern (*Cyathea* and *Dicksonsia*) pollen spectra. The area of the circles is proportional to the percentage of pollen.

ever, in mountainous regions a reduction in winter rainfall would lead to drier soil conditions, and moister, cloudier summers would severely restrict subalpine vegetation. Rogers and McGlone (1989) suggested that the mild, cloudy, and moist conditions that may have prevailed throughout the axial mountain chain in the early postglacial were the primary reason that low scrubland was abundant and *Nothofagus* forest scarce. *Nothofagus*-free regions, such as Mount Egmont, southern Ruahines, central Westland, and Stewart Island, all have high precipitation, persistent cloudiness, freedom from severe drought, and mild winters.

In eastern districts soil moisture would have been reduced during the early postglacial because of lack of winter recharge; hence the dominance of vegetation characteristic of dry sites and the absence of lakes and bogs. However, moist summers without severe foehn winds would have curtailed outbreaks of fire.

After about 7000 yr B.P. the forest cover throughout New Zealand went through a series of shifts in composition and structure. Although these shifts were more or less contemporaneous, they differed greatly among regions both in nature and in the species involved. We recognize four major shifts: (1) low forest replaced shrubland-grassland in central Otago; (2) *Nothofagus* forest spread in many upland areas, especially in the northeastern sector of South Island (Moar, 1971; Burrows and Russell, 1990) (Figs. 12.12 and 12.15)

and in southern North Island (Mildenhall and Moore, 1983; Rogers and McGlone, 1989); (3) a group of conifer and hardwood species (*Agathis australis, Libocedrus plumosa, Phyllocladus trichomanoides, Prumnopitys-Podocarpus, Knightia excelsa,* and *Nestegis*) began to increase in North Island (McGlone *et al.,* 1984; Newnham *et al.,* 1989); and (4) *Dacrydium cupressinum* and *Nothofagus menziesii* began to spread in coastal districts of South Island (McGlone and Bathgate, 1983) (Figs. 12.10a and 12.13).

At the same time as these new plant associations were forming, important taxa of the previous forests were declining. *Ascarina lucida* in particular went from being perhaps one of the most ubiquitous understory trees of lowland and montane forests to being rare or absent over great areas of its former range (Fig. 12.9a,b) (McGlone and Moar, 1977; McGlone, 1988). *Dacrydium cupressinum* declined gradually in many western and northern sites (Figs. 12.9a, 12.10b, and 12.11).

It is important to note that these changes cannot be attributed to migration rather than to environmental change. Rather, it appears that species initially at low population densities expanded in response to both climatic and edaphic change (McGlone, 1985).

Many of the vegetation changes after 7000 yr B.P. are related to alterations in precipitation patterns. In northern and western locations summers appear to have be-

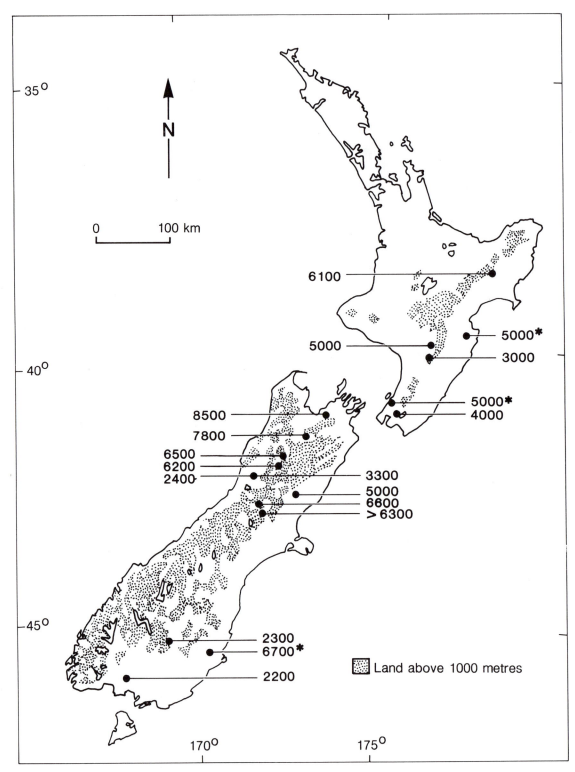

Fig. 12.15. Dates (yr B.P.) for the first rapid and major rise of *Nothofagus fusca*-type pollen curves. Asterisks indicate sites where the *Nothofagus* pollen was derived from distant sources. From McGlone (1988); reprinted by permission from Kluwer Academic Publishers, The Hague.

come drier. Outside the very high rainfall zone, summer water deficits, rather than an occasional feature of the climate, became a regular part of the annual cycle, as they are at present. Winters, on the other hand, became wetter, and lakes, swamps, and peat bogs began to form in dry eastern districts. The spread of cold-tolerant species in the uplands and the reduction of frost-sensitive species in the lowlands point to cooler winter conditions. The suite of plants that expanded in the late postglacial share an ability to tolerate frequent disturbance and poor soils, and these characteristics may also have assisted the spread of taxa such as *Agathis australis, Nothofagus,* and *Quintinia.*

Beginning around 2500 yr B.P. forests dominated by trees of the *Nothofagus fusca* group expanded in the southern and western regions of South Island (Figs. 12.10a, 12.12, and 12.15). At the same time, natural fires began to burn through large areas of inland southeastern South Island (McGlone, 1973 and unpublished data). The expansion of intense westerly wind flow onto southern New Zealand probably explains both these events. Increased disturbance and cooler conditions associated with stronger winds certainly would have favored the expansion of *N. fusca*-group forest in the southwest, and the prolonged droughts and strong foehn winds associated with intense westerly airflow in this region predispose forest vegetation to fire.

Natural changes since 2500 yr B.P. have been relatively minor. From around 1000 yr B.P. Maori settlers began to clear very large areas of lowland and montane forest by fire (McGlone, 1983, 1989), and this clearance was continued by European settlers in the 19th century. These anthropogenic changes in the environment have tended to obscure any evidence for recent adjustment of the vegetation to climatic change.

Pattern of Climatic Change

Paleovegetation data from New Zealand give no more than a guide to the magnitude of temperature changes. However, other proxy evidence clearly demonstrates significant fluctuations in temperature. Oxygen-isotope data from speleothems in North Island (Hendy and Wilson, 1968), the magnitude and timing of glacier advances and retreats in South Island (Suggate and Moar, 1970; Burrows, 1979; Gellatly *et al.,* 1988; Suggate, 1990), and microfossil analyses from a deep-sea core close to the east coast of South Island (W. Prell, personal communication, 1987) show that annual temperatures rose sharply around 14,000 yr B.P., when mountain glaciers retreated rapidly, and around 10,000 yr B.P. These sources collectively suggest that annual temperatures were coldest (4.5-5.0°C below present) between 20,000 and 15,000 yr B.P. and highest (1.0-2.0°C above present) between 10,000 and

8000 yr B.P. Average annual temperatures have declined since about 7500 yr B.P.

The pollen, glacier, and deep-sea records (Stewart and Neall, 1984) all show that the late-glacial amelioration occurred in a steplike fashion. Stewart and Neall (1984) hypothesized that the ocean off the east coast of North Island warmed abruptly about 15,000 yr B.P. Pollen results from their deep-sea core show that changes in vegetation on the adjacent mainland were synchronous (M. S. McGlone, unpublished data).

Thiede (1979) suggested on the basis of wind patterns inferred from analysis of quartz in ocean cores that the subtropical convergence lay to the north of New Zealand at the glacial maximum. If so, a sudden movement southward of the subtropical convergence to approximately latitude 39°S at about 15,000 yr B.P. may explain the rapid reforestation of the northern two-thirds of North Island. A further rapid southward movement of the subtropical convergence at around 10,000-10,500 yr B.P. to its present position around the south of South Island would likewise have led to the reforestation of South Island. However, the subtropical convergence east of New Zealand seems to be physically tied to the Chatham Rise, and there is no evidence that it has moved substantially over the last 20,000 yr (J. Fenner, personal communication, 1988). In the Tasman Sea west of New Zealand, there is no equivalent physical barrier to the movement of water masses, and it is possible that cold water extended far northward. CLIMAP Project Members (1981) showed an offset of sea-surface temperatures between western and eastern New Zealand of about 2°C for summer and winter at the glacial maximum, which could relate to differential movement of the subtropical convergence. The persistence of shrub grassland in southwestern North Island for some 2000 yr after reforestation by podocarp-hardwood forest in the east may be related to this temperature asymmetry.

A much colder water mass in the Tasman Sea, in conjunction with stronger zonal westerly-southwesterly airflow, may have controlled the general nature of the vegetation cover on land by facilitating the frequent intrusion of cold high-latitude maritime airmasses over New Zealand. Lower rainfall over New Zealand during the last glacial maximum may have been related to this cooler water in the Tasman Sea, but isolation from tropical sources of precipitation was probably as important. Rapid southward movement of warm water in the Tasman Sea could have provided the mechanism for abrupt climatic change on land during the late glacial.

Warmer ocean temperatures in the South Pacific (Hays *et al.,* 1976) around 9000 yr B.P. would have raised the moisture content of the overlying atmosphere, and we therefore could expect higher precipitation in the New Zealand region. However, as we have seen, total precipitation over New Zealand probably was diminished compared to the present, although it may have had a summer maximum rather than the present winter maximum over large regions. We suggest that at 9000 yr B.P. total wind flow over New Zealand was less than at present and that the reduced westerly circulation was more northerly in direction, unlike the mean west-southwest direction of the present circulation. The enhanced northerly component in the airflow would have increased the incidence of air of subtropical origin over the country, resulting in milder and more equable temperatures. Less vigorous cyclogenesis is probably the main cause of this relatively low early-postglacial rainfall in the Zealand region. Increasing rainfall during the late postglacial is therefore probably linked with increasing westerly wind flow and more vigorous cyclogenesis.

We suggest that the westerlies continued in this weakened state at 6000 yr B.P. but that airflow from southerly and easterly quarters was more frequent than at present. Such a circulation pattern would have brought higher precipitation to the east of both islands and resulted in more frequent incursion of cool Southern Ocean airmasses and thus cooler average temperatures overall.

Since 3000–2500 yr B.P. the increased incidence of fire, the establishment of the modern pattern of high-altitude vegetation dominated by *Nothofagus* and *Libocedrus*, the further restriction of frost-sensitive plants, and the spread of lowland trees favored by disturbance, seasonal climates, and edaphic extremes indicate a westerly and southwesterly airflow pattern similar to that of the present.

We may summarize paleovegetation results and their climatic interpretation as follows:

18,000 yr B.P.: Grassland and shrubland dominant throughout, except for closed forest at the northern tip of North Island. Forest rare elsewhere, especially on flat or rolling terrain. *Nothofagus* the most common forest tree. Glaciers at full extent. Climate much cooler than at present (-5°C), windier, and drier.

14,000 yr B.P.: Forest spreads in the northern two-thirds of North Island; glaciers retreat. Climate still cool but markedly warmer than at 18,000 yr B.P.; precipitation also significantly greater.

12,000 yr B.P.: All of North Island except the southern tip is forested. South Island lacks tall forest, but scrub and low forest increase. Temperatures 2°C lower than at present but rising; rainfall increasing; influence of cold airmasses still strong in south.

9000 yr B.P.: Podocarp-hardwood forest cover largely complete, but central Otago remains shrubland-grassland for 1500 more years. Moist forests in the west; drier, drought-tolerant forests in the east. *Nothofagus* forest rare nearly everywhere. Climate warmer than at present; overall rainfall probably lower, but much lower in the east; lesser windiness, seasonality, drought, and frost.

6000 yr B.P.: Spread of *Nothofagus, Agathis,* and associated species has begun; species favored by moist, mild environments decline; drought- and frost-tolerant species spread. *Dacrydium cupressinum* and *Nothofagus menziesii* spread in coastal regions of southern South Island. Central Otago gained forest at 7500 yr B.P. Climate cooler and windier. Southeast winds increase rainfall in the east of both islands, and total rainfall increases. Shift to winter rainfall maximum begins, but drought becomes more common.

3000 yr B.P.: Final expansion of *Nothofagus,* mainly in southern South Island. Deforestation by fire of large areas of the interior of South Island from 2500 yr B.P. on. Climatic conditions close to those of the present.

Comparisons with Model Results

New Zealand's small size excludes it from occupying even one land grid point in the Community Climate Model of the National Center for Atmospheric Research. Model results for the New Zealand region are derived from ocean grids only. Boundary conditions set for the ocean grid points are therefore extremely important. Sea-surface temperatures (SSTs) are fixed in model simulations. The prescribed SSTs for 18,000 yr B.P. are estimates (based on planktonic microfossils) made by CLIMAP (1981). For the 12,000-yr B.P. simulation the SST anomalies are reduced to half of their 18,000-yr B.P. values, and at 9000 yr B.P. and after SSTs are set at modern values. Sea-ice boundaries, another profound influence on New Zealand paleoclimates, are set at CLIMAP-specified positions for 18,000 yr B.P. in the model, kept there at 15,000 and 12,000 yr B.P., and thereafter placed at the modern positions.

Comparing model results with data from the New Zealand region is difficult. The first question is whether the New Zealand landmass has closely tracked oceanic conditions or whether it has responded more as a continental landmass. Modern New Zealand mean annual temperatures at sea level are closely comparable to those of the adjacent ocean, but most pollen sites are inland and at moderate altitude. The lowering of sea level by about 120 m at the glacial maximum increased the surface area of New Zealand by a third (Fig. 12.6) and thereby would have created a more continental climate in the interior. We therefore compare the New Zealand data with both the CLIMAP SSTs and the model temperature averages for land between 30 and 45°S.

Rind and Peteet (1985) questioned the accuracy of CLIMAP SST reconstructions for the glacial maximum for tropical and subtropical regions by showing that estimates of mean surface-air temperatures derived from terrestrial snowline depressions and pollen studies were 4°C lower. They found that uniform lowering of the CLIMAP SSTs by 2°C gave a better fit to the terrestrial data. The New Zealand data provide a check on the CLIMAP SST estimates (discussed below).

Table 12.1. Selected CLIMAP (1981) sea-surface temperatures (°C) for the New Zealand region and Patagonia, 18,000 yr B.P. and present

Region	18,000 yr B.P.		Present		Anomaly (0 yr B.P.–18,000 yr B.P.)	
Month	West	East	West	East	West	East
New Zealand mainland, 39°S						
February	18.0	19.5	18.5	18.5	-0.5	+1.0
August	11.0	13.0	13.5	13.5	-2.5	-0.5
Annual	14.5	16.0	16.0	16.0	-1.5	+0.75
New Zealand mainland, 45°S						
February	13.5	14.5	15.5	14.5	-2.0	0.0
August	8.5	9.5	11.0	11.0	-2.5	-1.5
Annual	11.0	12.0	13.3	12.7	-2.3	-0.75
Chatham Islands, 44°S						
February		15.0		15.0		
August		10.0		10.0		
Annual		12.3		12.3		
Patagonia,[a] 46-52°S						
February			11.5	8.0		
August			8.5	6.5		
Annual			10.0	7.3		

[a]Present Patagonian temperatures relate to 46°S (first column) and 52°S (second column).

A second difficulty in data-model comparisons arises because New Zealand precipitation patterns are largely controlled by the orientation of the high axial mountain chain. Because changes in the average strength and direction of wind flow produce major shifts in these patterns, it is difficult to estimate changes in total precipitation over the region. On the other hand, if precipitation patterns are well expressed in the data, they provide information on wind strength and direction.

18,000 YR B.P.

CLIMAP August (winter) SSTs for 18,000 yr B.P. are 2-3°C cooler than at present to the west of New Zealand and 1-2°C cooler to the east (Table 12.1). February (summer) SSTs are surprisingly close to those of the present, with the west only 0-2°C cooler and the east slightly warmer than or the same as at present. Model land-surface temperature averages for 30-45°S for July and January are about 3°C below present values (Fig. 12.16). The predominant westerly surface wind flow over New Zealand in July is strengthened by at least 5 m/s in the model, and the core of the strongest winds is south of its present position by 5° (50°S compared to 45°S). January surface winds show little change from the present. Annual precipitation changes are slight in the model.

New Zealand data strongly support the strengthened westerly airflow. Thick deposits of loess extend throughout New Zealand in eastern districts from the extreme south to about latitude 37°S and are particularly extensive south of 39°S (McCraw, 1975). Vegeta-

tion was sparse and low east of the axial mountain ranges. This evidence points to vigorous airflow from the west that produced dry foehn winds in the east.

The unchanged precipitation level simulated in the model is probably not in accord with the data. Rainfall was undoubtedly lower than at present in the east of both North and South islands, as dry lake basins and sparse vegetation both make clear. In the west and north lower lake levels are suggested (Kennedy et al., 1978; Green and Lowe, 1985), and the prevalence of open grassland and shrubland lends further support to the argument for a drier climate. The whole New Zealand region was probably somewhat drier than at present, but the confounding effects of both higher wind speeds and cooler temperatures on soil moisture make it difficult to be more specific.

In a narrow, highly oceanic landmass such as New Zealand, ocean temperatures are strongly correlated with land temperatures. Regional snowlines at 18,000 yr B.P. were lower than at present throughout New Zealand by about 800-850 m (Porter, 1975; Soons, 1979; Chinn, 1981), suggesting an average annual temperature depression of 4-5°C. However, CLIMAP (1981) annual average SSTs at latitudes 39 and 45°S were depressed by 1.5 and 2.3°C, respectively, in the west, while in the east they were slightly warmer (by 0.75°C) at latitude 39°S and slightly cooler at 45°S (Table 12.1). With the predominant wind flow from the west, SSTs in the west would have had a major influence on paleosnowlines, but even so the discrepancy is marked, especially in North Island. The average surface temperature depression of about 3°C on the

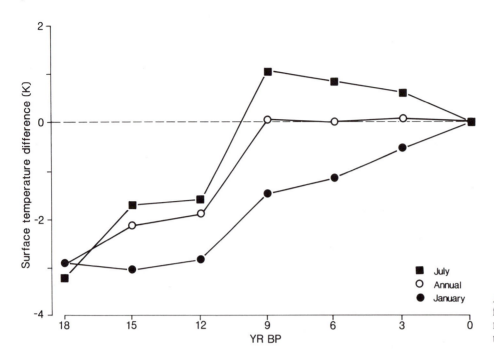

Fig. 12.16. Model temperature differences (°K) for 18,000 to 0 yr B.P. for land grid squares between 30 and 45°S.

model land grids (30–45°S) is much closer to the paleo-snowline temperature estimates.

The vegetation reconstructions are consistent with paleosnowline evidence in that they indicate a considerable depression of surface temperatures. They give no hint of any more complete vegetation cover in the east than in the west, as the CLIMAP SSTs would suggest.

The fact that New Zealand data suggest that cooling of the land surface was more in line with the model predictions for continental landmasses in the same latitudes than with the model SSTs indicates either that New Zealand behaved more like a continental land area than like ocean or that the CLIMAP SSTs underestimate the actual temperature depression. Some evidence suggests that the problem lies with the CLIMAP SSTs.

The Chatham Islands (44°S, 176°30'W) east of New Zealand are a small, isolated, low-lying group of islands. These highly oceanic islands therefore essentially record sea-surface conditions and can directly indicate changes in oceanic temperatures in their region. SSTs at 18,000 yr B.P. close to the Chathams were at present values according to the CLIMAP (1981) reconstructions (Table 12.1). Vegetation cover at the glacial maximum consisted of open grassland in place of the Holocene low forest and tall shrubland (D. C. Mildenhall, personal communication, 1986). Such a striking change in the vegetative cover seems improbable without a lowering of SSTs of the surrounding ocean. Although the larger landmass of New Zealand could conceivably have substantially modified its local climate, this is not possible for the much smaller Chatham Islands.

The present-day Patagonian ice cap, lying between 46 and 52°S at the southern tip of South America, provides an instructive analogy to the New Zealand glacial maximum. The Patagonian topography resembles that of southwestern New Zealand: a high coastal chain of mountains intercepts a strong, moist, westerly airflow. Present Patagonian winter SSTs are 2–4.5°C cooler than CLIMAP (1981) values for southwestern New Zealand at the glacial maximum (Table 12.1). It seems unlikely that the Patagonian ice cap could have withstood winter SSTs as warm as the CLIMAP SSTs for glacial-maximum New Zealand, which in turn suggests that the extensive valley and piedmont glaciers of the Southern Alps that formed during the last glacial maximum would not have formed under the same conditions.

We tentatively conclude that the CLIMAP SSTs for the New Zealand region underestimate the actual temperature depression at 18,000 yr B.P. by about 2°C. However, the resulting temperature depression of 4–5°C alone is insufficient to explain the largely treeless lowlands. We must also postulate frequent outbreaks of cold maritime polar air and a strengthened west-southwest circulation located north of its present position.

15,000–10,000 YR B.P.

The model results indicate little change between 18,000 and 12,000 yr B.P. Wind speeds and precipitation remain much the same. SSTs change little, and model land grid points between 30 and 45°S (Fig. 12.16) show January (summer) temperatures as cold at 12,000 yr B.P. as at 18,000 yr B.P., although annual and July (winter) temperatures rise by about 1.0 and 1.5°C, respectively.

Vegetation data indicate that New Zealand before 15,000 yr B.P. was little different from 18,000 yr B.P. A few northern sites show some climatic amelioration before this time, but it is slight. Around 14,000 yr B.P., however, the landscape began to alter significantly, with rapid afforestation in the north and central North Island and major and rapid retreat of the ice in South Island (Suggate and Moar, 1970). Thus the data and the model predictions for this time period are widely divergent.

9000 YR B.P.

Model ocean boundary conditions are the same at 9000 yr B.P. as at present. Radiation was above present levels in July and below them in January. The model shows westerly wind flow decreased from the 18,000-yr B.P. levels but still significantly stronger than at present, with a greater northerly component than at 12,000 yr B.P. Little or no change in precipitation is indicated. Model land grid surface temperatures for 30-45°S (Fig. 12.16) show annual averages close to those of the present, but with January (summer) temperatures still at the 15,000- to 12,000-yr B.P. level (about 1.5°C below present) and July (winter) temperatures 1°C higher than at present.

Physical evidence from the New Zealand region from speleothems, deep-sea cores, and glaciers puts annual temperatures at 9000 yr B.P. somewhat higher than at present, perhaps by as much as 1-2°C. Paleovegetation records do not give unequivocal evidence of warmer annual average temperatures. For instance, they do not indicate that treelines have ever been higher than they are now. However, the vegetation evidence points strongly toward mild, less seasonal climates. Thus the model results suggesting warmer winters, cooler summers, and consequently lower seasonality agree well with the data. Assuming that summers at 9000 yr B.P. were slightly cooler than at present would enable us to reconcile warmer annual temperatures with apparently unchanging treelines.

Evidence from past vegetation suggests increased northerly wind flow and less intense westerly and southwesterly flow at 9000 yr B.P. relative to the present. The model simulates changes in mean wind flow consistent with the vegetation evidence, but the model's overall wind speeds seem to differ little from those of the present, which is not supported in the data.

6000-0 YR B.P.

Model results for the last 6000 yr show increasing seasonality and decreasing westerly wind flow in winter. The vegetation data amply demonstrate increasing seasonality but also indicate increased windiness and an increased southerly wind flow, which are not reflected in the model.

Conclusions

The model simulations capture only some of the climatic changes of the last 18,000 yr in the New Zealand region. Because the New Zealand results deal with rather small changes in temperature and precipitation in a highly oceanic setting, they are as much a test of the accuracy of the boundary conditions as of the functioning of the model.

The temperature changes suggested by the model for the land grids for the New Zealand latitudes are reasonable approximations to the data, except that they seem to underestimate the severity of the last glacial maximum and do not reproduce the early-postglacial warm episode. The SSTs, on the other hand, suggest an amplitude of only 2°C for the glacial-interglacial cycle, whereas the data indicate something close to 6°C. In the model simulations seasonality is low at 9000 yr B.P. and increases toward the present, and the data strongly support this trend.

Precipitation is virtually unchanging in the model, in contrast to indications in the New Zealand data of substantial alterations. However, determining absolute precipitation from the data is difficult because of the strong influence of other factors, such as seasonality and wind speed.

Increased windiness in the New Zealand region at 18,000 yr B.P. is well indicated in the model results and is supported by the data. However, the lower wind speeds in the early postglacial suggested by the data are not demonstrated in the model, possibly because the boundary conditions are set at present values.

The model appears to underestimate climatic change in oceanic regions. Accurate climatic reconstructions for oceanic regions will probably not be possible until a fully interactive ocean is incorporated into the models.

Acknowledgments

We are grateful to M. K. Macphail for providing unpublished pollen results for Figures 12.12 and 12.13, W. Prell for SST results from core P69, and D. C. Mildenhall for Chatham Islands pollen evidence. We thank H. E. Wright, Jr., T. Webb III, and W. F. Ruddiman for comments on a draft manuscript.

References

Burrows, C. J. (1979). A chronology for cool-climate episodes in the Southern Hemisphere 12,000-1000 yr B.P. *Palaeogeography, Palaeoclimatology, Palaeoecology* 27, 287-347.

Burrows, C. J., and Russell, J. B. (1990). Aranuian vegetation history of the Arrowsmith Range, Canterbury. 1. Pollen diagrams, plant macrofossils, and buried soils from Prospect Hill. *New Zealand Journal of Botany* 28, 323-345.

Chinn, T. J. H. (1981). Late Quaternary glacial episodes in New Zealand. Report WS364. Ministry of Works and Development, Christchurch.

CLIMAP Project Members (1981). Seasonal reconstructions of the earth's surface at the last glacial maximum. *Geological Society of America Map and Chart Series* MC-36.

Froggatt, P. C., and Lowe, D. J. (1990). A review of late Quaternary silicic and some other tephra formations from New Zealand: Their stratigraphy, nomenclature, distribution, volume, and age. *New Zealand Journal of Geology and Geophysics* 33, 89-109.

Froggatt, P. C., and Rogers, G. M. (1990). Tephrostratigraphy of high-altitude peat bogs along the axial ranges, North Island, New Zealand. *New Zealand Journal of Geology and Geophysics* 33, 111-124.

Gellatly, A. F., Chinn, T. J. H., and Röthlisberger, F. (1988). Holocene glacier variations in New Zealand: A review. *Quaternary Science Reviews* 7, 227-242.

Green, J. D., and Lowe, D. J. (1985). Stratigraphy and development of c. 17,000 yr old Lake Maratoto, North Island, New Zealand, with some inferences about postglacial climatic change. *New Zealand Journal of Geology and Geophysics* 28, 675-699.

Griffiths, G. A., and McSaveney, M. J. (1983). Distribution of mean annual precipitation across some steepland regions of New Zealand. *New Zealand Journal of Science* 26, 197-209.

Harris, W. F. (1963). Paleo-ecological evidence from pollen and spores. *Proceedings of the New Zealand Ecological Society* 10, 38-44.

Hays, J. D., Imbrie, J., and Shackleton, N. J. (1976). Variations in the earth's orbit: Pacemaker of the ice ages. *Science* 194, 1121-1132.

Hendy, C. H., and Wilson, A. T. (1968). Palaeoclimatic data from speleothems. *Nature* 219, 48-51.

Kennedy, N. M., Pullar, W. A., and Pain, C. F. (1978). Late Quaternary land surfaces and geomorphic changes in Rotorua Basin, North Island, New Zealand. *New Zealand Journal of Science* 21, 249-264.

Lewis, K. B., and Mildenhall, D. C. (1985). The late Quaternary seismic sedimentary and palynological stratigraphy beneath Evans Bay, Wellington Harbour. *New Zealand Journal of Geology and Geophysics* 28, 129-152.

McCraw, J. D. (1975). Quaternary airfall deposits in New Zealand. *In* "Quaternary Studies" (R. P. Suggate and M. M. Cresswell, Eds.), pp. 35-44. Bulletin 13. Royal Society of New Zealand, Wellington.

McGlone, M. S. (1973). Pollen analysis. Appendix to "Relict periglacial landforms at Clarks Junction, Otago" by D. M. Leslie. *New Zealand Journal of Geology and Geophysics* 16, 575-578.

——. (1982). Modern pollen rain, Egmont National Park, New Zealand. *New Zealand Journal of Botany* 20, 253-262.

——. (1983). Polynesian deforestation of New Zealand: A preliminary synthesis. *Archaeology in Oceania* 18, 11-25.

——. (1985). Plant biogeography and the late Cenozoic history of New Zealand. *New Zealand Journal of Botany* 23, 723-749.

——. (1988). New Zealand. *In* "Vegetation History" (B. Huntley and T. Webb III, Eds.), pp. 557-599. Kluwer Academic Publishers, The Hague.

——. (1989). The Polynesian settlement of New Zealand in relation to environmental and biotic changes. *New Zealand Journal of Ecology* 12 (Supplement), 115-129.

McGlone, M. S., and Bathgate, J. L. (1983). Vegetation and climate history of the Longwood Range, South Island, New Zealand, 12,000 yr B.P. to the present. *New Zealand Journal of Botany* 21, 293-315.

McGlone, M. S., and Moar, N. T. (1977). The *Ascarina* decline and post-glacial climatic change in New Zealand. *New Zealand Journal of Botany* 15, 485-489.

McGlone, M. S., and Topping, W. W. (1977). Aranuian (post-glacial) pollen diagrams from the Tongariro region, North Island, New Zealand. *New Zealand Journal of Botany* 15, 749-760.

——. (1983). Late Quaternary vegetation, Tongariro region, central North Island, New Zealand. *New Zealand Journal of Botany* 21, 53-76.

McGlone, M. S., Nelson, C. S., and Todd, A. J. (1984). Vegetation history and environmental significance of pre-peat and surficial peat deposits at Ohinewai, Lower Waikato lowland. *Journal of the Royal Society of New Zealand* 14, 233-244.

McIntyre, D. J., and McKellar, I. C. (1970). A radiocarbon dated post glacial pollen profile from Swampy Hill, Dunedin, New Zealand. *New Zealand Journal of Geology and Geophysics* 13, 346-349.

Macphail, M. K. (1980). Fossil and modern *Beilschmiedia* (Lauraceae) pollen in New Zealand. *New Zealand Journal of Botany* 18, 453-457.

Macphail, M. K., and McQueen, D. R. (1983). The value of New Zealand pollen and spores as indicators of Cenozoic vegetation and climates. *Tuatara* 26, 37-59.

Maunder, W. J. (1971). The climate of New Zealand—Physical and dynamical features. *In* "World Survey of Climatology" (J. Gentilli, Ed.), Vol. 13, pp. 213-227. Elsevier, Amsterdam.

Mildenhall, D. C., and Moore, P. R. (1983). A late Holocene sequence at Turakirae Head, and climatic and vegetational change in the Wellington area in the last 10,000 years. *New Zealand Journal of Science* 26, 447-459.

Moar, N. T. (1970). Recent pollen spectra from three localities in the South Island, New Zealand. *New Zealand Journal of Botany* 8, 210-211.

——. (1971). Contributions to the Quaternary history of the New Zealand flora. 6. Aranuian pollen diagrams from Canterbury, Nelson and north Westland, South Island. *New Zealand Journal of Botany* 9, 80-145.

——. (1973). Late Pleistocene vegetation and environment in southern New Zealand. *In* "Palaeoecology of Africa and the Surrounding Islands and Antarctica" (E. M. van Zinderen Bakker, Ed.), Vol. 3, pp. 179-198. Balkema, Cape Town.

——. (1980). Late Otiran and early Aranuian grassland in central South Island. *New Zealand Journal of Ecology* 3, 4-12.

Newnham, R. M., Lowe, D. J., and Green, J. D. (1989). Palynology, vegetation and climate of the Waikato lowlands, North Island, New Zealand, since c. 18,000 years ago. *Journal of the Royal Society of New Zealand* 19, 127-150.

Nicholls, J. L. (1983). The extent and variability of the native lowland forests. *In* "Lowland Forests in New Zealand" (K. Thompson, A. P. H. Hodder, and A. S. Edmonds, Eds.), pp. 79-92. University of Waikato, Hamilton.

Norton, D. A., McGlone, M. S., and Wigley, T. M. L. (1986). Quantitative analyses of modern pollen-climate relationships in New Zealand indigenous forests. *New Zealand Journal of Botany* 24, 331-342.

Porter, S. C. (1975). Equilibrium line of late Quaternary glaciers in Southern Alps, New Zealand. *Quaternary Research* 5, 27-48.

Rind, D., and Peteet, D. (1985). Terrestrial conditions at the last glacial maximum and CLIMAP sea-surface temperature estimates: Are they consistent? *Quaternary Research* 24, 1-22.

Rogers, G. M., and McGlone, M. S. (1989). A postglacial vegetation history of the southern-central uplands of North Island, New Zealand. *Journal of the Royal Society of New Zealand* 19, 229-248.

Salinger, M. J. (1980a). New Zealand climate: I. Precipitation patterns. *Monthly Weather Review* 108, 1892-1904.

——. (1980b). New Zealand climate: II. Temperature patterns. *Monthly Weather Review* 108, 1905-1912.

Soons, J. M. (1979). Late Quaternary environments in the central South Island of New Zealand. *New Zealand Geographer* 35, 16-23.

Soons, J. M., and Burrows, C. J. (1978). Dates for Otiran deposits, including plant microfossils and macrofossils, from Rakaia Valley. *New Zealand Journal of Geology and Geophysics* 21, 607-615.

Stewart, R. B., and Neall, V. E. (1984). Chronology of palaeoclimatic change at the end of the last glaciation. *Nature* 311, 47-48.

Suggate, R. P. (1990). Late Pliocene and Quaternary glaciations of New Zealand. *Quaternary Science Reviews* 9, 175-197.

Suggate, R. P., and Moar, N. T. (1970). Revision of the chronology of the late Otira glacial. *New Zealand Journal of Geology and Geophysics* 13, 742-746.

Thiede, J. (1979). Wind regimes over the late Quaternary southwest Pacific Ocean. *Geology* 7, 259-262.

Wardle, J. (1984). "The New Zealand Beeches." New Zealand Forest Service, Christchurch.

CHAPTER 13

Holocene Vegetation, Lake Levels, and Climate of Africa

F. Alayne Street-Perrott and R. A. Perrott

Africa, the world's second largest landmass, sits astride the equator. It spans a range of latitudes from 37°N near Tunis to 34.5°S at Cape Agulhas and is topographically more uniform than any other continent. Apart from restricted highland areas in the Atlas Mountains (highest peak 4165 m above sea level), the central Sahara (Tibesti Mountains, 3415 m), West Africa (Mount Cameroon, 4095 m), Ethiopia (Ras Dashan, 4543 m), East Africa (Mount Kilimanjaro, 5895 m), and the eastern escarpment of southern Africa (Thabana Ntlenyana, 3484 m), the vast majority of the continent lies below 2000 m. Because of the restricted extent of continental shelf, the continent's area has fluctuated little in response to Quaternary sea-level fluctuations (CLIMAP Project Members, 1981).

As a result of this physiographic simplicity, the distribution of modern climates is more or less symmetric about the equator (Thompson, 1965; Leroux, 1983) (Fig. 13.1). The northern and southern extremities of the continent, which project into the belts of midlatitude westerlies, experience Mediterranean summer-dry climates, receiving most of their precipitation in winter from westerly cyclonic disturbances. These temperate areas are bordered on the equatorward side by subtropical deserts: the very extensive and extremely arid Sahara Desert north of the equator, and the much smaller Namib coastal desert in southwestern Africa. A wide belt of tropical climates separates the two arid zones. The meteorological equator—the confluence of the surface airflows from the two hemispheres—migrates northward to about 15-24°N in June-August and southward to about 8°N (West Africa) to 16°S (eastern Africa) in December-February (Leroux, 1983), giving rise to an equatorial zone of humid climates with a double rainfall maximum, flanked on the north and south by broad belts of

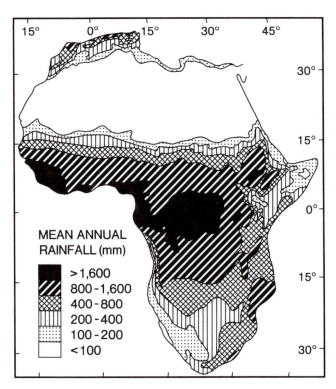

Fig. 13.1. Rainfall map of Africa. Redrawn from Nicholson (1986b); used with permission from *Journal of Climate and Applied Meteorology* and the American Meteorological Society.

monsoonal climates characterized by summer rains and winter drought (Fig. 13.2).

In mountainous areas this broadly zonal pattern of climates is greatly modified by altitude and aspect. In East Africa, for example, a marked maximum of cloud cover and rainfall is usually observed on the windward slopes of the mountains at about 2000–

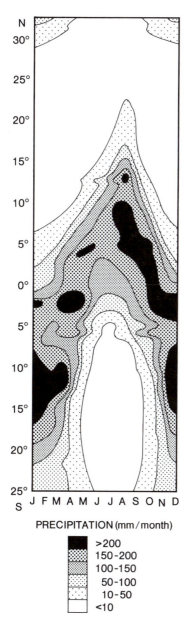

Fig. 13.2. Seasonal distribution of African rainfall as a function of latitude. Redrawn from Nicholson and Chervin (1983).

3000 m, with a drier zone above (Hastenrath, 1984). The average lapse rate of air temperatures recorded at surface stations ranges from 5 to 6.75°C/10³ m (Street, 1979; Lancaster, 1980; Hamilton, 1982; Maley and Livingstone, 1983), although the local temperature distribution is greatly affected by radiation and cloudiness.

Much of the rainfall in the tropics originates from organized disturbances such as easterly waves that are steered by the easterly winds aloft. A well-developed tropical easterly jet is present at 200–100 mb at 5–15°N in June-August; a weaker easterly wind maximum is found at 5–10°S in December-February (Leroux, 1983; Hastenrath, 1984). These jets are simulated by general circulation models such as the Communi-

ty Climate Model (CCM) of the National Center for Atmospheric Research (NCAR). In both northern and southern intertropical Africa, reduced upper-tropospheric easterly flow appears to be associated with drought (Newell and Kidson, 1984; Nicholson, 1986b).

The causes and spatial patterns of short-term climatic variations in Africa are discussed in Newell and Kidson (1984), Lamb (1985), Tyson (1986), and Nicholson (1986a,b). The distribution of rainfall over the continent has been found to be very responsive not only to shifts in the large-scale wind fields at different levels in the troposphere but also to sea-surface temperature variations (Folland et al., 1986; Palmer, 1986), atmospheric aerosols (Coakley and Cess, 1985), and changes in surface boundary conditions such as albedo, soil moisture, and surface roughness (Mintz, 1984).

The principal controls on the distribution of vegetation in Africa are total annual rainfall and the timing, duration, and intensity of the dry season(s). The vegetation classification used here (White, 1983), which is partly floristic and partly physiognomic, bears a very close relationship to large-scale climate (Fig. 13.3). The primary mapping units (phytochoria) are arranged in a broadly zonal pattern, except for the mountains of tropical and southern Africa, where extra altitudinal categories are introduced.

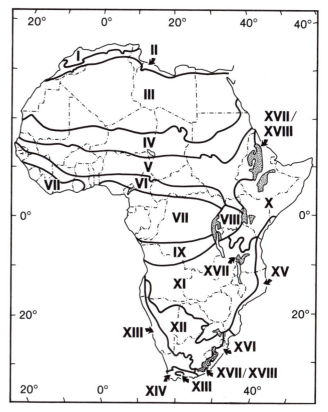

Fig. 13.3. Major floristic regions of Africa. Redrawn from White (1983) (regions renumbered); used with permission of Kluwer Academic Publishers.

The major floristic regions of Africa, with principal vegetation types and examples of characteristic pollen taxa, are as follows (after the UNESCO/AETFAT/UNSO 1:5,000,000 vegetation map of Africa [White, 1983], with minor modifications):

(I) Mediterranean: broad-leaved sclerophyllous forest with evergreen *Quercus* (oak); coniferous forest with *Pinus* (pine), *Cedrus* (cedar), etc.; deciduous *Quercus* forest.

(II) Pre-Saharan: sclerophyllous forest with *Juniperus* (juniper), *Olea* (olive), *Pistacia* (pistachio), and others; scrub forest, bushland, and shrubland; tussock grasslands (steppes) with *Artemisia* (sagebrush), *Lygeum spartum*, and abundant dryland taxa.

(III) Saharan: desert (very restricted pollen flora); certain plants characteristic of wadis (e.g., *Tamarix* [tamarisk]) and of sandy, rocky, gravelly, or saline surfaces (e.g., *Cornulaca, Calligonum, Fagonia*); Saharomontane vegetation (Ahaggar, Tibesti) includes Mediterranean/pre-Saharan elements such as *Olea laperrini, Artemisia, Ephedra*, and *Erica arborea* (giant heather).

(IV) Sahelian: semidesert grassland and thorny shrubland (north) to wooded grassland and bushland (south), with *Acacia* spp., *Commiphora africana, Balanites aegyptiaca*, Euphorbiaceae, and abundant dryland taxa.

(V) Sudanian: woodland and dry forest, with *Celtis integrifolia, Hymenocardia acida, Lannea, Mitragyna inermis, Prosopis africana*, etc.

(VI) Sudano-Guinean: mosaic of dry, peripheral, semievergreen rainforest and woodland or secondary grassland, transitional between regions V and VII.

(VII) Guineo-Congolian: lowland rainforest and swamp forest with very diverse endemic flora including *Chlorophora, Holoptelea, Uapaca, Musanga*, and *Elaeis guineensis* (oil palm); montane rainforest and grassland (above 1000 m in elevation) with *Olea hochstetteri, Podocarpus*, and *Ilex.*

(VIII) Lake Victoria regional mosaic: similar to region VI; dry, peripheral, semievergreen rainforest and scrub forest with *Celtis* spp., *Holoptelea*, and *Rhus natalensis*; wooded grassland with *Acacia* and palms (*Borassus*).

(IX) Zambezo-Congolian: similar to region VI.

(X) Somali-Masai: deciduous bushland and thicket similar to region IV, grading upward into region XVIII through semievergreen to evergreen bushland and thicket with *Cordia, Croton*, etc.

(XI) Zambezian: dry forest and woodland (bushveld) with *Brachystegia* (miombo), *Burkea africana*, Combretaceae, and Proteaceae (*Protea + Faurea*).

(XII) Kalahari-Highveld: wooded grassland and bushland (Kalahari thornveld) with *Acacia* spp., Capparidaceae, Tarchonantheae (Compositae), and *Aloe*; upland grassland (highveld) with abundant Gramineae (grasses), Cyperaceae (sedges), Compositae (daisy family), and Ericaceae (heaths).

(XIII) Karoo-Namib: semidesert shrubland and desert; few relevant pollen data.

(XIV) Cape: sclerophyllous shrubland (*fynbos*, macchia) with Asteraceae, Restionaceae, Proteaceae, Ericaceae, *Artemisia, Cliffortia, Myrica, Stoebe*, etc.

(XV) Zanzibar-Inhambane and (XVI) Tongaland-Pongaland: not discussed.

(XVII) Afromontane: dry montane forest with *Podocarpus, Juniperus*, and *Olea*; wet montane forest with *Afrocrania, Macaranga, Neoboutonia, Prunus, Ilex*, etc.; montane veld with *Celtis* and *Rhus* (southern Africa).

(XVIII) Afroalpine and Austroafroalpine: ericaceous bushland and shrubland with *Cliffortia, Erica arborea*, and other giant heathers (lower part, eastern Africa); high-altitude shrubland and grassland, similar to paramo in South America, with *Dendrosenecio* (giant groundsel), *Lobelia, Artemisia afra, Alchemilla*, and *Helichrysum* (everlasting flowers) (upper part, eastern Africa); alpine veld with *fynbos* elements (Ericaceae, *Passerina, Cliffortia*, etc.) (southern Africa).

The extensive literature on late-Quaternary climatic changes in Africa has been summarized by van Zinderen Bakker (1967, 1976, 1978, 1983), Faure (1969), Grove and Goudie (1971), Butzer *et al.* (1972), Hamilton (1974, 1982), Livingstone (1975), Rognon (1976, 1981, 1987), Street and Grove (1976, 1979), Rognon and Williams (1977), Sarnthein (1978), Nicholson and Flohn (1980), Street (1981), Street-Perrott and Roberts (1983), and Deacon and Lancaster (1988). Useful collections of papers can be found in Williams and Faure (1980) and Vogel (1984), as well as in recent volumes of *Palaeoecology of Africa*. For historical, political, or environmental reasons, most of the available paleoclimatic data come from sites in East Africa, South Africa, or north of the equator.

In this chapter we review the pollen and lake-level evidence relating to 9000 and 6000 yr B.P. Initially, we discuss each data set independently in order to examine its strengths and weaknesses. Although both types of data strongly reflect hydrologic changes, they are largely complementary. The paleoclimatic reconstructions derived from the two types of evidence are then compared and finally merged to facilitate comparisons with the NCAR CCM simulations for 9000 and 6000 yr B.P.

Pollen Data

A total of 32 sites, including four Atlantic marine cores obtained close to the African continent, yielded usable pollen data for 9000 or 6000 yr B.P. or both (Table 13.1). Since this data set was compiled, several important new diagrams have been presented (Lézine, 1987; Ritchie and Haynes, 1987; Scott, 1987b; Bonnefille and Riollet, 1988; Vincens, 1989a, b; Maley *et al.*, 1990; Taylor, 1990; Maitima, 1991; Meadows and Sugden, 1991; Scott *et al.*, 1991; Elmoutake *et al.*, 1992; Hooghiemstra

Table 13.1. African pollen data set for 9000 and 6000 yr B.P.

Site no.	Site name	Latitude	Longitude	Altitude (m)	¹⁴C dates[a] (no.)	Dating control[b] 9000 yr B.P.	Dating control[b] 6000 yr B.P.	Reference
1	Oumm el-Khaled, Algeria	35° 05'N	7° 36'E	~500	2	ND	7	Ritchie (1984)
2	Meteor core 80-17B, Atlantic	33° 37'N	9° 24'W	-3016	5	1	ND	Agwu (1979), Agwu and Beug (1982), H. Hooghiemstra (personal communication)
3	Tigalmamine, Morocco	32° 55'N	5° 21'W	1626	9	6	3	Lamb et al. (1989)
4	Ait Blal, Morocco	32° 00'N	5° 36'W	~1100	1	7	4	Ballouche et al. (1986)
5	Taoudenni, Mali	22° 30'N	4° 00'W	120	1	ND	7	Cour and Duzer (1976)
6	Mouskorbe, Tibesti	21° 22'N	18° 32'E	~2600	3	2	3	Maley (1981, 1983)
7	Enneri Tabi 4, Chad	21° 20'N	17° 03'E	~1100	2	1	ND	Schulz (1980b)
8	Meteor core 16017-2, Atlantic	21° 15'N	17° 48'W	-800	8	2	2	Hooghiemstra (1988b)
9	Oyo, Sudan	19° 16'N	26° 11'E	510	4	7	1	Ritchie et al. (1985)
10	Thiaye, Senegal	15° 55'N	17° 04'W	~0	4	ND	7	Lézine et al. (1985)
11	Meteor core 12345-5, Atlantic	15° 29'N	17° 22'W	-945	0	7	7	Rossignol-Strick and Duzer (1979a,b)
12	Tanma, Senegal	14° 54'N	17° 04'W	~0	8	2	5	Médus (1984a,b)
13	Tjeri, Chad	13° 44'N	16° 30'E	~300	2	1	4	Maley (1981, 1983)
14	Badda, Ethiopia	7° 52'N	39° 22'E	4000	5	2	2	Hamilton (1982), Bonnefille and Hamilton (1986)
15	Abiyata, Ethiopia	7° 42'N	38° 36'E	1585	6	2	7	Lézine (1981), Lézine and Bonnefille (1982)
16	Danka, Ethiopia	6° 58'N	38° 47'E	3830	1	ND	5	Hamilton (1982), Bonnefille and Hamilton (1986)
17	Bosumtwi, Ghana	6° 30'N	1° 25'W	100	9	2	2	Maley and Livingstone (1983), Talbot et al. (1984)
18	Niger Delta, Nigeria	4° 33'N	6° 26'E	~0	5	ND	4	Sowunmi (1981a,b)
19	Koitoboss, Kenya	1° 08'N	34° 36'E	3940	1	ND	4	Hamilton (1982)
20	Kimilili, Kenya	1° 06'N	34° 34'E	4150	9	1	1	Hamilton (1982)
24	Mahoma, Uganda	0° 21'N	29° 58'E	2960	3	4	6	Livingstone (1967)
26	Victoria, Uganda	0° 18'N	33° 20'E	1134	28	1	1	Kendall (1969), Laseski (1983)
28	Sacred Lake, Kenya	0° 02'S	37° 28'E	2400	4	4	3	Coetzee (1964, 1967), van Zinderen Bakker and Coetzee (1972)
29	Rutundu, Kenya	0° 02'S	37° 22'E	3140	4	1	1	Coetzee (1964, 1967), van Zinderen Bakker and Coetzee (1972)
30	Satima, Kenya	0° 18'S	36° 35'E	3670	21	1	2	Perrott (1982, 1987, unpublished), Perrott and Street-Perrott (1987)
31	Karimu, Kenya	0° 30'S	36° 41'E	3040	4	6	5	Perrott and Street-Perrott (1982); R. A. Perrott (unpublished)
32	Muchoya, Uganda	1° 16'S	29° 48'E	2256	3	2	4	Morrison (1968)
34	Ishiba Ngandu, Zambia	11° 12'S	31° 50'E	1400	1	7	7	Livingstone (1971)
35	Inyanga, Zimbabwe	18° 17'S	32° 45'E	~2000	1	6	6	Tomlinson (1974)
36	Wonderkrater, South Africa	24° 24'S	28° 46'E	1100	5	1	3	Scott (1982b)
37	Groenvlei, South Africa	33° 48'S	22° 52'E	~0	2	ND	3	Martin (1968)
38	Hangklip, South Africa	34° 20'S	18° 54'E	45	3	ND	3	Schalke (1973)

[a]Number of ¹⁴C dates obtained from the profile(s) containing the pollen spectra for 9000 and 6000 yr B.P.
[b]The dating control is coded as follows: 1, bracketing dates within 2000 yr of selected level; 2, one bracketing date within 2000 and one within 4000 yr; 3, bracketing dates within 4000 yr; 4, one bracketing date within 4000 and one within 6000 yr; 5, bracketing dates within 6000 yr; 6, one bracketing date within 6000 and one within 8000 yr or one within 4000 and one within 10,000 yr; 7, no dates in core, no bracketing dates, or bracketing dates more than 14,000 yr apart; ND, no data. Note that the top of the profile, if not disturbed, can be counted as one bracketing date with an assumed age of 0 yr B.P. Additional sources of uncertainty (e.g., inverted dates) are acknowledged by downgrading the ranking.

et al., 1992; and volumes 18-22 of *Palaeoecology of Africa*). These data, however, do not affect our main conclusions.

Where possible, our analysis is based on the raw counts extracted from published and unpublished sources. Following the practice established by French and German palynologists, we used the vegetation classification summarized above to group the pollen taxa according to their most probable region of origin. This seemed the most practical method in view of the huge number of taxa involved (227 in the Sahara-Sahel-Sudan zone alone). In northern and western Africa, an additional "dryland" group was created to include families with wide distributions that are nevertheless indicative of dry, open habitats, namely, Chenopodiaceae-Amaranthaceae, Compositae, Cruciferae, Gramineae, and Plumbaginaceae (compare Agwu [1979]). Ubiquitous taxa (including Cyperaceae), plants of localized moist or aquatic habitats, coastal vegetation (including mangrove), and exotic taxa were excluded from the pollen sum to emphasize changes in regional vegetation.

Because the ecology of individual pollen taxa often differs in different parts of the continent, we compiled separate plant lists for northern and western Africa, eastern Africa, and southern Africa. When different authorities conflicted, we assigned taxa to floristic regions in the way that yielded the most conservative climatic interpretation of past vegetation changes.

NORTHERN AND WESTERN AFRICA

Pollen taxa in northern and western Africa were classified based mainly on Schulz (1980b, 1984, 1986b), Agwu (1979), Agwu and Beug (1982), and Sowunmi (1981b), supplemented by White (1983). The classification differs slightly but not significantly from the groupings proposed more recently by Hooghiemstra *et al.* (1986) and Lézine (1987). For convenience the pollen diagrams from North Africa (floristic regions I and II), the Sahara-Sahel-Sudan region (III-V), and West Africa (VI and VII) are discussed separately, because the interpretation of pollen diagrams presents rather different problems in each of these three areas. Marine cores are included when sufficiently well dated in order to supplement the patchy continental record.

North Africa

Most of North Africa is characterized by a winter precipitation maximum, which is most pronounced on the north coast and in the Atlas Mountains. Frost and snow are common in winter at higher elevations.

Although French and North African palynologists have conducted a considerable number of studies in Morocco and Tunisia (Reille, 1976, 1977; Brun, 1979, 1983, 1985; Ben Tiba and Reille, 1982), the scarcity of ^{14}C dates and the large variations in sedimentation rates make it impossible to identify the 9000- and 6000-yr B.P. levels in their cores with any certainty. These stud-

ies demonstrate, however, that the late-Quaternary pollen record from North Africa consists of alternating phases dominated by either Mediterranean forest or pre-Saharan steppe and dryland taxa.

The best-dated diagram currently available comes from Tigalmamine (Fig. 13.4, site 3), a karstic depression at an elevation of 1626 m in the Middle Atlas range (Lamb *et al.,* 1989). Today the lake is surrounded by subhumid evergreen *Quercus* forest with *Cedrus* and occasional deciduous *Quercus*. Between about 18,000 and 8500 yr B.P. the pollen assemblages were dominated by Chenopodiaceae, *Artemisia*, and Gramineae, with small traces of evergreen *Quercus*, especially between 14,000 and 11,800 yr B.P. They suggest an open, steppelike environment, with a climate too dry or too cold or both to support more than scattered stands of trees. Around 8500 yr B.P. both types of *Quercus* increased sharply and herbaceous taxa decreased, indicating a shift to a much more humid climate with significant summer rainfall. From 4000 yr B.P. onward, deciduous *Quercus* declined and *Cedrus* became more abundant, suggesting that a drier and cooler regime had set in.

A similar change just before 8260 ± 180 yr B.P. is evident at site 4 (Fig. 13.5). Here zone A represents a local pre-Saharan steppe rich in Chenopodiaceae and Cruciferae, with *Cedrus* forest on the Atlas Mountains. In zone B *Pinus* and evergreen *Quercus* increase in the regional pollen rain, while greater percentages of *Artemisia*, *Typha*, and mosses indicate more abundant local moisture. A decrease in *Pinus* and *Cedrus* in zone C, accompanied by a rise in *Olea* and Gramineae, may indicate drying (Ballouche *et al.,* 1986).

Several other cores also show an important change between 5000 and 4000 yr B.P. that can be attributed to increasing summer drought. At site 1 arboreal taxa (*Pinus, Quercus,* and Cupressaceae) decreased around 4100 ± 80 yr B.P., with an increase in Gramineae, Compositae, *Artemisia* and, slightly later, *Olea*. In the Kroumirie Mountains of northwest Tunisia, evergreen *Quercus* and *Erica arborea* increased at the expense of deciduous *Quercus* around 4630 ± 580 yr B.P. (Ben Tiba and Reille, 1982). This major change, which can be interpreted as the onset of the modern Mediterranean summer-dry regime, is also evident in pollen records from southern Europe (Figs. 6.40-6.42 in Huntley and Birks [1983]).

Sahara-Sahel-Sudan Region

A strong climatic gradient is evident in the Sahara-Sahel-Sudan region, from subtropical desert with low, irregular rainfall in the north to a tropical climate with markedly dry winters in the south. The Sahara Desert presents greater difficulties for palynology than almost any other region of comparable size. Accessibility is a major problem. Pollen-bearing deposits are rare and are difficult to date precisely because of a scarcity of organic matter. Many sequences have been truncat-

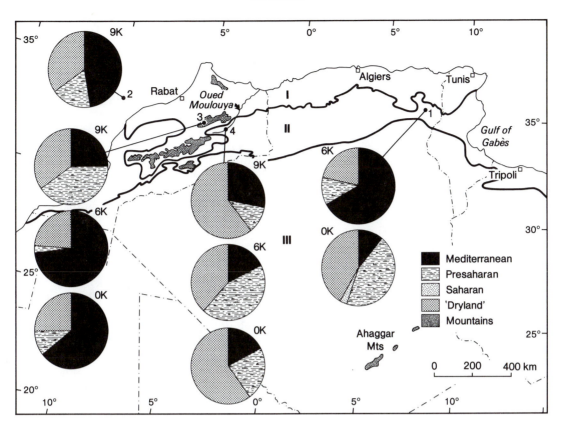

Fig. 13.4. Pollen spectra for 9000, 6000, and 0 yr B.P. from North Africa.

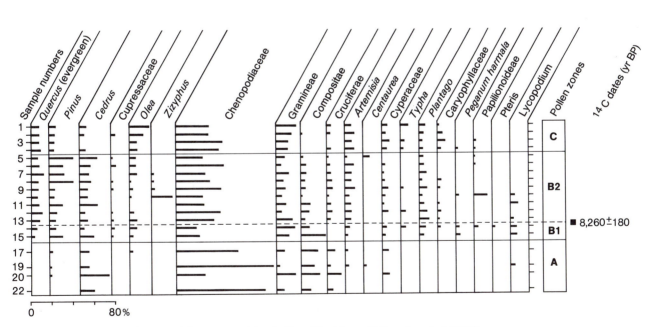

Fig. 13.5. Pollen diagram for Ait Blal, Morocco (site 4). Simplified from Ballouche *et al.* (1986).

ed by deflation or oxidation. Furthermore, pollen may be transported over immense distances by the wind, as shown by the occurrence of European forest taxa in both modern (Ritchie, 1986) and fossil spectra.

In view of the scarcity of polliniferous late-Holocene deposits in arid and semiarid areas, in situ modern samples that can be compared directly with early- and mid-Holocene spectra are extremely rare. In an attempt to overcome this problem, researchers have made extensive use of pollen counts on dust collected by filters mounted on vehicles or ships (Cour and Duzer, 1976; Caratini and Cour, 1980; Schulz, 1980a,b, 1984, 1986b; Melia, 1984). Hooghiemstra and co-workers also did a detailed survey of marine core-top samples (Hooghiemstra and Agwu, 1986; Hooghiemstra et al., 1986). All these workers found that the pollen contained in modern dust varies systematically with latitude. However, as noted by Schulz (1980a, p. 792), Mediterranean elements are continuously present south to 22°N, and tropical and Sahelian elements are similarly present north to 27°N. The values for Gramineae are highest in areas free of vegetation.

To make matters worse, modern filter samples are often not qualitatively or even quantitatively very different from early- or mid-Holocene spectra (Cour and Duzer, 1976; Schulz, 1980a,b; Petit-Maire and Schulz, 1981; Baumhauer and Schulz, 1984). These findings have led some palynologists to conclude that regional pollen deposition in the desert is swamped by long-distance eolian transport and that the regional vegetation has not changed significantly during the Holocene. For example, in the hyperarid Sebkha de Taoudenni in Mali (site 5) modern dust yields spectra almost identical to those from mid-Holocene muds containing *Acacia* wood and *Hippopotamus* bones (Cour and Duzer, 1976), a patently absurd situation. In contrast, palynologists working on continuous deep-sea records have detected significant vegetational shifts during the Holocene (Agwu, 1979; Caratini et al., 1979; Rossignol-Strick and Duzer, 1979a,b; Agwu and Beug, 1982; Hooghiemstra, 1988a,b).

This divergence of views may be explained by the fact that the "modern" desert dust contains abundant freshwater diatoms and phytoliths, which suggests that at least part of it originates from the deflation of Holocene or older lake and swamp deposits (Pokras and Mix, 1985, 1987). This hypothesis could be tested by ultraviolet-fluorescence analysis (compare Maley [1981], p. 374) or better yet by accelerator dating of the pollen in a large sample of airborne dust. In view of our skepticism about the origin of filter samples, we have not used them to represent the modern vegetation.

The pollen diagram from Oyo, in hyperarid northern Sudan (site 9, Fig. 13.6), provides convincing evidence for Holocene vegetational change in the Sahara. During the period 8500–6100 yr B.P. the lake was surrounded by a deciduous wooded grassland with a strong Sudanian element, similar to that found today

about 500 km to the south. Many of these tropical taxa produce little pollen or have large pollen grains with low dispersal capability (Ritchie et al., 1985). Between 6100 and 4500 yr B.P. the wooded grassland was replaced by Sahelian thorn scrub and semidesert grassland. Finally, around 4500 yr B.P., the lake dried out and desert conditions set in.

Ritchie et al. (1985) attributed the early-Holocene wet phase recorded at Oyo to a northward displacement of the African summer-monsoon isohyets. A longer monsoon record is provided by two deep-sea cores from sites 8 and 11, which show that arid conditions prevailed on the southern margin of the Sahara during and after the last glacial maximum. According to Hooghiemstra (1988b), the northward penetration of the summer rains reached a minimum in the interval 16,700–14,300 yr B.P. (site 8). The resurgence of the monsoon, marked by a strong increase in pollen from tropical forest and woodland and coastal mangrove, is dated at 14,300 yr B.P. (site 8) and 12,500 yr B.P. (site 11). Maximum percentages of Sudano-Guinean and Sudanian types were attained before 8800 and 7000 yr B.P., respectively. At site 11, where the late-Holocene record is more nearly complete, Sahelian taxa peaked around 5500 yr B.P. and then decreased slightly. The subsequent increase in Saharan and Mediterranean/pre-Saharan elements suggests a recession of the summer rains after 5500 yr B.P.

The pollen spectra corresponding to 9000, 6000, and 0 yr B.P. are mapped in Figure 13.7. Because the high percentages of dryland taxa (66–96%) obscure the changes in the climatically more diagnostic groups, the latter are shown separately in Figure 13.8. The following summary is based mainly on the spectra in Figure 13.8, in which the spatial and temporal trends are more obvious.

In all three time slices the pollen spectra exhibit large-scale spatial patterns. For example, the Mediterranean/pre-Saharan group decreases in abundance with distance from its source areas in North Africa and the Saharan mountains. Similarly, Sudano-Guinean and Guinean elements decline rapidly northward from the southern boundary of the region.

The percentages of Sudanian, Sudano-Guinean, and Guinean pollen decreased significantly from 9000 to 6000 to 0 yr B.P. at sites 8 and 11, suggesting that tropical summer rains were most abundant during the early Holocene. The same trend has been identified in shorter sequences from peat deposits along the coast of Senegal (sites 10 and 12, Fig. 13.7) (Médus, 1984a,b; see also newer data in Lézine, 1987). Relict gallery forests of Sudano-Guinean and Guinean affinities can still be found in moist interdune depressions (*niayes*) (Lézine, 1986). Unfortunately, however, the overwhelming dominance of coastal mangrove pollen in some of the Senegalese diagrams makes it hard to reconstruct the changes in terrestrial vegetation in any

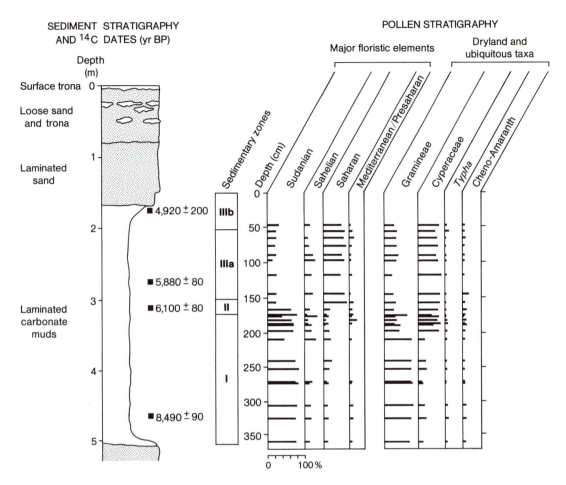

Fig. 13.6. Summary pollen diagram for Oyo, Sudan (site 9). The major floristic elements and the dryland and ubiquitous taxa are expressed as percentages of their respective pollen sums. Modified from Ritchie *et al.* (1985); used with permission from *Nature* (314, 352-355); copyright 1985 Macmillan Magazines Limited.

detail (Lézine, 1985). Hence, sites 10 and 12 are omitted from Figure 13.8.

The Tibesti Mountains (central Sahara) are represented by only a few well-dated fossil spectra (sites 6 and 7). Mediterranean/pre-Saharan elements and fern spores are abundant in high-altitude samples (site 6). According to Schulz (1980a), the early-Holocene vegetation of the mountains consisted of a dry, open woodland of Mediterranean character. Lower down (e.g., site 7), gallery forest consisting of *Acacia, Tamarix,* and other woody species of Sahelian affinity, with a dense understory of grasses and chenopods, extended much farther out into the piedmont than today. A similar altitudinal zonation probably existed in other Saharan massifs, but with the abundance of Mediterranean/pre-Saharan elements decreasing toward the south (Schulz, 1986a). The two spectra from site 6 are compatible with the trend toward drier conditions at 6000 yr B.P. observed in the eastern Sahara (site 9, Fig. 13.6).

One site stands out from the general temporal pattern described above: Tjeri (site 13) in the Lake Chad basin. Here the pollen assemblage for 9000 yr B.P. is dominated by Sahelian elements and *Tamarix,* which implies that the climate was still relatively dry at that time. Sudano-Guinean taxa rise to a peak between about 8000 and 7000 yr B.P., after which Sudanian elements increase (Maley, 1983). This site is unusual, however, in that today a significant amount of pollen is transported into Lake Chad by large rivers (the Chari and the Logone) that rise in zone VI, far to the south (Maley, 1981, p. 45). Hence it is not surprising that the Holocene pollen sequence shows considerable similarity to diagrams from the forested regions of West Africa (see below).

West Africa

In West Africa, the wettest area discussed so far, mean annual rainfall increases from the interior toward the Gulf of Guinea. Totals in excess of 3000 mm are received in two coastal areas within the disjunct Guinean and Congolian rainforest blocks, situated

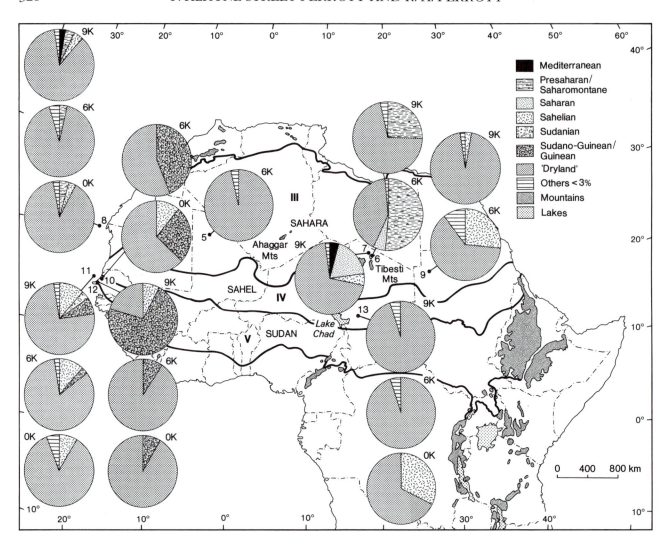

Fig. 13.7. Pollen spectra for 9000, 6000, and 0 yr B.P. from the Sahara-Sahel-Sudan region, including dryland taxa.

west and east, respectively, of the relatively dry Da-homey Gap (Fig. 13.9). Very wet climates also occur on the flanks of the mountains, especially Mount Ca-meroon. A double (equatorial) rainfall regime charac-terizes the coastal zone between Ivory Coast (e.g. Abidjan) and the Niger Delta.

Summary pollen diagrams covering the Holocene are available for two sites in West Africa (sites 17 and 18, Fig. 13.9). The main trends recorded are confirmed by discontinuous spectra from marine cores recov-ered off Abidjan (Assémien *et al.,* 1971) and Pointe Noire (Caratini and Giresse, 1979). So far, however, no data at all are available from the interior of the Con-golian forest block, a very significant gap in view of the biogeographic debate over the extent of lowland forest during the last glacial stage (Hamilton, 1976).

Lake Bosumtwi (site 17), 65 km south of the forest-savanna (VII-VI) boundary, is bordered today by relict patches of dry, semievergreen rainforest (Fig.

13.9) (Hall *et al.,* 1978). During and after the last glacial maximum the pollen assemblages were dominated by Gramineae and Cyperaceae, indicating the absence of a continuous forest cover (Fig. 13.10) (Talbot *et al.,* 1984). After about 8500 yr B.P. arboreal taxa and ferns became dominant. A rise in the abundance of *Elaeis guineensis* from 3500-3000 yr B.P. onward probably re-flects the spread of agriculture.

The pollen spectra for the period 14,500-8500 yr B.P. are particularly interesting (Maley and Livingstone, 1983). They contain a significant proportion (2.5-7.9%) of the wild olive *Olea capensis,* which grows today in West African cloud forests at elevations above about 1000-1200 m. Wild olive pollen reached its maximum percentage around 9000 yr B.P. The pollen evidence is confirmed by the presence in the lake sediments of abundant pooid grass cuticles (Fig. 13.10) and leaf fos-sils (Hall *et al.,* 1978; Talbot and Hall, 1981), which shows that the Bosumtwi region supported a tropical

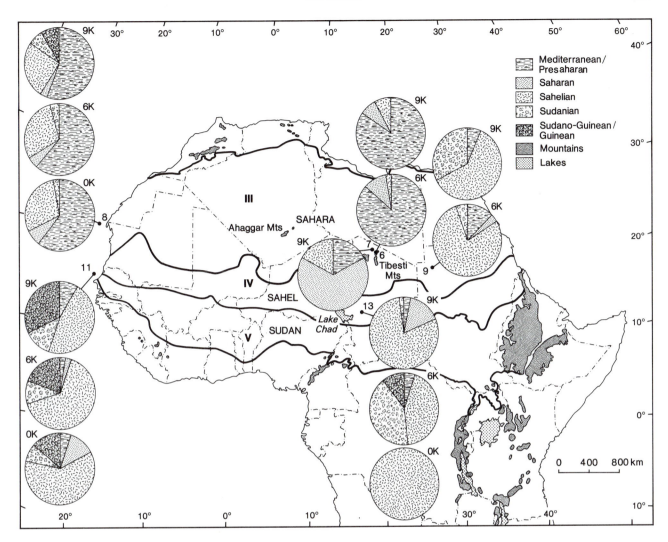

Fig. 13.8. Pollen spectra for 9000, 6000, and 0 yr B.P. from the Sahara-Sahel-Sudan region, excluding dryland taxa.

montane grassland with patches of dry semiever-green forest, but with a relatively high representation of trees characteristic of secondary forest, such as *Canarium schweinfurthii*. Many of the arboreal taxa are common today in upland environments (White, 1983). In tropical Africa pooid grasses are now confined to montane areas above 1500 m (Talbot *et al.,* 1984). Maley and Livingstone (1983) inferred a regional temperature lowering of at least 2–3°C, with an increase in mist and low cloud. Grass fires may have played a role in limiting the extent of trees (Talbot and Hall, 1981). The limited extent of the Guinean forests during the Pleistocene-Holocene transition is confirmed by the pollen spectrum from a submerged mangrove peat near Abidjan dated at 11,900 ± 250 yr B.P. (Assémien *et al.,* 1971), in which rainforest trees are poorly represented.

The Niger Delta core (site 18) from the western end of the main Congolian forest block (Fig. 13.9) unfortunately exhibits a substantial hiatus, resulting from the glacial lowering of sea level, as well as high percentages of mangrove pollen. Nevertheless, it is possible to infer a "drastic reduction" in lowland rainforest and swamp forest, accompanied by an expansion of savanna, during the period of sea-level rise before 7600 yr B.P. (Sowunmi, 1981a,b). Moist forest is strongly represented after that time. The main exception is the section dated 6800–5600 yr B.P., in which a peak of grass pollen suggests a dry climatic oscillation, although a marine invasion of the delta would be another possible explanation. *Elaeis guineensis* and weed pollen increase markedly around 2800 yr B.P. Forest clearance beginning about 2500–2000 yr B.P. is also evident in two short cores from the Cameroon highlands (Richards, 1986; Tamura, 1986).

Farther south, marine core C237 from just north of the mouth of the Zaire River provides a schematic outline of vegetation changes since about 20,000 yr B.P. (Caratini and Giresse, 1979). Gramineae and Cyperaceae are abundant at the base but decline to very

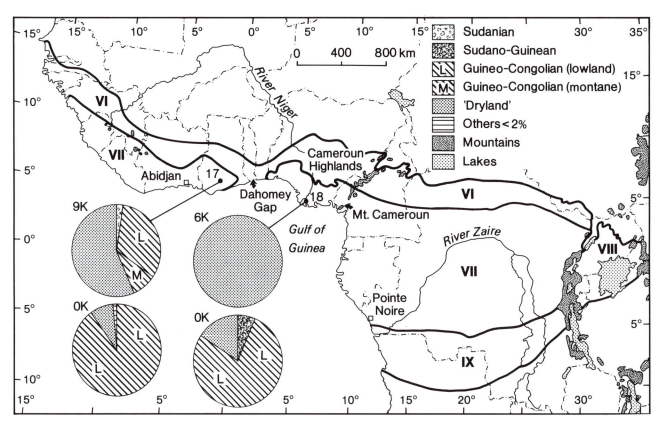

Fig. 13.9. Pollen spectra for 9000, 6000, and 0 yr B.P. from West Africa.

low levels from 12,000 yr B.P. onward. In contrast, the percentage of tropical forest pollen increases in the 12,000-yr B.P. spectrum and reaches its maximum around 3000 yr B.P.

In searching for an explanation for the cool, cloudy climate recorded at Lake Bosumtwi around 9000 yr B.P., Maley (1987) raised an intriguing possibility that would repay further investigation. At present such conditions are associated with strong coastal up-welling, which is characteristic along the northern shores of the Gulf of Guinea between 8°W and 2°E in July-September (Houghton, 1983). We suspect that the presence of this strip of cold water (below 24°C) off-shore during the summer monsoon season may be partly responsible for the existence of the Dahomey Gap, which today separates the Guinean and Congolian forest blocks. A comparison of the sea-surface temperature anomalies corresponding to wet and dry years in the Sahel (Lamb, 1978) suggests that the en-hanced summer monsoon circulation at 9000 yr B.P. may have increased coastal upwelling, thus explain-ing how conditions unfavorable for forest persisted at sites 17 and 18 into the early Holocene.

EASTERN AFRICA

Unlike other equatorial areas, most of tropical and subtropical eastern Africa has a dry-subhumid or semiarid climate (Trewartha, 1981), with the excep-tion of the wetter highland regions, where our pollen sites are located. These sites show an altitudinal maxi-mum of cloud cover and precipitation corresponding to the Afromontane forest belt (see earlier discussion). Other major climatic features of the region include a general decrease in the duration of the rainy seasons and in the total amount of rainfall from west to east and with distance from the equator (Nieuwolt, 1977).

In general eastern Africa has two rainy seasons, the more important one centering on March-May and the lesser one on September-November. At the most northerly sites (14-16), however, the minor rains occur in March-April and the main rains during the North-ern Hemisphere summer monsoon season (June-Sep-tember) (Degefu, 1987), when Ethiopia comes under the influence of the tropical easterly jet over the northern Indian Ocean (Hastenrath, 1985).

The distribution of natural vegetation types in east-ern Africa is as complex as the rainfall pattern. White (1983) provided the basis for grouping the pollen taxa into three lowland categories (Somali-Masai, Lake Vic-toria mosaic forest, and Lake Victoria mosaic grass-land) and three highland categories (wet Afromon-tane forest, dry Afromontane forest, and Afroalpine). The Afroalpine category includes both the ericaceous and the Afroalpine belts of Hamilton (1982) and earli-

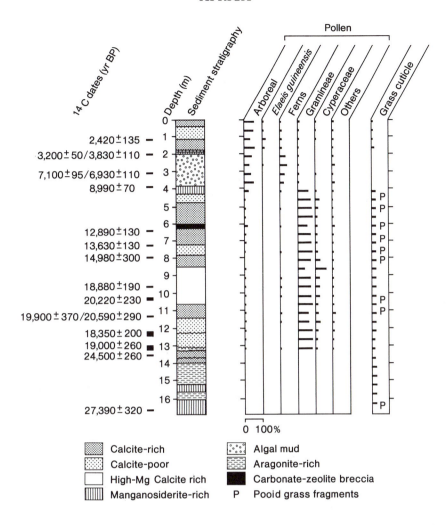

Fig. 13.10. Summary pollen diagram for Bosumtwi, Ghana (site 17). Redrawn from Talbot *et al.* (1984); used with permission from *Palaeoecology of Africa* (16, 173-192) and A. A. Balkema, Rotterdam, Netherlands.

er authors. Because Gramineae pollen reaches high percentages in both dry lowland and Afroalpine communities today, no separate "dryland" group was identified in this region. Instead, Gramineae, like Cyperaceae, was classified as ubiquitous and was excluded from the pollen sum.

Data are available from 12 sites in eastern Africa with adequate dating control. One of the earliest and most informative cores was collected from Sacred Lake, Mount Kenya (site 28), at an altitude of 2400 m. The lake is situated in a small crater surrounded by humid montane forest. The pollen diagram (Fig. 13.11) was derived from a 10.90-m core collected from the center of the lake. Because the top meter of sediment was not recovered, the exact age of the core top is uncertain.

The pollen diagram (Fig. 13.11) is divided into nine zones (R-Z) (Coetzee, 1967). The basal zone R, with its possibly infinite radiocarbon date of 33,350 ± 1000 yr B.P., shows a strong representation of Ericaceae and

Compositae, suggesting that the Afroalpine belt was depressed to an altitude close to the lake. The beginning of zone S is marked by a dramatic decrease in *Cliffortia*, which, together with an increase in taxa of the montane forest belt, suggests an upward migration of the forest in response to a climatic amelioration. The sediment stratigraphy suggests that the climate at this time was unstable but was warmer and wetter than in zone R.

The lower part of zone T is characterized by rapid shifts in the major pollen types: *Podocarpus* increases and *Hagenia* decreases, then *Podocarpus* declines and a high percentage of *Cliffortia* reappears. These changes could signal the beginning of a cold period, with initially dry conditions that allowed *Podocarpus* to expand at the expense of *Hagenia*. As the cold conditions intensified, the treeline was depressed and *Cliffortia* was once again able to grow near the lake. This cool, dry climate continued to the top of the zone at around 14,050 ± 360 yr B.P.

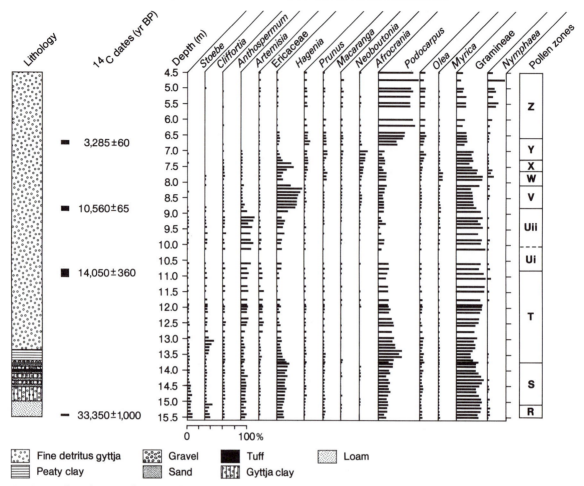

Fig. 13.11. Pollen diagram for Sacred Lake, Kenya (site 28). Simplified from Coetzee (1967); used with permission from *Palaeoecology of Africa* (3, 1-146) and A. A. Balkema, Rotterdam, Netherlands.

Zone U differs little from zone T, although *Artemisia* reaches slightly higher levels as *Podocarpus* gradually declines. Pollen of *Hagenia* and other montane forest species increases slightly. At present zone U and the underlying zone T are climatically indistinguishable.

Zone V, which is dated at about 10,000 yr B.P. at the base, shows dramatic changes in pollen representation. *Artemisia,* ericaceous shrubs, and Compositae all decline abruptly, while *Hagenia* and other montane forest trees increase. This suggests a rapid change to warmer, wetter conditions. By 9000 yr B.P. humid montane forest pollen types were dominant, indicating that this forest type was growing near the lake, if not surrounding it.

The transition from zone V to zone W is abrupt; Gramineae expands to levels higher than 40% as *Hagenia* declines. Small peaks of *Macaranga, Myrica, Juniperus,* and *Nymphaea* show up in zone W. Coetzee (1967) interpreted these changes as indications that the treeline had migrated up above the lake and suggested that the abundance of large grains of Gramineae pollen indicated the presence of mountain

bamboo (*Arundinaria alpina*) around the lake. However, the ratio of large to small grains of Gramineae is virtually identical to that of the rest of the profile. Hamilton (1982) pointed out that large grains of Gramineae are not a reliable indicator of bamboo. Moreover, *Arundinaria* produces little pollen (Osmaston, 1958; Hamilton, 1972; Hamilton and Perrott, 1980). Studies of modern pollen rain on Mount Elgon have demonstrated that percentages of Gramineae pollen in excess of 30% occur only in grassland and rocky-ground communities (Hamilton and Perrott, 1980). Therefore it can be argued that at the boundary between zones V and W the onset of cool, dry conditions once again depressed the treeline to some distance below the lake, which was surrounded by Afroalpine grassland. The montane forest indicators present in zone W, such as *Podocarpus, Olea, Macaranga, Myrica,* and *Juniperus,* all have high or moderate export ability and could have been transported upslope by the mountain wind systems. Such a cool, dry event would correlate with the similar event recorded between 8500 and 6500 yr B.P. at site 30 (Perrott, 1987) and with the dramatic fall in lake levels recorded in eastern Africa at this time (Street, 1979; Gillespie *et al.,* 1983).

In zone X Gramineae values fall and *Hagenia* values rise once again. Coetzee (1967) concluded that the treeline was forced downward by cold, dry conditions. It seems more probable, however, that the treeline was readvancing up the mountain with a return to warmer, wetter conditions. This trend continues into zone Y, in which taxa of the wetter montane forest, such as *Afrocrania, Prunus,* and *Neoboutonia,* gain in strength. The midpoint of zone Y probably marks the warmest, wettest time of the Holocene, with high percentages of *Afrocrania, Prunus, Olea, Macaranga,* and *Neoboutonia.* Toward the top of zone Y, a major change in forest type occurs; a large increase in the representation of *Podocarpus* indicates an abrupt shift to drier conditions, which is maintained to the top of the core.

The pollen spectrum for 9000 yr B.P. (Fig. 13.12a) is dominated by the wetter montane forest, although the strong representation of Afroalpine elements indicates that the climate was still cool enough to depress ericaceous vegetation to the vicinity of the lake. By 6000 yr B.P. (Fig. 13.12b) the wetter montane forest had expanded farther as Afroalpine elements decreased, indicating that warm, wet conditions had allowed forest to surround the lake. At the top of the core (Fig. 13.12c), the situation is completely different; the pollen spectra are dominated by taxa of the dry montane forest, reflecting the major change to drier conditions.

The second set of cores from Mount Kenya was recovered from site 29 (3140 m), about 740 m upslope from Sacred Lake. Lake Rutundu occupies a deeper crater in the lower part of the Afroalpine belt, about 90 m above the treeline of the dry montane forest. Two cores were collected near the northern shore of the lake. Coetzee (1967) perceived a similarity among a few widely spaced pollen counts and assumed an overlap between the cores. Because we do not find the pollen evidence for an overlap convincing, and because an overlap is not consistent with the lithology, we used only core B in deriving the position of the 9000-yr B.P. level.

Site 29 shows no significant change between 9000 and 6000 yr B.P. (Fig. 13.12a,b); the pollen spectra are dominated by Afroalpine elements. At both times the dry-forest representation was slightly stronger than that of the wet forest, but because the dominant dry-forest taxa have pollen of high to moderate export ability (Hamilton and Perrott, 1982), both spectra are taken to indicate wet conditions. Today (Fig. 13.12c) the dry montane forest elements have greatly increased, again reflecting the change to a drier climate on the mountain.

The pollen sites closest to Mount Kenya are Karimu mire (site 31) (3040 m) and Mount Satima mire (site 30) (3670 m) on the Aberdare Range. Karimu mire covers a very extensive area of the Aberdare Plateau. The area selected for study was a southeastern embayment of the mire. Three cores were analyzed for pollen; we selected core P as the most suitable for this study. At 9000 yr B.P. (Fig. 13.12a) the local pollen was dominated by Afroalpine elements, while pollen from the wet montane forest dominated the pollen transported from afar, indicating cool, wet conditions on the Aberdare range. By 6000 yr B.P. (Fig. 13.12b) dry montane forest elements had increased, probably indicating that a change to warmer conditions had allowed the montane forest to expand toward the site. However, when we take into account the underrepresentation of the wet montane forest in the pollen rain at high-altitude sites, the strong showing of the wet taxa indicates that the climate on the range continued to be wet. The modern pollen spectrum (Fig. 13.12c) is completely different; the dominance of dry montane forest taxa reflects conditions drier than those experienced at 6000 and 9000 yr B.P.

The pollen spectrum for 9000 yr B.P. at site 31, located in a glaciated valley high in the Afroalpine belt, is almost totally dominated by Afroalpine taxa, indicating that the treeline was severely depressed by cool conditions and/or that the montane forest was slow to expand from its glacial refugia (Fig. 13.12a). The small but almost equal representation of taxa from the wet and dry montane forests indicates wet conditions at this time. At 6000 yr B.P. (Fig. 13.12b) montane forest elements had expanded considerably, especially taxa of the wet montane forest, indicating warmer, wetter conditions. As at most East African pollen sites, the modern spectrum (Fig. 13.12c) is dominated by taxa of the dry montane forest, which have increased in representation by 50% relative to 6000 yr B.P., confirming the change to much drier conditions. It is important to note that the high-resolution diagram from Mount Satima records an extreme cool, dry "event" between 8500 and 6500 yr B.P.

The other pollen diagrams from eastern Africa can be divided into those from the drier region to the north and those from the wetter region to the west. To the north, pollen diagrams have been obtained from the Balé and Arussi mountains of Ethiopia (sites 14 and 16) and from Lake Abiyata in Ethiopia (site 15). The important diagrams from the Cherangani Hills in Kenya (van Zinderen Bakker, 1962, 1964; Coetzee, 1967) could not be included in this study because of the major dating problems encountered.

The Ethiopian sites also have dating control problems. The Lake Abiyata core (site 15) has inverted dates at the base of the Holocene, which made the selection of the 9000-yr B.P. level difficult. Because the core top is of mid-Holocene age (Street, 1979), the present-day pollen spectrum was derived from a nearby surface sample (Lézine, 1981; Lézine and Bonnefille, 1982). The Danka bog core (site 16) has only one date at the base, leading to uncertainties in the selection of the 6000-yr B.P. level, especially in view of the widespread evi-

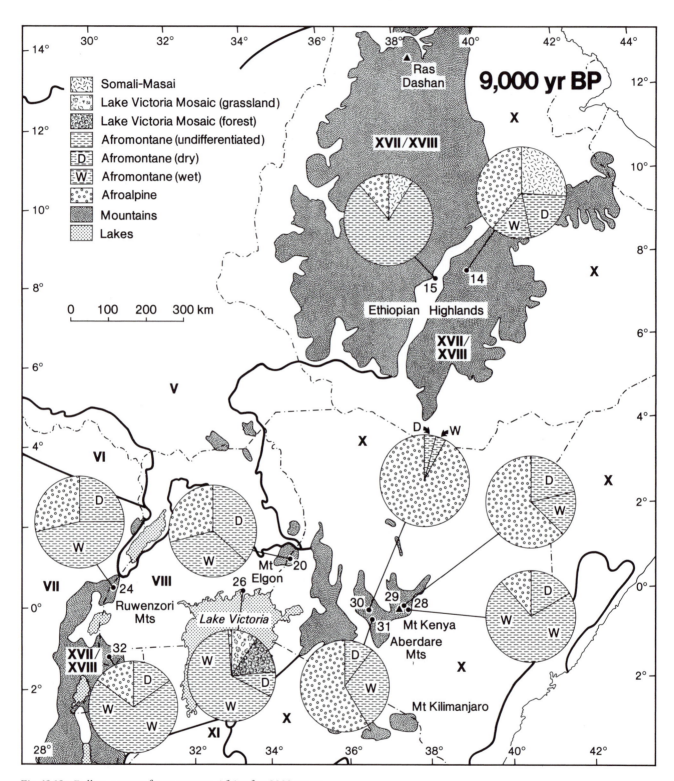

Fig. 13.12a. Pollen spectra from eastern Africa for 9000 yr B.P.

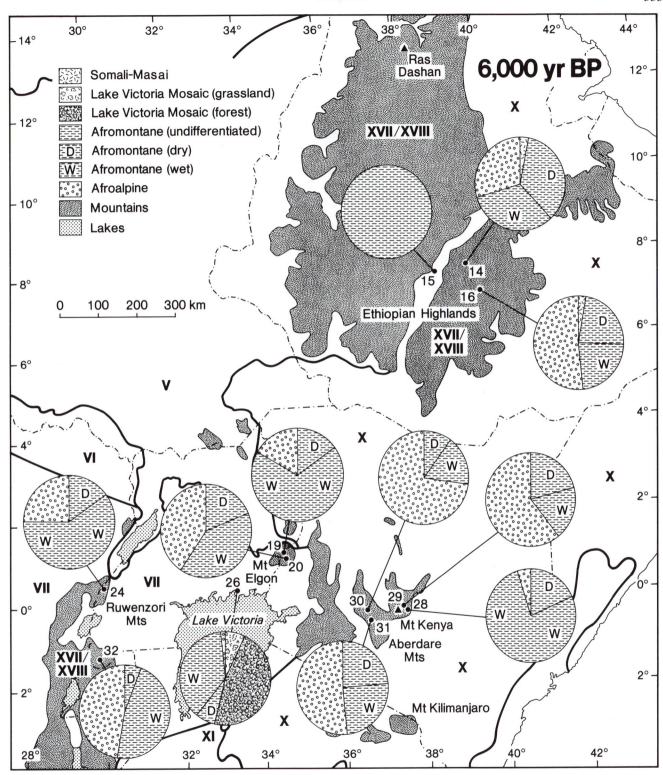

Fig. 13.12b. Pollen spectra from eastern Africa for 6000 yr B.P.

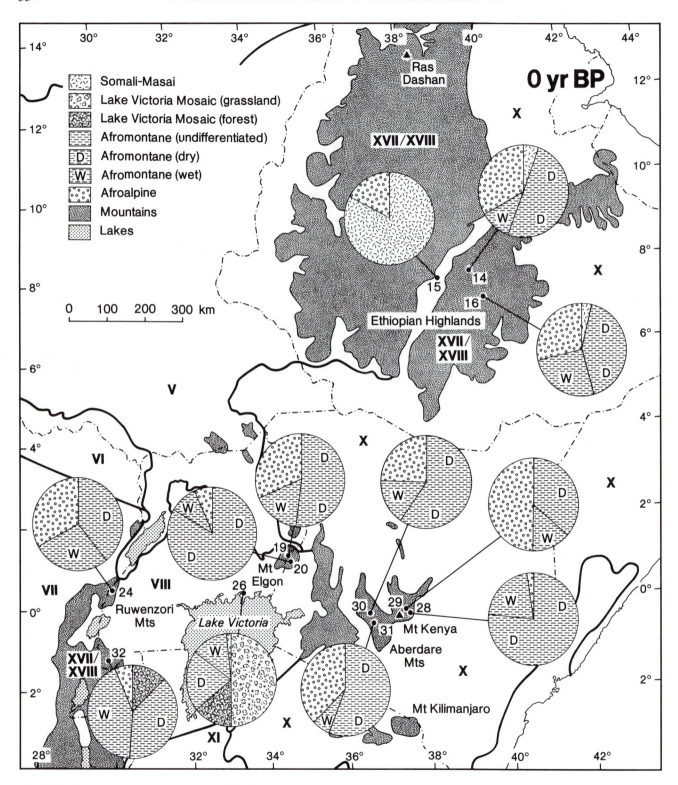

Fig. 13.12c. Pollen spectra from eastern Africa for 0 yr B.P.

dence for changes in sedimentation rates in East African mires during the Holocene (Hamilton, 1982; Perrott, 1982). The representativeness of the present-day pollen sample is also in question, because the ericaceous vegetation at this site is regularly burned over by pastoralists. The Mount Badda core (site 14) is relatively well dated and confirms that sedimentation rates varied widely, which introduces some uncertainty into the selection of the 9000- and 6000-yr B.P. levels.

Site 15 (1585 m), on the floor of the Rift Valley, is the second lowest site examined in eastern Africa. The pollen spectrum for 9000 yr B.P. (Fig. 13.12a) is dominated by Afromontane taxa, indicating that the climate was wet enough to allow montane forest to expand on the Ethiopian Rift escarpments and adjacent plateaus. This conclusion is supported by the low representation of pollen from the dry Somali-Masai vegetation (*Acacia* bushland and wooded grassland) that occupies the Rift floor today. The relatively large percentage of Afroalpine elements suggests that conditions were still cool. The 6000-yr B.P. spectrum (Fig. 13.12b) is almost exclusively montane forest taxa, indicating wetter and warmer conditions. The modern spectrum (Fig. 13.12c) shows a dramatic reversal, with Somali-Masai elements dominating the diagram. Although the decline of the montane forest may be attributed in part to anthropogenic impact, there is also clear evidence for drier conditions, not least in the low lake levels (Gillespie *et al.,* 1983).

Site 16 (3830 m) lies in a much wetter area than the other high-altitude site (site 14). At 6000 yr B.P. the spectrum for site 16 (Fig. 13.12b) is dominated by pollen from the surrounding Afroalpine vegetation, with equal representation of wet and dry Afromontane forest types. However, given the lower transport ability of the wetter taxa (Hamilton and Perrott, 1982), the equal representation suggests that moist conditions had allowed wetter montane forest to spread widely in the highlands. Drier conditions since 6000 yr B.P. are associated with much larger percentages of dry montane forest elements (Fig. 13.12c).

At site 14 (4000 m) the spectrum for 9000 yr B.P. (Fig. 13.12a) is difficult to interpret. The strong representation of Afroalpine taxa suggests that cool conditions were keeping the upper treeline altitudinally depressed, whereas the long-distance pollen indicates that the montane forest, although fairly wet in character, was still relatively restricted in extent. The surprisingly large proportion of Somali-Masai elements in fact consists mainly of chenopods, which were probably living on areas of exposed lake floor not yet covered by the early-Holocene transgression (Gillespie *et al.,* 1983). The spectrum for 6000 yr B.P. (Fig. 13.12b) suggests that in response to wetter conditions the montane forest had expanded its range at the expense of the Afroalpine belt. The modern pollen spectrum (Fig. 13.12c) is dominated by dry montane forest. This change was almost certainly caused by a drier climate,

an inference supported by the slight increase in Somali-Masai elements.

In the wetter area to the west, the major diagrams are derived from Mount Elgon (sites 19 and 20), Lake Victoria (site 26), the Rukiga highlands (site 32), and the Ruwenzori Mountains (site 24). Site 20 (4150 m) is a cirque lake in the southern wall of the caldera of Mount Elgon at the head of a glaciated valley. The pollen spectra (Fig. 13.12a-c) are very difficult to interpret. At 9000 yr B.P., as could be expected for a site at this altitude, Afroalpine elements were well represented, but long-distance pollen from both wet and dry Afromontane forest types was surprisingly abundant. This presents a major problem in interpretation, for a pollen diagram from Laboot Swamp (Hamilton, 1982), located within the montane forest belt on the same aspect of the mountain at an altitude of 2280 m, implies that montane forest of the types encompassed by our Afromontane categories did not exist at or near the site until some time after 5800 yr B.P. Although Laboot Swamp was not included in this study because of its poor radiocarbon control, its evidence in this respect is very clear. The only explanation suggested for this anomaly is that before 5800 yr B.P. environmental conditions favored the spread of mountain bamboo (*Arundinaria alpina*) throughout the niche available to montane forest on the southern and western aspects of the mountain. However, the bamboo cover may have been incomplete in the north and east, allowing pollen of dry and wet montane taxa to be transported to the summit area and deposited at site 20.

At 6000 yr B.P. wet montane forest dominated the spectrum at site 20. Afroalpine representation remained virtually the same as at 9000 yr B.P., and dry montane forest elements were much reduced, suggesting not only that 6000 yr B.P. was wetter than 9000 yr B.P. but also that the cool, dry event discussed above (8500–6500 yr B.P.) had reduced the domination of *A. alpina,* allowing wet montane forest to expand. The subsequent return of dry conditions has resulted in almost total domination of the modern spectrum by pollen from the dry montane forest.

Site 26 (1130 m) is the lowest site studied in eastern Africa. In contrast to most of our sites, Lake Victoria is vast, in terms of both its own area (68,800 km^2) and the size of its catchment. This, together with the diversity of vegetation types within region VIII, makes its pollen spectra the most complex of those shown in Figure 13.12a-c. Although many wet-forest genera (e.g., *Alchornea, Myrica, Macaranga, Acalypha*) occur in this area from the water's edge to the flanks of the nearby mountains, they have been included in the wet Afromontane forest category. The Lake Victoria mosaic elements have been divided into those commonly found in dry lowland forest and in wooded grassland (see earlier descriptions).

At 9000 yr B.P. the spectrum (Fig. 13.12a) was dominated by pollen of wet-forest taxa, indicating a much moister climate than today in the lake basin. By 6000 yr B.P. (Fig. 13.12b) the dry lowland forest had become the most prominent group, suggesting that this site was drier at 6000 yr B.P. than at 9000 yr B.P. An important consideration in the records for both 9000 and 6000 yr B.P. is the very low representation of dry montane forest elements, especially *Podocarpus*. Because the levels seen are consistent with long-distance transport, questions arise as to where *Podocarpus* was growing at this time and where the refugia were located from which it spread to dominate the modern pollen spectra throughout East Africa. At present wooded grassland and dry montane forest elements dominate the pollen rain, reflecting the ubiquitous change to drier conditions in eastern Africa.

The core from site 32 (2256 m) has only two radiocarbon dates within the Holocene. It also has a "floating" chronology, as the top meter of sediment was not sampled. However, a surface sample from the mire was available (Hamilton, 1972) to represent the present. A further problem arises because the large samples used for dating, comprising 50 and 100 cm of core material, make the selection of the 6000- and 9000-yr B.P. levels problematic. Care must be taken in interpreting the pollen spectra from site 32 (Fig. 13.12a-c) because the "Afroalpine" elements recorded at both 9000 and 6000 yr B.P. consist entirely of Ericaceae and *Alchemilla*, which probably grew on the swamp surface (Morrison, 1968), whereas the modern spectrum contains *Anthospermum*, which is often found on disturbed land (Hamilton, 1972).

At 9000 yr B.P. (Fig. 13.12a) conditions moister than at present are implied by the dominance of wet Afromontane forest. The high proportion of *Hagenia* (58%), however, suggests that the treeline was depressed by lower temperatures. At 6000 yr B.P. (Fig. 13.12b) the wetter Afromontane elements were still dominant, but the forest had made a major shift to a more diverse composition with *Macaranga*, *Myrica*, *Anthocleista*, *Hagenia*, and *Hypericum*. Morrison (1968) noted a change from lake mud to herbaceous woody peat just below this level, which probably allowed an open swamp forest to develop on the drier mire surface. We cannot be certain whether this progression was caused by a climatic change or by a natural infilling of the lake. Between 6000 yr B.P. and the present a clear shift to drier conditions is reflected in a large increase in dry-forest taxa.

Site 24 (2960 m), the lowest known glacial lake in East Africa, lies at the boundary between the bamboo zone of the upper montane forest belt and the Afroalpine belt. This core must be interpreted with caution, as the counts are small. At 9000 yr B.P. (Fig. 13.12a) the pollen spectrum is dominated by wet montane taxa, with a fairly strong Afroalpine representation, suggesting cool, wet conditions. The proportion of dry montane taxa is also relatively large but appears to be made up solely of Oleaceae. This suggests

that the montane forest had not achieved its full extent or stability and that *Olea* was exploiting its potential as a colonizing species. At 6000 yr B.P. (Fig. 13.12b) the increased representation of the wet montane forest suggests that 6000 yr B.P. was wetter than 9000 yr B.P., while the reduction in dry montane forest elements, especially Oleaceae, implies that the potential for colonization was now limited. Today (Fig. 13.12c) dry montane forest taxa dominate the spectrum, reflecting the general trend to drier conditions.

SOUTHERN AFRICA

[14]C-dated pollen spectra representing 9000 yr B.P., 6000 yr B.P., or both are currently available for five sites in southern Africa, which spans floristic regions XI-XIV (Fig. 13.3). Three of these diagrams represent the Zambezian region (XI) and two represent the Cape (XIV). The lack of information from the intervening dry zone reflects a general scarcity of lake deposits and peat. Although pollen diagrams have been obtained from a number of alluvial, spring, and cave sites in regions XII and XIII (van Zinderen Bakker, 1957, 1982, 1984; Coetzee, 1967; Scott, 1982a, 1984, 1986; Scott and Vogel, 1983), they terminate before 9000 yr B.P., are highly discontinuous, or exhibit stratigraphically inconsistent ^{14}C dates that make it hard to locate the 9000- and 6000-yr B.P. levels with any confidence.

The groupings of pollen taxa used here for southern Africa are based primarily on Werger (1978) and White (1983), supplemented by the modern pollen spectra and ecological interpretations reported by Coetzee (1967), Martin (1968), Schalke (1973), Scott (1982c), Meadows (1984b), and van Zinderen Bakker and Müller (1987). Since this data set was prepared, additional modern spectra have been published by Cooremans (1989), Scott (1989b), and Meadows and Sugden (1991). As in eastern Africa, no separate group of dryland taxa has been created because of the difficulty of interpreting pollen types like Gramineae and Compositae (undifferentiated), which are treated as ubiquitous.

Zambezian and Kalahari-Highveld Regions

The Zambezian phytogeographic region, the largest in Africa after the Sahara, comprises not only the richest and most diverse flora (about 8500 species) but also the widest range of vegetation types (White, 1983). It is characterized by a tropical summer-rain regime. Total annual precipitation decreases from the equatorward margin of the region toward the borders of the Kalahari-Highveld region. Highland areas with cooler climates are found on its northeastern, eastern, and southeastern flanks.

The dry deciduous forests and savanna woodlands that cover vast expanses of region XI are very inadequately portrayed by the regional pollen rain. Many of the dominant tree species are insect-pollinated legumes with very low pollen production, such as

Baikaea, Brachystegia (miombo), *Julbernardia, Iso-berlinia,* and *Colophospermum* (mopane). Of these, only *Brachystegia* even appears in the pollen diagrams discussed here, and it is seriously underrepresented (Livingstone, 1971). As a result, variations in rainfall as large as ±50% may not be palynologically detectable in the interior of the Zambezian region, a situation in which "the worst fears of palynologists from higher latitudes seem amply justified" (Livingstone, 1971).

The core from Lake Ishiba Ngandu (site 34), which is situated in miombo woodland, exemplifies these problems. Gramineae and Cyperaceae dominate the pollen assemblages for the last 21,000 yr. The remaining taxa vary little in abundance. Afromontane and Afroalpine elements are better represented at 9000 than at 6000 or 0 yr B.P. (Fig. 13.13), and Zambezian taxa that could reflect either forest clearance (Livingstone, 1971) or climatic drying (Hamilton, 1982, p. 197) increase after about 3000 yr B.P. These changes are almost certainly not statistically significant.

The montane grasslands of Malawi and Zimbabwe (site 35) are also relatively complacent palynologically, given the dominance of Gramineae, Cyperaceae, and Compositae in modern and fossil spectra (Tomlinson, 1974; Meadows, 1984a,b) (Fig. 13.13). Although the appearance of small percentages of *Brachystegia* suggests that the modern climate may be slightly drier

than the climates at 9000 and 6000 yr B.P., the overall impression is one of stability.

Fortunately the bushveld region of the Transvaal, South Africa, near the boundary with region XII, has proved much more sensitive to climatic change. The Wonderkrater site (site 36) is situated on a strong climatic gradient between the warm, semiarid Kalahari thornveld in the west and the cool, moist highveld and Afromontane environments to the south and east. From before 14,180 yr B.P. to around 11,000 yr B.P., this site was surrounded by open grassland with Austroafroalpine elements such as Ericaceae, *Cliffortia,* and *Passerina* (Fig. 13.14), indicative of a cool to cold-temperate subhumid climate (Scott, 1982b). Peaks of *Anthospermum, Artemisia,* and *Stoebe* are compatible with this picture. Distant pockets of montane forest are recorded by small percentages of *Podocarpus* and *Myrica.* Between about 11,000 and 6000 yr B.P. the pollen of Afromontane trees fell to very low levels. Austroafroalpine taxa also declined, while Zambezian (*Burkea africana,* Combretaceae-type) and Kalahari (Tarchonantheae, Capparidaceae, *Aloe*-type) elements increased, suggesting greater warmth and dryness. This drying trend culminated between about 9500 and 6000 yr B.P., when a Kalahari thornveld vegetation characteristic of warm-temperate, semiarid conditions became established in the local area. A rise in Zambezian taxa after 6000 yr B.P. may reflect an increase in

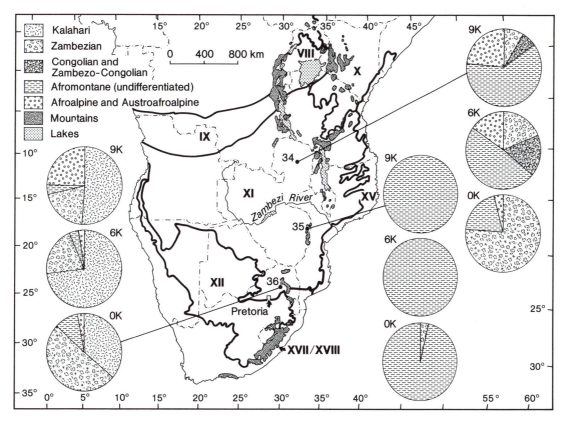

Fig. 13.13. Pollen spectra for 9000, 6000, and 0 yr B.P. from the Zambezian region.

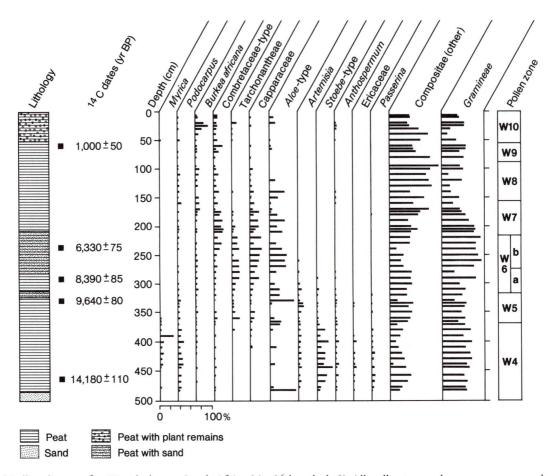

Fig. 13.14. Pollen diagram for Wonderkrater, South Africa (site 36, borehole 3). All pollen types shown are expressed as percentages of total nonlocal pollen, except that Cyperaceae and Gramineae are shown as percentages of the local pollen sum, which includes aquatic taxa, swamp taxa, and grasses. Simplified from Scott (1982b); used with permission from Academic Press.

moisture. Since about 4000 yr B.P. Afromontane tree pollen has shown a slight resurgence, and Zambezian bushveld has surrounded the site.

The indications of early-Holocene aridity at Wonderkrater are supported by three less well-dated diagrams from region XII: Rietvlei in the highveld (Scott and Vogel, 1983) and Equus and Wonderwerk caves in the Kalahari thornveld (van Zinderen Bakker, 1982; Scott, 1987a). Coetzee (1967) also inferred an oscillatory transition from cold, humid Austroafroalpine to warm, dry Karoo-like conditions between 12,600 and 9650 yr B.P. from short cores collected at Aliwal North (region XII).

The pie diagrams for the three Zambezian sites are shown in Figure 13.13. No compelling palynologic evidence exists for an enhanced early-Holocene monsoon in the summer-rainfall area of southern Africa (van Zinderen Bakker, 1976). The most sensitive and well-dated site, site 36, as well as several sites in the adjacent Kalahari-Highveld region registered conditions drier than at present in the interval 11,000–6000 yr B.P.

Cape Region

The Cape is a small area at the southwest tip of Africa with a "Mediterranean" climate. Annual precipitation increases from 250 mm on the boundary with the semiarid Karoo (region XIII) to more than 1750 mm in the mountains. The western part receives 60–80% of its rain in winter, and the severity of summer drought decreases eastward. Snow is common on high ground, and frost occurs in inland areas.

Although the Cape region has a large endemic flora, it exhibits significant affinities with the Afroalpine and Austroafroalpine regions of eastern and southern Africa (Taylor, 1978). The characteristic sclerophyllous *fynbos* vegetation, which is adapted to summer drought, is palynologically fairly distinct from the Karoo shrubland and Afromontane forest that border it to the north and southeast, respectively.

Two pollen diagrams are available from region XIV. Groenvlei (site 37) is located in sandy heathland (coastal *fynbos*) just outside the Knysna temperate rainforest enclave (which is classified as part of the Afromontane region [White, 1978]). The pollen dia-

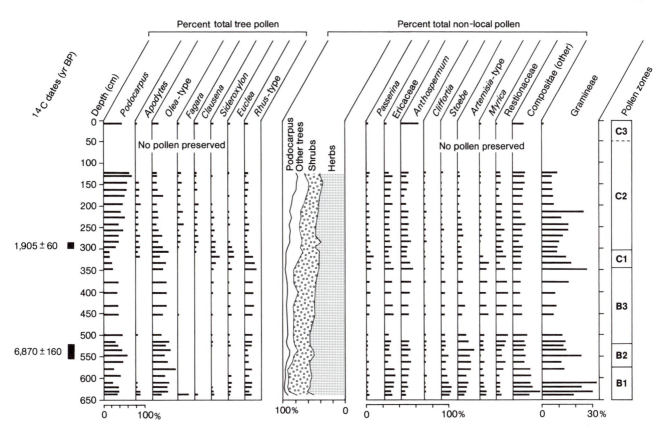

Fig. 13.15. Pollen diagram for Groenvlei, South Africa (site 37). Simplified from Martin (1968); used with permission from Elsevier Science Publishers.

gram from this site has an estimated basal age of 8000 yr B.P. (Fig. 13.15). At that time the local area appears to have been occupied by a dry, open sclerophyllous heath dominated by Asteraceae (Compositae). Forest was much less extensive than now and may have been limited by fire, for charcoal fragments are found in the core. Martin (1968) suggested that the 400-mm isohyet had retreated 60-100 km to the east of its present position and that the summer drought period was more pronounced.

From about 7000 yr B.P. (zone B2) onward, the diagram shows a general increase in arboreal taxa and a decrease in shrubs and herbs, reflecting an overall trend toward moister conditions, with only minor superimposed oscillations. The maximum spread of trees is recorded in zone C2 (Fig. 13.15) between about 2000 and 1000 yr B.P. Although the site was never invaded by forest, woodland became more extensive in the local area at that time. Its subsequent decline resulted at least in part from clearance by European settlers (zone C3).

Site 38 is also located in the coastal *fynbos* but is much farther from areas of contrasting vegetation. Although slight oscillations in moisture may have been recorded between 7700 yr B.P. and the present, they were not large enough, or the site was not sensitive

enough, for any major changes in vegetation to be registered (Fig. 13.16).

The pollen evidence from the Cape region agrees well with the data from the summer-rainfall area discussed above. Sites far from major floristic boundaries (34, 35, and 38) show little significant variation during the Holocene, whereas sensitive ecotonal sites (36 and 37) record conditions drier than today before 7000-6000 yr B.P. Newer Holocene records from Madagascar (about 20°S), summarized by Burney (1987a,b), confirm the trend seen in the African sites. Open grassland and ericaceous bushland, swept by frequent fires, gave way to wooded savanna after 4000 yr B.P. The dryness during the early Holocene here is attributed to increased summer drought—in other words, to a slightly weaker summer monsoon over southern Africa. This hypothesis is discussed in more detail below.

PALEOCLIMATIC MAPS BASED ON POLLEN DATA

We used the changing representation of the different phytogeographic elements in the pollen spectra from each site to evaluate past changes in temperature and effective moisture on a five-point relative scale (Figs. 13.17 and 13.18). We evaluated moisture conditions at 9000 and 6000 yr B.P. at almost all sites for

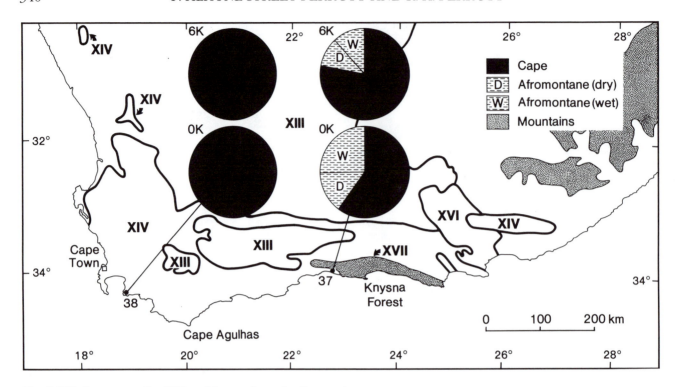

Fig. 13.16. Pollen spectra for 6000 and 0 yr B.P. from the Cape region.

which data were available. We evaluated temperatures at 9000 yr B.P. in highland areas with well-developed altitudinal zonation based either on the relative proportions of lowland, montane, and alpine taxa in the pollen spectra or, at some sites in East Africa, on the representation of trees, such as *Hagenia,* that are characteristically found near treeline, within the wet montane forest element. No map is presented for temperatures at 6000 yr B.P. because the abundances of high-altitude taxa were not very different from those of today.

Temperatures at 9000 yr B.P.

Figure 13.17 shows the paleotemperature reconstruction for 9000 yr B.P. No data points are shown for the Atlas or Tibesti mountains, because the pollen spectra from those areas are not yet readily interpretable in terms of temperature.

At all the sites in tropical and southern Africa where it was possible to code the data, pollen types characteristic of modern high-altitude samples were more strongly represented at 9000 yr B.P. than at either 6000 or 0 yr B.P.; that is, the vegetation was reacting as if temperatures were cooler. It is important to note, however, that our data set has no seasonal resolution. Moreover, temperatures on the tropical high mountains are partly controlled by cloudiness and hence by precipitation and prevailing winds (Hastenrath, 1984, chap. 2).

Modeling studies suggest that the Laurentide ice-sheet remnant had little impact on temperatures in

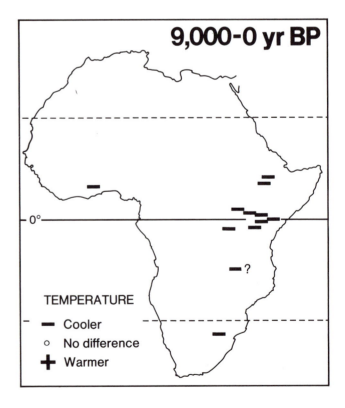

Fig. 13.17. Paleotemperature anomalies at 9000 yr B.P. reconstructed from pollen evidence.

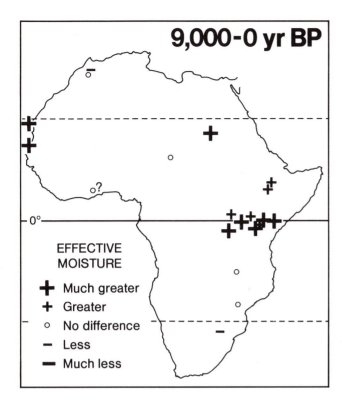

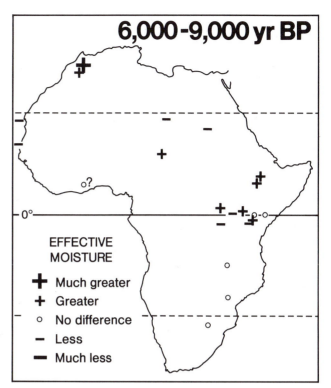

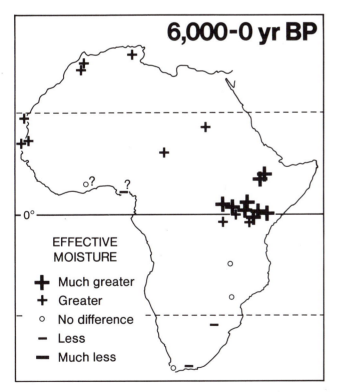

Fig. 13.18. Effective moisture anomalies reconstructed from pollen evidence for 9000-0 yr B.P., 6000-9000 yr B.P., and 6000-0 yr B.P. Some closely spaced sites have been shifted slightly to avoid superposition.

Africa at 9000 yr B.P. (Kutzbach and Otto-Bliesner, 1982; Mitchell *et al.,* 1988). A stronger Northern Hemisphere summer monsoon, however, would probably have led to increased cloud and fog and thus to cooler temperatures at high elevations. The correlation between cloudiness and coastal upwelling in West Africa has already been stressed; it is possible that the increased coastal upwelling in northern summer off the Horn of Africa and Arabia (Prell *et al.,* 1990) contributed to the relatively cool climate of the high mountains in Ethiopia and East Africa at 9000 yr B.P.

An additional reason for the generally cooler temperatures throughout Africa at 9000 yr B.P. is provided by the earth's orbital configuration. Mitchell *et al.* (1988) showed that the zonal-mean incoming solar radiation, averaged over the *entire* year, was less than today between about 40°N and S. This currently appears to be the only plausible explanation for the occurrence of cooler temperatures in southern Africa.

Effective Moisture at 9000 and 6000 yr B.P.

Figure 13.18 shows the reconstructions of effective moisture anomalies at 9000 and 6000 yr B.P. based on the pollen data set, as well as the net change between these two time periods. At 9000 yr B.P. North Africa and the Zambezian and Kalahari-Highveld regions of South Africa appear to have experienced conditions similar to or drier than today. Most of the sites in sub-Saharan and eastern Africa registered greater moisture than at present except for site 13, as discussed previously.

The pattern of moisture anomalies at 6000 yr B.P. (Fig. 13.18) differs subtly from that at 9000 yr B.P. The sites in North Africa had become wetter, whereas those in the eastern Sahara and sub-Saharan Africa were drier except for site 13. In eastern Africa, while all the diagrams suggest that moist forest types were more extensive at 6000 yr B.P. than at present, the distribution of maximum effective moisture apparently shifted northward between 9000 and 6000 yr B.P.

The moisture reconstructions in Figure 13.18 based on pollen data are compared with moisture reconstructions based on lake-level data in a later section.

Lake-Level Data

The Holocene record of water-level fluctuations in lake basins in Africa has been studied in more detail than that of any other continent. Past lake levels have been reconstructed from a variety of sources, including old shorelines; the stratigraphy, geochemistry, and paleoecology of sediment cores; and the distribution of lakeside archaeological sites (Street-Perrott and Harrison, 1985).

The data set used here was extracted from the November 1985 version of the Oxford lake-level data bank. It comprises 42 basins with evidence for 9000 ±

500 yr B.P. and 52 for 6000 ± 500 yr B.P. (Table 13.2). Only lakes that are believed to have been closed during all or part of the late Quaternary are included. The method of compilation is summarized in Street-Perrott and Roberts (1983) and Street-Perrott *et al.* (1985). Briefly, the water level in each basin at each time period was assessed on a relative scale divided into the following classes, each with a roughly similar frequency of occurrence in the data set: low, 0–15% of the total vertical range of fluctuation, including dry lakes; intermediate, 15–70%; and high, 70–100%, including overflowing lakes. The direction of change in lake level was also evaluated.

Fluctuations in lake area and lake depth are caused by variations in net water balance. For a closed, sealed lake, equilibrium with the prevailing climate is reached when

$$A_L/A_B = (P_B - E_B)/(E_L - P_L),$$

where A is area, P is annual precipitation, E is annual evaporative loss, and the subscripts L and B refer to the lake and its drainage basin, respectively. The lakes most sensitive to climatic fluctuations are closed lakes fed predominantly by river runoff rather than by direct precipitation or groundwater; these are known as amplifier lakes (Street-Perrott and Harrison, 1985). Many of the basins situated along the East African Rift Valley are of this type. Their equilibrium response time for a step change in water balance is given by

$$\tau_{EQ} = A_L/(dA_L/dD)(E_L - P_L),$$

where τ_{EQ} is the time taken to accomplish $(1 - 1/e)$ (i.e. 63%) of the equilibrium change in area and D is water depth (Mason *et al.,* in press). τ_{EQ} probably has an upper limit of 10^3 yr.

Many of the lakes in the Sahara and Namib deserts and adjacent areas occupy interdune basins fed primarily by groundwater. In such cases, the response time of the aquifer to climatic change may introduce a large and variable lag into the lake-level response (Gasse *et al.,* 1987). The groundwater imprint, however, seems unlikely to have obscured climatic trends on the broad spatial and temporal scale discussed in this chapter.

Street-Perrott and Roberts (1983, Fig. 2d) illustrated the relationship between the modern distribution of lake levels and climate. Most African lakes are low at present except for a small number of lakes in near-equatorial areas with relatively humid climates (e.g., sites 16, 17, 65, 75, 82, and 110) and site 23, which is fed by runoff from the wetter, forested regions of West Africa.

In view of the large size of the full African data set (113 basins) and the extensive literature dealing with individual regions, we will not attempt here to summarize "typical" late-Quaternary lake-level records. In-

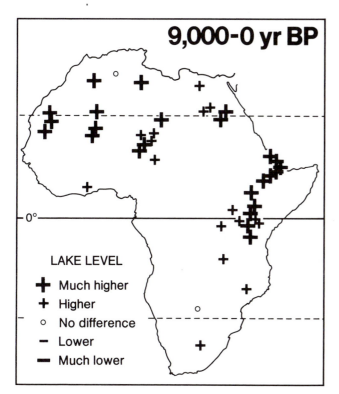

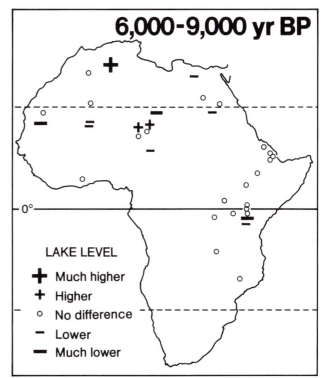

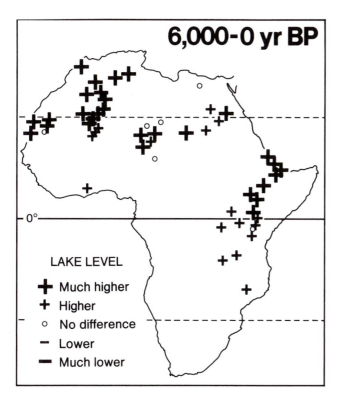

Fig. 13.19. Lake-level anomalies for 9000–0 yr B.P., 6000–9000 yr B.P., and 6000–0 yr B.P. Some closely spaced sites have been shifted slightly to avoid superposition.

Table 13.2. *African lake-level data set for 9000 and 6000 yr* B.P.

Site no.	Site name	Latitude	Longitude	Altitude (m)	^{14}C dates[a] (no.)	Dating control[b] 9000 yr B.P.	6000 yr B.P.	References[c]
1	Abhe, Ethiopia-Djibouti	11° 15'N	42° 00'E	240	73	1	1	119, 120, 130, 291, 312, 421, 446
2	Abu Ballas, Egypt	24° 14'N	27° 25'E	506	7	2	2	252
4	Afrera, Ethiopia	13° 25'N	40° 50'E	-82	9	1	1	13, 96, 130
5	Agadem, Niger	16° 50'N	13° 20'E	350	6	1	1	68, 96, 112, 293, 294
7	Alexandersfontein, S. Africa	28° 50'S	24° 48'E	1119	35	2	ND	51, 53, 409
8	Asal, Djibouti	11° 36'N	42° 30'E	-155	35	2	1	107, 120, 128, 130, 448
13	Besaka, Ethiopia	8° 53'N	39° 52'E	1000	7	2	ND	362, 366
14	Bilma, Niger	18° 45'N	13° 00'E	310	10	2	4	96, 112, 113, 294
16	Bogoria, Kenya	0° 18'N	36° 06'E	990	19	4	3	318, 371, 410, 411, 412
17	Bosumtwi, Ghana	6° 30'N	1° 25'W	100	51	1	2	313–316
18	Bou Ali, Algeria	27° 10'N	0° 15'E	350	2	ND	2	70
19	Bou Bernous, Algeria	27° 20'N	3° 04'W	200	1	ND	2	69, 70
20	Bouloum Gana, Niger	15° 01'N	10° 37'E	340	1	1	ND	112
23	Chad, Chad-Nigeria-Niger	13° 00'N	14° 00'E	282	84	1	3	46, 74, 134, 142, 227, 293, 294, 361, 418–420
25	Chemchane-Aderg, Mauritania	21° 00'N	12° 07'W	260	6	1	6	54, 56
26	Cheshi, Zambia	8° 55'S	29° 43'E	928	4	4	3	694
27	Chew Bahir, Ethiopia	4° 45'N	37° 00'E	500	1	ND	2	143
28	Chilwa, Malawi	15° 30'S	35° 30'E	622	4	1	2	213, 214, 405, 406
29	Dayet Aaoua, Morocco	35° 21'N	5° 02'W	1180	2	ND	1	416
31	Dobi-Hanle, Ethiopia	11° 30'N	42° 00'E	120	11	1	3	120, 130, 421
33	Enneri Bardague, Chad	21° 30'N	17° 00'E	1025	44	1	4	31, 133, 134, 194, 253
35	Erg Ine Sakane, Mali	20° 48'N	0° 42'W	320	2	ND	2	175, 423
36	Erg Ine Sakane 20, Mali	20° 50'N	0° 38'W	320	4	ND	2	175, 257, 423
38	Erg Ine Sakane 57-19, Mali	20° 57'N	0° 32'W	320	2	ND	1	175, 423
41	Fachi, Niger	18° 07'N	11° 40'E	275	9	1	3	94, 112, 113, 293, 294
42	Fderick, Mauritania	22° 40'N	12° 42'W	350	1	1	ND	109
44	Ghadames, Libya	30° 10'N	10° 06'E	300	7	2	ND	68, 258
45	Gilf Kebir, Egypt	23° 30'N	26° 00'E	750	5	2	ND	
48	Hassei Gaboun, Mauritania	18° 18'N	15° 49'W	150	1	ND	1	68, 161
51	Hassi El Mejna, Algeria	31° 30'N	3° 20'E	500	5	1	1	422
52	Hassi Messouad, Algeria	32° 00'N	5° 51'E	470	3	ND	1	95
55	Ichourad Well, Mali	20° 48'N	0° 42'W	300	2	1	2	175, 423
56	Ine Kousamene, Mali	20° 40'N	0° 35'W	330	14	1	1	175, 423
60	Kadda, Algeria	26° 12'N	0° 53'E	250	3	ND	3	70
64	Khat, Mauritania	19° 10'N	12° 30'W	100	4	2	6	68, 319
65	Kivu, Rwanda-Burundi-Zaire	2° 00'S	29° 00'E	1462	15	6	4	90, 162, 248, 265
70	Magadi, Kenya	1° 53'S	36° 18'E	604	1	1	7	50
71	Makgadikgadi, Botswana	20° 24'S	24° 25'E	890	39	2	ND	71, 72, 145, 163-165, 384, 415
72	Manyara, Tanzania	3° 37'S	35° 49'E	945	14	6	2	177, 447
75	Mobutu Sese Seko, Uganda-Zaire	1° 30'N	36° 00'E	619	7	1	4	149, 303
80	Nabta, Egypt	23° 00'N	31° 00'E	250	33	1	1	353–355
82	Naivasha, Kenya	1° 15'S	36° 20'E	1890	5	2	2	9, 50, 426
83	Nakuru-Elmenteita, Kenya	0° 25'N	36° 10'E	1750	22	1	3	50, 62, 191, 271, 345
84	Oukechert, Mali	20° 40'N	0° 45'W	320	7	3	2	175, 423
85	Oum Arouaba, Mauritania	20° 52'N ·	12° 52'W	150	3	ND	1	55, 56

stead, we refer readers to the discussions in Butzer *et al.* (1972), Hecky and Degens (1973), Livingstone (1975), Rognon and Williams (1977), Servant and Servant-Vildary (1980), Gasse *et al.* (1980, 1987), Wendorf and Schild (1980), Petit-Maire and Riser (1983), Street-Perrott and Roberts (1983), Street-Perrott *et al.* (1985), and Gasse (1987). To date, however, no detailed and consistent picture has emerged from southern Africa.

Figure 13.19 summarizes the distribution of lake-level anomalies at 9000 and 6000 yr B.P., as well as the net changes between these two time periods. The data coverage for the whole area south of 9° is unfortunately still very sparse.

At 9000 yr B.P. all the sites from 15°30'S northward registered higher water levels than at present except the most northerly one (site 51, 31°30'N), where conditions were similar to today (Fig. 13.19). There are still no data from the Atlas to confirm this hint of a north-

Site no.	Site name	Latitude	Longitude	Altitude (m)	¹⁴C dates[a] (no.)	Dating control[b] 9000 yr B.P.	6000 yr B.P.	References[c]
86	Ounianga Kebir, Chad	19° 03'N	20° 30'E	310	2	ND	1	133, 134
87	Oyo, Sudan	19° 16'N	26° 11'E	510	8	1	2	399
90	Rukwa, Tanzania	8° 00'S	32° 30'E	793	4	ND	2	59, 144, 156
91	Saoura, Algeria	30° 00'N	2° 00'W	400	20	1	1	5, 70, 94, 432
92	Seguidine, Niger	20°24'N	12° 48'E	570	2	ND	7	402
93	Selima, Sudan	21° 19'N	29° 20'E	200	2	7	1	157, 428, 429
98	Siwa, Egypt	29° 13'N	25° 21'E	-30	4	1	6	150
99	Suguta, Kenya	1° 55'N	36° 28'E	900	5	2	ND	28
100	Tagnout-Chaggaret, Mali	21° 02'N	0° 45'W	310	8	ND	2	175, 423
101	Taoudenni, Mali	22° 30'N	4° 00'W	120	1	ND	2	75
103	Termit Ouest, Niger	16° 05'N	11° 15'E	300	6	2	2	112, 293, 294
105	Tirersioum, Mauritania	21° 22'N	16° 41'W	100	4	ND	1	97
108	Turkana, Kenya	5° 00'N	36° 00'E	375	51	1	1	50, 52, 262, 343, 370, 373, 374, 444
110	Victoria, Uganda-Tanzania-Kenya	1° 00'S	33° 00'E	1134	34	1	1	202, 308, 417
113	Ziway-Shala, Ethiopia	7° 45'N	38° 40'E	1558	52	2	1	136, 140, 143, 153, 304, 425, 427, 433

[a]Total number of ¹⁴C dates obtained from each lake basin.

[b]The dating control is coded as in Table 13.1 (for continuous sequences) and as follows for discontinuous samples such as shoreline dates: single radiocarbon date within 250 yr of the level being mapped, code 1; within 500 yr, code 2; within 750 yr, code 3; within 1000 yr, code 4; within 1250 yr, code 5; within 2000 yr, code 6; more than 2000 yr from the level being mapped, code 7. ND = no data.

[c]1, Alimen (1976a); 9, Ambrose (1984); 13, Bannert et al. (1970); 28, Bishop (1975); 31, Böttcher et al. (1972); 46, Buckley and Willis (1970); 50, Butzer et al. (1972); 51, Butzer et al. (1973); 52, Butzer (1980); 53, K. W. Butzer (personal communication, 1983); 54, Chamard (1973); 55, Chamard et al. (1970); 56, Chamard (1972); 59, Clark et al. (1970); 62, Cohen et al. (1983); 68, Commelin et al. (1980); 69, Conrad and Conrad (1965); 70, Conrad (1969); 71, Cooke (1984); 72, H. J. Cooke and K. Heine (personal communication, 1983); 74, Copens et al. (1968); 75, Cour and Duzer (1976); 90, Degens and Hecky (1974); 94, Delibrias et al. (1966); 95, Delibrias et al. (1971); 96, Delibrias et al. (1974); 97, Delibrias et al. (1976); 107, Eldridge et al. (1975); 109, Evin et al. (1975); 112, Faure et al. (1963); 113, Faure (1966); 119, Fontes and Pouchan (1975); 120, Fontes et al. (1973); 128, Gasse and Stieltjes (1973); 130, Gasse (1975); 133, Geyh and Jäkel (1974a); 134, Geyh and Jäkel (1974b); 136, Geze (1975); 140, Gillespie et al. (1983); 142, Grove and Warren (1968); 143, Grove et al. (1975); 144, A. T. Grove (unpublished); 145, Grove (1969); 149, Harvey (1976); 150, Hassan (1978); 153, Haynes and Haas (1974); 156, Haynes et al. (1971); 157, Haynes et al. (1979); 161, Hebrard (1978); 162, Hecky and Degens (1973); 163, Heine (1978); 164, Heine (1979); 165, Heine (1982); 175, Hillaire-Marcel et al. (1983); 177, Holdship (1976); 191, Isaac (1976); 194, Jäkel (1979); 202, Kendall (1969); 213, Lancaster (1979a); 214, Lancaster (1979b); 227, Maley (1973); 248, Olsson and Broecker (1959); 252, Pachur and Braun (1980); 253, Pachur (1975); 257, Petit-Maire and Riser (1981); 258, Petit-Maire and Delibrias (1979); 262, Phillipson (1977); 265, Pouclet (1975); 271, Richardson (1972); 291, Semmel (1971); 293, Servant and Servant-Vildary (1980); 294, Servant (1973); 303, Stoffers and Singer (1979); 304, Street (1980); 308, Stuiver et al. (1960); 312, Taieb (1971); 313, Talbot and Delibrias (1980); 314, Talbot and Delibrias (1977); 315, Talbot (1976); 316, Talbot et al. (1984); 318, Tiercelin et al. (1981); 319, Trompette and Manguin (1968); 343, Vondra et al. (1971); 345, Washbourn-Kamau (1972); 353, Wendorf and Schild (1980); 354, Wendorf et al. (1976); 355, Wendorf and the Members of the Combined Prehistoric Expedition (1977); 361, Williams and Adamson (1976); 362, Williams et al. (1977); 366, Williams et al. (1981); 370, Yamasaki et al. (1972); 371, Young and Renaut (1979); 373, Harvey and Grove (1982); 374, Owen et al. (1982); 384, Grove (1978); 399, Ritchie et al. (1985); 402, Baumhauer and Schulz (1984); 405, Lancaster (1981); 406, Shaw et al. (1984); 409, Butzer (1984); 410, Renaut (1982); 411, Vincens et al. (1986); 412, Carbonel et al. (1983); 415, Cooke and Verstappen (1984); 416, R. A. Perrott (personal communication, 1985); 417, Stager (1982); 418, Durand (1982); 419, Durand and Mathieu (1980); 420, Durand et al. (1984); 421, Gasse and Delibrias (1976); 422, Fontes et al. (1985); 423, Riser et al. (1983); 425, Lézine and Bonnefille (1982); 426, Richardson and Richardson (1972); 427, Bonnefille et al. (1986); 428, Haynes (1982); 429, Mehringer (1982); 432, Alimen (1976b); 433, Perrott (1979); 444, Meulengracht et al. (1981); 446, Gasse and Descourtieux (1979); 447, Keller et al. (1975); 448, Gasse and Fournier (1983); 694, Stager (1984).

ward decrease in moisture. South of 15°S one site (site 7) records slightly greater wetness and one (site 71) records conditions similar to the present. However, there is a notable lack of sites in southern Africa with well-preserved shorelines similar to those that surround expanded early-Holocene lakes north of the equator.

At 6000 yr B.P. water levels higher than at present prevailed in the Sahara and along the East African Rift Valley (Fig. 13.19). The most southerly sites in the data set (sites 90 at 8°S, 26 at 8°55'S, and 28 at 15°30'S) record slightly greater wetness than today.

Some interesting regional changes in lake status took place between 9000 and 6000 yr B.P. (Fig. 13.19). Apart from two interdune basins in Niger with a strong groundwater influence (sites 14 and 41), the lakes in the southern and northeastern Sahara either showed no net change or had begun to desiccate (compare with Fig. 13.6). In contrast, lake levels generally rose in the northwestern Sahara, including basins at least as far north as site 29 (35°N) in the Atlas. Two sites just south of the equator (sites 70 and 72) also experienced a net drop in lake level.

These large-scale trends in lake status, especially the contrasting behavior of the basins in the northeastern and northwestern Sahara, cannot be attributed merely to local hydrogeologic factors. We interpret them below as the result of changes in the atmospheric circulation between 9000 and 6000 yr B.P.

Comparison of the Pollen and Lake-Level Data

The paleoclimatic reconstructions based on pollen and lake-level data (Figs. 13.18 and 13.19) reveal strong general similarities, despite some minor differences attributable either to variable data quality or to differences in the climatic response of the two indicators. The most striking feature of the maps for 9000 yr B.P. is the broad belt of sites with greater effective moisture than today; the belt extends from the western Sahara across to eastern Africa and from about 30°N to at least 9°S (Lancaster, 1981; Stager, 1984; Haberyan, 1987; Finney and Johnson, 1991). Despite the limited data available, it seems clear that conditions similar to or drier than those of today prevailed in the Atlas Mountains and the northernmost Sahara (31-32°N) at the beginning of the Holocene. This pattern is most readily explained by a greater northward migration and intensification of the Northern Hemisphere summer monsoon. Pollen studies in the eastern Sahara (Ritchie and Haynes, 1987) suggest that the rainfall isohyets may have advanced about 5° north of their present positions.

South of 10°S the existing data are scanty and rather inconclusive, but they give no clear-cut indications of a stronger Southern Hemisphere summer monsoon. Greatest weight should probably be placed on the consistent palynologic evidence for drier early-Holocene climates in South Africa and Madagascar (van Zinderen Bakker, 1983; Burney 1987a,b; Scott, 1989a).

The two data sets differ in their climatic implications for 9000 yr B.P. in Ethiopia and the northern part of East Africa, where the lakes stood at or near their highest levels, whereas the pollen evidence suggests that the climate was cooler and slightly wetter than today. Two explanations for this discrepancy can be proposed. First, lower temperatures may have reduced evaporation rates and reinforced the effects of increased precipitation on the water budget of the lakes. Alternatively, the montane forest niche may have been occupied by a cloud forest consisting mainly of palynologically undetectable bamboo, while forest trees, which had been restricted by glacial aridity

to small refugia in the highlands (Hamilton, 1976), were gradually expanding their ranges. This hypothesis could readily be tested by searching for bamboo cuticle in early-Holocene lake sediments from the modern forest belt.

At 6000 yr B.P. a broad belt of moister conditions was still present in the Sahara and eastern Africa (Figs. 13.18 and 13.19), but its configuration had changed slightly. The northeastern and southern Sahara were drier than at 9000 yr B.P., whereas the northwestern Sahara and Atlas Mountains had become significantly wetter. Some of the more southerly sites in eastern Africa were also drier. South of 10°S data are insufficient to make comparisons between 9000 and 6000 yr B.P.

One may speculate about the seasonal distribution and synoptic origin of the rains that fell in the northwest Sahara and Atlas Mountains at 6000 yr B.P. The palynologic evidence reviewed above—for example, the greater abundance of deciduous relative to evergreen oaks—plus a limited number of stable-isotope measurements on ^{14}C-dated groundwaters (e.g., Edmunds and Wright, 1979) suggest that convective summer precipitation was involved. Could the monsoon have reached sites as far north as 35°N? On the face of it, this seems unlikely, for the maximum northward penetration of the moisture-bearing winds at the surface today averages 24°N in August (Leroux, 1983).

An alternative explanation is provided by paleoceanographic evidence (Mix *et al.,* 1986) suggesting that sea-surface temperatures off the coast of northwest Africa warmed by about 4°C between 8000 and 6000 yr B.P. This warming indicates that advection of cool water in the Canaries Current and coastal upwelling decreased markedly during this interval, probably as a result of weaker trade winds and a greater tendency for zonal (westerly) airflow in summer (compare Petit-Maire, 1980; Sarnthein *et al.,* 1981). The warmer nearshore waters would have provided a moisture source for summer thunderstorms in the Atlas and the northern Sahara. This scenario is consistent with the occurrence of cooler summer temperatures in the Mediterranean (Huntley and Prentice, 1988), more frequent summer rains in Crete and Israel (Roberts and Wright, this vol.), and semiarid conditions in the eastern Sahara, which was remote from the Atlantic moisture source.

Comparison of the Paleoclimatic Data with Model Simulations

The paleoclimatic reconstructions for 9000 and 6000 yr B.P. are in excellent agreement with the results of numerical modeling experiments of Kutzbach and Otto-Bliesner (1982) and Kutzbach and Guetter (1986). Because a detailed comparison of the hydrologic

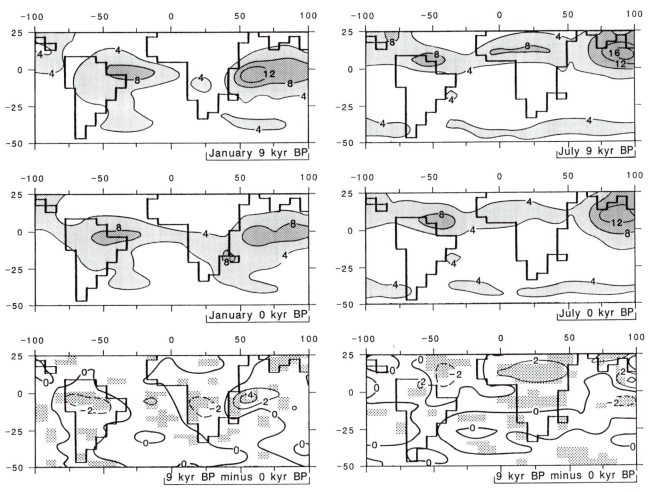

Fig. 13.20. Simulated precipitation (mm/day) over Africa and adjacent areas in the Community Climate Model experiments for January and July, 9000 and 0 yr B.P., together with difference maps for both seasons. Increases of more than 4 mm/day are shaded. Dotted grid squares indicate differences significant at the 5% level.

cycle for the last 18,000 yr for the latitude belt 8.9–26.6°N has already been published (Kutzbach and Street-Perrott, 1985), the present discussion focuses on the spatial pattern of precipitation anomalies at 9000 yr B.P. (Fig. 13.20) and its causes.

At 9000 yr B.P. perihelion occurred in late July; that is, at the height of the summer monsoon. The earth's axial tilt was 24.2°, compared with 23.4° today (Kutzbach and Guetter, 1986, Table 2). As a direct consequence, the insolation received at the top of the atmosphere in July between the equator and 30°N was 7% greater at 9000 yr B.P. than at present. According to the experiments of Kutzbach and Guetter (1986, Table 3) with the NCAR CCM, the increase in net radiation received at the land surface in these latitudes in July averaged 18 W m^{-2} (11%) and resulted in a considerable strengthening (2.6 mb) of the pressure gradient between the heated continents and the oceans. This was in turn associated with stronger onshore flow into northern and eastern Africa (as far south as about 16°S), which enhanced cloudiness and moisture con-

vergence over these areas. The tropical easterly jet at 190 mb over northern intertropical Africa also intensified by 10 m/s in July compared with the modern control case. Between 7 and 20°N, the simulated increase in rainfall exceeded 2 mm/day and was highly significant (Fig. 13.20). Despite a small increase in evaporation, the overall result was a large increase in effective moisture (P-E).

During the Northern Hemisphere winter (the dry season), reduced insolation resulted in lower evaporation rates but little change in precipitation. So, paradoxically, the net outcome of the compensating seasonal *radiation* anomalies was a large increase in effective *moisture* at 9000 yr B.P. South of the equator, however, the radiation anomalies at 9000 yr B.P. were smaller but of opposite sign. In the CCM experiment the lower insolation during the summer monsoon season resulted in reduced low-level convergence over land and lower pressure over the adjacent oceans. Simulated precipitation decreased by more than 2 mm/day in southern Africa and increased by more than 4 mm/day over the western Indian Ocean (Fig. 13.20).

Numerical experiments with general circulation models that incorporate a mixed-layer ocean (Kutzbach and Gallimore, 1988; Mitchell *et al.,* 1988) confirm the conclusions of the simulations described above, in which sea-surface temperatures were prescribed. None of the models employed so far is capable of computing changes in oceanic circulation, although they can be predicted in general terms from the anomalies of the low-level wind field (Kutzbach, Guetter, *et al.,* this vol.). The intensification of the North Atlantic high in July would lead to slightly enhanced upwelling in the Canaries Current off northwest Africa. Coastal upwelling driven by the strengthened southwest monsoon flows would also have increased in summer along the northern shores of the Gulf of Guinea, as well as off the Horn of Africa and Arabia. In contrast, the weakening of the South Atlantic high associated with the insolation minimum during southern summer at 12,000–9000 yr B.P. may have reduced upwelling intensity off southwest Africa and resulted in more frequent rain showers in the Namib coastal desert.

By 6000 yr B.P. the seasonal insolation anomalies in both hemispheres had declined considerably (Kutzbach and Guetter, 1986), weakening the pressure and circulation patterns described above. However, effective moisture (P–E) in the latitude belt 8.9–26.6°N was still significantly greater than today (Kutzbach and Street-Perrott, 1985).

Conclusions

The African pollen and lake-level data described above are in excellent agreement with the results of numerical simulations based on the astronomical theory of climatic change. The insolation maximum in northern summer at 9000 yr B.P. led to stronger monsoon rains right across northern Africa, extending along the East African Rift Valley down to about 9°S. Slightly stronger upwelling and more anticyclonic conditions than today along the northwest continental margin gave rise to dry conditions in the Atlas Mountains.

In the Southern Hemisphere the CCM results suggest that both the summer monsoon and the adjacent oceanic anticyclones were weaker at 9000 yr B.P. than at present because of the decreased insolation and reduced land-sea contrast. Early-Holocene pollen spectra from the Zambezian, Kalahari-Highveld, and Cape regions of South Africa as well as from Madagascar record dry conditions. This area is about 15° farther south than the area of maximum dryness in the model (Fig. 13.20). Slightly more frequent rains in the Namib coastal desert would be expected from the model results.

By 6000 yr B.P. the insolation anomalies had begun to decay. The Northern Hemisphere summer monsoon remained stronger than today, although its northern limit had retreated southward in the central and eastern Sahara. Wetter conditions in the northwestern Sahara and Atlas Mountains were linked with decreased upwelling and warmer sea-surface temperatures offshore. The interior of South Africa, the Cape region, and Madagascar became wetter after 6000 yr B.P. as summer radiation began to increase, whereas the opposite trend may have occurred in the Namib Desert.

The preceding discussion ignored the shorter (10^2–10^3 yr) climatic oscillations superimposed on the broad trends attributable to orbital forcing (Street-Perrott and Roberts, 1983). These smaller-scale fluctuations appear to reflect a different set of causal mechanisms. No description of Holocene climates on the African continent will be complete until these shorter-term oscillations are understood.

Acknowledgments

We are indebted to A. Ballouche, R. Bonnefille, J. A. Coetzee, H. Hooghiemstra, H. F. Lamb, A.-M. Lézine, and J. C. Ritchie for supplying unpublished pollen counts and to A. H. Hamilton, H. F. Lamb, J. C. Ritchie, and E. M. van Zinderen Bakker for reviewing the manuscript. N. Roberts, S. P. Harrison, and D. A. Baines assisted with the compilation of the lake-level data and T. Webb III and his assistants with the plotting of the pollen and lake-level maps. We also thank Mrs. A. Newman for drafting the figures.

References

Agwu, C. O. C. (1979). "Vegetations- und Klimageschichtliche Untersuchung an marinen Sedimenten vor der westafrikanischen Küste." Unpublished dissertation, University of Göttingen.

Agwu, C. O. C., and Beug, H.-J. (1982). Palynological studies of marine sediments off the West African coast. *"Meteor" Forschung-Ergebnisse* C36, 1–30.

Alimen, H. (1976a). Alternances "pluvial-aride" et "érosion-sédimentation" au Sahara nord-occidental. *Revue de géographie physique et de géologie dynamique* 18, 301–311.

——. (1976b). Variations climatiques dans les zones désertiques de l'Afrique nord-equatoriale durant les quarantes derniers millénaires. *In* "Proceedings of the VII Pan-African Congress on Prehistory and Quaternary Studies" (B. Abebe, J. Chavaillon, and J. E. G. Sutton, Eds.), pp. 337–370. Ethiopian Ministry of Culture, Addis Ababa.

Ambrose, S. H. (1984). "Holocene Environments and Human Adaptations in the Central Rift Valley, Kenya." Unpublished Ph.D. thesis, University of California, Berkeley.

Assémien, P., Filleron, J. C., Martin, L., and Tastet, J. P. (1971). Le Quaternaire de la zone littorale de Côte d'Ivoire. *Quaternaria* 15, 305–316.

Ballouche, A., Lefèvre, D., Carruesco, C. V., Raynal, J. F., and Texier, J. P. (1986). Holocene environments of coastal and continental Morocco. *In* "Quaternary Climate in Western

Mediterranean" (F. López-Vera, Ed.), pp. 517-531. Universidad Autónoma, Madrid.

Bannert, D., Brinkmann, J., Kading, K.-C., Knetsch, G., Kursten, M., and Mayrhofer, H. (1970). Zur Geologie der Danakil-Senke. *Geologische Rundschau* 59, 409-443.

Baumhauer, R., and Schulz, E. (1984). The Holocene lake of Seguedine, Kaouar, NE Niger. *Palaeoecology of Africa* 16, 283-290.

Ben Tiba, B., and Reille, M. (1982). Recherches pollenanalytiques dans les montagnes de Kroumirie (Tunisie septentrionale): Premiers résultats. *Ecologia Mediterranea* 8, 75-86.

Bishop, W. W. (1975). Geological reconnaissance of the lower Suguta Valley in northern Kenya. *In* "Cambridge Meeting on Desertification" (A. T. Grove, Ed.), p. 62. Department of Geography, Cambridge University, Cambridge.

Bonnefille, R., and Hamilton, A. S. (1986). Quaternary and late Tertiary history of Ethiopian vegetation. *Acta Universitatis Upsaliensis Symb. Bot. Ups.* 26, 48-63.

Bonnefille, R., and Riollet, G. (1988). The Kashiru pollen sequence (Burundi): Palaeoclimatic implications for the last 40,000 yr B.P. in tropical Africa. *Quaternary Research* 30, 19-35.

Bonnefille, R., Robert, C., Delibrias, G., Elenga, C., Herbin, J. P., Lézine, A. M., Périnet, G., and Tiercelin, J. J. (1986). Palaeoenvironment of Lake Abiyata, late Holocene, Ethiopia, during the past 2000 years. *In* "Sedimentation in the African Rift" (L. E. Frostick, R. W. Renaut, I. Reid, and J.-J. Tiercelin, Eds.), pp. 253-265. Geological Society of London Special Publication 25. Blackwell Scientific Publications, Oxford.

Böttcher, U., Ergenzinger, P.-J., Jaeckel, S. H., and Kaiser, K. (1972). Quartäre Seebildungen und ihre Mollusken Inhalte im Tibesti-Gebirge. *Zeitschrift für Geomorphologie* 16, 182-234.

Brun, A. (1979). Recherches palynologiques sur les sédiments marins du golfe de Gabès. *Revue de géologie méditerranéenne* 6, 247-264.

——. (1983). Etude palynologique des sédiments marins holocènes de 5000 B.P. à l'Actuel dans le golfe de Gabès (Mer pélagienne). *Pollen et spores* 25, 437-460.

——. (1985). La couverture steppique en Tunisie au Quaternaire supérieur. *Comptes rendus hebdomadaires de l'Académie des sciences, Paris, série II* 14, 1085-1090.

Buckley, J. D., and Willis, E. H. (1970). Isotopes' radiocarbon measurements VIII. *Radiocarbon* 12, 105.

Burney, D. A. (1987a). Late Quaternary stratigraphic charcoal records from Madagascar. *Quaternary Research* 28, 274-280.

——. (1987b). Pre-settlement vegetation changes at Lake Tritivakely, Madagascar. *Palaeoecology of Africa* 18, 357-381.

Butzer, K. W. (1980). The Holocene lake plain of North Rudolph, East Africa. *Physical Geography* 1, 42-58.

——. (1984). Late Quaternary environments in South Africa. *In* "Late Cainozoic Palaeoclimates of the Southern Hemisphere" (J. C. Vogel, Ed.), pp. 235-264. Balkema, Rotterdam.

Butzer, K. W., Isaac, G. L., Richardson, J. L., and Washbourn-Kamau, C. (1972). Radiocarbon dating of East African lake levels. *Science* 175, 1069-1076.

Butzer, K. W., Fock, G. J., Stuckenrath, R., and Zilch, A. (1973). Palaeohydrology of late Pleistocene Lake Alexandersfontein, Kimberley, South Africa. *Nature* 243, 328-330.

Caratini, C., and Cour, P. (1980). Aéropalynologie en Atlantique oriental au large de la Mauritanie, du Sénégal et de la Gambie. *Pollen et spores* 22, 245-256.

Caratini, C., and Giresse, P. (1979). Contribution palynologique à la connaissance des environnements continentaux et marines à la fin du Quaternaire. *Comptes rendus hebdomadaires de l'Académie des sciences, Paris, série D* 289, 379-382.

Caratini, C., Bellet, J., and Tissot, C. (1979). Etude microscopique de la matière organique: Palynologie et palynofaciès. *In* "Orgon III—Mauritanie-Sénégal-Iles du Cap Vert, Octobre 1976" (M. Arnould and R. Pelet, Eds.), pp. 215-247. Centre national de la recherche scientifique, Paris.

Carbonel, P., Grosdidier, E., Peypouquet, J.-P., and Tiercelin, J.-J. (1983). Les ostracodes, temoins de l'évolution hydrologique d'un lac de rift. Exemple du lac Bogoria, Rift Gregory, Kenya. *Bulletin Centres recherche exploration-production Elf-Aquitaine* 7, 301-313.

Chamard, P. C. (1972). Les lacs holocènes de l'Adrar de Mauritanie et peuplements préhistoriques. *Notes africaines, Institut français d'Afrique noire* 133, 1-8.

——. (1973). Monographie d'une sebkha continentale du sud-ouest saharien: la sebkha de Chemchane (Adrar de Mauritanie). *Bulletin de l'Institut français d'Afrique noire, Sénégal* 35A, 207-243.

Chamard, P. C., Guitat, R., and Thilmans, G. (1970). Le lac holocène et le gisement neolithique de l'Oum Arouaba (Adrar de Mauritanie). *Bulletin de l'Institut français d'Afrique noire, Sénégal* 32, 688-723.

Clark, J. D., Haynes, C. V., Jr., Mawby, J. E., and Gautier, A. (1970). Interim report on palaeo-anthropological investigations in the Lake Malawi Rift. *Quaternaria* 13, 305-354.

CLIMAP Project Members (1981). Seasonal reconstructions of the earth's surface at the last glacial maximum. *Geological Society of America Map and Chart Series* MC-36.

Coakley, J. A., and Cess, R. D. (1985). Response of the NCAR Community Climate Model to the radiative forcing by the naturally occurring tropospheric aerosol. *Journal of the Atmospheric Sciences* 42, 1677-1692.

Coetzee, J. A. (1964). Evidence for a considerable depression of the vegetation belts during the Upper Pleistocene on the East African mountains. *Nature* 204, 564-566.

——. (1967). Pollen analytical studies in East and southern Africa. *Palaeoecology of Africa* 3, 1-146.

Cohen, A. S., Dussinger, R., and Richardson, J. (1983). Lacustrine paleochemical interpretations based on eastern and southern African ostracodes. *Palaeogeography, Palaeoclimatology, Palaeoecology* 43, 129-151.

Commelin, D., Petit-Maire, N., and Casanova, J. (1980). Chronologie isotopique saharienne des derniers 10,000 ans. *Bulletin du Musée d'anthropologie préhistorique de Monaco* 23, 37-88.

Conrad, G. (1969). "L'évolution continentale post-hercynienne du Sahara algérien." Série géologique 10, Centre national de la recherche scientifique-Centre de recherche sur les zones arides, Paris.

Conrad, G., and Conrad, J. (1965). Précisions stratigraphiques sur les depôts holocènes du Sahara occidental grâce à la geochronologie absolue. *Comptes rendus hebdomadaires des séances de l'Académie des sciences* 7, 234-236.

Cooke, H. J. (1984). The evidence from northern Botswana of late Quaternary climatic change. *In* "Late Cainozoic Palaeoclimates of the Southern Hemisphere" (J. C. Vogel, Ed.), pp. 265-278. Balkema, Rotterdam.

Cooke, H. J., and Verstappen, H. T. (1984). The landforms of the western Makgadikgadi basin in northern Botswana, with a consideration of the chronology of the evolution of Lake Palaeo-Makgadikgadi. *Zeitschrift für Geomorphologie* 28, 1-19.

Cooremans, B. (1989). Pollen production in central South Africa. *Pollen et spores* 31, 61-78.

Coppens, R., Durand, G. L. A., and Guillet, B. (1968). Nancy natural radiocarbon measurements I. *Radiocarbon* 10, 119-123.

Cour, P., and Duzer, D. (1976). Persistance d'un climat hyperaride au Sahara central et méridional au cours de l'Holocène. *Revue de géographie physique et de géologie dynamique* 18, 175-198.

Deacon, J., and Lancaster, N. (1988). "Late Quaternary Palaeoenvironments of Southern Africa." Clarendon Press, Oxford.

Degefu, W. (1987). Some aspects of meteorological drought in Ethiopia. *In* "Drought and Hunger in Africa" (M. H. Glantz, Ed.), pp. 23-36. Cambridge University Press, Cambridge.

Degens, E. T., and Hecky, R. E. (1974). Paleoclimatic reconstruction of late Pleistocene and Holocene based on biogenic sediments from the Black Sea and a tropical African lake. *Colloques internationaux du C.N.R.S.* 219, 1-11.

Delibrias, G., Guillier, M. T., and Labeyrie, J. (1966). Gif natural radiocarbon measurements II. *Radiocarbon* 8, 92.

——. (1971). Gif natural radiocarbon measurements VI. *Radiocarbon* 13, 223.

——. (1974). Gif natural radiocarbon measurements VIII. *Radiocarbon* 16, 15-94.

Delibrias, G., Ortlieb, L., and Petit-Maire, N. (1976). New ^{14}C data for the Atlantic Sahara (Holocene): Tentative interpretations. *Journal of Human Evolution* 5, 535-546.

Durand, A. (1982). Oscillations of Lake Chad over the past 50,000 years: New data and new hypothesis. *Palaeogeography, Palaeoclimatology, Palaeoecology* 39, 37-53.

Durand, A., and Mathieu, P. (1980). Evolution paléogéographique et paléoclimatique du bassin tchadien au Pléistocène supérieur. *Revue de géologie dynamique et de géographie physique* 22, 329-341.

Durand, A., Fontes, J.-C., Gasse, F., Icole, M., and Lang, J. (1984). Le nord-ouest du lac Tchad au Quaternaire: Etude de paléoenvironnements alluviaux, éoliens, palustres et lacustres. *Palaeoecology of Africa* 16, 215-243.

Edmunds, W. M., and Wright, E. F. (1979). Groundwater recharge and palaeoclimate in the Sirte and Kufra basins, Libya. *Journal of Hydrology* 40, 215-241.

Eldridge, K. L., Stipp, J. J., and Cohen, S. J. (1975). University of Miami radiocarbon dates III. *Radiocarbon* 17, 239-246.

Elmoutaki, S., Lézine, A. M., and Thomassin, B. A. (1992). Mayotte (canal de Mozambique). Evolution de la végétation et du climat au cours de la dernière transition glaciaire-interglaciaire et de l'Holocène. *Comptes rendus de l'Académie des sciences de Paris* 314, sér. III, 237-244.

Evin, J., Marien, G., and Pachiaudi, C. (1975). Lyon natural radiocarbon measurements V. *Radiocarbon* 17, 23.

Faure, H. (1966). "Reconnaissance géologique des formations sédimentaires post-paléozoïques du Niger oriental." Mémoires 47, Bureau de recherches géologiques et minières, Paris.

——. (1969). Lacs quaternaires du Sahara. *Mitteilungen der internationale Vereinigung für theoretische und angewandte Limnologie* 17, 131-146.

Faure, H., Manguin, E., and Nydal, R. (1963). Formations lacustres du Quaternaire supérieur du Niger oriental: diatomites et âges absolus. *Bulletin du Bureau de recherches géologiques et minières, Paris* 3, 41-63.

Finney, B. P., and Johnson, T. C. (1991). Sedimentation in Lake Malawi (East Africa) during the past 10,000 years: A continuous paleoclimatic record from the southern tropics. *Palaeogeography, Palaeoclimatology, Palaeoecology* 85, 351-366.

Folland, C. K., Palmer, T. N., and Parker, D. E. (1986). Sahel rainfall and worldwide sea temperatures, 1901-1985. *Nature* 320, 602-607.

Fontes, J.-C., and Pouchan, P. (1975). Les cheminées du lac Abhé (T.F.A.I.): stations hydroclimatiques de l'Holocène. *Comptes rendus hebdomadaires de l'Académie des sciences, Paris* 280D, 383-386.

Fontes, J.-C., Moussié, C., Pouchan, P., and Weidmann, P. (1973). Phases humides au Pléistocène supérieur et à l'Holocène dans le sud de l'Afar (T.F.A.I.). *Comptes rendus hebdomadaires de l'Académie des sciences, Paris* 277, 1973-1976.

Fontes, J.-C., Gasse, F., Callot, Y., Plaziat, J.-C., Carbonel, P., Dupeuble, P. A., and Kaczmarska, I. (1985). Freshwater to marine-like environments from Holocene lakes in northern Sahara. *Nature* 317, 608-610.

Gasse, F. (1975). "L'évolution des lacs de l'Afar Central (Ethiopie et T.F.A.I.) du Plio-Pléistocène à l'Actuel: reconstitution des paléomilieux lacustres à partir de l'étude des diatomées." Unpublished thesis, University of Paris VI.

——. (1987). Diatoms for reconstructing palaeoenvironments and palaeohydrology in tropical semi-arid zones: Example of some lakes from Niger since 12,000 yr B.P. *Hydrobiologia* 154, 127-163.

Gasse, F., and Delibrias, G. (1976). Les lacs de l'Afar Central (Ethiopie et T.F.A.I.) au Pléistocène supérieur. *Palaeolimnology of Lake Biwa and the Japanese Pleistocene* 4, 529-575.

Gasse, F., and Descourtieux, C. (1979). Diatomées et évolution de trois milieux éthiopiens d'altitude différente, au cours du Quaternaire supérieur. *Palaeoecology of Africa* 11, 117-134.

Gasse, F., and Fournier, M. (1983). Sédiments plio-quaternaires et tectonique en bordure du golfe de Tadjoura (République de Djibouti). *Bulletin Elf-Aquitaine* 7, 285-300.

Gasse, F., and Stieltjes, L. (1973). Les sédiments du Quaternaire récent du lac Asal (Afar Central, Territoire français des Afars et des Issas). *Bulletin du Bureau de recherches géologiques et minières, Paris (deuxième série)* 4, 229-245.

Gasse, F., Rognon, P., and Street, F. A. (1980). Quaternary history of the Afar and Ethiopian Rift lakes. *In* "The Sahara and the Nile: Quaternary Environments and Prehistoric Occupation in Northern Africa," pp. 361–400. A. A. Balkema, Rotterdam.

Gasse, F., Fontes, J.-C., Plaziat, J.-C., Carbonel, P., Kaczmarska, I., De Deckker, P., Soulié-Marsche, I., Callot, Y., and Dupeuble, P. A. (1987). Biological remains, geochemistry and stable isotopes for the reconstruction of environmental and hydrological changes in the Holocene lakes from north Sahara. *Palaeogeography, Palaeoclimatology, Palaeoecology* 60, 1-46.

Geyh, M. A., and Jäkel, D. (1974a). Spätpleistozäne und holozäne Klima-geschichte der Sahara aufgrund zugänglicher ^{14}C-Daten. *Zeitschrift für Geomorphologie* NF 18, 82-98.

——. (1974b). ^{14}C-Altersbestimmungen im Rahmen der Forschungsarbeiten der Aussenstelle Bardai/Tibesti der Freien Universität Berlin. *Pressedienst Wissenschaft der Freien Universität Berlin* 5, 107-117.

Geze, F. (1975). New dates on ancient Galla lake levels (Ethiopian Rift Valley). *Bulletin of the Geophysical Observatory, Addis Ababa* 15, 119-124.

Gillespie, R., Street-Perrott, F. A., and Switsur, V. R. (1983). Postglacial arid episodes in Ethiopia have implications for climate prediction. *Nature* 306, 680-683.

Grove, A. T. (1969). Landforms and climatic change in the Kalahari and Ngamiland. *Geographical Journal* 135, 191-212.

——. (1978). Late Quaternary climatic change and the conditions for current erosion in Africa. *Geo-Eco-Trop* 2, 291-300.

Grove, A. T., and Goudie, A. S. (1971). Late Quaternary lake levels in the Rift Valley of Ethiopia and elsewhere in tropical Africa. *Nature* 234, 403-405.

Grove, A. T., and Warren, A. (1968). Quaternary landforms and climate on the south side of the Sahara. *Geographical Journal* 134, 194-208.

Grove, A. T., Street, F. A., and Goudie, A. S. (1975). Former lake levels and climatic change in the Rift Valley of southern Ethiopia. *Geographical Journal* 141, 177-202.

Haberyan, K. A. (1987). Fossil diatoms and the paleolimnology of Lake Rukwa, Tanzania. *Freshwater Biology* 17, 429-436.

Hall, J. B., Swaine, M. D., and Talbot, M. R. (1978). An early Holocene leaf flora from Lake Bosumtwi, Ghana. *Palaeogeography, Palaeoclimatology, Palaeoecology* 24, 247-261.

Hamilton, A. C. (1972). The interpretation of pollen diagrams from highland Uganda. *Palaeoecology of Africa* 7, 45-49.

——. (1974). The history of the vegetation. In "East African Vegetation" (E. M. Lind and M. E. S. Morrison, Eds.), pp. 188-209. Longman, London.

——. (1976). The significance of patterns of distribution shown by forest plants and animals in tropical Africa for the reconstruction of Upper Pleistocene palaeoenvironments: A review. *Palaeoecology of Africa* 9, 63-97.

——. (1982). "Environmental History of East Africa." Academic Press, London.

Hamilton, A. C., and Perrott, R. A. (1980). Modern pollen deposition on a tropical African mountain. *Pollen et spores* 22, 437-468.

Harvey, P., and Grove, A. T. (1982). A prehistoric source for the Nile. *Geographical Journal* 148, 327-336.

Harvey, T. J. (1976). "The Paleolimnology of Lake Mobutu Sese Seko, Uganda-Zaire: The Last 28,000 Years." Unpublished Ph.D. thesis, Duke University, Durham, N.C.

Hassan, F. A. (1978). Archaeological explorations of the Siwa Oasis region, Egypt. *Current Anthropology* 19, 146-148.

Hastenrath, S. (1984). "The Glaciers of Equatorial East Africa." D. Reidel, Dordrecht.

——. (1985). "Climate and Circulation of the Tropics." D. Reidel, Dordrecht.

Haynes, C. V., Jr. (1982). Lacustrine chronology and geomorphology of Selima Oasis, northern Sudan. In "Abstracts with Programs, 95th Annual Meeting, October 1982," p. 511. Geological Society of America, Boulder, Colo.

Haynes, C. V., Jr., and Haas, H. (1974). Southern Methodist University radiocarbon date list I. *Radiocarbon* 16, 368-380.

Haynes, C. V., Jr., Grey, D. C., and Long, A. (1971). Arizona radiocarbon dates VIII. *Radiocarbon* 13, 1-18.

Haynes, C. V., Jr., Mehringer, P. J., and Zaghloul, E. L. A. (1979). Pluvial lakes of north-western Sudan. *Geographical Journal* 145, 437-445.

Hebrard, L. (1978). "Contribution à l'étude géologique du Quaternaire du littoral mauritanien entre Nouakchott et Nouadhibou (18-21 N)." Unpublished thesis, Laboratoire de géologie, Faculté des sciences, Université de Lyon.

Hecky, R. E., and Degens, E. T. (1973). "Late Pleistocene-Holocene Chemical Stratigraphy and Paleolimnology of the Rift Valley Lakes of Central Africa." Technical Report WHOI-73-28. Woods Hole Oceanographic Insitution, Woods Hole, Mass.

Heine, K. (1978). Radiocarbon chronology of late Quaternary lakes in the Kalahari, southern Africa. *Catena* 5, 145-149.

——. (1979). Reply to Cooke's discussion of: K. Heine: Radiocarbon chronology of late Quaternary lakes in the Kalahari, southern Africa. *Catena* 6, 259-266.

——. (1982). The main stages of the late Quaternary evolution of the Kalahari region, southern Africa. *Palaeoecology of Africa* 15, 53-76.

Hillaire-Marcel, C., Riser, J., Rognon, P., Petit-Maire, N., Rosso, J. C., and Soulié-Marsche, I. (1983). Radiocarbon chronology of Holocene hydrologic changes in northeastern Mali. *Quaternary Research* 20, 145-164.

Holdship, S. A. (1976). "The Paleolimnology of Lake Manyara, Tanzania: A Diatom Analysis of a 56 Meter Sediment Core." Unpublished Ph.D. thesis, Duke University, Durham, N.C.

Hooghiemstra, H. (1988a). Palynological records from northwest African margin marine sediments: A general outline of the interpretation of the pollen signal. *Philosophical Transactions of the Royal Society of London* B318, 431-449.

——. (1988b). Changes of major wind belts and vegetation zones in NW Africa 28,000-5000 yr B.P., as deduced from a marine pollen record near Cap Blanc. *Review of Palaeobotany and Palynology* 55, 101-140.

Hooghiemstra, H., and Agwu, C. O. C. (1986). Distribution of palynomorphs in marine sediments: A record for seasonal wind patterns over NW Africa and adjacent Atlantic. *Geologische Rundschau* 75, 81-95.

Hooghiemstra, H., Agwu, C. O. C., and Beug, H.-J. (1986). Pollen and spore distribution in recent marine sediments: A record of NW-African seasonal wind patterns and vegetation belts. *"Meteor" Forschung-Ergebnisse* C40, 87-135.

Hooghiemstra, H., Stalling, H., Agwu, C. O. C., and Dupont, L. M. (1992). Vegetational and climatic changes at the northern fringe of the Sahara 250,000-5000 years B.P.: Evidence from 4 marine pollen records located between Portugal and the Canary Islands. *Review of Palynology and Palaeobotany* 74, 1-53.

Houghton, R. W. (1983). Seasonal variations of the subsurface thermal structure of the Gulf of Guinea. *Journal of Physical Oceanography* 13, 2070-2081.

Huntley, B., and Birks, H. J. B. (1983). "An Atlas of Past and Present Pollen Maps for Europe 0-13,000 Years Ago." Cambridge University Press, Cambridge.

Huntley, B., and Prentice, I. C. (1988). July temperatures in Europe from pollen data, 6000 years before present. *Science* 241, 687-690.

Isaac, G. H. (1976). A preliminary report on stratigraphic studies in the Nakuru Basin, Kenya. In "Proceedings of the VII Pan-African Congress on Prehistory and Quaternary Studies" (B. Abebe, J. Chavaillon, and J. E. G. Sutton, Eds.), pp. 409-411. Ethiopian Ministry of Culture, Addis Ababa.

Jäkel, D. (1979). Run-off and fluvial formation processes in the Tibesti Mountains as indicators of climatic history in the central Sahara during the late Pleistocene and Holocene. *Palaeoecology of Africa* 11, 13–44.

Keller, C. M., Hansen, C., and Alexander, C. S. (1975). Archaeology and palaeoenvironments in the Manyara and Engaruka basins, Tanzania. *Geographical Review* 65, 3364–3376.

Kendall, R. L. (1969). An ecological history of the Lake Victoria basin. *Ecological Monographs* 39, 121–176.

Kutzbach, J. E., and Gallimore, R. G. (1988). Sensitivity of a coupled atmosphere/mixed-layer ocean model to changes in orbital forcing at 9000 yr B.P. *Journal of Geophysical Research* 93, 803–821.

Kutzbach, J. E., and Guetter, P. J. (1986). The influence of changing orbital parameters and surface boundary conditions on climate simulations for the past 18,000 years. *Journal of the Atmospheric Sciences* 43, 1726–1759.

Kutzbach, J. E., and Otto-Bliesner, B. (1982). The sensitivity of the African-Asian monsoonal climate to orbital parameter changes for 9000 years B.P. in a low-resolution general circulation model. *Journal of the Atmospheric Sciences* 39, 1177–1188.

Kutzbach, J. E., and Street-Perrott, F. A. (1985). Milankovitch forcing of fluctuations in the level of tropical lakes from 18 to 0 kyr B.P. *Nature* 317, 130–134.

Lamb, H. F., Eicher, U., and Switsur, V. R. (1989). An 18,000-year record of vegetational, lake-level and climatic change from the Middle Atlas, Morocco. *Journal of Biogeography* 16, 65–74.

Lamb, P. J. (1978). Case studies of tropical Atlantic surface circulation patterns during recent sub-Saharan weather anomalies: 1967 and 1968. *Monthly Weather Review* 106, 482–491.

——. (1985). Rainfall in Subsaharan West Africa during 1941–1983. *Zeitschrift für Gletscherkunde und Glazialgeologie* 21, 131–139.

Lancaster, N. (1979a). Late Quaternary events in the Chilwa basin, Malawi. *Palaeoecology of Africa* 11, 233–235.

——. (1979b). The changes in lake level. *In* "Lake Chilwa: Studies of Change in a Tropical Ecosystem" (M. Kalk, A. J. McLachlan, and C. Howard-Williams, Eds.), pp. 43–58. Monographiae Biologicae 35. Junk, The Hague.

——. (1980). Relationships between altitude and temperature in Malawi. *South African Geographical Journal* 62, 89–97.

——. (1981). Formation of the Holocene Lake Chilwa sandbar, southern Malawi. *Catena* 8, 369–382.

Laseski, R. A. (1983). "Modern Pollen Data and Holocene Climate Change in Eastern Africa." Unpublished Ph.D. dissertation, Brown University, Providence, R.I.

Leroux, M. (1983). "Le Climat de l'Afrique tropicale," 2 vols. Editions Champion, Paris.

Lézine, A.-M. (1981). "Le Lac Abiyata (Ethiopie)—Palynologie et paléoclimatologie du Quaternaire récent," 2 vols. Unpublished thesis, Université de Bordeaux I.

——. (1985). Commentaire sur "L'essai de reconstitution de la végétation et du climat holocènes sur la côte septentrionale de Sénégal" de J. Médus (*Review of Palaeobotany and Palynology* 41, 31–38). *Review of Palaeobotany and Palynology* 45, 373–376.

——. (1986). Environnement et paléoenvironnement des niayes depuis 12,000 B.P. *In* "Changements globaux en Afrique durant le Quaternaire: Passé-présent-futur" (H. Faure, L. Faure, and E. S. Diop, Eds.), pp. 261–263. Travaux et documents 197, Editions de l'ORSTOM, Paris.

——. (1987). "Paléoenvironnements végétaux d'Afrique nord-tropicale depuis 12,000 B.P.: Analyse pollinique de séries

sédimentaires continentales (Sénégal-Mauritanie)," 2 vols. Unpublished thesis, Université d'Aix-Marseille II.

Lézine, A.-M., and Bonnefille, R. (1982). Diagramme pollinique holocène d'un sondage du lac Abiyata (Ethiopie, 7°42' Nord). *Pollen et spores* 24, 463–480.

Lézine, A.-M., Bieda, S., Faure, H., and Saos, J.-L. (1985). Etude palynologique d'un milieu margino-littoral: La tourbière de Thiaye (Sénégal). *Sciences géologiques bulletin* 38, 79–89.

Livingstone, D. A. (1967). Postglacial vegetation of the Ruwenzori Mountains in Equatorial Africa. *Ecological Monographs* 37, 25–52.

——. (1971). A 22,000-year pollen record from the plateau of Zambia. *Limnology and Oceanography* 16, 349–356.

——. (1975). Late Quaternary climatic change in Africa. *Annual Reviews of Ecology and Systematics* 6, 249–280.

Maitima, J. M. (1991). Vegetation response to climatic change in Central Rift Valley, Kenya. *Quaternary Research* 35, 234–245.

Maley, J. (1973). Mécanisme des changements climatiques aux basses latitudes. *Palaeogeography, Palaeoclimatology, Palaeoecology* 14, 193–227.

——. (1981). "Etudes palynologiques dans le bassin du Tchad et paléoclimatologie de l'Afrique nord-tropical de 30,000 ans à l'époque actuelle." Thesis, Université de sciences et techniques du Languedoc. Travaux et documents 127, Editions de l'ORSTOM, Paris.

——. (1983). Histoire de la végétation et du climat de l'Afrique nord-tropicale au Quaternaire récent. *Bothalia* 14, 377–389.

——. (1987). Fragmentation de la forêt dense humide africaine et extension des biotopes montagnards au Quaternaire récent: Nouvelles données polliniques et chronologiques. Implications paléoclimatiques et biogéographiques. *Palaeoecology of Africa* 18, 307–334.

Maley, J., and Livingstone, D. A. (1983). Extension d'un élément montagnard dans le sud du Ghana (Afrique de l'Ouest) au Pléistocène et à l'Holocène inférieur: premières données polliniques. *Comptes rendus hebdomadaires de l'Académie des sciences, Paris, série II* 296, 1287–1292.

Maley, J., Livingstone, D. A., Giresse, P., Thouveny, N., Brenac, P., Kelts, K., Kling, G., Stager, C., Haag, M., Fournier, M., Bandet, Y., and Zogning, A. (1990). Lithostratigraphy, volcanism, palaeomagnetism and palynology of Quaternary lacustrine deposits from Barombi Mbo (West Cameroon): Preliminary results. *Journal of Volcanic and Geothermal Research* 42, 319–335.

Martin, A. R. H. (1968). Pollen analysis of Groenvlei Lake sediments, Knysna (South Africa). *Review of Palaeobotany and Palynology* 7, 107–144.

Mason, I. M., Guzkowska, M. A. J., Rapley, C. G., and Street-Perrott, F. A. (In press). The response of lake levels and areas to climatic change. *Climatic Change*.

Meadows, M. E. (1984a). Late Quaternary vegetation history of the Nyika Platau, Malawi. *Journal of Biogeography* 11, 209–222.

——. (1984b). Contemporary pollen spectra and vegetation of the Nyika Plateau, Malawi. *Journal of Biogeography* 11, 223–233.

Meadows, M. E., and Sugden, J. M. (1991). A vegetation history of the last 14,000 years on the Cederberg, south-western Cape Province. *South African Journal of Science* 87, 33–43.

Médus, J. (1984a). Analyse pollinique des sédiments holocènes du lac Tanma, Sénégal. *Palaeoecology of Africa* 16, 255–264.

——. (1984b). Essai de reconstitution de la végétation et du climat holocènes de la côte septentrionale du Sénégal. *Review of Palaeobotany and Palynology* 41, 31-38.

Mehringer, P. J. (1982). Early Holocene climate and vegetation in the eastern Sahara: The evidence from Selima Oasis, Sudan. *In* "Abstracts with Programs, 95th Annual Meeting, October 1982," p. 564. Geological Society of America, Boulder, Colo.

Melia, M. B. (1984). The distribution and relationship between palynomorphs in aerosols and deep-sea sediments off the coast of north-west Africa. *Marine Geology* 58, 345-371.

Meulengracht, A., McGovern, P., and Lawn, B. (1981). University of Pennsylvania radiocarbon dates XXI. *Radiocarbon* 23, 227-240.

Mintz, Y. (1984). The sensitivity of numerically simulated climates to land-surface boundary conditions. *In* "The Global Climate" (J. T. Houghton, Ed.), pp. 79-105. Cambridge University Press, Cambridge.

Mitchell, J. F. B., Grahame, N. S., and Needham, K. J. (1988). Climate simulations for 9000 years before present: Seasonal variations and effect of the Laurentide ice sheet. *Journal of Geophysical Research* 93, 8283-8303.

Mix, A. C., Ruddiman, W. F., and McIntyre, A. (1986). Late Quaternary paleoceanography of the tropical Atlantic. 1: Spatial variability of annual mean sea-surface temperatures, 0-20,000 years B.P. *Paleoceanography* 1, 43-66.

Morrison, M. E. S. (1968). Vegetation and climate in the uplands of southwestern Uganda during the later Pleistocene period, 1: Muchoya Swamp, Kigezi District. *Journal of Ecology* 56, 363-384.

Newell, R. E., and Kidson, J. W. (1984). African mean wind changes between Sahelian wet and dry periods. *Journal of Climatology* 4, 27-33.

Nicholson, S. E. (1986a). The nature of rainfall variability in Africa south of the equator. *Journal of Climatology* 6, 516-530.

——. (1986b). The spatial coherence of African rainfall anomalies: Interhemispheric teleconnections. *Journal of Climate and Applied Meteorology* 25, 1365-1381.

Nicholson, S. E., and Chervin, R. M. (1983). Recent rainfall fluctuations in Africa—Interhemispheric teleconnections. *In* "Variations in the Global Water Budget" (A. Street-Perrott, M. Beran, and R. Ratcliffe, Eds.), pp. 221-238. D. Reidel, Dordrecht.

Nicholson, S. E., and Flohn, H. (1980). African environmental and climatic changes and the general atmospheric circulation in late Pleistocene and Holocene. *Climatic Change* 2, 313-348.

Nieuwolt, S. (1977). "Tropical Climatology: An Introduction to the Climates of Low Latitudes." Wiley, London.

Olsson, E. A., and Broecker, W. S. (1959). Lamont natural radiocarbon measurements V. *Radiocarbon* 2, 13-14.

Osmaston, H. A. (1958). "Pollen Analysis in the Study of the Past Vegetation and Climate of the Ruwenzori and Its Neighborhood." Uganda Forest Department, Kampala.

Owen, R. B., Barthelme, J. W., Renaut, R. W., and Vincens, A. (1982). Palaeolimnology and archaeology of Holocene deposits north-east of Lake Turkana, Kenya. *Nature* 298, 523-529.

Pachur, H.-J. (1975). Zur spätpleistozäne und holozänen Formung auf der Nordabdachung des Tibestigebirges. *Die Erde* 106, 21-46.

Pachur, H.-J., and Braun, G. (1980). The paleoclimate of the central Sahara, Libya and the Libyan desert. *Palaeoecology of Africa* 12, 351-364.

Palmer, T. N. (1986). Influence of the Atlantic, Pacific and Indian oceans on Sahel rainfall. *Nature* 322, 251-253.

Perrott, F. A. (1979). "Late Quaternary Lakes in the Ziway-Shala Basin, Southern Ethiopia." Unpublished Ph.D. thesis, University of Cambridge.

Perrott, R. A. (1982). A high altitude pollen diagram from Mount Kenya: Its implications for the history of glaciation. *Palaeoecology of Africa* 14, 77-83.

——. (1987). Postglacial climatic events in eastern Africa. *In* "Abstracts XII INQUA Congress, Ottawa," p. 241. International Union for Quaternary Research.

Perrott, R. A., and Street-Perrott, F. A. (1982). New evidence for a late Pleistocene wet phase in northern intertropical Africa. *Palaeoecology of Africa* 14, 57-75.

——. (1987). Postglacial climatic events in the tropics: Possible mechanisms. *Terra Cognita* 7, 214-215.

Petit-Maire, N. (1980). Holocene biogeographical variations along the northwestern African coast (28°-19°N): Paleoclimatic implications. *Palaeoecology of Africa* 12, 365-377.

Petit-Maire, N., and Delibrias, G. (1979). Late Holocene palaeoenvironment in the Ghadamès area. *Maghreb Review* 4, 138-139.

Petit-Maire, N., and Riser, J. (1981). Holocene lake deposits and palaeoenvironments in central Sahara, northeastern Mali. *Palaeogeography, Palaeoclimatology, Palaeoecology* 35, 45-61.

——, Eds. (1983). "Sahara ou Sahel?" Laboratoire de géologie du Quaternaire, Centre national de la recherche scientifique, Marseille.

Petit-Maire, N., and Schulz, E. (1981). Data on Holocene vegetation in the Atlantic Sahara. *Palaeoecology of Africa* 13, 199-203.

Phillipson, D. W. (1977). Lowasera. *Azania* 12, 1-32.

Pokras, E. M., and Mix, A. C. (1985). Eolian evidence for spatial variability of late Quaternary climates in tropical Africa. *Quaternary Research* 24, 137-149.

——. (1987). Earth's precession cycle and Quaternary climatic change in tropical Africa. *Nature* 326, 486-487.

Pouclet, A. (1975). Histoire des grands lacs de l'Afrique centrale mise au point des connaissances actuelles. *Revue de géographie physique et de géologie dynamique* 2, 475-482.

Prell, W. L., Marvil, R. E., and Luther, M. E. (1990). Variability in upwelling fields in the northwestern Indian Ocean 2. Data-model comparison at 9000 years B.P. *Paleoceanography* 5, 447-457.

Reille, M. (1976). Analyse pollinique de sédiments postglaciaires dans le Moyen Atlas et le Haut Atlas marocains: Premiers résultats. *Ecologia Mediterranea* 2, 153-170.

——. (1977). Contribution pollenanalytique à l'histoire holocène de la végétation du Rif (Maroc septentrional). *Bulletin de l'Association française pour l'étude du Quaternaire, supplément* 50, 53-76.

Renaut, R. W. (1982). "Late Quaternary Geology of the Lake Bogoria Fault-Trough, Kenya Rift Valley." Unpublished Ph.D. thesis, University of London.

Richards, K. (1986). Preliminary results of pollen analysis of a 6000 year core from Mboandong, a crater lake in Cameroun. *In* "The Hull University Cameroun Expedition 1981-82 Final Report" (R. G. E. Baker, K. Richards, and C. A. Rimes, Eds.), pp. 14-28. Miscellaneous Series 30, Department of Geography, University of Hull.

Richardson, J. L. (1972). Palaeolimnological records from Rift lakes in central Kenya. *Palaeoecology of Africa* 6, 131-136.

Richardson, J. L., and Richardson, A. E. (1972). History of an African Rift lake and its climatic implications. *Ecological Monographs* 42, 499-534.

Riser, J., Hillaire-Marcel, C., and Rognon, P. (1983). Les phases lacustres holocènes. *In* "Sahara ou Sahel? Quaternaire récent du bassin de Taoudenni (Mali)" (N. Petit-Maire and J. Riser, Eds.), pp. 65-86. Laboratoire de géologie du Quaternaire, Centre national de la recherche scientifique, Marseille.

Ritchie, J. C. (1984). Analyse pollinique de sédiments holocènes supérieurs des hauts plateaux du Maghreb oriental. *Pollen et spores* 26, 489-496.

——. (1986). Modern pollen spectra from Dakhleh Oasis, western Egyptian desert. *Grana* 25, 177-182.

Ritchie, J. C., and Haynes, C. V., Jr. (1987). Holocene vegetation zonation in the eastern Sahara. *Nature* 330, 645-647.

Ritchie, J. C., Eyles, C. H., and Haynes, C. V., Jr. (1985). Sediment and pollen evidence for an early to mid-Holocene humid period in the eastern Sahara. *Nature* 314, 352-355.

Rognon, P. (1976). Essai d'interprétation des variations climatiques au Sahara depuis 40,000 ans. *Revue de géographie physique et géologie dynamique, série 2* 18, 251-282.

——. (1981). Une extension des déserts (Sahara et Moyen-Orient) au cours du Tardiglaciaire (18,000-10,000 ans B.P.). *Revue de géologie dynamique et de géographie physique* 22, 313-328.

——. (1987). Late Quaternary climatic reconstruction for the Magreb (North Africa). *Palaeogeography, Palaeoclimatology, Palaeoecology* 58, 11-34.

Rognon, P., and Williams, M. A. J. (1977). Late Quaternary climatic changes in Australia and North Africa: A preliminary investigation. *Palaeogeography, Palaeoclimatology, Palaeoecology* 21, 285-327.

Rossignol-Strick, M., and Duzer, D. (1979a). Late Quaternary pollen and dinoflagellate cysts in marine cores off West Africa. *"Meteor" Forschung-Ergebnisse* C30, 1-14.

——. (1979b). West African vegetation and climate since 22,500 B.P. from deep-sea cores palynology. *Pollen et spores* 21, 105-134.

Sarnthein, M. (1978). Sand deserts during glacial maximum and climatic optimum. *Nature* 272, 43-45.

Sarnthein, M., Tetzleff, G., Koopman, B., Wolter, K., and Pflaumann, U. (1981). Glacial and interglacial wind regimes over the eastern subtropical Atlantic and northwest Africa. *Nature* 293, 193-196.

Schalke, H. J. W. G. (1973). The Upper Quaternary of the Cape Flats area (Cape Province, South Africa). *Scripta Geologica* 15, 1-57.

Schulz, E. (1980a). An investigation of current geological processes and an interpretation of Saharan palaeoenvironments. *In* "The Geology of Libya," Vol. III (M. J. Salem and M. T. Busrewil, Eds.), pp. 791-796. Academic Press, New York.

——. (1980b). Zur Vegetation der östlichen zentralen Sahara und zu ihrer Entwicklung im Holozän. *Würzburger Geographische Arbeiten* 51, 1-194.

——. (1984). The recent pollen rain in the eastern central Sahara—A transect between northern Libya and southern Niger. *Palaeoecology of Africa* 16, 245-253.

——. (1986a). Holocene vegetation in central Sahara (eastern Niger and southern Libya). *In* "Changements globaux en Afrique durant le Quaternaire: Passé-présent-futur" (H.

Faure, L. Faure, and E. S. Diop, Eds.), pp. 427-429. Travaux et documents 197, Editions de l'ORSTOM, Paris.

——. (1986b). Present pollen rain between the Mediterranean and the Atlantic Ocean. A Libya-Togo transect. *In* "Changements globaux en Afrique durant le Quaternaire: Passé-présent-futur" (H. Faure, L. Faure, and E. S. Diop, Eds.), pp. 431-433. Travaux et documents 197, Editions de l'ORSTOM, Paris.

Scott, L. (1982a). A 5000-year old pollen sequence from spring deposits in the bushveld at the north of the Soutpansberg, South Africa. *Palaeoecology of Africa* 14, 45-55.

——. (1982b). A late Quaternary glacial record from the Transvaal bushveld, South Africa. *Quaternary Research* 17, 339-370.

——. (1982c). Late Quaternary fossil pollen grains from the Transvaal, South Africa. *Review of Palaeobotany and Palynology* 36, 241-278.

——. (1984). Late Quaternary paleoenvironments in the Transvaal on the basis of palynological evidence. *In* "Late Cainozoic Palaeoclimates of the Southern Hemisphere" (J. C. Vogel, Ed.), pp. 317-327. Balkema, Rotterdam.

——. (1986). Pollen analysis and paleoenvironmental interpretation of late Quaternary sediment exposures in the eastern Orange Free State, South Africa. *Palaeoecology of Africa* 17, 113-122.

——. (1987a). Pollen analysis of hyena coprolites and sediments from Equus Cave, Taung, southern Kalahari (South Africa). *Quaternary Research* 28, 144-156.

——. (1987b). Late Quaternary forest history in Venda, South Africa. *Review of Palaeobotany and Palynology* 53, 1-10.

——. (1989a). Climatic conditions in southern Africa since the last glacial maximum, inferred from pollen analysis. *Palaeogeography, Palaeoclimatology, Palaeoecology* 70, 345-353.

——. (1989b). Late Quaternary vegetation history and climatic change in the eastern Orange Free State, South Africa. *South African Journal of Botany* 55, 107-116.

Scott, L., and Vogel, J. C. (1983). A late Quaternary pollen profile from the Transvaal highveld, South Africa. *South African Journal of Science* 79, 266-272.

Scott, L., Cooremans, B., de Wet, J. S., and Vogel, J. C. (1991). Holocene environmental changes in Namibia inferred from pollen analyses of swamp and lake deposits. *The Holocene* 1, 8-13.

Semmel, A. (1971). Zur jungquartären Klima- und Reliefentwicklung in der Danakilwuste (Athiopien) und ihren westlichen Randgebieten. *Erdkunde* 25, 199-208.

Servant, M. (1973). "Sequences continentales et variations climatiques: Evolution du bassin du Tchad au Cenozoïque supérieur." Unpublished thesis, University of Paris.

Servant, M., and Servant-Vildary, S. (1980). L'environnement quaternaire du bassin du Tchad. *In* "The Sahara and the Nile: Quaternary Environments in Northern Africa" (M. A. J. Williams and H. Faure, Eds.), pp. 133-162. Balkema, Rotterdam.

Shaw, P., Crossley, R., and Davison-Hirschmann, S. (1984). A major fluctuation in the level of Lake Chilwa, Malawi, during the Iron Age. *Palaeoecology of Africa* 16, 391-395.

Sowunmi, M. A. (1981a). Aspects of late Quaternary vegetational changes in West Africa. *Journal of Biogeography* 8, 457-474.

——. (1981b). Palynological indications of late Quaternary environmental changes in Nigeria. *Pollen et spores* 23, 125-148.

Stager, J. C. (1982). The diatom record of Lake Victoria (East Africa): The last 17,000 years. *In* "Proceedings of the VII International Symposium on Living and Fossil Diatoms, August 1982," pp. 455-476.

——. (1984). "Environmental Changes at Lake Cheshi, Zambia since 40,000 Years B.P. with Additional Information from Lake Victoria, East Africa." Unpublished Ph.D. thesis, Duke University, Durham, N.C.

Stoffers, P., and Singer, A. (1979). Clay minerals in Lake Mobutu Sese Seko (Lake Albert)—Their diagenetic changes as an indicator of the paleoclimate. *Geologische Rundschau* 68, 1009-1024.

Street, F. A. (1980). Chronology of late Pleistocene and Holocene lake-level fluctuations, Ziway-Shala Basin, Ethiopia. *In* "Proceedings of the VIII Pan-African Congress on Prehistory and Quaternary Studies" (R. E. Leakey and B. A. Ogot, Eds.), pp. 143-146. International Louis Leakey Memorial Institute for African Prehistory, Nairobi.

——. (1981). Tropical palaeoenvironments. *Progress in Physical Geography* 5, 157-185.

Street, F. A., and Grove, A. T. (1976). Environmental and climatic implications of late Quaternary lake-level fluctuations in Africa. *Nature* 261, 385-390.

——. (1979). Global maps of lake-level fluctuations since 30,000 yr B.P. *Quaternary Research* 12, 83-118.

Street-Perrott, F. A., and Harrison, S. P. (1985). Lake levels and climate reconstruction. *In* "Paleoclimate Analysis and Modeling" (A. D. Hecht, Ed.), pp. 291-340. Wiley, New York.

Street-Perrott, F. A., and Roberts, N. (1983). Fluctuations in closed-basin lakes as an indicator of past atmospheric circulation patterns. *In* "Variations in the Global Water Budget" (A. Street-Perrott, M. Beran, and R. Ratcliffe, Eds.), pp. 331-345. D. Reidel, Dordrecht.

Street-Perrott, F. A., Roberts, N., and Metcalfe, S. E. (1985). Geomorphic implications of late Quaternary hydrological and climatic changes in the Northern Hemisphere tropics. *In* "Environmental Change and Tropical Geomorphology" (I. Douglas and T. Spencer, Eds.), pp. 165-183. George Allen and Unwin, London.

Stuiver, M., Deevey, E. S., and Gralenski, L. J. (1960). Yale natural radiocarbon measurements V. *Radiocarbon* 2, 55-56.

Taieb, M. (1971). Aperçus sur les formations quaternaires et la néotectonique de la basse vallée de l'Aouache. *Comptes rendus de la Société géologique de France* 13, 63-65.

Talbot, M. R. (1976). "Late Quaternary Sedimentation in Lake Bosumtwi, Ghana." 20th Annual Report, Research Institute of African Geology, University of Leeds.

Talbot, M. R., and Delibrias, G. (1977). Holocene variations in the level of Lake Bosumtwi, Ghana. *Nature* 268, 722-724.

——. (1980). A new late Pleistocene-Holocene water-level curve for Lake Bosumtwi, Ghana. *Earth and Planetary Science Letters* 47, 336-344.

Talbot, M. R., and Hall, J. B. (1981). Further Quaternary leaf fossils from Lake Bosumtwi, Ghana. *Palaeoecology of Africa* 13, 83-92.

Talbot, M. R., Livingstone, D. A., Palmer, P. G., Maley, J., Melack, J. M., Delibrias, G., and Gulliksen, S. (1984). Preliminary results from sediment cores from Lake Bosumtwi, Ghana. *Palaeoecology of Africa* 16, 173-192.

Tamura, T. (1986). Regolith-stratigraphic study of late Quaternary environmental history in the West Cameroon highlands. *In* "Geomorphology and Environmental Changes in Tropical Africa: Case Studies in Cameroon and Kenya—A Preliminary Report of the Tropical African Geomorphology and Late-Quaternary Palaeoenvironments Research

Project 1984/85" (H. Kadomura, Ed.), pp. 63-93. Hokkaido University, Sapporo, Japan.

Taylor, D. M. (1990). Late Quaternary pollen records from two Ugandan mires: Evidence for environmental change in the Rukiga Highlands of southwest Uganda. *Palaeogeography, Palaeoclimatology, Palaeoecology* 80, 283-300.

Taylor, H. C. (1978). Capensis. *In* "Biogeography and Ecology of Southern Africa," Vol. 1 (M. J. A. Werger, Ed.), pp. 170-229. Monographiae Biologicae 31. Junk, The Hague.

Thompson, B. W. (1965). "The Climate of Africa." Oxford University Press, Oxford.

Tiercelin, J.-J., Renaut, R. W., Delibrias, G., LeFournier, J., and Bieda, S. (1981). Late Pleistocene and Holocene lake level fluctuations in the Lake Bogoria basin, northern Kenya Rift Valley. *Palaeoecology of Africa* 13, 105-120.

Tomlinson, R. W. (1974). Preliminary biogeographical studies on the Inyanga Mountains, Rhodesia. *South African Geographical Journal* 56, 15-26.

Trewartha, G. T. (1981). "The Earth's Problem Climates." University of Wisconsin Press, Madison.

Trompette, R., and Manguin, E. (1968). Nouvelles observations sur le Quaternaire lacustre de l'extremité sud-est de l'Adrar de Mauritanie (Sahara occidental). *Annales de la Faculté des sciences, Université de Dakar, série sciences de la terre* 22, 151-162.

Tyson, P. D. (1986). "Climatic Change and Variability in Southern Africa." Oxford University Press, Cape Town.

van Zinderen Bakker, E. M. (1957). A pollen analytical investigation of the Florisbad deposits (South Africa). *In* "Proceedings of the 3rd Pan-African Congress of Prehistory, Livingstone, 1955" (J. D. Clark, Ed.), pp. 56-67. Chatto and Windus, London.

——. (1962). A late-glacial and post-glacial climatic correlation between East Africa and Europe. *Nature* 194, 201-203.

——. (1964). A pollen diagram from equatorial Africa, Cherangani, Kenya. *Geologie en Mijnbouw* 43, 123-128.

——. (1967). Upper Pleistocene and Holocene stratigraphy and ecology on the basis of vegetation changes in sub-Saharan Africa. *In* "Background to Evolution in Africa" (W. W. Bishop and J. D. Clark, Eds.), pp. 125-147. University of Chicago Press, Chicago.

——. (1976). The evolution of late Quaternary palaeoclimates of southern Africa. *Palaeoecology of Africa* 9, 160-202.

——. (1978). Quaternary vegetation changes in southern Africa. *In* "Biogeography and Ecology of Southern Africa" (M. J. A. Werger, Ed.), pp. 131-143. Monographiae Biologicae 31. Junk, The Hague.

——. (1982). Pollen analytical studies of the Wonderwerk Cave, South Africa. *Pollen et spores* 24, 235-250.

——. (1983). The late Quaternary history of climate and vegetation in East and southern Africa. *Bothalia* 14, 369-375.

——. (1984). A late- and post-glacial pollen record from the Namib Desert. *Palaeoecology of Africa* 16, 421-428.

van Zinderen Bakker, E. M., Sr. and Coetzee, J. A. (1972). A reappraisal of late Quaternary climatic evidence from tropical Africa. *Palaeoecology of Africa* 7, 151-181.

van Zinderen Bakker, E. M., and Müller, M. (1987). Pollen studies in the Namib Desert. *Pollen et spores* 29, 185-206.

Vincens, A. (1989a). Les forêts claires zambéziennes du bassin Sud-Tanganyika. Evolution entre 25,000 et 6000 ans B.P. *Comptes rendus de l'Académie des sciences de Paris* 308, sér. II, 809-814.

——. (1989b). Paléoenvironnements du bassin Nord-Tanganyika (Zaire, Burundi, Tanzanie) au cours des 13 derniers mille ans: apport de la palynologie. *Reveiw of Palaeobotany and Palynology* 61, 69-88.

Vincens, A., Casanova, J., and Tiercelin, J.-J. (1986). Paleolimnology of Lake Bogoria (Kenya) during the 4500 B.P. high lacustrine phase. *In* "Sedimentation in the African Rift" (L. E. Frostick, R. W. Renaut, I. Reid, and J.-J. Tiercelin, Eds.), pp. 323-330. Geological Society of London Special Publication 25. Blackwell Scientific Publications, Oxford.

Vogel, J. C., Ed. (1984). "Late Cainozoic Palaeoclimates of the Southern Hemisphere." Balkema, Rotterdam.

Vondra, G. F., Johnson, G. D., Bowen, B. E., and Behrensmeyer, A. K. (1971). Preliminary stratigraphical studies of the east Rudolf basin, Kenya. *Nature* 231, 245.

Washbourn-Kamau, C. (1972). Studies on former lake levels in the Nakuru-Elmenteita and the Naivasha basins. *Palaeoecology of Africa* 6, 138.

Wendorf, F., and Schild, R. (1980). "Prehistory of the Eastern Sahara." Academic Press, New York.

Wendorf, F., Schild, R., Said, R., Haynes, C. V., Gautier, A., and Kobusiewicz, M. (1976). The prehistory of the Egyptian Sahara. *Science* 193, 103-114.

Wendorf, F., and the Members of the Combined Prehistoric Expedition (1977). Late Pleistocene and recent climatic changes in the Egyptian Sahara. *Geographical Journal* 143, 211-234.

Werger, M. J. A., Ed. (1978). "Biogeography and Ecology of Southern Africa," 2 vols. Monographiae Biologicae 31. Junk, The Hague.

White, F. (1978). The Afromontane region. *In* "Biogeography and Ecology of Southern Africa," Vol. 1 (M. J. A. Werger, Ed.), pp. 465-511. Monographiae Biologicae 31. Junk, The Hague.

——. (1983). "The Vegetation of Africa; A Descriptive Memoir to Accompany the UNESCO/AETFAT/UNSO Vegetation Map of Africa." UNESCO Natural Resources Research 20. UNESCO, Paris.

Williams, M. A. J., and Adamson, D. (1976). "The Origins of the Soils between the Blue and White Nile Rivers, Central Sudan, with Some Agricultural and Climatological Implications." Occasional Paper 6, Economic and Social Research Council, Khartoum.

Williams, M. A. J., and Faure, H., Eds. (1980). "The Sahara and the Nile: Quaternary Environments in Northern Africa." Balkema, Rotterdam.

Williams, M. A. J., Bishop, P. M., Dakin, F. M., and Gillespie, R. (1977). Late Quaternary lake levels in southern Afar and the adjacent Ethiopian Rift. *Nature* 267, 690-693.

Williams, M. A. J., Williams, F. M., and Bishop, P. M. (1981). Late Quaternary history of Lake Besaka, Ethiopia. *Palaeoecology of Africa* 13, 93-104.

Yamasaki, F., Hamada, C., and Hamada, T. (1972). Riken natural radiocarbon measurements VII. *Radiocarbon* 14, 223-238.

Young, J. A. T., and Renaut, R. W. (1979). A radiocarbon date from Lake Bogoria, Kenya Rift Valley. *Nature* 278, 243-245.

CHAPTER 14

Climatic History of Central and South America since 18,000 yr B.P.: Comparison of Pollen Records and Model Simulations

Vera Markgraf

Overview

This chapter documents the paleoclimatic history of Central and South America derived from pollen and lake-level records and compares it with results from paleoclimatic modeling experiments based on the general circulation model (GCM) of the National Center for Atmospheric Research (NCAR). Because the GCM results are analyzed at 3000-yr intervals over the last 18,000 yr, I discuss the pollen and lake-level data in the same time resolution.

Sixty pollen records (up to 1987) fulfill the requirements of continuity and chronological control imposed in this synthesis: nine from tropical and subtropical lowlands in Central and South America; five from the Central American highlands; 25 from the high Andes both north and south of the equator; three from the southern subtropics; and 18 from the southern Andes and subantarctic islands. Paleoclimatic interpretation of the pollen records is semiquantitative and is based on bioclimatic parameters such as elevational temperature gradients or latitudinal precipitation gradients that can be related to vegetation zonation. Additional paleoclimatic information is gleaned from lake-level records, derived either from dated shorelines (eight additional records) or from the limnological history of lake sediment sections for which a pollen record also exists (seven records).

Large-scale paleoclimatic evaluation of the paleoenvironmental records indicates that full- and late-glacial climates throughout Central and South America were cooler than the modern climate by about 4-5°C. The overall lower temperatures reflect the influence of the globally increased terrestrial and marine ice cover, which also resulted in lower sea levels and cooler sea-surface temperatures. Although these parameters led to an overall decrease in ocean-land moisture trans-

port and hence overall drier terrestrial climates, in some regions full- and late-glacial climates were as moist as or moister than today, such as in the northern and southern subtropics of the Americas. In paleocirculation terms, this specific scenario can be interpreted as a weakening of the subtropical high-pressure cells and an intensification of the midlatitude westerly circulation. This implies an equatorward shift of the northern westerlies and a poleward shift of the southern westerlies, as also suggested by paleoclimatic simulation results from GCMs.

The GCM simulations also predict an enhanced early-Holocene summer monsoon in the northern subtropics, contrasting with the suppressed monsoon in the southern subtropics. Although paleoenvironmental records from the southern subtropics of the Americas are few, they indicate a period of aridity during that time that supports the model results. Although the Northern Hemisphere subtropics of the Americas are only partially covered in this review with records from Central America, neither the paleoenvironmental data nor the model results clearly signal an enhanced monsoon during the early Holocene, in contrast to the greatly enhanced monsoon activity in Africa and India.

One of the biggest controversies in South American paleoclimatic interpretations concerns the existence of a Younger Dryas equivalent, a cold period between 11,000 and 10,000 yr B.P. Available data from South America show either no change or numerous paleoenvironmental fluctuations during the transition period from late-glacial to postglacial times (Markgraf, 1991). Thus, the concept of an interval characterized by the same climatic signal documented for the North Atlantic seems too simplistic to explain the South American paleoclimatic changes. Moreover, at this point only a few records have been analyzed and dated in sufficient detail to yield unambiguous climatic information for the interval in question.

The principal outcomes of this paleoenvironmental and paleoclimatic synthesis are a better understanding of the type and quality of data available at present and the identification of regions and time periods for which a much greater effort is needed to gather data suitable for testing the validity of paleoclimatic model results. The value of this exercise of comparing paleoclimatic data with paleoclimatic model results, however, extends beyond the aspect of validating the model simulations. Such comparisons can deepen our understanding of the dynamics of climate and climate change and of the forcing of the specific boundary conditions on regional paleoclimates.

Background

Interactive atmospheric GCMs specify a set of relevant boundary conditions, such as solar radiation, sea-surface temperatures, and ice sheets. When it became feasible to use these models to simulate global paleoclimates, the need arose to test the models' predictions, for example by comparing them with paleoclimatic proxy data. Among the many types of terrestrial regional paleoclimatic proxy data, pollen and lake-level sequences stand out as the most geographically widespread and numerically abundant sets of records that can be considered comparable to paleoclimatic model simulations. In an attempt to elucidate the mechanisms of climatic change that led from a glacial to an interglacial environment during the last 18,000 yr, COHMAP (Cooperative Holocene Mapping Project) proposed comparing paleoclimatic model results computed at 3000-yr intervals for the last 18,000 yr to a regional and global set of paleoenvironmental data.

For this comparison to be realistic, the geographic and temporal resolution of the model and the paleoenvironmental data sets had to be similar, and the paleoclimatic parameters also had to be comparable. European and eastern North American paleoenvironmental data sets have been quantified in terms of paleoclimatic parameters such as temperature and precipitation via modern pollen-climate transfer functions (Prentice, 1985; Webb *et al.,* 1987; Huntley and Prentice, 1988). Because paleoenvironmental records from South America lack the calibration data necessary to establish a quantitative relation between pollen assemblages and specific climatic parameters, fossil pollen records from South America cannot be quantified in climatic terms to the same degree as the paleoclimatic model. The approach followed in this analysis is therefore semiquantitative: biogeographic information on specific taxa or vegetation zones and their relation to specific climatic parameters is used to reconstruct past patterns of precipitation, temperature, and atmospheric circulation. From a network of modern climate data in Patagonia, for example, either

mean annual precipitation or seasonal effective moisture defined according to Walter's (1973) climadiagram approach (see Fig. 14.2) can be derived from the steppe-forest boundary, which is an ecotone that can be recognized quite easily by its characteristic pollen assemblage. The paramo of the northern Andes, also palynologically well differentiated, can be defined by certain values of temperature or summer temperature that delimit it from other altitudinal vegetation zones, especially the Andean forest. Paleoclimatic changes can thus be inferred from the records on the basis of shifts in vegetation types and boundaries and the climatic parameters they represent. At times the fossil pollen assemblage does not appear to resemble any modern assemblage. In most instances, however, this apparent lack of analogue has been shown to be the result of insufficient modern pollen data.

South America and Central America south of the Tropic of Cancer together cover a latitudinal range of 79° (from 23°N to 56°S) and encompass a correspondingly wide variety of biogeographic and climatic zones. Physiographically, the region comprises the Precambrian shields of Guyana and Brazil, the numerous primarily crystalline rocks of varying tectonic and structural origins in the Andes between latitudes 10°N and 56°S, and the sedimentary deposits east of the ranges, including Mesozoic and Cenozoic marine, lacustrine, fluvial, glacial, and eolian deposits. Climatic belts include the easterly trade wind zones between latitudes 10 and 30° north and south of the equator, the subtropical high-pressure areas, the westerlies between 30 and 65°S, the Antarctic high-pressure area, and the polar easterlies south of latitude 65°S. Bioclimatic zones include the tropics, subtropics, and temperate, subantarctic, and high-elevation zones. Finally, vegetation types include tropical rainforests, subtropical semideciduous or deciduous forests and savannas, tropical and subtropical deserts, temperate grasslands and semideserts, temperate and subantarctic evergreen rainforests and deciduous forests, sclerophyllous woodlands, Andean grasslands, moorland, and tundra (Fig. 14.1).

The described multitude of environments is not matched by a comparable number of palynologic and lake-level records covering the late Quaternary and especially the Holocene. Although tropical and subtropical forests and savannas represent the largest proportion of surface area in the region, only nine pollen and lake records fulfill the continuity and dating control requirements for inclusion in this paleoclimatic analysis. On the other hand, the environmentally more uniform high-elevation zone of the Andes above 2000 m is much better represented, with 28 records from sites in Costa Rica, Venezuela, Colombia, Peru, Bolivia, and Argentina. Equally well represented are the southern temperate forest, steppe, and tundra environments, with 20 records from Argentina, Chile, and the subantarctic islands.

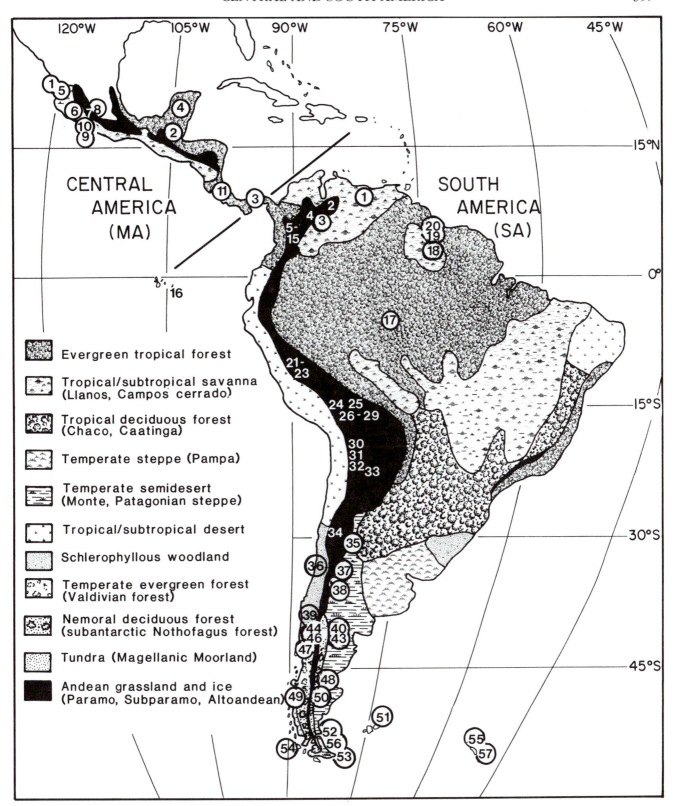

Fig. 14.1. Map of major vegetation types in Central and South America (according to Hueck and Seibert, 1981), showing the location of the sites in Central America (MA 1–11) and South America and the subantarctic islands (SA 1–57) reviewed in this chapter (see also Table 14.1). Reprinted from Markgraf (1989a), with permission from *Quaternary Science Reviews* and Pergamon Press.

Because of the uneven geographic coverage of the records, the paleoclimatic interpretations proposed in this chapter must be considered a first approximation. Because of our still limited understanding of the modern distribution and ecological significance of many of the principal plant taxa as well as the modern relationship between plant frequency and pollen frequency and the modern climate patterns and their variability, any calibration of pollen frequencies in terms of paleoclimatic parameters may lead to unrealistic conclusions. The first such calibration effort for South America was undertaken for southern Chile (Heusser and Streeter, 1980; Heusser *et al.,* 1981; Heusser, 1984b). Because the calibration data and procedures were not reported, the results are difficult to assess. However, incomplete coverage of the entire environmental gradient in the modern data set and inconsistencies between timing and amplitude of changes in fossil pollen assemblages and changes in inferred climates suggest problems in the approach (see discussion in Markgraf [1989a] and Ashworth and Hoganson [1984]).

The following paleoclimatic overview for Central and South America illustrates the type and quality of paleoclimatic proxy records available at present. It proposes large-scale paleoclimatic correlations that can be compared with the equally large-scale paleoclimatic model results, points out regions that should receive more attention in future paleoclimatic research, and documents the limits of paleoclimatic interpretation in view of the limited knowledge of modern vegetation, pollen dispersal, and climate mechanisms.

Methods

Of the fully published late-Quaternary records from Central and South America that were screened for continuity and chronologic control, 61 pollen and 15 lake-level records (up to 1987) were selected for analysis (Fig. 14.1). Table 14.1 lists site-specific information, including location, number of radiocarbon dates, ranking of dates with respect to target dates, and literature references. The records are grouped into seven major plant-geographic units (Fig. 14.1): (1) tropical-subtropical lowland forest and savannas north of the equator, (2) high-elevation montane woodland and grassland in Central America, (3) high- elevation paramo and puna grasslands in the northern and central Andes, (4) upper-Andean shrubland and woodland of the northern Andes (subparamo-subpuna), (5) altoandean grassland and shrubland of the central and southern Andes, (6) subtropical woodlands and desert lowlands south of the equator, and (7) temperate evergreen rainforests and deciduous forests of the southern Andes, along with Patagonian steppe, moorland, and tundra.

To allow comparison with the paleoclimatic model results, which are computed at 3000-yr intervals from 18,000 yr B.P. to today, target dates in the paleoenvironmental records were selected for the same intervals. Data for 15,000 yr B.P. appear similar to data for 18,000 yr B.P. in most records and are not discussed here. For the target dates of 18,000, 12,000, 9000, 6000, 3000, and 0 yr B.P., interpolated from each record's radiocarbon-dated chronology, the original pollen percentages are combined in pie diagrams showing the major vegetation associations distinguished in each region. In cases where the original percentages were based on a pollen sum that excluded taxa of regional relevance (such as herbs), the values were recalculated to allow comparison with other records. Aquatic taxa and fern spores are excluded except in the Bolivian records (Graf, 1979, 1981), where they occur in exceptionally high numbers and are considered environmentally significant. Cyperaceae percentages are included or excluded from the sum following each author's treatment of the data.

Because the 3000-yr interval between target dates precludes documentation of some paleoenvironmental changes, the overall paleoenvironmental sequence is described as well. Paleoclimatic interpretations follow those of the respective authors; in some instances, however, an alternative interpretation is proposed. Paleoclimatic inferences from lake-level records derived from changes in relative water depth and collected in the Oxford data bank are presented diagrammatically to allow comparison with the pollen records.

The paleoenvironmental history of each region is discussed in detail in Markgraf (1989a). Thus only the major paleoenvironmental features and their paleoclimatic implications are presented here to provide the background information necessary for comparison with paleoclimatic modeling results.

Regional Paleoenvironmental Changes

Tropical and Subtropical Lowlands

Except for the coastal sites in Guyana (Wijmstra and van der Hammen, 1966) and the Amazon Basin (Absy, 1979), none of the nine records (MA 1-3, SA 1 and 17-20, and one lake-level record [MA 4]) from the tropical and subtropical lowlands (Fig. 14.1) is located in the true tropical rainforest (Terra Firme forest). Instead the records are from sites in semievergreen forests, deciduous forests, or savanna, all of which are characterized by one to several months of moisture deficit (Fig. 14.2). Regional paleoclimatic interpretation of sites in tropical rainforest is complicated by the fact that many sites are from coastal or fluvial floodplains and reveal primarily a record of ocean or river-level changes.

With the exception of one record from the coastal plain (SA 20), none of the available continuous re-

Table 14.1. Site-specific information[a]

Site no.	Site name	Latitude	Longitude	Elevation (m)	Country	Pollen (P), lake level (L)	References	¹⁴C dates (no.)	Ranking[b] 18	12	9	6	3 ka B.P.
MA 1	Sinaloa	22°40′N	105°42′W	10	Mexico	P	Sirkin, 1985	11	–	–	–	1	1
MA 2	Peten	17°N	89°20′W	200	Guatemala	P	Leyden, 1985	2	–	–	1	4	4
MA 3	Panama Canal	9°10′N	79°52′W	35	Panama	P	Bartlett and Barghoorn, 1973	14	–	2	1	1	1
MA 4	Chichencanab	19°50′N	88°45′W	200	Guatemala	L	Covich and Stuiver, 1974	4	–	–	2	1	2
MA 5	S. Nicolas	10°23′N	101°17′W	1700	Mexico	P	Brown, 1985	15	–	–	1	2	1
MA 6	Patzcuaro	19°35′N	101°35′W	2044	Mexico	P	Watts and Bradbury, 1982	9	2	4	2	3	1
MA 7	Texcoco	19°27′N	99°W	2022	Mexico	L/P	Watts and Bradbury, 1982; Bradbury, 1989	4	2	2	2	4	4
MA 8	Chalco	19°27′N	99°W	2040	Mexico	L	Watts and Bradbury, 1982; Bradbury, 1989	3	–	3	2	2	4
MA 9	Malinche	19°15′N	98°W	3100	Mexico	P	Ohngemach and Straka, 1978	1	–	–	2	2	4
MA 10	Lerma/Zacapu	19°13′N	99°W	1988	Mexico	L	Metcalfe, 1985	10	–	–	2	1	1
MA 11	V. Lachner	9°43′N	83°56′W	2400	Costa Rica	P	Martin, 1964	3	4	4	2	4	4
SA 1	L. Valencia	10°16′N	67°45′W	403	Venezuela	L/P	Salgado-Labouriau,1980; Bradbury et al, 1981; Leyden, 1984	27	–	1	1	1	1
SA 2	Paramo Culata	8°45′N	71°4′W	3800	Venezuela	P	Salgado-Labouriau and Schubert, 1976	5	–	1	1	–	–
SA 3	Cienaga/Lagunillas	6°30′N	72°18′W	3510	Colombia	P	van der Hammen and Gonzalez, 1965; van der Hammen et al, 1981	3/9	4	1	1	6	4
SA 4	Cien. Visitador	6°8′N	72°47′W	3300	Colombia	P	van der Hammen and Gonzalez, 1965	2	–	2	2	4	4
SA 5	Lag. Fuquene	5°30′N	73°45′W	2580	Colombia	L/P	van Geel and van der Hammen, 1973	2	4	2	2	6	4
SA 6	El Abra Valley	5°1′N	73°57′W	2570	Colombia	P	Schreve-Brinkman, 1978	9	4	2	4	6	4
SA 7	Billar II	4°50′N	75°51′W	3600	Colombia	P	Melief, 1985	11	–	1	2	1	2
SA 8	P. Palacios	4°46′N	73°51′W	3500	Colombia	P	van der Hammen and Gonzalez, 1960a	1	–	–	2	4	4
SA 9	Sabana de Bogotá	4°38′N	74°5′W	2560	Colombia	L/P	van der Hammen and Gonzalez, 1960b	5	6	4	2	4	4
SA 10	Lag. Amer. I/II	4°15′N	74°W	3550	Colombia	P	van der Hammen and Gonzalez, 1960a	4	–	–	1	1	1
SA 11	Andabobos	4°5′N	74°10′W	3570	Colombia	P	Melief, 1985	2	–	2	2	5	4
SA 12	Lag. Primavera	4°N	74°10′W	3510	Colombia	P	Melief, 1985	6	–	2	1	1	1
SA 13	Alsacia	4°N	74°W	3100	Colombia	P	Melief, 1985	3	3	4	2	2	4
SA 14	Guitarra	4°N	74°W	3600	Colombia	P	Melief, 1985	3	–	2	2	2	2
SA 15	Gobernador	4°N	74°10′W	3815	Colombia	P	Melief, 1985	2	–	–	2	2	2
SA 16	Junco/Galapagos	0°55′S	89°30′W	500	Ecuador	L/P	Colinvaux and Schofield, 1976	23	–	–	1	1	1
SA 17	Lago Surara	4°9′S	61°46′W	76	Brazil	L/P	Absy, 1979	2	–	–	2	2	3
SA 18	L. Moriru/Moreiru	4°S	59°W	110	Guyana	P	Wijmstra and van der Hammen, 1966	2	–	6	2	1	3
SA 19	Kwakwani	5°15′N	58°3′W	150	Guyana	P	van der Hammen, 1963	1	–	–	–	2	4
SA 20	Ogle Bridge	6°50′N	58°10′W	0	Guyana	P	Wijmstra and van der Hammen, 1966	2	7	4	2	4	4
SA 21	L. Huatacocha	10°43′S	76°35′W	4500	Peru	P	Hansen et al, 1984	3	–	–	2	2	2
SA 22	Rio Blanca	10°50′S	75°20′W	4270	Peru	P	Hansen et al, 1984	3	–	2	2	5	2
SA 23	Laguna Junin	11°3′S	76°7′W	4100	Peru	P	Hansen et al, 1984	8	1	1	2	4	2
SA 24	Lag. Katantica	14°48′S	69°11′W	4820	Bolivia	P	Graf, 1979, 1981	3	–	–	–	1	2

Table 14.1. Site-specific information[a] (continued)

Site no.	Site name	Latitude	Longitude	Elevation (m)	Country	Pollen (P), lake, level (L)	References	¹⁴C dates (no.)	Ranking[b] 18	12	9	6	3 ka B.P.
SA 25	Cotapampa	15°13S	69° 6W	4450	Bolivia	P	Graf, 1981	5	–	–	1	2	1
SA 26	E. Cumbre Unduavi	16°21S	68° 2W	4620	Bolivia	P	Graf, 1979, 1981	5	–	–	2	2	1
SA 27	Chacaltaya C	16°22S	68° 2W	4570	Bolivia	P	Graf, 1979, 1981	5	–	–	–	1	1
SA 28	Chacaltaya B	16°22S	68° 9W	4570	Bolivia	P	Graf, 1979, 1981	5	–	–	1	1	1
SA 29	Rio Kaluyo	16°26S	68° 8W	4070	Bolivia	P	Graf, 1979, 1981	3	–	–	2	3	1
SA 30	Monte Blanco	17° 1S	67°21W	4780	Bolivia	P	Graf, 1979, 1981	6	–	–	2	1	1
SA 31	Tauca	19°30S	68°W	3800	Bolivia	L	Servant and Fontes, 1978	8	–	1	–	–	–
SA 32	Khota	21°36S	68°W	3800	Bolivia	L	Servant and Fontes, 1978	2	–	1	–	–	–
SA 33	Aguilar	23° 5S	65°45W	4000	Argentina	P	Markgraf, 1985a	3	–	–	2	2	2
SA 34	Salina 2	32°15S	69°20W	2000	Argentina	P	Markgraf, 1983	2	–	–	2	1	2
SA 35	L. Bebedero	33°20S	66°45W	380	Argentina	L	Gonzalez et al, 1981	15	1	2	2	2	2
SA 36	L. Tagua Tagua	34°30S	71°10W	200	Chile	L/P	Heusser, 1983	9	3	4	6	7	4
SA 37	Gruta Indio	34°45S	68°22W	600	Argentina	P	D'Antoni, 1983	7	7	2	4	4	4
SA 38	Vaca Lauquen	36°50S	71° 5W	1550	Argentina	P	Markgraf, 1987	3	–	–	1	1	2
SA 39	Rucañancu	39°33S	72°18W	290	Chile	P	Heusser, 1984a	7	–	–	1	2	2
SA 40	Lago Moreno	41° 3S	71°31W	800	Argentina	P	Markgraf, 1984	3	–	2	2	1	2
SA 41	L. Cari Laufquen G.	41°10S	69°34W	800	Argentina	L	Galloway et al, 1988	3	1	4	2	–	–
SA 42	L. Mascardi-Gut.	41°15S	71°28W	800	Argentina	P	Markgraf, 1983	3	–	–	2	2	1
SA 43	M. Book	41°20S	71°35W	800	Argentina	P	Markgraf, 1983	9	–	1	2	1	1
SA 44	Alerce I	41°24S	72°54W	100	Chile	P	Heusser, 1966	6	–	1	1	1	2
SA 45	Alerce III	41°25S	72°52W	100	Chile	P	Heusser, 1966	5	–	1	1	4	4
SA 46	Calbuco	41°44S	73°12W	100	Chile	P	Heusser, 1966	4	–	2	2	2	4
SA 47	L. Pastahue	42°22S	73°49W	150	Chile	P	Villagran, 1985	4	–	2	1	2	2
SA 48	L. Cardiel	48°55S	71°14W	276	Argentina	L	Galloway et al, 1988; Stine and Stine, 1990	2	–	–	2	–	2
SA 49	Puerto Eden	49° 8S	74°28W	10	Chile	P	Heusser, 1973	2	–	–	2	2	2
SA 50	Moreno Glacier	50°27S	73°W	231	Argentina	P	Mercer and Ager, 1983	2	–	–	2	3	2
SA 51	West Falkland	51°38S	59°34W	100	United Kingdom	P	Barrow, 1978	2	–	–	2	4	2
SA 52	Fells Cave	52° 4S	69° 7W	200	Chile	P	Markgraf, 1988	6	–	–	1	2	4
SA 53	La Misión	53°30S	67°50W	5	Argentina	P	Markgraf, 1980, 1985b	4	–	–	1	4	4
SA 54	Isla Clarence	54°12S	71°14W	21	Chile	P	Auer, 1974	3	–	–	2	2	1
SA 55	Sphagnum Valley	54°16S	36°35W	48	United Kingdom	P	Barrow, 1978	2	–	–	–	2	4
SA 56	Lago Yehuin	54°20S	67°45W	100	Argentina	P	Markgraf, 1983	5	–	–	1	1	1
SA 57	Gun Hut Valley	54°33S	36°28W	21	United Kingdom	P	Barrow, 1978	4	–	–	1	2	1

[a]Updated from Markgraf (1989a); used with permission from *Quaternary Science Reviews* and Pergamon Press.
[b]1, + and –1000 yr; 2, ±1000 yr; 3, + and –2000 yr; 4, ±2000 yr; 5, + and –3000 yr; 6, ±3000 yr.

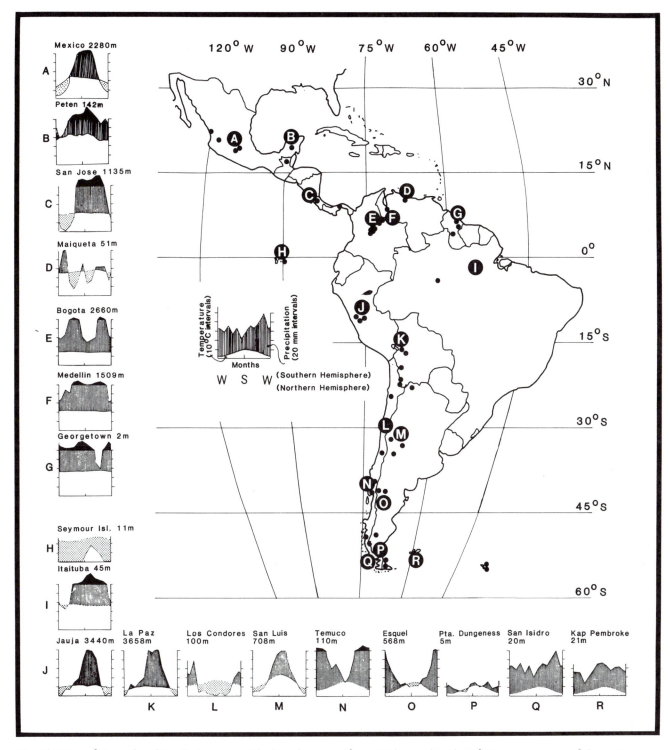

Fig. 14.2. Map of Central and South America with climadiagrams (from Walter and Lieth, 1960) representative of the major vegetation types illustrated in Figure 14.1. Dotted area indicates season with moisture deficit; shaded area indicates season with moisture surplus. Reprinted from Markgraf (1989a), with permission from *Quaternary Science Reviews* and Pergamon Press.

cords dates back before 13,000 yr B.P. Records that apparently span the entire last glacial period have recently been recovered from Panama and Brazil, and paleoclimatic analyses of these records may change the interpretations proposed herein (Bush and Colinvaux, 1990; Absy *et al.*, 1991; Ledru, 1993).

From 13,000 to about 10,000 yr B.P. all records are dominated by nonarboreal pollen, interpreted as representing savanna vegetation (Fig. 14.3) and thus suggesting substantially less moisture than today. Additional evidence of late-Pleistocene aridity in the tropical-subtropical lowlands of Central and South America before 10,000 yr B.P. comes from limnological and geomorphic studies, especially from the Amazon and Orinoco drainages. Instead of lakes, marshes and playas characterized the late-glacial landscapes (Absy, 1979; Bradbury *et al.*, 1981; Schubert and Fritz, 1985; Colinvaux, 1987). In the Amazon and Orinoco drainages, large areas with dune fields, alluvial fans, and river terraces that could not have been deposited under the modern forested conditions reportedly date to the full-glacial interval (Bigarella, 1975; Tricart, 1975, 1977).

Between 10,000 and 8500 yr B.P. those records with sufficient paleoenvironmental detail show a rapid stepwise succession of vegetation types, replacing the late-Pleistocene grasslands (Fig. 14.3). The early-Holocene mixed assemblages of rainforest and montane forest elements were replaced around 9000 yr B.P. at first by semievergreen rainforest elements and later by woodland savanna elements. This succession suggests a stepwise warming and increase in precipitation.

With one exception (Covich and Stuiver, 1974) early-Holocene (10,000–8500 yr B.P.) limnological changes (Fig. 14.4) parallel the vegetation succession: marshes and playas became ephemeral lakes or saline lakes and later large freshwater lakes. Early-Holocene climates must have been moist and cool at first and later moist and warm; that is, with a dry season shorter than the modern climadiagrams indicate (Fig. 14.2). Also sea level appears to have risen to the modern level during this time. Except for low-lying areas where rising sea level would have altered the groundwater tables (and lake and river levels), sea-level fluctuations probably contributed only indirectly to regional climate change.

After about 8500 yr B.P. semievergreen forest or savanna woodland elements were replaced by deciduous forest or savanna elements. Lake levels were already low or were still falling, and increased sedimentation rates in lake sediments indicate an increase in erosion (Liu and Colinvaux, 1988). In paleoclimatic terms this implies an increase in seasonally dry periods. In contrast, the Galapagos record at that time shows an expansion of tree elements simultaneously with rising water levels. This is interpreted as the onset of modern climatic conditions, perhaps even slightly warmer ("Hypsithermal") than today (Colinvaux and Schofield, 1976).

For the remaining 5000 yr (the late Holocene) environmental changes lacked the synchroneity shown earlier. More mesic forest and higher lake levels returned in some records after 5000 yr B.P., in others only after 3000 yr B.P. To what extent the difference in climatic signal during the middle and late Holocene is related to the difference in resolution of the records, differential environmental responses of local biota, or actual climatic differences is unclear. The fact that the climatic history inferred from the Galapagos record appears out of phase with other records could reflect a real difference, as suggested by modern climate data: today the Galapagos Islands receive precipitation only between January and May, when precipitation is at its lowest in the other regions discussed.

Part of the environmental heterogeneity noted during the late Holocene may be related to human impact. Bush *et al.* (1989) registered continuous presence of maize (*Zea*) since 6000 yr B.P. in a record from the Amazon lowlands. The regional differentiation is even more striking during the last 2500 yr, when all records from the tropical and subtropical lowlands show great environmental variability, primarily declines of forest taxa coupled with increases in savanna or woodland elements (Fig. 14.3). At the same time indicators of human disturbance (e.g., *Zea* pollen and sedimentologic changes) appear in the records. This implies that the trend resembling increased aridity reflects prehistoric land use (Binford *et al.*, 1987; Liu and Colinvaux, 1988).

In summary, the absence of pollen records older than 13,000 yr B.P. preempts inferences on full-glacial regional vegetation in the tropical and subtropical lowlands. Neither the location nor the composition of potential late-Pleistocene rainforest refugia (see Leyden, 1984; Bush and Colinvaux, 1990) can be identified. Geomorphic evidence, however, suggests that the climate must have been substantially drier than at present. The lowlands at 12,000 yr B.P. were cool and dry, in a continuation of the glacial mode. By 9000 yr B.P. moisture levels had reached or even surpassed modern levels, but temperatures were probably still lower than today. After 8500 yr B.P. moisture levels fell, reaching a minimum at 6000 yr B.P. After 5000 yr B.P. regional paleoenvironmental history became very variable, making interregional correlations difficult. Agricultural evidence is documented after 6000 yr B.P., and the impact of land use was widespread throughout the lowlands after 2500 yr B.P.

HIGH-ELEVATION SITES IN THE ANDES

Twenty-eight pollen and six lake-level records from sites ranging from the central Mexican plateau to the altiplano (puna) of Argentina provide paleoclimatic information on high-elevation environments. These records are discussed in the following groups: the montane Central American forest sites (pollen records MA 5-9 and 11; lake-level record MA 10), the

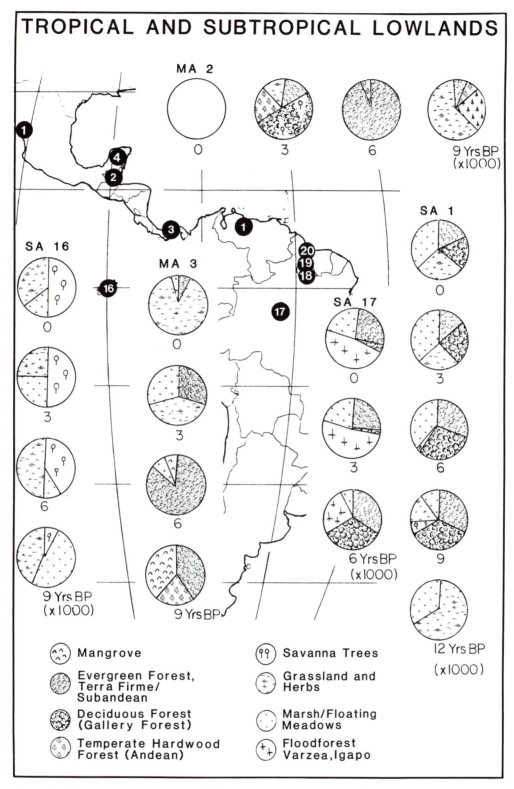

Fig. 143. Palynologic "snapshots" at 3000-yr intervals based on pollen percentage data from records from the tropical and subtropical lowlands in Central and South America, showing location of sites. Reprinted from Markgraf (1989a), with permission from *Quaternary Science Reviews* and Pergamon Press.

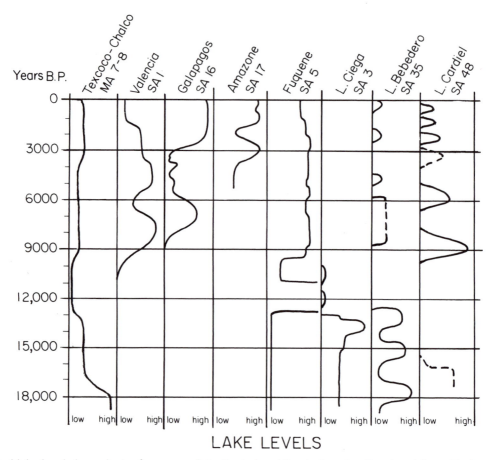

Fig. 14.4. Selected lake-level chronologies from records in Central and South America. Reprinted from Markgraf (1989a), with permission from *Quaternary Science Reviews* and Pergamon Press.

paramo sites of the northern and central Andes at elevations above 3300 m (SA 2-4, 7, 8, 10-15, and 21-23), the subparamo sites of the northern and central Andes at elevations between 2000 and 3300 m (SA 5, 6, and 9), and the altoandean sites of the central Andes above 2000 m (pollen records SA 24-30 and 33; lake-level records SA 31 and 32).

Along the whole north-south transect the high-elevation climate is characterized by summer precipitation, with little or no rainfall during the cold months. Toward higher latitudes south of the equator in Peru, Bolivia, and northern Argentina the number of dry months exceeds the number of months with precipitation (Fig. 14.2), so the more xeric puna is more important than the more humid paramo of the northern Andes and Central America.

Montane Central America

Four of the six paramo and upper montane records from Central America (Fig. 14.5) date back to the full-glacial period (assumed to be 18,000 yr B.P.). The records from sites above 2400 m in elevation (MA 9 and 11) are dominated by paramo taxa before 10,000 yr B.P., implying a lowering of the treeline by 600-800

m and a temperature about 4°C below the present level (Martin, 1964; Heine, 1984). Records from below 2400 m, on the other hand, are dominated by arboreal taxa before 10,000 or 11,000 yr B.P., just as today. The proportions, however, are different from those of today and suggest that full- and late-glacial climates were drier as well as cooler than at present. Late-Pleistocene aridity is also suggested by lake-level data based on ostracod and diatom analyses of lake cores from the Zacapu Basin (MA 10) and the Cuenca of Mexico (MA 8) (Bradbury, 1989) (Fig. 14.4).

Essentially modern vegetation became established throughout the higher elevations in Central America between 10,000 and 9000 yr B.P. (Fig. 14.5). Differentiating between early- and mid-Holocene environments is difficult, in part because of inadequate dating control and in part because of the small scale of changes. Several records suggest that the early Holocene might have been somewhat moister than the middle Holocene, but lake-level data from the Cuenca of Mexico suggest that the early Holocene was arid, and only during the late Holocene does increased spring activity indicate greater moisture (Bradbury, 1989).

After 6000 yr B.P. paleoenvironmental changes in

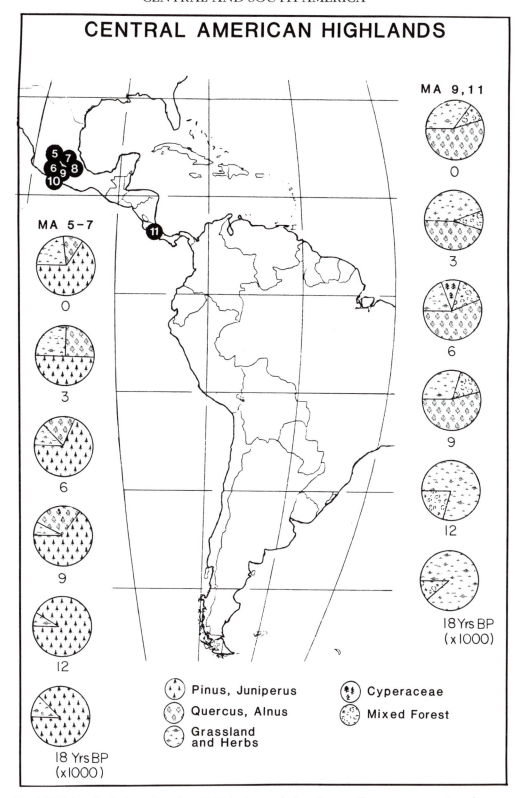

Fig. 14.5. Palynologic "snapshots" at 3000-yr intervals based on pollen percentage data from records from the Central American highlands, showing location of sites. Reprinted from Markgraf (1989a), with permission from *Quaternary Science Reviews* and Pergamon Press.

Central America can no longer be ascribed to climatic changes alone, because the first traces of corn pollen (*Zea*) appear in the records from the Valley of Mexico (Niederberger, 1979; Gonzalez-Quintero and Fuentes Mata, 1980). Even the shift in Lake Chalco (Cuenca of Mexico) from a saline to a freshwater marsh at 6000 yr B.P. is probably not natural but rather related to human attempts to control water levels by engineering (Bradbury, 1971, 1989). Evidence of forest clearing, burning, weed proliferation, and agriculture is widespread throughout Central America by 3500 yr B.P. These paleoenvironmental changes are so prominent that any paleoclimatic interpretation for that time is ambiguous at best.

Paramos of the Northern and Central Andes

Paleoclimatic trends for the paramo sites in Venezuela, Colombia, and Peru are interpreted from changes in the pollen proportions of Gramineae, Compositae, other herbaceous taxa, and Andean forest taxa. Most of these records are from Colombia and are the result of more than 30 yr of palynologic research by T. van der Hammen and co-workers, culminating in the detailed analysis of the record from the Sabana de Bogotá, which is 3.5 million years old (van der Hammen and Gonzalez, 1960b; van der Hammen, 1974; Hooghiemstra, 1984). At least 26 couplets of glacial and interglacial periods can be recognized in this paleoenvironmentally highly sensitive record, providing answers to questions about the evolution and migration of Andean biota since the late Tertiary (Simpson, 1975; Hooghiemstra, 1984). This record is discussed further in the subparamo section.

Only three paramo records (SA 3, 13, and 23) extend back continuously through the full-glacial interval (named the Fuquene stadial), estimated to date between 20,000 and 14,000 yr B.P. At that time paramo (puna) pollen types dominated the spectra, and subparamo (subpuna) and Andean forest types were present in minor amounts. Proportions of *Isoetes* and other water plants and algae imply that the lakes at that time were shallow in Colombia but deep in Peru.

The majority of the high-elevation records from the paramo date back only to 12,000–14,000 yr B.P., after the glaciers had receded from these elevations. After about 14,000 yr B.P. (Fig. 14.6) subparamo taxa (Compositae) replaced the paramo taxa, and Andean forest taxa increased as well. Several short-term paleoenvironmental intervals have been distinguished between 14,000 and 10,000 yr B.P.: the Susaca interstadial, Ciega stadial, Guantiva interstadial, and El Abra stadial, successively. Although the chronological control is inadequate to define the exact timing of these intervals, they are assumed to be contemporaneous with the classical European late-glacial stages of Bölling, Older Dryas, Alleröd, and Younger Dryas, respectively (van der Hammen *et al.*, 1981). The distinction between the warm (interstadial) and the cool (stadial) intervals in

Colombia is based on shifts of the treeline, interpreted from shifts between taxa representative of the paramo versus taxa representative of the subparamo and the Andean forest. The Guantiva interstadial (between 12,000 and 11,000 yr B.P.) is believed to have been as warm as or even warmer than today. Stadials, on the other hand, are believed to have been as cold as the full glacial (Salgado-Labouriau *et al.*, 1977; van der Hammen *et al.*, 1981; Melief, 1985). Dating of moraine sequences in the Eastern and Central Cordillera in Colombia suggests that the stadials correlate with periods of glacier readvance (Helmens, 1988).

Changes in lake levels during the late glacial paralleled the regional vegetation changes (Fig. 14.4). Lake stands were high during interstadials and low during stadials, including the full glacial (Melief, 1985). Radiocarbon dates from high lake terraces in the Bolivian and Peruvian altiplano (Laguna Tauca and Laguna Khota) (Servant and Fontes, 1978) suggest lakes larger than today between 12,000 and 11,000 yr B.P. At that time glaciers had retreated behind their subsequent neoglacial limits, and the climate must have been as warm as at present. The increased warmth could have been related to increased precipitation, as suggested by the climatic model results of Hastenrath and Kutzbach (1985). The postulated 50-75% increase in precipitation needed to explain the high lake levels seems unlikely, however, in view of the absence of a comparable vegetational change at that time.

Depending on the resolution of the record, the onset of Holocene-type climates is dated at 12,000 yr B.P. in Peru (Hansen *et al.*, 1984) or at 10,000 yr B.P. in Colombia (van der Hammen *et al.*, 1981) (Fig. 14.6). In the paramo records an increase in Andean forest taxa between 10,000 and 7000 yr B.P. is interpreted as an upslope shift of the Andean forest zone and hence warmer and moister climates. Curiously, the upper and lower Andean forest elements appear in the same proportions, despite the fact that the upper forest shifted closer to the sites and thus should have provided substantially more pollen than the lower forest. This problem is addressed again in the context of the subparamo records.

A further upslope shift of the forest belt between 7000 and 5000 yr B.P. is interpreted from increases in lower Andean forest taxa and Compositae. Therefore temperatures are thought to have been higher than before ("Hypsithermal"). At that time *Plantago* increased substantially in several records (SA 2–4). Although *Plantago* could indicate local bog conditions (Melief, 1985), it is generally thought to be an indicator of open paramo vegetation (Hooghiemstra, 1984). Modern pollen assemblages from treeless paramo vegetation are mixed; they contain up to 40% tree pollen in addition to herbaceous taxa (Grabandt, 1980). The resemblance between the mid-Holocene assemblage and this modern mixed assemblage suggests that instead of an upslope shift of the forest, paramo

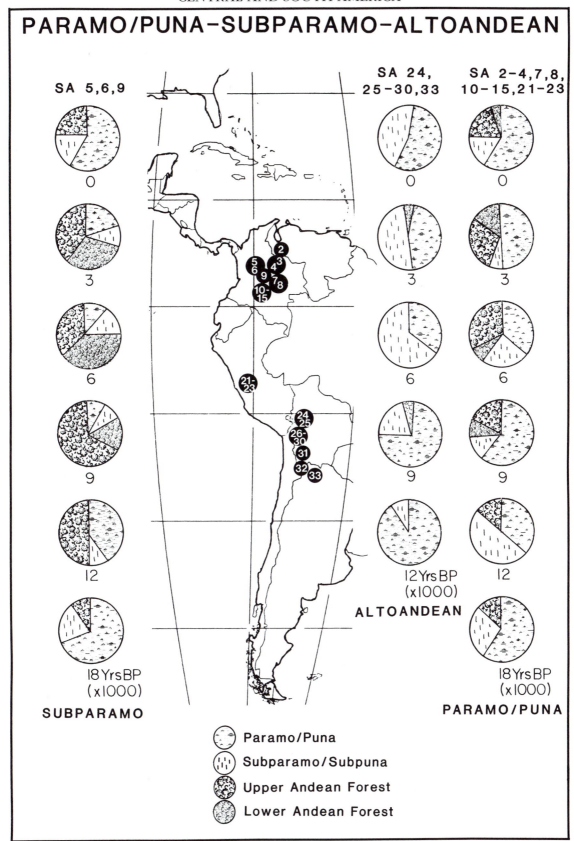

Fig. 14.6. Palynologic "snapshots" at 3000-yr intervals based on pollen percentage data from records from the paramo-puna, subparamo, and altoandean regions in South America, showing location of sites. Reprinted from Markgraf (1989a), with permission from *Quaternary Science Reviews* and Pergamon Press.

vegetation expanded at that time. Comparison with the lower-elevation (subparamo) records should help resolve the question of whether the forest shifted up-slope or sparser local paramo vegetation expanded.

After 5000 yr B.P. upper Andean forest elements gradually replaced the lower forest elements, and Compositae decreased as well. This is interpreted as a gradual downslope movement of the treeline in response to a temperature decrease. By 3000 yr B.P. essentially modern vegetation conditions had become established. An overall forest decline shortly afterward coupled with an increase in herbaceous taxa most likely reflects human impact on the environment, probably primarily intensified livestock grazing.

Subparamo of the Northern and Central Andes

Because they are closer to the Andean forest belt, the subparamo sites (SA 5, 6, and 9) (about 2570 m in elevation) yield records that show the paleoenvironmental changes more distinctly than do the paramo records (Fig. 14.6). During full-glacial times (Fuquene stadial) these elevations were dominated by paramo vegetation only. Between 12,000 and 11,000 yr B.P. (Guantiva interstadial) tree pollen increased substantially, suggesting that an open Andean forest had become established at these elevations (Fig. 14.6). After a brief interval (11,000–10,000 yr B.P., El Abra stadial) when nonarboreal taxa reexpanded and the climate is thought to have been far cooler than during the previous interval, Andean forest elements returned and began to dominate. This is interpreted as a local establishment of the Andean forest in response to higher temperatures.

After about 8000 yr B.P. lower Andean forest elements replaced the upper Andean forest taxa. These lower forest elements are recorded in equally high proportions at the subparamo and paramo sites. If the forest belts had shifted upslope, as is generally interpreted, then pollen assemblages from subparamo sites should show a greater proportion of lower Andean forest taxa than samples from paramo sites, reflecting the difference in proximity to the Andean forest. Furthermore, the subparamo records at that time also show high values for Compositae, a local subparamo element. Thus, an alternative interpretation to an upslope shift of the forest belts would be more open local vegetation, with shrubs (Compositae) replacing the upper Andean forest elements and allowing greater input of long-distance pollen from the lower Andean forest. This hypothesis contradicts the generally accepted interpretation but does explain the mixed assemblages in both paramo and subparamo records and the paleoenvironmental changes in both types of records in a paleoclimatically consistent manner. The paleoclimatic interpretation of this "Hypsithermal" interval between 8000 and 3000 yr B.P. would

call primarily for greater aridity (expressed mainly in the subparamo records) coupled with lower rather than higher temperatures (expressed in the paramo records). This scenario is supported by indications of lower lake levels during that time (Melief, 1985).

After 3000 yr B.P. lower Andean forest elements and Compositae decreased, while upper forest elements increased. Instead of a downslope shift of the forest belt, this change could imply a more dense forest due to increased precipitation. The abrupt decline in tree pollen after about 2000 yr B.P., coupled with an increase in disturbance and weed pollen (Dodonaea, Ambrosia, Polygonum, and Gramineae), suggests human impact.

Because of the records' location the described paleoenvironmental changes are interpreted as treeline shifts related to shifts of the 9.5°C isotherm (Hooghiemstra, 1984). During full-glacial times (Fuquene stadial) the treeline is supposed to have been lower than at present by about 1000 m, at an elevation of about 2000 m, corresponding to a temperature decrease of 8°C. During the Guantiva interstadial (12,000–11,000 yr B.P.) the treeline must have been close to the modern level at about 3000 m. During the following El Abra stadial, the treeline descended to 2500 m, implying a temperature decrease of about 3°C. According to van der Hammen (1974), the treeline reached its highest elevation at 3600 m during the "Hypsithermal" interval (7000–5000 yr B.P.), corresponding to a temperature increase of 6°C between the early and middle Holocene. The treeline descended to modern levels after 5000 yr B.P. If, on the other hand, the middle Holocene is interpreted as a period of lower precipitation rather than higher temperature, the inferred Holocene temperature fluctuations would be less extreme, which perhaps is more realistic.

Except at Sabana de Bogotá, where water-level changes in the last 30,000 yr were tectonically controlled, lake-level records from paramo and subparamo sites show a similar history. In general lake levels were low during cold stadial periods and high during warm interstadial periods. Changes during the Holocene were minor but seem to indicate lower water levels between 7000 and 5000 yr B.P. (Fig. 14.4).

Altoandean of the Central Andes

Altoandean records from Bolivia (SA 24–30) and Argentina (SA 33) do not present as much of a problem with long-distance transport of forest pollen as the paramo and subparamo records discussed above. Except for the great abundance of fern spores in the Bolivian records, assumed to represent the cloud forest vegetation downslope (Graf, 1985), none of the records contains substantial amounts of forest pollen; the source area of forests in Bolivia and Argentina is smaller, and the sites are much farther from the source than sites farther north. On the other hand, the absence of Andean forest pollen also diminishes the diversity in the pollen spectra, rendering paleoclimatic interpretation of the altoandean pollen records difficult. The paleoclimatic interpretation is based on changes in the proportions of Gramineae and herbaceous taxa, representing the altoandean (puna) vegetation, and the proportions of Compositae, *Ephedra*, and Chenopodiaceae, representing the subpuna vegetation.

None of the records extends back beyond 10,000 yr B.P., but preliminary pollen data from a section that also contains extinct fauna and is dated between 12,000 and 10,000 yr B.P. suggest that paramo grasslands expanded downslope during the late glacial, replacing puna environments (Fernandez *et al.*, 1991). This in turn suggests substantially greater effective moisture than today, probably seasonally quite uniform. The existence of glaciers in the region indicates that temperatures must have been lower than today (Markgraf, 1985a) and that the moisture for glacier growth originated primarily from the east; that is, as summer precipitation (Strecker *et al.*, 1986).

Between 10,000 and 8000 or 7000 yr B.P. temperatures must have risen throughout the altoandean region to explain the increase in Compositae, but in addition the relatively high proportions of moist paramo indicators suggest that moisture was still greater than today. Long-distance pollen, although in low amounts, is represented by taxa from the eastern lowlands, suggesting an easterly wind component, which today is related to summer rains.

After 8000 (7000) yr B.P. greater proportions of subpuna vegetation indicate increased aridity, probably due to higher temperatures. The increase in fern spores in the Bolivian records suggests regional expansion of cloud forest.

After 4000 yr B.P. altoandean elements increased again, but puna elements remained relatively high. Long-distance pollen at that time is primarily from the southwest (*Nothofagus*, *Podocarpus*), suggesting westerly winds. Sedimentologic data suggest channel filling at that time (Fernandez, 1986), which implies also primarily winter precipitation.

Repeated high-amplitude changes in pollen assemblages starting after 2000 yr B.P. indicate human impact, probably onset of overgrazing in the high-elevation environments. Unquestionable evidence of human impact is found after 500 yr B.P. everywhere in the altiplano and in Peru (Hansen *et al.*, 1984), Colombia (Kuhry *et al.*, 1983), and Argentina (Markgraf, 1985a).

SUBTROPICAL WOODLAND AND DESERT (LATITUDES 30–34°S)

The three pollen records at latitude 34°S come from both sides of the Andes and are divided by mountain ranges over 6000 m in elevation (Fig. 14.7). Laguna Tagua Tagua in Chile (SA 36) is located in sclerophyllous *Nothofagus* woodland and receives mean annual

precipitation of 800 mm, primarily during winter. The other two pollen sites are in Argentina: Gruta del Indio (SA 37) in the Monte Desert, with mean annual precipitation of 350 mm, primarily in summer, and Uspallata Valley (Salina, SA 34), located in the transition (puneña) vegetation, with mean annual precipitation of about 100 mm. There is also one sequence of lake terrace dates, from Laguna Bebedero (SA 35) in Argentina (Gonzalez *et al.*, 1981). The winter rains in these latitudes are linked to the westerly circulation, which shifts northward (equatorward) from its main position at latitude 40°S, primarily at the beginning of winter. The summer rains, on the other hand, depend on the seasonal southward (poleward) shift of the subtropical high-pressure system (Pittock, 1980a,b). Two pollen records extend back apparently without interruption to beyond radiocarbon dating limit. However, the long records have no dates during the last 10,000 yr B.P., which complicates Holocene intersite correlation.

High proportions of *Nothofagus dombeyi*-type and *Prumnopitys* (*Podocarpus*) *andina* pollen in the Chilean record (SA 36) and of Gramineae and herbaceous pollen in the Argentine record (SA 37) (Fig. 14.7) were originally interpreted as representing a northward shift of the Valdivian forest and Patagonian steppe, respectively. Because these taxa today occur at the same latitudes at elevations 600–800 m higher, it is more likely that during the glacial period they shifted downslope, suggesting cooler climates than today (Markgraf, 1989b).

Laguna Tagua Tagua, which is not in a closed basin, was a large lake until about 10,000 yr B.P. Lake terrace dates from Laguna Bebedero, a closed basin (SA 35), reveal high stands at 18,600, 15,600, and 13,000 yr B.P., separated by intervals of low stands (Fig. 14.4). Whereas the Laguna Bebedero high stands were interpreted as recording ablation phases of the Pleistocene Andean glaciers (Gonzalez, 1981), Heusser (1983) interpreted the increased late-Pleistocene moisture in the Chilean lowlands as an increase in winter rains caused by an equatorward shift of the westerlies. However, if a cooling of at least 4°C is assumed, increased precipitation may not be needed to explain the greater effective moisture (Markgraf, 1989b).

Water levels fell rapidly in both lake basins after 12,000 yr B.P. Laguna Tagua Tagua turned into a marsh, while Laguna Bebedero became a playa lake. At the same time in the Argentine record, grassland vegetation was replaced by desert scrub, and in Chile the montane forest taxa disappeared from the lowlands (Fig. 14.7). Unfortunately, from that time on local salt-flat or marsh pollen dominates the Chilean record, and no inferences can be made about the Holocene regional forest vegetation.

On the basis of plant growth form and its climatic implications, D'Antoni (1983) argued that the Argentine pollen record implies precipitation seasonality.

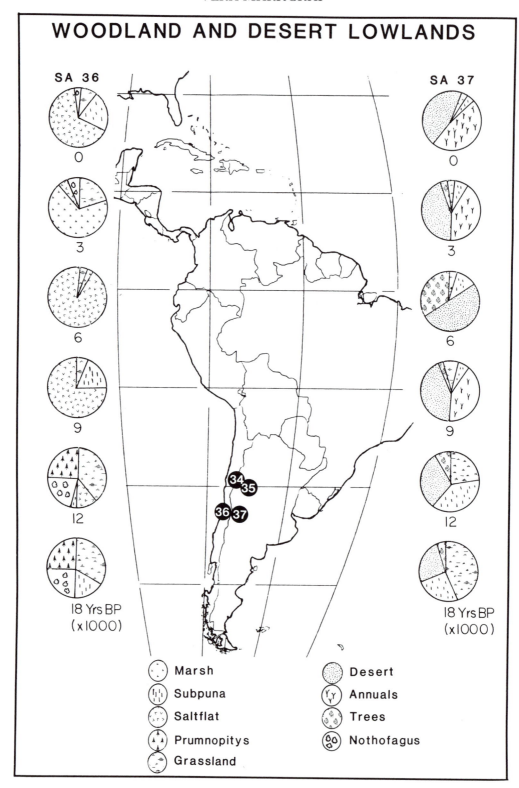

Fig. 14.7. Palynologic "snapshots" at 3000-yr intervals based on pollen percentage data from records from woodland and desert lowland regions in South America, showing location of sites. Reprinted from Markgraf (1989a), with permission from *Quaternary Science Reviews* and Pergamon Press.

The increase in summer-rain desert taxa, especially annuals (therophytes), after 12,000 yr B.P. suggests a substantial temperature increase, coupled with a shift to summer precipitation. After 9000 yr B.P. even the annuals declined and only geophytes remained, suggesting further reduction in effective moisture; that is, absence of summer rains. This arid phase lasted until about 5000 yr B.P., when tree taxa increased in the Argentine lowlands and Andean taxa increased in the eastern Andean foothills (SA 34), implying most likely an increase in winter rains. Mid-Holocene lake levels above the modern playa floor in Laguna Bebedero support the increase in precipitation at 5000 yr B.P. suggested by the pollen evidence. The modern desert-scrub vegetation began to dominate after 3000 yr B.P., suggesting the onset of the modern mixed winter-summer rain pattern. Effective moisture was apparently the greatest during the late Holocene, as confirmed by recently analyzed records from the northernmost Valdivian relict forests at latitude 32°S (Villagran and Varela, 1990).

In summary, the paleoclimate in the southern subtropics was characterized by greater effective moisture during full- and late-glacial times. Lowering of higher-elevation vegetation by at least 600 m suggests temperatures were at least 4°C cooler than today, which alone could explain the increased moisture. Aridity was greatest by 9000 yr B.P. and was reduced after 5000 yr B.P. and especially after 3000 yr B.P., when today's mixed summer-winter precipitation regime became established.

Temperate Nemoral and Evergreen *Nothofagus* Forests of the Southern Andes

South of latitude 36°S pollen records have been analyzed from the Valdivian and Patagonian rainforest in Chile (SA 39, 44–47, 49, and 54), the mixed *Nothofagus* forest environments in Argentina (SA 40, 42, 43, and 50), the steppe-*Nothofagus* forest ecotone in Argentina (SA 38, 52, 53, and 56), and the Magellanic moorland or tundra in the subantarctic islands (SA 51, 55, and 57).

Paleoenvironmental reconstruction based on pollen records from the Valdivian rainforest region in Chile was pioneered by Heusser (1966). Understanding of the relation established by Heusser (1974) between modern plant distribution and pollen assemblages was greatly advanced by detailed phytosociologic analysis of vegetation zonation along elevational gradients and its relation to pollen assemblages and climatic parameters (Villagran, 1980, 1985).

Heusser and Streeter (1980) proposed another approach to paleoclimatic reconstruction of past pollen assemblages. They applied the relation between modern pollen and climate data (mean annual precipitation and January [summer] temperatures) derived from a longitudinal transect in lowland southern Chile between latitudes 33 and 55°S to the fossil pollen

assemblages. The unrealistic results obtained, especially relating to temperature (Heusser and Streeter, 1980; Heusser et al., 1981; Heusser, 1984b), suggest that the modern pollen and climate data used did not provide adequate analogues to relate to the fossil assemblages. The fossil assemblages, especially during pre-Holocene times, contained substantial amounts of both steppe and high-elevation taxa, which today are not represented in the Chilean lowlands. Thus this approach cannot be accepted uncritically, especially because neither the data nor the procedures have been fully published.

Data for full-glacial (18,000 yr B.P.) environments from the Valdivian region are based on several long records from Isla Chiloé (Heusser and Flint, 1977; Villagran, 1987, 1988) and several discontinuous records from the Chilean lake district (Puerto Octay, Puerto Varas, and Rupanco) (Heusser, 1974). Valdivian rainforest environments in the lowlands of the Chilean lake district during full-glacial times were characterized by 50–75% nonarboreal taxa (Fig. 14.8). The arboreal taxa were primarily *Nothofagus dombeyi* type, which includes evergreen species (*N. dombeyi, N. betuloides*) and deciduous taxa (*N. pumilio, N. antarctica*) with different ecological ranges. The openness of the forest suggests a climate with seasonal moisture stress. A decrease in annual precipitation by 1500 mm and a drop in January temperatures by 4°C inferred by Heusser and Streeter (1980) fall short of explaining the full-glacial environment, because a closed Valdivian rainforest could exist even today under such climatic conditions. On Isla Chiloé, on the other hand, dominance of Magellanic moorland elements suggests a climate substantially windier and cooler than today (Villagran, 1987, 1988). Both paleoclimatic interpretations—less seasonal moisture in the lake district and more moisture and windiness in Chiloé—reflect the same circulation change: intensification of the westerlies and a poleward shift of the storm tracks. This result has also been inferred from pollen records from Argentina (Markgraf, 1983, 1987, 1989a,b).

After 12,000 yr B.P. temperate, cool-summer forest taxa became dominant in the pollen records from the Valdivian region (Villagran, 1985) (Fig. 14.8). These taxa suggest high effective moisture, with temperatures cooler than today. Compared to late-glacial conditions, this implies an increase in *precipitation* and for Isla Chiloé an increase in *temperatures* coupled with a decrease in windiness. Although a doubling of precipitation between 12,000 and 11,000 yr B.P., proposed by the pollen-climate transfer function approach, seems plausible, a simultaneous increase in January (summer) temperature to 6°C above the modern value, followed by a decrease of 11°C after 11,000 yr B.P. (Heusser and Streeter, 1980), is not consistent with the recorded vegetation changes.

The case for a short-term environmental change between 11,000 and 10,000 yr B.P. correlated with the European Younger Dryas interval is strongly debated

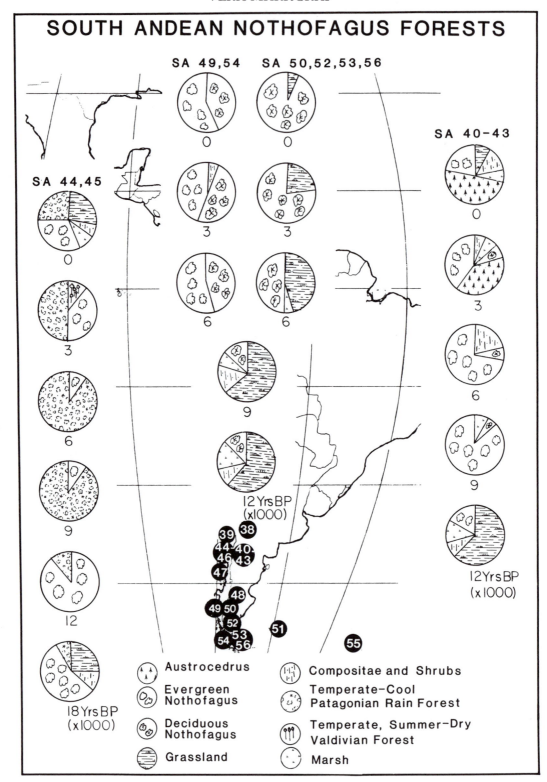

Fig. 14.8. Palynologic "snapshots" at 3000-yr intervals based on pollen percentage data from records from the South Andean *Nothofagus* forest region, showing location of sites. Reprinted from Markgraf (1989a), with permission from *Quaternary Science Reviews* and Pergamon Press.

(Markgraf, 1991). First, dating control of this well-defined short-term interval is inadequate for most records. Second, many pollen records do not document any change during the interval in question (Ashworth and Hoganson, 1984). Third, pollen data that do show changes roughly at that time (although generally of longer duration) do not show a reversal to glacial-type conditions but rather a shift to a different assemblage, primarily a decrease in *Nothofagus dombeyi* type and an increase in Myrtaceae pollen. This environmental change could be explained by a local edaphic change at the site rather than by a regional climatic alteration.

After 10,000 yr B.P. elements of the warm-temperate Valdivian rainforest (at first *Weinmannia*) increased, suggesting warmer temperatures than before. After 8000 yr B.P. elements of the dry-summer Valdivian forest (e.g. *Eucryphia*) became more prominent. This change is especially well documented in records from the northern limit of the Valdivian rainforest, where increased aridity would be expected to have had the greatest effect. By 6000 yr B.P. subantarctic Patagonian rainforest elements (especially conifers) expanded. The last 6000 yr have seen several short-term environmental shifts between warm-temperate and cool-temperate forest elements. Correlation of these short-term phases among records is difficult, however, because of either dating problems or differences in resolution. At 3000 yr B.P. the climate is assumed to have been somewhat warmer than today. Unambiguous evidence of human disturbance appears only during the last 500 yr (Veblen and Markgraf, 1988).

Records from the mixed deciduous *Nothofagus* forest environments east of the Andes range from latitude 36 to 54°S. Within this wide latitudinal range many taxa reach either their northern or southern distributional limits, complicating detailed paleoenvironmental comparisons. The major vegetation types, however, are sufficiently similar to justify joint treatment of the data.

Before 12,000 yr B.P. the forest area was greatly reduced all along this latitudinal stretch. Full-glacial dominance of steppe-scrub and small-tree taxa that today grow as understory trees at the steppe-forest limit (V. Markgraf, unpublished) and are highly resistant to drought and frost (Weinberger, 1974; Steubing *et al.,* 1983) suggests that between 18,000 and 14,000 yr B.P. summers especially must have been dry and winters cold. Lake levels in central Patagonia apparently were high, indicating a positive water balance (Galloway *et al.,* 1988).

By 14,000 yr B.P. Gramineae began to dominate, suggesting that effective moisture had increased, although not enough to result in general forest expansion. Some tree taxa, however, did expand their territory during late-glacial times. *Prumnopitys* (*Podocarpus*) *andina* is documented during late-glacial times from both sides of the Andes at lower elevations than today (SA 36 and 38), and *Nothofagus*

alpina type (including *N. obliqua*) expanded farther south than today (SA 40).

After 12,000 yr B.P. the vegetational history at middle versus high latitudes developed separately. The low-elevation records from latitude 41°S document a rapid forest expansion shortly after 12,000 yr B.P., primarily involving the evergreen *Nothofagus dombeyi.* Its codominant today, *Austrocedrus,* was only a minor component at that time. At high latitudes (51–54°S, but also at 46°S [SA 49]), the forest expansion did not begin until after 9500 yr B.P. Because the modern steppe-forest limit appears related to a minimum annual precipitation of 600–800 mm, the paleoclimatic implication is that precipitation reached this value nearly 3000 yr earlier at midlatitudes than at high latitudes. This suggests that the high latitudes continued under the direct influence of Antarctica, which emerged from its glacial mode only after 9000 yr B.P.

Information on past temperature changes comes from a high-elevation record at latitude 37°S (SA 38) and from high latitudes (SA 52 and 53). In both cases an environmental shift at about 10,000 yr B.P. suggests increased temperatures.

After 9500 yr B.P. when the forest began to expand at high latitudes, midlatitude sites showed an increase in shrub and understory tree elements instead, suggesting open forest conditions. Precipitation apparently increased at high latitudes and simultaneously decreased at midlatitudes. This opposition of climatic signals continues through the remainder of the Holocene. Between 6000 and 5000 yr B.P. midlatitude records show a return to more closed forests, suggesting increased precipitation, whereas high-latitude records indicate steppe and heath expansion, suggesting a decrease in precipitation.

After 5000 yr B.P. all records began to resemble the modern environments. Records from low elevations at midlatitudes suggest climates relatively cooler and drier than before. Records from high elevations and high latitudes south of 46°S suggest cooler and moister conditions than before. Further evidence for this climatic opposition comes from lake-level records. Lakes at latitude 49°S were apparently highest after 9000 yr B.P. and again during the last 3000 yr (Galloway *et al.,* 1988; Stine and Stine, 1990), whereas midlatitude (41°S) lake levels have been dropping ever since the late-glacial high stands (Galloway *et al.,* 1988).

This latitudinal divergence in the paleoclimatic signal has also been documented for modern climates of southern South America and has been linked with atmospheric circulation anomalies (Pittock, 1980a,b). When the modern climatic scenarios are used to interpret the past patterns (Markgraf, 1983), the widespread late-Pleistocene aridity is explained by intensification and a poleward shift of the westerly storm tracks coupled with a transpolar shift of the circumpolar trough toward the Australian sector. Whereas by 12,000 yr B.P. the westerly storm tracks presumably had shifted back north to their modern latitudinal position between latitudes 40 and 50°S, the circumpolar trough remained in the Australian sector, resulting in continued aridity in the high southern latitudes of South America. This independent behavior of the two principal circulation systems is also the most likely cause of the paleoclimatic opposition between middle and high latitudes during the Holocene (Markgraf, 1983).

Paleoclimatic reinterpretation of Tasmanian pollen records provided the first intrahemispheric test of the paleocirculation shifts proposed for southern South America (Markgraf *et al.,* 1986). Analysis of the paleoclimatic similarity of records from South America, Australia, and New Zealand (Markgraf *et al.,* 1992) suggests that in fact the same large-scale mechanism is at work in the circum-South Pacific (see also Webb, Ruddiman, *et al.,* this vol.).

In contrast to the substantial paleoclimatic variability documented in the records from mainland southern South America, little paleoenvironmental change is visible in records from the Falklands and South Georgia. The proportions of the major pollen taxa represented, Gramineae and *Acaena,* change over time, but the changes are not contemporaneous among the records. This suggests that the changes reflect local

(perhaps edaphic) differences rather than regional climate variations. Long-distance pollen from forest vegetation, which has been used in records from high northern latitudes to interpret past changes in wind direction and circulation (Nichols *et al.,* 1978; Barry *et al.,* 1981), apparently cannot be used for this purpose in the high southern latitudes because it is present in extremely low and erratic amounts.

Discussion

This section describes the patterns of late-Pleistocene and Holocene paleoenvironmental changes specific to each plant-geographic region in Central and South America and their inferred paleoclimates. The discussion focuses on the trends and differences at 18,000, 12,000, 9000, 6000, 3000, and 0 yr B.P. Given the still limited ecological and climatological information, quantification of the paleoclimatic inferences is unrealistic. The plotted temperature and precipitation curves based on environmental and lake-level changes (Figs. 14.4 and 14.9) are semiquantitative at best. These curves primarily illustrate the frequency and relative scale of change through time in the specific regions.

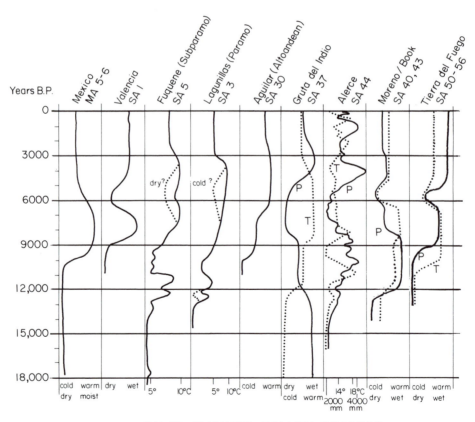

PALEO CLIMATE - POLLEN RECORDS

Fig. 14.9. Selected paleoclimatic curves interpreted from pollen records in Central and South America. Dotted lines refer to temperature (T), solid lines to precipitation (P). Reprinted from Markgraf (1989a), with permission from *Quaternary Science Reviews* and Pergamon Press..

According to the pollen records, late-Pleistocene climates before 12,000 yr B.P. were colder than today everywhere in Central and South America, in the high Andes as well as in the tropical, subtropical, and temperate lowlands. Inferred mean annual temperatures range from 5°C lower than at present for the paramo and subparamo sites in Colombia (van der Hammen et al., 1981; Hooghiemstra, 1984; Melief, 1985) to 9°C colder than at present for the mountains in Central America (Costa Rica [Martin, 1964]) and the southern Andes at midlatitudes (Markgraf, 1989b). The cooler temperatures probably occurred primarily during winter, while summers may have been somewhat warmer than at present, especially east of the Andes in areas of precipitation shadow (Markgraf, 1984).

In contrast to the more uniform temperature patterns, full- and late-glacial moisture patterns in Central and South America show significant regional differentiation. The southern subtropical latitudes (sites from about 20 to 38°S latitude) and the southern temperate latitudes on the west coast of Chile (from 43°S to probably as far south as 45°S) had more effective moisture before 14,000 yr B.P. than at present. All other regions, at high and low elevations, were far drier than today and probably had much higher winds. Estimates of the decrease in precipitation range from 50% less than modern levels for the tropical lowlands (Leyden, 1985) and the southern temperate forests (Markgraf, 1983) to 25% less for the Valdivian rainforest at midlatitudes (Heusser and Streeter, 1980; Villagran, 1985).

The late-Pleistocene increase in effective moisture at subtropical latitudes was previously ascribed to an equatorward shift of the westerly circulation (Heusser, 1983) or to a poleward shift through intensification (Markgraf, 1983, 1987, 1989b; Servant, 1984; Kessler, 1985). Reinterpretation of earlier records and the development of higher-resolution paleoenvironmental data from Chiloé and the Chilean lake district indicate that the latter mechanism is more probable (Markgraf, 1989b).

After 14,000 and especially 12,000 yr B.P. regional patterns of climatic change developed on smaller geographic scales than seen earlier. This regionalization, which began at low and middle latitudes, may reflect weakening of the extreme full-glacial forcing of global scale. As a consequence, regional climatic forcing and its respective regional climatic patterns began to emerge, starting in the subtropics and subsequently expanding poleward. Although climatic patterns from that time onward may differ, the changes appear synchronous under the existing dating control, indicating linkage of the climate system.

By 12,000 yr B.P. temperatures appear to have reached near-modern levels. The temperature rose either in a series of steps, with (e.g., in Colombia and Chile) or without (e.g., in Peru and southern temperate latitudes) intervening reversals, or in a single step

(e.g., in the tropics, Mexican highlands, and southern Andes). This difference cannot be explained solely on the basis of different resolution of the respective records but more likely reflects regional climatic responses of the different biota. For example, the "complacent" pollen records from the Mexican highlands (Watts and Bradbury, 1982) contrast with the records from the Colombian Andes, which document great variability at the end of the Pleistocene and beginning of the Holocene (van der Hammen et al., 1981).

The existence of a Younger Dryas equivalent in southern South America, a supposedly cold interval between 11,000 and 10,000 yr B.P., has been vigorously debated. South American glacial evidence dated to this interval is ambiguous (Clapperton, 1983, 1987). Even in the Colombian Andes, where many pollen records show environmental reversals at that time, evidence for glacier readvance is scarce (van der Hammen et al., 1981; Helmens, 1988). Records that were interpreted as documenting a Younger Dryas or equivalent cool phase in middle and high latitudes (Heusser, 1966; Heusser and Rabassa, 1987) have been critically reevaluated (Markgraf, 1989a, 1991). Resolving the Younger Dryas question will require a concerted effort to study both glacial and paleoenvironmental data that satisfy the same resolution and dating control criteria.

The regionalization of paleoclimatic patterns that produced diverse climatic conditions developed throughout the Holocene and reached its peak during the late Holocene. For example, the late Holocene was cool and moist at temperate midlatitudes in Chile and Argentina but cool and dry at high latitudes between 12,000 and 9000 yr B.P. (Markgraf, 1983), and the reverse was true between 9000 and 6000 yr B.P.

The middle Holocene, represented by the 6000-yr B.P. target date, is reflected in many records by a short-term environmental change lasting only about 1000 yr or by culmination of a trend that started after 8000 yr B.P. The paleoclimatic interpretation in general suggests a period of moisture stress for both the shorter and the longer interval.

ATMOSPHERIC CIRCULATION PATTERNS

During full- and late-glacial times, the modern summer-rain pattern in the northern subtropics and tropics must have been greatly reduced, according to the paleoclimatic interpretation of the palynologic data. This suggests that the subtropical high-pressure system bringing easterly moisture to Central and northern South America was weaker than today and perhaps also shifted somewhat closer to the equator. The upwelling history reconstructed from an ocean sediment record off the coast of Venezuela supports this full-glacial circulation scenario. Foraminifera data indicate that before 11,000 yr B.P. there was essentially no upwelling, and thus trade wind intensity must have been greatly reduced (Overpeck *et al.,* 1989). Another line of support could be interpreted from data on the historical "little ice age" interval that could serve as ice-age analogue. In the Quelccaya ice core the "little ice age" was recognized as a cold and arid interval with greatly reduced easterly (summer) precipitation and enhanced westerly winds in winter, characterized by high proportions of windblown dust (Thompson *et al.,* 1984).

The slightly greater effective moisture during full-glacial times compared to the present along the northern margin of the northern subtropics (Mexico) is probably related to proportionally greater winter rains, a prominent full- and late-glacial feature in the American Southwest related to the equatorward displacement of the northern westerlies (Thompson *et al.,* this vol.). Full-glacial atmospheric circulation has been far more difficult to assess in the southern subtropics because of the paucity of data and the ambiguity of their interpretation (see Markgraf, 1989b). The unequivocal evidence in the records for greater effective moisture during the full glacial than at present has been previously ascribed either to an equatorward shift of the southern westerlies (Heusser, 1983) or to a poleward shift of the subtropical high-pressure system and the southern westerlies (Markgraf, 1983). Preference for either interpretation hinges not only on a more critical analysis of the available records from the subtropics as well as from adjacent higher latitudes (38–42°S) but also on a better understanding of the relationship between precipitation patterns and the dynamics of the associated circulation systems.

Modern climates toward the northern margin of the westerlies between latitudes 36 and 38°S are characterized by a gradually shortened winter-rain season, caused by the seasonal shift of the westerly storm tracks. This shift—poleward in summer and equatorward in winter—is a consequence of the seasonally changing pole-equator pressure and temperature gradients, because the storm tracks are located where the gradient is steepest (Aceituno, 1987). The other source of winter precipitation in the subtropical region is related to warm El Niño/Southern Oscillation (ENSO) events and to times when the southeast Pacific anticyclone is weak (Rutllant and Fuenzalida, 1991).

Full-glacial evidence suggests that the zone of increased seasonal moisture stress, comparable to the northern margin of the westerlies, had shifted south to about latitude 42°S to encompass the Chilean lake region. Thus the westerly storm tracks must have been positioned farther south and could not have been responsible for the greater moisture at subtropical latitudes. The increased moisture and windiness south of latitude 42°S then represents the northern limit of the full- and late-glacial location of the westerly storm tracks. Hence the westerlies were more zonal than today, and the precipitation gradient at its northern border was far more compressed than today. A steepening of the gradient that would compress the westerlies and the frontal activity has been proposed by CLIMAP Project Members (1981) on the basis of full-glacial sea-surface temperatures (SSTs) along the west coast of Chile. On the other hand, the increased moisture in the subtropics could have been a combination of reduced evaporation resulting from lower temperatures and higher precipitation in response to a weaker subtropical high-pressure zone positioned toward the pole (Pittock, 1980a,b; Markgraf, 1983, 1989b; Servant, 1984; Kessler, 1985). The westerlies weakened during late-glacial times, between 14,000 and 12,000 yr B.P., and the storm tracks could not be traced until after 9500 yr B.P.

Because most land west of the Andes south of latitude 45°S was covered by ice, records from sites in that region generally start after 14,000 yr B.P. The paleoenvironmental evidence shows a period of markedly lower precipitation at high latitudes before 12,000 yr B.P. (Markgraf, 1983; Ashworth and Markgraf, 1989). At these high southern latitudes Antarctica dominates the climate. The location of the circum-Antarctic low-pressure belt determines precipitation: if it is positioned toward the Australian sector, southern South America is relatively dry (Pittock, 1980a,b); if it is positioned toward South America, the high latitudes receive more precipitation (Markgraf, 1983).

The principal shift of the atmospheric circulation systems to their modern positions and the onset of seasonal variability occurred in Central and South America at 12,000 yr B.P. for the southern low and middle latitudes and at 9500 yr B.P. for the northern tropical and subtropical and southern high latitudes. For the circum-Antarctic region glacial-type conditions continued until 9500 yr B.P, suggesting that the circum-Antarctic low-pressure belt probably continued in its glacial mode into the early Holocene. The delayed emergence of the northern subtropics and tropics from their glacial mode could have been influenced by the slow disappearance of the Laurentide ice and its long-distance role in lower latitudes' atmospheric circulation systems (see also Thompson *et al.,* this vol.).

One interesting paleocirculation feature concerns the ENSO events. Sedimentologic analyses of flood deposits in coastal northern Peru (Wells, 1987) suggest that the frequency of extreme flood events was one every 1000 yr during the Holocene and that the frequency was higher during the late Holocene than earlier. The greater precipitation variability and overall greater moisture at low latitudes during the last 5000–3000 yr suggest that ENSO events became far more common during the late Holocene than they had been earlier (McGlone *et al.,* 1992).

Model Results and Paleoclimatic Data

Although discussions of such large-scale paleoenvironmental and paleoclimatic successions necessarily brush over much temporal and regional detail, they provide an excellent framework for global comparison of paleoclimates, which is one approach to improving our understanding of the dynamics of climatic change. With the more recent development of paleoclimatic modeling and experimental "prediction" of specific regional paleoclimatic patterns has come the need to validate the models' results with paleoclimatic evidence (Kutzbach, 1981; Kutzbach and Guetter, 1984a; Schneider, 1986; Webb, 1988). Although testing models by comparing their results with paleoclimatic evidence has limitations (see discussion by Rind and Peteet, 1986), the exercise of this joint COHMAP venture (COHMAP Members, 1988) has proved very rewarding.

For this review, results from global paleoclimatic "snapshot" model experiments with the NCAR Community Climate Model, an atmospheric GCM (Kutzbach and Guetter, 1986), were compared with paleoclimatic data from Central and South American records, analyzed at the same geographic and temporal resolution as the model experiments. The GCM data for the South American sector were calculated as area averages of four to 12 grid points in latitudinal bands that correspond more or less to the specific biogeographic zones selected for the records (Fig. 14.10). This procedure was chosen to illustrate the regional climatic patterns on a comparable scale. Kutzbach and Guetter (1986) discussed in detail the model's boundary conditions, which are related to changes in solar seasonal radiation, terrestrial- and sea-ice cover, and SSTs.

Seasonal insolation in the Southern Hemisphere at 18,000 yr B.P. was close to modern values. Solar radiation subsequently decreased by 6% in summer and increased by 5% in winter. This trend culminated between 12,000 and 9000 yr B.P. The insolation variations were proportionally stronger at lower latitudes than at higher latitudes. The immediate result was overall decreased seasonality in the early Holocene (Kutzbach and Guetter, 1986).

The predominance of ocean surface in the Southern Hemisphere explains the great climatic influence, especially during the late Pleistocene, of the other two boundary parameters, SSTs and sea-ice extent. The SST values (difference from modern values) based on the CLIMAP (1981) reconstruction for 18,000 yr B.P. were continued through 15,000 yr B.P., reduced to one-half for 12,000 yr B.P., and assumed to be near modern values at and after 9000 yr B.P. The full-glacial extent of Antarctic sea ice was continued until 15,000 yr B.P. and reduced to its modern extent by 12,000 yr B.P. Full- and late-glacial climates in the model predictions for South America are primarily influenced by (1) SSTs in the subtropical South Pacific as warm as today, (2) SSTs at midlatitudes in the southern oceans 4-6°C colder than today, and (3) a maximum equatorward expansion of Antarctic sea ice.

The lower terrestrial surface temperature predictions of the GCM for Central and South America (Fig. 14.10) agree with the paleoclimatic evidence of temperatures lower than today before 12,000 yr B.P., especially in winter (Kutzbach and Guetter, 1986). The model predicts an interesting temperature anomaly for latitudes 47-56°S, namely summers warmer than today in full- and late-glacial times (Fig. 14.10). In the model this anomaly is related to the warm subtropical South Pacific and a more northerly wind direction in the high latitudes (Fig. 7 in Kutzbach and Guetter, 1986). Late-glacial summers warmer than today are in fact suggested in records east of the Andes at latitude 41°S (Markgraf, 1984). At high latitudes a high-elevation record (Markgraf, 1990), probably sensitive to summer temperatures, shows instead that temperatures were lower than at present until about 10,000 yr B.P.

In contrast to the more uniform temperature patterns, full- and late-glacial moisture patterns in Central and South America show significant regional differentiation, both in the model and in the paleoclimatic data. Moisture results from the model are plotted as annual values of precipitation minus evaporation (P-E) in terms of the difference from today (Fig. 14.10).

The GCM results for Central and South America poleward to 38°S show substantially less moisture in full- and late-glacial times than today. In contrast, the model proposes greater effective moisture than at present south of latitude 38°S (Fig. 14.10). Paleoclimatic data that can be interpreted in terms of moisture show comparable moisture patterns; that is, drier conditions, especially in the tropics and to a lesser degree in the subtropics. The data also support greater effective moisture during full- and late-glacial times between latitudes 38 and 47°S, although only on the oceanic west coast of Chile. Records east of the Andes instead show substantially less moisture, most likely related to the increased rainshadow effect of the enlarged ice barrier of the Andes. At high latitudes the data again disagree with the model predictions: instead of the higher effective moisture simulated in the model, the data indicate strongly decreased moisture in glacial times until after 10,000 yr B.P.

The reason for the only partial agreement between the moisture results of the model and the data is probably related to the inability of the model to account for elevational gradients, which are a major influence on the distribution of precipitation in South America today. The discrepancies at high latitudes, on the other hand, probably stem from the model's sea-ice boundary conditions, which assign the modern value by 12,000 yr B.P. Analyses of deep-sea cores suggest that Antarctica's ice shelves and sea ice did not begin

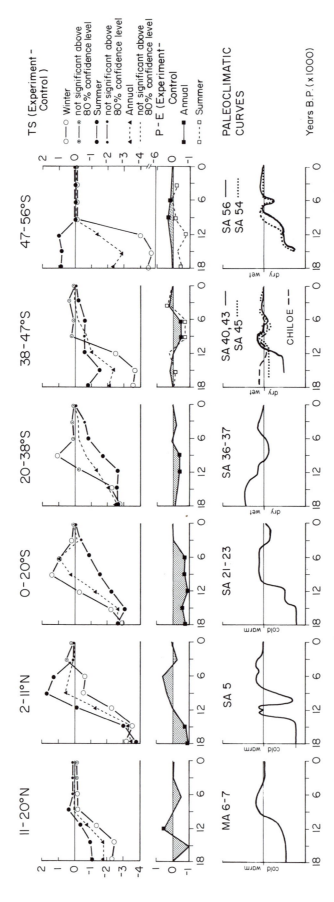

Fig. 14.10. General circulation model results for latitudinal grid averages for Central and South America for surface temperature (TS) and precipitation minus evaporation (P-E) at 3000-yr intervals for the last 18,000 yr B.P. compared with paleoclimatic curves interpreted from pollen records (see also Fig. 14.9). Reprinted from Markgraf (1989a), with permission from *Quaternary Science Reviews* and Pergamon Press.

to break up until about 9000 yr B.P. (Herron and Anderson, 1990).

Of great interest in this comparison are the wind data. During full- and late-glacial times, especially in winter, winds were stronger than at present by about 10 m/s at 50°S, and the westerlies were shifted year-round south over the northward-extended sea-ice limit (Kutzbach and Guetter, 1986). This scenario is supported by the paleoclimatic data from middle and high latitudes that had been interpreted as the result of a poleward shift of the westerlies (Markgraf, 1983). The wind information combined with pressure data allows us to determine the position of the moisture-bringing storm tracks and thus the location of precipitation. For full-glacial times the model predicts a year-round steepened poleward pressure gradient, steepest at about 40°S (Kutzbach and Guetter, 1986). If so, the storm tracks then should be located south of latitude 40°S. Data from Isla Chiloé at latitude 43°S (Villagran, 1985, 1988) in fact suggest that this region was substantially moister and windier during full-glacial times than today. In that case the westerly storm tracks could not have been responsible for the higher effective moisture farther equatorward at latitudes 32-36°S, which must have been caused instead by substantially lower temperatures.

After about 12,000 yr B.P. paleoclimatic data indicate a regionalization of climatic patterns on geographic scales much smaller than earlier. This regionalization probably reflects the weakening of global climatic forcing. Because of its global scale and the chosen time intervals, the GCM fails to simulate the detail seen in paleoclimatic records for the Holocene. Nevertheless, some climatic trends in the model simulations are worth comparing with the data.

Evidence of decreased moisture and higher summer temperatures in the southern subtropics (20-38°S) in the early Holocene can be seen in both the model and the data (Fig. 14.10). The insolation trend toward generally cooler summers coupled to warmer winters predicts such a "suppressed" summer monsoon for the southern subtropics. In contrast, the Northern Hemisphere tropics and subtropics (2-11°N), with their enhanced seasonality contrast of warmer summers and colder winters than today, received more moisture 9000 yr ago than at present, according to both the model and the data (Fig. 14.10). This enhanced monsoon effect, however, was much smaller in the Americas than in the Asian-African sector (Kutzbach and Guetter, 1984b; Kutzbach and Street-Perrott, 1985), perhaps because of the dampening effect of the lingering melting of the Laurentide ice sheet (see Thompson et al., this vol.).

Decreased effective moisture is also predicted in the model for the early Holocene between latitudes 38 and 47°S. Pollen data support this result. In the model this moisture deficit is related to a weaker trough and thus less convergence at these latitudes (Kutzbach and Guetter, 1986), which in turn is related to the lessened seasonal contrast in insolation.

To summarize, because many of the primary aspects of paleoclimatic patterns simulated by the model are comparable to the paleoclimatic data, we can use the experimental results to learn more about the mechanisms that underlie climatic change. Instances where model results and paleoenvironmental data disagree seem to be linked to (1) the model's inability to account for the elevational climatic gradient, which is a major factor in South American climate, (2) the specification of the SST and sea-ice boundary conditions in the model, and (3) the grid scale of the model cells, which is too large to register the paleoclimatic detail, especially during the Holocene. Once the model addresses these three concerns, its predictions will probably relate much better to the real world of paleoclimatic data.

Acknowledgments

This work is the result of several years of interaction with the members of COHMAP, especially Tom Webb III and John E. Kutzbach, whose help and never-ceasing interest and enthusiasm are greatly appreciated. I am greatly indebted to Tom Webb and Herb Wright for editorial help with the manuscript. The first part of the chapter is an abridged version of Markgraf (1989a). The illustrations and table are reproduced with permission of the editor and publisher of *Quaternary Science Reviews*. Financial support was provided in part by National Science Foundation grant ATM-8618217.

References

Absy, M. L. (1979). "A Palynological Study of Holocene Sediments in the Amazon Basin." Unpublished Ph.D. dissertation, University of Amsterdam.

Absy, M. L., Cleef, A., Fournier, M., Martin, L., Servant, M., Sifedine, A., Ferreira de Silva, M., Soubies, F., Suguio, K., Turcq, B., and van der Hammen, T. 1991. Mise en évidence de quatre phases d'ouverture de la forêt dense dans le sud-est de l'Amazonie au cours des 60,000 dernières années. Première comparaison avec d'autres régions tropicales. *Comptes rendues de l'Academie des sciences, Paris, séries II* 312, 673-678.

Aceituno, P. (1987). "On the Interannual Variability of South American Climate and the Southern Oscillation." Unpublished Ph.D. dissertation, University of Wisconsin, Madison.

Ashworth, A. C., and Hoganson, J. W. (1984). Testing the late Quaternary climatic record of southern Chile with evidence from fossil Coleoptera. *In* "Late Cainozoic Palaeoclimates of the Southern Hemisphere" (J. C. Vogel, Ed.), pp. 85-102. Balkema, Rotterdam.

Ashworth, A. C., and Markgraf, V. (1989). Climate of the Chilean Channels between 11,000 to 10,000 yr B.P. based on fossil beetle and pollen analyses. *Revista Chilena de Historia Natural* 62, 61-74.

Auer, V. (1974). The isorhythmicity subsequent to the Fuego-Patagonian and Fennoscandian ocean level transgressions and regressions of the latest glaciation. *Annales Academiae Scientiarum Fennicae A* 115, 1-88.

Barrow, C. J. (1978). Postglacial pollen diagrams from South Georgia (sub-Antarctic) and West Falkland Island (south Atlantic). *Journal of Biogeography* 5, 251-274. (Reprinted in *British Antarctic Survey Bulletin* 58, 15-42, 43-60.)

Barry, R. G., Elliott, D. L., and Crane, R. G. (1981). The palaeoclimatic interpretation of exotic pollen peaks in Holocene records from the eastern Canadian Arctic: A discussion. *Review of Palaeobotany and Palynology* 33, 153-167.

Bartlett, A. S., and Barghoorn, E. (1973). Phytogeographic history of the Isthmus of Panama during the past 12,000 years (a history of vegetation, climate, and sea-level change). *In* "Vegetation and Vegetational History of Northern Latin America" (A. Graham, Ed.), pp. 203-300. Elsevier, Amsterdam.

Bigarella, J. J. (1975). Lagoa dune field (State of Santa Catarina, Brazil), a model of eolian and pluvial activity. *Boletim Paranaense Geociencias* 33, 133-168.

Binford, M. W., Brenner, M., Whitmore, T. J., Higuera-Gundy, A., Deevey, E. S., and Leyden, B. (1987). Ecosystems, palaeoecology and human disturbance in subtropical and tropical America. *Quaternary Science Reviews* 6, 115-128.

Bradbury, J. P. (1971). Paleolimnology of Lake Texcoco, Mexico: Evidence from diatoms. *Limnology and Oceanography* 66, 385-416.

——. (1989). Late Quaternary lacustrine palaeoenvironments in the Cuenca de Mexico. *Quaternary Science Reviews* 8, 75-100.

Bradbury, J. P., Leyden, B., Salgado-Labouriau, M. L., Lewis, W. M., Schubert, C., Binford, M. W., Frey, D. G., Whitehead, D. R., and Weibezahn, F. H. (1981). Late Quaternary environmental history of Lake Valencia, Venezuela. *Science* 214, 1299-1305.

Brown, R. B. (1985). A summary of late-Quaternary pollen records from Mexico west of the Isthmus of Tehuantepec. *In* "Pollen Records of Late-Quaternary North American Sediments" (V. M. Bryant and R. G. Holloway, Eds.), pp. 71-94. American Association of Stratigraphic Palynologists Foundation, Dallas.

Bush, M. B., and Colinvaux, P. A. 1990. A pollen record of a complete glacial cycle from lowland Panama. *Journal of Vegetation Science* 1, 105-118.

Bush, M. B., Piperno, D. T. C., and Colinvaux, P. A. 1989. A 6000 year history of Amazonian maize cultivation. *Nature* 340, 303-305.

Clapperton, C. M. (1983). The glaciation of the Andes. *Quaternary Science Reviews* 2, 83-155.

——. (1987). Glacial geomorphology, Quaternary glacial sequence and paleoclimatic inferences in the Ecuadorian Andes. *In* "International Geomorphology 1986, Part II" (V. Gardiner, Ed.). Wiley, New York.

CLIMAP Project Members (1981). Seasonal reconstructions of the earth's surface at the last glacial maximum. *Geological Society of America Map and Chart Series* MC-36.

COHMAP Members (1988). Climatic changes of the last 18,000 years: Observations and model simulations. *Science* 241, 1043-1052.

Colinvaux, P. (1987). Environmental history of the Amazon Basin. *Quaternary of South America and Antarctic Peninsula* 5, 223-237.

Colinvaux, P. A., and Schofield, E. K. (1976). Historical ecology in the Galapagos Islands. *Journal of Ecology* 64, 989-1012.

Covich, A., and Stuiver, M. (1974). Changes in oxygen 18 as a measure of long-term fluctuations in tropical lake levels and molluscan populations. *Limnology and Oceanography* 19, 682-691.

D'Antoni, H. L. (1983). Pollen analysis of Gruta del Indio. *Quaternary of South America and Antarctic Peninsula* 1, 83-104.

Fernandez, J. (1986). *Hippidion:* New data on the new extinct American horse in northwest Argentina: Palaeoenvironmental and palaeoclimatic implications. *Current Research in the Pleistocene* 3, 64-67.

Fernandez, J., Markgraf, V., Panarello, H. O., Albero, M., Angiolini, F. E., Valencio, S., and Arriaga, M. (1991). Paleoenvironments and paleoclimates, extinct and extant fauna, and human occupation in the altiplano of northwestern Argentina during late Pleistocene/Holocene transition. *Geoarchaeology* 6, 251-272.

Galloway, R. W., Markgraf, V., and Bradbury, J. P. (1988). Dating shorelines of lakes in Patagonia, Argentina. *Journal of South American Earth Sciences* 1, 195-198.

Gonzalez, M. A. (1981). Evidencias palaeoclimaticas en la Salina del Bebedero (San Luis). *Actas VIII Congreso Geologico Argentino* 3, 411-438.

Gonzalez, M. A., Musacchio, E. A., Garcia, A., Pascual, R., and Corte, A. E. (1981). Las lineas de costa holoceno de la Salina del Bebedero (San Luis, Argentina): Implicancias palaeoambientales de sus microfosiles. *Actas VIII Congreso Geologico Argentino* 3, 617-628.

Gonzalez-Quintero, L., and Fuentes Mata, L. (1980). El Holoceno de la porción central de la Cuenca del Valle de Mexico. *Memorias Instituto Nacional de Antropología e Historia, III Coloquio Paleobotanica y Palinología, Mexico City, 1977* 86, 113-132.

Grabandt, R. A. J. (1980). Pollen rain in relation to arboreal vegetation in the Colombian Cordillera Oriental. *Review of Palaeobotany and Palynology* 29, 65-147.

Graf, K. J. (1979). "Untersuchungen zur rezenten Pollen- and Sporenflora in der nördlichen Zentralkordillere Boliviens und Versuch einer Auswertung von Profilen aus postglazialen Torfmooren." Ph.D. dissertation, University of Zürich, Switzerland.

——. (1981). Palynological investigations of two post-glacial peat bogs near the boundary of Bolivia and Peru. *Journal of Biogeography* 8, 353-369.

——. (1985). Der Rezentpollenflug als Klimaindikator in den bolivianischen Anden. *Zentralblatt für Geologie und Paläontologie* 11/12 (1984), 1679-1689.

Hansen, B. C. S., Wright, H. E., Jr., and Bradbury, J. P. (1984). Pollen studies in the Junin area, central Peruvian Andes. *Bulletin of the Geological Society of America* 95, 1454-1465.

Hastenrath, S., and Kutzbach, J. (1985). Late Pleistocene climate and water budget of the South American altiplano. *Quaternary Research* 24, 249-256.

Heine, K. (1984). The classical late Weichselian climatic fluctuations in Mexico. *In* "Climatic Changes on a Yearly to Millennial Basis" (N. A. Mörner and W. Karlén, Eds.), pp. 95-115. Reidel, Dordrecht.

Helmens, K. F. (1988). Late Pleistocene glacial sequence in the area of the high plain of Bogotá (eastern Cordillera, Colombia). *Palaeogeography, Palaeoclimatology, Palaeoecology* 67, 263-283.

Herron, M. J., and Anderson, J. B. (1990). Late Quaternary glacial history of the South Orkney Plateau, Antarctica. *Quaternary Research* 33, 265-275.

Heusser, C. J. (1966). Polar hemispheric correlation: Palynological evidence from Chile and the Pacific Northwest of America. In "World Climate from 8000 to 0 B.C." (J. S. Swayer, Ed.), pp. 124-142. Royal Meteorological Society, London.

——. (1973). An additional post-glacial pollen diagram from Patagonia occidental. Pollen et spores 14, 157-167.

——. (1974). Vegetation and climate of the southern Chilean lake district during and since the last interglaciation. Quaternary Research 4, 293-321.

——. (1983). Quaternary pollen record from Laguna Tagua Tagua, Chile. Science 220, 1429-1432.

——. (1984a). Late-glacial-Holocene pollen sequence in the northern lake district in Chile. Quaternary Research 22, 77-90.

——. (1984b). Late Quaternary climates of Chile. In "Late Cainozoic Palaeoclimates of the Southern Hemisphere" (J. C. Vogel, Ed.), pp. 59-84. Balkema, Rotterdam.

Heusser, C. J., and Flint, R. F. (1977). Quaternary glaciations and environment of northern Isla Chiloé, Chile. Geology 5, 305-308.

Heusser, C. J., and Rabassa, J. (1987). Cold climatic episode of Younger Dryas age in Tierra del Fuego. Nature 328, 609-611.

Heusser, C. J., and Streeter, S. S. (1980). A temperature and precipitation record of the past 16,000 years in southern Chile. Quaternary Research 16, 293-321.

Heusser, C. J., Streeter, S. S., and Stuiver, M. (1981). Temperature and precipitation record in southern Chile extended to -43,000 yr ago. Nature 294, 65-67.

Hooghiemstra, H. (1984). Vegetational and climatic history of the high plain of Bogotá, Colombia: A continuous record of the last 3.5 million years. Dissertationes Botanicae 79, 1-368.

Hueck, K., and Seibert, P. (1981). "Vegetationskarte von Südamerika," 2nd ed. G. Fischer, Stuttgart and New York.

Huntley, B., and Prentice, I. C. (1988). July temperatures in Europe from pollen data, 6000 years before present. Science 241, 687-690.

Kessler, A. (1985). Zur Rekonstruktion von spätglazialem Klima und Wasserhaushalt auf dem peruanisch-bolivianischen Altiplano. Zeitschrift für Gletscherkunde 21, 107-114.

Kuhry, P., Salomons, B., Riezebos, P. A., and van der Hammen, T. (1983). Paleoecologia de los últimos 6000 años en el area de la Laguna de Otun-El Bosque. In "Studies on Tropical Andean Ecosystems, Vol. 1: La Cordillera Central Colombiana transecto Parque Los Nevados" (T. van der Hammen, A. Perez-P., and P. Pinto-E., Eds.), pp. 227-261. Cramer, Vaduz, Liechtenstein.

Kutzbach, J. E. (1981). Monsoon climate of the early Holocene: Climatic experiment using the earth's orbital parameters for 9000 years ago. Science 214, 59-61.

Kutzbach, J. E., and Guetter, P. J. (1984a). Sensitivity of late-glacial and Holocene climates to the combined effects of orbital parameter changes and lower boundary condition changes: "Snapshot" simulations with a general circulation model for 18, 9, and 6 ka B.P. Annals of Glaciology 5, 85-87.

——. (1984b). The sensitivity of monsoon climates to orbital parameter changes for 9000 years B.P.: Experiments with the NCAR general circulation model. In "Milankovitch and Climate" (A. K. Berger, J. Imbrie, J. Hays, G. Kukla, and B. Saltzman, Eds.), part II, pp. 801-820. D. Reidel, Dordrecht, The Netherlands.

——. (1986). The influence of changing orbital parameters and surface boundary conditions on climate simulations for the past 18,000 years. Journal of the Atmospheric Sciences 43, 1726-1759.

Kutzbach, J. E., and Street-Perrott, A. F. (1985). Milankovitch forcing of fluctuations in the level of tropical lakes from 18 to 0 kyr B.P. Nature 317, 130-134.

Ledru, M. P. (1993). "Late Quaternary environmental and climatic changes in central Brazil. Quaternary Research 39, 90-98.

Leyden, B. W. (1984). Guatemalan forest synthesis after Pleistocene aridity. Proceedings of the National Academy Sciences 81, 4856-4859.

——. (1985). Late Quaternary aridity and Holocene moisture fluctuations in the Lake Valencia Basin, Venezuela. Ecology 66, 1279-1295.

Liu, K., and Colinvaux, P. (1988). A 5200-year history of Amazon rain forest. Journal of Biogeography 15, 231-248.

Markgraf, V. (1980). New data on the late and postglacial vegetational history of La Misión, Tierra del Fuego, Argentina. Proceedings, IV International Palynological Conference 1976-1977 3, 68-74.

——. (1983). Late and postglacial vegetational and paleoclimatic changes in subantarctic, temperate, and arid environments in Argentina. Palynology 7, 43-70.

——. (1984). Late Pleistocene and Holocene vegetation history of temperate Argentina: Lago Morenito, Bariloche. Dissertationes Botanicae 72, 235-254.

——. (1985a). Paleoenvironmental history of the last 10,000 years in northwestern Argentina. Zentralblatt für Geologie und Paläontologie 11/12, 1739-1749.

——. (1985b). Late Pleistocene faunal extinctions in southern Patagonia. Science 228, 1110-1112.

——. (1987). Paleoenvironmental changes at the northern limit of the subantarctic Nothofagus forest. Quaternary Research 28, 119-129.

——. (1988). Fells Cave: 11,000 years of changes in paleoenvironments, fauna, and human occupation. In "Travels and Archaeology in South Chile" by J. B. Bird, pp. 196-201. University of Iowa Press, Iowa City.

——. (1989a). Palaeoclimates in Central and South America since 18,000 B.P. based on pollen and lake-level records. Quaternary Science Reviews 8, 1-24.

——. (1989b). The southern westerlies during the last glacial maximum: Reply to C. J. Heusser. Quaternary Research 31, 426-432.

——. (1990). Paleoclimates in Tierra del Fuego. In "Programme and Abstracts, CANQUA/AMQUA 1990," p. 24. First joint meeting, Canadian Quaternary Association/American Quaternary Association, Waterloo, Canada.

——. (1991). Younger Dryas in southern South America? Boreas 20, 63-69.

Markgraf, V., Bradbury, J. P., and Busby, J. (1986). Paleoclimates of the last 13,000 years in southwestern Tasmania. Palaios 1, 368-380.

Markgraf, V., Dodson, J. R., Kershaw, A. P., McGlone, M. S., and Nicholls, N. (1992). Evolution of late Pleistocene and Holocene climates in the circum-South Pacific land areas. Climate Dynamics 6, 193-211.

Martin, P. S. (1964). Paleoclimatology and a tropical pollen profile. In "Proceedings of the VI International Quaternary Congress, International Quaternary Association (INQUA), Warsaw, 1961," pp. 319-323.

McGlone, M. S., Kershaw, A. P., and Markgraf, V. (1992). El Niño/Southern Oscillation climatic variability in Australasian and South American paleoenvironmental records. *In* "El Niño: Historical and Paleoclimatic Aspects of the Southern Oscillation" (H. F. Diaz and V. Markgraf, Eds.), pp. 435-462. Cambridge University Press, Cambridge.

Melief, A. B. M. (1985). "Late Quaternary Paleoecology of the Parque Nacional Natural los Nevados (Cordillera Central) and Sumapaz (Cordillera Oriental) areas, Colombia." Unpublished Ph.D. dissertation, University of Amsterdam.

Mercer, J. H., and Ager, T. (1983). Glacial and floral changes in southern Argentina since 14,000 years ago. *National Geographic Society Research Reports* 15, 457-477.

Metcalfe, S. E. (1985). "Late Quaternary Environments of Central Mexico: A Diatom Record." Unpublished Ph.D. dissertation, Oxford University.

Nichols, H., Kelly, P. M., and Andrews, J. T. (1978). Holocene palaeo-wind evidence from palynology in Baffin Island. *Nature* 273, 140-142.

Niederberger, C. (1979). Early sedentary economy in the basin of Mexico. *Science* 203, 131-142.

Ohngemach, D., and Straka, H. (1978). La historia de la vegetación en la region Puebla Tlaxcala durante el cuaternario tardio. *Comunicaciones Proyecto Puebla-Tlaxcala* 15, 189-203.

Overpeck, J. T., Peterson, L. C., Kipp, N., Imbrie, J., and Rind, D. (1989). Climate change in the circum-North Atlantic region during the last deglaciation. *Nature* 338, 553-557.

Pittock, A. B. (1980a). Patterns of climatic variation in Argentina and Chile, I. Precipitation 1931-1960. *Monthly Weather Review* 108, 1347-1362.

——. (1980b). Patterns of climatic variation in Argentina and Chile, II. Temperature 1931-1960. *Monthly Weather Review* 108, 1362-1369.

Prentice, I. C. (1985). Forest-composition calibration of pollen data. *In* "Handbook of Paleoecology" (B. E. Berglund, Ed.), pp. 799-816. Wiley, New York.

Rind, D., and Peteet, D. (1986). Comment on S. H. Schneider's editorial "Can modeling of the ancient past verify prediction of future climates?" (*Climatic Change* 8, 117-119). *Climatic Change* 9, 357-360.

Rutllant, J., and Fuenzalida, H. (1991). Synoptic aspects of the central Chile rainfall variability associated with the Southern Oscillation. *International Journal of Climatology* 11, 63-76.

Salgado-Labouriau, M. L. (1980). A pollen diagram of the Pleistocene-Holocene boundary of Lake Valencia, Venezuela. *Review of Palaeobotany and Palynology* 30, 297-312.

Salgado-Labouriau, M. L., and Schubert, C. (1976). Palynology of Holocene peat bogs from the central Venezuelan Andes. *Palaeogeography, Palaeoclimatology, Palaeoecology* 19, 147-156.

Salgado-Labouriau, M. L., Schubert, C., and Valastro, S. (1977). Palaeoecologic analysis of a late Quaternary terrace from Mucubaji, Venezuelan Andes. *Journal of Biogeography* 4, 313-325.

Schneider, S. H. (1986). Can modeling of the ancient past verify prediction of future climates? An editorial. *Climatic Change* 8, 117-119.

Schreve-Brinkman, E. J. (1978). A palynological study of the upper Quaternary sequence in the El Abra corridor and rock shelters (Colombia). *Palaeogeography, Palaeoclimatology, Palaeoecology* 25, 1-109.

Schubert, C., and Fritz, P. (1985). Radiocarbon ages of peat. Guayana Highlands (Venezuela). *Naturwissenschaften* 72, 427-429.

Servant, M. (1984). Climatic variations in the low continental latitudes during the last 30,000 years. *In* "Climatic Changes on a Yearly to Millennial Basis" (N. A. Mörner and W. Karlén, Eds.), pp. 117-120. Reidel, Dordrecht.

Servant, M., and Fontes, J. C. (1978). Les lacs Quaternaires des hauts plateaux des Andes boliviénnes: premières interpretations paléoclimatiques. *Cahier ORSTROM, séries géologie* 10, 5-23.

Simpson, B. B. (1975). Pleistocene changes in the flora of the high tropical Andes. *Paleobiology* 1, 273-294.

Sirkin, L. (1985). Late Quaternary stratigraphy and environments of the west Mexican coastal plain. *Palynology* 9, 3-26.

Steubing, L., Alberdi, M., and Wenzel, H. (1983). Seasonal changes of cold resistance of Proteaceae of the south Chilean laurel forest. *Vegetatio* 52, 35-44.

Stine, S., and Stine, M. (1990). A record from Lake Cardiel of climate change in southern South America. *Nature* 345, 705-708.

Strecker, M., Bloom, A. L., and Cahill, T. (1986). Pleistocene (PSL) and modern (MSL) snowline trends in the northern Sierra Pampeanas and southern puna, Argentine Andes (25° to 28°S lat., 65° to 69°W long.). *Geological Society of America, Abstracts and Programs* 18, 765.

Thompson, L. G., Mosley-Thompson, E., and Morales Arnao, B. (1984). El Niño-Southern Oscillation events recorded in the stratigraphy of the tropical Quelccaya ice cap, Peru. *Science* 226, 50-53.

Tricart, J. (1975). Influence des oscillations climatiques récentes sur le modèle en Amazonie Orientale (région de Santarem d'après les images radar lateral). *Zeitschrift für Geomorphologie* 19, 140-163.

——. (1977). Aperçus sur le Quaternaire Amazonien. *Supplément, Bulletin Association française pour l'étude du Quaternaire* 50, 265-271.

van der Hammen, T. (1963). A palynological study on the Quaternary of British Guiana. *Leidse Geologische Mededelingen* 29, 125-180.

——. (1974). The Pleistocene changes of vegetation and climate in tropical South America. *Journal of Biogeography* 1, 3-26.

van der Hammen, T., and Gonzalez, A. E. (1960a). Climate and vegetation history of the late glacial and Holocene of the Paramos de Palacio (Eastern Cordillera, Colombia, South America). *Geologie en Mijnbow* 39, 737-746.

——. (1960b). Upper Pleistocene and Holocene climate and vegetation of the Sabana de Bogotá, Colombia, South America. *Leidse Geologische Mededelingen* 25, 261-315.

——. (1965). Late glacial and Holocene pollen diagram from Cienaga del Visitador (dep. Boyaca, Colombia). *Leidse Geologische Mededelingen* 32, 193-201.

van der Hammen, T., Barelds, J., De Jong, H., and De Veer, A. A. (1981). Glacial sequence and environmental history in the Sierra Nevada del Cocuy (Colombia). *Palaeogeography, Palaeoclimatology, Palaeoecology* 32, 247-340.

van Geel, B., and van der Hammen, T. (1973). Upper Quaternary vegetational climate sequence of the Fuquene area (Eastern Cordillera, Colombia). *Palaeogeography, Palaeoclimatology, Palaeoecology* 4, 9-92.

Veblen, T. T., and Markgraf, V. (1988). Steppe expansion in Patagonia? *Quaternary Research* 30, 331-338.

Villagran, C. (1980). Vegetationsgeschichtliche und pflanzensoziologische Untersuchung im Vicente Perez Rosales Nationalpark (Chile). *Dissertationes Botanicae* 54, 1-165.

——. (1985). Analisis palinologico de los cambios vegetacionales durante el Tardiglacial y Postglacial en Chiloé, Chile. *Revista Chilena Historia Natural* 58, 57-69.

——. (1987). Historia de la vegetación de la Isla Grande de Chiloé, Chile. *Actas VII Simposio Argentino de Palaeobotanica y Palinologia, Buenos Aires, 1987* 133-137.

——. (1988). Expansion of Magellanic moorland during the last glaciation: Palynological evidence from northern Isla de Chiloé, Chile. *Quaternary Research* 30, 304-324.

Villagran, C., and Varela, J. (1990). Palynological evidence for increased aridity on the central Chilean coast during the Holocene. *Quaternary Research* 34, 198-207.

Walter, H. (1973). "Vegetation of the Earth in Relation to Climate and the Eco-physiological Conditions." Springer, New York.

Walter, H., and Lieth, H. (1960). "Klimadiagramm-Weltatlas." G. Fischer, Stuttgart and Jena.

Watts, W. A., and Bradbury, J. P. (1982). Paleoecological studies at Lake Patzcuaro on the west-central Mexican plateau and at Chalco in the Basin of Mexico. *Quaternary Research* 17, 56-70.

Webb, T. III (1988). Eastern North America. *In* "Vegetation-History" (B. Huntley and T. Webb III, Eds.), pp. 385-414. Kluwer Academic Publishers, Dordrecht, The Netherlands.

Webb, T. III, Bartlein, P. J., and Kutzbach, J. E. (1987). Climatic change in eastern North America during the past 18,000 years: Comparisons of pollen data with model results. *In* "North America and Adjacent Oceans during the Last Deglaciation" (W. F. Ruddiman and H. E. Wright, Jr., Eds.), pp. 447-462. The Geology of North America, Vol. K-3. Geological Society of America, Boulder, Colo.

Weinberger, P. (1974). Verbreitung und Wasserhaushalt araukanopatagonischer Proteaceen in Beziehung zu mikroklimatischen Faktoren. *Flora* 163, 251-264.

Wells, L. E. (1987). An alluvial record of El Niño events from northern coastal Peru. *Journal of Geophysical Research* 92, 14463-14470.

Wijmstra, T. A., and van der Hammen, T. (1966). Palynological data on the history of tropical savannas in northern South America. *Leidse Geologische Mededelingen* 38, 71-90.

CHAPTER 15

Holocene Vegetation and Climate Histories of Alaska

Patricia M. Anderson and Linda B. Brubaker

The history of Beringia, comprising northeasternmost Asia, northwesternmost North America, and the land bridge joining the two continents, has sparked the imagination of Quaternary scientists for decades. This history is a central theme of several books (Hopkins, 1967; Hopkins *et al.*, 1982; Kontrimavichus, 1984), which focus primarily on reconstructing late-Pleistocene environments and presenting guidelines for subsequent research in Alaska. Interest in the Alaskan Holocene was not pronounced until the 1970s, and regional syntheses have only recently been published (e.g., Ager, 1983; Ager and Brubaker, 1985; Barnosky *et al.*, 1987). The data and ideas in this chapter are current to 1986.

Many kinds of evidence have been used for interpreting Holocene paleoenvironments, but palynologic data have probably increased the most dramatically among types of fossil data during the past 15 yr. The results of recent studies, coupled with the pioneering work of Heusser (1952, 1957, 1960), Livingstone (1955, 1957), and Colinvaux (1964a,b, 1967), have greatly expanded our understanding of the Alaskan past. However, the history of Holocene landscapes and climatic conditions, whether based on palynologic or other data, remains sketchy and regionally incomplete. Because of the massive size of Alaska, the remoteness of most areas, and the emphasis in many Quaternary studies on the Pleistocene, interpretations of the Alaskan Holocene are less detailed than those for other areas of North America. Despite these limitations, the available fossil data show great potential for yielding detailed reconstructions of Alaskan paleoenvironments, particularly past vegetation and climate.

This chapter describes the vegetation and climate histories of Alaska over the past 12,000 yr and compares these fossil-based interpretations with model simulations of past climates. Vegetation history is summarized with isochrone maps for northern Alaska, where there is an adequate grid of fossil sites. In southern Alaska, where sites are fewer and less well dated, representative pollen diagrams are used to illustrate vegetational changes. We also discuss climatic interpretations for 12,000, 9000, and 6000 yr B.P., with emphasis on the Holocene times.

The Setting

Alaska, located at the far northwestern tip of North America, covers 20° of latitude and 31° of longitude. The state has been divided into four physiographic provinces (Wahrhaftig, 1965), four climatic zones (Watson, 1959), and three major vegetational formations (Viereck and Little, 1975) (Fig. 15.1). Alaska's highest peaks and largest glaciers are in the southern ranges of the Pacific Mountain system, which stretches from the Alaskan panhandle to the Aleutian Islands. The rugged terrain of these mountains changes inland to a gentler topography of alternating plateaus and lowlands that characterize the Intermontane region of interior Alaska. The northern border of the Intermontane region is formed by the Brooks Range, which marks the northernmost extent of the Rocky Mountain system. The Brooks Range is generally lower in elevation than the Pacific Mountain system. It gives way on the north to low rolling foothills and the flat Arctic coastal plain.

The four climatic zones of Alaska are maritime, continental, transitional maritime-continental, and arctic (Fig. 15.1 and Table 15.1). The modern climate reflects the interaction of Arctic, cool Pacific, and mild Pacific airmasses. The maritime climate of Alaska's Pacific coast is the mildest of the four regimes. Autumn and winter typically are very wet and cloudy as a result of frequent cyclonic storms over the Pacific

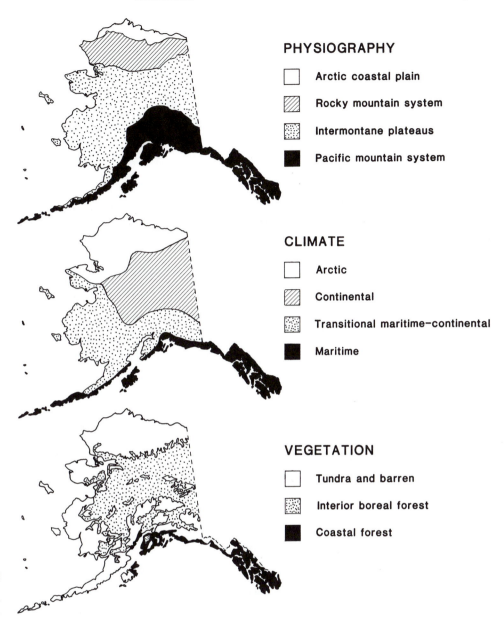

PHYSIOGRAPHY

☐ Arctic coastal plain

▨ Rocky mountain system

▦ Intermontane plateaus

■ Pacific mountain system

CLIMATE

☐ Arctic

▨ Continental

▦ Transitional maritime–continental

■ Maritime

VEGETATION

☐ Tundra and barren

▦ Interior boreal forest

■ Coastal forest

Fig. 15.1. Physiography, climate, and vegetation of Alaska. Adapted from Viereck and Little (1975).

Ocean. During these seasons the Pacific storms regularly move onshore because of increased zonal circulation and the presence of the Arctic front along the coastal ranges. The onshore airmasses, while causing heavy precipitation, also make winter temperatures along Alaska's southern coast relatively mild. By spring the zonal circulation and baroclinicity decrease, and rainfall, though still substantial, also decreases. Cyclogenesis occurs farther north in summer, with the breakdown of the Aleutian low and northward movement of the subtropical high. Airflow is less zonal, and storm intensity decreases, especially in southeastern Alaska.

The winter climate of interior and northern Alaska is dominated by Arctic airmasses that bring cold, dry conditions. The northward movement of the Arctic front in summer to a position over the Brooks Range permits mild, moist Pacific air to flow into interior Alaska. Weak cyclonic systems that originate in Siberia and the Arctic Ocean move along the front and are additional sources of summer precipitation. Summers are relatively warm south of the Arctic front except along the Bering coast, where temperatures remain cool because of the proximity of chilly seas. North of the Arctic front summers are cool as a result of continued dominance of Arctic airmasses and the coldness of the nearby Beaufort and Chukchi Seas.

The major Alaskan vegetation formations are tundra, boreal forest, and coastal forest (Viereck and Little, 1972, 1975) (Fig. 15.1). Tundra is found along the coasts of northern and western Alaska, on the islands of the Bering Strait, and at high elevations in the inte-

Table 15.1. Mean annual and seasonal temperature and precipitation ranges for Alaskan climate zones

Climate variable[a]	Arctic	Continental	Transitional	Maritime
MAT (°C)	-12.5 to -7	-7.5 to -2.5	-5 to +2.5	2.5-5
JANT (°C)	-27.5 to -20	-27.5 to -20	-20 to -10	-10 to -2.5
JULT (°C)	5-12.5	12.5-15	10-12.5	10-12.5
MAP (mm)	<8 to 16	<8 to 16	16-24	24 to >160
JANP (mm)	<10 to 25	<10 to 25	25-100	100 to >200
JULP (mm)	25-75	25-100	50-100	100-150

[a]MAT, mean annual temperature; JANT, mean January temperature; JULT, mean July temperature; MAP, mean annual precipitation; JANP, mean January precipitation; JULP, mean July precipitation.

rior. Tundra in low-lying wet areas along the coast is a mixture of meadowlike vegetation, with abundant Cyperaceae (sedges) and Gramineae (grasses) (Viereck and Dryness, 1980). Better-drained low- to middle-elevation sites support a shrub-tussock tundra, of which *Betula nana* (dwarf birch), *B. glandulosa* (shrub birch), *Salix* (willow), *Alnus* (alder), Ericales (heaths), Cyperaceae, and Gramineae are important components. Alpine tundra is found on exposed slopes at high elevation. This vegetation usually is sparse and dominated by herbs (e.g., *Dryas octopetala, D. integrifolia, Oxyria digyna*) and *Salix* spp.

The boreal forest, which covers low- to middle-elevation sites of interior Alaska, is the dominant vegetation of the Intermontane system. *Picea glauca* (white spruce), *P. mariana* (black spruce), *Betula papyrifera* (paper birch), *Populus balsamifera* (balsam poplar), *Populus tremuloides* (quaking aspen), and *Larix laricina* (tamarack) characterize this forest type. Latitudinal treeline follows the southern flanks of the Brooks Range and extends from about 67°N in northwestern Alaska to about 69°N in the northeast. North-central and northeastern sections of the Intermontane system support relatively dense stands of trees, but the closed forest thins to large areas of mixed open forest-tundra in far northwestern Alaska.

The coastal forest is confined to a relatively narrow strip of land that borders the North Pacific coast and Gulf of Alaska. *Tsuga heterophylla* (western hemlock) and *Picea sitchensis* (Sitka spruce) are the predominant tree species, with scattered growth of *T. mertensiana* (mountain hemlock), *Pinus contorta* (lodgepole pine), *Chamaecyparis nootkatensis* (Alaska cedar), *Populus trichocarpa* (black cottonwood), and *Alnus rubra* (red alder). *Vaccinium* spp. (blueberry), *V. parvifolium* (huckleberry), *Oplopanax horridus* (devil's club), and *Gaultheria shallon* (salal) commonly form a dense understory.

The Data

Fossil pollen sites are distributed unevenly throughout the state, with major concentrations in the central Brooks Range, Tanana River Valley-Alaska Range foothills, Prince William Sound, and the Alaska panhandle (Fig. 15.2 and Table 15.2). The sites in north-central Alaska have the best combination of site distribu-

tion, temporal control, and length of continuous record. Many of the sites in southern Alaska are of limited use for this study because of poor dating control and/or short records.

DATING CONTROL

We evaluated the quality of dating control for pollen diagrams from 124 sites (Table 15.3). We arbitrarily assigned all core tops an age of 0 yr B.P. This age assignment may be unwarranted because of the extremely slow sedimentation rates in many Arctic lakes. The number of moderately well-dated sites (rank 4) may be overstated as a result of the assumption of a 0-yr B.P. bracket date.

In general, the Alaskan pollen spectra at 6000 and 9000 yr B.P. are poorly dated. Records from northern Alaska (i.e., sites north of the Alaska Range) usually are dated more accurately at 6000 and 9000 yr B.P. than are records from the south. Dating is poorer in the south primarily because of the predominance of sites analyzed before radiocarbon dating and the large number of records with only a single radiocarbon date (e.g., Heusser, 1952, 1955, 1957, 1960; Ager and Brubaker, 1985). Almost all northern sites are lakes that have complete Holocene sections and multiple radiocarbon dates.

Southern Alaska—Level 1 Data

VEGETATION HISTORY

Most of the available pollen data for southern Alaska are the result of C. J. Heusser's long-term and intensive study of coastal areas of the Gulf of Alaska and the Aleutian Islands (Heusser, 1952, 1955, 1957, 1960, 1965, 1973, 1978, 1983a,b, 1985). Sirkin and Tuthill (1969), Mckenzie (1970), Ager (1983), Mann (1983), and Peteet (1986) have supplemented Heusser's work, and sufficient data are now available to describe the postglacial vegetational history of southern Alaska. The number of adequately dated pollen records is not sufficient for isopoll mapping, so the vegetational history must be interpreted from a small number of diagrams.

The oldest continuous pollen sequence for southern Alaska is from central Kenai Peninsula and dates to about 14,000 yr B.P. (Ager, 1983; Ager and Brubaker, 1985) (Fig. 15.3). The earliest pollen assemblages at this

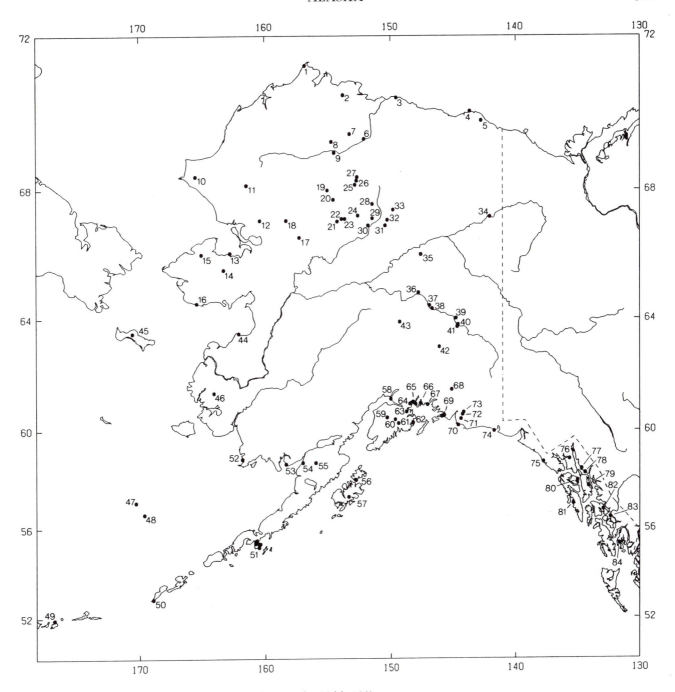

Fig. 15.2. Location of sites reviewed for this study (see also Table 15.2).

site indicate that an herb-willow tundra was replaced by birch shrub tundra around 13,400 yr B.P. Between 11,000 and 8000 yr B.P. poplar-willow scrub vegetation occupied areas of central Kenai Peninsula, the northern Chugach Mountains, and northern Cook Inlet. Alder appeared in this region between 9000 and 8000 yr B.P., apparently arriving first on the coast and spreading rapidly to the north and west. Spruce (presumably white spruce) appeared at about 9100 yr B.P. in the Gulkana uplands, 8700 yr B.P. in the northern

Chugach Mountains, and 8000 yr B.P. in Kenai Peninsula and the Anchorage area. These dates suggest that spruce migrated southward from interior Alaska to the Copper River Valley and Cook Inlet area (Ager and Brubaker, 1985). Sitka spruce, mountain hemlock, and western hemlock did not appear in south-central Alaska until about 4000–3000 yr B.P. (Heusser, 1983a; Ager and Brubaker, 1985).

A lodgepole pine parkland existed in southeastern Alaska at approximately 11,000 yr B.P. (Heusser, 1965;

Table 15.2. Fossil pollen sites (see also Fig. 15.2)

Site number	Site name	Reference
1	Point Barrow	Colinvaux, 1964b
2	Teshekpuk Lake	Anderson, 1982
3	Milne Point	Wilson, 1984
4	Barter Island	Wilson, 1984
5	Niguanak	Wilson, 1984
6	Umiat	Livingstone, 1957
7	Square Lake	Anderson, 1982
8	Ikpikpuk	Colinvaux, 1964b
9	Killik	Colinvaux, 1964b
10	Ogotoruk Creek	Heusser, 1963b
11	Kaiyak Lake[a]	Anderson, 1985
12	Squirrel Lake[a]	Anderson, 1985
13	Cape Deceit[a]	Matthews, 1974a
14	Imuruk Lake	Colinvaux, 1964a
15	Whitefish Lake	Shackleton, 1982
16	Nome[a]	Hopkins et al, 1960
17	Kiliovilik Lake	Anderson, 1982
18	Epiguruk[a]	Schweger, 1982
19	Headwaters Lake[a]	Brubaker et al, 1983
20	Redondo Lake[a]	Brubaker et al, 1983
21	Ruppert Lake[a]	Brubaker et al, 1983
22	Angal Lake[a]	Brubaker et al, 1983
23	Ranger Lake[a]	Brubaker et al, 1983
24	Redstone Lake[a]	Edwards et al, 1985
25	Eight Lake[a]	Livingstone, 1955
26	Chandler Lake[a]	Livingstone, 1955
27	Lake A[a]	Livingstone, 1955
28	Screaming Yellowlegs Pond[a]	Edwards et al, 1985
29	Death Valley Lake[a]	Livingstone, 1955
30	John River sites	Schweger, 1982
31	Tramway Bar Bog	Baker, 1984
32	Grayling Bog	Baker, 1984
33	Rebel Lake[a]	Edwards et al, 1985
34	Ped Pond[a]	Edwards and Brubaker, 1986
35	Sands of Time Lake[a]	Lamb and Edwards, 1988
36	Isabella Basin[a]	Matthews, 1974b
37	Harding Lake	Ager, 1983
38	Birch Lake[a]	Ager, 1975
39	Healy, Hidden Lakes	Ager, 1975; Anderson, 1975
40	George Lake[a]	Ager, 1975
41	Johnson River	Ager, 1975; Anderson, 1975
42	Tangle Lake[a]	Ager and Sims, 1981
	Rock Creek	Schweger, 1981
43	Eightmile Lake	Ager, 1983
44	Puyuk, Zagoskin Lakes[a]	Ager, 1982, 1983
45	St. Lawrence Island	Colinvaux, 1967
46	Tungak Lake[a]	Ager, 1982
47	St. Paul Island	Colinvaux, 1964a, 1981
48	St. George Island	Parrish, 1980
49	Adak Island	Heusser, 1978
50	Umnak Island	Heusser, 1973
51	Ungak Island	Heusser, 1983b
52	Goodnews Bay	Colinvaux, 1967
53	Bristol Bay	Ager, 1982
54	Naknek River	Heusser, 1963a
55	Brooks River	Heusser, 1963a
56	Afognak Island	Heusser, 1960
57	Kodiak Island	Heusser, 1960
58	Point Woronzof	Ager and Brubaker, 1985
59	Hidden Lake	Ager, 1983
60	Kenai	Heusser, 1955
61	Divide	Heusser, 1955
62	Verdant Island	Heusser, 1983a
63	Blackstone Bay	Heusser, 1983a
64	Harriman, Lower Barry Arm, Point Quick	Heusser, 1983a
65	Golden	Heusser, 1983a
66	Jonah Bay	Heusser, 1983a
67	Emerald Cove	Heusser, 1983a
68	70 Mile	Ager and Brubaker, 1985
69	Eyak, Orca Inlet	Heusser, 1955
70	Whale Island, Cape Martin	Sirkin and Tuthill, 1969
71	Katalla sites A	Sirkin and Tuthill, 1969
72	Ragged Mountain	Sirkin and Tuthill, 1969
73	Katalla sites B	Sirkin and Tuthill, 1969
74	Icy Cape	Heusser, 1960
75	Muskeg Cirque	Mann, 1983
76	Adams Inlet	Mckenzie, 1970
77	Montana Creek	Heusser, 1960
78	Juneau sites	Heusser, 1952
79	Hasselborg	Heusser, 1960
80	Whitestone Harbor	Heusser, 1960
81	Sitka sites	Heusser, 1952
82	Petersburg sites	Heusser, 1952
83	Wrangell sites	Heusser, 1952
84	Ketchikan sites	Heusser, 1952

[a]Sites used in mapping of northern Alaskan pollen data (Figs. 15.5 and 15.6).

Table 153. Dating control for Alaskan pollen records

| Region | Dating rank[a] (no. of sites) | | | | | | | |
Time (yr B.P.)	1	2	3	4	5	6	7	No data[b]
Northern Alaska								
0	3	8	6	6	3	4	24	8
6000	6	8	3	8	0	1	18	18
9000	10	4	5	2	1	1	22	17
Southern Alaska								
0	4	0	6	2	0	2	48	0
6000	2	0	2	3	0	1	37	17
9000	4	0	1	0	0	0	10	47

[a]Ranks were assigned as follows: bracketing radiocarbon dates both within 2000 yr (rank 1), one within 2000 yr (rank 2), both within 4000 yr (rank 3), one within 4000 yr and one within 6000 yr (rank 4), both within 6000 yr (rank 5), one within 6000 yr and one within 8000 yr (rank 6). Rank 7 means that the record is poorly dated, meeting none of the ranking criteria or lacking radiocarbon dates.
[b]Records with levels missing or that do not extend back to 6000 or 9000 yr B.P.

Mckenzie, 1970; Mann, 1983). Alder was present in the Alaska panhandle at the same time. Sitka spruce grew in Lituya Bay (Mann, 1983) by 10,500 yr B.P. but was not present at Icy Cape until 7600 yr B.P. (Peteet, 1986). Western hemlock and mountain hemlock migrated into Lituya Bay by 7100 yr B.P. and 4900 yr B.P., respectively (Mann, 1983). These hemlock species were absent from Icy Cape until 3500 yr B.P. (Peteet, 1986). Sitka spruce, western hemlock, and mountain hemlock appear to have moved at different rates and, at least for the area between Lituya Bay and Icy Cape, at different times in the Holocene, although all migration routes have a south-to-north pattern.

Pollen diagrams from the Aleutian Islands indicate tundra vegetation for the past 10,000 yr (Heusser, 1973, 1978). Vegetation was a maritime willow shrub tundra throughout the Holocene; crowberry (*Empetrum*) became a more important component of the vegetation after 3000 yr B.P.

CLIMATE HISTORY

According to initial reconstructions, southeastern Alaska was cool and moist between about 10,000 and 6000 yr B.P. (e.g., Heusser, 1952, 1960). The climate warmed and became drier between 6000 and 2000 yr B.P., with a time of maximum warmth and minimum precipitation between 5000 and 2000 yr B.P. Temperatures once again cooled and precipitation increased between 2000 and 200 yr B.P. After 200 yr B.P. precipitation fell and temperature rose.

More recent quantitative reconstructions differ from the earlier qualitative interpretations (Heusser *et al.*, 1985). They show that July temperatures were warmer than at present between 10,000 and 8000 yr B.P. Temperatures gradually declined to modern values by about 5000 yr B.P. A precipitation minimum is evident at 8000 yr B.P., and the highest values inferred for precipitation were at approximately 4000 yr B.P.

On the Aleutian Islands, the climate was cool and moist between 10,000 and 8500 yr B.P. and became warmer and drier between 8500 and 3000 yr B.P. Cool, moist conditions returned after 3000 yr B.P. and have continued to the present (Heusser, 1973, 1978).

SUMMARY: SOUTHERN ALASKA AT 9000 AND 6000 YR B.P.

At 9000 yr B.P. conifer-dominated forests covered much of southeastern Alaska, vegetation in south-central Alaska was a mixture of birch-willow shrub tundra and scattered stands of poplars, and farther west in the Aleutian Islands willow shrub tundra grew. The climate in southeastern and south-central Alaska probably was warmer and drier than at present, whereas conditions in the southwest evidently were cool and moist.

By 6000 yr B.P. hemlock and/or spruce species were established in southeastern and much of south-central Alaska, and the climate was probably cooler and moister than at 9000 yr B.P. Minor changes in tundra composition inferred for the Aleutian Islands suggest that the climate there was warmer and drier at 6000 yr B.P. than at 9000 yr B.P.

Northern Alaska—Level 1 Data

VEGETATION HISTORY

Recent work in the Brooks Range (Brubaker *et al.*, 1983; Anderson, 1985; Edwards *et al.*, 1985; Edwards and Brubaker, 1986) and in the Intermontane region (Ager, 1982, 1983; Ager and Brubaker, 1985) has provided a grid of comparatively well-dated lacustrine pollen sequences. These data, coupled with information from previous studies in the area (Livingstone, 1955, 1957; Colinvaux, 1964a,b, 1967; Ager, 1975, 1982,

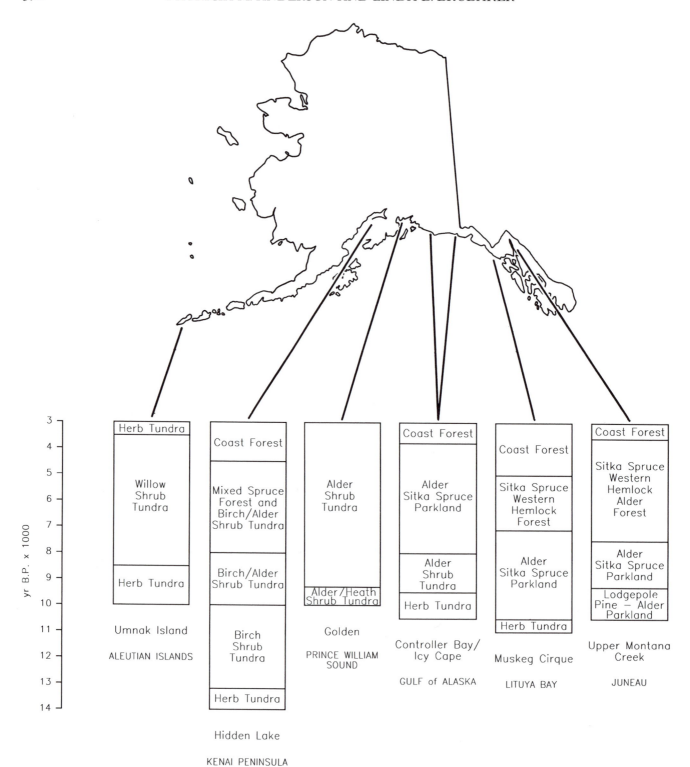

Fig. 153. Summary of vegetation history of southern Alaska.

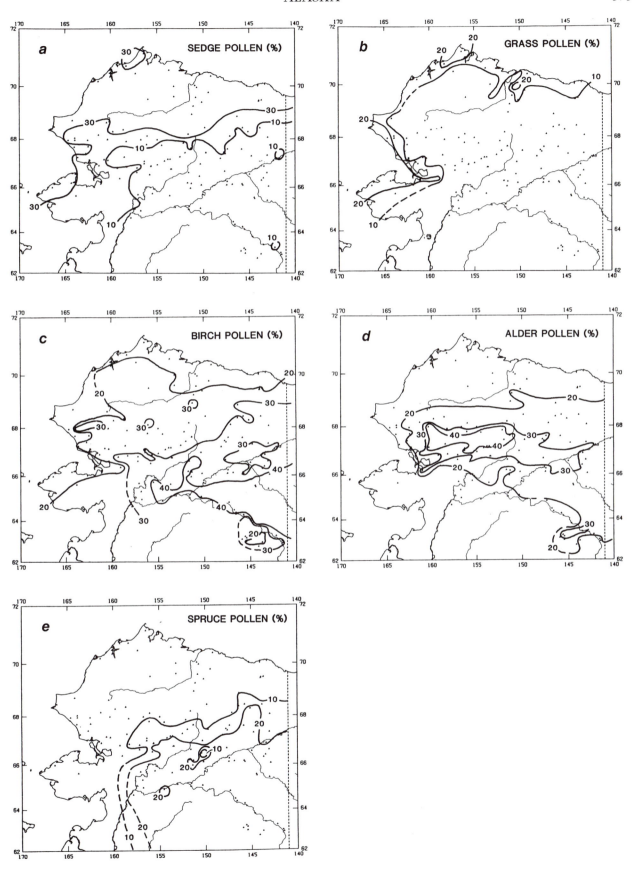

Fig. 15.4. Isopoll maps of modern pollen percentages for (a) sedges, (b) grasses, (c) birch, (d) alder, and (e) spruce.

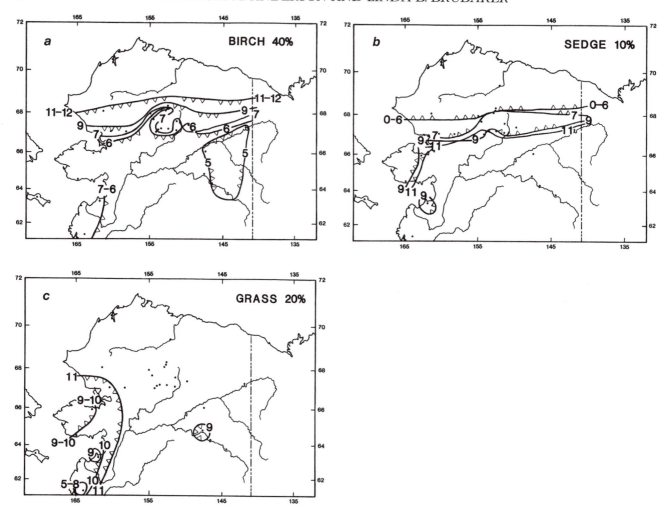

Fig. 15.5. Isochrone maps of (a) 40% birch isopoll, (b) 10% sedge isopoll, and (c) 20% grass isopoll.

1983; Schweger, 1976, 1982), permit a more detailed inspection of the vegetational history of northern Alaska through isopoll maps than is possible with simple stratigraphic correlations.

Isochrone maps of spruce, birch, alder, sedge, and grass pollen from 25 sites (Table 15.2) trace the major changes in paleovegetation of northern Alaska (Anderson, 1982). The maps depict a more complex Holocene vegetation history than was previously recognized for Arctic Alaska. Isolines chosen for use as isochrones reflect range boundaries or formation boundaries as defined in the modern pollen-vegetation study of Anderson and Brubaker (1986) (Fig. 15.4). For the sake of map clarity, questionable isopoll locations (i.e., areas where sites do not exist to constrain isopolls) are not distinguished graphically from more certain locations. The tentative locations, however, are obvious from site distribution.

Birch was the predominant pollen type over most of the region at 12,000 yr B.P. (Fig. 15.5a). The vegetation at this time is difficult to reconstruct in detail, because birch pollen percentages are generally much higher than those found in modern tundra samples. Nonetheless, the pollen spectra from 12,000 yr B.P. have been interpreted as representing a birch shrub tundra that was similar throughout northern Alaska and northwestern Canada, perhaps with some minor altitudinal variation (Ritchie, 1984). The isochrone maps, however, show some regional variation in pollen, with areas of far western Alaska having higher percentages of grass pollen (Fig. 15.5c). By 11,000 yr B.P. the Brooks Range sites were characterized by high percentages of both birch and sedge pollen (Fig. 15.5a,b). Interior lowland sites, in contrast, were dominated by birch pollen alone (Fig. 15.5a).

Isochrone maps of birch, sedge, and grass pollen show little change from 11,000 to 9000 yr B.P. The only important difference between the two times is the apparent postglacial reestablishment of spruce and alder populations in far northeastern and western Alaska, respectively (Fig. 15.6). Birch shrub tundra, however, remained the predominant vegetation throughout the region at 9000 yr B.P. Temperature and precipitation probably increased between 12,000 and 9000 yr B.P., fi-

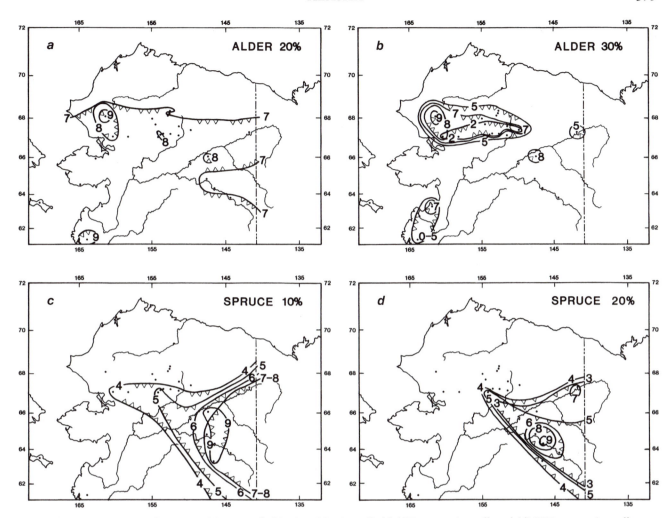

Fig. 15.6. Isochrone maps of (a) 20% alder isopoll, (b) 30% alder isopoll, (c) 10% spruce isopoll, and (d) 20% spruce isopoll.

nally achieving a threshold between 10,000 and 9000 yr B.P. that permitted the initial expansion of spruce and alder populations.

Many sites in the area show an increase in poplar pollen (probably balsam poplar) between 11,000 and 9000 yr B.P. (Ager and Brubaker, 1985; Barnosky *et al.,* 1987). Not all sites in north-central Alaska have a peak in poplar pollen (Anderson *et al.,* 1988), however, and in some cases the magnitude of the peak differs between nearby sites (e.g., Angal and Ruppert lakes [Brubaker *et al.,* 1983]). The significance of the poplar subzone is poorly understood. The rise and decline of poplar pollen, and presumably the density of trees, have been related to climate (e.g., Brubaker *et al.,* 1983), edaphic factors (e.g., Cwynar, 1982), and community dynamics (e.g., Ritchie, 1984).

By 6000 yr B.P. areas with high percentages of grass pollen were gone, and the 10% sedge isopoll was close to its modern position (compare Figs. 15.4a and 15.5b). Birch pollen percentages decreased in Brooks Range sites, and by 5000 yr B.P. high frequencies of birch

pollen were restricted to areas of the Yukon Flats and Tanana River Valley (compare Figs. 15.4c and 15.5a).

Both spruce and alder maps show major changes between 6000 and 9000 yr B.P. Alder values of 20% or greater were found at only one site in northern Alaska at 9000 yr B.P., but by 7000 yr B.P. all Brooks Range sites and many sites farther south had high percentages of alder pollen (Fig. 15.6a). Alder apparently colonized all of northern Alaska in a span of 2000 yr or less if we assume that, like today, the 20% contour represents the regional presence of alder shrubs (Anderson and Brubaker, 1986) (Fig. 15.4d). Site density is not high enough to determine whether alder plants moved quickly from west to east or spread from several refugia in the region.

The Holocene history of alder, as reflected in the 30% isochrone map, shows steady increases in the percentage of alder pollen in the western and central Brooks Range until around 5000 yr B.P. (Fig. 15.6b). The area with more than 30% alder pollen gradually contracted until 2000 yr B.P., when it approximated the present location of the 40% alder isopoll. This decline reflects in part the addition of spruce pollen to the fossil spectra of northwestern Alaska. However, the input of spruce pollen to most of these sites is negligible (7% or less) and is not sufficient to account for the "collapse" of the alder area. Alder shrubs in the central and western Brooks Range therefore were either more abundant or more productive between 7000 and 5000 yr B.P.

Spruce dispersed over large areas of northeastern and north-central Alaska between 6000 and 9000 yr B.P. but seems to have migrated at a slower rate than alder (Fig. 15.6c,d). The locations of glacial refugia for spruce are unknown, but both the 10% (mixed forest-tundra) and 20% (main body of forest) isolines show distinct east-to-west trends. Current data suggest that spruce forests were first established in Alaska in the Yukon-Tanana valleys and spread to the north, south, and west. The 10% and 20% isopolls indicate that between 9000 and 6000 yr B.P. dense boreal forest was limited in area and probably was surrounded by a broad band of open forest-tundra.

The vegetation of northern Alaska at 6000 yr B.P. was similar to the modern vegetation in that the same major plants were present in the region, but the distribution of those plants as reflected in the isopoll maps appears to have differed from that observed today. In the Brooks Range and lower elevations in the northwest, the vegetation was an alder-birch shrub tundra, possibly with isolated stands of spruce on valley slopes and along river courses. Boreal forest was limited to the northeast, but it was probably bordered by a broad band of open woodland extending into surrounding shrub tundra. After a period of relative stability between 9000 and 6000 yr B.P., spruce populations continued their westward expansion between 6000 and 4000 yr B.P., as shown by both the 10% and 20% spruce isopolls. Spruce did not move down the Kobuk River Valley until the mid-Holocene, approximating its modern distribution in northwestern Alaska at about 4000 yr B.P. (Fig. 15.6c).

CLIMATE HISTORY

Although the climate history of northern Alaska is not understood in great detail, some generalities about climate over the past 14,000 yr can be drawn from both pollen and nonpollen fossil data. The impact of an abrupt climatic change beginning at about 14,000 yr B.P. is recorded most clearly in the widespread and almost synchronous change from an herb-dominated to a shrub-dominated tundra. Conditions were warmer and moister than those of the Late Wisconsinan maximum (Hopkins et al., 1981, 1982; Ager, 1983; Carter et al., 1984). The high percentages of sedge and grass pollen in the Brooks Range and western Alaska may be related to the dominance in summer of cool, moist air as the seaway in the Bering Strait began to open with the rise in sea level. The increased moisture availability may also account for a glacial advance

at about 12,800 yr B.P. in the western Brooks Range (Hamilton, 1986). Evidence of warm conditions throughout northern Alaska from 11,000 to 8000 yr B.P. includes increased eolian activity (Carter and Hopkins, 1982), melting of ice wedges in northwestern Alaska (McCulloch and Hopkins, 1966), range extensions of cattail and balsam poplar (McCulloch and Hopkins, 1966; Hopkins et al., 1981), regionwide peaks in poplar and juniper pollen (Ritchie, 1984; Ager and Brubaker, 1985), and localized peaks in spruce pollen (Brubaker et al., 1983). Fossil invertebrate and plant assemblages from alluvial terraces north of the Brooks Range indicate at least a 2°C increase in summer temperature. These data also suggest that winter snowfall was greater than at present (Carter et al., 1984).

Other arguments have been made for the presence of a deep winter snowpack for the period 11,000–8000 yr B.P. A deeper snowpack would result in greater spring floods, which could account for observed downcutting of streams (Carter and Hopkins, 1982), increased slope runoff, as evidenced by higher percentages of Polypodiaceae spores in pollen records (Brubaker et al., 1983), and higher poplar populations (probably restricted to gravel bars and disturbed areas bordering rivers). Growth of ice wedges on the Arctic coastal plain also suggests increased snow cover. Deeper snow, and the resulting increase in insulation, may have reduced ice thickness in shallow lakes and permitted beavers to survive beyond their current limits (D. M. Hopkins, personal communication).

Evidence for a "Hypsithermal," or mid-Holocene (about 7000–5000 yr B.P.) thermal maximum, in northern Alaska is slight. Pollen data from individual sites are ambiguous, and mid-Holocene range extensions seem to be limited to a single find of alder twigs and catkins in the central Brooks Range (Tedrow and Walton, 1964). This paucity of data in northern Alaska is in strong contrast to north-central and northwestern Canada (Nichols, 1975; Ritchie et al., 1983) and western Siberia (Kind, 1967), where good evidence documents the northward extension of forest between about 9000 and 5000 yr B.P.

Indications of increased precipitation in Alaska during the mid-Holocene are stronger than those for temperature but are still not definitive. Fluctuations in alder pollen percentages in the north-central and northwestern Brooks Range have been linked to changes in precipitation (Anderson, 1982). Hopkins et al. (1981) suggested that the mid-Holocene migration of spruce was related to heavier precipitation and moister soils. On the basis of spruce macrofossil data, they inferred a "stall" in spruce migration between 8000 and 5000 yr B.P. after the initial migration into northeastern Alaska between 9500 and 8000 yr B.P. The boreal forest once again moved westward between 5000 and 3000 yr B.P. This trend is also evident in the pollen data (Fig. 15.6c,d). Moister conditions in north-central Alaska also are suggested by increased peat accumulation (Hamilton and Robinson, 1977).

SUMMARY: NORTHERN ALASKA AT 9000 AND 6000 YR B.P.

Major changes in vegetation took place between 9000 and 6000 yr B.P. in northern Alaska. By 9000 yr B.P. a gallery spruce forest was established in the Tanana Valley, with perhaps a birch shrub tundra-spruce woodland in the northern foothills of the Alaska Range. The rest of northern Alaska was covered by shrub tundra. Vegetation in the Brooks Range was predominantly a birch-sedge tundra, although high percentages of alder pollen in the Kotzebue Sound drainage indicate that alder was also a component of the vegetation in some areas. Coastal areas retained high percentages of grass and sedge pollen, suggesting the continued importance of these elements in the vegetation. Poplar was probably the only tree species growing in areas of northern Alaska beyond spruce treeline.

The pollen data indicate that the climate at 9000 yr B.P. had become warmer and wetter than the full-glacial conditions of 15,000 or 18,000 yr B.P. However, unlike in northwestern Canada (Ritchie *et al.,* 1983), the pollen data in Alaska do not give any indication of northward extension of range boundaries beyond the present limits, with the exception of cattail (*Typha latifolia*) at Ped Pond in northeastern Alaska (Edwards and Brubaker, 1986). The qualitative interpretations of the pollen diagrams, while indicating a general climatic amelioration, do not specify the extent of the improvement. Other types of proxy data, as discussed above, indicate more clearly than the pollen records the occurrence of a postglacial thermal maximum in northern Alaska.

The interior boreal forest probably had achieved its modern composition and distribution over much of interior Alaska by 6000 yr B.P. (Ager, 1975). Spruce had migrated to the north and west of the Tanana Valley, but most of north-central Alaska remained a mixed open forest-shrub tundra. By 6000 yr B.P. alder had become an important component of the vegetation throughout northern Alaska. Qualitative interpretations of pollen data do not indicate a decline in temperature. If anything they suggest either that temperatures were similar to those at 9000 yr B.P. (with a delayed spruce migration to the west) or that temperatures in northwestern and north-central Alaska increased during the mid-Holocene, which allowed spruce to grow in these areas. Other data, as described previously, indicate a decrease in temperature at 6000 yr B.P. compared to 9000 yr B.P. Increased percentages of alder pollen in the western and central Brooks Range might indicate a Holocene maximum for precipitation at about 6000 yr B.P. Evidence for such an event in other fossil data is ambiguous.

Comparison with Climate Simulations for 9000 and 6000 yr B.P.

Paleoclimatic temperature and precipitation trends as interpreted from fossil data are consistent among regions of northern Alaska but show more local variation in southeastern, south-central, and southwestern Alaska (Tables 15.4 and 15.5). The inconsistency in the

Table 15.4. Temperature (T) and precipitation (P) reconstructions for Alaska (relative to modern values)

Time (yr B.P.)	Region, northern Alaska					
	North Slope		Brooks Range		Intermontane	
	T	P	T	P	T	P
6000	Warmer[a]	Much wetter;[a,b] same[c]	Warmer[a]	Much wetter;[a,b] same[c]	Warmer[a]	Drier;[c] same[d]
9000	Much warmer[a]	Wetter[e]	Much warmer[a]	Wetter[e]	Much warmer[a]	Wetter[e]
12,000	Cooler	Drier	Cooler	Drier	Cooler	Drier

Time (yr B.P.)	Region, southern Alaska					
	Southeastern		South-central		Southwestern	
	T	P	T	P	T	P
6000	Warmer[a]	Much drier	Same	Same	Warmer	Drier
9000	Much warmer[a]	Drier	Warmer[a]	Wetter	Cooler (?)	Wetter (?)
12,000	Cooler	Drier	Cooler	Drier	Glaciated	

[a]Probable summer condition.
[b]Western and central portions of region.
[c]Eastern portion of region.
[d]Western portion of region.
[e]Probable winter condition.

Table 15.5. Reconstructed temperature (T) and precipitation (P) trends in Alaska based on fossil data

	Region, northern Alaska					
	North Slope		Brooks Range		Intermontane	
Interval (yr B.P.)	T	P	T	P	T	P
6000-0	Cooling	Drier;[a] no change[b]	Cooling	Drier;[a] no change[b]	Cooling	Wetter;[b] no change[c]
9000-6000	Cooling	Wetter	Cooling	Wetter	Cooling	Wetter
12,000-9000	Warming	Wetter	Warming	Wetter	Warming	Wetter
18,000-12,000	Warming	Wetter	Warming	Wetter	Warming	Wetter

	Region, southern Alaska					
	Southeastern		South-central		Southwestern	
	T	P	T	P	T	P
6000-0	Cooling	Wetter	No change	No change	Cooling	Wetter
9000-6000	Cooling	Wetter	Cooling	Drier	Warming	Drier
12,000-9000	Warming	Wetter	Warming	Wetter	Warming	Wetter
18,000-12,000	Warming	Wetter	Warming	Wetter	Glaciated	

[a]Western and central portions of region.
[b]Eastern portion of region.
[c]Western portion of region.

south may reflect the nature of the data (fewer sites and fewer types of paleoenvironmental data used for interpretation) or the complexity of mesoscale or microscale climate as recorded at sites in a variety of topographic and edaphic settings.

Fossil data indicate that both temperature and precipitation increased between 12,000 and 9000 yr B.P. in all regions of Alaska and that summer temperatures reached a maximum at 9000 yr B.P. in northern and southeastern areas of the state. Winter precipitation was greater at 9000 yr B.P. than at present in northern Alaska. South-central and southwestern Alaska also received more moisture, although the seasonal distribution of precipitation is not clear. In southeastern Alaska mean annual precipitation was less than today at both 9000 and 6000 yr B.P., whereas in north-central and northwestern Alaska 6000 yr B.P. was the time of greatest summer precipitation. Fossil data indicate a general decrease in temperatures over the past 6000 yr for the entire state. In some areas of northern Alaska this cooling was paralleled by reduced precipitation, but in southern Alaska conditions became wetter over the same period.

Model simulations indicate warm summers at 9000 and 6000 yr B.P. (Table 15.6). Fossil data from both southern and northern Alaska generally support the model's trends in paleotemperatures, suggesting maximum summer warmth at 9000 yr B.P. Model simulations indicate no significant differences in seasonal or mean annual precipitation at 9000 or 6000 yr B.P. Fossil data, however, suggest an increase in snow cover and/or snow depth (i.e., an increase in winter precipitation) for northern portions of Alaska and wetter than modern conditions in most of southern Alaska at 9000 yr B.P. Fossil-based interpretations for 6000 yr B.P. suggest wet summers in northern Alaska, whereas areas of southern Alaska were drier than they are today.

Table 15.6. Model simulations of seasonal and annual climatic anomalies for Alaska[a]

Time (yr B.P.)	January	July	Annual
	Surface temperature (°C)		
3000	-6.4	+1.7*	-2.4
6000	-3.7	+2.2*	-1.2
9000	-3.4	+2.0*	-1.6
12,000	-5.1	+0.4	-3.4
	Precipitation (mm/day)		
3000	-0.33	-0.61	-0.48
6000	-0.19	-0.41	-0.32
9000	-0.09	-0.56	-0.35
12,000	-0.35	+0.20	-0.11
	Precipitation - evaporation (mm/day)		
3000	-0.35	-0.62	-0.49
6000	-0.22	-0.51	-0.37
9000	-0.11	-0.70	-0.41
12,000	-0.44	+0.07	-0.20

[a]Anomalies are expressed as differences from the modern values. An asterisk indicates that the anomaly is statistically significant.

Acknowledgments

This research was funded by the National Science Foundation, Division of Polar Programs and Office of Climate Dynamics. We have benefited from discussions with all COHMAP members but especially wish to acknowledge the input of P. Bartlein, K. Gajewski, J. Ritchie, and T. Webb III. R. Reanier drafted the figures.

References

Ager, T. A. (1975). "Late Quaternary Environmental History of the Tanana Valley, Alaska." Institute for Polar Studies Report 54. Ohio State University, Columbus.

——. (1982). Vegetational history of western Alaska during the Wisconsin glacial interval and the Holocene. In "Paleoecology of Beringia" (D. M. Hopkins, J. V. Matthews, Jr., C. E. Schweger, and S. B. Young, Eds.), pp. 75-93. Academic Press, New York.

——. (1983). Holocene vegetational history of Alaska. In "Late Quaternary Environments of the United States, Vol. 2: The Holocene" (H. E. Wright, Jr., Ed.), pp. 128-141. University of Minnesota Press, Minneapolis.

Ager, T. A., and Brubaker, L. B. (1985). Quaternary palynology and vegetational history of Alaska. In "Pollen Records of Late Quaternary North American Sediments" (V. M. Bryant, Jr., and R. G. Holloway, Eds.), pp. 353-384. American Association of Stratigraphic Palynologists Foundation, Dallas.

Ager, T. A., and Sims, J. D. (1981). Holocene pollen and sediment record from the Tangle Lakes area, central Alaska. Palynology 5, 85-98.

Anderson, J. H. (1975). A palynological study of late Holocene vegetation and climate in the Healy Lake area, Alaska. Arctic 28, 29-62.

Anderson, P. M. (1982). "Reconstructing the Past: The Synthesis of Archaeological and Palynological Data, Northern Alaska and Northwestern Canada." Unpublished Ph.D. dissertation, Brown University, Providence, R.I.

——. (1985). Late Quaternary vegetational change in the Kotzebue Sound area, northwestern Alaska. Quaternary Research 24, 307-321.

Anderson, P. M., and Brubaker, L. B. (1986). Modern pollen assemblages from northern Alaska. Review of Palaeobotany and Palynology 46, 273-291.

Anderson, P. M., Reanier, R. E., and Brubaker, L. B. (1988). Late Quaternary vegetational history of the Black River region in northeastern Alaska. Canadian Journal of Earth Sciences 25, 84-94.

Baker, S. A. (1984). "Holocene Palynology and Reconstruction of Paleoclimates in North-Central Alaska." Unpublished M.S. thesis, University of Colorado, Boulder.

Barnosky, C. W., Anderson, P. M., and Bartlein, P. J. (1987). The northwestern U.S. during deglaciation: Vegetational history and paleoclimatic implications. In "North America and Adjacent Oceans during the Last Deglaciation" (W. F. Ruddiman and H. E. Wright, Jr., Eds.), pp. 289-321. The Geology of North America, Vol. K-3. Geological Society of America, Boulder, Colo.

Brubaker, L. B., Garfinkel, H. L., and Edwards, M. E. (1983). A late Wisconsin and Holocene vegetation history from the central Brooks Range: Implications for Alaskan paleoecology. Quaternary Research 20, 194-214.

Carter, L. D., and Hopkins, D. M. (1982). Late Wisconsinan winter snow cover and sand-moving winds on the Arctic coastal plains of Alaska. In "Eleventh Annual Arctic Workshop Abstracts," p. 8. University of Colorado, Boulder.

Carter, L. D., Nelson, R. E., and Galloway, J. P. (1984). Evidence for early Holocene increased precipitation and summer

warmth in Arctic Alaska. Abstracts of the Geological Society of America Cordilleran Section Meeting 80, 274.

Colinvaux, P. A. (1964a). The environment of the Bering land bridge. Ecological Monographs 34, 297-329.

——. (1964b). Origin of ice ages: Pollen evidence from Arctic Alaska. Science 145, 707-708.

——. (1967). A long record from St. Lawrence Island, Bering Sea (Alaska). Palaeogeography, Palaeoclimatology, Palaeoecology 3, 29-48.

——. (1981). Historical ecology in Beringia: The south land bridge coast at St. Paul Island. Quaternary Research 16, 18-36.

Cwynar, L. C. (1982). A late-Quaternary vegetation history from Hanging Lake, northern Yukon. Ecological Monographs 52, 1-24.

Edwards, M. E., and Brubaker, L. B. (1986). Late Quaternary environmental history of the Fishhook Bend area, Porcupine River, Alaska. Canadian Journal of Earth Sciences 23, 1765-1773.

Edwards, M. E., Anderson, P. M., Garfinkel, H. L., and Brubaker, L. B. (1985). Late Wisconsin and Holocene vegetational history of the upper Koyukuk region, Brooks Range, Alaska. Canadian Journal of Botany 63, 616-626.

Hamilton, T. D. (1986). Late Cenozoic glaciation of the central Brooks Range. In "Glaciation in Alaska: The Geologic Record" (T. D. Hamilton, K. M. Reed, and R. M. Thorson, Eds.), pp. 9-50. Alaska Geological Society, Anchorage.

Hamilton, T. D., and Robinson, S. (1977). Late Holocene (Neoglacial) environmental changes in central Alaska. Geogical Society of America Abstracts with Programs 9, 1003.

Heusser, C. J. (1952). Pollen profiles from southeastern Alaska. Ecological Monographs 22, 331-352.

——. (1955). Pollen profiles from Prince William Sound and southeastern Kenai Peninsula, Alaska. Ecology 36, 185-202.

——. (1957). Pleistocene and postglacial vegetation of Alaska and the Yukon Territory. In "Arctic Biology" (H. P. Hansen, Ed.), pp. 131-151. Oregon State College, Corvallis.

——. (1960). "Late-Pleistocene Environments of North Pacific North America." Special Publication 35. American Geographical Society, New York.

——. (1963a). Postglacial palynology and archeology in the Naknek River drainage area, Alaska. American Antiquity 29, 74-81.

——. (1963b). Pollen diagrams from Ogotoruk Creek, Cape Thompson, Alaska. Grana Palynologica 4, 149-159.

——. (1965). A Pleistocene phytogeographical sketch of the Pacific Northwest and Alaska. In "The Quaternary of the United States" (H. E. Wright, Jr. and D. G. Frey, Eds.), pp. 469-483. Princeton University Press, Princeton, N.J.

——. (1973). Postglacial palynology of Umnak Island, Aleutian Islands, Alaska. Review of Palaeobotany and Palynology 15, 277-285.

——. (1978). Postglacial vegetation of Adak Island, Aleutian Islands, Alaska. Bulletin of the Torrey Botanical Club 15, 18-23.

——. (1983a). Holocene vegetation history of the Prince William Sound region, south-central Alaska. Quaternary Research 19, 337-355.

——. (1983b). Pollen diagrams from the Shumagin Islands and adjacent Alaska Peninsula, southwestern Alaska. Boreas 12, 279-295.

——. (1985). Quaternary pollen records from interior Pacific Northwest coast: Aleutians to the Oregon-California boundary. *In* "Pollen Records of Late Quaternary North American Sediments" (V. M. Bryant, Jr., and R. G. Holloway, Eds.), pp. 141-166. American Association of Stratigraphic Palynologists Foundation, Dallas.

Heusser, C. J., Heusser, L. E., and Peteet, D. M. (1985). Late-Quaternary climatic change on the American North Pacific coast. *Nature* 315, 485-487.

Hopkins, D. M., Ed. (1967). "The Bering Land Bridge." Stanford University Press, Stanford.

Hopkins, D. M., MacNeil, F. S., and Leopold, E. B. (1960). The coastal plain at Nome, Alaska: A late Cenozoic type section for the Bering Strait region. *Report of the 21st International Geological Congress* Part 4, 44-57.

Hopkins, D. M., Smith, P. A., and Matthews, J. V., Jr. (1981). Dated wood from Alaska and the Yukon: Implications for forest refugia in Beringia. *Quaternary Research* 15, 217-249.

Hopkins, D. M., Matthews, J. V., Jr., Schweger, C. E., and Young, S. B., Eds. (1982). "Paleoecology of Beringia." Academic Press, New York.

Kind, N. V. (1967). Radiocarbon chronology in Siberia. *In* "The Bering Land Bridge" (D. M. Hopkins, Ed.), pp. 172-192. Stanford University Press, Stanford.

Kontrimavichus, V. L., Ed. (1984). "Beringia in the Cenozoic Era." (Translated from the Russian [1976] by R. Chakravarty.) Amerind Publishing Co. Pvt. Ltd., New Delhi.

Lamb, H. F., and Edwards, M. E. (1988). The Arctic. *In* "Vegetation History" (B. Huntley and T. Webb III, Eds.), pp. 519-555. Kluwer Academic Publishers, Boston.

Livingstone, D. A. (1955). Some pollen profiles from Arctic Alaska. *Ecology* 36, 587-600.

——. (1957). Pollen analysis of a valley fill near Umiat, Alaska. *American Journal of Science* 255, 254-260.

Mann, D. H. (1983). "The Quaternary History of the Lituya Glacial Refugium, Alaska." Unpublished Ph.D. dissertation, University of Washington, Seattle.

Matthews, J. V., Jr. (1974a). Quaternary environments at Cape Deceit (Seward Peninsula, Alaska): Evolution of a tundra ecosystem. *Geological Society of America Bulletin* 85, 1353-1384.

——. (1974b). Wisconsin environment of interior Alaska: Pollen and macrofossil analysis of a 27-meter core from the Isabella basin (Fairbanks, Alaska). *Canadian Journal of Earth Sciences* 11, 828-841.

McCulloch, D. S., and Hopkins, D. M. (1966). Evidence for an early Recent warm interval in northwestern Alaska. *Geological Society of America Bulletin* 77, 1089-1108.

Mckenzie, G. D. (1970). "Glacial Geology of Adams Inlet, Southeastern Alaska." Institute of Polar Studies Report 25. Ohio State University, Columbus.

Nichols, H. (1975). "Palynological and Paleoclimatic Study of the Late Quaternary Displacement of the Boreal Forest-Tundra Ecotone in Keewatin and Mackenzie, N.W.T." Institute of Arctic and Alpine Research Occasional Paper 15. University of Colorado, Boulder.

Parrish, L. (1980). "A Record of Holocene Climate Changes from St. George Island, Pribilofs, Alaska." Institute of Polar Studies Report 75. Ohio State University, Columbus.

Peteet, D. M. (1986). Modern pollen rain and vegetational history of the Malaspina glacier district, Alaska. *Quaternary Research* 25, 100-120.

Ritchie, J. C. (1984). "Past and Present Vegetation of the Far Northwest of Canada." University of Toronto Press, Toronto.

Ritchie, J. C., Cwynar, L. C., and Spear, R. W. (1983). Evidence from northwest Canada for an early Holocene Milankovitch thermal maximum. *Nature* 305, 126-128.

Schweger, C. E. (1976). "Late Quaternary Paleoecology of the Onion Portage Region, Northwestern Alaska." Unpublished Ph.D. dissertation, University of Alberta, Edmonton.

——. (1981). Chronology of late glacial events from the Tangle Lakes, Alaska Range. *Arctic Anthropology* 18, 97-101.

——. (1982). Late Pleistocene vegetation of eastern Beringia: Pollen analysis of dated alluvium. *In* "Paleoecology of Beringia" (D. M. Hopkins, J. V. Matthews, Jr., C. E. Schweger, and S. B. Young, Eds.), pp. 95-112. Academic Press, New York.

Shackleton, J. (1982). "Environmental Histories from Whitefish and Imuruk lakes, Seward Peninsula, Alaska." Institute of Polar Studies Report 76. Ohio State University, Columbus.

Sirkin, L. A., and Tuthill, S. (1969). Late Pleistocene palynology and stratigraphy of Controller Bay region, Gulf of Alaska. *In* "Etudes sur le Quaternaire dans le Monde" (M. Ters, Ed.), pp. 197-208. Eighth Congress of the International Association for Quaternary Research, Paris.

Tedrow, J. C. F., and Walton, G. F. (1964). Some Quaternary events of northern Alaska. *Arctic* 17, 268-271.

Viereck, L. A., and Dryness, C. T. (1980). "A Preliminary Classification System for Vegetation of Alaska." U.S. Department of Agriculture, Washington, D.C.

Viereck, L. A., and Little, E. L. (1972). "Alaska Trees and Shrubs." U.S. Department of Agriculture, Washington, D.C.

——. (1975). "Atlas of United States Trees, Vol. 2: Alaska Trees and Common Shrubs." U.S. Department of Agriculture, Washington, D.C.

Wahrhaftig, C. (1965). "Physiographic Divisions of Alaska." Paper 482, U.S. Geological Survey, Washington, D.C.

Watson, C. E. (1959). "Climates of the States, Alaska." Climatography of the United States, No. 60-49. U.S. Weather Bureau, Washington, D.C.

Wilson, M. J. (1984). "Modern and Holocene Environments of the North Slope of Alaska." Unpublished M.S. thesis, University of Colorado, Boulder.

CHAPTER *16*

Vegetation, Lake Levels, and Climate in Western Canada during the Holocene

J. C. Ritchie and S. P. Harrison

The western interior of Canada extends from the Ontario-Manitoba boundary and the west coast of Hudson Bay to roughly the Rocky Mountain foothills and includes most of Manitoba, Saskatchewan, Alberta, and the Northwest Territories (mainland). This large landmass (about 3 million km²) has the lowest density of Holocene pollen sites in North America, in part because of difficult access. Similarly, we found useful data in the literature for only 12 lake basins.

The region consists of a vast plain sloping toward the east, from the Cordilleran foothills at about 2000 m to sea level at Hudson Bay. It is divisible into three physiographic regions (Fig. 16.1). The Interior Plain has a thick mantle of glacial till and descends in three irregular levels: the high plains of Alberta at about 1200 m, the Saskatchewan plains between roughly 1000 and 400 m, and the Manitoba lowlands at 300 m. In the east is the Shield, a peneplain of predominantly Archean and Proterozoic rocks scoured by Laurentide ice, with extensive irregular glacial deposits alternating with vast expanses of rough bedrock topography. In the west is the Cordilleran region, comprising the Rocky Mountain foothills, the Liard plateau, and the Mackenzie and Richardson mountains, alternating in the northwest with extensive unglaciated intermontane plains.

The climate of the western interior is determined on the one hand by the interaction of the midlatitude westerly air currents with the massive north-south barrier of the Cordillera, and on the other hand by the unimpeded expanses of the plains, which permit "great meridional excursions of both arctic and tropical air masses" (Bryson and Hare, 1974, p. 1). In addition the general circulation patterns (standing disturbances) are affected by the two subpolar low-pressure cells: the Aleutian low, which occupies the Gulf of Alaska and the North Pacific for all except the sum-

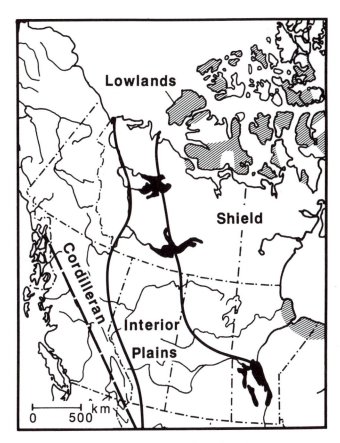

Fig. 16.1. The main physiographic features of western Canada.

mer months, when it penetrates Alaska and the Yukon, and the Icelandic low, which influences the circulation of northern Hudson Bay and the eastern Canadian Arctic. The resulting bioclimate is typically continental, characterized by long, very cold winters

401

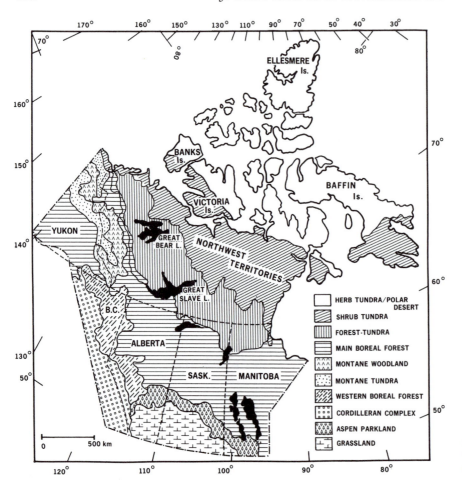

Fig. 16.2. Schematic map of the major vegetation zones of the western interior of Canada.

and short, warm, dry summers (Hare and Thomas, 1979). The isotherms and isohyets trend northwest to southeast and, because of the general uniformity of topography and surface materials, the western interior has a correspondingly regular zonal pattern of plant communities (Rowe, 1972).

For the purposes of this chapter a simple grouping of these vegetation zones will suffice (Fig. 16.2). A large central forested core is bounded by a northern treeless zone (the shrub tundra) and a southern treeless zone (the grasslands). The forest core is differentiated into forest-tundra, main boreal forest, montane boreal forest, and aspen parkland. These zones are described in detail in Rowe (1972). Further information on the bioclimates, vegetation, and modern pollen spectra of the entire region can be found in Ritchie (1987).

The ecotone from northern boreal forest to tundra coincides with a relatively steep gradient in summer temperatures (Hare and Ritchie, 1972), but winter temperatures vary little across the same transition. The northern limits of *Picea*, tree *Betula*, and *Populus* are probably determined by summer temperatures (Black and Bliss, 1980) (Table 16.1). The northern limit of *Pinus banksiana* (jack pine) is several tens of kilometers south of the limits of spruce, birch, and poplar

in the eastern portion of the study area and several hundred kilometers south of these boundaries in the western segment (Porsild and Cody, 1980). In the Mackenzie Valley region the abundance of *P. banksiana* decreases sharply near Great Slave Lake, and the tree's northern distributional limit lies near Great Bear Lake. If we assume that jack pine has reached its northern limit there, we can only speculate that the controlling factors are related to summer thermal conditions and the length of the growing season (Table 16.1). Frequent fires and the availability of freely drained habitats are regarded as important requirements for abundant jack pine stands (Johnson and Rowe, 1975), but there is no clear evidence that these factors vary across the present limits of the tree.

The transition from the southern boreal forest to the aspen parkland and prairies coincides more or less with moderate gradients of increasing summer temperature and decreasing effective moisture (Table 16.1). The boreal zone is characterized by markedly lower mean January temperatures compared to stations to the south, but it is not obvious that this factor would limit tree growth.

Almost the entire area was under Laurentide or Cordilleran ice until about 15,000 yr B.P., and the region

Table 16.1. Mean seasonal and annual (1930-1980) thermal and moisture variables along two north-south transects across the western interior of Canada and the approximate positions of the main phytogeographic boundaries

Location	Mean January temperature (°C)	Mean July temperature (°C)	Total degree-days	Frost-free days (no.)	Precipi-tation (mm)	Important phytogeographic boundaries
Tuktoyaktuk (69°20'N, 133°W)	-27	10	600	50	229	Tundra; northern limits of *Picea, Populus,* and arboreal *Betula*
Inuvik (68°20'N, 134°W)	-28	14	1016	70	317	
Yellowknife (62°30'N, 144°W)	-28	16	1586	110	250	Boreal forest; northern limit of *Pinus banksiana*
Edmonton (53°19'N, 118°35'W)	-15	17	2198	127	446	Parkland; southern limit of *Picea* and *Betula*
Lethbridge (49°38'N, 112°48'W)	-9.7	18	2286	175	310	Grassland
Churchill (58°45'N, 94°04'W)	-28	12	837	81	397	Forest-tundra; northern limit of *Picea*
Flin Flon (54°46'N, 101°51'W)	-22	17	1600	91	405	Boreal forest
Prince Albert (53°13'N, 105°41'W)	-21	18	1726	96	389	Southern limit of *Picea*
Regina (50°26'N, 104°40'W)	-17	19	2105	107	398	Grassland

did not become completely ice-free until 6000 yr B.P. Thus the history of postglacial reestablishment of vegetation must have involved substantial migrations of the dominant taxa from distant source areas. This contrasts with such unglaciated regions as the southern and central United States and southern Europe, where small plant populations that apparently survived in scattered (usually montane) microhabitats spread over relatively short distances to occupy their modern areas (Huntley and Birks, 1983; Delcourt *et al.,* 1984; Bennett, 1985).

Pollen Data and General Stratigraphy

Holocene pollen data for 9000 and 6000 yr B.P. are available from 51 sites in the region (Table 16.2 and Fig. 16.3). Forty-one are lake sites, and the others are either mires or pingos. We present summary percentage pollen diagrams for eight lake sites chosen as representative regional records (Fig. 16.4).

We used the following changes in pollen spectra and their reconstructions (Table 16.2 and Fig. 16.5) as the criteria for inferred changes in temperature and moisture for the periods we considered: (1) at all northern sites and some others, nonarboreal pollen (NAP) assemblages with high frequencies of dwarf birch, sedges, and herbs were taken to represent tundra vegetation, and a change to a tree assemblage was interpreted as indicating increased temperature; (2) at

southern sites, replacement of a tree assemblage by an NAP assemblage with high percentages of *Artemisia,* Gramineae, and Chenopodiaceae-Amaranthaceae pollen was taken to indicate an increase in summer temperature and/or a decrease in effective moisture; (3) at southern sites, a change from NAP-dominated spectra interpreted as prairie to assemblages dominated by arboreal pollen was taken to mean a lower summer temperature, increased moisture, or both; (4) at northern sites, replacement of conifer-dominated assemblages by NAP-dominated spectra was interpreted as a retreat of the treeline in response to lower temperature.

Before we examine the evidence for 9000 and 6000 yr B.P., a brief review of general trends will serve to provide a frame of reference. Southern sites exposed by the melting of ice cover before 12,000 yr B.P. and now located around the broad ecotones between the grasslands, aspen parkland, and southern boreal forest show a *Picea*-dominated assemblage, associated at some sites with *Populus.* Nonarboreal taxa are abundant. Spruce frequencies decline sharply to less than 2% by about 9000 yr B.P., and spectra dominated by Gramineae, Ambrosieae, and Chenopodiaceae-Amaranthaceae increase (representative pollen diagrams are sites 27 and 14 in Fig. 16.4). Subsequent changes vary with the distance of the site from the modern southern limit of the boreal forest. At sites south of the forest border (e.g., site 14, Fig. 16.4), the nonarboreal assemblage (prairie or grassland) persists until about 3000 yr B.P., when *Quercus* increases. At sites

Table 16.2. Location,[a] literature source, dating control (DC) ranking,[b] and temperature (T) and moisture (M) reconstructions[c] for pollen sites in this study

Site no.	Name	Latitude (°N)	Longitude (°W)	Reference[d]	12,000–9000 DC T M	9000 DC T M	6000 DC T M	3000 DC T M	3000–0 DC T M
1	Alpen Siding	54 27	113 00	37b					
2	Antifreeze Pond	62 21	140 50	50	2 +	2 ++	3 +	4 0 −	5 0
3	Belmont Lake	49 26	99 26	63	3 +	6 +	3 ++	2 0 +	2 0 0
4	Boone Lake	55 34	119 24	80					
5	Chalmers Bog	50 37	114 25	45	5	4 +	4 0	6 0	6 0
6	Clearwater Lake	50 52	107 56	44		5 +	5 0	5 0	6 0
7	Crestwynd	50 10	105 40	54		6	3		
8	Cycloid Lake	55 16	105 16	44		3 +	5 0 0	6 0 0	6 0 0
9	Eaglenest Lake	57 46	112 06	74	3	2 + −	4 0 0	4 0 0	4 0 0
10	Eildun Lake	63 08	122 46	68	3	1 +	3 0	1 0	1 0
11	Eskimo Lakes	69 24	131 40	28		2 +	2 0	4 0	4 0
12	Fiddler's Pond	56 15	121 20	81		2	2	3	2
13	Flin Flon	55 45	102 05	54	6 +	6 +	1 0	1 0 0	1 0 0
14	Glenboro Lake	49 26	99 17	63	2 +	3 +	2 ++	1 0 +	1 0 0
15	Grand Rapids	53 02	99 43	61		3 +	1 0	2 0	3 0
16	Hafichuk	50 20	105 48	64	1	1 +			
17	Hanging Lake 2	68 23	138 23	9	1 +	1 ++	1 0	1 0	1 0
18	Hendrickson Island	69 32	133 30	28		2 +	3 0	4 0	4 0
19	John Klondike	60 21	123 39	43		1 0	1 0	2 0	2 0
20	Kananaskis	51 04	115 03	38	2 −	1 +	2 0	4 0	4 0
21	Kate's Pond	68 22	133 20						
22	Lac Ciel Blanc	59 33	122 10	40	2	1 +	2 0	2 0	2 0
23	Lac Demain	62 03	118 42	40	2	1 +	2 0	2 0	1 0
24	Lac Mélèze	65 13	126 07	40	1 +	2 +	2 0	1 0	1 0
25	Lake A	53 14	105 43	44	6 +	6 ++	6 0	6 − +	6 0 0
26	Lake B	53 48	106 05	44	6 +	6 ++	6 0	6 − +	6 0 0
27	Riding Mountain	50 43	99 39	51,52	2 +	1 ++	1 +	2 0 +	2 0 0
28	Lake M	68 06	133 28	55	2	1 +	2 0	1 0	1 0
29	Lake Manitoba	50 15	98 25	71	2	1	2 + −	1 0 0	1 0 0
30	Lateral Pond	65 57	135 31	57	3 +	2	2	3	4

that today are within the boreal forest (sites 31 and 27, Fig. 16.4), prairie spectra are replaced by boreal taxa (spruce, pine, birch) earlier, at about 6000 yr B.P. The broad vegetation reconstruction suggested is an early-Holocene version of the spruce-dominated boreal forest, which was replaced at its southern margin by grassland or prairie at about 9000 yr B.P., followed by a later southward shift of spruce and pine.

By contrast, northern sites near the modern tundra-forest transition zone (sites 41 and 21, Fig. 16.4) show an early-Holocene (about 11,000 yr B.P.) assemblage dominated by dwarf birch pollen replaced by a *Picea* assemblage with variable amounts of poplar and juniper. The period of relatively rapid changes in pollen frequency lasts from 11,000 to 6000 yr B.P., by which time pollen spectra similar to modern ones are first established. These spectra remain little changed to the present.

Sites in the interior boreal zone lie closer to the position of the late-Holocene Laurentide ice margins and therefore have shorter histories (rarely longer than 11,000 yr). Early nonarboreal assemblages (dwarf birch and herbs) are replaced at about 9000 yr B.P. by *Picea*-dominated spectra, which in turn are later mod-

ified by the addition of arboreal *Betula, Alnus,* and finally *Pinus*. The timing of the increase in pine frequency varies with latitude from about 7000 yr B.P. at Lone Fox Lake (site 32, Fig. 16.4; after MacDonald [1987]) to 2500 yr B.P. at Eildun Lake (Slater, 1985). Interior sites (e.g., sites 24 and 32, Fig. 16.4) show no evidence of late-Holocene pollen changes similar to those found near the northern and southern margins.

Lake-Level Records—Sites and Methods

Data from 12 basins in western Canada (Table 16.3) can be used to reconstruct the sequence of hydrologic fluctuations in this region over the last 10,000 yr. This data base, compiled from published literature, consists of records of lake-level changes at basins that are closed or have been closed at some stage of their late-Quaternary history. We excluded from the study records or parts of records where water depth was primarily controlled by tectonic uplift, sea-level changes, or glacier fluctuations. The chronology of

Site no.	Name	Latitude (°N)	Longitude (°W)	Reference[d]	12,000–9000 DC T M			9000 DC T M			6000 DC T M			3000 DC T M			3000–0 DC T M		
31	Lofty Lake	54 44	112 29	37a	2			1	+	−	1	++	−	2	0	0	2	0	0
32	Lone Fox Lake	56 43	119 43	40	2			1	+		2	0		2	0		2	0	
33	Lynn Lake	56 50	101 03	46							1			1			1		
34	Natla Bog	63 00	129 05	39				1			1	+		1	0		1	0	
35	Polybog	67 48	139 48	48	1			1			1			1			1		
36	Porcupine Mt.	52 31	101 15	47							1	+		1			1		
37	Porter Lake	61 45	108 00	56							1			3	0		4	0	
38	Ra Lake	65 14	126 25	40							3			3	0		4	0	
39	Reindeer Lake	56 40	102 35	54							3			2	0		2	0	
40	Sewell Lake	49 35	99 15	54	6	+		7			7			7			7		
41	Sleet Lake	69 17	133 35	69	2	+		2	++		2	+		2	0		3	0	
42	Snowshoe Lake	57 27	120 40	40	2			1	+		2	0		2	0		2	0	
43	Spring Lake	55 31	119 35	80															
44	Sweet Little Lake	67 38	132 00	58				1	+		2	0		2	0		1	0	
45	Thompson	56 10	97 50	54							6	0		6			6		
46	Tuktoyaktuk 5	69 03	133 27	62	1	+		2	++		1	+		1	0		1	0	
47	Tuktoyaktuk 6	69 06	133 25	53	7			7			7			7			7		
48	Twin Tamarack	68 18	133 25	59	1			2	+		1	0		1	0		1	0	
49	Tyrrell Lake	66 03	135 39	57	6	+		6	+		2	0		7			7		
50	Wild Spear Lake	59 15	114 09	40	2			1	+		1	0		1	0		1	0	
51	Yesterday Lake	56 46	119 29	40	2			2	+		3	0		4	0		4	0	

[a]See also Figure 16.3.

[b]According to the scheme adopted by COHMAP (Webb, 1985, pp. 68–69).

[c]+ = warmer (T) or wetter (M) than at present, (+) = probably warmer or wetter than at present, 0 = no difference, (−) = probably colder (T) or drier (M) than at present, − = colder or drier than at present.

[d]9, Cwynar (1982); 28, Hyvarinen and Ritchie (1975); 37a, Lichti-Federovich (1970); 37b, Lichti-Federovich (1972); 38, MacDonald (1982); 39, MacDonald (1983); 40, MacDonald (1984); 43, Matthews (1980); 44, Mott (1973); 45, Mott and Jackson (1982); 46, Nichols (1967); 47, Nichols (1969); 48, Ovenden (1982); 50, Rampton (1971); 51, Ritchie (1964); 52, Ritchie (1969); 53, Ritchie (1972); 54, Ritchie (1976); 55, Ritchie (1977); 56, Ritchie (1980); 57, Ritchie (1982); 58, Ritchie (1984); 59, Ritchie (1985); 61, Ritchie and Hadden (1975); 62, Ritchie and Hare (1971); 63, Ritchie and Lichti-Federovich (1968); 64, Ritchie and de Vries (1964); 68, Slater (1985); 69, Spear (1983); 71, Teller and Last (1981); 74, Vance (1986); 80, White (1983); 81, White and Mathewes (1982).

lake-level changes was established by radiocarbon dating and, in the case of Lake Isle, Wedge Lake, and Phair Lake, by the presence of tephra units within the lake sediments. A detailed description of the primary data from each site and their interpretation in terms of lake-level changes is given in Harrison (1988).

Lake sites in western Canada are spread over a latitudinal range from 50 to 63°N and a longitudinal range from 98 to 123°W (Fig. 16.3). The sites are not distributed evenly, however; seven lakes are clustered within Alberta, with five more isolated sites. The temporal distribution is also uneven; only five sites have records extending back to 10,000 yr B.P., and records at three sites cover less than the last 5000 yr.

The types of data used to reconstruct changes in water level at the lake sites are shown in Table 16.3. All the data from western Canada are derived from cores of lake sediment. Changes in sediment lithology are the major source of information for most of the basins. Additional information is provided by marked changes in reconstructed sedimentation rates, where extremely low rates indicate a hiatus in deposition and hence a lowered lake level. Lake-level changes are also reconstructed from changes in pollen preserva-tion, the presence of reworked pollen within the sediments, geochemical data, and paleoecological indicators such as aquatic pollen and diatom assemblages. The interpretation of these data is discussed in Harrison (1988) and is not repeated here.

Because lake-level fluctuations are inferred from several different kinds of information, the data must be standardized for intersite comparison. The maximum and minimum water depths registered within the basin are identified on the basis of all available data and are used to define high and low status classes, respectively. Transitional periods are then assigned intermediate status. Assessments of lake status have been coded at 1000-yr intervals (Table 16.4). This tripartite classification necessarily involves the loss of a certain amount of information about lake behavior. It is a slightly modified version of the method used in previous compilations of the lake data from western Canada (e.g., Harrison and Metcalfe, 1985a,b).

To focus attention on changes through time, we used the data to derive the difference in "effective moisture" between selected time periods (Table 16.4). This index, which is independent of the lake status measures, is a record of whether the lake data indicate

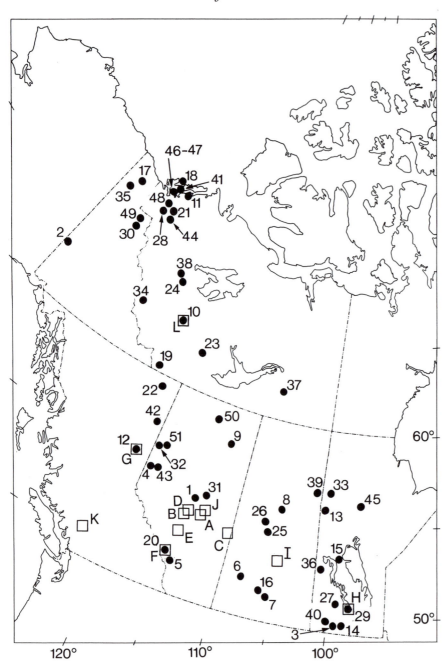

Fig. 16.3. Pollen (●) and lake-level (□) sites from western Canada. The site numbers and letters correspond to those listed in Tables 16.2 and 16.3, respectively.

a more or less positive water balance between two times, regardless of the absolute scale of the change in water level. The effective moisture index has been calculated for the intervals 9000–6000 and 6000–3000 yr B.P. as well as between each of these times and the present.

Pollen Spectra, Vegetation, and Climates at 9000 and 6000 yr B.P.

A major change in pollen stratigraphy at roughly 9000 yr B.P. can be seen in all diagrams that have adequate radiocarbon control (Table 16.2). At southern sites *Picea* frequencies decline as taxa associated with grasslands or deciduous-tree woodlands increase. Spectra older than 9000 yr B.P. have high percentages of *Picea* (30–60%) and smaller amounts of *Populus*, *Juniperus*, *Shepherdia canadensis*, Gramineae, *Artemisia*, and Ambrosieae. Later this assemblage is replaced by one dominated by nonarboreal taxa (Gramineae, *Artemisia*, and Ambrosieae), associated with low frequencies of such characteristic prairie taxa as *Amorpha*, *Petalostemum*, *Phlox hoodii*, and *Sphaeralcea*. The vegetation represented by the latter assemblage is almost certainly a grassland or prairie, similar

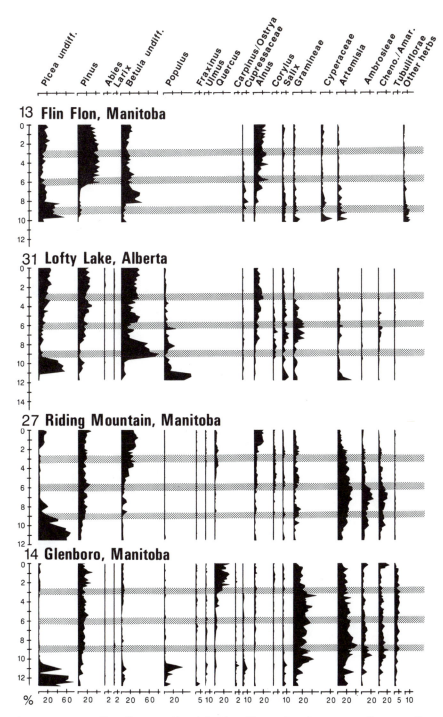

Fig. 16.4. Representative summary pollen diagrams for eight sites, illustrating the patterns from southern boreal forest sites (13, 31, and 27, this page), an aspen-parkland site (14, this page), northern treeline sites (41 and 21, p. 408), and northern boreal forest sites (24 and 32, p. 408). The vertical scale is in years B.P. × 10³.

to the modern plant communities found at undisturbed sites in southern Manitoba, Saskatchewan, and Alberta. The earlier spruce assemblage is interpreted as a closed spruce forest on mesic sites, with local shrub and herb communities on the most xeric habitats (Ritchie and Lichti-Federovich, 1968).

Conditions at 9000 yr B.P. were warmer than at present at almost all sites (Fig. 16.5), but by 6000 yr B.P. all sites except those at the northern and the southeastern extremities showed temperatures similar to those of the present day. Modern thermal conditions were established by 3000 yr B.P.

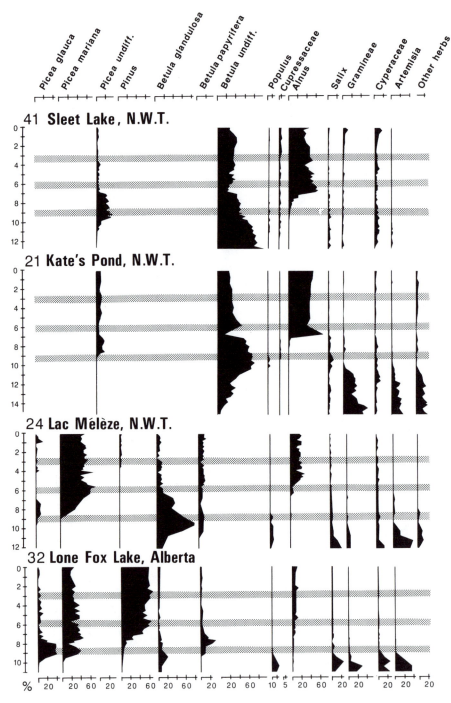

Fig. 16.4, continued from p. 413

In several early-Holocene sequences from the grassland region, a pollen zone of short duration between the spruce zone and the prairie zone is marked by significant proportions of *Larix*, *Fraxinus*, *Ulmus*, and *Carpinus-Ostrya*, with occasional grains of *Carya* and *Tsuga*. At least some of these trees (*Larix*, *Fraxinus*, *Ulmus*) may have been present at low abundances in the early-Holocene forests near these sites. Watts (1983) reviewed a similar transitional phase with deciduous trees between *Picea* forest and prairie at sites in the Great Plains region of the United States and noted that the transition from forest to prairie was time-transgressive from approximately 12,000 yr B.P. in South Dakota to between 10,000 and 9000 yr B.P. near the Canadian border.

Northern sites at and beyond the Arctic treeline show a maximum northward extension of spruce centered at 9000 yr B.P. (Ritchie *et al.*, 1983). The three

Table 16.3. The western Canada lake-level data bank

Site letter[a]	Basin	Latitude (°N)	Longitude (°W)	Elevation (m)	Maximum depth (m)	Area (ha)	^{14}C dates[b] (no.)	Length of record (yr)	Data sources[c]: L	SR	G	A	D	PP	References[d]
A	Hastings, Alberta	53 30	113 00	735	8.0	844.0	5	4800	*		*				1, 2
B	Isle, Alberta	52 37	114 26	700	8.0	1340.0	3[T]	10,300	*	*	*		*		3, 4
C	Moore, Alberta	53 00	110 30	500	30.0		6	12,000	*	*	*	*	*		5, 6
D	Smallboy, Alberta	53 35	114 08	762	7.5	1.0	5	7400	*		*	*			2
E	Wabamun, Alberta	53 30	114 25	723	10.0	8288.0	10	9800	*		*	*	*	*	7-10
F	Wedge, Alberta	50 52	115 10	1500	2.0		2[T]	10,500	*		*				11
G	Fiddler's Pond, B.C.	56 15	121 20	630	1.0	1.8	3	7500	*	*		*		*	12
H	Manitoba, Manitoba	51 00	98 00	248	6.3		9	10,000	*						13-16
I	Waldsea, Sask.	52 17	105 12	530	14.5	470.0	4	4000	*		*				17
J	E.I. Pond, Alberta	53 38	112 51	724	2.5	0.8	2	4000	*		*	*			2
K	Phair, B. C.	50 34	122 05	716		3.0	4[T]	7000	*						18
L	Eildun, N.W.T.	63 08	122 46	302	3.0	50.0	6	11,000	*	*					19

[a]See also Figure 16.3.
[b]T indicates additional tephra dating.
[c]L = lithology-stratigraphy, SR = sedimentation rates, G = geochemistry, A = aquatic pollen, D = diatoms, PP = pollen preservation or pollen reworking.
[d]1, Forbes and Hickman (1981); 2, Vance *et al.* (1983); 3, Hickman and Klarer (1981); 4, Hickman (personal communication, 1983); 5, Schweger and Hickman (1980); 6, Schweger *et al.* (1983); 7, Fritz and Krouse (1973); 8, Holloway (1980); 9, Holloway *et al.* (1981); 10, Hickman *et al.* (1984); 11, MacDonald (1982); 12, White and Mathewes (1982); 13, Fenton *et al.* (1983); 14, Last and Teller (1983); 15, Teller and Last (1981); 16, Teller and Last (1982); 17, Last and Schweyen (1985); 18, Mathewes and Westgate (1980); 19, Slater (1985).

Table 16.4. Summary of the lake-level data from western Canada

Basin	Change in effective moisture[a]: 9000-0 yr B.P.	6000-0 yr B.P.	3000-0 yr B.P.	9000-6000 yr B.P.	6000-3000 yr B.P.	Status[b]: 10	9	8	7	6	5	4	3	2	1	0	Dating control[c]: 9	6	3
Hastings, Alberta		D	0		D						3	2	1	1	1	1		7	1
Isle, Alberta	D	D	0	(D)	D	2	3	3	3	3	2	1	1	1	1	1	2	1	1
Moore, Alberta	D	D	0	(W)	D	2	3	3	3	3	2	1	1	1	1	1	2	2	1
Smallboy, Alberta	(D)	D	0	(D)	D				3	3	2	2	1	1	1	1	7	1	1
Wabamun, Alberta	(W)	D	0	W	D		1	2	3	3	3	2	2	2	2	2	2	2	3
Wedge, Alberta	D	0	0	D	0	3	3	2	2	1	1	1	1	1	1	1	2	2	3
Fiddler's Pond, B.C.		D	0		D				3	3	2	2	1	1	1	1		1	2
Manitoba, Manitoba	D	D	0	(W)	D												2	2	1
Waldsea, Saskatchewan			0									3	1	1	1	1			1
E.I. Pond, Alberta		(D)	0		(D)							2	2	1	1	1			1
Phair, British Columbia		D	D		D				1	3	2	2	2	1	1	1		1	2
Eildun, N.W.T.	D	0	0	D	0												1	1	1

[a] For each period, D = less effective moisture, (D) = probably less effective moisture, 0 = no difference, (W) = probably more effective moisture, W = more effective moisture.
[b]For each time (yr B.P. × 10^3), 1 = high, 2 = intermediate, 3 = low.
[c]Dating control for each time (yr B.P. × 10^3) according to the scheme adopted by COHMAP (Webb, 1985, pp. 68-69).

diagrams currently available with values for pollen accumulation rates show a steady increase in total pollen from about 15,000 to 9000 yr B.P., suggesting a gradual increase in summer warmth (Cwynar, 1982; Ritchie, 1982, 1985) culminating at 10,000-9000 yr B.P. in maximum values for the indicator taxa *Picea, Populus, Typha,* and *Myrica.* Ritchie (1984, p. 154) suggested that the dominant plant cover in northwest Canada shifted at about 10,000 yr B.P. from tundra to spruce forest and inferred that "the regional climate, particularly in the growing season, warmed slowly between 15,000 and 12,000 yr B.P. to conditions warmer than present by not later than 10,000 yr B.P. Mean July temperatures were 3-5°C higher than modern values, with proportional increases in the length of the growing season, degree day totals, and the length of the frost-free period." Evidence from treeline sites, particularly those on the Tuktoyaktuk Peninsula (e.g., Sleet Lake, site 41, Fig. 16.4; from Spear [1983]), suggests that "the period of warmer-than-modern conditions lasted until

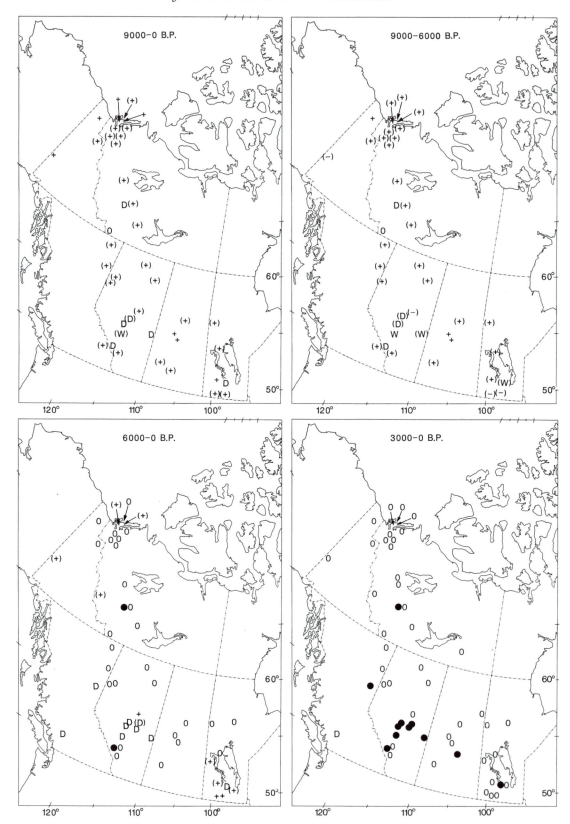

Fig. 16.5. Mapped summaries of the effective moisture index reconstructed from lake-level data and of temperature changes reconstructed from pollen data. W = more effective moisture, (W) = probably more effective moisture, ● = no change in effective moisture, (D) = probably less effective moisture, D = less effective moisture; + = warmer, (+) = probably warmer, 0 = no change in temperature, (−) = probably colder, and − = colder.

4500 yr B.P. but a slow cooling began at 8000 yr B.P." (Ritchie, 1984, p. 155).

The pollen evidence centered at 6000 yr B.P. is more variable in its possible significance. Among the southern sites the most sensitive appear to be sites 25–27 (Table 16.2) at the southern limit of the boreal forest of Saskatchewan and Manitoba. The southernmost stands of the boreal forest often occur on uplands surrounded by lowlands occupied by aspen parkland and prairie vegetation. The pollen assemblages at site 27 (Fig. 16.4) change at roughly 6000 yr B.P., indicating a decrease in prairie vegetation on the uplands and its replacement by forests dominated by deciduous trees, chiefly oak and birch. At Lofty Lake (site 31, Fig. 16.4) a similar but less well-defined change involves increases in birch pollen percentages and decreases in some of the herb taxa. However, in the Tiger Hills area (sites 3 and 14, the latter shown in Fig. 16.4) no significant changes were registered at 6000 yr B.P. The mid- to late-Holocene record at many southern sites may be misleading because of the familiar problem of the erratic preservation of poplar pollen. Poplar (*Populus*) is the dominant tree of the modern aspen parkland, an ecotonal zone between modern grassland and boreal forest, and it is likely, though seldom shown in the pollen record, that slight shifts in summer temperatures and effective moisture would have produced significant responses in the abundance of the several species of *Populus*.

Significant changes in pollen frequency at northern and central sites for this time interval can be seen in some but not all diagrams. The major change is an increase in alder percentages from less than 1% to about 40%, with a corresponding decrease in the frequencies of the other important taxa. Estimates of concentration and accumulation rate show that the increase in alder is real, although the overrepresentation of alder in the pollen sum must be considered in any reconstructions. However, the rise in alder pollen varies among sites in both timing and magnitude, with the highest percentages at sites beyond 62°N. The available data provide no support for a clear time-transgressive trend reflecting a directional migration of alder. Possible alternative explanations have been proposed elsewhere (Ritchie, 1984, 1985), but the sparseness of the data network precludes a critical analysis of these speculations. The only convincing pollen evidence for climatic change in the late Holocene comes later, at about 3000 yr B.P., when it appears that the plains became cooler and/or wetter, promoting a southward extension of the boreal forests.

Lake-Level Results

The categorized lake-level data from western Canada are summarized in Table 16.4. Given the small number

of sites and the paucity of data from the early Holocene, the interpretations and conclusions put forward here must be regarded as tentative.

The temporal sequence of changes in lake status during the Holocene (Fig. 16.6) shows that the early to middle Holocene was a time of dryness, with peak aridity centered on 6000 yr B.P. The return to the high lake levels characteristic of today started after 5000 yr B.P. and was completed by 2000 yr B.P. Since most of this area was ice-free before the beginning of the Holocene (Denton and Hughes, 1981), the fact that so many of the records start in the later part of the Holocene is consistent with aridity during the early to middle Holocene.

The spatial patterns in the lake records are illustrated by maps of the reconstructed differences in effective moisture between critical times (Fig. 16.5). At 9000 yr B.P. conditions were drier than at present throughout the area, and they remained drier at 6000 yr B.P. except in the extreme north, where the one available record (Eildun Lake) suggests that conditions were already similar to those of the present by 6000 yr B.P. The effective moisture differences between 9000 and 6000 yr B.P. suggest that peak aridity occurred during the early phase in the west and later in the eastern part of the region. By 3000 yr B.P. conditions were similar to those of the present everywhere except in the extreme southwest (Phair Lake), which continued drier than at present. This latitudinal pattern in lake

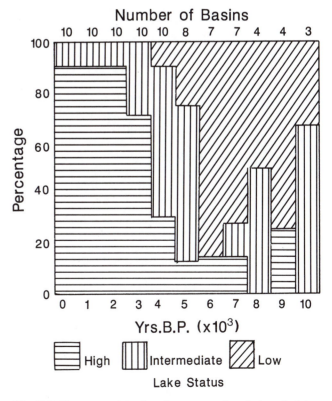

Fig. 16.6. Histogram showing the temporal variations in lake status during the Holocene in western Canada.

behavior in western Canada suggests that the return to moister conditions after the early- to mid-Holocene arid phase occurred earliest in the north and latest in the southwest. The time of peak aridity varied along an east-west gradient.

Discussion and Conclusions

These general paleoclimatic reconstructions based on both pollen and lake-level data conform with the Milankovitch proposition of maximum summer insolation at high latitudes around 9000 yr B.P. Results of general circulation models of Holocene global climate based on the same theory (Kutzbach, Guetter, et al., this vol.) are also consistent with our interpretations, for they suggest that July surface temperatures were 2.5°C higher than at present in both the northwest and the central plains regions of Canada at this time.

The lake-level record clearly shows that drier conditions prevailed over this region during the early Holocene. It seems likely that this aridity resulted from the suppression of the westerly flow of moist Pacific air into the Canadian interior. The Laurentide ice sheet could have acted as a geographic barrier, displacing westerly flow south of the continental ice and southeasterly anticyclonic flow over the deglaciated areas at the margin of the ice sheet (Kutzbach, Guetter, et al., this vol.). Studies of stabilized dune ridges in northern Saskatchewan, which apparently formed between 10,000 and 8800 yr B.P., indicate that this area was subject to southeasterly winds at this time (David, 1981).

Harrison and Metcalfe (1985b) suggested that Holocene aridity in western Canada reflected poleward migration of the frontal zone between Arctic and dry Pacific airmasses, possibly to north of 56°N. They argued that this shift culminated at around 6000 yr B.P., when the Arctic front appeared to have reached its most northerly position. The increase in effective moisture reflected by the record from Eildun Lake at 6000 yr B.P., however, suggests that the southward migration of the Arctic front had already begun by this stage. High lake levels after 4000 yr B.P. probably reflect the continued southward migration of this zone to its present position. Model results for 3000 yr B.P. (Kutzbach, Guetter, et al., this vol.) suggest that the winter westerlies were stronger than at present, which would have increased precipitation in the region. The lake records provide no evidence of increased precipitation, but this may simply reflect the fact that the lakes were open systems and that a more positive water balance was compensated by increased outflow.

The western interior of Canada provides sensitive and direct pollen and lake-level registrations of the Holocene climatic changes predicted by and modeled on the basis of the Milankovitch theory (Kutzbach, Guetter, et al., this vol.). The most important factors that determined this response to changes in climate

were the absence of major topographic contrasts and the locations of the sites west or southwest of the Laurentide ice dome. The most important climatic changes appear to have been those that affected the thermal and moisture conditions of the growing season.

By contrast, data from pollen sites farther northwest, in Alaska, although correlated closely with those reported here, show considerable regional variation explained in part by topographic diversity and in part by the importance of coastal migration routes for some tree and shrub taxa (Ager, 1983; Anderson and Brubaker, this vol.).

Pollen data from sites in the area adjacent to the east and southeast of our study area show some similar responses and trends but also some differences, probably because of major floristic changes and because of the "upstream" climatic effects of the Laurentide ice during the first half of the Holocene (Webb et al., 1983; Thompson et al., this vol.; Webb, Bartlein, et al., this vol.).

References

Ager, T. (1983). Holocene vegetation history of Alaska. *In* "Late Quaternary Environments of the United States, Vol. 2: The Holocene" (H. E. Wright, Jr., Ed.), pp. 128-141. University of Minnesota Press, Minneapolis.

Bennett, K. D. (1985). The spread of *Fagus grandifolia* across eastern North America during the last 18,000 years. *Journal of Biogeography* 12, 147-164.

Black, R. A., and Bliss, L. C. (1980). Reproductive ecology of *Picea mariana* (Mill.) BSP, at treeline near Inuvik, Northwest Territories, Canada. *Ecological Monographs* 50, 331-354.

Bryson, R. A., and Hare, F. K. (1974). "The Climates of North America." Vol. VII, World Survey of Climatology (H. Landsberg, Ed.). Elsevier, Amsterdam.

Cwynar, L. C. (1982). A late-Quaternary vegetation history from Hanging Lake, northern Yukon. *Ecological Monographs* 52, 1-24.

David, P. P. (1981). Stabilized dune ridges in northern Saskatchewan. *Canadian Journal of Earth Sciences* 18, 286-310.

Delcourt, P. A., Delcourt, H. R., and Webb, T. III (1984). Atlas of paired isophyte and isopoll maps for important tree taxa of eastern North America. *American Association of Stratigraphic Palynologists Contribution Series* 14, 1-131.

Denton, G. H., and Hughes, T. J. (1981). "The Last Great Ice Sheets." Wiley-Interscience, New York.

Fenton, M. M., Moran, S. R., Teller, J. T., and Clayton, L. (1983). Quaternary stratigraphy and history in the southern part of the Lake Agassiz basin. *In* "Glacial Lake Agassiz" (J. T. Teller and L. Clayton, Eds.), pp. 49-74. Special Paper 26, Geological Association of Canada.

Forbes, J. R., and Hickman, M. (1981). Paleolimnology of two shallow lakes in central Alberta, Canada. *Internationale Revue der Gesamten Hydrobiologie* 66, 863-888.

Fritz, P., and Krouse, H. R. (1973). Wabamun Lake past and present, an isotopic study of the water budget. *In* "Proceedings of the Symposium on the Lakes of Western Canada" (E. R. Reinelt, A. H. Laycock, and W. M. Schultz,

Eds.), pp. 244-258. Water Resources Centre, University of Alberta, Edmonton.

Hare, F. K., and Ritchie, J. C. (1972). The boreal bioclimates. *Geographical Review* 62, 333-365.

Hare, F. K., and Thomas, M. K. (1979). "Climate Canada," 2nd ed. Wiley, Toronto.

Harrison, S. P. (1988). Lake-level records from Canada and the eastern U.S.A. *Lundqua Report* 29. Department of Quaternary Geology, Lund University, Lund.

Harrison, S. P., and Metcalfe, S. E. (1985a). Spatial variations in lake levels since the last glacial maximum in the Americas north of the equator. *Zeitschrift für Gletscherkunde und Glazialgeologie* 21, 1-15.

——. (1985b). Variations in lake levels during the Holocene in North America: An indicator of changes in atmospheric circulation patterns. *Geographie physique et Quaternaire* 39, 141-150.

Hickman, M., and Klarer, D. M. (1981). Paleolimnology of Lake Isle, Alberta, Canada. *Archiwum Hydrobiologii* 91, 490-508.

Hickman, M., Schweger, C. E., and Habgood, T. (1984). Lake Wabamun, Alta.: A paleoenvironmental study. *Canadian Journal of Botany* 62, 1438-1465.

Holloway, R. G. (1980). Absolute pollen analysis of Lake Wabamun, Alberta, Canada. *Palynology* 4, 243-244.

Holloway, R. G., Bryant, V. M., and Valastro, S. (1981). A 16,000 year pollen record from Lake Wabamun, Alberta, Canada. *Palynology* 5, 195-208.

Huntley, B., and Birks, H. J. B. (1983). "An Atlas of Past and Present Pollen Maps for Europe: 0-13,000 Years Ago." Cambridge University Press, Cambridge.

Hyvarinen, H., and Ritchie, J. C. (1975). Pollen stratigraphy of Mackenzie pingo sediments, N.W.T., Canada. *Arctic and Alpine Research* 7, 261-272.

Johnson, E. A., and Rowe, J. S. (1975). Fire in the wintering grounds of the Beverley caribou herd. *American Midland Naturalist* 94, 1-14.

Last, W. M., and Schweyen, T. H. (1985). Late Holocene history of Waldsea Lake, Saskatchewan, Canada. *Quaternary Research* 24, 219-234.

Last, W. M., and Teller, J. T. (1983). Holocene climate and hydrology of the Lake Manitoba basin. *In* "Glacial Lake Agassiz" (J. T. Teller and L. Clayton, Eds.), pp. 333-353. Special Paper 26, Geological Association of Canada.

Lichti-Federovich, S. (1970). The pollen stratigraphy of a dated section of late Pleistocene lake sediment from central Alberta. *Canadian Journal of Earth Sciences* 7, 938-945.

——. (1972). "Pollen Stratigraphy of a Sediment Core from Alpen Siding, Alberta," pp. 113-115. Report of Activities, Paper 72-1B, Geological Survey of Canada, Ottawa.

MacDonald, G. M. (1982). Late Quaternary paleoenvironments of the Morley Flats and Kananaskis Valley of southwestern Alberta. *Canadian Journal of Earth Sciences* 19, 23-35.

——. (1983). Holocene vegetation history of the upper Natla River area, Northwest Territories, Canada. *Arctic and Alpine Research* 15, 169-180.

——. (1984). "Postglacial Plant Migration and Vegetation Development in the Western Canadian Boreal Forest." Unpublished Ph.D. thesis, University of Toronto.

——. (1987). Postglacial development of the subalpine-boreal transition forest in western Canada. *Journal of Ecology* 75, 303.

Mathewes, R. W., and Westgate, J. A. (1980). Bridge River tephra: Revised distribution and significance for detecting old carbon errors in radiocarbon dates of limnic sediments in southern British Columbia. *Canadian Journal of Earth Sciences* 17, 1454-1461.

Matthews, J. V., Jr. (1980). "Paleoecology of John Klondike Bog, Fisherman Lake Region, Southwest District of Mackenzie." Paper 80-22, Geological Survey of Canada, Ottawa.

Mott, R. J. (1973). "Palynological Studies in Central Saskatchewan: Pollen Stratigraphy from Lake Sediment Sequences." Paper 72-49, Geological Survey of Canada, Ottawa.

Mott, R. J., and Jackson, L. E., Jr. (1982). An 18,000 year palynological record from the southern Alberta segment of the Classical Wisconsinan, "Ice-free Corridor." *Canadian Journal of Earth Sciences* 19, 504-513.

Nichols, H. (1967). The post-glacial history of vegetation and climate at Ennadai Lake, Keewatin, and Lynn Lake, Manitoba, Canada. *Eiszeitalter und Gegenwart* 18, 176-197.

——. (1969). The late Quaternary history of vegetation and climate at Porcupine Mountain and Clearwater Bog, Manitoba. *Arctic and Alpine Research* 1, 155-167.

Ovenden, L. E. (1982). Vegetation history of a polygonal peatland, northern Yukon. *Boreas* 11, 209-224.

Porsild, A. J., and Cody, W. J. (1980). "Vascular Plants of Continental Northwest Territories." National Museums of Canada, Ottawa.

Rampton, V. N. (1971). Late Quaternary vegetational and climatic history of the Snag-Klutlan area, southwestern Yukon Territory, Canada. *Geological Society of America Bulletin* 82, 959-978.

Ritchie, J. C. (1964). Contributions to the Holocene paleoecology of west-central Canada. 1. The Riding Mountain area. *Canadian Journal of Botany* 42, 181-196.

——. (1969). Absolute pollen frequencies and carbon-14 age of a section of Holocene lake sediment from the Riding Mountain area of Manitoba. *Canadian Journal of Botany* 47, 1345-1349.

——. (1972). Pollen analysis of late-Quaternary sediments from the Arctic treeline of the Mackenzie River Delta region, Northwest Territories, Canada. *In* "Climatic Changes in Arctic Areas during the Last Ten-Thousand Years" (Y. Vasari, H. Hyvarinen, and S. Hicks, Eds.), pp. 253-271. Acta Universitatis Ouluensis, Series A, Scientiae Rerum Naturalium No. 3, Geologica No. 1. Oulu University, Oulu, Finland.

——. (1976). The late-Quaternary vegetational history of the western interior of Canada. *Canadian Journal of Botany* 54, 1793-1818.

——. (1977). The modern and late Quaternary vegetation of the Campbell-Dolomite uplands near Inuvik, N.W.T., Canada. *Ecological Monographs* 47, 401-423.

——. (1980). Towards a late-Quaternary palaeoecology of the ice-free corridor. *Canadian Journal of Anthropology* 1, 15-28.

——. (1982). The modern and late-Quaternary vegetation of the Doll Creek area, north Yukon, Canada. *New Phytologist* 90, 563-603.

——. (1984). "Past and Present Vegetation of the Far Northwest of Canada." University of Toronto Press, Toronto.

——. (1985). Late-Quaternary climatic and vegetational change in the lower Mackenzie Basin, northwest Canada. *Ecology* 66, 612-621.

—. (1987). "Postglacial Vegetation of Canada." Cambridge University Press, Cambridge.

Ritchie, J. C., and de Vries, B. (1964). Contributions to the Holocene paleoecology of west-central Canada: A late-glacial deposit from the Missouri Coteau. *Canadian Journal of Botany* 42, 677–692.

Ritchie, J. C., and Hadden, K. A. (1975). Pollen stratigraphy of Holocene sediments from the Grand Rapids area, Manitoba, Canada. *Review of Palaeobotany and Palynology* 19, 193–202.

Ritchie, J. C., and Hare, F. K. (1971). Late-Quaternary vegetation and climate near the Arctic tree line of northwestern North America. *Quaternary Research* 1, 331–342.

Ritchie, J. C., and Lichti-Federovich, S. (1968). Holocene pollen assemblages from the Tiger Hills, Manitoba. *Canadian Journal of Earth Sciences* 5, 873–880.

Ritchie, J. C., Cwynar, L. C., and Spear, R. W. (1983). Evidence from northwest Canada for an early Holocene Milankovitch thermal maximum. *Nature* 305, 126–128.

Rowe, J. S. (1972). "Forest Regions of Canada." Publication 1300, Canadian Forestry Service, Department of the Environment, Ottawa.

Schweger, C., and Hickman, M. (1980). Postglacial palynology and paleolimnology, Alberta, western Canada. *In* "Abstracts, Fifth International Palynological Conference, Cambridge," p. 357.

Schweger, C., Habgood, T., and Hickman, M. (1983). Late-glacial Holocene climatic changes of Alberta: The record from lake sediment studies. *In* "The Impacts of Climatic Fluctuations on Alberta's Resources and Environment," pp. 47–60. Edmonton Report, WAES-1-81. Atmospheric Environment Service, Environment Canada, Ottawa.Slater, D. S. (1985). Pollen analysis of postglacial sediments from Eildun Lake, District of Mackenzie, N.W.T., Canada. *Canadian Journal of Earth Sciences* 22, 663–674.

Spear, R. W. (1983). Paleoecological approaches to a study of treeline fluctuations in the Mackenzie Delta region, Northwest Territories: Preliminary results. *Collection Nordicana* 47, 61–72.

Teller, J. T., and Last, W. M. (1981). Late Quaternary history of Lake Manitoba, Canada. *Quaternary Research* 16, 97–116.

—. (1982). Pedogenic zones in postglacial sediment of Lake Manitoba, Canada. *Earth Surface Processes and Landforms* 7, 367–379.

Vance, R. E. (1986). Pollen stratigraphy of Eaglenest Lake, northeastern Alberta. *Canadian Journal of Earth Sciences* 23, 11–20.

Vance, R. E., Emerson, D., and Habgood, T. (1983). A mid-Holocene record of vegetative change in central Alberta. *Canadian Journal of Earth Sciences* 20, 364–376.

Watts, W. A. (1983). Vegetational history of the eastern United States, 25,000 to 10,000 years ago. *In* "Late Quaternary Environments of the United States, Vol. 1: The Late Pleistocene" (S. C. Porter, Ed.), pp. 294–310. University of Minnesota Press, Minneapolis.

Webb, T. III (1985). A global paleoclimatic data base for 6000 yr B.P. Technical Report DOE/EV/10097-6, TRO18. U.S. Department of Energy, Washington, D.C.

Webb, T. III, Cushing, E. J., and Wright, H. E., Jr. (1983). Holocene changes in the vegetation of the Midwest. *In* "Late Quaternary Environments of the United States, Vol. 2: The Holocene" (H. E. Wright, Jr., Ed.), pp. 142–165. University of Minnesota Press, Minneapolis.

White, J. M. (1983). "Late Quaternary Geochronology and Palaeoecology of the Upper Peace River District, Canada." Unpublished Ph.D. thesis, Simon Fraser University, Burnaby, B.C.

White, J. M., and Mathewes, R. W. (1982). Holocene vegetation and climatic change in the Peace River district, Canada. *Canadian Journal of Earth Sciences* 19, 555–570.

Vegetation, Lake Levels, and Climate in Eastern North America for the Past 18,000 Years

Thompson Webb III, Patrick J. Bartlein, Sandy P. Harrison, and Katherine H. Anderson

Climatologists face the challenge of understanding regional-scale climate change as the outcome of an interplay of global-scale controls, whereas palynologists and geomorphologists face the problem of extracting a regional-scale signal from data that often record local events. Climatologists, therefore, often work top-down and use global climate models to simulate regional patterns. In contrast, geologists proceed from the bottom up and use statistical analyses and maps of their data to synthesize regional patterns from irregularly spaced individual records. Such simulations and analyses have been completed for eastern North America, and we compare the results.

During the past 18,000 yr the major global climate controls have included changes in insolation, glaciation, atmospheric composition, and oceanic circulation and temperatures (Bartlein, 1988). The importance of these controls varies over space and time, so that each region has a unique climate history. Since 18 ka the critical controls or "boundary conditions" for eastern North American climates have been the changing intensity of seasonal insolation, the retreat of the Laurentide ice sheet, and changing sea-surface temperatures, especially in the Gulf of Mexico and western North Atlantic (Kutzbach and Wright, 1985; Kutzbach, 1987; Webb *et al.*, 1987; COHMAP Members, 1988).

The interaction among these controls, especially between the direct and indirect effects of changing insolation and the decay of the Laurentide ice sheet, has produced a sequence of climatic changes in eastern North America (Kutzbach, 1987; Webb *et al.*, 1987). The sequence begins with an initial "glacial" period (18 and 15 ka), when the influence of the ice sheet predominated and temperatures were much lower than at present. A "transitional" period (12 and 9 ka) follows, during which the influence of the ice sheet

waned while the effects of insolation changes increased. Finally, an "interglacial" period begins after 9 ka, when insolation changes became the predominant influence on climate (Webb *et al.*, 1987).

The regional imprint of this sequence arises from the unique interaction among controls in eastern North America. For example, the residual Laurentide ice sheet delayed the full thermal response to increased summer insolation until about 6 ka, considerably later than the hemispheric radiation maximum (COHMAP Members, 1988).

When we shift our perspective from climate to vegetation and landscape dynamics, the focus on regional changes becomes even more important, because regional—not global—climatic changes directly influence biotic and physical systems. Changes in regional patterns of temperature and precipitation affect the composition, structure, and function of vegetation as well as the regional water balance and hence the extent of lakes.

Climatic gradients across eastern North America today produce striking regional patterns in vegetation and lakes: the decrease in prairie and locally saline lakes and the increase in forests and freshwater lakes from midcontinent eastward and the gradational sequence of forest regions from Florida north to the Arctic treeline. These patterns are clear enough that past climates can be reconstructed from measures of past vegetation cover and regional water balance, such as pollen and lake-level data (e.g., Bartlein *et al.*, 1984; Harrison and Metcalfe, 1985a,b; Webb *et al.*, 1987; Harrison, 1989). Because each type of paleoenvironmental data records a somewhat different aspect of climate, comparing climate reconstructions based on two (or more) environmental sensors can broaden and deepen our understanding of past climate changes.

In this chapter we present mapped temporal syntheses of the available fossil pollen and lake-level data from eastern North America. By concentrating on the long-term, broad-scale patterns in these data, we reconstruct past changes in the regional patterns of vegetation and lake levels. We then use response surfaces (Bartlein *et al.*, 1986) to estimate temperature and precipitation from the pollen data, and we interpret the lake-level reconstructions to estimate moisture budgets. We systematically compare the patterns of climate changes inferred from the pollen and lake-level data. Finally, we compare these climate reconstructions with the results of paleoclimatic modeling experiments made with the Community Climate Model (CCM) of the National Center for Atmospheric Research (NCAR) (Kutzbach and Guetter, 1986; Kutzbach, 1987; Kutzbach, Guetter, *et al.*, this vol.) to check the model results. We also use the model results to help explain the observed changes in climate. Figure 17.1 illustrates our overall interpretive scheme. Our overarching aim is to describe and explain the nature of the environmental and climatic changes in eastern North America over the last 18,000 yr and to relate these changes to their ultimate causes, namely changes in the global controls on climate.

Previous broad-scale paleoclimatic interpretations for eastern North America include those by Bryson and Wendland (1967), Lamb and Woodroffe (1970), Bartlein *et al.* (1984), Delcourt and Delcourt (1984), Wright (1984), Bartlein and Webb (1985), Kutzbach and Wright (1985), Harrison and Metcalfe (1985a,b), Kutzbach (1987), and Webb *et al.* (1987). Our study builds on these syntheses by combining the results from model simulations with those from both pollen and lake-level data analyses.

The Region: Topography, Climate, and Vegetation

For the purposes of this chapter, eastern North America extends from 25 to 70°N and from 55 to 105°W. It is a region of relatively moderate relief (Fig. 17.2). The main mountain chain, the Appalachians, extends from Georgia in the southeastern United States northeast to the Maritime Provinces of Canada and has few elevations above 1500 m. Other highlands include the Ozarks of Missouri and northern Arkansas and a large highland plateau with elevations above 500 m in Quebec and Labrador. Lowlands occur around Hudson Bay, along the coastal plain from New York City to Alabama and Texas, and along the Mississippi River from Illinois to the Gulf of Mexico. In the central region, elevations increase northward to 300-500 m near the Great Lakes and westward to over 1000 m near the Rocky Mountains.

In general, mean summer and winter temperatures trend south to north (Fig. 17.2). Mean January temperatures range from 20°C in southern Florida to -35°C in Keewatin at 70°N; mean July temperatures range from 29 to 5°C. Summer temperatures decrease along a southwest-to-northeast gradient in Canada, and winter temperatures decrease to the northwest from Florida to Nebraska. Annual precipitation east of 90°W also decreases south to north from over 1500 mm along the Gulf coast to less than 200 mm at 70°N. However, annual precipitation exceeds 1000 mm over most of the coastal plain and in parts of Nova Scotia and Newfoundland because of their proximity to the ocean. West of 90°W the main precipitation gradient is east-west; annual rainfall ranges from more than 1000 mm at 90°W to less than 400 mm at 105°W. The east-

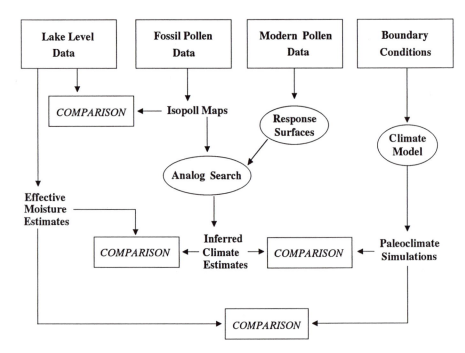

Fig. 17.1. Research design for comparison of data and climate estimates based on pollen data, lake-level data, and climate-model results.

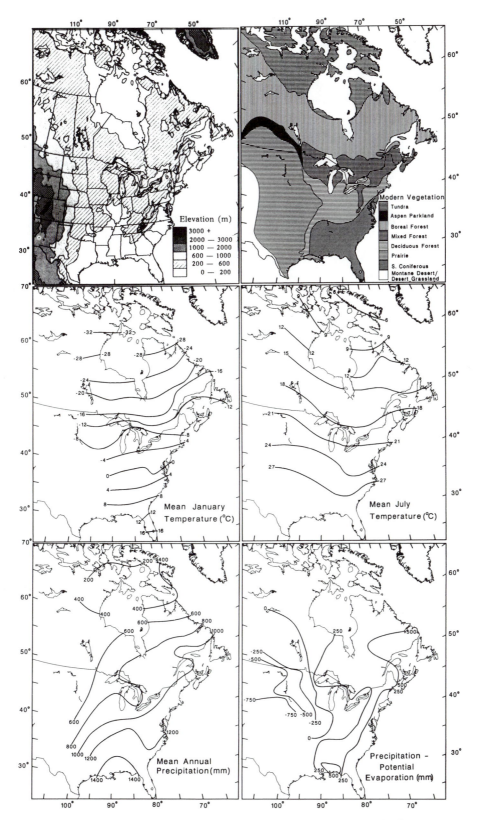

Fig. 17.2. Topography, vegetation, and climate variables (January and July temperatures, annual precipitation, and precipitation minus potential evaporation) in eastern North America. Climate data are 1941–1970 normals.

west gradient is even more pronounced in the map of moisture balance as represented by precipitation minus potential evaporation (Fig. 17.2). The steepest moisture gradient is east-northeast to west-southwest from Newfoundland to Kansas. Continental-scale vegetation patterns reflect these variations in climate and topography (Bryson, 1966; Bryson and Hare, 1974; Bartlein *et al.,* 1986), with prairie in the dry west, tundra in the cold north, and broad north-to-south gradients in the distribution of forest assemblages (Fig. 17.2).

The Data

POLLEN DATA

Pollen data are a well-proven source of distributional and stratigraphic information about past vegetation and climates. Davis (1983), Jacobson *et al.* (1987), Ritchie (1987), Watts (1983), and Webb (1988) have summarized past vegetational changes in eastern North America.

The modern data set is documented in Avizinis and Webb (1983) and Delcourt *et al.* (1984); we used the subset of 851 samples selected by Bartlein *et al.* (1986). The fossil pollen data come from paleoecologic studies of radiocarbon-dated sediment cores from lakes and wetlands (Fig. 17.3 and Table 17.1) and represent the work of more than 100 palynologists. The network of 328 sites with fossil pollen data includes 19 sites with data for 18 ka, more than 80 sites with data for 12 ka, and more than 270 sites with data for most of the Holocene (i.e., the past 10,000 yr) (Fig. 17.4). The ranking of the dating control for each mapped date at each site is given in Table 17.1 (see footnote in table for definition of ranks). Twenty new sites not included in Webb (1988) were added from New Jersey, Iowa, and the South and improve the coverage, especially for 18 and 15 ka. They are site numbers 40, 53, 54, 66, 89, 114, 119, 121, 133, 165, 166, 211, 219, 224, 229, 233, 279, 290, 313, and 328 in Table 17.1.

LAKE-LEVEL DATA

Changes in the depth and extent of a lake provide a record of changes in the mean annual water budget. Lake-level changes have been used to reconstruct geographic patterns of change in the hydrologic balance and thus to infer paleoclimatic changes at a continental scale (Street and Grove, 1976, 1979; Street-Perrott and Roberts, 1983; Street-Perrott and Harrison, 1984, 1985). This approach has been applied only recently in temperate regions, probably because many temperate lakes are overflowing today and the existence of an outlet moderates the response to hydrologic change, and because most temperate lakes are near their highest levels at present, which makes the evidence for water-level changes harder to find. These conditions,

however, do not mean that temperate lakes are insensitive or unsuitable for paleoclimatic reconstructions, as is well documented in Harrison and Metcalfe (1985a,b), Street-Perrott and Harrison (1985), and Harrison (1989). Syntheses of lake-level data from temperate regions (e.g., Gaillard, 1985; Harrison and Metcalfe, 1985a,b; Street-Perrott, 1986; Winkler *et al.,* 1986; Harrison, 1989; Harrison *et al.,* 1991; Harrison and Digerfeldt, in press) in fact are a rich source of data for reconstructing paleoclimatic changes.

We extracted lake-level information from published stratigraphic literature for 25 basins in eastern North America (Fig. 17.5 and Table 17.2). Most of the sites were originally studied to reconstruct regional vegetation history, and we reinterpreted the published information to infer changes in water levels. Harrison (1988) discussed the evidence and the detailed record of changing lake level at each site. Records where water depth is primarily controlled by crustal movement, sea-level change, or glacier fluctuations were excluded.

The chronology of lake-level change at all of the sites except Elk Lake, Minnesota, and Sunfish Lake, Ontario, was established by radiocarbon dating (Table 17.2). The Holocene sediments from Elk Lake are annually laminated (Dean *et al.,* 1984) and provide a precise chronology for the changes in water level. The onset of lacustrine sedimentation at Sunfish Lake has been radiocarbon-dated to about 10.5 ka (Sreenivasa and Duthie, 1973), and the timing of water-level changes during the Holocene was established by pollen correlation with a nearby site (Ballycroy Bog) (Anderson, 1971).

The basins with lake-level evidence lie between 27 and 52°N, and 64% (16 basins) lie between 40 and 45°N (Fig. 17.5). The spatial distribution of lake-level information is thus uneven. The temporal range of the data is from 12 ka to the present. The data are distributed more uniformly in time than in space because most are derived from lacustrine cores, which provide quasi-continuous records.

Methods

POLLEN DATA

Choice of a Pollen Sum

We selected a pollen sum of 24 pollen types by excluding infrequently observed taxa, indicators of human disturbance (*Ambrosia,* Gramineae, *Plantago,* and *Rumex*), and several wetland taxa, which are often locally overrepresented (e.g., *Nyssa, Larix, Salix,* and Ericaceae). The sum included 12 major types (*Picea, Betula, Pinus, Quercus,* forbs [Chenopodiaceae-Amaranthaceae, *Artemisia,* and other Compositae, excluding *Ambrosia*], Cyperaceae, *Alnus, Abies, Tsuga, Fagus, Carya,* and *Ulmus*) and 12 other taxa

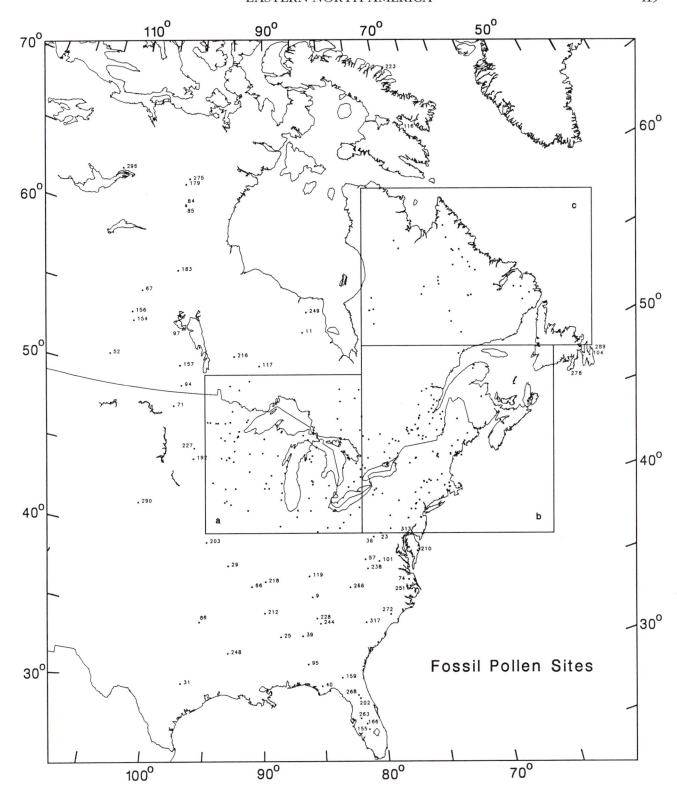

Fig. 17.3. Sites with fossil pollen data in eastern North America (see also Table 17.1, pp. 421-27). Boxed areas shown in expanded view on pp. 420–21.

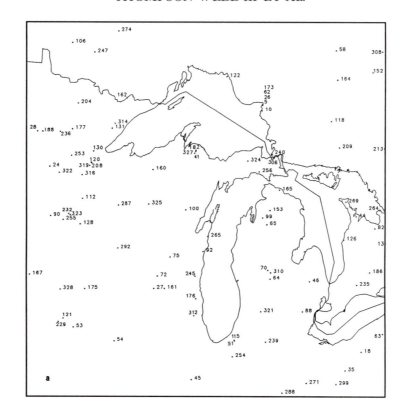

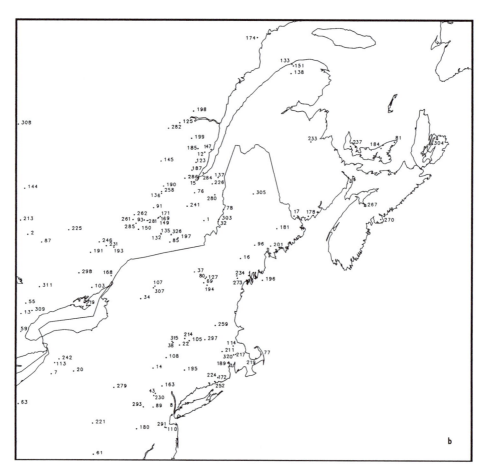

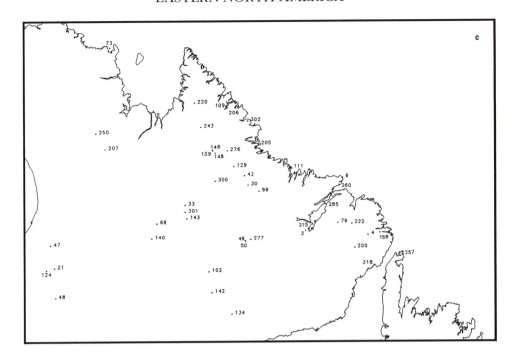

Table 17.1. Sites with fossil pollen data in eastern North America (see also Fig. 173)

Site number and name[a]	State or province[b]	Latitude (N)	Longitude (W)	Site type[c]	Elevation (m)	Dating control rank[d] 3	6	9	12	15	18 ka	Source
*1 Albion	PQ	45°40'	71°19'	L	320	2	1	1				Richard, 1977
2 Alderdale	ON	46° 3'	79°12'	B	240	3	3					Terasmae, 1968
*3 Alexander Lake	LB	53°20'	60°35'	L	143	2	1					Jordan, 1975
*4 Alexis Lake	LB	52°31'	57° 2'	L	200	3	2	2				Lamb, 1978
*5 Alfies Lake	ON	47°53'	84°52'	L	288	3	3	1				Saarnisto, 1974
*6 Aliuk Pond	LB	54°35'	57°22'	L	25	3	2					Jordan, 1975
*7 Allenberg	NY	42°15'	78°52'	B	494	6	7	7	7			Miller, 1973
8 Alpine Peat Bog	NJ	40°57'	73°54'	B	30	3	2	1	1			Peteet et al., 1990
9 Anderson Pond	TN	36° 2'	85°30'	L	305	3	2		2	2	1	Delcourt, 1979
10 Antoine	ON	47°44'	84°54'	L	271			3				Saarnisto, 1975
11 Attawapisket	ON	53° 0'	85°10'	B	100	3	3					Terasmae, 1968
*12 Baie St. Paul—Ange	PQ	47°28'	70°41'	L	640	1	1	2				Labelle and Richard, 1981
13 Ballycroy Bog	ON	53°57'	79°52'	B	297	2	5	5				Karrow et al., 1975
14 Balsam Lake	NY	42° 1'	74°36'	L	762	1	1	3	1			Ibe, 1985
15 Base de Plein Air de St. Foy	PQ	46°47'	71°20'	B	16	2	3					A. Larouche and P. Richard, unpublished
16 Basin Pond	ME	44°28'	70° 3'	L	124							Gajewski et al., 1987
*17 Basswood Road Lake	NB	45°15'	67°20'	L	106	2	1	2	1			Mott, 1975
18 Battaglia Bog	OH	41° 8'	81°19'	B	318				1	1		Shane, 1975
19 Bay of Quinte	ON	44° 2'	77° 5'	L	75	2	1					T. W. Anderson, unpublished
*20 Belmont Bog	NY	42°15'	77°55'	B	497	2	4	4	1			Spear and Miller, 1976
*21 Bereziuk	PQ	54° 3'	76° 7'	L	205	2	1					Richard, 1979
*22 Berry Pond	MA	42°30'	73°19'	L	600	1	2	1	2			Whitehead, 1979
*23 Big Pond	PA	39°46'	78°33'	L	634	5	5	5				Watts, 1979
24 Billy's Lake	MN	46°16'	94°33'	L	383	1	1	2				Jacobson, 1979
*25 BL-Tombigbee	AL	33°33'	88°28'	L	49	3	1					Whitehead and Sheehan, 1985
*26 Blackington Lake	ON	47°54'	84°52'	L	261	4	4					Saarnisto, 1975
*27 Blue Mound	WI	43° 5'	89°52'	B	335	2	1	1				Davis, 1977
*28 Bog D	MN	47°11'	95°10'	L	457	1	3	1				McAndrews, 1966
29 Boney Springs (terrace sample)	MO	38° 6'	93°22'	B	61						2	King, 1973
30 Border Beacon	LB	55°20'	63°12'	L	470	1	1					Lamb, 1982

Table 17.1. (continued)

Site number and name[a]	State or province[b]	Latitude (N)	Longitude (W)	Site type[c]	Elevation (m)	Dating control rank[d] 3	6	9	12	15	18 ka	Source
31 Boriack Bog (core A)	TX	30°21'	97° 7'	B	143	2	3	3	2	1		Bryant, 1977
*32 Boundary	ME	45°34'	70°41'	L	603	1	1	2				Mott, 1977
*33 Boundary Lake	PQ	55°15'	67°24'	L	525	1	1					Stravers, 1981
*34 Brandreth Lake	NY	43°55'	74°41'	B	583	2	1	1				Overpeck, 1985
*35 Brown's Lake Bog	OH	40°41'	82° 3'	B	290	1	2	1				J. G. Ogden III, unpublished
36 Buckle's Bog	MD	39°34'	79°16'	B	814	6			1	2	1	Maxwell and Davis, 1972
37 Bugbee Bog	VT	44°22'	72° 9'	B	398	1	2	3				McDowell et al., 1971
*38 Burden Lake	NY	42°36'	73°34'	L	192	1	1					Gaudreau, 1986
*39 Cahaba Pond	AL	33°34'	86°31'	L	204	1	2	1	1			Delcourt et al., 1983
40 Camel Lake	FL	30°16'	85° 1'	L	20	6	6	6	5	5	6	Watts et al., 1992
*41 Camp 11 Lake	MI	46°40'	88° 1'	L	549	1	1	1				Brubaker, 1975
42 Caribou Hill	LB	55°40'	63°15'	L	475	1	1					Lamb, 1982
43 Cedar Bog	NJ	41°20'	74°40'	B	137	6	7	7				Niering, 1953
*44 Charles Lake	ON	44°44'	81° 1'	L	219	4	4	3				R. E. Bailey, unpublished
*45 Chatsworth Bog	IL	40°40'	88°20'	B	219	2	1	1	2			King, 1981
*46 Chippewa Bog	MI	43° 7'	83°15'	B	262	1	2	1				Bailey and Ahearn, 1981
*47 Chism-I	PQ	54°48'	76° 9'	L	340	3	3					Richard, 1979
*48 Chism-II	PQ	53° 5'	76°19'	L	273	3	3					Richard, 1979
49 Churchill Falls North	PQ	53°36'	64°19'	B	398	3						Morrison, 1970
50 Churchill Falls South	PQ	53°35'	64°18'	B	398	3						Morrison, 1970
*51 Clear Lake	IN	41°39'	86°32'	L	244	6	3	6	2			Bailey, 1972
52 Clearwater	SK	50°52'	107°56'	L	686	3	4					Mott, 1973
53 Colorado Marsh	IA	42° 1'	93°16'	B	329	2	1	1	3			Kim, 1986
54 Conklin Quarry	IA	41°41'	91°33'	L	213						3	Baker et al., 1986
55 Cookstown Bog	ON	44°13'	79°37'	B	234		4	5				Karrow et al., 1975
56 Coppermine Saddleback	NT	67°50'	115°19'	B	43	1						Nichols, 1975
*57 Cranberry Glades	WV	38°12'	80°17'	B	1029	2	1					Watts, 1979
58 Crates Lake	ON	49°11'	81°16'	L	259	2	1	3				Liu, 1982
59 Crawford Bog	ON	43°28'	79°57'	B	279	6	7	7				J. H. McAndrews, unpublished
60 Crewsell Bay	NT	72°52'	93°37'	B	160	2	1					S. K. Short, unpublished
*61 Crider's Pond	PA	39°58'	77°33'	L	290	6	5	6	1	1		Watts, 1979
62 Crozier	ON	47°54'	84°52'	L	223			5				Saarnisto, 1975
*63 Crystal Lake	PA	41°33'	80°22'	L	260	6	4	3				Walker and Hartman, 1960
*64 Crystal Lake	MI	43°15'	84°55'	L	313	1	6	7				R. O. Kapp and D. Lay, unpublished
65 Cub Lake	MI	44°42'	84°57'	L	370	1	2	6				J. Rasmussen and R. E. Bailey, unpublished
66 Cupola Pond	MO	36°48'	91° 6'	L	244	3	4	2	1	1		Smith, 1984
*67 Cycloid Lake	SK	55°16'	105°16'	L	369	3	1					Mott, 1973
*68 Daumont Lake	PQ	54°52'	69°24'	L	600	1	3					Richard et al., 1982
*69 Deer Lake Bog	NH	44° 2'	71°50'	B	1325	2	1	1	1			Spear, 1989
70 Demont Lake	MI	43°29'	85°0'	L	248	2	2	1				Kapp, 1977
71 Devils Lake	ND	48° 5'	99°55'	L	448	3	3					J. H. McAndrews, unpublished
72 Devils Lake	WI	43°25'	89°43'	L	294	1	1	1	1			Maher, 1982
*73 Diana Island	PQ	60°59'	69°57'	L	110	2	2					Richard, 1977
*74 Dismal Swamp (core 1)	VA	36°35'	76°27'	B	6	2	3	3				Whitehead, 1972
*75 Disterhaft Farm Bog	WI	43°55'	89°10'	B	329	2	2	1	1			Baker, 1970 (Webb and Bryson, 1972)
76 Dosquet	PQ	46°27'	71°30'	B	140	4	4	3				Richard, 1977
77 Duck Pond (core b)	MA	41°50'	70° 0'	L	2	1	2	1	3			Winkler, 1985
*78 Dufresne	PQ	45°51'	70°21'	L	650	1	2	1				Mott, 1977
*79 Eagle Lake	LB	53°14'	58°33'	L	400	1	1	2				Lamb, 1980
*80 Eagle Lake Bog	NH	44°10'	71°40'	B	1275	2	2	1				Spear, 1989
81 East Baltic Bog	PEI	46°26'	62° 7'	B	45	2	1					Anderson, 1980
*82 Edward Lake	ON	44°22'	80°15'	L	518	2	1	1				McAndrews, 1981
83 Ennadai	NWT	61°10'	100°55'	B	168	1						Nichols, 1967
84 Ennadai 72	NWT	61°14'	100°57'	B	325	1						Nichols, 1975
85 Farnham Bog	PQ	45°17'	72°59'	B	55	2	1					P. J. H. Richard, unpublished

Table 17.1. (continued)

Site number and name[a]	State or province[b]	Latitude (N)	Longitude (W)	Site type[c]	Elevation (m)	Dating control rank[d]						Source
						3	6	9	12	15	18 ka	
86 Ferndale Bog (core IV)	OK	34°24'	95°48'	B	263	2						Albert and Wyckoff, 1981
*87 Found Lake	ON	45°48'	78°38'	L	488	2	1	1				McAndrews, 1981
*88 Frains Lake	MI	42°20'	83°38'	L	271	1	1	1	2			Kerfoot, 1974
89 Francis Lake II	NJ	40°59'	74°51'	L	189	6	5	6	1	1	1	Cotter, 1984
90 French Lake	MN	44°57'	94°25'	L	327	1						Grimm, 1983
*91 Gabriel	PQ	46°16'	73°28'	L	250	1	1					Richard, 1977
92 Gass Lake	WI	44° 3'	87°44'	L	211	2	1	2	3			Webb, 1983
*93 Gea-I—St. Hippolyte	PQ	45°59'	73°59'	L	365	6	4	5				P. J. H. Richard, unpublished
*94 Glenboro Lake	MB	49°26'	99°17'	L	450	1	3	3	3			Ritchie and Lichti-Federovich, 1968
*95 Goshen Springs	AL	31°43'	86° 8'	L	105	2	2	3				Delcourt, 1980
96 Gould Pond	ME	44°44'	69°19'	L	89	1	1	1	1			Jacobson et al., 1987
*97 Grand Rapids	MB	53° 2'	99°43'	L	350	2	1					Ritchie and Hadden, 1975
98 Gravel Ridge	LB	55° 2'	62°38'	L	559	1	1					Lamb, 1982
*99 Green Lake	MI	44°53'	85° 7'	L	277	2	4	4	2			Lawrenz, 1975
100 Green Lake	WI	45°10'	88°27'	L	305	2	2	1				A. M. Swain, unpublished
*101 Hack Pond	VA	37°59'	79° 0'	L	451	6	4	3	2			Craig, 1969
102 Harrie Lake	LB	52°56'	66°57'	L	534	1	1					King, 1987
103 Harrowsmith	ON	44°25'	76°42'	B	145	6	5	5				Terasmae, 1968
104 Hawke Hills Kettle	NF	47°19'	53° 8'	L	220	2	1					MacPherson, 1982
*105 Hawley Bog Pond	MA	42°34'	72°53'	L	549	5	4	1	1			W. A. Patterson, unpublished
*106 Hayes Lake	ON	49°35'	93°45'	L	391	2	1	2				McAndrews, 1982
*107 Heart Lake	NY	44°11'	73°58'	L	664	1	1	1				Whitehead and Jackson, 1990
108 Heart's Content Bog	NY	42°14'	73°58'	B	107	6	7	7	7			Ibe, 1982
109 Hebron Lake	LB	58°12'	63° 2'	L	170	3	1					Lamb, 1982
*110 Helmetta Bog	NJ	40°23'	74°26'	B	15	6	4	3				Watts, 1979
111 Hopedale Pond	LB	55°28'	60°17'	L	76	1						Short and Nichols, 1977
*112 Horseshoe Lake	MN	45°27'	93° 3'	L	331	1	1	1				E. J. Cushing, unpublished
*113 Houghton Bog	NY	42°32'	78°40'	L	428	6	5	6				Miller, 1973
*114 Houghton Pond	MA	42°12'	71° 5'	L	48	2	2	3				P. C. Newby, unpublished
*115 Hudson Lake	IN	41°40'	86°32'	L	239	1	3	4	3			Bailey, 1972
*116 Iglutalik Lake	BF	66° 8'	66° 5'	L	90	1	1					Davis, 1980
117 Indian Lake	ON	50°55'	90°27'	L	383	3	3	1				Björck, 1985
118 Jack Lake	ON	47°19'	81°46'	L	430	2	2	1				Liu and Lam, 1985
119 Jackson Pond	KY	37°24'	85°42'	L	212	6	5	3	3	3	2	Wilkins, 1985
*120 Jacobson Lake	MN	46°25'	92°43'	L	324	2	2	1				Wright and Watts, 1969
121 Jewell site	IA	42°15'	93°42'	B	317	3	1	1	2			Kim, 1986
*122 Jock Lake	ON	48°41'	86°27'	L	290	1	1					Saarnisto, 1975
*123 Joncas Lake	PQ	47°15'	71°10'	L	780	4	4					P. J. H. Richard, unpublished
*124 Kanaaupscow	PQ	54° 1'	76°38'	L	200	3	3					Richard, 1979
125 Kenogami	PQ	48°22'	71°34'	B	166	2	1					Richard, 1977
126 Kincardine Bog	ON	44° 9'	81°39'	B	198	4	4	1				Karrow et al., 1975
*127 Kinsman Pond	NH	44° 8'	71°44'	L	1140	4	4	3				Spear, 1989
*128 Kirchner Marsh	MN	44°46'	93° 7'	B	254	2	1	1	1			Wright et al., 1963
*129 Kogaluk Plateau Lake	LB	56° 4'	63°45'	L	530	1	2					Short and Nichols, 1977
*130 Kotiranta	MN	46°43'	92°37'	L	386	6	5	6	1	1		Wright and Watts, 1969
131 Kylen Lake	MN	47°20'	91°48'	L	485				1	1	1	Birks, 1981
*132 Lac à la Tortue	PQ	45°32'	73°19'	L	137	2	2	2				Gauthier, 1981
133 Lac à Leonard, Mont St.-Pierre	PQ	49°12'	65°48'	L	17	2	1	1				Labelle and Richard, 1984
134 Lac au Sable	PQ	51°24'	66°13'	L	534	2	1					King, 1987
*135 Lac Bouleaux	PQ	45°33'	73°19'	L	126	6	7	7				R. J. Mott, unpublished
*136 Lac Castor	PQ	46°36'	72°59'	L	220	1	1	1				P. J. H. Richard, unpublished
*137 Lac Colin	PQ	46°43'	70°18'	L	658	1	1	1				Mott, 1977
138 Lac Cote	PQ	48°58'	65°57'	L	96	6	4	3				P. J. H. Richard, unpublished

Table 17.1. (continued)

Site number and name[a]	State or province[b]	Latitude (N)	Longitude (W)	Site type[c]	Elevation (m)	Dating control rank[d]						Source
						3	6	9	12	15	18 ka	
139 Lac de Roches Moutonnees	PQ	56°46'	64°49'	L	410	1						Samson and McAndrews, 1977
*140 Lac Delorme II	PQ	54°25'	69°55'	L	538	1	1					Richard et al., 1982
*141 Lac des Atocas	PQ	45°32'	73°18'	L	120	2	1	1				Gauthier, 1981
142 Lac Gras	PQ	52°15'	67° 4'	L	564	1	2					King, 1987
*143 Lac Hamard	PQ	54°48'	67°30'	L	564	2	2					Stravers, 1981
*144 Lac Louis	PQ	47°17'	79° 7'	L	300	2	1	1				Vincent, 1973
*145 Lac Martini	PQ	47°28'	72°45'	L	242	2	1	3				P. J. H. Richard, unpublished
*146 Lac Martyne	PQ	56°47'	64°50'	L	365	1	1					J. H. McAndrews, unpublished
*147 Lac Mimi	PQ	47°30'	70°22'	L	411	2	4	1				Richard, 1977
148 Lac Patricia	PQ	56°40'	64°42'	L	540	2						J. H. McAndrews, unpublished
*149 Lac Romer	PQ	45°58'	73°20'	L	20	2	1					Comtois, 1982
*150 Lac Tania	PQ	45°46'	74°18'	L	305	1	1	1				P. J. H. Richard, unpublished
151 Lac Turcotte	PQ	49° 9'	65°45'	L	457	2	1	1				Labelle and Richard, 1984
*152 Lac Yelle	PQ	48°30'	79°38'	L	355	2	2	3				Richard, 1980
153 Lake 27	MI	45° 4'	84°47'	L	378	1						Bernabo, 1981
154 Lake A	SK	53°14'	105°44'	L	440	6	5	6				Mott, 1973
155 Lake Annie	FL	27°12'	81°21'	L	37	1	4	4	1			Watts, 1975a
156 Lake B	SK	53°48'	106° 4'	L	553	6	5	5				Mott, 1973
*157 Lake E	MB	50°43'	99°39'	L	724	2	1	2				Ritchie, 1969
158 Lake Hope Simpson	LB	52°27'	56°20'	L	295	2	1	1				Engstrom and Hansen, 1985
*159 Lake Louise	GA	30°43'	83°15'	L	61	3	4	3				Watts, 1971
*160 Lake Mary	WI	46°15'	89°54'	L	488	2	3	1				Webb, 1974
161 Lake Mendota	WI	43° 5'	89°25'	L	257	1	2	1	4			Winkler et al., 1986
162 Lake of the Clouds	MN	48° 9'	91° 7'	L	453	1	1	1				Craig, 1972
*163 Lake Rogerine	NJ	41°30'	74°20'	L	137	6	4	3				Nicholas, 1968
164 Lake Six	ON	48°24'	81°19'	L	305	2	1					Liu, 1982
165 Lake Sixteen	MI	45°36'	84°19'	L	216	1	1	1				Futyma and Miller, 1986
166 Lake Tulane	FL	27°35'	81°30'	L	36	2	1	1	1	2	2	Watts and Hansen, 1988
*167 Lake West Okoboji	IA	43°22'	95°11'	L	415	1	1	2	1			Van Zant, 1979
168 Lambs Pond	ON	44°39'	75°48'	L	107			1	1			T. W. Anderson, unpublished
*169 Lanoraie, St. Henri Bog	PQ	45°59'	73°18'	B	18	1	1					Comtois, 1982
*170 Lanoraie, St. Joseph Bog	PQ	45°59'	73°18'	B	18	1						Comtois, 1982
171 Lanoraie, St. Jean Bog	PQ	46° 0'	73°13'	B	20	1						Comtois, 1982
*172 Lantern Hill Pond	CT	41°27'	71°57'	L	36	1	1	2				K. M. Trent, unpublished
173 Last Lake	ON	47°53'	84°52'	L	265		1					Saarnisto, 1975
*174 "LD" Lake	PQ	50° 8'	67° 7'	L	122	2	2					Mott, 1976
175 Lily Lake	MN	43° 3'	92°50'	L	258	1	1	1	3			Brugam et al., 1988
176 Lima Bog	WI	42°48'	88° 5'	B	238	2	2	1	4			Baker et al., 1992
*177 Little Bass Lake	MN	47°17'	93°36'	L	391	1	1	1				Swain, 1979
*178 Little Lake	NB	45° 9'	66°43'	L	64	3	1	2	3			Mott, 1975
179 Long Lake	NWT	62°38'	101°14'	L	259	2						Kay, 1979
180 Longswamp	PA	40°29'	75°40'	B	192			3	1			Watts, 1979
181 Loon Pond	ME	45° 2'	68°12'	L	110	1	1	1	1			G. L. Jacobson, Jr., unpublished
*182 Lost Lake	MI	46°43'	87°58'	L	500	2	3	3				Brubaker, 1975
*183 Lynn Lake	MB	56°50'	101° 3'	B	340	2	1					Nichols, 1967
*184 Maclaughlin Pond	PEI	46°23'	62°47'	L	24	1	2	1				T. W. Anderson, unpublished
*185 Malbaie	PQ	47°36'	70°58'	B	800	2	3					Richard, 1977
*186 Maplehurst	ON	43°13'	80°39'	L	300	3	1	1	2			Mott and Farley-Gill, 1978
*187 Marcotte	PQ	47° 4'	71°25'	L	503	2	1	1				Labelle and Richard, 1981
*188 Martin Pond	MN	47°11'	94°56'	L	429	6	7	7				McAndrews, 1966
189 Mashapaug Pond	RI	41°47'	71°26'	L	12	2						Bernabo, 1977
*190 Mauricie	PQ	46°47'	72°50'	L	270	3	2	1				Richard, 1977
191 McLaughlan Lake	ON	45°21'	76°33'	L	196		5	5				T. W. Anderson, unpublished

Table 17.1. (continued)

Site number and name[a]	State or province[b]	Latitude (N)	Longitude (W)	Site type[c]	Elevation (m)	Dating control rank[d]						Source
						3	6	9	12	15	18 ka	
192 Medicine Lake	SD	44°49'	97°21'	L	519	1	2	1				Radle, 1981
193 Mer Bleue	ON	45°24'	75°30'	B	69	3	4					Mott and Camfield, 1969
194 Mirror Lake	NH	43°57'	71°42'	L	258	2	1	2				Likens and Davis, 1975
195 Mohawk Pond	CT	41°49'	73°17'	L	360	1	1	1	2			Gaudreau, 1986
*196 Monhegan Island Meadow	ME	43°46'	69°18'	B	3	1	1	1				Bostwick, 1978
*197 Mont Shefford	PQ	45°21'	72°35'	B	282	1	1	2				Richard, 1977
*198 Mont Valin	PQ	48°36'	70°50'	L	891	2	1	1				P. J. H. Richard, unpublished
199 Montagnais	PQ	47°54'	71°10'	B	800	4	4	3				Richard, 1977
200 Moraine Lake	LB	52°16'	58° 3'	L	0	1	1	2				Engstrom and Hansen, 1985
*201 Moulton Pond	ME	44°37'	68°38'	L	143	1	1	1	1			Davis et al., 1975
202 Mud Lake	FL	29°18'	81°52'	L	15	2	2	3				Watts, 1969
*203 Muscotah	KS	39°32'	95°31'	B	320		3		3	7	6	Grüger, 1973
*204 Myrtle Lake	MN	47°59'	93°23'	L	393	1	1	1				Janssen, 1968
*205 Nain Pond	LB	56°32'	61°49'	L	80	3	2	3				Short and Nichols, 1977
*206 Napaktok Lake	LB	57°55'	62°34'	L	143	1	3					S. K. Short, unpublished
207 Nedlouc	PQ	57°39'	71°39'	L	330	1						Richard, 1981
*208 Nelson Pond	MN	46°24'	92°41'	L	335		1					Jacobson, 1979
209 Nina Lake	ON	46°36'	81°30'	L	380	2	2	1				Liu, 1982
210 Ninepin 24	MD	38°18'	75°17'	B	15						2	Sirkin et al., 1977
211 Nipmuck Pond	MA	42° 7'	71°35'	L	121	1	2	1	1			P. Tzedakis, unpublished
212 Nonconnah Creek-1	TN	35° 5'	89°55'	B	79					2	2	Delcourt et al., 1980
*213 North Bay	ON	46°27'	79°28'	B	369	6	4	3				Terasmae, 1968
214 North Pond	MA	42°39'	73° 3'	L	586	1	1	1				Whitehead and Crisman, 1978
215 Northwest River Pond	LB	53°31'	60°10'	L	29	2						Jordan, 1975
216 Nungesser	ON	51°30'	93°24'	B	116		4	3				Terasmae, 1968
*217 Nunkets Pond	MA	41°58'	71° 3'	L	18	2	2	2				R. H. W. Bradshaw, unpublished
*218 Old Field	MO	37° 7'	89°50'	B	98		1	3				King and Allen, 1979
219 Owl Pond	MA	41°45'	70°15'	L	7	1	2	2				Tzedakis, 1992
*220 Palsa Lake	PQ	58°28'	65°10'	L	143	1						S. K. Short, unpublished
*221 Panther Run Pond	PA	40°48'	77°25'	L	634	2	3		3			Watts, 1979
*222 Paradise Lake	LB	53° 3'	57°45'	L	180	3	2	2				Lamb, 1980
*223 Patricia Bay Lake	BF	70°28'	68°30'	L	11	2	1					Mode, 1980
224 Pequot Cedar Swamp	CT	41°27'	71°57'	B	37				1	1		Webb, 1990
*225 Perch Lake	ON	46° 2'	77°22'	L	160	3	4	3				J. H. McAndrews, unpublished
*226 Petit Lac Terrien	PQ	46°35'	70°37'	L	404	6	6	7	3			Mott, 1977
*227 Pickerel Lake	SD	45°30'	97°20'	L	395	1	2	3				Watts and Bright, 1968
228 Pigeon Marsh	GA	34°40'	85°10'	B	630	6	5	5		5	6	Watts, 1975b
229 Pilot Mound Site	IA	42° 9'	93°54'	B	352	1	2	4	7			Kim, 1986
230 Pine Bog	NJ	41°17'	74°45'	B	137	6	7	7	7			Niering, 1953
*231 Pink Lake	PQ	45°28'	75°49'	L	162	2	4	4				Mott and Farley-Gill, 1981
232 Pogonia Bog Pond	MN	45° 2'	93°38'	L	292	1	2	2				Swain, 1979
233 Point Escuminac	NB	47° 4'	65°49'	B	76	1	1	1				Warner et al., 1991
*234 Poland Spring Pond	ME	44° 2'	70°21'	L	94	1	1	1	1			G. L. Jacobson, unpublished
*235 Pond Mills Pond	ON	42°55'	81°15'	L	274	6	4	5				McAndrews, 1981
*236 Portage	MN	47°12'	94° 9'	L	396	2	2	1				J. H. McAndrews, unpublished
*237 Portage Bog	PEI	46°40'	64° 6'	B	8	2	4					Anderson, 1980
*238 Potts Mountain Pond	VA	37°36'	80° 8'	L	840	4	4	3				Watts, 1979
*239 Pretty Lake	IN	41°35'	85°15'	L	294	1	1	1	1			Williams, 1974
240 Prince	ON	46°34'	84°33'	L	290			1				Saarnisto, 1974
*241 Princeville	PQ	46° 8'	71°56'	L	135			7	7			Richard, 1977
242 Protection Bog	NY	42°37'	78°28'	B	430	1	2	3	7			Miller, 1973
*243 Pyramid Hills Lake	LB	57°38'	65°10'	L	381	1	1					Short and Nichols, 1977
*244 Quicksand	GA	34°19'	84°52'	L	285	6	6	7	5	4	4	Watts, 1970
245 Radtke Lake	WI	43°24'	88° 6'	L	274	1	1	1				Webb, 1983
*246 Ramsay Lake	PQ	45°36'	76° 6'	L	200	2	2	2				Mott and Farley-Gill, 1981
247 Rattle Lake	ON	49°21'	92°42'	L	460	3	3	1				Björck, 1985

Table 17.1. (continued)

Site number and name[a]	State or province[b]	Latitude (N)	Longitude (W)	Site type[c]	Elevation (m)	Dating control rank[d] 3	6	9	12	15	18 ka	Source
248 Rayburn Salt Dome	LA	32°28'	93°10'	B	61					3	7	Kolb and Fredlund, 1981
*249 Riley Lake	ON	54°19'	84°33'	L	142	2	2					McAndrews et al., 1982
250 Rivière and Feuilles	PQ	58°14'	72° 4'	L	200	2						Richard, 1981
251 Rockyhock Bay	NC	36°10'	76°41'	B	6	4	4	2	1	1	1	Whitehead, 1981
*252 Rogers Lake	CT	41°22'	72° 7'	L	91	1	1	1	1	1		Davis, 1969
253 Rossburg Bog	MN	46°35'	93°36'	B	372	1	2	1				Wright and Watts, 1969
*254 Round Lake	IN	41°14'	86°38'	L	216	2	3	4				R. E. Bailey, unpublished
*255 Rutz Lake	MN	44°52'	93°52'	L	314	1	1	1	1			Waddington, 1969
*256 Ryerse Lake	MI	46° 7'	85°10'	L	259	1	1					Futyma, 1982
257 Saddle Hill Pond	NF	51°30'	55°31'	L		2	2	2				A. M. Davis and J. H. McAndrews, unpublished
*258 Sam	PQ	46°39'	72°58'	L	240	1	2	1				P. J. H. Richard, unpublished
259 Sandogardy Pond	NH	42°50'	71°40'	L	100	1	2	3	2			Davis, 1978
260 Sandy Cove Pond	LB	54°24'	57°43'	L	100	1						Jordan, 1975
*261 Sav-I—Ste. Agathe	PQ	46° 3'	74°28'	L	454	2	2	4				Savoie and Richard, 1979
*262 Sav-II—Lac aux Quenoilles	PQ	46°10'	74°23'	L	403	4	4	1				Savoie and Richard, 1979
263 Scott Lake	FL	27°58'	81°57'	L	51	2						Watts, 1971
*264 Second Lake	ON	44°50'	79°59'	L	196	2	5	7				J. H. McAndrews, unpublished
265 Seidel 1	WI	44°27'	87°31'	L	219			1	5			J. C. B. Waddington, unpublished
266 Shady Valley Peat	TN	36°31'	81°56'	B	383	4	4	3				Barclay, 1957
*267 Shaw's Bog	NS	45° 1'	64°11'	B	30	2	1	1				Hadden, 1975
268 Sheelar Lake	FL	29°31'	82° 0'	L	51	4	4	1	1		3	Watts and Stuiver, 1980
*269 Shouldice Lake	ON	45° 9'	81°25'	L	177	4	4	1				J. H. McAndrews, unpublished
270 Silver Lake	NS	44°33'	63°38'	L	69	2	1	1				Livingstone, 1968
*271 Silver Lake	OH	40°26'	83°40'	L	341	1	1	2				Ogden, 1966
272 Singletary Lake	NC	34°30'	78°30'	L	18		3	2	4	2	5	Whitehead, 1967
*273 Sinkhole Pond	ME	43°58'	70°21'	L	95	1	1		1			G. L. Jacobson, Jr., unpublished
274 Sioux Lake	ON	49°56'	91°34'	L	410	3	1	2				Björck, 1985
275 Slow River	NWT	63° 2'	100°45'	B	265	2						Kay, 1979
276 Snow Lake	LB	56°38'	63°53'	L	535	1						Lamb, 1982
277 Sona Lake West	PQ	53°35'	63°57'	B	429	3						Morrison, 1970
278 South Burin Peninsula	NF	46°55'	55°36'	L	114				1			Anderson, 1983
279 Spring Lake	PA	41°40'	76°21'	L	342	4	4	3	1			Barnosky et al., 1988
*280 St. Benjamin	PQ	46°17'	70°36'	L	330	6	5	3				Richard, 1977
*281 St. Calixte	PQ	45°57'	73°52'	L	261	2	2	1				P. J. H. Richard, unpublished
*282 St. Francois de Sales	PQ	48°17'	72° 8'	L	358	3	2	1				P. J. H. Richard, unpublished
*283 St. Germain	PQ	45°56'	74°22'	L	473	2	1	1				Savoie and Richard, 1979
*284 St. Jean, Ile d'Orleans	PQ	46°56'	70°56'	B	68	1	2					Richard, 1977
*285 St. John's Island Pond	LB	53°57'	58°55'	L	137	3	3					Jordan, 1975
*286 St. Raymond	PQ	46°53'	71°48'	B	160	2	2					Richard, 1977
*287 Stewart's Dark Lake	WI	45°18'	91°27'	L	335	1	1	1				Peters and Webb, 1979
288 Stotzel-Leis site	OH	40°13'	84°41'	B	320		1	1	1	2		Shane, 1987
289 Sugarloaf Pond	NF	47°37'	52°40'	L	100	4	4	1				MacPherson, 1982
290 Swan Lake	NE	41°43'	102°30'	L	1161	2	3	3				Wright et al., 1985
*291 Szabo Pond	NJ	40°24'	74°29'	L	29	4						Watts, 1979
292 Tamarack Creek	WI	44° 9'	91°27'	B	244	1						Davis, 1979
293 Tannersville Bog	PA	41° 2'	75°16'	B	277	2	2	1	1			Watts, 1979
*294 Terhell Pond	MN	47°12'	95°47'	L	442	2	5	7				McAndrews, 1966
*295 Thompson	MN	47°12'	96° 5'	L	370	5	7	7				McAndrews, 1966
296 Thompson Landing	NWT	63° 4'	110°47'	L	180	1	1					Nichols, 1975
297 Tom Swamp	MA	42°31'	72°13'	B	229			1	1			Gaudreau, 1986
*298 Tonawa Lake	ON	44°51'	77°10'	L	305	2	2	2				J. H. McAndrews, unpublished
299 Torrens Bog	OH	40°21'	82°28'	B	302	1	1	1				Ogden and Hay, 1967
300 Track Lake	LB	55°46'	65°10'	L	440	1						Short and Nichols, 1977

Table 17.1. (continued)

Site number and name[a]	State or province[b]	Latitude (N)	Longitude (W)	Site type[c]	Elevation (m)	Dating control rank[d] 3	6	9	12	15	18 ka	Source
*301 Tunturi Lake	PQ	55° 1'	67°30'	L	610	2	2					Stravers, 1981
*302 Ublik Lake	LB	57°23'	62° 3'	L	122	1	1	2				Short and Nichols, 1977
303 Unknown Lake	ME	45°37'	70°38'	L	489	2	5	4				Mott, 1977
304 Upper Gillies Lake	NS	46° 4'	60°30'	L	81	6	7	7				Livingstone, 1968
*305 Upper South Branch Pond	ME	46° 5'	68°54'	L	300	2	2	2				Anderson, 1979
306 Upper Twin	ON	46°33'	84°35'	L	302			1				Saarnisto, 1974
*307 Upper Wallface Pond	NY	44° 9'	74° 3'	L	945	2	2	2	1			Whitehead and Jackson, 1990
*308 Val St. Gilles Bog	PQ	49° 1'	79° 5'	B	290	3	1					Terasmae and Anderson, 1970
*309 Van Nostrand Lake	ON	44° 0'	79°23'	L	297	2	1	1				McAndrews, 1970
*310 Vestaburg	MI	43°25'	84°53'	L	255	3	2	2				Gilliam et al, 1967
*311 Victoria Road Bog	ON	44°37'	78°57'	B	274	6	4	6				Terasmae, 1968
312 Volo Bog	IL	42°21'	88°11'	B	229	1	1	2				King, 1981
313 Walters Puddle	DE	39°24'	75°41'	L	17	3			2			P. C. Newby, unpublished
*314 Weber Lake	MN	47°28'	91°40'	L	567	4	4	1	2	3		Fries, 1962
315 West Sand Lake Peat Bog	NY	42°38'	73°36'	B	170		4	3	6			Gaudreau, 1986
316 White Lily	MN	46° 5'	93° 6'	L	344			7	7			Cushing, 1967
*317 White Pond	SC	34°10'	80°46'	L	90	6	4	3	2	4	4	Watts, 1980a
*318 Whitney's Gulch	LB	51°31'	57°18'	L	98	1	2	1				Lamb, 1980
*319 Willow River Pond	MN	46°20'	92°47'	L	314		1					Jacobson, 1979
320 Winneconnet Pond	MA	41°58'	71° 8'	L	20	1	1	1	1			Suter, 1985
*321 Wintergreen Lake	MI	42°24'	85°23'	L	283	1	2	1	1			Manny et al, 1978
322 Wolf Creek	MN	46° 7'	94° 7'	L	375			3	1	3	3	Birks, 1976
323 Wolsfeld Lake	MN	45° 0'	93°34'	L	291	1	1	1				Grimm, 1983
*324 Wolverine Lake	MI	46°25'	85°39'	L	259	2	1	1				Futyma, 1982
*325 Wood Lake	WI	45°20'	90° 5'	L	464	1	1	1	1			Heide, 1984
*326 Yamaska	PQ	45°28'	72°52'	L	265	1	1	1				P. J. H. Richard, unpublished
*327 Yellow Dog Lake	MI	46°45'	87°57'	L	445	1	1	1				Brubaker, 1975
328 Zuehl Farm site	IA	43° 1'	93°52'	B	356	1	4	4	1			Kim, 1986

[a]Asterisks indicate sites used in Webb (1985b).

[b]In addition to states and provinces, LB = Labrador, NWT = Northwest Territories, and BF = Baffin Island.

[c]L = lake, B = bog or fen.

[d]Rank criteria (based on dating information in Table 17.4 and in other sources cited in text and in that table): two bracketing dates (1) both within 2000 yr of the mapped date; (2) one date within 2000 yr and one within 4000 yr; (3) both dates within 4000 yr or just one date within 1000 yr; (4) one date within 4000 yr and one within 6000 yr; (5) both dates within 6000 yr or just one date within 2000 yr; (6) one date within 6000 yr and one within 8000 yr or just one date within 3000 yr; (7) no radiocarbon dates or or both bracketing dates more than 6000 yr older or younger than the mapped date. Uncorrected radiocarbon dates and stratigraphic dates for the top of the core were the main sources of information used in setting the ranks.

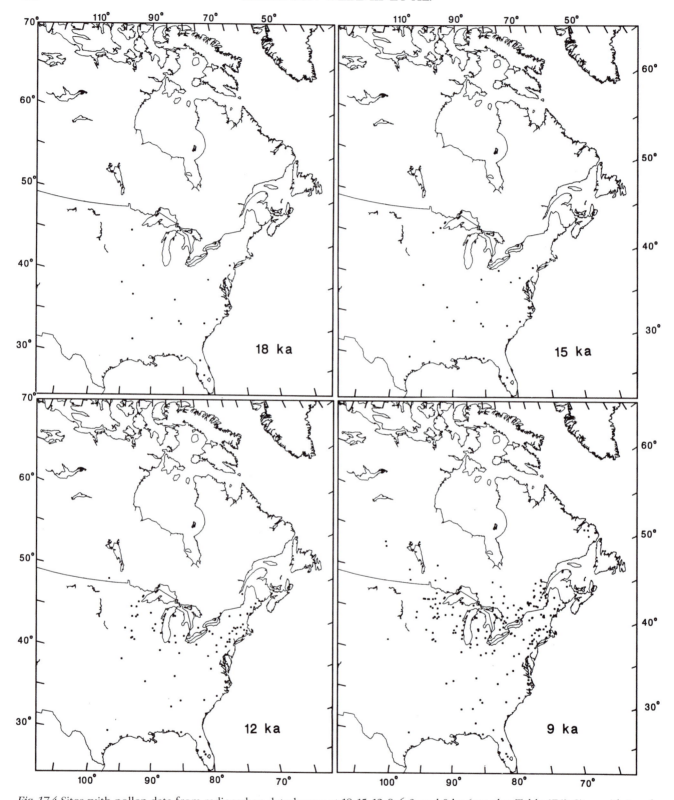

Fig. 17.4. Sites with pollen data from radiocarbon-dated cores at 18, 15, 12, 9, 6, 3, and 0 ka (see also Table 17.1). Sites with "modern" data were used to calculate response surfaces. Table 17.1 gives a ranking of the dating control at each site for each date from 18 to 3 ka.

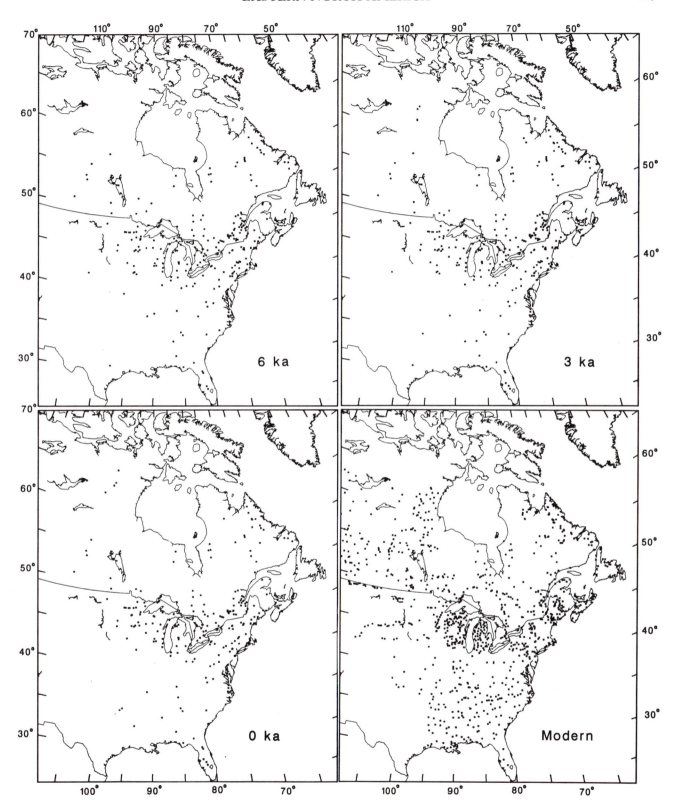

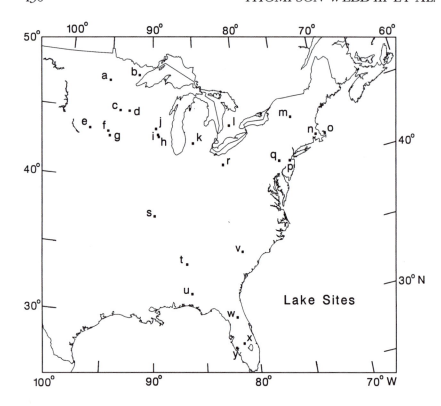

Fig. 17.5. Sites with lake-level data in eastern North America (see also Table 17.2).

Table 17.2. Sites with lake-level data in eastern North America (see also Fig. 17.5)

Basin code and name	State or province	Latitude (N)	Longitude (W)	Elevation (m)	Maximum depth[a,b] (m)	Area[b] (ha)	[14]C dates[c] (no.)	Length of record (yr B.P.)	L	SR	PP	G	A	D	M	O	AR
a Elk Lake	MN	47°13'	95°13'	453	29.60	113.00	0	10,400	X			X		X			
b Weber Lake	MN	47°28'	91°39'	559	2.00	3.00	4	11,000	X				X				
c Rutz Lake	MN	44°52'	93°52'	314	4.00	13.00	8	12,000	X		X		X				
d Kirchner Marsh	MN	44°50'	92°46'	275	S	NA	12	13,000	X	X		X	X	X			
e Pickerel Lake	SD	43°30'	97°20'	395	13.40	387.00	6	11,000	X		X		X	X	X		
f Lake West Okoboji	IA	43°20'	95°12'	425	41.00	NA	14	14,000	X				X	X			
g Kettle Hole Lake	IA	43° 0'	95° 0'	350	NA	NA	4	12,000	X				X				
h Hook Lake Bog	WI	42°57'	89°20'	260	S	100.00	11	13,000	X	X			X				
i Lake Mendota	WI	43° 6'	89°25'	257	24.00	3490.00	14	12,000	X				X				
j Washburn Bog	WI	43°32'	89°39'	248	S	6.00	6	11,000	X	X			X				
k Wintergreen Lake	MI	42°24'	85°23'	271	6.30	14.60	10	13,000	X			X	X	X			
l Sunfish Lake	ON	43°30'	81° 0'	365	20.00	8.30	1	12,000	X				X				
m Lake George	NY	43°31'	73°39'	96	59.40	11,396.00	2	12,000	X				X				
n Titicut Swamp	MA	41°57'	71° 2'	20	S	NA	3	12,000	X								
o Duck Pond	MA	41°56'	70° 0'	3	18.50	5.10	9	12,000	X	X			X				
p Szabo Pond	NJ	40°24'	74°29'	29	1.00	NA	3	12,000	X		X		X				
q Longswamp	PA	40°29'	75°40'	192	S	NA	6	15,000	X				X				
r Brown's Lake	OH	40°41'	82° 4'	340	6.00	2.60	2	17,000	X		X						
s Old Field Swamp	MO	37° 7'	89°50'	97	S	NA	4	9000					X				
t Cahaba Pond	AL	33°30'	86°32'	210	NA	0.20	13	12,000	X	X			X				
u Goshen Springs	AL	31°43'	86° 8'	105	NA	NA	8	26,000	X	X	X		X				
v White Pond	SC	34°10'	80°45'	90	NA	NA	3	19,000	X				X				
w Mud Lake	FL	29°18'	81°52'	15	1.00	NA	5	35,000	X				X	X		X	
x Lake Annie	FL	27°18'	81°24'	36	18.50	0.20	9	44,000	X	X							
y Little Salt Spring	FL	27° 0'	82°10'	5	12.00	0.60	18	12,000	X				X				X

[a]S = swamp.

[b]NA = not available.

[c]0 indicates varve chronology only; 1 indicates pollen correlation.

[d]L = lithology-stratigraphy, SR = sedimentation rates, PP = pollen preservation/reworking, G = geochemistry, A = aquatics, D = diatoms, M = mollusks, O = ostracods, AR = archaeological evidence.

Table 17.3. Adjustments to the pollen sum[a] for selected sites and taxa[b]

Site number and name	Minimum date[c] (yr B.P.)	Maximum date[c] (yr B.P.)	Correction[d]
9 Anderson Pond		<9100	*Alnus* cap 1%, *Acer* cap 4%, *Ostrya-Carpinus* cap 10%
13 Ballycroy Bog		<4200	*Acer* cap 8%
		<4200	Cyperaceae cap 4%
14 Balsam Lake	9040	11,140	*Abies* cap 10%
17 Basswood Road Lake			*Populus* cap 6%
20 Belmont Bog			*Alnus* cap 6%
21 Bereziuk			*Populus* cap 6%
25 BL-Tombigbee			*Alnus* cap 6%
27 Blue Mound		7829	*Alnus* cap 2%
	4500	9580	Cyperaceae cap 10%
	>9580		Cyperaceae cap 6%
31 Boriack Bog (core A)			*Alnus* cap 0%; Cyperaceae cap 2%
34 Brandreth Lake		<1066	*Picea* cap 9%
35 Brown's Lake Bog			*Corylus* cap 2%; *Acer* cap 3%
55 Cookstown Bog			Cyperaceae cap 10%
57 Cranberry Glades			*Alnus* cap 6%
59 Crawford Bog			Cyperaceae cap 5%
61 Crider's Pond			*Abies* cap 12%
63 Crystal Lake			*Alnus* cap 10%
72 Devils Lake (WI)			Delete 584, 588 cm (*Populus* spike)
75 Disterhaft Farm Bog			Cyperaceae cap 10%
93 Gea-I—St. Hippolyte			Cupressaceae cap 15%; *Populus* cap 6%
96 Gould Pond			Delete 980 cm (*Alnus* spike)
108 Heart's Content Bog			*Abies* cap 17%; *Acer* cap 7%; *Alnus* cap 7%; Cyperaceae cap 5%
119 Jackson Pond			Delete 9-ka sample (*Picea* spike)
124 Kanaaupscow			*Populus* cap 10%; Cupressaceae cap 10%
133 Lac à Leonard, Mont St.-Pierre		<3279	Cyperaceae cap 3%; *Alnus* cap 5%
157 Lake E			*Corylus* cap 7%
202 Mud Lake			Forbs cap 8%
203 Muscotah			Cyperaceae cap 2%
206 Napaktok Lake			Delete 140, 145, 150, 160, 165, 170 cm (Cyperaceae spike)
228 Pigeon Marsh			Cyperaceae cap 5%
266 Shady Valley Peat			*Alnus* cap 7%
290 Swan Lake			Cyperaceae cap 2%
292 Tamarack Creek			Cyperaceae cap 5%
293 Tannersville Bog			Delete 980, 998 cm (*Tsuga* spike)
304 Upper Gillies Lake			Delete 480 cm (*Picea, Abies* spike)
317 White Pond			Cyperaceae cap 5%
320 Winneconnet Pond			*Alnus* cap 7%

[a]A pollen sum of 25 types was used for adjustments: *Pinus* (northern and southern), *Betula, Picea, Quercus, Carya,* Cupressaceae, *Alnus,* Cyperaceae, *Tsuga, Ulmus, Abies, Ostrya-Carpinus,* forbs (*Artemisia* + Chenopodiaceae-Amaranthaceae + other Compositae, excluding *Ambrosia*), *Fagus, Acer, Fraxinus, Populus, Corylus, Juglans, Liquidambar, Castanea, Tilia, Celtis,* and *Platanus.*

[b]Adjustments were made to avoid problems of local overrepresentation. In addition to the site-specific corrections listed in the table, the following general correction was made: total Cupressaceae was set to 0 for (1) latitude <40°N and date ≤4.5 ka, (2) latitude <38°N and date >4.5 ka and ≤9.5 ka, (3) latitude <37°N and date >9.5 ka and ≤13.5 ka, (4) latitude <35°N and date >13.5 ka and ≤15.5 ka, and (5) latitude <33°N and date >15.5 ka and ≤18.5 ka.

[c]Minimum and maximum dates indicate upper and lower dates, respectively, in which correction applies. If no dates are listed, then correction was applied to entire core.

[d]Cap is defined as the upper limit that the taxa was allowed to reach for the time period indicated. At some sites, anomalous samples with locally overrepresented taxa were deleted.

(*Acer, Ostrya-Carpinus, Fraxinus,* Cupressaceae, *Populus, Corylus, Celtis, Liquidambar, Castanea, Juglans, Tilia,* and *Platanus*).

At a few sites where exceedingly high values of Cyperaceae, *Alnus,* and *Acer* pollen indicated local growth, we capped their values at 10% or less (Table 17.3). Peak values for *Abies, Ostrya-Carpinus, Populus,* forbs, Cupressaceae, *Picea,* and *Corylus* were also capped at 12 sites, and a few isolated unusual samples were excluded at six sites. We also applied a general correction to exclude the influence of *Taxodium* on the Cupressaceae counts (Table 17.3, footnote b).

The isopoll maps (Fig. 17.6) show the 12 major taxa as percentages of the sum of all 24 pollen types, but we restricted the sum to just the 12 taxa for the response functions and the climate reconstructions. Isopoll maps for these 12 taxa based on a sum of the same 12 taxa closely resemble the maps in Figure 17.6.

Subdivision of Pinus *and* Betula

To facilitate use of the response functions, we subdivided *Pinus* and *Betula,* two bimodally distributed pollen types, into separate unimodal groups. We designated a northern and a southern group for *Pinus* pollen: *Pinus resinosa, P. banksiana, P. strobus,* and *P. rigida* are the dominant species producing northern *Pinus* pollen, and *P. taeda, P. echinata, P. serotina, P. palustris, P. elliottii, P. clausa,* and *P. virginiana* are the dominant species producing southern *Pinus* pollen. We defined northern and southern *Pinus* pollen in the modern data set as all *Pinus* north and south, respectively, of 39°N (Webb *et al.,* 1987). Some of the species in the two groups intermingle in the Appalachians, so our split of these types is somewhat arbitrary. Evidence from pollen morphological studies and plant macrofossils (Whitehead, 1964, 1981; Watts, 1970, 1980b; Delcourt, 1979) and from associated taxa (e.g. *Picea*) indicates that the location of this division in *Pinus* types was near the Gulf Coast at 18 and 15 ka and moved north from 15 to 9 ka. We therefore assumed that all *Pinus* pollen south of 30°N at 18 and 15 ka was southern and all *Pinus* pollen north of 34°30'N and at Rayburn Salt Dome was northern. Between these latitudes we designated a zone with both types of pines. We moved the zone borders north to 30°30' and 36°N at 12 ka, and we assumed that the modern situation prevailed after 9 ka. More research is needed to establish how the abundance of different species of *Pinus* has varied over time.

We subdivided the bimodal distribution of *Betula* pollen today into "shrub" birch north of 55°N in Quebec and Labrador and "tree" birch elsewhere. On the basis of pollen stratigraphy, we interpreted shrub birch pollen as the dominant type at sites in central Quebec north of 50°30'N at 6 ka. We judged that tree birch pollen was dominant at 12 and 9 ka except at a few sites and therefore did not apply the distinction to the data for these dates.

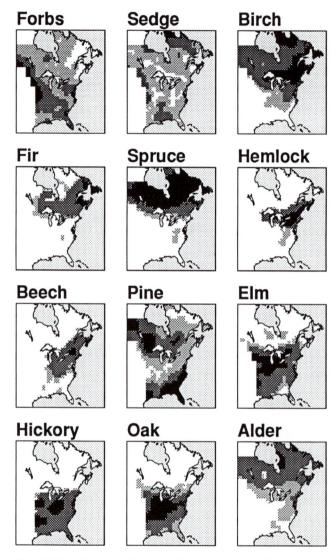

Fig. 17.6. Modern distribution of the 12 major pollen groups used in the analyses. The shading indicates levels of abundance for each pollen type, with black the highest and white the lowest. For forbs, or leafy herbs (*Artemisia* + Chenopodiaceae-Amaranthaceae + other Compositae types, excluding *Ambrosia*), sedges (Cyperaceae), birch (*Betula*), spruce (*Picea*), and alder (*Alnus*), white is <1.0%, light gray 1-5%, dark gray 5-20%, and black >20%. For oak (*Quercus*) and pine (*Pinus*), white is <5%, light gray 5-20%, dark gray 20-40%, and black >40%. For fir (*Abies*), hemlock (*Tsuga*), beech (*Fagus*), elm (*Ulmus*), and hickory (*Carya*), the ranges are <0.5%, 0.5-1%, 1-5%, and >5.0%.

Our experience in using response surfaces for climatic calibration led us to make these subdivisions of the modern and fossil data for *Pinus* and *Betula.* For fossil samples that were dominated by either of these taxa, the dissimilarity measure (squared chord distance) used to assign analogues sometimes gave ambiguous results by assigning southern sites when northern sites were indicated and vice versa. Subdividing the taxa was one way to solve this problem.

Isopoll Maps

We used radiocarbon and stratigraphic dates to estimate the age for each pollen sample at each site, and the pollen percentages from adjacent samples were linearly interpolated to estimate pollen percentages at the time intervals to be mapped. In general, no data were mapped at a site for dates more than 500 yr older than the oldest radiocarbon date at that site (Table 17.4). We analyzed preliminary maps for anomalous values and then examined the data at individual sites to identify the source of the anomalies. We made corrections to dates when such were plausible (documented in Table 17.4 and tables in Webb *et al.* [1983a,b] and Gaudreau and Webb [1985]). Stratigraphic dates were also added when they were justified by the pollen data (Table 17.4). Two commonly used stratigraphic dates were the culturally induced increase in *Ambrosia* pollen percentages (the ragweed rise) and the more or less synchronous decline in *Tsuga* pollen percentages (the hemlock decline) at 4700 ± 200 yr B.P. (Davis, 1981; Webb, 1982; McAndrews, 1988).

The data were machine-contoured by interpolation of the fossil pollen data onto an equal-area grid with a cell size of 10,000 km². First a distance-weighted average of the data was calculated within a window of 200 km centered on each grid point. In regions with few samples, adjacent values were averaged to fill in the grid. Finally, the gridded values were slightly smoothed. The grid used extended farther west (to 109°W) than that used by Webb *et al.* (1987) or Prentice *et al.* (1991). The grid for 18 and 15 ka was more restricted than for the other dates to allow for the smaller number of sites at these earlier dates.

The interpolation procedure resulted in some spatial smoothing of the patterns. The combined smoothing and generalization reduce the effect of small-scale spatial and temporal variability unrelated to climatic controls and also reduce the discrepancy between the coarse spatial scale of the climate model and the generally finer scale of the data. The resulting maps approximate the hand-contoured isopoll maps in Jacobson *et al.* (1987) and Webb (1987, 1988).

The pollen percentages were recoded to four levels: <5, 5–20, 20–40, and >40% for the most abundant taxa (*Quercus* and *Pinus*); <1, 1–5, 5–20, and >20% for the other abundant taxa (forbs, *Alnus*, *Betula*, Cyperaceae, and *Picea*); and <0.5, 0.5–1, 1–5, and >5% for less abundant taxa (*Abies*, *Fagus*, *Tsuga*, *Carya*, and *Ulmus*). For most taxa the contour with the lowest value lies near or outside the range boundary for the plants, the intermediate contour marks areas where the plants are found in low to moderate numbers, and the contour with the highest value identifies the region of high abundance (i.e., the density center) for the taxon. Studies of how the pollen percentages in recently deposited sediments match the contemporary abundances of plants growing near the sediment samples (Webb *et al.*, 1981; Bradshaw and Webb, 1985;

Gaudreau, 1986; Prentice and Webb, 1986; Prentice, 1988; Jackson, 1989) were useful in choosing the contours for each taxon.

Response Surfaces

Response surfaces quantify and display the relationships between individual pollen types and certain climate variables (Bartlein *et al.*, 1986). They illustrate the multivariate nature of this relationship for each pollen type by showing the pollen data as a function of two or more climate variables. Response surfaces thus depict in a specific form the general relationships that can be inferred from isotherm, isohyet, and isopoll maps (Figs. 17.2 and 17.6–17.9). In the terminology of Imbrie and Kipp (1971), response surfaces are "ecological equations," because they estimate the percentages of various pollen types from values for mean July temperature, mean January temperature, and annual precipitation. We choose these three particular climate variables to represent the general control of plant distribution by summer warmth, extreme winter cold, and moisture availability. We recognize that other variables could be used. We assume that linear combinations of these three variables represent most of the variation in the actual climate variables that influence or limit plant growth and reproduction at specific locations. We also recognize that response surfaces are based on the modern correlation between the distribution of pollen and climatic variables and therefore do not necessarily constitute a direct cause-and-effect model for this relationship in the past.

Bartlein *et al.* (1986), Huntley *et al.* (1989), Prentice *et al.* (1991), and P. J. Bartlein, T. Webb III, K. H. Anderson, R. S. Webb, and B. Lipsitz (unpublished) describe the methods used to calculate the response surfaces. The fitting method we used approximates the smoothing of a multidimensional scatter diagram by the robust, locally weighted regression method (LOWESS) (see Cleveland and Devlin, 1988). This approach involves moving a window over the climate space defined by the three climate variables and calculating the expected value of a pollen type at the center of the window (Fig. 17.7). The resulting surface can then be depicted by interpolation among these "fitted values."

For display purposes, we present a set of response surfaces that illustrate the abundance of the different pollen types as a function of two climate variables at a time. Response surfaces depicting the relative abundance of the pollen types as a function of all three climate variables can be found in Prentice *et al.* (1991).

Climatic Calibration of Pollen Data: Inferred Climate Estimates

In previous work, transfer functions or calibration functions derived by multiple regression and other multivariate procedures were used to interpret pollen data in climatic terms (Webb and Bryson, 1972; Webb

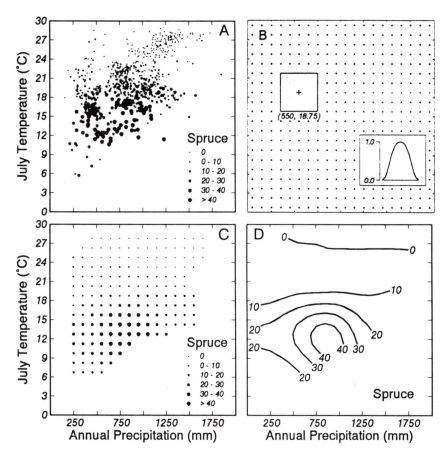

Fig. 17.7. (A) Percentage of spruce (*Picea*) pollen at individual sites plotted in climate space along axes for mean July temperature and annual precipitation. (B) Grid laid over the climate data to which the pollen percentages are fitted by local-area regression. The box with the plus sign is the window used for local-area regression. (C) Spruce pollen percentages fitted onto the grid. (D) Contours representing the response surface and pollen percentages shown in part C.

and Clark, 1977; Howe and Webb, 1983; Bartlein *et al.,* 1984; Bartlein and Webb, 1985; Huntley and Prentice, 1988). These methods are cumbersome to use at a subcontinental scale (Bartlein and Webb, 1985).

We have developed a new interpretive method that uses response surfaces, fossil pollen data, and dissimilarity coefficients (Bartlein *et al.,* 1986; Prentice *et al.,* 1991). We used the temporally interpolated pollen values for 14 pollen categories (the 12 major pollen types, with *Pinus* split into northern and southern *Pinus* and *Betula* split into shrub and tree *Betula*) to define a fossil pollen assemblage at each site at 3000-yr intervals from 18 to 0 ka. Chord distances (Overpeck *et al.,* 1985) are then used to compare each temporally interpolated fossil pollen assemblage with "modern" pollen assemblages formed by the set of fitted values that define the 14 response surfaces and thereby to identify the closest modern analogues of the fossil assemblage. The climate values associated with the analogue assemblages thus provide an estimate of the climate associated with the fossil assemblage. These estimates are called "inferred" climatic values because they are inferred from the pollen values (Fig. 17.1). The inferred temperature and precipitation values provide estimates of the climatic values most compatible with the observed past distribution of the 14 pollen categories.

The critical value delimiting "good" analogues can

be determined by comparing the modern assemblages (Overpeck *et al.,* 1985; Anderson *et al.,* 1989). We used squared chord distances to compare modern observed spectra with assemblages estimated from the 14 response functions and thus established a criterion for analogue matches between pollen spectra (Overpeck *et al.,* 1985). We chose 0.20 as the critical value for analogue matches because more than 97% of the observed spectra matched the response surface spectra with a squared chord distance of 0.20 or less (Table 17.5). This value is slightly larger than that used by Overpeck *et al.* (1985) and Anderson *et al.* (1989). This increase arises because the chord distances between observed assemblages and those generated by the fitted values from the response surfaces are inevitably greater than the chord distances between two observed pollen assemblages. This solution arises because the fitted-value assemblages contain more nonzero pollen types than do the observed. We generated inferred climate estimates only for fossil samples with analogue matches; that is, with squared chord distances less than 0.20.

Our application of response surfaces to fossil data for climatic calibration assumes that changes in the pollen data over time result from pollen-climate relationships that are comparable to those represented by the response functions and therefore by the modern geographic patterns of vegetation and climate (Pren-

Table 17.4. Dating control and adjustments for sites with pollen data

Site number and name[a]	Date for mapping (yr B.P.) Youngest	Date for mapping (yr B.P.) Oldest	Youngest date assigned to core	Ragweed rise (R) date inserted[b]	Oldest 14C date[c] (yr B.P.)	14C dates (no.)	14C dates used (no.)	Total dates used (no.)	Stratigraphic dates used[d] (no.)	Hemlock decline date (H) inserted	Changes and additions to dates used in age model[e]
2 Alderdale	0	6000	0		6090	1	1	3	2		ad: 0.1 ka (NAP rise)
3 Alexander Lake	2000	6000	2000		5985	3	3	4	1		
4 Alexis Lake	0	11,000	30		20,520	4	4	5	1		Samples date >11 ka not used for maps; dates of 20 and 14 ka ignored
5 Alfies Lake	0	10,000	0	R	9210	3	3	5	2		xd: 10 ka
6 Aliuk Pond	3000	7000	1000		7170	2	2	3	1		de: 0.1 ka
9 Anderson Pond	0	18000	0	R	25,000	10	9	11	2		ad: 7.2 ka (based on Alfies Lake)
10 Antoine	7500	9000	7200		8830	1	1	2	1		
11 Attawapisket	0	6000	0		5670	1	1	2	1		ad: 7.3 ka (based on beech rise)
13 Ballycroy Bog	0	11000	0	R	10,900	2	2	5	3		
15 Base de Plein Air de St. Foy	0	6000	0	R	5540	2	2	4	2		
17 Basswood Road Lake	0	12,500	0	R	12,250	6	6	9	3	H	corr: 0.35 ka at all dates (based on hemlock decline)
18 Battaglia Bog	10,000	16,000	10,060		15,570	3	3	3	0		
19 Bay of Quinte	100	8000	100		7920	4	4	5	1		
21 Bereziuk	0	6630	20		6630	3	3	4	1		
25 BL-Tombigbee	0	10,000	0		9765	7	5	6	1		de: 3.98, 8.7 ka (contaminants); ha: 3.0–8.0 ka
26 Blackington Lake	0	8500	20		8640	2	2	3	1		ad: top at 1.7 ka (from average of four wood dates just above top sample)
29 Boney Springs (terrace sample)	17,000	18,000	17,000		27,480	4	4	5	1		
30 Border Beacon	0	6500	0		6500	5	5	6	1		
31 Boriack Bog (core A)	0	16,000	0		15,460	4	4	5	1		
35 Brown's Lake Bog	0	11,000	10	R	10,595	11	11	13	2		ha: 6–9 ka
36 Buckle's Bog	0	18,000	0	R	18,550	6	6	9	3	H	ad: 0.65 ka (top sample based on constant sedimentation rate)
39 Cahaba Pond	650	12000	650		11,960	13	13	14	1		
40 Camel Lake	100	18,000	100		31,610	4	4	5	1		
42 Caribou Hill	0	6000	0		6120	5	5	6	1		
44 Charles Lake	0	9000	0	R	8763	1	1	4	3	H	xd: 6 ka
47 Chism-I	0	6000	20		5465	1	1	2	1		
48 Chism-II	0	6500	20		6430	1	1	2	1		
49 Churchill Falls North	200	5000	200		5255	1	1	2	1		ad: 0.2 ka (for top)
50 Churchill Falls South	200	5000	200		5450	1	1	2	1		ad: 0.2 ka (for top)
52 Clearwater	0	6000	0		6480	2	2	3	1		corr: 1.12 ka at 0.05 ka (based on chenopod rise), 1.79 ka at 6.48 ka (allowing for difference between inorganic and organic dates)
53 Colorado Marsh	100	13,500	100		(13,500)	5	5	7	2		ad: 13.5 ka (for glacial retreat)

Table 17.4. Dating control and adjustments for sites with pollen data (continued)

Site number and name[a]	Date for mapping (yr B.P.) Youngest	Date for mapping (yr B.P.) Oldest	Youngest date assigned to core	Ragweed rise (R) date inserted[b]	Oldest ^{14}C date[c] (yr B.P.)	^{14}C dates (no.)	^{14}C dates used (no.)	Total dates used (no.)	Stratigraphic dates used[d] (no.)	Hemlock decline (H) date inserted	Changes and additions to dates used in age model[e]
54 Conklin Quarry	17,500	18,000	17,900		18,090	1	1	2	1		ad: 17.9-ka date inserted and pollen sample duplicated to allow interpolation to 18 ka
55 Cookstown Bog	5000	10,500	20	R	10,200	1	1	4	3		ha: 0–5 ka
56 Coppermine Saddleback	0	3700	0		3715	5	4	5	1	H	de: 2.92 ka (reversal)
57 Cranberry Glades	0	12,500	20	R	12,185	3	3	5	2		ha: 7.3–11.9 ka
58 Crates Lake	0	9000	0		8800	6	5	6	1		de: 5.189 ka (reversal)
59 Crawford Bog	0	11,500	0	R	(11,300)	0	0	6	6	H	ad: 7.2 ka (beech rise, Van Nostrand Lake), 10.2 ka, 11.3 ka spruce decline and max (from Ballycroy Bog and Maplehurst)
60 Creswell Bay	1000	7500	1320		7590	4	4	4	0		ad: 6.8, 9.0 ka
62 Crozier	6500	9000	6800		(9000)	1	1	3	2		corr: all dates by 1.0 ka (based on hemlock decline); xd: 9.3 ka
65 Cub Lake	50	9000	20	R	8220	5	5	7	2		
66 Cupola Pond	100	17,000	100		17,000	10	8	9	1		
67 Cycloid Lake	0	8500	0		8520	2	2	3	1		
68 Daumont Lake	500	6000	100		5490	4	4	5	1		
71 Devils Lake (ND)	0	6000	20		6120	1	1	2	1		
72 Devils Lake (WI)	0	13000	0	R	12,550	12	10	12	2		de: 12.52 ka (reversal); av: one date pair at 12.55 ka
73 Diana Island	0	7000	10		6820	3	3	4	1		
74 Dismal Swamp (core 1)	0	9000	0	R	8900	2	2	4	2		
79 Eagle Lake (LB)	0	10,000	20		10,550	5	5	6	1		
81 East Baltic Bog	0	8500	0	R	8430	3	3	5	2		corr: 0.03 and 0.04 ka at 4.02 and 7.0 ka, respectively
82 Edward Lake	0	10,000	0	R	9750	3	3	5	2		corr: 0.8 ka at all dates (based on hemlock decline
83 Ennadai	600	5500	600		5780	12	9	10	1		ad: top date 0.6 ka; av: one date pair (three dates) at 1.37 ka; de: 5.72 ka
84 Ennadai 72	800	5000	870		4690	7	5	5	0		de: 0.97 and 4.52 ka (reversals), 0.87 ka at 3.8 cm; ha: 3–4 ka
86 Ferndale Bog (core IV)	0	5000	0		5170	5	5	6	1		
87 Found Lake	0	10,000	1		9999	7	0	Varved	0		
89 Francis Lake II	3000	18,000	2600		18,570	4	4	5	1		ad: 1.07 ka for topmost pollen sample at 60 cm in core (based on constant sedimentation rate [Cotter, 1984])
90 French Lake	0	3500	0	R	3465	4	3	5	2		corr: 0.32 ka at all dates (based on ragweed rise)
92 Gass Lake	0	12,000	0	R	11,760	6	6	8	2		ad: 0.2 ka (Gramineae rise)
93 Gea-I–St. Hippolyte	0	10,000	50		10,140	1	1	4	3	H	ad: 9.43 ka (spruce decline); corr: 1.0 ka at 11.8 ka (based on ice cover at 13 ka [Denton and Hughes, 1981])
94 Glenboro Lake	0	12,000	0	R	11,800	5	5	8	3		

Site number and name[a]	Date for mapping (yr B.P.) Youngest	Oldest	Youngest date assigned to core	Ragweed rise (R) date inserted[b]	Oldest ^{14}C date[c] (yr B.P.)	^{14}C dates (no.)	^{14}C dates used (no.)	Total dates used (no.)	Stratigraphic dates used[d] (no.)	Hemlock decline date (H) inserted	Changes and additions to dates used in age model[e]
95 Goshen Springs	0	9000	0		33,000	5	5	7	2		ha: 9.1–26 ka
97 Grand Rapids	300	7000	0	R	7220	3	3	4	1		First sample at 5 cm (300 yr); 0 yr = 0 cm
98 Gravel Ridge	0	6000	0		6470	9	4	5	1		de: 202, 337, 3.96, 53, and 6.51 ka (Lamb, 1982)
100 Green Lake (WI)	0	10,000	0	R	10,410	6	5	7	2		de: 1.09 ka (replaced with ragweed rise)
101 Hack Pond	0	13,000	0		12,720	2	2	3	1		
102 Harrie Lake	0	6000	0		6030	6	6	7	1		
103 Harrowsmith	0	10,500	0	R	10,390	1	1	4	3	H	
104 Hawke Hills Kettle	0	7500	0		7290	3	3	4	1		
106 Hayes Lake	100	10,000	0		(9800)	3	3	5	2		ad: 9.8 ka (J. H. McAndrews, personal communication)
109 Hebron Lake	0	7000	0		8350	3	3	4	1		
111 Hopedale Pond	200	5500	200		5440	5	3	4	1		de: 3.865, 4.24 ka (Short and Nichols, 1977)
112 Horseshoe Lake	0	11,000	−24	R	13,530	12	12	14	2		corr: 0.45 ka at 0.46, 1.84, 3.255, 4.2, and 5.715 ka
114 Houghton Pond	0	9000	0	R	8930	2	2	5	3	H	
116 Iglutalik Lake	0	8400	0		8815	5	5	6	1		
117 Indian Lake	5500	10,000	0		9140	3	3	4	1		
118 Jack Lake	0	9200	0		9270	3	3	4	1		
119 Jackson Pond	0	18,000	0	R	20,330	6	6	8	2		ad: 9-ka date at 200 cm based on spruce decline
121 Jewell site	1000	13,500	0		14,450	7	6	8	2		de: 14.45 ka; ad: 13.5 ka (for glacial retreat)
122 Jock Lake	1000	6000	0		9060	9	9	10	1		
124 Kanaaupscow	0	7000	20		6450	1	1	2	1		xd: 8.043 ka
126 Kincardine Bog	0	11,000	0	R	11,200	4	4	7	3	H	
129 Kogaluk Plateau Lake	2500	8500	2500		8610	5	3	4	1		ad: 25 ka (top sample based on constant sedimentation rate); de: 4.36 and 4.655 ka
130 Kotiranta	200	16,000	200		16,150	3	3	7	4		ad: 4, 7.5, and 10.5 ka (from Jacobson Lake)
131 Kylen Lake	8500	15,500	8410		15,850	9	9	9	0		de: 8.62 ka (reversal [P. J. H. Richard, personal communication]; xd: 12,769 ka
132 Lac à la Tortue	0	10,000	20	R	9215	4	3	6	3	H	
133 Lac à Leonard, Mont St-Pierre	0	9000	0		9040	5	5	6	1		
134 Lac au Sable	0	7000	0		6860	3	3	4	1		ad: 7 and 9.6 ka (based on pine and spruce decline at Yamaska)
135 Lac Bouleaux	200	11,000	150		13,000	2	2	6	4	H	
136 Lac Castor	0	10,000	20	R	9540	6	6	8	2		

Table 17.4. Dating control and adjustments for sites with pollen data (continued)

Site number and name[a]	Date for mapping (yr B.P.) Youngest	Oldest	Youngest date assigned to core	Ragweed rise (R) date inserted[b]	Oldest 14C date[c] (yr B.P.)	14C dates (no.)	14C dates used (no.)	Total dates used (no.)	Stratigraphic dates used[d] (no.)	Hemlock decline date (H) inserted	Changes and additions to dates used in age model[e]
137 Lac Colin	0	11,000	20	R	11,100	8	8	10	2		
138 Lac Cote	0	9500	0		9810	1	1	2	1		
139 Lac de Roches Moutonnees	500	5000	510		5185	5	4	4	1		de: 4.09 ka (reversal)
140 Lac Delorme II	0	6000	0		6320	5	5	6	1		
141 Lac des Atocas	100	10,000	100		10,250	4	4	5	1		
142 Lac Gras	0	7000	0		6510	5	5	6	1		
143 Lac Hamard	0	6500	0		16,975	4	1	4	3		de: 24, 6345, and 16.975 ka; ad: 55 and 6.2 ka (spruce rise and alder decline at Lac Delorme)
144 Lac Louis	0	9000	20	R	9090	3	3	5	2		
145 Lac Martini	0	9000	50		8740	4	4	6	2	H	
146 Lac Martyne	0	6000	20		5980	3	3	4	1		
148 Lac Patricia	0	5000	20		4855	2	2	3	1		
149 Lac Romer	0	7000	0	R	6920	4	4	6	2		
150 Lac Tania	200	10,000	50		10,000	6	6	8	2	H	
151 Lac Turcotte	500	10,500	0		10,360	5	5	6	1		ad: 0.5 ka (calculated on constant sedimentation rate to top sample)
152 Lac Yelle	0	9000	20	R	8900	3	3	5	2		
154 Lake A	0	11,000	0		11,560	1	1	3	2		ad: 0.05 ka (chenopod rise)
155 Lake Annie	0	13,000	0		44,300	8	7	8	1		de: >25 ka; ha: 13–33 ka
156 Lake B	0	10,000	0		10,260	1	1	3	2		ad: 0.05 ka (chenopod rise)
157 Lake E	100	11,000	100		11,140	8	8	9	1		
158 Lake Hope Simpson	200	10,500	0		10,400	6	6	7	1		
159 Lake Louise	0	10,000	0		49,000	3	3	5	2		ad: 0.17 ka inserted at 25 cm even though no clear ragweed rise; ha: ca. 8.5–49 ka; hiatus led to restriction on oldest date mapped
161 Lake Mendota	0	14,000	0	R	16,440	6	5	8	3		de: 16.44 ka; ad: 14 ka (Devils Lake, WI); ha: 3.2–6.9 ka; hiatus led to restriction on dates used
164 Lake Six	0	7000	0		6970	3	3	4	1		
165 Lake Sixteen	2500	11,000	0		10,690	5	5	6	1		
166 Lake Tulane	0	18,000	0	R	39,600	12	12	14	2		
168 Lambs Pond	8000	12,000	8320		12,300	4	4	4	0		
169 Lanoraie, St. Henri Bog	0	6000	0	R	5960	6	4	7	3	H	de: 3.82 and 4.65 ka (replaced with hemlock decline)
170 Lanoraie, St. Joseph Bog	0	5000	0	R	4790	3	3	5	2		
173 Last Lake	8500	9500	9000		9220	2	2	2	0		
174 "LD" Lake	0	7000	50		6960	2	2	3	1		
175 Lily Lake	0	12,000	0	R	11,770	7	7	9	2		
176 Lima Bog	0	14,000	50		25,680	8	8	10	2		ad: 14 ka (for ice retreat based on Devils Lake, WI, and other evidence)

Site number and name[a]	Date for mapping (yr B.P.) Youngest	Date for mapping (yr B.P.) Oldest	Youngest date assigned to core	Ragweed rise (R) date inserted[b]	Oldest 14C date[c] (yr B.P.)	14C dates (no.)	14C dates used (no.)	Total dates used (no.)	Stratigraphic dates used[d] (no.)	Hemlock decline date (H) inserted	Changes and additions to dates used in age model[e]
177 Little Bass Lake	0	10,500	0	R	10,660	5	5	7	2		de: 6.44, 9.14, 14.3, and 16.5 ka; ad: 3.02, 5.12, 6.58, 9.46, 11.3, and 12.6 ka (based on Basswood Road Lake)
178 Little Lake	0	12,500	0	R	16,500	4	0	8	8		
179 Long Lake	0	5550	0		5550	2	2	3	1		av: two date pairs at 5135 and 6.015 ka
183 Lynn Lake	2000	6500	2170		6530	6	4	4	0		
184 Maclaughlin Pond	50	10,000	50		9670	5	5	6	1		corr: 0.48 ka at 0.1 ka (based on 14C dated ragweed rise of 0.58 ka)
186 Maplehurst	0	12,500	0	R	12,500	5	5	6	1		de: 10.7 ka; ad: 6.5, 8, and 10 ka (based on hemlock, spruce, pine at Ramsay Lake)
191 McLaughlan Lake	6000	10,000	6500		10,700	1	0	3	3		
192 Medicine Lake	0	10,500	0		10,940	6	5	6	1		corr: 1.05 ka at 0.1 ka; 1.35 ka at 1.35, 3.7, and 7.4 ka; 1.67 ka at 10.94 ka (E. C. Grimm, personal communication)
193 Mer Bleue	300	7500	0	R	7650	1	1	4	3		
197 Mont Shefford		11,500	200		11,400	10	10	11	1	H	ad: 0.2 ka for top (guess), first sample at 0.254 ka
200 Moraine Lake	0	10,000	0		9640	6	6	7	1		
201 Moulton Pond	0	13,500	0	R	13,510	16	11	13	2		de: 0.1, 0.72, 0.1, 5.56, and 3.075 ka (reversals)
202 Mud Lake	0	9000	0	R	34,500	5	5	7	2		
203 Muscotah	0	18,000	0	R	23,040	4	4	6	2		
205 Nain Pond	0	9000	50		(9000)	8	2	5	3		ad: 4.5 and 9 ka (based on spruce rise and sedge decline at Ublik Lake); de: 1.655, 5.385, 3.645, 13.235, 8.725, and 7.195 ka (Short and Nichols, 1977)
206 Napaktok Lake	0	6500	0		8735	5	4	5	1		de: 10.2 ka (reversal)
207 Nedlouc	0	4000	10		3950	7	5	6	1		de: 1.78 and 3.92 ka (reversals)
209 Nina Lake	0	9500	0		9510	3	3	4	1		
211 Nipmuck Pond	0	12,750	0		12,750	6	6	7	1		
212 Nonconnah Creek-1	13,000	18,000	13,000		22,305	1	1	4	3		ad: 13 ka (P. A. Delcourt, personal communication estimate), 17 ka, 20 ka
213 North Bay	0	9500	0		9575	1	1	3	2		
215 Northwest River Pond	0	5000	30		4805	1	1	2	1		ad: 0.07 ka (settlement horizon)
216 Nungesser	6000	9000	0		8860	1	1	2	1		
218 Old Field	3500	9000	3500		8810	4	4	5	1		ad: 3.5 ka (top sample based on constant sedimentation rate)
219 Owl Pond	0	10,400	0	R	10,270	4	4	6	2		
220 Palsa Lake	0	6000	0		5825	6	4	5	1		de: 6.935 ka (reversal), 16.8 ka (too old)
221 Panther Run Pond	0	12,500	0	R	12,610	2	2	5	3	H	ha: 7.1–10.9 ka

Table 17.4. Dating control and adjustments for sites with pollen data (continued)

Site number and name[a]	Date for mapping (yr B.P.) Youngest	Date for mapping (yr B.P.) Oldest	Youngest date assigned to core	Ragweed rise (R) date inserted[b]	Oldest 14C date[c] (yr B.P.)	14C dates (no.)	14C dates used (no.)	Total dates used (no.)	Stratigraphic dates used[d] (no.)	Hemlock decline date (H) inserted	Changes and additions to dates used in age model[e]
222 Paradise Lake	0	9500	0		9810	3	2	3	1		de: 6.075 ka (reversal)
223 Patricia Bay Lake	2600	6500	2600		6320	3	3	3	0		ad: 2.6 ka (for top of core, guess)
224 Pequot Cedar Swamp	7500	13,000	0		12,690	6	6	7	1		ha: 0-7.5 ka
225 Perch Lake	0	9500	0	R	9430	1	1	4	3	H	
226 Petit Lac Terrien	0	12,000	0	R	12,640	1	1	7	6		ad: 4.9, 6.3, 9.1, and 11.2 ka (based on Lac Colin)
228 Pigeon Marsh	0	18,000	0	R*	19,520	2	2	3	1		ha: 11.1-14.9 ka
229 Pilot Mound Site	350	13,500	0		13,500	5	4	7	3		de: 9.3 ka; ad: 11.5 and 13.5 ka (based on spruce decline, glacial retreat)
231 Pink Lake	0	10,500	0	R	10,600	3	2	5	3	H	de: 7.8 ka (too old based on hemlock decline)
232 Pogonia Bog Pond	0	11,000	0	R	11,190	7	7	9	2		av: 10.21 and 10.61 ka; de: 1.83, 2.26, and 9.11 ka (B. Warner, personal communication)
233 Point Escuminac	0	11,000	0	R	10,755	23	19	21	2		
234 Poland Spring Pond	0	13,000	0	R	12,860	11	10	12	2		de: 0.135 ka
235 Pond Mills Pond	0	10,500	0	R	(11,400)	1	0	7	7	H	ad: 7.1, 8, 9.6, and 11.4 ka (based on Maplehurst)
237 Portage Bog	0	8000	0	R	9880	3	3	6	3	H	corr: -27 per mil (for 13C correction) at 0.49 and 3.43 ka
238 Potts Mountain Pond	0	11,111	0	R	11,140	2	2	6	4	H	ad: 6 ka to bracket hiatus (65-68 cm); ha: 6.1-8.9 ka
240 Prince	6500	11,000	6800		10,800	2	2	3	1		ad: 6.8 ka (for upper sample, Alfies Lake)
242 Protection Bog	100	12,500	0	R	(12,400)	3	3	6	3		ad: 12.4 ka (basal date, Belmont Bog)
243 Pyramid Hills Lake	100	7000	100		6815	7	5	6	1		de: two dates based on sedimentation rates
244 Quicksand	0	18,000	0	R	20,100	2	2	4	2		
245 Radtke Lake	0	11,500	0	R	11,460	7	7	9	2		
246 Ramsay Lake	0	10,700	0	R	10,800	5	4	6	2		
247 Rattle Lake	0	11,000	0		11,110	6	6	7	1		de: 0.54 ka (reversal)
248 Rayburn Salt Dome	15,000	18,000	0		(22,000)	3	3	7	4		ad: 6, 19, and 22 ka (based on southern evergreen, spruce, pine [P. and H. Delcourt, personal communication]); ha: 0.2-13.ka
249 Riley Lake	0	7000	0	R	6490	2	2	4	2		
250 Rivière aux Feuilles	500	5500	10		5235	3	3	4	1		
251 Rockyhock Bay	0	18,000	0	R	27,700	13	6	8	2		de: 5.82 ka (reversal), all six dates from Ohio Wesleyan University (six dates used from Indiana University)
256 Ryerse Lake	500	9000	870		8960	6	6	6	0		
257 Saddle Hill Pond	0	10,000	0		9430	4	4	5	1		
260 Sandy Cove Pond	0	4500	0		4555	3	3	4	1		

Site number and name[a]	Date for mapping (yr B.P.) Youngest	Date for mapping (yr B.P.) Oldest	Youngest date assigned to core	Ragweed rise (R) date inserted[b]	Oldest 14C date[c] (yr B.P.)	14C dates (no.)	14C dates used (no.)	Total dates used (no.)	Stratigraphic dates used[d] (no.)	Hemlock decline date (H) inserted	Changes and additions to dates used in age model[e]
261 Sav-I–Ste. Agathe	0	10,000	0	R	10,170	3	3	6	3		
262 Sav-II–Lac aux Quenoilles	0	11,000	20	R	10,820	3	3	6	3	H	
263 Scott Lake	0	4500	0		4360	1	1	2	1	H	
264 Second Lake	0	10,000	0	R	(10,000)	1	1	5	4		corr: 0.975 ka at 4 ka (hemlock decline); ad: 10 ka (based on spruce decline at Edward Lake)
266 Shady Valley Peat	300	10,000	0		9500	1	1	3	2	H	
267 Shaw's Bog	0	9180	0	R	9180	4	4	6	2		
268 Sheelar Lake	150	18,000	0		23,880	8	8	10	2		ad: 18 ka (guess) data at 18.5 ka close enough; ha: 14.6–18.5 ka
269 Shouldice Lake	0	9500	0	R	9390	2	2	5	3	H	
270 Silver Lake (NS)	0	10,000	0	R	9650	3	3	5	2		
272 Singletary Lake	6000	18,000	0		40,000	5	5	6	1		ha: 0–6 ka
274 Sioux Lake	0	10,000	0		9740	3	3	4	1		
275 Slow River	0	3805	0		3805	1	1	2	1		
276 Snow Lake	0	5000	0		5270	3	3	4	1		
277 Sona Lake West	300	5500	200		5575	1	1	2	1		
278 South Burin Peninsula	10,500	13,500	10,700		13,400	3	3	3	0		
279 Spring Lake	0	13,000	0	R	12,500	4	4	7	3	H	
280 St. Benjamin	0	9000	20	R	9100	1	1	4	3	H	Hiatus 690–615 cm
285 St. John's Island Pond	0	7000	30		(6550)	2	1	3	2		ad: 6.55 ka (based on alder peak on second core)
289 Sugarloaf Pond	0	9500	0		9270	2	2	3	1		ha: 6–9 ka (inclusive)
290 Swan Lake	0	9200	0	R	8950	2	2	3	1		
291 Szabo Pond	0	11,500	0	R	11,400	2	2	4	2		ad: 2.98, 3.38, 43, and 6.17 ka (Short and Nichols, 1977)
296 Thompson Landing	3000	6000	2980		(6170)	0	0	4	4		corr: 1.0 ka at all dates (based on hemlock decline)
298 Tonawa Lake	0	9500	0	R	9030	2	2	4	2		
299 Torrens Bog	0	10,500	0	R	(10,400)	0	0	9	9		ad: 10.4-ka date from Stotzel-Leis and Silver Lake for pine peak; other stratigraphic dates from J. G. Ogden (personal communication); 13 14C dates from another core used
300 Track Lake	0	5000	50		4755	7	3	4	1		de: 3.06, 3.275, 2.845, and 4.28 ka
301 Tunturi Lake	0	6500	0		14,040	3	1	4	3		de: 5.99 and 14.04 ka; ad: 5.5 and 6.2 ka (based on spruce rise and alder decline at Lac Delorme)
302 Ublik Lake	100	9000	100		10,260	4	4	4	0		
304 Upper Gillies Lake	0	10,500	0		(10,400)	0	0	5	5	H	ad: 0.15 ka (herb rise), 9 ka, 10.4 ka (based on sedge decline, birch rise, and pine rise at Gillis Lake)
306 Upper Twin	8500	10,500	8760		10,650	3	3	3	0		

Table 17.4. Dating control and adjustments for sites with pollen data (continued)

Site number and name[a]	Date for mapping (yr B.P.) Youngest	Date for mapping (yr B.P.) Oldest	Youngest date assigned to core	Ragweed rise (R) date inserted[b]	Oldest 14C date[c] (yr B.P.)	14C dates (no.)	14C dates used (no.)	Total dates used (no.)	Stratigraphic dates used[d] (no.)	Hemlock decline date (H) inserted	Changes and additions to dates used in age model[e]
308 Val St. Gilles Bog	500	6500	50		6460	2	2	3	1		
309 Van Nostrand Lake	0	11,000	0	R	(10,750)	2	1	6	5	H	de: 5.7 ka; ad: 10.75 ka (J. H. McAndrews, personal communication), 7.35 ka (hemlock rise at Kincardine Bog and Maplehurst)
311 Victoria Road Bog	0	10,000	0	R	9600	1	1	3	2		
313 Walters Puddle	100	14,000	100		14,400	3	3	6	3		ad: 5.82 and 11.88 ka (to bracket hiatus); hiatus 582-1188 ka
316 White Lily	8500	12,000	8500		(11,400)	0	0	3	3		ad: 11.4, 95, and 85 ka (based on spruce, birch, pine at Horseshoe)
317 White Pond	0	18,000	0	R	19,110	3	3	5	2		
318 Whitney's Gulch	20	9500	20		9820	7	5	6	1		de: 6.275 and 9.940 ka (reversals)
323 Wolsfeld Lake	0	11,500	0	R	12,060	11	11	13	2		
324 Wolverine Lake	100	10,000	100		9980	5	5	6	1		
328 Zuehl Farm site	1000	13,000	0		(13,000)	4	4	6	2		ad: 13 ka (for glacial retreat)

[a]Sites not listed are in Table 1 in Gaudreau and Webb (1985), Table 10-3 in Webb et al. (1983a), or Table 1 in Webb et al. (1983b).

[b]R* indicates ragweed rise observed but not used in dating.

[c]Dates in parentheses are stratigraphic dates.

[d]Number of stratigraphic dates used to set the chronology. These may include the date assigned to the uppermost sample, the ragweed rise date, the hemlock decline date (4.7 ± 0.2 ka), and any other dates added (see footnote e).

[e]Dates used in age models were the published radiocarbon dates and any of the standard stratigraphic dates listed (see footnote d). All modifications to this practice are listed here: ad = stratigraphic dates added; de = dates deleted; ha = bracketing dates for hiatuses within time interval of mapping; xd = bottom dates extrapolated based on constant sedimentation rate from above; av = average of two or more dates. NAP = nonarboreal pollen.

Table 17.5. Summary of analogue matches: Number and percentage of samples that match at least one (other) contemporary sample with squared chord distances below selected values for the modern data set and for selected dates in the past

Date (yr B.P.)	Squared chord distance										Total no. of samples
	0.15		0.20		0.30		0.4		>0.4		
	No.	%	No.	%	No.	%	No.	%	No.	%	
Modern	895	94	920	97	944	99	949	99.8	951	100	951
3000	252	85	277	94	292	99	296	100	296	100	296
6000	195	71	234	85	261	95	273	99	276	100	276
9000	179	80	204	91	220	99	222	99.5	223	100	223
12,000	33	39	45	53	67	79	79	93	85	100	85
15,000	5	20	14	56	20	80	21	84	25	100	25
18,000	7	37	13	68	18	95	18	95	19	100	19

tice *et al.,* 1991). By illustrating the multivariate response of pollen taxa to climate, the response surfaces show that no one climate variable controls the whole range of a taxon. Therefore no simple vegetation response to climate changes should be anticipated, because changes in climate are by definition multivariate, and the vegetational response to such changes involves nonlinear (compounded or sometimes compensating) effects.

LAKE-LEVEL DATA

Estimation of Lake-Level Status

Because lake-level changes are derived from many different kinds of geomorphologic, sedimentologic, and biostratigraphic information (see Harrison, 1988), the data must be standardized for the purposes of intersite comparison (Street and Grove, 1976). In previous compilations of lake-level data from North America (e.g., Harrison and Metcalfe, 1985a) based on information in the Oxford lake-level data bank (Street and Grove, 1976, 1979; Street-Perrott and Roberts, 1983; Street-Perrott and Harrison, 1984, 1985; Street-Perrott, 1986), the information was standardized to yield estimates of lake status. The total range of altitudinal changes in lake level within each basin was divided into three status classes: low, representing the bottom 15% of the range and including dry lakes; intermediate, the middle 15–70% of the range; and high, the upper 30% of the range, including overflowing lakes. The boundaries between status classes were originally chosen so that each class had a similar frequency of occurrence within the global data set.

This classification scheme was designed for use with closed-basin lakes in semiarid regions and works well when the absolute magnitude of lake-level changes at a site can be reconstructed, for example, from shore elevations. It can also be applied to some temperate lake basins in eastern North America, notably Little Salt Spring, Lake Mendota, and Duck Pond, where the data are derived from shoreline elevations or from a transect of cores. The scheme is not well suited, however, for use at sites where it is only possible to derive relative measures of changes in water depth, as at most sites in temperate regions. The records from such basins can usually be categorized by identifying the maximum and minimum water depths registered within the basin on the basis of all available data and using these to define high and low status classes, respectively. Periods when the lake was in transition between these states are assigned to an intermediate status category. Lake status has been assessed in this way at 1000-yr intervals (Table 17.6). This tripartite ordinal classification is roughly comparable to the methods employed in the Oxford lake-level data bank.

Estimation of Effective Moisture from Lake-Level Data

The use of status categories involves the loss of some information about lake behavior and thus may obscure fundamental changes in the hydrologic regime as monitored by lakes, particularly where the response is categorized according to the maximum differences in behavior observed within the system regardless of the length of the record. Lowered lake levels in response to a given hydrologic stress, for example, may be recorded as a transition from high to intermediate status in one basin but may not be large enough to register as a change in the status category at another site.

To focus on changes in water balance through time, we used the primary data to derive a second measure of the difference in effective moisture conditions between selected time periods. This index records whether the lakes indicate a more or less positive water balance between two intervals, regardless of the absolute change in water level (see Thompson *et al.,* this vol.). We used this index to compare lake levels at each 3000-yr interval from 12 to 3 ka with the present as well as between adjacent dates (Table 17.6).

CLIMATE-MODEL RESULTS

Interpolation of Model Simulations to Fossil Pollen Sites

We followed the procedure described by Webb *et al.* (1987) to convert the model results of Kutzbach

Table 17.6. Lake status at 1000-yr intervals and effective moisture differences between each past date and today and between adjacent pairs of dates[a]

Basin code and name	State or province	Lake status[b] 0	1	2	3	4	5	6	7	8	9	10	11	12 ka	Diff 12 -0	9 -0	6 -0	3 -0 ka	12 -9	9 -6	6 -3	3 -0 ka
a Elk Lake	MN	1	1	1	1	0	1	3	3	3	1	1	0	0		S	D	S		W	D	S
b Weber Lake	MN	1	1	1	3	3	3	3	2	2	2	1	1	0		D	D	D		W	S	D
c Rutz Lake	MN	2	2	2	2	3	3	3	2	2	2	1	1	1		S	D	S		W	D	S
d Kirchner Marsh	MN	2	2	2	2	2	2	3	3	2	2	1	1	1	W	S	D	S	W	W	D	S
e Pickerel Lake	SD	1	1	1	1	1	2	3	3	3	3	2	0	0		(D)	D	S		(S)	D	S
f Lake West Okoboji	IA	1	1	1	1	1	2	3	3	3	2	1	1	1	S	D	D	S	W	W	D	S
g Kettle Hole Lake	IA														W	S	W	S	W	D	W	S
h Hook Lake Bog	WI	2	2	2	2	3	3	3	2	2	2	2	1	1	W	S	D	S	W	W	D	S
i Lake Mendota	WI	2	2	2	2	3	3	3	2	1	1	1	1	2	(S)	(W)	D	S	(D)	W	D	S
j Washburn Bog	WI	2	2	2	2	3	3	3	2	2	1	1	0	0		S	D	S		W	D	S
k Wintergreen Lake	MI	2	2	2	3	3	2	2	2	2	2	2	1	1	W	S	S	D	W	S	S	D
l Sunfish Lake	ON															S	D	D		W	W	D
m Lake George	NY	1	1	1	1	1	2	2	3	3	3	3	1	1	S	D	(D)	S	W	D	(D)	S
n Titicut Swamp	MA																			D		
o Duck Pond	MA	2	2	2	2	2	3	3	3	2	2	1	2	2	S	S	D	S	S	W	D	S
p Szabo Pond	NJ	1	0	0	0	0	0	0	0	0	0	0	3	1	(W)	(D)	(D)	S	W	(W)	(D)	S
q Longswamp	PA	2	0	0	0	0	0	0	0	0	3	1	1	1	W	(D)	(D)	S	W	S	D	S
r Brown's Lake	OH	0	0	0	0	0	0	0	0	0	0	2	1	1	W				W			
s Old Field Swamp	MO																			W	D	
t Cahaba Pond	AL	2	2	2	3	3	3	3	3	3	3	1	1	1	W	D	D	S	W	S	D	S
u Goshen Springs	AL	2	2	2	2	2	1	1	1	2	2	0	0	0		D	S	S		D	S	S
v White Pond	SC	2	3	3	3	3	3	3	3	3	3	1	1	1	W	D	D	D	W	S	S	D
w Mud Lake	FL														D	D	D	S	S	D	D	S
x Lake Annie	FL	1	1	1	1	1	1	2	2	2	2	2	2	2	D	D	D	S	S	S	D	S
y Little Salt Spring	FL	1	1	1	1	1	2	2	2	1	1	2	3	3	D	S	D	S	D	W	D	S

[a]This table includes corrections to errors in the summary table in Harrison (1988).

[b]1 = high, 2 = intermediate, 3 = low, 0 = no data.

[c]W = wetter than at younger date, D = drier than at younger date, S = similar at both dates. Parentheses indicate uncertainty in the estimate.

and Guetter (1986) and Kutzbach, Guetter, *et al.* (this vol.) into estimates of the simulated climatic values at each pollen site. We used an inverse distance-weighting method to average the model estimates of particular climate variables for the several grid points adjacent to each fossil pollen site. Because ocean and ice-covered grid points were excluded from this procedure, the number of grid points that contributed to the average for each site ranged from one to three in coastal areas and along the ice sheet to four or five in the interior. This interpolation procedure does not introduce any new features into the resulting data that were not present in the original grid point data, and features in the original data that are evident across several grid points are preserved in the interpolated data.

The reliability of general circulation models is generally proportional to the size of the area considered. Moreover, the inherent variability of the simulated climate, which in turn affects the statistical significance of simulated climatic changes, decreases with the size of the area average. Because of these sampling considerations, simulations and observations are often compared in terms of broad zonal averages or large-area averages (Kutzbach and Street-Perrott, 1985). We fol-

lowed that practice by using area averages to compare the simulated (model-derived) and inferred (pollen-derived) climate estimates. When we compare maps of simulated and inferred climate estimates, however, we are working near the lower limit of the potential resolving power of the NCAR CCM, and our study provides a test of new procedures for comparing simulations with available data and with climatic estimates derived from these data.

General circulation models do not reproduce modern climates exactly but exhibit some systematic biases. These biases are likely to occur also in the paleoclimatic simulations, and they pose a major problem in making direct comparisons between model simulations and climate reconstructions inferred from data. Webb *et al.* (1987) eliminated this source of difficulty by using the anomaly values simulated by the model; that is, the differences between the paleoclimatic experiment and the control, or modern, run. We have followed their practice.

Area Averaging of Climate-Model Results

To summarize the data and climate-model results and to facilitate comparison of the simulations and reconstructions of climate values, we calculated areal averages for three regions (southeastern, northeastern, and north-central) within which the temporal changes in lake levels were generally similar. These regions differ from those used by Kutzbach (1987) and Webb *et al.* (1987). The model simulations of the climate variables at individual grid points were averaged as follows: for the southeastern region, the four grid points at 33.3°N, 82.5°W; 37.8°N, 82.5°W; 33.3°N, 90°W; and 37.8°N, 90°W; for the northeastern region, the five grid points at 42.2°N, 75°W; 42.2°N, 82.5°W; 46.7°N, 67.5°W; 46.7°N, 75°W; and 46.7°N, 82.5°W; and for the north-central region, the four grid points at 42.2°N, 90°W; 42.2°N, 97.5°W; 46.7°N, 90°W; and 46.7°N, 97.5°W. Averages were constructed only for ice-free grid points. In the construction of the averages, the individual values simulated by the model were weighted by the area of the cells defined by the grid points.

Area Averaging of Climate Estimates Inferred from the Data

We calculated area averages of the climate values inferred from the pollen data by first interpolating the values onto an equal-area grid and then averaging them for the three regions used for the simulations. Because the latitude and longitude of each model grid point are at the center of the grid points, the coverage of each region in the data is actually as follows: for the southeastern region, data between 26.6 and 40°N and between 78.75 and 101.25°W; for the northeastern region, data between 40 and 44.4°N and between 71.5 and 86.25°W, plus that between 44.4 and 48.9°N and between 63.75 and 86.25°W; and for the north-central region, data between 40 and 48.9°N and between 86.25 and 101.25°W.

Area averages of the "aridity index" for lakes were constructed for the three regions by simple averaging. The northeastern region includes nine lakes with status data, the north-central region six lakes, and the southeastern region five lakes. To allow comparison with the climate-model results, the lake-status data were converted to an "aridity index" by averaging the status values for all lakes in a region, where the status was coded as 1 for high, 2 for intermediate, and 3 for low. The model results are the average annual water balance (precipitation minus evaporation, or P–E) for all the grid squares in a region. The aridity index is presented as differences from today and the simulated P–E values as differences from the 0-ka control simulation scaled to millimeters per day.

Results

POLLEN DATA

Modern Distribution and Response Surfaces

Maps of modern pollen data show that patterns in the abundance of major pollen types parallel the patterns of modern plant formations (Webb, 1988), which in turn are aligned along climatic gradients. The modern distribution of pollen values must therefore ultimately reflect these climatic gradients (Figs. 17.2 and 17.6). For example, along the East Coast the north-south gradient from high values of Cyperaceae pollen in the north to southern *Pinus* pollen in Florida reflects the general north-south increase in temperature for both January and July. The northeast-to-southwest gradient from Cyperaceae pollen on Baffin Island to forb pollen in Nebraska reflects the increase in mean July temperature, the northwest-to-southeast gradient from forb pollen in Nebraska to southern *Pinus* pollen in Florida reflects the increase in mean January temperature, and the west-east gradient in the Midwest from prairie to forest reflects the increase in precipitation (Fig. 17.6).

Pollen data record changes in broad-scale vegetational patterns during the late Quaternary (Jacobson *et al.*, 1987). Assuming that the vegetation has stayed in dynamic equilibrium with climate (Webb, 1986), we used changes in the distribution and gradients of pollen percentages to reconstruct past climatic gradients.

The modern geographic covariations of pollen types and climate variables (Figs. 17.2 and 17.6) provide the information for defining response surfaces, and the shapes of the response surfaces reflect how the pollen abundances vary in climate space, along axes representing climate variables (Figs. 17.8 and 17.9). For example, along a north-south transect from cold, dry conditions in north-central Canada to warmer but still relatively dry conditions in the Great Plains, the pollen assemblages change from a maximum of Cyperaceae pollen to maxima of *Picea*, northern pines, and then forb pollen (Figs. 17.2 and 17.6). On the response surfaces (Figs. 17.8 and 17.9) the highest values of forb pollen are associated with the combination of warm and dry conditions (the upper left corner of Fig. 17.8), whereas the highest values of *Picea* pollen lie in the portion of the figure that represents cooler and wetter conditions (Fig. 17.8).

The shapes of the response surfaces are useful in interpreting observed changes in pollen abundance in climatic terms. They show how taxa differ in climatic preference and sensitivity. For example, the horizontal orientation of the isopolls for *Picea* (Fig. 17.9) indicates that changes in *Picea* pollen abundance are mainly influenced by mean July temperature rather than mean January temperature. The isopolls for northern *Pinus* for July temperatures below 15°C are similarly oriented. In contrast, the isopolls for southern *Pinus* have a vertical orientation, indicating primary sensitivity to mean January temperature. The diagonal orientation of the isopolls for *Quercus* pollen indicates sensitivity to both January and July temperature.

Sensitivity to annual precipitation is evident in the response surfaces for many pollen types, including forbs, southern and northern *Pinus*, *Ulmus*, *Quercus*,

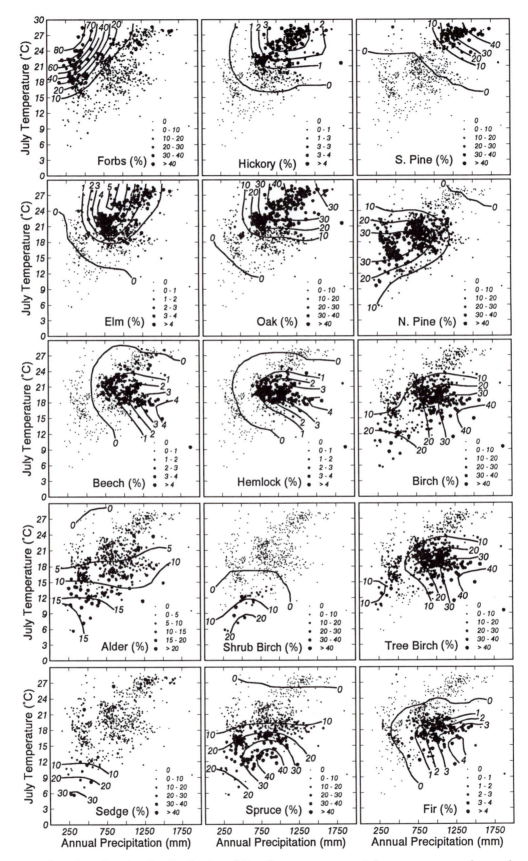

Fig. 17.8. Response functions showing the distribution of 15 pollen groups versus July temperature and annual precipitation.

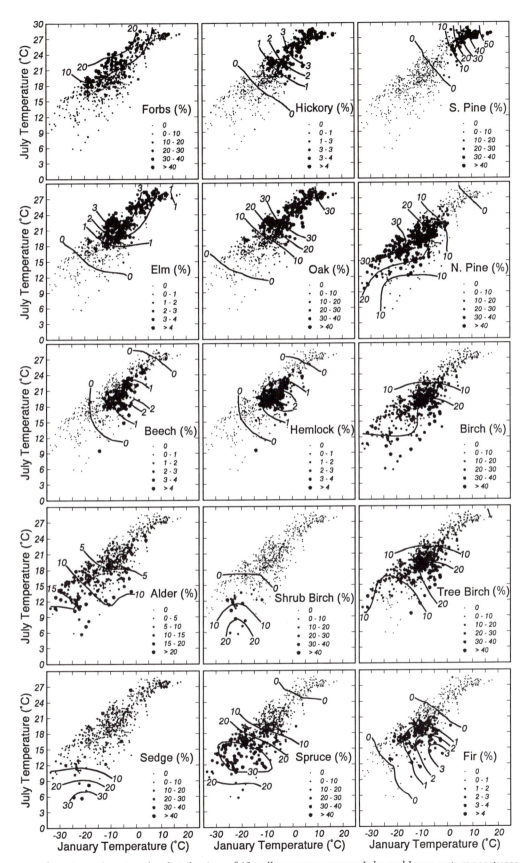

Fig. 17.9. Response functions showing the distribution of 15 pollen groups versus July and January temperatures.

Fagus, Tsuga, Betula, Abies, and *Picea* (Fig. 17.8). Among mixed-forest taxa, northern *Pinus* pollen dominates at lower values of annual precipitation than does pollen of *Fagus, Tsuga,* and *Betula.* These differences in the response of pollen types from place to place in climate space (defined by the three climate variables) provide the basis for interpreting observed temporal changes in pollen abundances in climatic terms (Fig. 17.10) and for transforming temperature and precipitation estimates into pollen abundance for different taxa (Webb *et al.,* 1987; Prentice *et al.,* 1991).

Mapped Temporal Variation in Fossil Pollen Data

The isopoll maps for the 12 pollen types (Fig. 17.10) show how the pollen distributions changed between 18 ka and today. The modern patterns for these pollen types did not develop until after 9 ka (Webb, 1988). One of the most obvious changes is the northward movement of many taxa in a manner suggestive of the general warming of eastern North America (Ritchie, 1987). The movement was gradual before 12 ka, rapid from 12 to 9 ka, and then gradual again. Most taxa reached their maximum northward extent at 6 ka. After 6 ka several taxa, including *Picea* and *Abies,* expanded in abundance southward, implying a reversal in climatic trends.

Before 12 ka *Picea,* Cyperaceae, northern *Pinus, Quercus,* and *Carya* maps show a general north-south distribution, with *Picea* and Cyperaceae in the north and northern *Pinus, Quercus,* and *Carya* in the south. Other pollen types, including *Betula, Abies, Fagus,* and *Tsuga,* are not abundant. By 12 ka this pattern had begun to change; the abundance maximum for northern *Pinus* pollen was confined to the East Coast, and *Fagus, Abies,* and *Ulmus* pollen was more abundant than before.

The period between 12 and 9 ka was marked by rapid changes. The modern spatial patterns in vegetation first appeared at 9 ka. A major northeast-to-southwest vegetational gradient developed between central Canada and Nebraska. This gradient is represented by a sequence of peak values for *Picea, Pinus, Quercus,* and forb pollen (Fig. 17.10) that reflects a sequence of vegetational assemblages from *Picea*-dominated forest to *Pinus* forest, *Quercus* forest, and finally prairie. This pattern later extended to northeastern Canada after the shrinkage and final disappearance of the Laurentide ice sheet. High values of Cyperaceae pollen indicate that tundra then grew in the north. This vegetation gradient parallels the modern gradient in summer isotherms (Fig. 17.2).

After 9 ka a second major gradient became pronounced, as values of southern *Pinus* pollen increased in the southern United States and the modern southern conifer forest developed (Fig. 17.10). This gradient extended southeastward from Nebraska to Florida and probably indicates both the increase in

winter temperature in the middle to late Holocene and the development of the modern gradient in winter temperatures (Webb *et al.,* 1987).

Other changes during the last 18 ka include the switch in alignment of the gradient in herb pollen taxa from its original north-south orientation (see forb and Cyperaceae maps for 18-12 ka) to an east-west orientation with the formation of the modern prairie after 12 ka (Fig. 17.10). During the Holocene the prairie-forest border moved eastward in the northern Midwest from 10 to 6 ka (or 3 ka, depending on location [Webb *et al.,* 1983a; Winkler *et al.,* 1986; Baker *et al.,* 1992]) and then retreated westward. In the east, tree *Betula, Fagus,* and *Tsuga* pollen increased at the expense of northern *Pinus* pollen after 9 ka (Fig. 17.10). Northern *Pinus* pollen was dominant from Minnesota to Nova Scotia by 9 ka.

Climatic Interpretation and Inferred Climate Estimates

The response surfaces provide an initial qualitative interpretation of the changing patterns in the isopoll maps (Figs. 17.8-17.10). The northward movement of spruce and oak populations indicates that temperatures increased, particularly summer temperatures; then, after 6 ka, spruce populations moved south as summer temperatures decreased (Figs. 17.8 and 17.10). During the time of warming in the early Holocene, the western Midwest dried out, and the prairie formed and moved eastward. Southern *Pinus* pollen became more abundant in the southeast as winter temperatures rose after 9 ka and contributed to the formation of the northwest-to-southeast vegetational gradient. Northern *Pinus* pollen was replaced by tree *Betula* and *Fagus* pollen in the northeastern United States after 9 ka as this region became moister and summer temperatures fell (Figs. 17.8-17.10).

The maps of our inferred climate estimates show patterns that are consistent with these qualitative interpretations (Figs. 17.11-17.13). The maps of mean January and July temperature illustrate the general warming as the ice sheet retreated, as well as the delay in peak warmth until after 9 ka. Both sets of isotherm maps show a big increase in temperature between 12 and 9 ka, and mean January temperatures also increased significantly from 15 to 12 ka. The warming continued until 6 ka; after that mean July temperatures fell somewhat as the 20°C isotherm shifted south of the Canadian border. Mean January temperatures continued to increase from 9 ka to the present, especially in the southeast (Fig. 17.11). These changes indicate a decrease in seasonality since 9 ka and the development of the modern temperature gradients during the Holocene.

The maps of inferred precipitation estimates (Fig. 17.13) show the development of the modern east-to-west precipitation decrease in the midcontinent beginning at 9 ka and strengthening by 6 ka, when con-

Forbs *(Artemisia + Chenopodiaceae/Amaranthaceae + other Compositae)*

Sedge *(Cyperaceae)*

Spruce *(Picea)*

Birch *(Betula)*

Pine *(Pinus)*

Oak *(Quercus)*

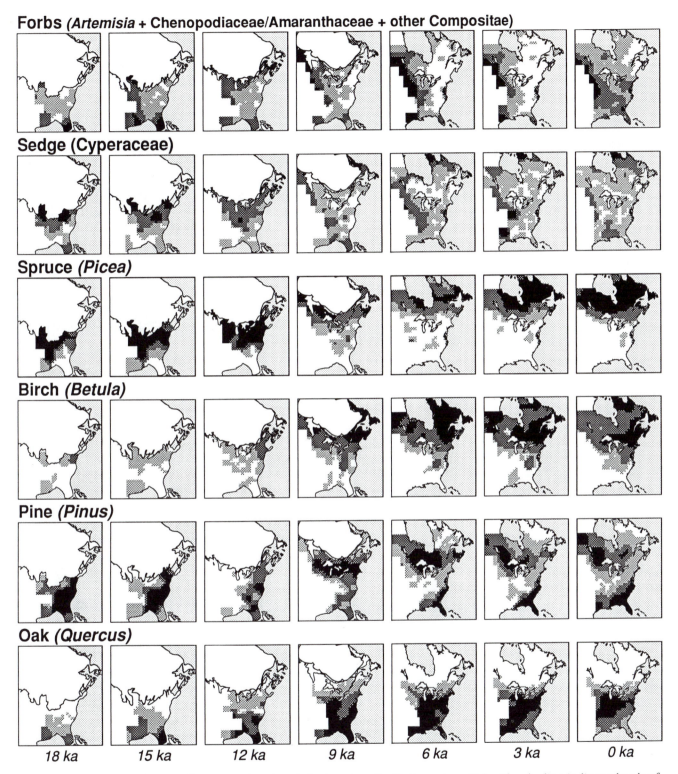

18 ka   15 ka   12 ka   9 ka   6 ka   3 ka   0 ka

Fig. 17.10. Isopoll maps for 12 pollen types showing their changing distributions since 18 ka. The shading indicates levels of abundance for each pollen type, with black the highest and white the lowest, as in Figure 11.6. The percentages are based on a sum of 24 pollen categories. The white ice sheet is shown in the north from 18 to 9 ka.

Fir *(Abies)*

Hemlock *(Tsuga)*

Beech *(Fagus)*

Elm *(Ulmus)*

Hickory *(Carya)*

Alder *(Alnus)*

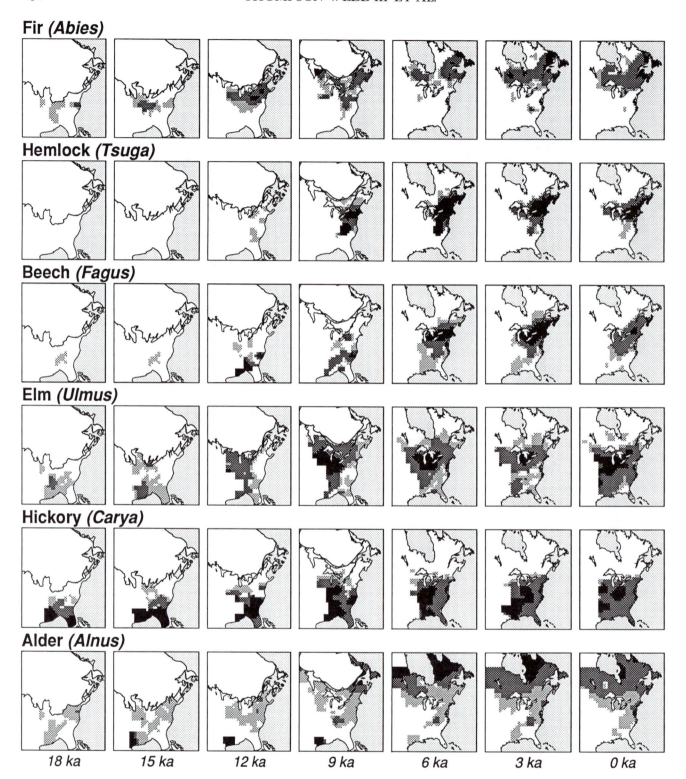

| 18 ka | 15 ka | 12 ka | 9 ka | 6 ka | 3 ka | 0 ka |

Fig. 17.10 (continued)

Mean January Temperature (°C)

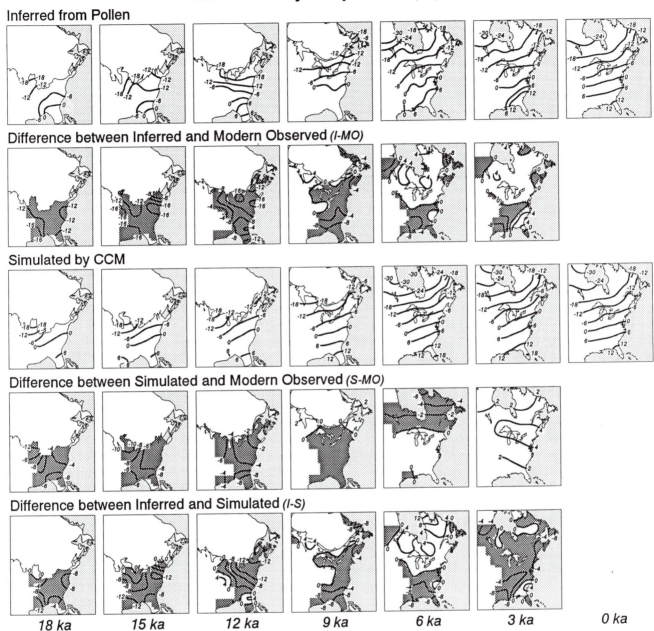

Fig. 17.11. Maps of (1) the estimates of mean January temperature from 18 to 3 ka inferred from the pollen data, (2) the differences between the estimates for each date and modern observed values, (3) the mean January temperatures simulated by the climate model, (4) the differences between the simulated values and modern observed values, and (5) the differences between the inferred and simulated values. Negative differences are shaded gray.

Mean July Temperature (°C)

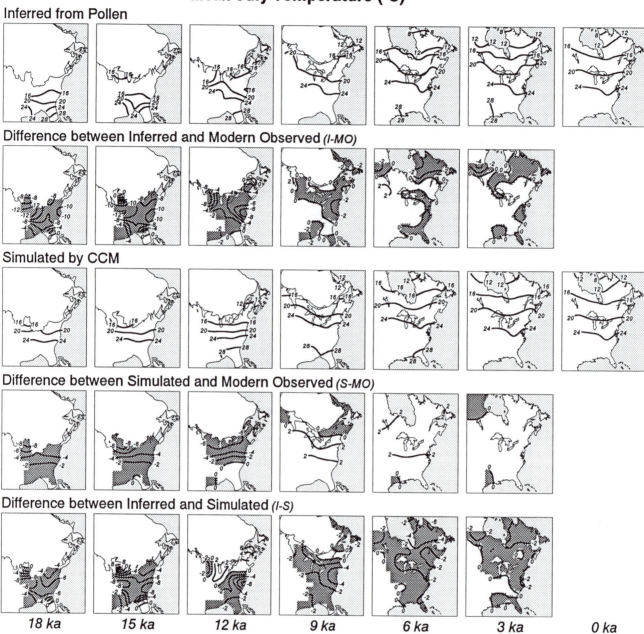

Fig. 17.12. Maps of (1) the estimates of mean July temperature from 18 to 3 ka inferred from the pollen data, (2) the differences between the estimates for each date and modern observed values, (3) the mean July temperatures simulated by the climate model, (4) the differences between the simulated values and modern observed values, and (5) the differences between the inferred and simulated values. Negative differences are shaded gray.

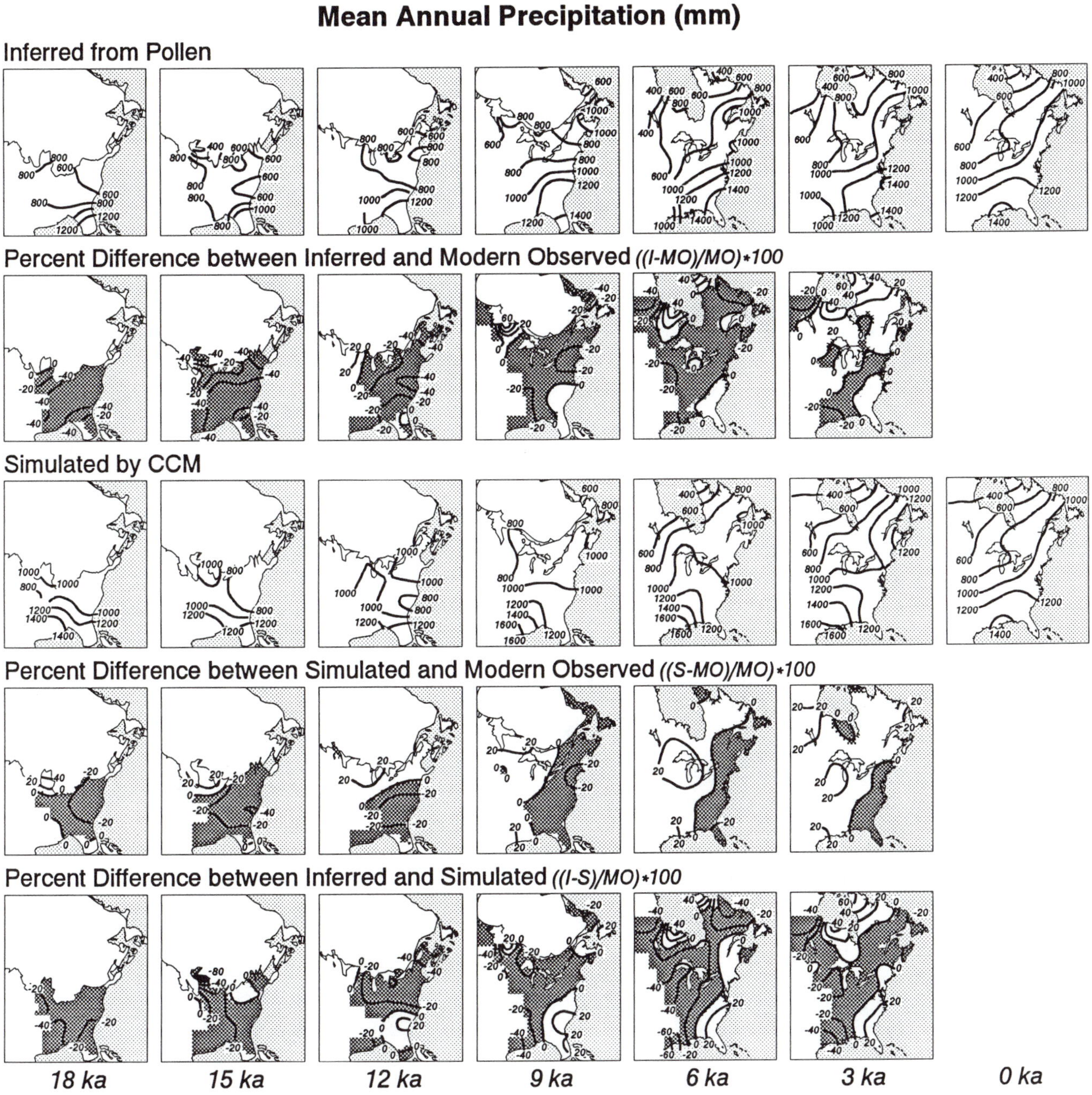

Mean Annual Precipitation (mm)

Inferred from Pollen

Percent Difference between Inferred and Modern Observed ((I-MO)/MO)*100

Simulated by CCM

Percent Difference between Simulated and Modern Observed ((S-MO)/MO)*100

Percent Difference between Inferred and Simulated ((I-S)/MO)*100

| 18 ka | 15 ka | 12 ka | 9 ka | 6 ka | 3 ka | 0 ka |

Fig. 17.13. Maps of (1) the estimates of annual precipitation from 18 to 3 ka inferred from the pollen data, (2) the percentage differences between the estimates for each date and modern observed values, (3) the annual precipitation simulated by the climate model, (4) the percentage differences between the simulated values and modern observed values, and (5) the percentage differences between the inferred and simulated values. Negative differences are shaded gray.

ditions were driest there. The 800-mm contours in the northeast at 9 ka show that this area was drier than at present by 10% or more. The inferred estimates for 18–12 ka indicate that most of the area south of the ice sheet received less precipitation than today, but the lower temperatures at that time made conditions moister than at present (Figs. 17.11–17.13).

These major trends in the climate estimates inferred from the pollen data are based on analyses of fossil samples with modern analogues. The time with the most no-analogue samples (48%) was 12 ka, when the pollen spectra from North Carolina to Minnesota did not match any modern spectra (Table 17.7). Several sites also lacked analogues at 15 ka (40%) and 18 ka (26%), but almost all pollen spectra from 9 to 3 ka (>85%) closely resembled one or several modern samples.

LAKE-LEVEL DATA

Lake-level data are a source of climatic information independent of pollen data and therefore provide additional evidence of the timing, pattern, and nature of climatic changes. A histogram of the aggregated lake-level data for eastern North America (Fig. 17.14) shows that conditions were moister at 12 ka than they are today and then became rapidly drier after 10 ka. Peak aridity was centered at 6 ka, when almost all lakes in the Midwest and two lakes in the Southeast were low (Fig. 17.15). The period after about 4 ka was marked by a return to more humid conditions, which have persisted to the present, although lakes do not appear to have returned to the high levels characteristic of the early part of the record.

At 12 ka most of the lakes in eastern North America were high or intermediate in status (Fig. 17.15 and Table 17.6), indicating that effective moisture was greater then than today or at 9 ka (Figs. 17.16 and 17.17). At 9 ka (Fig. 17.15) lakes in the Southeast and along the eastern seaboard were either intermediate or low in status, and several lakes in the Midwest that had been high at 12 ka had become intermediate. The Midwest remained relatively moist at 9 ka, however (Figs. 17.16 and 17.17).

By 6 ka the data from almost all lakes indicate conditions drier than at present (Fig. 17.16), and water lev-

els were low in all lakes in the northwestern Midwest (Fig. 17.15). This pronounced change in the pattern of lake status and effective moisture in this region indicates increased aridity during the early to middle Holocene (Figs. 17.15–17.17).

The present-day pattern of lake status became established between 5 and 2 ka (Table 17.6). By 3 ka most of the lake sites in eastern North America registered the high or intermediate water levels characteristic of today (Fig. 17.15) and were similar to the present in terms of effective moisture (Figs. 17.16 and 17.17; Table 17.6).

COMPARISON OF MAPPED PATTERNS OF POLLEN AND LAKE-LEVEL DATA

The patterns and types of change in the lake-level data are similar to those in the pollen data. The overall pattern and status of lake levels changed the most between 12 and 9 ka (Fig. 17.14), also the interval of greatest change in vegetation patterns (Fig. 17.10). In the Southeast north of Florida, the high lake levels and higher-than-present effective moisture at 12 ka (Figs. 17.15, 17.16, and 17.18) occurred while the temperatures inferred from the pollen data were lower than at present by 6°C or more in January and 1–4°C in July (Figs. 17.11, 17.12, and 17.19). The low effective moisture in this area at 9 and 6 ka (Figs. 17.15, 17.16, and 17.18) occurred when inferred July temperatures were similar to those of today (Fig. 17.19) but inferred annual pre-

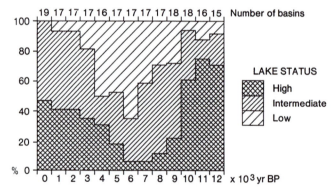

Fig. 17.14. Histogram showing the temporal variations in lake-level status for all sites in eastern North America.

Table 17.7. Number of fossil samples with analogue matches (squared chord distances <0.20) with one or more modern gridded samples[a]

| Date (yr B.P.) | Number of gridded samples | | | | | | | | | | | Total no. of fossil samples |
	0	1	2	3	4	5	6	7	8	9	10	
3000	19	0	1	0	1	2	0	0	3	1	269	296
6000	42	1	5	4	1	1	2	2	1	3	214	276
9000	19	1	1	0	1	2	2	1	3	1	192	223
12,000	41	0	0	1	0	1	0	1	1	0	40	85
15,000	10	1	1	0	0	3	0	0	0	0	10	25
18,000	5	0	0	0	0	3	1	0	0	0	10	19

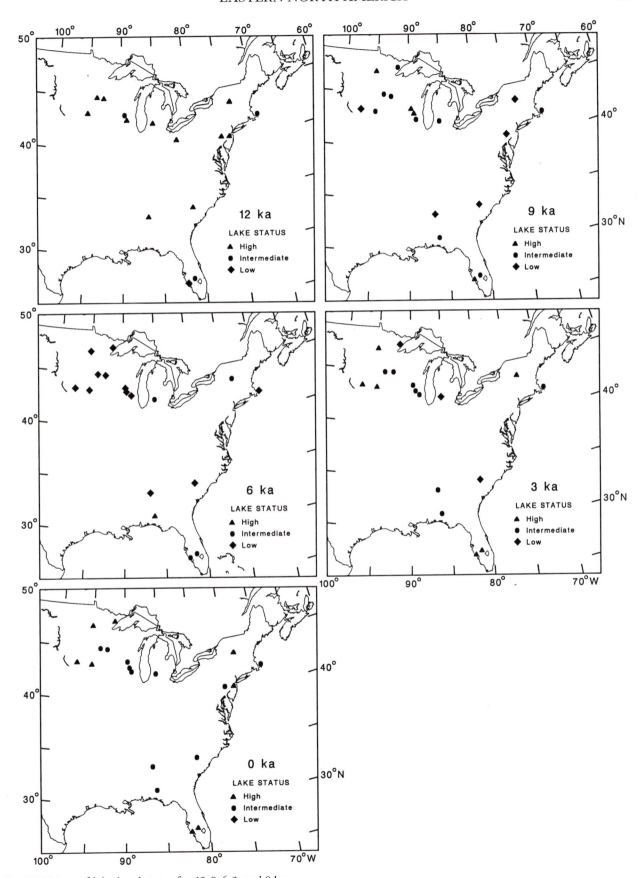

Fig. 17.15. Maps of lake-level status for 12, 9, 6, 3, and 0 ka.

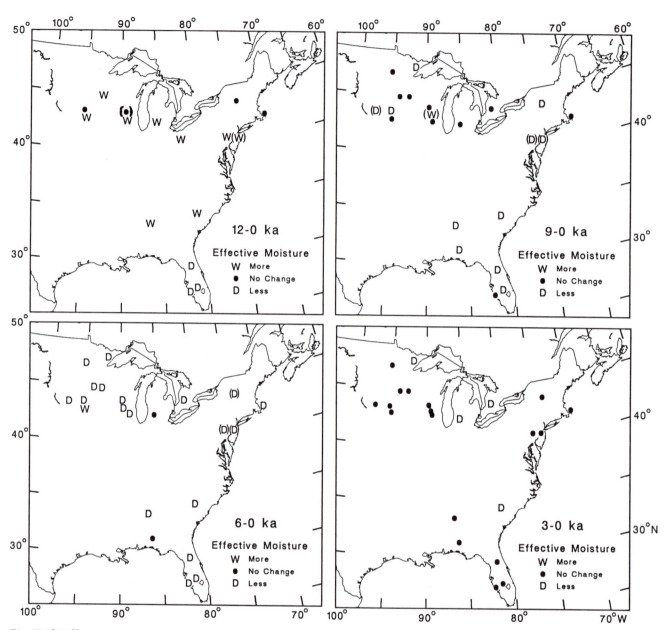

Fig. 17.16. Differences between the effective moisture index at 12, 9, 6, and 3 ka and at present. W indicates wetter conditions, with more effective moisture than at present; D indicates drier conditions, with less effective moisture than at present; and ● indicates no difference in effective moisture. Parentheses indicate uncertainty in the estimate.

cipitation was as much as 20% lower than at present (Fig. 17.19). By 3 ka effective moisture conditions in the Southeast were similar to those of today, as were the summer-season climatic conditions inferred from the pollen data (Fig. 17.19).

In the north-central region, the history of the lake-level data matches the well-studied history of the prairie-forest border (Fig. 17.10) (Wright *et al.*, 1963; McAndrews, 1966; Webb *et al.*, 1983a). The lake data indicate wetter-than-present conditions at 12 ka followed by a decline in water levels to near present values by 9 ka (Figs. 17.16 and 17.17), when the prairie-for-

est border had moved eastward into Minnesota and Iowa (Fig. 17.10). Inferred estimates of July temperature in the Midwest were much lower than present values at 12 ka but were approaching present values by 9 ka (Fig. 17.19). Inferred annual precipitation was close to modern levels at 12 and 9 ka (Fig. 17.19), and the well-marked east-west precipitation gradient had not yet formed (Fig. 17.13). By 6 ka, when lake levels were at their lowest in the Midwest (Figs. 17.15-17.18), the prairie-forest border (i.e., the eastern edge of black shading in Fig. 17.10) was farthest east, inferred precipitation was at its lowest point in the northwestern

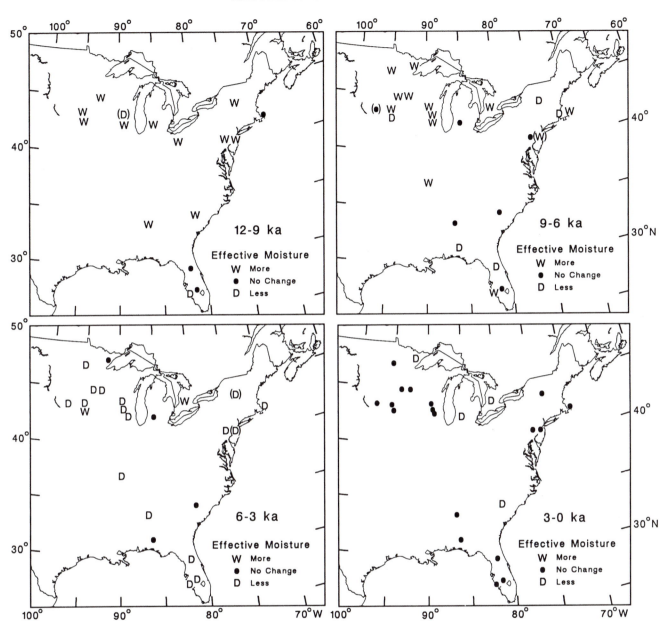

Fig. 17.17. Differences in the effective moisture index between 12 and 9 ka, 9 and 6 ka, 6 and 3 ka, and 3 ka and the present. W indicates wetter conditions, or more effective moisture, at the older date; D indicates drier conditions, or less effective moisture, at the older date; and ● indicates no difference in effective moisture. Parentheses indicate uncertainty in the estimate.

Midwest (Figs. 17.13 and 17.19), and inferred July temperatures were slightly higher than at present (Fig. 17.19). After 6 ka lake levels returned to present levels, and inferred precipitation and temperature conditions became similar to present conditions (Figs. 17.15–17.19).

The trends in the lake data are less marked in the Northeast than in the other regions, but the somewhat lower lake levels at 9 ka (Figs. 17.15–17.18) coincide with the period when northern *Pinus* populations were most abundant in the east and inferred precipitation was lower than at present (Figs. 17.10,

17.13, and 17.19). After 9 ka *Fagus, Tsuga,* and *Betula* populations replaced the populations of northern *Pinus* (Fig. 17.10), inferred precipitation values increased (Fig. 17.13), and inferred temperatures first rose and then fell (Fig. 17.19). Lake levels in general had risen to their present levels by 3 ka (Fig. 17.16).

In summary the general trends in the lake-level data are consistent with the temporal patterns of geographic change evident in the pollen data. The lake-level data are also consistent with the changes in climate inferred from the pollen data.

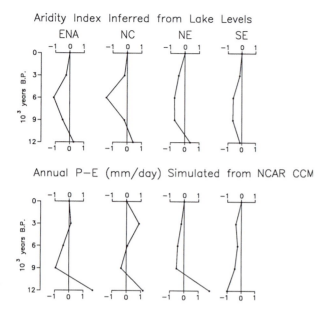

Fig. 17.18. Time series of the aridity index derived from lake-level data (difference of average lake status from present status) and of the model-simulated annual moisture balance (precipitation minus evaporation [P-E] scaled as differences from today in millimeters per day) for eastern North America (ENA) and for the north-central (NC), northeastern (NE), and southeastern (SE) regions.

CLIMATIC CHANGES INDICATED BY THE POLLEN AND LAKE-LEVEL DATA

The pollen and lake-level data indicate the following major changes in climate: (1) increasing mean January and July temperatures as ice retreated from 18 to 6 ka, with changes in the range of 5-10°C and a larger increase in winter than in summer (Fig. 17.19); (2) development of major modern gradients in summer and winter temperatures beginning by 9 ka and continuing during the Holocene (Figs. 17.11 and 17.12); (3) maximum warmth, mainly summer warmth, centered at 6 ka, with mean July temperatures 1-2°C higher than today (Fig. 17.19); (4) increased winter temperatures in the Southeast after 9 ka (Fig. 17.19); (5) maximum seasonality in temperature at 9 and 6 ka, although this is not a strong feature in the inferred climate maps; (6) moister conditions (P-E) than today at 12 ka, even though annual precipitation was less than today in most regions (Figs. 17.18 and 17.19); (7) development of an east-to-west decline in precipitation in the Midwest, with maximum aridity centered at 6 ka (Fig. 17.13); and (8) lower precipitation and moisture balance than today in the northeastern United States at 9 ka (Figs. 17.13 and 17.18).

COMPARISON OF DATA WITH CLIMATE-MODEL RESULTS

The maps and the area-average plots of the simulated temperature and precipitation estimates show many of the same general patterns and trends that are

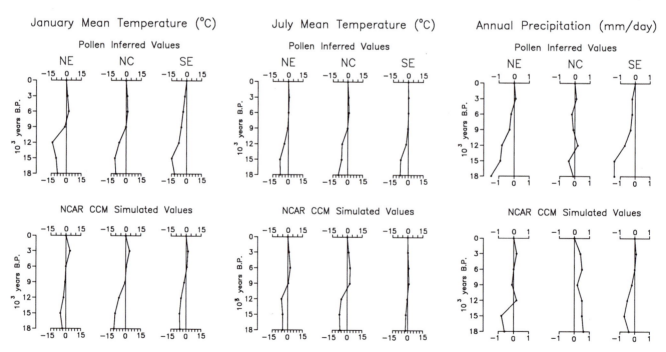

Fig. 17.19. Time series of area-average values (northeastern [NE], north-central [NC], and southeastern [SE] regions) for pollen-inferred and model-simulated January and July temperatures and annual precipitation from 18 ka to the present expressed as differences from today's values.

evident in the maps of the climate estimates inferred from the pollen data (Figs. 17.11–17.13 and 17.19). The simulated temperatures are lower than at present at 18 and 15 ka, increase slightly at 12 ka, and show the largest increase between 12 and 9 ka (Figs. 17.11, 17.12, and 17.19). The temperature estimates then change relatively little during the Holocene, but July temperatures in Canada decrease after 6 ka, and January temperature in the Southeast increases somewhat after 9 ka (Figs. 17.11 and 17.12). The modern gradients in summer and winter temperatures emerge by 9 ka, and an east-to-west gradient in precipitation is evident in the Midwest by 9 ka.

Despite the general agreement in pattern, there are some specific differences between the simulated and inferred climates. The simulated temperatures at 18 and 15 ka are much higher than the inferred temperatures (by 2–10°C for January and July), especially just south of the ice sheet (Figs. 17.11 and 17.12). The difference between inferred and simulated July temperatures is still pronounced at 12 ka, but the two sets of temperature estimates are generally within 1–3°C of each other for Holocene dates. The differences between inferred and simulated January temperatures are large, reflecting the greater overall range of this variable as well as the greater uncertainty in the inferred values (relative to those for July) (Prentice et al., 1991).

The inferred estimates for annual precipitation also are 10–20% lower than simulated values at 18 and 15 ka but are generally within 10% of the simulated values from 12 ka onward (Figs. 17.13 and 17.19). One key discrepancy is in the northern Midwest at 6 ka, where the inferred precipitation values are 20% less than the simulated values (Figs. 17.13 and 17.19).

The broad-scale pattern of hydrologic changes over eastern North America reconstructed from lake-level data is in general agreement with subcontinent-wide changes in the annual water budget simulated by the model (Fig. 17.18). Conditions overall were wetter than or as wet as today at 12 ka, became drier than today at 9 and 6 ka, and were similar to today by 3 ka. Like the pollen data, however, the lake data indicate much drier conditions at 12 ka than those simulated by the model, particularly in the northeastern and north-central regions (Figs. 17.13, 17.18, and 17.19). Climatic reconstructions from lake-level and pollen data suggest that the model overestimates moisture and precipitation at 12 ka.

A comparison for the whole region suggests a discrepancy between the data and model results in the timing of maximum aridity. The model simulates maximum aridity at 9 ka, whereas the lake-level data indicate peak aridity at 6 ka (Fig. 17.18). The model results for 6 ka are influenced by its simulation of much higher precipitation in the northern Midwest than is compatible with our interpretation of the pollen and lake-level data (Figs. 17.10, 17.13, and 17.16).

In the northeastern region both the model and the data show a more positive water balance at 12 ka than at present and a more negative water balance during the Holocene (Fig. 17.18). The trend of changes during the Holocene is also similar for both. The agreement between the lake data and the model results is good, even though the model simulated much wetter conditions at 12 ka than seem compatible with the data. The temporal trends for simulated and inferred temperature and precipitation are also similar, even though the inferred temperatures for January were much lower than the simulated values from 18 to 12 ka (Fig. 17.19).

In the north-central region, both the model and the lake-level data show a positive water balance at 12 ka (Fig. 17.18). The two sets of results, however, do not agree for the Holocene. The model simulated more precipitation in the region at 6 and 3 ka than the lake data indicate. Climate reconstructions from pollen data also suggest that the model overestimated precipitation in the western Great Lakes region at 6 ka (Figs. 17.13 and 17.19).

In the southeastern region, both the model and the lake data show more arid conditions (more negative water balance) than today at 12 ka and throughout most of the Holocene (Fig. 17.18). The model, however, predicts maximum aridity at 12 ka, whereas the data suggest that aridity was maximum at 9 and 6 ka. The temporal trends for inferred and simulated temperature and annual precipitation are similar, but the range of the simulated values is much less than that of the inferred values (Fig. 17.19).

Overall climatic inferences from the lake-level data agree with the model results more than they disagree. Simulations of precipitation and water balance are acknowledged to be key areas in the model that require improvement. This degree of agreement is therefore a significant finding. The lake-level data also reinforce the pollen data by indicating certain discrepancies with the model results. For example, the region of relatively high simulated precipitation around Lake Superior at 6 ka is not supported by either data set (Figs. 17.13, 17.15, 17.18, and 17.19).

Discussion

The set of climatic estimates inferred from the pollen data by the use of response surfaces and dissimilarity coefficients is unique to our study. Prentice et al. (1991) also calculated a set of these estimates but used only six pollen types. For this study we expanded the set of response surfaces to 14 pollen types. Theoretically the use of additional response surfaces should improve the climate estimates. Our analysis of the results suggests as much, but we have reserved the detailed comparison for another study (P. J. Bartlein, T. Webb III, K. H. Anderson, R. S. Webb, and B. Lipsitz, unpublished).

Another key contribution of our study is the combination and comparison of results for eastern North America that were previously presented separately. Our primary focus here has been to identify areas of agreement and disagreement between the pollen data and the lake-level data and among the data, the climate estimates inferred from them, and the climate-model results (Fig. 17.1). The comparison of the data sets has strengthened our confidence both in certain inferences about past climates and in our judgment of the model results. The comparison of the data, climate estimates, and model results has also supported our identification of the factors that varied among model runs as the controls of paleoclimatic variations in eastern North America.

The pollen and lake-level data and the climatic estimates inferred from them highlight six major features of the past 18,000 yr of climatic history in eastern North America: (1) the climate changed continuously from 18 ka to the present at the temporal and spatial scales that we focused on; (2) the major change was the temperature increase between 15 and 9 ka; (3) the greatest climatic changes took place between 12 and 9 ka; (4) the modern spatial gradients for July temperature and annual precipitation first appeared at 9 ka (Figs. 17.2, 17.12, and 17.13); (5) the time of maximum warmth was about 6 ka, when temperatures were not more than 1-2°C higher than at present; and (6) the time of maximum dryness in the last 12 ka was 6 ka in the Midwest and 9 ka in the northeastern United States.

The model simulations generally agree with these results (Figs. 17.11-17.13, 17.18, and 17.19). They show continuous changes in climate, with warming predominant from 15 to 6 ka and the largest change between 12 and 9 ka. The model results also indicate that the northeast-to-southwest gradient for July temperatures and the north-south and east-west gradients in precipitation first appeared at 9 ka (Figs. 17.12 and 17.13) (Kutzbach, Guetter, *et al.,* this vol.). Simulated July temperatures show a maximum at about 6 ka (Fig. 17.19), and simulated precipitation and moisture balance in the northeastern United States at 9 ka were less than at present (Figs. 17.18 and 17.19).

The main differences between the data and the model results are (1) that the model simulated temperatures south of the ice sheet much higher than those inferred from the pollen data from 18 to 12 ka and (2) that the model simulated too much rainfall and a positive moisture balance in the northern Midwest at 6 ka (Figs. 17.18 and 17.19). Both the pollen and the lake-level data differ from these model results.

Despite these differences, the simulated climatic values reproduce the patterns on the isopoll maps reasonably well (Webb *et al.,* 1987). This result shows that the direction, scale, and magnitude of changes in the isopoll maps are consistent with the types of change that might be attributable to climate. The strong tie between the pollen and lake-level results helps to confirm this result. The good agreement between the pollen and lake-level data would be hard to explain if the variations in these data sets were not both generated by climate variations. Our results, along with those of Webb *et al.* (1987) and Prentice *et al.* (1991), support the view that the vegetation was in equilibrium with the climate on the time and space scales shown on our maps (Webb, 1986). Because the climate changes continuously, vegetation is perpetually chasing climate, and despite lags in the response, vegetation tracks the major variations over time scales of 10^3-10^5 yr.

The general similarity between the data and the model results also implies that changes in the boundary conditions used in the climate simulations induced the climatic changes inferred from the data. The two key factors for eastern North America are the retreat and disappearance of the Laurentide ice sheet and the orbitally induced changes in insolation. These factors, combined with the changes in sea-surface temperatures, forced the sequence of regional climate change across eastern North America, which in turn caused the patterned sequence of changes in vegetation and regional water budgets.

The principal sequence of climate changes, described by Webb *et al.* (1987), began with the "glacial" period at 18 and 15 ka, when the influence of the Laurentide ice sheet dominated. The large ice sheet split the jet stream in winter, and storms tracked to the south of their positions today. Glacial anticyclonic flow predominated over the ice sheet and along its margins, and the winds in these areas had a greater easterly component than the winds today. A "transition" period followed in which increased seasonality in insolation combined with the still significant Laurentide ice sheet to create combinations of climate variables that do not exist today. The ice sheet was no longer high enough to split the winter jet stream, but glacial anticyclonic flow still influenced airflow just south of the ice sheet. The added summer radiation heated the southern United States and set up a sharp temperature gradient between there and the ice sheet. The seasonal contrast was pronounced at this time when the largest number of pollen samples differ significantly from the modern samples (i.e., have no modern analogues; see Table 17.7). The climatic patterns were changing relatively rapidly from their glacial orientation to their modern configuration. "Interglacial" conditions developed shortly after 9 ka, and maximum summer warmth came at 6 ka, when the Laurentide ice sheet was gone and July insolation was still 5% greater than today. The results of our study strengthen our confidence in this general description and explanation of the prominent climate changes across eastern North America during the past 18,000 yr.

Acknowledgments

Grants from the National Science Foundation Climate Dynamics Program to COHMAP (ATM-8406832 and ATM-8713981) and from the Department of Energy Carbon Dioxide Research Program (DE-FG02-85ER60304) supported this research. We thank J. Avizinis, P. Greenberg, P. Jaumann, T. Judd, P. Klinkman, S. Klinkman, B. Lipsitz, M. Mack, P. C. Newby, R. S. Webb, and L. Sheehan for technical assistance and H. E. Wright, Jr., and J. C. Ritchie for critical review of the manuscript. T. W. Anderson, R. E. Bailey, R. G. Baker, E. J. Cushing, A. M. Davis, R. P. Futyma, D. C. Gaudreau, B. C. S. Hansen, G. L. Jacobson, Jr., R. O. Kapp, G. King, H. F. Lamb, A. Larouche, J. H. McAndrews, R. J. Mott, P. C. Newby, R. J. Nickmann, J. G. Ogden III, W. A. Patterson III, P. J. H. Richard, L. C. K. Shane, S. K. Short, E. N. Smith, R. W. Spear, A. M. Swain, K. M. Trent, P. Tzedakis, K. L. Van Zant, J. C. B. Waddington, W. A. Watts, and D. R. Whitehead kindly contributed unpublished pollen data. We thank J. E. Kutzbach, P. Guetter, and P. Behling for supplying the model results and J. E. Kutzbach for discussions concerning them.

References

Albert, L. E., and Wyckoff, D. G. (1981). "Ferndale Bog and Natural Lake: Five Thousand Years of Environmental Change in Southeastern Oklahoma." Oklahoma Archaeological Survey, Studies in Oklahoma's Past, Vol. 7. University of Oklahoma, Norman.

Anderson, P. M., Bartlein, P. J., Brubaker, L. B., Gajewski, K., and Ritchie, J. C. (1989). Modern analogues of late-Quaternary pollen spectra from the western interior of North America. *Journal of Biogeography* 16, 573-596.

Anderson, R. S. (1979). "A Holocene Record of Vegetation and Fire at Upper South Branch Pond in Northern Maine." Unpublished M.S. thesis, University of Maine, Orono.

Anderson, T. W. (1971). "Post-glacial Vegetative Changes in the Lake Huron-Lake Simcoe District, Ontario, with Special Reference to Glacial Lake Algonquin." Unpublished Ph.D. thesis, University of Waterloo, Waterloo, Ontario.

——. (1980). Holocene vegetation and climatic history of Prince Edward Island, Canada. *Canadian Journal of Earth Sciences* 17, 1152-1165.

——. (1983). "Preliminary Evidence for Late Wisconsinian Climatic Fluctuations from Pollen Stratigraphy in Burin Peninsula, Newfoundland." Current Research, Part B, Paper 83-1B. Geological Survey of Canada, Ottawa.

Avizinis, J., and Webb, T. III (1983). The computer file of modern pollen and climatic data at Brown University. Unpublished manuscript, Department of Geological Sciences, Brown University.

Bailey, R. E. (1972). "Late- and Postglacial Environmental Changes in Northwestern Indiana." Unpublished Ph.D. thesis, Indiana University, Bloomington.

Bailey, R. E., and Ahearn, P. J. (1981). A late- and postglacial pollen record from Chippewa Bog, Lapeer Co., MI: Further examination of white pine and beech immigration into the central Great Lakes region. *In* "Geobotany II" (R. C. Romans, Ed.), pp. 53-74. Plenum Press, New York.

Baker, R. G. (1970). A radiocarbon-dated pollen chronology for Wisconsin: Disterhaft Farm Bog revisited. *Geological Society of America Abstracts* 2, 488.

Baker, R. G., Rhodes, R. S. II, Schwert, D. P., Ashworth, A. C., Frest, R. J., Hallberg, G. R., and Janssen, J. A. (1986). A full-glacial biota from southeastern Iowa, USA. *Journal of Quaternary Science* 1, 91-107.

Baker, R. G., Maher, L. J., Chumbley, C. A., and Van Zant, K. L. (1992). Patterns of Holocene environmental change in the midwestern United States. *Quaternary Research* 37, 379-389.

Barclay, F. H. (1957). "The Natural Vegetation of Johnson County, Tennessee, Past and Present." Unpublished Ph.D. thesis, University of Tennessee, Knoxville.

Barnosky, A. D., Barnosky, C. W., Nickmann, R. J., Ashworth, A. C., Schwert, D. P., and Lantz, S. W. (1988). Late Quaternary paleoecology at the Newton site, Bradford Co., northeastern Pennsylvania: *Mammuthus columbi*, palynology and fossil insects. *In* "Late Pleistocene and Early Holocene Paleoecology and Archaeology of the Eastern Great Lakes Region" (R. S. Laub, N. G. Miller, and D. W. Steadman, Eds.), pp. 173-184. Bulletin 33, Buffalo Society of Natural Sciences, Buffalo, N.Y.

Bartlein, P. J. (1988). Late Tertiary and Quaternary paleoenvironments. *In* "Vegetation History" (B. Huntley and T. Webb III, Eds.), pp. 113-152. Kluwer Academic Publishers, Dordrecht, The Netherlands.

Bartlein, P. J., and Webb, T. III (1985). Mean July temperature at 6000 yr B.P. in eastern North America: Regression equations for estimates from fossil-pollen data. *Syllogeus* 55, 301-342.

Bartlein, P. J., Webb, T. III, and Fleri, E. C. (1984). Holocene climatic change in the northern Midwest: Pollen-derived estimates. *Quaternary Research* 22, 361-374.

Bartlein, P. J., Prentice, I. C., and Webb, T. III (1986). Climatic response surfaces from pollen data for some eastern North American taxa. *Journal of Biogeography* 13, 35-57.

Bernabo, J. C. (1977). "Sensing Climatically and Culturally Induced Environmental Changes with Palynological Data." Unpublished Ph.D. dissertation, Brown University.

——. (1981). Quantitative estimates of temperature changes over the last 2700 years in Michigan based on pollen data. *Quaternary Research* 15, 143-159.

Birks, H. J. B. (1976). Late-Wisconsin vegetational history at Wolf Creek, central Minnesota. *Ecological Monographs* 46, 395-429.

——. (1981). Late Wisconsin vegetation and climatic history at Kylen Lake, northeastern Minnesota. *Quaternary Research* 16, 322-355.

Björck, S. (1985). Deglaciation chronology and revegetation in northwestern Ontario. *Canadian Journal of Botany* 22, 850-871.

Bostwick, L. K. (1978). "An Environmental Framework for Cultural Change in Maine: Pollen Influx and Percentage Diagrams from Monhegan Island." Unpublished M.S. thesis, University of Maine, Orono.

Bradshaw, R. H. W., and Webb, T. III (1985). Relationships between contemporary pollen and vegetation data from Wisconsin and Michigan, U.S.A. *Ecology* 66, 721-737.

Brubaker, L. B. (1975). Postglacial forest patterns associated with till and outwash in northcentral upper Michigan. *Quaternary Research* 5, 499-527.

Brugam, R. B., Grimm, E. C., and Eyster-Smith, N. M. (1988). Holocene environmental changes in Lily Lake, Minnesota inferred from fossil diatom and pollen assemblages. *Quaternary Research* 30, 53-66.

Bryant, V. M., Jr. (1977). A 16,000 year pollen record of vegetational change in central Texas. *Palynology* 1, 143-156.

Bryson, R. A. (1966). Air masses, streamlines, and the boreal forest. *Geographical Bulletin* 8, 228-269.

Bryson, R. A., and Hare, F. K. (1974). The climates of North America. In "Climates of North America" (R. A. Bryson and F. K. Hare, Eds.), pp. 1-47. Elsevier Scientific Publishing Company, Amsterdam.

Bryson, R. A., and Wendland, W. M. (1967). Tentative climatic patterns for some late glacial and postglacial episodes in central North America. *In* "Life, Land and Water" (W. J. Meyer-Oakes, Ed.), pp. 271-289. University of Manitoba Press, Winnipeg.

Cleveland, W. S., and Devlin, S. J. (1988). Locally weighted regression: An approach to regression analysis by local fitting. *Journal of the American Statistical Association* 83, 596-610.

COHMAP Members (1988). Climatic changes of the last 18,000 years: Observations and model simulations. *Science* 241, 1043-1052.

Comtois, P. (1982). Histoire holocène du climat et de la végétation à Lanoraie (Québec). *Canadian Journal of Earth Sciences* 19, 1938-1952.

Cotter, J. F. P. (1984). "The Minimum Age of the Woodfordian Deglaciation of Northeastern Pennsylvania and Northwestern New Jersey." Unpublished Ph.D. dissertation, Lehigh University.

Craig, A. J. (1969). Vegetational history of the Shenandoah Valley, Virginia. *Geological Society of America Special Paper* 123, 283-296.

——. (1972). Pollen influx to laminated sediments: A pollen diagram from northeastern Minnesota. *Ecology* 53, 46-57.

Cushing, E. J. (1967). Late-Wisconsin pollen stratigraphy and the glacial sequence in Minnesota. *In* "Quaternary Paleoecology" (E. J. Cushing and H. E. Wright, Jr., Eds.), pp. 59-88. Yale University Press, New Haven.

Davis, A. M. (1977). The prairie-deciduous forest ecotone in the upper Middle West. *Annals of the Association of American Geographers* 67, 204-213.

——. (1979). Wetland succession, fire and pollen record: A midwestern example. *American Midland Naturalist* 102, 86-94.

Davis, M. B. (1969). Climatic changes in southern Connecticut recorded by pollen deposition at Rogers Lake. *Ecology* 50, 409-422.

——. (1978). Climatic interpretation of pollen in Quaternary sediments. *In* "Biology and Quaternary Environments" (D. Walker and J. C. Guppy, Eds.), pp. 35-51. Australian Academy of Science, Canberra.

——. (1981). Outbreaks of forest pathogens in Quaternary history. *In* "Fourth International Palynological Conference, Proceedings, Vol. III" (D. C. Bharadwaj, Vishnu-Mittre, and H. K. Maheshwari, Eds.), pp. 216-227. Birbal Sahni Institute of Palaeobotany, Lucknow, India.

——. (1983). Holocene vegetation history of the eastern United States. *In* "Late Quaternary Environments of the United States, Vol. 2: The Holocene" (H. E. Wright, Jr., Ed.), pp. 166-181. University of Minnesota Press, Minneapolis.

Davis, P. T. (1980). "Late Holocene Glacial, Vegetational, and Climatic History of Pangnirtung and Kingnait Fiord Area, Baffin Island, Canada." Unpublished Ph.D. thesis, University of Colorado, Boulder.

Davis, R. B., Bradstreet, T. E., Stuckenrath, R., Jr., and Borns, H. W., Jr. (1975). Vegetation and associated environments during the past 14,000 years near Moulton Pond, Maine. *Quaternary Research* 5, 436-465.

Dean, W. E., Bradbury, J. P., Anderson, R. Y., and Barnosky, C. W. (1984). The variability of Holocene climatic change: Evidence from varved lake sediments. *Science* 226, 1191-1194.

Delcourt, H. R. (1979). Late Quaternary vegetation history of the eastern Highland Rim and adjacent Cumberland Plateau of Tennessee. *Ecological Monographs* 49, 255-280.

Delcourt, H. R., Delcourt, P. A., and Spiker, E. C. (1983). A 12,000-year record of forest history from Cahaba Pond, St. Clair County, Alabama. *Ecology* 64, 871-887.

Delcourt, P. A. (1980). Goshen Springs: Late Quaternary vegetation record for southern Alabama. *Ecology* 61, 371-386.

Delcourt, P. A., and Delcourt, H. R. (1984). Late Quaternary paleoclimates and biotic responses in eastern North America and the western North Atlantic Ocean. *Palaeogeography, Palaeoclimatology, Palaeoecology* 48, 263-284.

Delcourt, P. A., Delcourt, H. R., Brister, R. C., and Lackey, L. E. (1980). Quaternary vegetation history of the Mississippi embayment. *Quaternary Research* 13, 111-132.

Delcourt, P. A., Delcourt, H. R., and Webb, T. III (1984). Atlas of paired isophyte and isopoll maps for important tree taxa of eastern North America. *American Association of Stratigraphic Palynologists, Contribution Series* 14, 1-131.

Denton, G. H., and Hughes, T. J., Eds. (1981). "The Last Great Ice Sheets." Wiley-Interscience, New York.

Engstrom, D., and Hansen, B. (1985). Vegetational change and soil development in southeastern Labrador as inferred from pollen and chemical stratigraphy. *Canadian Journal of Botany* 63, 543-561.

Fries, M. (1962). Pollen profiles of late Pleistocene and recent sediments at Weber Lake, northeastern Minnesota. *Ecology* 43, 295-308.

Futyma, R. P. (1982). "Postglacial Vegetation of Eastern Upper Michigan." Unpublished Ph.D. thesis, University of Michigan, Ann Arbor.

Futyma, R. P., and Miller, N. G. (1986). Stratigraphy and genesis of the Lake Sixteen peatland, northern Michigan. *Canadian Journal of Botany* 64, 3008-3019.

Gaillard, M.-J. (1985). Postglacial paleoclimatic changes in Scandinavia and central Europe: A tentative correlation based on studies of lake-level fluctuations. *Ecologia Mediterranea* 11, 159-175.

Gajewski, K., Swain, A. M., and Peterson, G. M. (1987). Late Holocene pollen stratigraphy in four northeastern United States lakes. *Géographie physique et Quaternaire* 41, 377-386.

Gaudreau, D. C. (1986). "Late-Quaternary Vegetational History of the Northeast: Paleoecological Implications of Topographic Patterns in Pollen Distributions." Unpublished Ph.D. thesis, Yale University.

Gaudreau, D. C., and Webb, T. III (1985). Late-Quaternary pollen stratigraphy and isochrone maps for the northeastern United States. *In* "Pollen Records of Late-Quaternary North American Sediments" (V. M. Bryant, Jr., and R. G. Holloway, Eds.), pp. 247-280. American Association of Stratigraphic Palynologists Foundation, Dallas.

Gauthier, R. (1981). "Histoire de la colonisation végétale postglaciaire des Monteregiennes: deux sites du Mont Saint-Bruno." Unpublished M.Sc. memoir, Université de Montréal.

Gilliam, J. A., Kapp, R. O., and Bogue, R. D. (1967). A post-Wisconsin pollen sequence from Vestaburg Bog, Moncalm County, Michigan. *Michigan Academy of Science, Arts and Letters* 52, 3–17.

Grimm, E. C. (1983). Chronology and dynamics of vegetation change in the prairie-woodland region of southern Minnesota, U.S.A. *New Phytologist* 93, 311–350.

Grüger, J. (1973). Studies on the late Quaternary vegetation history of northeastern Kansas. *Geological Society of America Bulletin* 84, 239–250.

Hadden, K. A. (1975). A pollen diagram from a postglacial peat bog in Hants County, Nova Scotia. *Canadian Journal of Botany* 53, 39–47.

Harrison, S. P. (1988). Lake level records from Canada and the eastern U.S.A. *Lundqua Report* 29. Department of Quaternary Geology, Lund University, Lund.

——. (1989). Lake levels and climatic change in eastern North America. *Climate Dynamics* 3, 157–167.

Harrison, S. P., and Digerfeldt, G. (in press). European lakes as palaeohydrological and palaeoclimatic indicators. *Quaternary Science Reviews*.

Harrison, S. P., and Metcalfe, S. E. (1985a). Variations in lake levels during the Holocene in North America: An indicator of changes in atmospheric circulation patterns. *Géographie physique et Quaternaire* 39, 141–150.

——. (1985b). Spatial variations in lake levels since the last glacial maximum in the Americas north of the equator. *Zeitschrift für Gletscherkunde und Glazialgeologie* 21, 1–15.

Harrison, S. P., Saarse, L., and Digerfeldt, G. (1991). Holocene changes in lake levels as climate proxy data in Europe. *Paläoklimaforschung* 6, 159–179.

Heide, K. (1984). Holocene pollen stratigraphy from a lake and small hollow in north-central Wisconsin, USA. *Palynology* 8, 3–19.

Howe, S. E., and Webb, T. III (1983). Calibrating pollen data in climatic terms: Improving the methods. *Quaternary Science Reviews* 2, 17–51.

Huntley, B., and Prentice, I. C. (1988). July temperatures in Europe from pollen data, 6000 years before present. *Science* 241, 687–690.

Huntley, B., Prentice, I. C., and Bartlein, P. J. (1989). Climatic control of the distribution and abundance of beech (*Fagus* L.) in Europe and North America. *Journal of Biogeography* 16, 551–560.

Ibe, R. A. (1982). "Quaternary Palynology of Five Lacustrine Deposits in the Catskill Mountain Region of New York." Unpublished Ph.D dissertation, New York University.

——. (1985). Postglacial montane vegetational history around Balsam Lake, Catskill Mountains, New York. *Bulletin of the Torrey Botanical Club* 112, 176–186.

Imbrie, J., and Kipp, N. G. (1971). A new micropaleontological method for quantitative paleoclimatology: Application to a late Pleistocene Caribbean core. *In* "The Late Cenozoic Glacial Ages" (K. K. Turekian, Ed.), pp. 71–181. Yale University Press, New Haven.

Jackson, S. T. (1989). Pollen source area and representation in small lakes of the northeastern United States. *Review of Palaeobotany and Palynology* 63, 53–76.

Jacobson, G. L., Jr. (1979). The paleoecology of white pine (*Pinus strobus*) in Minnesota. *Journal of Ecology* 67, 697–726.

Jacobson, G. L., Jr., Webb, T. III, and Grimm, E. C. (1987). Patterns and rates of vegetation change during the deglacia-

tion of eastern North America. *In* "North America and Adjacent Oceans during the Last Deglaciation" (W. F. Ruddiman and H. E. Wright, Jr., Eds.), pp. 277–288. Geology of North America, Vol. K-3. Geological Society of America, Boulder, Colo.

Janssen, C. R. (1968). Myrtle Lake: A late- and postglacial pollen diagram from northern Minnesota. *Canadian Journal of Botany* 46, 1397–1410.

Jordan, R. H. (1975). Pollen diagrams from Hamilton Inlet, central Labrador, and their environmental implications for the northern maritime Arctic. *Arctic Anthropology* 12, 92–116.

Kapp, R. O. (1977). "Paleoecology in Central Lower Michigan." Field Trip Guide, Paleoecology Section, Ecological Society of America. Michigan State University, East Lansing.

Karrow, P. F., Anderson, T. W., Clarke, A. H., Delorme, L. D., and Sreenivasa, M. R. (1975). Stratigraphy, paleontology, and age of Lake Algonquin sediments in southwestern Ontario, Canada. *Quaternary Research* 5, 49–87.

Kay, P. A. (1979). Multivariate statistical estimates of Holocene vegetation and climate change, forest-tundra transition zone, NWT, Canada. *Quaternary Research* 11, 125–140.

Kerfoot, W. C. (1974). Net accumulation rates and the history of cladoceran communities. *Ecology* 55, 51–61.

Kim, H. K. (1986). "Late-Glacial and Holocene Environment in Central Iowa: A Comparative Study of Pollen Data from Four Sites." Unpublished Ph.D. thesis, University of Iowa, Iowa City.

King, G. A. (1987). "Deglaciation and Vegetation History of Western Labrador and Adjacent Quebec." Unpublished Ph.D. thesis, University of Minnesota, Minneapolis.

King, J. E. (1973). Late Pleistocene palynology and biogeography of the western Missouri Ozarks. *Ecological Monographs* 43, 539–565.

——. (1981). Late Quaternary vegetational history of Illinois. *Ecological Monographs* 51, 43–62.

King, J. E., and Allen, W. H., Jr. (1979). A Holocene vegetation record from the Mississippi River valley, southeastern Missouri. *Quaternary Research* 8, 307–323.

Kolb, C. R., and Fredlund, G. G. (1981). "Palynological Studies, Vacherie and Rayburn's Domes, North Louisiana Salt Dome Basin." Topical Report E530-02200-T-2. Institute for Environmental Studies, Louisiana State University, Baton Rouge.

Kutzbach, J. E. (1987). Model simulations of the climatic patterns during the deglaciation of North America. *In* "North America and Adjacent Oceans during the Last Deglaciation" (W. F. Ruddiman and H. E. Wright, Jr., Eds.), pp. 425–446. Geology of North America, Vol. K-3. Geological Society of America, Boulder, Colo.

Kutzbach, J. E., and Guetter, P. J. (1986). The influence of changing orbital parameters and surface boundary conditions on climate simulations for the past 18,000 years. *Journal of the Atmospheric Sciences* 43, 1726–1759.

Kutzbach, J. E., and Street-Perrott, F. A. (1985). Milankovitch forcing of fluctuations in the level of tropical lakes from 18 to 0 kyr B.P. *Nature* 317, 130–134.

Kutzbach, J. E., and Wright, H. E., Jr. (1985). Simulation of the climate of 18,000 years B.P.: Results for the North American/North Atlantic/European sector and comparison with the geologic record of North America. *Quaternary Science Reviews* 4, 147–187.

Labelle, C., and Richard, P. J. H. (1981). Végétation

tardiglaciaire et postglaciaire au sud-est du Parc des Laurentides, Québec. *Géographie physique et Quaternaire* 35, 345-359.

——. (1984). Histoire postglaciaire de la végétation dans la région de Mont-Saint-Pierre Gaspesie, Québec. *Géographie physique et Quaternaire* 38, 257-274.

Lamb, H. F. (1978). "Post-glacial Vegetation Change in Southern Labrador." Unpublished M.S. thesis, University of Minnesota, Minneapolis.

——. (1980). Late Quaternary vegetational history of southeastern Labrador. *Arctic and Alpine Research* 12, 117-135.

——. (1982). "Late Quaternary Vegetational History of the Forest-Tundra Ecotone in North-Central Labrador." Unpublished Ph.D. dissertation, Cambridge University.

Lamb, H. H., and Woodroffe, A. (1970). Atmospheric circulation during the last ice age. *Quaternary Research* 1, 29-58.

Lawrenz, R. (1975). "Biostratigraphic Study of Green Lake, Michigan." Unpublished M.S. thesis, Central Michigan University, Mount Pleasant.

Likens, G. E., and Davis, M. B. (1975). Post-glacial history of Mirror Lake and its watershed in New Hampshire, U.S.A.: An initial report. *Verhandlungen der Internationalen Vereinigung für theoretische und angewandte Limnologie* 19, 982-993.

Liu, K.-b. (1982). "Postglacial Vegetational History of Northern Ontario: A Palynological Study." Unpublished Ph.D. dissertation, University of Toronto.

Liu, K.-b., and Lam, N. S.-N. (1985). Paleovegetational reconstruction based on modern and fossil pollen data: An application of discriminant analysis. *Annals of the Association of American Geographers* 75, 115-130.

Livingstone, D. A. (1968). Some interstadial and postglacial pollen diagrams from eastern Canada. *Ecological Monographs* 38, 87-125.

McAndrews, J. H. (1966). Postglacial history of prairie, savanna, and forest in northeastern Minnesota. *Torrey Botanical Club Memoirs* 22, 1-72.

——. (1970). Fossil pollen and our changing landscape and climate. *Rotunda* 3, 30-37.

——. (1981). Late Quaternary climate of Ontario: Temperature trends from the fossil pollen record. *In* "Quaternary Paleoclimate" (W. C. Mahaney, Ed.), pp. 319-333. Geo Abstracts, Ltd., Norwich, England.

——. (1982). Holocene environment of a fossil bison from Kenora, Ontario. *Ontario Archaeology* 37, 41-51.

——. (1984). Histoire postglaciaire de la végétation dans la région de Mont-Saint-Pierre Gaspesie, Québec. *Géographie physique et Quaternaire* 38, 257-274.

——. (1988). Human disturbance of North American forests and grasslands: The fossil pollen record. *In* "Vegetation History" (B. Huntley and T. Webb III, Eds.), pp. 673-697. Kluwer Academic Publishers, Dordrecht, The Netherlands.

McAndrews, J. H., Riley, J. L., and Davis, A. M. (1982). Vegetation history of the Hudson Bay lowland: A postglacial pollen diagram from Sutton Ridge. *Naturaliste canadien* 109, 597-608.

McDowell, L. L., Dole, R. M., Jr., Howard, M., Jr., and Farrington, R. A. (1971). Palynology and radiocarbon chronology of Bugbee Wildflower Sanctuary and Natural Area, Caledonia County, Vermont. *Pollen et spores* 13, 73-91.

MacPherson, J. B. (1982). Postglacial vegetation history of the eastern Avalon Peninsula, Newfoundland and Holocene climatic change along the eastern Canadian seaboard. *Géographie physique et Quaternaire* 36, 175-196.

Maher, L. (1982). The palynology of Devils Lake, Sauk County, Wisconsin. *In* "Quaternary History of the Driftless Area" (M. E. Ostrom, Ed.), pp. 119-135. Field Trip Guide Book 5. Geological and Natural History Survey, University of Wisconsin-Extension, Madison.

Manny, B. A., Wetzel, R. G., and Bailey, R. E. (1978). Paleolimnological sedimentation of organic carbon, nitrogen, phosphorus, fossil pigments, pollen and diatoms in a hypereutrophic hardwater lake: A case history for eutrophication. *Polskie Archiwum Hydrobiologii* 25, 243-267.

Maxwell, J. A., and Davis, M. B. (1972). Pollen evidence of Pleistocene and Holocene vegetation on the Allegheny Plateau, Maryland. *Quaternary Research* 2, 506-530.

Miller, N. G. (1973). Late-glacial plants and plant communities in northwestern New York State. *Journal of the Arnold Arboretum* 54, 123-159.

Mode, W. (1980). "Quaternary Stratigraphy and Palynology of the Clyde Foreland, Baffin Island, N.W.T., Canada." Unpublished Ph.D. thesis, University of Colorado, Boulder.

Morrison, A. (1970). "Late-Glacial and Postglacial Vegetation Change in Southwestern New York State." Bulletin 420, New York State Museum and Science Service, Albany.

Mott, R. J. (1973). "Palynological Studies in Central Saskatchewan: Pollen Stratigraphy from Lake Sediment Sequences." Paper 72-49, Geological Survey of Canada, Ottawa.

——. (1975). Palynological studies of lake sediment profiles from southwestern New Brunswick. *Canadian Journal of Earth Sciences* 12, 273-288.

——. (1976). A Holocene pollen profile from the Sept-Iles area, Quebec. *Naturaliste canadien* 103, 457-467.

——. (1977). Late Pleistocene and Holocene palynology in southeastern Quebec. *Géographie physique et Quaternaire* 31, 139-149.

Mott, R. J., and Camfield, M. (1969). "Palynological Studies in the Ottawa Area." Paper 69-38, Geological Survey of Canada, Ottawa.

Mott, R. J., and Farley-Gill, L. D. (1978). A late-Quaternary pollen profile from Woodstock, Ontario. *Canadian Journal of Earth Sciences* 15, 1101-1111.

——. (1981). "Two Late Quaternary Pollen Profiles from Gatineau Park, Quebec. Paper 80-41, Geological Survey of Canada, Ottawa.

Nicholas, J. (1968). "Late Pleistocene Palynology of Southeastern New York and Northern New Jersey." Unpublished Ph.D. dissertation, New York University.

Nichols, H. (1967). Central Canadian palynology and its relevance to northwestern Europe in the late Quaternary period. *Review of Palaeobotany and Palynology* 2, 231-243.

——. (1975). "Palynological and Paleoclimatic Study of the Late Quaternary Displacement of the Boreal Forest-Tundra Ecotone in Keewatin and Mackenzie, N.W.T., Canada." Occasional Paper 15, Institute of Arctic and Alpine Research. University of Colorado, Boulder.

Niering, W. A. (1953). The past and present vegetation of High Point State Park, New Jersey. *Ecological Monographs* 23, 127-148.

Ogden, J. G. III (1966). Forest history of Ohio: Radiocarbon dates and pollen stratigraphy of Silver Lake, Ohio. *Ohio Journal of Science* 66, 387-400.

Ogden, J. G. III, and Hay, R. J. (1967). Ohio Wesleyan University natural radiocarbon measurements III. *Radiocarbon* 9, 316-332.

Overpeck, J. T. (1985). A pollen study of a late-Quaternary

peat bog: South-central Adirondack Mountains, New York. *Geological Society of America Bulletin* 96, 145-154.

Overpeck, J. T., Webb, T. III, and Prentice, I. C. (1985). Quantitative interpretation of fossil pollen spectra: Dissimilarity coefficients and the method of modern analogs for pollen data. *Quaternary Research* 23, 87-108.

Peteet, D. M., Vogel, J. S., Nelson, D. E., Southon, J. R., Nickman, R. J., and Heusser, L. E. (1990). Younger Dryas climatic reversal in northeastern USA? AMS ages for an old problem. *Quaternary Research* 33, 219-230.

Peters, M. A., and Webb, T. III (1979). A radiocarbon-dated pollen diagram from west-central Wisconsin. *Bulletin of the Ecological Society of America* 60, 102.

Prentice, I. C. (1988). Records of vegetation in time and space: The principles of pollen analysis. *In* "Vegetation History" (B. Huntley and T. Webb III, Eds.), pp. 17-24. Kluwer Academic Publishers, Dordrecht, The Netherlands.

Prentice, I. C., and Webb, T. III (1986). Pollen percentages, tree abundances and the Fagerlind effect: A comparison of vegetational calibration methods for pollen spectra. *Journal of Quaternary Science* 1, 35-43.

Prentice, I. C., Bartlein, P. J., and Webb, T. III (1991). Vegetation and climate changes in eastern North America since the last glacial maximum: A response to continuous climatic forcing. *Ecology* 72, 2038-2056.

Radle, N. J. (1981). "Vegetation History and Lake-Level Changes at a Saline Lake in Northeastern South Dakota." Unpublished M.S. thesis, University of Minnesota, Minneapolis.

Richard, P. J. H. (1977). "Histoire post-Wisconsinienne de la végétation du Québec meridional par l'analyse pollinique," 2 vols. Service de la recherche, Direction générale des forêts, Ministère des terres et forêts, Québec.

——. (1979). Contribution à l'histoire postglaciaire de la végétation au nord-est de la Jamesie, Nouveau-Québec. *Géographie physique et Quaternaire* 33, 93-112.

——. (1980). Histoire postglaciaire de la végétation au sud du lac Abitibi, Ontario et Québec. *Géographie physique et Quaternaire* 34, 77-94.

——. (1981). Paleophytogéographie postglaciaire en Ungava par l'analyse pollinique. *Paleo-Québec* 13, 1-153.

Richard, P. J. H., Larouche, A., and Bouchard, M. (1982). Age de la deglaciation finale et histoire postglaciaire de la végétation dans la partie centrale du Nouveau-Québec. *Géographie physique et Quaternaire* 36, 63-90.

Ritchie, J. C. (1969). Absolute pollen frequencies and carbon-14 age of a section of Holocene lake sediment from the Riding Mountain area of Manitoba. *Canadian Journal of Botany* 47, 1345-1349.

——. (1987). "Postglacial Vegetation of Canada." Cambridge University Press, Cambridge.

Ritchie, J. C., and Hadden, K. A. (1975). Pollen stratigraphy of Holocene sediments from the Grand Rapids area, Manitoba, Canada. *Review of Palaeobotany and Palynology* 19, 193-202.

Ritchie, J. C., and Lichti-Federovich, S. (1968). Holocene pollen assemblages from the Tiger Hills, Manitoba. *Canadian Journal of Earth Sciences* 5, 873-880.

Saarnisto, M. (1974). The deglaciation history of the Lake Superior region and its climatic implications. *Quaternary Research* 4, 316-339.

——. (1975). Stratigraphic studies on the shoreline displacement of Lake Superior. *Canadian Journal of Earth Sciences* 12, 300-319.

Samson, G., and McAndrews, J. H. (1977). Analyse pollinique et implications archeologiques et geomorphologiques, lac de Hutte Sauvage (Mushuau Nipi), Nouveau-Québec. *Géographie physique et Quaternaire* 31, 177-183.

Savoie, L., and Richard, P. (1979). Paleophytogéographie de l'épisode de Saint-Narcisse dans la région de Sainte-Agathe, Québec. *Géographie physique et Quaternaire* 33, 175-188.

Shane, L. C. K. (1975). Palynology and radiocarbon chronology of Battaglia Bog, Portage County, Ohio. *Ohio Journal of Science* 75, 96-102.

——. (1987). Late-glacial climatic history of the Allegheny Plateau and the till plains of Indiana, U.S.A. *Boreas* 16, 1-20.

Short, S. K., and Nichols, H. (1977). Holocene pollen diagrams from subarctic Labrador-Ungava: Vegetational history and climatic change. *Arctic and Alpine Research* 9, 265-290.

Sirkin, L. A., Denny, C. S., and Rubin, M. (1977). Late Pleistocene environment of the central Delmarva Peninsula, Delaware-Maryland. *Geological Society of America Bulletin* 88, 139-142.

Smith, E. N., Jr. (1984). "Late-Quaternary Vegetational History at Cupola Pond, Ozark National Scenic Riverways, Southeastern Missouri." Unpublished M.S. thesis, University of Tennessee, Knoxville.

Spear, R. W. (1989). Late-Quaternary history of high-elevation vegetation in the White Mountains of New Hampshire. *Ecological Monographs* 59, 125-151.

Spear, R. W., and Miller, N. G. (1976). A radiocarbon dated pollen diagram from the Allegheny Plateau of New York State. *Journal of the Arnold Arboretum* 57, 369-403.

Sreenivasa, M. R., and Duthie, H. C. (1973). The postglacial diatom history of Sunfish Lake, southwestern Ontario. *Canadian Journal of Botany* 51, 1599-1609.

Stravers, L. K. S. (1981). "Palynology and Deglaciation History of the Central Labrador-Ungava Peninsula." Unpublished M.S. thesis, University of Colorado, Boulder.

Street, F. A., and Grove, A. T. (1976). Environmental and climatic implications of late Quaternary lake-level fluctuations in Africa. *Nature* 261, 385-390.

——. (1979). Global maps of lake-level fluctuations since 30,000 yr B.P. *Quaternary Research* 12, 83-118.

Street-Perrott, F. A. (1986). The response of lake levels to climatic change—Implications for the future. *In* "Climate-Vegetation Interactions" (C. Rosenzweig and R. Dickinson, Eds.), pp. 77-80. Report OIES-2. University Corporation for Atmospheric Research, Office of Interdisciplinary Earth Sciences, Boulder, Colo.

Street-Perrott, F. A., and Harrison, S. P. (1984). Temporal variations in lake levels since 30,000 yr B.P.—An index of the global hydrological cycle. *In* "Climate Processes and Climate Sensitivity" (J. E. Hansen and T. Takahashi, Eds.), pp. 118-129. Geophysical Monograph 29. American Geophysical Union, Washington, D.C.

——. (1985). Lake levels and climate reconstruction. *In* "Paleoclimate Analysis and Modeling" (A. D. Hecht, Ed.), pp. 291-340. John Wiley & Sons, New York.

Street-Perrott, F. A., and Roberts, N. (1983). Fluctuations in closed lakes as an indicator of past atmospheric circulation patterns. *In* "Variations in the Global Water Budget" (A. Street-Perrott, M. Beran, and R. Ratcliffe, Eds.), pp. 331-345. D. Reidel, Dordrecht.

Suter, S. M. (1985). Late-glacial and Holocene vegetational history in southeastern Massachusetts: A 14,000 year pollen record. *Current Research* 2, 87-88.

Swain, P. C. (1979). "The Development of Some Bogs in Eastern Minnesota." Unpublished Ph.D. thesis, University of Minnesota, Minneapolis.

Terasmae, J. (1968). A discussion of deglaciation and the boreal forest in the northern Great Lakes region. *Proceedings of the Entomological Society of Ontario* 99, 31-43.

Terasmae, J., and Anderson, T. W. (1970). Hypsithermal range extension of white pine (*Pinus strobus* L.) in Quebec, Canada. *Canadian Journal of Earth Sciences* 7, 406-413.

Tzedakis, P. (1992). Effects of soils on the Holocene history of forest communities. *Géographie physique et Quaternaire* 46, 113-124.

Van Zant, K. L. (1979). Late-glacial and postglacial pollen and plant macrofossils from Lake West Okoboji, northwestern Iowa. *Quaternary Research* 12, 358-380.

Vincent, J. S. (1973). A palynological study for the Little Clay Belt, northwestern Quebec. *Naturaliste canadien* 100, 59-69.

Waddington, J. C. B. (1969). A stratigraphic record of pollen influx to a lake in the Big Woods of Minnesota. *Geological Society of America Special Paper* 123, 263-282.

Walker, P. C., and Hartman, R. T. (1960). The forest sequence of the Hartstown Bog area in western Pennsylvania. *Ecology* 41, 461-474.

Warner, B. G., Tolonen, K., and Tolonen, M. (1991). A postglacial history of vegetation and bog formation at Point Escuminac, New Brunswick. *Canadian Journal of Earth Sciences* 28, 1572-1582.

Watts, W. A. (1969). A pollen diagram from Mud Lake, Marion County, north-central Florida. *Geological Society of America Bulletin* 80, 631-642.

——. (1970). The full-glacial vegetation of northwestern Georgia. *Ecology* 51, 17-33.

——. (1971). Postglacial and interglacial vegetation history of southern Georgia and central Florida. *Ecology* 52, 676-690.

——. (1975a). A late Quaternary record of vegetation from Lake Annie, south-central Florida. *Geology* 3, 344-346.

——. (1975b). Vegetation record for the last 20,000 years from a small marsh of Lookout Mountain, northwestern Georgia. *Geological Society of America Bulletin* 86, 287-291.

——. (1979). Late Quaternary vegetation of central Appalachia and the New Jersey coastal plain. *Ecological Monographs* 49, 427-469.

——. (1980a). Late-Quaternary vegetation history at White Pond on the inner coastal plain of South Carolina. *Quaternary Research* 13, 187-199.

——. (1980b). The late Quaternary vegetation history of the southeastern United States. *Annual Reviews of Ecology and Systematics* 11, 387-409.

——. (1983). Vegetation history of the eastern United States, 25,000 to 10,000 years ago. *In* "Late Quaternary Environments of the United States, Vol. 1: The Late Pleistocene" (S. C. Porter, Ed.), pp. 294-310. University of Minnesota Press, Minneapolis.

Watts, W. A., and Bright, R. C. (1968). Pollen, seed, and mollusk analysis of a sediment core from Pickerel Lake, northeastern South Dakota. *Geological Society of America Bulletin* 79, 855-876.

Watts, W. A., and Hansen, B. C. S. (1988). Environments of Florida in the Late Wisconsin and Holocene. *In* "Wet Site Archaeology" (B. Purdy, Ed.), pp. 307-323. Telford Press, Caldwell, N.J.

Watts, W. A., and Stuiver, M. (1980). Late Wisconsin climate of northern Florida and the origin of the species-rich deciduous forest. *Science* 210, 325-327.

Watts, W. A., Hansen, B. C. S., and Grimm, E. C. (1992). A 40,000-yr record of vegetational and forest history from northwest Florida. *Ecology* 73, 1056-1066.

Webb, R. S. (1990). "Late Quaternary Water-Level Fluctuations in the Northeastern United States." Unpublished Ph.D. thesis, Brown University.

Webb, S. L. (1983). "The Holocene Extension of the Range of American Beech (*Fagus grandifolia*) into Wisconsin: Paleoecological Evidence for Long-Distance Seed Dispersal." Unpublished M.S. thesis, University of Minnesota.

Webb, T. III (1974). A vegetational history from northern Wisconsin: Evidence from modern and fossil pollen. *American Midland Naturalist* 92, 12-32.

——. (1982). Temporal resolution in Holocene pollen data. *In* "Third North American Paleontological Convention, Proceedings, Vol. 2" (B. Mamet and M. J. Copeland, Eds.), pp. 569-572. Business and Economic Service Limited, Toronto.

——. (1986). Is vegetation in equilibrium with climate? How to interpret late-Quaternary pollen data. *Vegetatio* 67, 75-91.

——. (1987). The appearance and disappearance of major vegetational assemblages: Long-term vegetational dynamics in eastern North America. *Vegetatio* 69, 177-187.

——. (1988). Eastern North America. *In* "Vegetation History" (B. Huntley and T. Webb III, Eds.), pp. 385-414. Kluwer Academic Publishers, Dordrecht, The Netherlands.

Webb, T. III, and Bryson, R. A. (1972). Late- and post-glacial climatic change in the northern Midwest, U.S.A.: Quantitative estimates derived from fossil pollen spectra by multivariate statistical analysis. *Quaternary Research* 2, 70-115.

Webb, T. III, and Clark, D. R. (1977). Calibrating micropaleontological data in climatic terms: A critical review. *Annals of the New York Academy of Sciences* 288, 93-118.

Webb, T. III, Howe, S. E., Bradshaw, R. H. W., and Heide, K. M. (1981). Estimating plant abundances from pollen data: The use of regression analysis. *Review of Palaeobotany and Palynology* 34, 269-300.

Webb, T. III, Cushing, E. J., and Wright, H. E., Jr. (1983a). Holocene changes in the vegetation of the Midwest. *In* "Late Quaternary Environments of the United States, Vol. 2: The Holocene" (H. E. Wright, Jr., Ed.), pp. 142-165. University of Minnesota Press, Minneapolis.

Webb, T. III, Richard, P. J. H., and Mott, R. J. (1983b). A mapped history of Holocene vegetation in southern Quebec. *Syllogeus* 49, 273-336.

Webb, T. III, Bartlein, P. J., and Kutzbach, J. E. (1987). Climatic change in eastern North America during the past 18,000 years: Comparisons of pollen data with model results. *In* "North America and Adjacent Oceans during the Last Deglaciation" (W. F. Ruddiman and H. E. Wright, Jr., Eds.), pp. 447-462. Geology of North America, Vol. K-3. Geological Society of America, Boulder, Colo.

Whitehead, D. R. (1964). Fossil pine pollen and full-glacial vegetation in southeastern North Carolina. *Ecology* 45, 767-777.

——. (1967). Studies of full-glacial vegetation and climate in the southeastern United States. *In* "Quaternary Paleoecology" (E. J. Cushing and H. E. Wright, Jr., Eds.), pp. 237-248. Yale University Press, New Haven.

——. (1972). Developmental and environmental history of the Dismal Swamp. *Ecological Monographs* 42, 301-315.

——. (1979). Late-glacial and post-glacial vegetational history of the Berkshires, western Massachusetts. *Quaternary Research* 12, 333-357.

——. (1981). Late-Pleistocene vegetational changes in northeastern North Carolina. *Ecological Monographs* 51, 451-471.

Whitehead, D. R., and Crisman, T. (1978). Paleolimnological studies of small New England (U.S.A.) ponds. I: Late glacial and postglacial trophic oscillations. *Polskie Archiwum Hydrobiologii* 25, 471-481.

Whitehead, D. R., and Jackson, S. T. (1990). "The Regional Vegetational History of the High Peaks (Adirondack Mountains), New York." Bulletin 478, New York State Museum, Albany.

Whitehead, D. R., and Sheehan, M. C. (1985). Holocene vegetational changes in the Tombigbee River Valley, eastern Mississippi. *American Midland Naturalist* 113, 122-137.

Wilkins, G. R. (1985). "Late-Quaternary Vegetation History at Jackson Pond, Larue County, Kentucky." Unpublished M.S. thesis, University of Tennessee, Knoxville.

Williams, A. S. (1974). "Late-Glacial-Postglacial Vegetational History of the Pretty Lake Region, Northeastern Indiana."

Professional Paper 686, U.S. Geological Survey. U.S. Government Printing Office, Washington, D.C.

Winkler, M. G. (1985). A 12,000-year history of vegetation and climate for Cape Cod, Massachusetts. *Quaternary Research* 23, 301-312.

Winkler, M., Swain, A. M., and Kutzbach, J. E. (1986). Middle Holocene dry period in the northern midwestern United States: Lake levels and pollen stratigraphy. *Quaternary Research* 25, 235-250.

Wright, H. E., Jr. (1984). Sensitivity and response time of natural systems to climatic change in the late Quaternary. *Quaternary Science Reviews* 3, 91-131.

Wright, H. E., Jr., and Watts, W. A. (1969). "Glacial and Vegetational History of Northeastern Minnesota." SP-11, Minnesota Geological Survey, Minneapolis.

Wright, H. E., Jr., Winter, T. C., and Patten, H. L. (1963). Two pollen diagrams from southeastern Minnesota: Problems in the regional late-glacial and postglacial vegetational history. *Geological Society of America Bulletin* 74, 1371-1396.

Wright, H. E., Jr., Almendinger, J. C., and Grüger, J. (1985). Pollen diagram from the Nebraska sandhills and the age of dunes. *Quaternary Research* 24, 115-120.

CHAPTER *18*

Climatic Changes in the Western United States since 18,000 yr B.P.

Robert S. Thompson, Cathy Whitlock, Patrick J. Bartlein, Sandy P. Harrison, and W. Geoffrey Spaulding

The physiography of the western United States is characterized by high relief (Fig. 18.1) and a wide variety of landforms, including massive mountains, smaller fault-block ranges, and broad dissected plateaus. Today this region has a wide range of seasonal climates resulting from the overlap of climatic gradients related to North Pacific (winter) and subtropical (summer) moisture sources. The variety of environmental settings, coupled with the absence of a uniform source of paleoclimatic data, makes it difficult to reconstruct past climatic conditions in quantitative terms. It is possible, however, to identify large-scale qualitative patterns in western paleoclimates.

This chapter summarizes the available data and inferred paleoclimatic patterns for the western United States over the last 18,000 yr. Stratigraphic pollen records, plant macrofossil assemblages from packrat middens, and lake-level records are the primary sources of paleoenvironmental information and are augmented by isotopic, glacial, vertebrate, and insect data sets. The primary data sets are not evenly distributed with regard to geography or elevation in this mountainous terrain, nor do they record climatic changes with the same temporal or spatial resolution.

To circumvent problems inherent in comparing such divergent sources of information, paleoclimatic estimates based on each data set were converted into estimates of the difference in levels of effective moisture (and in some cases, temperature) between past periods and today. We mapped the inferred paleoclimatic conditions (relative to today) for 18 ka (the continental glacial maximum), 12 ka (the continental deglaciation phase), 9 ka (the period of maximum Northern Hemisphere July insolation), and 6 ka (the time of peak overall aridity in the West). The patterns in these maps can be compared against paleoclimatic predictions derived from general circulation model

simulations of climate based on a variety of past boundary conditions (Kutzbach, Guetter, *et al.,* this vol.) to test the appropriateness of the model and boundary conditions. The model simulations in turn

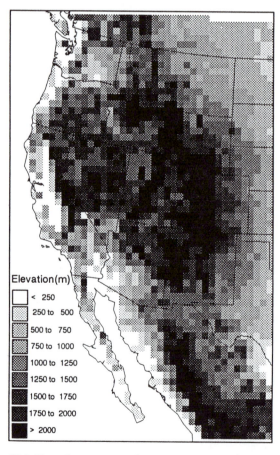

Fig. 18.1. Elevations across the western United States as viewed on a 50-km grid.

provide insights into the changes in atmospheric circulation that may have been responsible for the patterns observed in the paleoenvironmental data.

Modern Environment

CLIMATE

Regional climates in the American West are strongly influenced by the major mountain systems, which form rainshadows for westerly and, to a lesser degree, southerly flow and result in aridity in the interior of the region. Throughout the West, high relief creates strong elevational gradients in temperature and precipitation that greatly influence biotic distributions and local hydrologic conditions.

Temperature and seasonality patterns in the West are strongly correlated with latitude, elevation, and distance from the coast (Fig. 18.2). Precipitation in much of the region is associated with migratory low-pressure cells that follow westerly flow in the autumn, winter, and spring (Fig. 18.3). This flow is frequently blocked by the Cascade and Sierra Nevada ranges and is preferentially steered through lower-elevation divides among these major mountain systems (Bryson and Hare, 1974). Summer rainfall associated with southerly or southeasterly "monsoonal" flow from the Pacific or the Gulf of Mexico is abundant in the southeastern portion of the region (Bryson and Lowry, 1955; Hales, 1974; Huning, 1978; Tang and Reiter, 1984; Neilson, 1986). During this season the westerlies are weaker and are generally displaced far to the north, the eastern Pacific subtropical high-pressure system is expanded, and consequently the West Coast and much of the interior is dry.

Precipitation gradients show strong seasonal variations across the West as a result of contrasts between principally summer-dominant and winter-dominant precipitation regimes (Fig. 18.3). Instrumental data from this century (Sellers, 1968) and dendrochronologic evidence from the last 300 yr (LaMarche and Fritts, 1971) indicate a persistent opposition for many months of the year between precipitation regimes in the Northwest and the Southwest. This pattern appears to reflect in part the changing position of the westerlies. When the jet stream is shifted farther north than normal, the Northwest is moist and the Southwest is dry, whereas when the jet stream migrates farther south than normal, the reverse situation occurs. Strong contrasts between the Northwest and Southwest are evident in the changing patterns of effective moisture over much of the last 18,000 yr, although the causes of these contrasts have varied through time. Changes in temperature patterns across the West during this same time interval were influenced by latitude, elevation, and distance from the coast as well as by the same changes in circulation that affected moisture levels.

VEGETATION

Coniferous forests with pine, spruce, fir, and Douglas fir are dominant in the mountainous regions of most of the western United States. (Common names for plants and animals are used throughout this chapter; Table 18.1 provides the Latin equivalents. An exception is made for *Artemisia* pollen, which includes the widespread sagebrush as well as a number of other related shrubs and herbs.) Along the northwest coast these forests descend to sea level, whereas in many of the interior and southwestern ranges they are restricted to middle and upper elevations.

The western forests are not uniform in composition but rather vary with the climatic gradients described above. Along the Pacific coast is a zone of maritime coniferous forest growing in a climate of high winter precipitation and low seasonal contrasts. Included in these forests are hemlock, cedar, and Douglas fir as well as coastal species of spruce, pine, and fir. Where maritime air penetrates eastward into northern Idaho, northwestern Montana, and southeastern British Columbia, these coastal species intermingle with Rocky Mountain conifers. Alpine tundra and barren ground are found on high peaks throughout the West.

Woodland and chaparral communities occur below forests in California, the desert Southwest, and portions of the surrounding regions. Typical woodland dominants include junipers, piñon pines, and scrub oak. Evergreen, nonconiferous, sclerophyllous plants such as scrub oak, sumac, and buckthorn are typical of chaparral.

Grasslands cover the central valley of California, portions of the middle elevations in the Southwest, the Great Plains, and smaller areas of the Northwest. Steppe vegetation, dominated by sagebrush, shadscale, and/or other composite or chenopod shrubs, covers the valleys of the Great Basin, Colorado Plateau, and adjacent regions. To the south, in the Mojave, Sonoran, and Chihuahuan deserts, xerophytic shrubs such as creosote bush grow in hot and dry low-elevation landscapes. Subtropical succulents such as saguaro are present in the essentially frost-free Sonoran Desert but not in the colder winter climates of the Mojave and northern Chihuahuan deserts.

LAKES

Closed drainage basins are common in the arid interior of the western United States, and lakes in these settings provide sensitive records of changes in water balance (Street-Perrott and Harrison, 1985). Many western basins are now dry or contain lakes that are smaller than their maximum Pleistocene extents (Snyder *et al.*, 1964; Mifflin and Wheat, 1979; Smith and Street-Perrott, 1983). At present, permanent freshwater lakes occur only in the northern portion of the region or at high elevations, whereas the basins of the southwestern deserts contain dry or seasonally wet playas.

Temperature (°C)

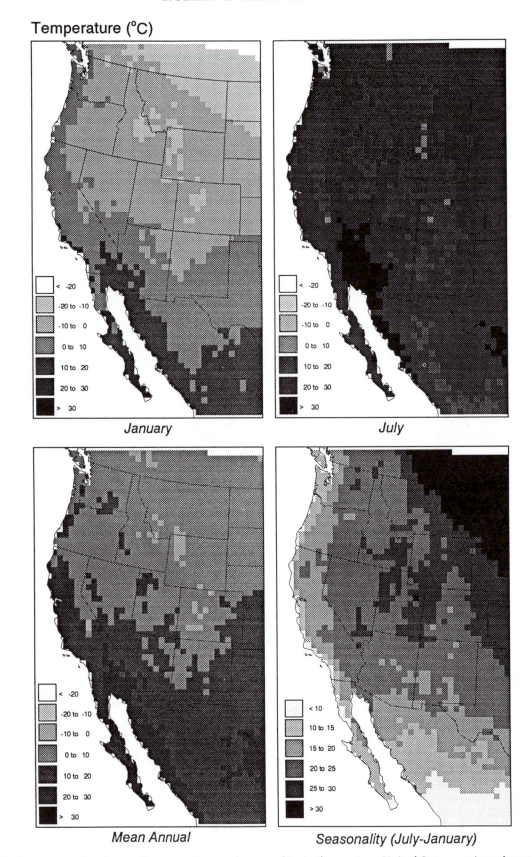

Fig. 18.2. Modern seasonal and annual temperatures and seasonality in the western United States as viewed on a 50-km grid. (See Lipsitz [1988] for a discussion of the techniques employed to estimate climatic parameters on this grid.)

Log of Precipitation and Months of Maximum Precipitation

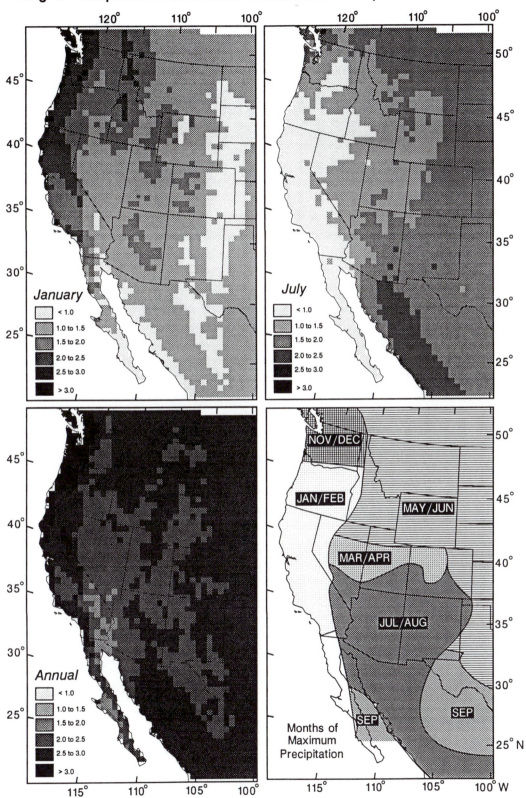

Fig. 183. Modern seasonal and annual precipitation in the western United States as viewed on a 50-km grid and (lower right) the months of maximum precipitation. Modified from Tang and Reiter (1984). (See Lipsitz [1988] for a discussion of the techniques employed to estimate climatic parameters on this grid.)

Table 18.1. Latin equivalents of plant and animal names used in this chapter

Common name	Latin equivalent	Common name	Latin equivalent
Plants		Palo verde, blue	*Cercidium floridum*
Alder	*Alnus*	Pine	*Pinus*
Alder, red	*Alnus rubra*	Pine, border piñon	*Pinus discolor*
Ash, single-leaf	*Fraxinus anomala*	Pine, bristlecone	*Pinus longaeva*
Barberry	*Berberis*	(intermountain)	
Blackbrush	*Coleogyne ramosisima*		
		Pine, Colorado piñon	*Pinus edulis*
Brittlebush	*Encelia farinosa*	Pine, Jeffrey	*Pinus jeffreyi*
Buckthorn	*Rhamnus*	Pine, limber	*Pinus flexilis*
Catclaw	*Acacia greggii*	Pine, lodgepole	*Pinus contorta*
Cedar	*Thuja* and/or	Pine, papershell piñon	*Pinus remota*
	Chamaecyparis		
Cedar, incense	*Calocedrus decurrens*	Pine, ponderosa	*Pinus ponderosa*
		Pine, single-needle piñon	*Pinus monophylla*
Cedar, pygmy	*Peucephyllum schottii*	Pine, sugar	*Pinus lambertiana*
Cedar, western red	*Thuja plicata*	Pine, whitebark	*Pinus albicaulis*
Chenopods	Chenopodiaceae	Prickly pear	*Opuntia* sec.
Cholla, teddy bear	*Opuntia thurberii*		*platyopuntia*
Composites	Compositae		
	(Asteraceae)	Rose	*Rosa*
		Sage, Mojave	*Salvia mohavensis*
Creosote bush	*Larrea divaricata*	Sagebrush	*Artemisia* sec.
Cypress, Arizona	*Cupressus arizonica*		*tridentatae*
Douglas fir	*Pseudotsuga menziesii*	Saguaro	*Cereus giganteus*
Fir	*Abies*	Saltbush	*Atriplex canescens*
Fir, grand	*Abies grandis*		
		Sequoia, giant	*Sequoiadendron*
Fir, red	*Abies magnifica*		*giganteum*
Fir, subalpine	*Abies lasiocarpa*	Shadscale	*Atriplex confertifolia*
Grasses	Gramineae (Poaceae)	Snowberry	*Symphoricarpos*
Hackberry	*Celtis*	Sotol	*Dasilyrion*
Hemlock	*Tsuga*	Spruce	*Picea*
Hemlock, mountain	*Tsuga mertensiana*	Spruce, blue	*Picea pungens*
Hemlock, western	*Tsuga heterophylla*	Spruce, Engelmann	*Picea engelmannii*
Hornwort	*Ceratophyllum*	Spruce, sitka	*Picea sitchensis*
	demersum	Sumac	*Rhus*
Joshua tree	*Yucca brevifolia*	Yucca	*Yucca*
Juniper	*Juniperus*	Yucca, whipple	*Yucca whippelii*
		Animals	
Juniper, California	*Juniperus californica*	Lemming	*Dicrostonyx* or
Juniper, prostrate	*Juniperus communis*		*Synaptomys*
Juniper, redberry	*Juniperus erythrocarpa*	Lemming, collared	*Dicrostonyx torquatus*
Juniper, Rocky Mountain	*Juniperus scopulorum*	Lemming, northern bog	*Synaptomys borealis*
Juniper, single-seed	*Juniperus monosperma*	Marmot	*Marmota*
		Packrat (woodrat)	*Neotoma*
Juniper, Utah	*Juniperus osteosperma*	Perch, tule	*Hysterocarpus traski*
Juniper, western	*Juniperus occidentalis*		
Mesquite	*Prosopis juliflora*	Pika	*Ochotona princeps*
Mormon-tea	*Ephedra*	Rabbit, pygmy	*Brachylagus idahoensis*
Mountain mahogany	*Cercocarpus*	Shrew, least	*Crytotis parva*
		Tortoise, desert	*Gopherus agassizi*
Mountain mahogany, little-leaf	*Cercocarpus intricatus*	Vole, heather	*Phenacomys*
Nutmeg, California	*Torreya california*		*intermedius*
Oak	*Quercus*		
Oak, Ajo	*Quercus ajoensis*	Vole, prairie	*Microtus ochrogaster*
Oak, Gambell	*Quercus gambellii*	Vole, sagebrush	*Lagurus curtatus*
		Woodrat, eastern	*Neotoma floridana*
Oak, Hinckley	*Quercus hinckleyii*		

Paleoenvironmental Data

VEGETATION DATA

Palynologic studies are the principal source of paleoclimatic data for the Pacific Northwest, California, and the Rocky Mountains, where lakes and wetlands are common. Such sites are rare at low elevations in the arid interior and southwestern deserts, and dry cave sediments and fluvial deposits provide the few stratigraphic pollen records from these areas. Chronologic controls for palynologic studies are provided by radiocarbon dates and, throughout the Pacific Northwest and northern Rocky Mountains, by volcanic ash layers.

Fossil pollen usually can be identified to the generic level. Some taxa can be identified at the subgeneric level, and plant macrofossils, where present, provide higher levels of taxonomic resolution than are possible with pollen data alone. The spatial coverage of pollen spectra varies with topographic setting, site type and size, amount of river input, and other factors. The temporal resolution varies with deposition rate and dating controls.

Most of the paleoclimatic information from the western deserts comes from plant macrofossil assemblages recovered from ancient packrat middens (e.g., Betancourt *et al.* [1990] and references therein). In a few cases, pollen and packrat midden records have been found together (Betancourt and Van Devender, 1981; Thompson and Kautz, 1983; Thompson, 1985; Mehringer and Wigand, 1987), but the arid conditions that favor midden preservation do not lend themselves to good stratigraphic pollen records, and vice versa.

The paleoecologic data gained from packrat middens are site-specific and are of high taxonomic resolution. Packrats forage generally within 50 m of their dens, collecting seeds, leaves, twigs, and spines from a wide range of plants. Modern midden contents correspond well with surrounding vegetation on a presence-absence basis (Cole, 1985; Spaulding, 1985). However, no measure yet devised can consistently relate the representation of a plant taxon in a midden assemblage to its abundance in the surrounding vegetation (e.g., Spaulding *et al.,* 1990). A midden may be built within a few months to a few years, but it is not known (given current dating techniques) whether a single stratum from a consolidated midden represents a similarly brief span. It is generally accepted that the standard deviation of a sample's radiocarbon date, although it does not define the period of accumulation, likely brackets the time interval of deposition. Our system for selecting dated assemblages for data-model comparisons dictated that the radiocarbon age (either single dates or weighted means [Long and Rippeteau, 1974] of several dates from one assemblage) be within 1000 yr of the "target date" (i.e., 6, 9, 12, or 18 ka).

In general the climatic signals in vegetation data are interpreted from past changes in the elevational or geographic limits of plant taxa. Upslope movement of upper range limits of plants may imply increased growing-season temperatures. Downslope movement of lower limits of plant species implies increased effective moisture at low elevations. Geographic changes in plant distributions reflect climate variations in a similar fashion; simply put, northward movement of a species over time may reflect rising temperatures and southward movement falling temperatures. Such range adjustments may also occur in response to changes in the amount, regional distribution, or seasonality of precipitation. Reliable regional climatic reconstructions are possible only when the effects of local variability can be overcome by comparison and corroboration of several records.

LAKE-LEVEL DATA

Two major approaches have been taken to the reconstruction of lake-level changes in the western United States: studies of shoreline features and studies of sediment stratigraphy. In the latter, fluctuations in lake level are recorded in sedimentological, geochemical, and paleontological data. Stratigraphic studies of cores or of lake-margin outcrops offer the best temporal resolution but generally yield estimates of relative changes in past lake depth only. Geomorphic studies of shorelines, on the other hand, permit more accurate reconstruction of past lakeshore elevations (and hence paleolake depth and area) but are more difficult to date. The most desirable approach combines both avenues of investigation.

Western lake basins vary greatly in size, bathymetry, and topographic setting, and these differences may greatly affect the hydrologic response of lakes to climatic changes. Lakes in small basins are highly sensitive to climatic changes and are potentially excellent recorders of high-frequency, low-amplitude events. Lakes in large basins respond more slowly and are likely to record only high-amplitude events. Deep, narrow basins such as fault-bounded troughs are highly sensitive to late-Quaternary climatic variations, whereas flat-floored basins are relatively insensitive. Lakes that lie in rainshadows of major mountain ranges, such as Pleistocene Lake Lahontan and smaller modern lakes in its basin, often are fed primarily by perennial rivers draining from the mountains and receive only small inputs from direct precipitation on the lake surface and immediate environs.

The quality of dating controls is central to the study of lake-level fluctuations because fluctuations in lake size not only cause variations in sedimentation rates, which make the interpolation of ages between chronologic markers unreliable, but also result in the superposition of shorelines of differing ages. Chronologic controls for western paleolimnological studies are provided by radiocarbon dating and, in some basins, tephrochronology. Accelerator mass spectrometer (AMS) radiocarbon dates may be more reliable than conventional dates, and dates on organic materials are probably more reliable than those on carbonates. Lake sediments from forest regions contain a variety of acceptable organic remains, whereas those from closed basins in arid regions may contain abundant carbonate but little organic material. Carbonate radiocarbon dates are often unreliable because of recrystallization and precipitation of secondary carbonates.

The basic paleoclimatic interpretation of lake-level changes is straightforward: rises in lake levels imply increased effective moisture, and decreases in lake levels imply decreased effective moisture. More detailed interpretations are difficult first because similar changes in lake levels may result from changes in precipitation, evaporation, or both and second because the controlling climatic parameter may vary in a given basin through time. Lake-level information is limited in one additional aspect—once a lake desiccates during an arid interval, it records no further information on the severity of the drought. As with vegetation data, reliable regional climatic reconstructions based on lake-level data are possible only when the effects of local factors can be evaluated by comparison of several records.

PALEOCLIMATIC MAPS

Collectively, the vegetation and lake-level data sets provide regional coverage of qualitative changes in climate since 18 ka. To facilitate comparisons of the data sets for mapping purposes, we interpreted the paleoenvironmental data on a three-part scale reflecting major variations in "effective moisture." (The term *effective moisture* is used to take into account the effect of variations in evaporation rates associated with temperature changes. For example, although portions of the West were cold and received relatively little precipitation during the Late Wisconsin, the reduced temperatures and increased cloud cover apparently lowered evaporation rates, which made these sites "effectively" more moist than they are today.) The reconstructed conditions at a particular site for a given time period were compared to modern conditions and ranked as (1) effectively moister than today (i.e., more precipitation or greatly reduced evaporation), (2) not unequivocally different from today, or (3) effectively drier than today (less precipitation or greatly increased evaporation). For the paleoclimatic "extremes" of 18 ka (cold) and 9 ka (warm), we also interpreted the selected paleoclimatic evidence in terms of qualitative departures from modern temperatures.

GEOGRAPHIC REGIONS

We used the following regional divisions (Fig. 18.4) as a framework for the paleoenvironmental discussions: (1) Northwest (the coastal and Cascade regions of Washington and Oregon, the Columbia Plateau,

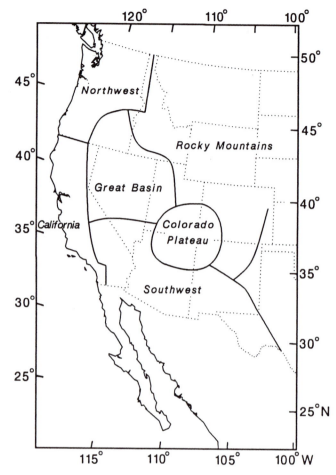

Fig. 18.4. Geographic regions discussed in text.

and parts of adjacent northern Idaho); (2) Rocky Mountains (northern and central Rocky Mountains, intervening valleys, and adjacent High Plains, covering portions of Idaho, Utah, and Colorado and all of Wyoming and Montana); (3) Great Basin (that portion of the basin-and-range physiographic province where sagebrush and shadscale dominate the valley-bottom vegetation, covering portions of Oregon, Idaho, California, Nevada, and Utah); (4) Colorado Plateau (the physiographic Colorado Plateau and associated mountain ranges, covering portions of Utah, Colorado, New Mexico, and Arizona); (5) Southwest (the Chihuahuan, Sonoran, and Mojave deserts and adjacent highlands, covering portions of west Texas, northern Mexico, New Mexico, Arizona, Nevada, and California); and (6) California (the Sierra Nevada, central valley, and coastal region of California).

Environmental Changes since the Last Glacial Maximum

In this section we characterize the nature, amplitude, and geographic patterning of the large-scale paleoenvironmental and paleoclimatic variations in the western United States at 3000-yr intervals centered on 18, 12, 9, and 6 ka inferred from fossil pollen data (Table 18.2), lake-level data (Table 18.3), and packrat midden assemblages (Table 18.4). This level of resolution and these time periods were chosen to facilitate comparisons with climate simulations from numerical models. Conditions at 15 ka are not a major focus in this discussion because neither the model simulations nor the paleoenvironmental data for 15 ka differ significantly from those for 18 ka.

18 KA: FULL-GLACIAL ENVIRONMENTS
(Figs. 18.5 and 18.6)

Northwest

In the Northwest alpine glaciers reached their maxima by 20 ka and retreated somewhat over the next two millennia (Porter *et al.*, 1983), as Cordilleran ice lobes moved southward into the Puget Trough (Booth, 1987). Undated alpine glacial and periglacial landforms of presumed full-glacial age exist throughout the high mountain ranges across the western United States as far south as 37°N latitude. Equilibrium-line altitudes descended more than 1000 m below modern levels, implying temperature depressions of more than 6°C (Porter *et al.*, 1983; Dohrenwend, 1984).

Between 20 and 16 ka pollen data from the Olympic Peninsula provide evidence of parkland communities of spruce, pine, mountain hemlock, and western hemlock. Tundra vegetation, inferred from a predominance of nonarboreal taxa, was confined to sites closest to alpine glacier margins. Modern analogues are the subalpine parkland and alpine tundra communities at high elevations in the Olympic Mountains (Heusser, 1973, 1977, 1983). The pollen data thus imply a lowering of treeline by 1000 m and temperatures 6°C lower than today.

The Puget Trough, located between the Coast Range and the Cascade Range, also supported subalpine parkland at 18 ka, although communities were more xerophytic than those along the coast. The presence of Engelmann spruce, lodgepole pine, *Artemisia*, and grasses and the absence of the coastal mountain hemlock most resemble the modern subalpine parkland of the northern Rocky Mountains (Barnosky, 1981, 1984, 1985a). Response surface analysis of the taxa present at 18 ka at Battle Ground Lake in the southern Puget Trough suggests that the mean annual temperature was 6-7°C colder than today and that annual precipitation (largely winter) was reduced by 700-1000 mm (Whitlock *et al.*, 1990).

Pollen data from Carp Lake, located east of the Cascade Range in the southeastern Columbia Basin, feature high percentages of *Artemisia*, grasses, and tundra herbs at 18 ka (Barnosky, 1985b). The vegetation is inferred to have been a sparsely vegetated periglacial steppe, indicative of cold and dry conditions. This site provides no record of montane conifers at 18 ka, which suggests that their range was either greatly compressed in the nearby Cascade Range or displaced an unknown distance into Oregon (Barnosky, 1985b). The water level at Carp Lake at 18 ka had dropped from an earlier high but may have been close to that of the present day.

Rocky Mountains

Only two pollen records from the Rocky Mountains date to 18 ka—Grays Lake in the Snake River Plain of southeastern Idaho (Beiswenger, 1987, 1991) and Devlins Park in the Colorado Front Range (Legg and Baker, 1980). Pollen data from Grays Lake suggest that a sparse *Artemisia* steppe covered the basin floor, while lodgepole pine and other conifers grew in the nearby foothills or other suitable habitats. The climate of the area was both cold and dry (although effectively moister than today). Data from Devlins Park, located in the mountains, record high percentages of pine and *Artemisia*, low pollen accumulation rates, and herb types characteristic of tundra, indicating a treeline at least 500 m below that of today (Legg and Baker, 1980).

Other proxy data also suggest cold and dry conditions in the basins. Permafrost features (such as frost wedges) of presumed full-glacial age suggest periglacial conditions in the valleys of central and southern Wyoming (Mears, 1981; Péwé, 1983). Fossil records of lemmings from Wyoming and Idaho (Mead and Mead, 1989) also may indicate cold full-glacial conditions. Full-glacial faunal assemblages from rock-shelters in the Bighorn Range in north-central Wyoming include a mixture of elements found also in contemporaneous eastern-midwestern faunas (collared lemming, heather vole, northern bog lemming) and western faunas (sagebrush vole, marmot, pika) (Graham and Mead, 1987). Open vegetation (but not northern tundra) with cool, dry climates was inferred. AMS radiocarbon-dated fossil insect assemblages from the High Plains near Denver contain nonthermophilic prairie taxa interpreted as indicating cold and dry conditions at 17.9 ka (Elias and Toolin, 1990).

Great Basin

The few packrat midden assemblages for 18 ka in the Great Basin indicate that cold-tolerant, xerophytic subalpine conifers descended to altitudes as much as 1000 m below their modern common lower limits (Thompson and Mead, 1982; Wells, 1983; Thompson, 1984, 1988, 1990). Pollen data from the Ruby Marshes in Nevada, a midvalley location, indicate that sagebrush steppe persisted in valleys and suggest that conifers were restricted to coarser substrates (Thompson, 1992). Hydrogen isotope ratios in cellulose from packrat middens in the eastern Great Basin indicate that temperatures were colder during this period than during the late glacial (Siegal, 1983; Long *et al.*, 1990),

Table 18.2. Fossil pollen localities that provided data for the paleoenvironmental maps

Site name	Reference	Latitude	Longitude	Effective moisture index[a]			
				6	9	12	18 ka
Alkali Lake	Markgraf and Scott, 1981	38°45'	106°50'	W	D	W	
Antelope Playa	Markgraf and Lennon, 1986	43°30'	105°27'	N	N		
Balsam Meadows	Davis et al., 1985	37°10'	119°30'	D	D		
Barrett Lake	Anderson, 1990	37°36'	119° 1'		D	D	
Battle Ground Lake	Barnosky, 1985a	45°40'	122°29'	D	D	W	D
Beaver Lake	Burkart, 1976	44°12'	107°15'		W		
Big Meadow	Mack et al., 1978c	48°55'	117°25'	D	D	W	
Blacktail Pond	Gennett, 1977	44°58'	110°36'	D	W	W	
Bogachiel River Site	Heusser, 1978	47°53'	124°20'	D	D	N	N
Bonaparte Meadows	Mack et al., 1979	48°45'	119° 5'	D	D		
Buckbean Fen	Baker, 1976	44°18'	110°15'	D	D	W	
Carp Lake	Barnosky, 1985b	45°55'	120°53'	D	D	D	D
Chewachan Lake	Hansen, 1947	42°30'	120°25'	D			
Clear Lake	Adam et al., 1981; Adam, 1988	39° 0'	122°45'	D	N	W	W
Coast Trail Pond, etc.	Rypins et al., 1989	37°59'	122°47'		N	W	
Creston Fen	Mack et al., 1976	47°35'	118°45'		W	W	
Cub Creek Pond	Waddington and Wright, 1974	45°10'	110°10'	D	D	W	
Cub Lake	Baker, 1983	44° 8'	111°11'	D	D	W	
Davis Lake	Barnosky, 1981	46°35'	122°15'	D	D	W	D
Dead Man Lake	Wright et al., 1973	36°15'	109° 0'				W
Devlins Park	Legg and Baker, 1980	40° 1'	105°33'				W
Diamond Pond	Wigand, 1987	43°15'	118°20'	D	D		
Exchequer Meadow	Davis and Moratto, 1988	37° 0'	119° 5'	N	D	D	
Fargher Lake	Heusser, 1983	45°53'	122°31'				N
Fish Lake	Mehringer, 1985	42°44'	118°38'	D	N		
Floating Island Lake	Burkart, 1976	44°33'	107°28'		W		
Forest Lake	Brant, 1980	45°30'	112°18'	N	W	W	
Gardiners Hole	Baker, 1983	44°55'	110°44'	N	W	W	
Gatecliff Shelter	Thompson and Kautz, 1983	39° 0'	116°47'	D			
Goose Lake	Nickmann and Leopold, 1984	48°20'	119°15'	D	D	W	
Grays Lake	Beiswenger, 1991	43° 0'	111°35'	D	D	W	W
Great Salt Lake	Mehringer, 1985	41° 0'	112°30'	D	W		
Guardipee Lake	Barnosky, 1989	48°33'	112°43'		N	N	
Hager Pond	Mack et al., 1978d	48°40'	116°55'	D	W		
Hall Lake	Tsukada et al., 1981	47°49'	122°18'	D	D	W	
Hay Lake	Fine Jacobs, 1983	34° 0'	109°30'			W	W
Hidden Cave	Wigand and Mehringer, 1985	39°20'	118°45'	D			
Hoh River Valley site 1	Heusser, 1974	47°50'	124°30'	D	D	N	N
Humptulips	Heusser, 1983	47°16'	123°54'				N
Ice Slough	Beiswenger, 1987	42°29'	107°54'	D	W		
Kalaloch	Heusser, 1972	47°33'	124°20'				N
Kelowna Bog	Alley, 1976	49°56'	119°23'	D	D		
Keystone Iron Bog	Fall, 1985, 1988	38°52'	107° 2'	W			
Kirk Lake	Cwynar, 1987	48° 7'	121°30'	D	D	W	
Lake Cleveland	Davis, 1981	42°19'	113°39'	D	D	W	
Lake Washington	Leopold et al., 1982	47°40'	122°13'	D	D	W	
Lilypad Pond	Baker, 1976	44°18'	110°15'	D	D	W	
Lost Lake	Barnosky, 1989	47°38'	110°20'	D	D		
Lost Trail Pass Bog	Mehringer et al., 1977	45°45'	113°58'	N	N	D	
Manis Mastodon	Petersen et al., 1983	48° 5'	123° 2'		N	N	
Marian and Surprise lakes	Mathewes, 1973	49°20'	123° 0'	D	D	W	
Mckillop Creek Pond	Mack et al., 1983	48°20'	115°27'	D	D		
Mineral Lake	Tsukada et al., 1981	46°44'	122°12'	N	D	D	D
Mission Cross Bog	Thompson, 1984	41°47'	115°29'	D	N		
Mosquito Lake	Hansen and Easterbrook, 1974	48°30'	122°10'	D	D	W	
Mud Lake	Mack et al., 1979	48°30'	119°45'	D	D	W	
Murphey's rock-shelter	Henry, 1984	43°12'	116°6'	D	D		
Nisqually Lake	Hibbert, 1979	47° 0'	122°13'	D	D	W	
Osgood Swamp	Adam, 1967	38°50'	120° 5'	D	W		
Pangborn Bog	Hansen and Easterbrook, 1974	48°50'	122°35'	D	D	W	
Pinecrest Lake	Mathewes and Rouse, 1975	50°30'	121°30'	D	D		

Table 18.2. Fossil pollen localities that provided data for the paleoenvironmental maps (continued)

Site name	Reference	Latitude	Longitude	Effective moisture index[a] 6	9	12	18 ka
Point Grenville	Heusser, 1983	47°19'	124°16'				N
Rattlesnake Cave	Davis, 1981	43°23'	112°38'	D			
Redrock Lake	Maher, 1972	40° 5'	105°33'	N	W		
Ruby Marshes	Thompson, 1992	40°35'	115°20'	D	W		W
San Agustin	Markgraf et al., 1984	33°50'	108°10'	W	W	W	W
Sherd Lake	Burkart, 1976	44°16'	107° 1'		W	W	
Simpson's Flats	Mack et al., 1978a	48°20'	118°35'	D	D		
Snowbird Bog	Madsen and Currey, 1979	40°35'	111°55'	W			
Soleduck Bog	Heusser, 1973	47°55'	124°28'	D	D		
Swan Lake	Bright, 1966	42°20'	112°25'	D	W	W	
Teepee Lake	Mack et al., 1983	48°20'	115°30'	D	D	W	
Telegraph Creek Marsh	Brant, 1980, 1982	46°30'	112°20'	N	W	W	
Tioga Pass Pond	Anderson, 1990	37°55'	119°16'	N	D		
Touchet Mammoth Site	Martin et al., 1982	46°20'	119°25'			W	
Tule Springs	Mehringer, 1967	36°19'	115°11'	N	W	W	W
Waits Lake	Mack et al., 1978b	48°10'	117°40'	D	D	W	
Wentworth Lake	Heusser, 1973	48°24'	124°30'	D	D		
Wessler Bog	Heusser, 1973	48°10'	124°30'	D	D	W	
Wildhorse Lake	Mehringer, 1985	42°44'	118°38'	D	D		
Williams Fen	Nickmann, 1979	47°20'	117°35'	D	W	W	
Zenkner Valley section	Heusser, 1977	46°45'	123° 0'	D	D	W	

[a]W indicates wetter conditions, or more effective moisture, than today; D indicates drier conditions, or less effective moisture, than today; N indicates no difference from today; blank indicates no data.

whereas those from the western Great Basin suggest that 18 ka was warmer than the late glacial (Flynn and Buchanan, 1990).

On the western edge of the Great Basin, alpine glaciers were near their Late Wisconsin maxima at 18 ka (Elliott-Fisk, 1987; Dorn et al., 1990). In contrast, major paleolakes in the Great Basin, including Bonneville (Currey and Oviatt, 1985; Thompson et al., 1990), Lahontan (Thompson et al., 1986; Benson and Thompson, 1987), and Russell (Lajoie, 1968), were filling by 18 ka but remained well below their Late Wisconsin maxima (Benson et al., 1990; Thompson et al., 1990). Stratigraphic studies and radiocarbon dating indicate that the lake maxima lagged behind the alpine glacial maxima by several thousand years in the Great Basin region (Gilbert, 1890; Putnam, 1950; Scott et al., 1983; Wayne, 1984). In addition, lacustrine and vegetational systems seem to differ in their sensitivity and time of response to paleoclimatic fluctuations (e.g., Thompson, 1992).

Table 18.3. Paleolacustrine localities that provided data used in the paleoenvironmental maps

Site name	Reference	Latitude	Longitude	Effective moisture index[a] 6	9	12	18 ka
San Agustin	Markgraf et al., 1984	33°50'	108°10'	W	W	W	W
Great Salt Lake (Lake Bonneville)	Currey and Oviatt, 1985; Thompson et al., 1990	40°30'	113° 0'	D	N	D	W
Sevier Lake	Oviatt, 1988	39° 0'	113°20'	N	N	W	W
Ruby Marshes	Thompson et al., 1990	40°35'	115°20'	D	W		W
Searles Lake	Smith and Street-Perrott, 1983; Benson et al., 1990	35°36'	117°42'	N	N	W	W
Mono Lake (Lake Russell)	Lajoie, 1968; Benson et al., 1990	38° 3'	118°46'	N	N	W	W
Harney Basin	Gehr, 1980	43°12'	119° 6'		W		
Lake Lahontan	Benson et al., 1990	40° 0'	119°30'	D		W	W
Fort Rock Basin	S. P. Harrison, this report	43°10'	120°45'	D	W		
Carp Lake	Barnosky, 1985b	45°55'	120°53'	D	D	D	D
Diamond Pond	Wigand, 1987	43°15'	118°20'	D			
Lake Cochise	Waters, 1989	32°10'	109°52'	W	W		
Lake Mojave	Wells et al., 1987	35°20'	116° 7'		W	W	
Las Vegas Valley	Quade, 1986	36°30'	115°25'		W	W	W

[a]W indicates wetter conditions, or more effective moisture, than today; D indicates drier conditions, or less effective moisture, than today; N indicates no difference from today; blank indicates no data.

Table 18.4. Packrat midden localities that provided data for the paleoenvironmental maps

Midden name	Reference	Latitude	Longitude	Effective moisture index[a]
6-ka middens				
Ajo Loop	Van Devender, 1987	31°58'	112°47'	W
Atlatl Cave	Betancourt and Van Devender, 1981	36° 2'	107°54'	N
Bida Cave	Cole, 1981	36° 0'	112° 0'	D
Carlins Cave	Thompson, 1990	38°18'	115° 2'	N
Chuar Valley	Cole, 1981	36°10'	111°55'	D
Desert View	Spaulding, 1981	36°38'	115° 2'	W
Etna	Madsen, 1973	37°33'	114°37'	N
Eureka View	Spaulding, 1980	37°20'	117°47'	W
Fishmouth Cave	Betancourt, 1984	37°25'	109°39'	W
Gatecliff/June Canyon	Thompson, 1990	39° 1'	116°45'	W
Grandview	Cole, 1981	36° 0'	111°59'	N
Hornaday Mountains	Van Devender et al., 1990	31°59'	113°36'	W
Lava beds	Mehringer and Wigand, 1986	41°45'	121°30'	N
Lucerne Valley	King, 1976	34°30'	117° 0'	N
Marble Canyon	Van Devender et al., 1987	32°50'	105°55'	W
McCullough Range	Spaulding, 1991	35°45'	115°10'	D
Rhodes Canyon	Van Devender and Toolin, 1983	33°11'	106°45'	W
Smith Creek Canyon	Thompson, 1984	39°20'	114° 6'	W
Valleyview	Thompson, 1984	39°30'	114°43'	W
Waterman Mountains	Anderson and Van Devender, 1991	32°24'	111°20'	W
Wellton Hills	Van Devender, 1973	32°36'	114° 7'	N
Wolcott Peak	Van Devender, 1973	32°27'	111°28'	N
9-ka middens				
Alamo Canyon	Van Devender, 1987	32° 7'	112°42'	W
Allen Canyon	Betancourt, 1984	37°47'	109°35'	W
Atlatl Cave	Betancourt and Van Devender, 1981	36° 2'	107°54'	W
Aysees Peak	Wells and Jorgensen, 1964	36°54'	115°48'	W
Basin Canyon	Spaulding, 1981	36°42'	115°16'	W
Bass Canyon	Cole, 1981	36°15'	112°30'	W
Bida Cave	Cole, 1981	36° 0'	112° 0'	W
Bloody Arm Cave	Thompson, 1984	39° 8'	114°25'	W
Brass Cap Point	Van Devender et al., 1985	33°26'	114° 5'	W
Carlins Cave	Thompson, 1990	38°18'	115° 2'	W
Cholla Pass	Van Devender, 1987	31°58'	112°47'	W
Chuar Valley	Cole, 1981	36°10'	111°55'	W
Clear Creek	Cole, 1981	36° 8'	112° 0'	W
Deadman	Spaulding, 1981	36°37'	115°17'	W
Death Valley	Wells and Woodcock, 1985	36°40'	117° 0'	W
Death Valley	Wells and Woodcock, 1985	36°47'	117°20'	W
Desert Almond	Phillips, 1977	36° 7'	113°56'	W
Esmeralda	Thompson, 1990	37°53'	117°14'	W
Etna	Madsen, 1973	37°33'	114°37'	W
Eureka View	Spaulding, 1980	37°20'	117°47'	W
Eyrie	Spaulding, 1981	36°38'	115°17'	W
Fishmouth Cave	Betancourt, 1984	37°25'	109°39'	W
Gatecliff Shelter	Thompson and Hattori, 1983	39° 0'	116°47'	W
Hornaday Mountains	Van Devender et al., 1990	31°59'	113°36'	W
Horseshoe Mesa	Cole, 1981	36° 2'	111°59'	W
Hueco Mountains	Van Devender and Riskind, 1979	31°54'	106° 9'	W
Hueco Tanks	Van Devender and Riskind, 1979	31°55'	106° 3'	W
Last Chance Range	Spaulding, 1985	36°17'	116° 8'	W
Lucerne Valley	King, 1976	34°30'	116°30'	W
Marble Canyon	Van Devender et al., 1984	32°50'	105°55'	W
Marble Mountains	Spaulding, 1980	34°40'	115°35'	W
Mercury Ridge	Wells and Berger, 1967	36°42'	115°53'	W
Navar Ranch	Van Devender and Wiseman, 1977	31°54'	106° 9'	W
Needle Eye Canyon	Phillips, 1977	36° 7'	113°56'	W
Negro Butte	Wells and Berger, 1967	34°29'	116°49'	W

Table 18.4. (continued)

Midden name	Reference	Latitude	Longitude	Effective moisture index[a]
Newberry Mountains	Leskinen, 1975	35°16'	114°37'	W
North Muddy Mountain	Wells and Hunziker, 1976	36°28'	114°37'	W
Old Man Cave	Thompson, 1984	39°10'	114° 8'	W
Penthouse	Spaulding, 1981	36°28'	115°15'	W
Picacho Peak (Arizona)	Van Devender and Spaulding, 1979	32°38'	111°24'	W
Picacho Peak (California)	Cole, 1986, 1990b	32°58'	114°50'	W
Point of Rocks	Spaulding, 1985	36°34'	116° 5'	W
Rampart Cave / Vulture Canyon	Phillips, 1977	36° 6'	113°56'	W
Redtail Peaks	Van Devender, 1977; Mead *et al.*, 1978; Van Devender *et al.*, 1985	34°16'	114°25'	W
Rhodes Canyon	Van Devender and Toolin, 1983	33°11'	106°45'	W
Sacatone Wash	Leskinen, 1975	35°15'	114°37'	W
San Andres	Van Devender *et al.*, 1984	32°50'	105°55'	W
Silver Peak	Thompson, 1990	37°45'	117°40'	W
Smith Creek Canyon	Thompson, 1984	39°21'	114° 5'	W
Spires	Spaulding, 1981	36°34'	115°18'	W
Spotted Range	Wells and Berger, 1967	36°38'	115°56'	W
Tinajas Altas	Van Devender *et al.*, 1985	32°15'	114°10'	W
Titus Canyon	Van Devender, 1977	36°45'	116°55'	W
Twin Peaks	Van Devender, 1987	31°58'	112°47'	W
Waterman Mountains	Anderson and Van Devender, 1991	32°24'	111°20'	W
Wells #25	Wells, 1983	39°20'	114° 7'	W
Wells #26	Wells, 1983	38°51'	114°13'	W
Wells #39	Wells, 1983	39°5'	113°34'	W
Wells #77	Wells, 1983	36°35'	114°32'	W
Wellton Hills	Van Devender, 1973	32°36'	114° 7'	W
Whipple Mountains	Van Devender, 1977	34°14'	114°22'	W
Willow Wash	Spaulding, 1981	36°28'	115°15'	W
12-ka middens				
Allen Canyon	Betancourt, 1984	37°47'	109°35'	W
Bennett Ranch	Van Devender and Spaulding, 1979	30°37'	104°59'	W
Bida Cave	Cole, 1981	36° 0'	112° 0'	W
Brass Cap Point	Van Devender, 1973	33°26'	114° 5'	W
Butler Mountains	Van Devender *et al.*, 1985	32°15'	114°15'	W
Canyon del Muerto	Betancourt and Davis, 1984	36° 9'	109°25'	W
Carlins Cave	Thompson, 1990	38°18'	115° 2'	W
Chuar Valley	Cole, 1981	36°10'	111°55'	W
Clark Mountain	Mehringer and Ferguson, 1969	35°33'	115°37'	W
Cottonwood Canyon	Cole, 1981	36° 3'	112° 0'	W
Death Valley	Wells and Woodcock, 1985	36°40'	117° 0'	W
Death Valley	Wells and Woodcock, 1985	36°47'	117°20'	W
Desert Almond	Phillips, 1977; Van Devender and Spaulding, 1979	36° 7'	113°54'	W
Eleana Range	Spaulding, 1985	37° 7'	116°14'	W
Esmeralda	Thompson, 1990	37°53'	117°14'	W
Falcon Hill	Thompson *et al.*, 1986	40°16'	119°20'	W
Falling Arches	Mead *et al.*, 1978	34°13'	114°22'	W
Fishmouth Cave	Betancourt, 1984	37°25'	109°39'	W
Funeral Range	Wells and Berger, 1967	36°23'	116°36'	W
Garrison	Thompson, 1984	38°57'	114° 3'	W
Granite Wash	Thompson, 1984	39°40'	113°50'	W
Guadalupe	Van Devender, 1977	31°55'	104°50'	W
Hance Canyon	Cole, 1981	36° 2'	111°58'	W
Hidden Forest	Spaulding, 1981	36°34'	115° 6'	W
Iceberg Canyon	Phillips, 1977	36°11'	114° 3'	W
Kings Canyon	Cole, 1983	36°48'	118°48'	W
Last Chance Range	Spaulding, 1985	36°17'	116° 8'	W
Lucerne Valley	King, 1976	34°30'	116°30'	W
Maravillas Canyon	Wells, 1966	29°32'	102°49'	W
Mercury Ridge	Wells and Berger, 1967	36°42'	115°53'	W

Table 18.4. Packrat midden localities that provide data for the paleoenvironmental maps (continued)

Midden name	Reference	Latitude	Longitude	Effective moisture index[a]
Nankoweap	Cole, 1981	36°15'	111°57'	W
Navar Ranch	Van Devender and Wiseman, 1977	31°54'	106° 9'	W
New Water Mountains	Mead *et al.*, 1978	33°36'	113°55'	W
Ord Mountain	King, 1976	34°40'	116°50'	W
Penthouse	Van Devender and Spaulding, 1979	36°28'	115°15'	W
Picacho Peak (Arizona)	Van Devender and Spaulding, 1979	32°38'	111°24'	W
Picacho Peak (California)	Cole, 1986, 1990b	32°58'	114°50'	W
Picture Cave	Van Devender and Riskind, 1979	31°53'	106° 9'	W
Point of Rocks	Spaulding, 1985	36°34'	116° 5'	W
Pontatoc Ridge		32°21'	110°53'	W
Potosi Mountain		36° 0'	115°30'	W
Puerto de Ventanillas	Van Devender and Burgess, 1985	26° 2'	102°45'	W
Quitman Mountains	Van Devender and Spaulding, 1979	31° 8'	105°24'	W
Rampart Cave/Vulture Canyon	Phillips, 1977; Mead and Phillips, 1981	36° 6'	113°56'	W
Redtail Peaks	Van Devender and Spaulding, 1979; Mead *et al.*, 1978	34°16'	114°25'	W
Robber's Roost	Van Devender and Spaulding, 1979	35°36'	117°57'	W
Shelter Cave	Thompson *et al.*, 1980; Van Devender and Spaulding, 1979	32°11'	106°11'	W
Sierra de la Misericordia	Van Devender and Burgess, 1985	25°56'	103°45'	W
Smith Creek Canyon	Thompson, 1984	39°21'	114° 5'	W
Tip Top	Van Devender and Toolin, 1983	33°11'	106°45'	W
Tucson Mountains	Van Devender, 1973	32°19'	111°12'	W
Tunnel Ridge	Mead *et al.*, 1978	34°13'	114°22'	W
Waterman Mountains	Anderson and Van Devender, 1991	32°24'	111°20'	W
Wells #10, 11	Wells, 1983	39°20'	114° 6'	W
Wells #12, 14, 15, 16	Wells, 1983	38°58'	114° 8'	W
Wells #13	Wells, 1983	38°58'	114°10'	W
Wells #17	Wells, 1983	39°20'	114° 7'	W
Wells #30	Wells, 1983	38°34'	113°35'	W
Wells #31	Wells, 1983	38°32'	113°31'	W
Wells #32	Wells, 1983	38°30'	113°32'	W
Wells #33	Wells, 1983	39° 1'	113°36'	W
Wells #34	Wells, 1983	39° 5'	113°34'	W
Wells #35	Wells, 1983	38°32'	113°31'	W
Wells #43	Wells, 1983	37°34'	114°32'	W
Wells #76	Wells, 1983	36°35'	114°32'	W
Wells #83	Wells, 1983	35°43'	114°38'	W
Wells #84	Wells, 1983	35°16'	114°41'	W
Williams Cave	Van Devender and Spaulding, 1979	31°54'	104°50'	W
Winnemucca Caves	Thompson *et al.*, 1986	40°12'	119°16'	W
Wolcott Peak	Van Devender and Spaulding, 1979	32°27'	111°28'	W
18-ka middens				
Artillery Mountains	Van Devender and Spaulding, 1979	34°20'	113°35'	W
Bennett Ranch	Van Devender and Spaulding, 1979	30°37'	104°59'	W
Big Boy	Van Devender *et al.*, 1984	32°50'	105°55'	W
Burro Mesa	Wells, 1966	29°16'	103°23'	W
Carlins Cave	Thompson, 1990	38°18'	115°2'	W
Cave of the Early Morning Light	Van Devender and Spaulding, 1979	35°43'	113°23'	W
Chuar Valley	Cole, 1981	36°10'	111°55'	W
Deadman	Spaulding, 1981	36°37'	115°17'	W
Death Valley	Wells and Woodcock, 1985	36°40'	117° 0'	W
Death Valley	Wells and Woodcock, 1985	36°47'	117°20'	W
Eleana Range	Spaulding, 1985	37° 7'	116°14'	W
Eyrie	Spaulding, 1981	36°38'	115°17'	W
Flaherty Mesa	Spaulding, 1981	36°29'	115°15'	W
Hance Canyon	Cole, 1981	36° 2'	111°58'	W
Horseshoe Mesa	Cole, 1981	36° 2'	111°59'	W
Kings Canyon	Cole, 1983	36°48'	118°48'	W
Montezuma Head	Van Devender and Spaulding, 1979	32° 7'	112°42'	W

Table 18.4. (continued)

Midden name	Reference	Latitude	Longitude	Effective moisture index[a]
Nankoweap	Cole, 1981	36°15'	111°57'	W
Pontatoc Ridge	Van Devender *et al.*, 1987	32°21'	110°53'	W
Rampart Cave/Vulture Canyon	Phillips, 1977; Mead and Phillips, 1981	36° 6'	113°56'	W
Ranger Mountains	Wells and Jorgensen, 1964	36°46'	115°51'	W
Smith Creek Canyon	Thompson, 1984	39°21'	114° 5'	W
Specter Range	Spaulding, 1985	36°40'	116°13'	W
Spires	Spaulding, 1981	36°34'	115°18'	W
Streerwitz Hills	Van Devender and Spaulding, 1979	31° 7'	105° 9'	W
Tinajas Altas	Van Devender *et al.*, 1985	32°15'	114°10'	W
Waterman Mountains	Anderson and Van Devender, 1991	32°24'	111°20'	W
Willow Wash	Spaulding, 1981	36°28'	115°15'	W

[a]W indicates wetter conditions, or more effective moisture, than today; D indicates drier conditions, or less effective moisture, than today; N indicates no difference from today; blank indicates no data.

Colorado Plateau

Pollen data of apparent full-glacial age from the Chuska Mountains of New Mexico (Wright *et al.*, 1973) indicate spruce parkland or alpine tundra with nearby pockets of spruce and fir in place of the modern ponderosa pine forest. In the eastern Grand Canyon packrat midden data indicate that many plant species lived 600–1000 m below their modern limits: juniper and shadscale grew in the modern range of desert scrub, fir and Douglas fir in the lower portion of piñon pine-juniper woodland, and limber pine and spruce in the upper range of woodland (Cole, 1981, 1982, 1990a,b). On the Colorado Plateau in Utah, packrat midden data indicate that limber pine, Rocky Mountain juniper, and sagebrush coexisted with xerophytes such as saltbush, little-leaf mountain mahogany, and prickly pear (Betancourt, 1990), indicating cool, dry conditions. A 17.4-ka midden assemblage from this area contains no conifers but instead has sagebrush, rose, prickly pear, and other steppe shrubs, again reflecting relatively dry conditions (but still moister than today).

Pollen spectra from the White Mountains of eastern Arizona (Fine Jacobs, 1983) indicate that the Pleistocene vegetation was spruce parkland near the upper treeline in what is now ponderosa pine forest. This suggests that the upper treeline was depressed no more than about 575 m below modern levels. At Walker Lake in north-central Arizona, pollen data suggest an expansion of boreal taxa at the expense of pine (Hevly, 1985). Pollen spectra dated near 18 ka from paleolacustrine sediments from the San Agustin Plains of southeastern New Mexico indicate a regional expansion of spruce, pine, sagebrush, grasses, and composites to low elevations (Markgraf *et al.*, 1984). Stable isotope and noble gas data from radiocarbon-dated groundwater from this basin suggest that full-glacial temperatures were 5–7°C cooler than today with a higher proportion of winter rainfall (Phillips *et al.*, 1986).

Although dating controls are poor for many of the lakes in the southwestern desert region, it appears that most of these lakes were relatively deep, or at least fresh, during the full glacial. One exception is

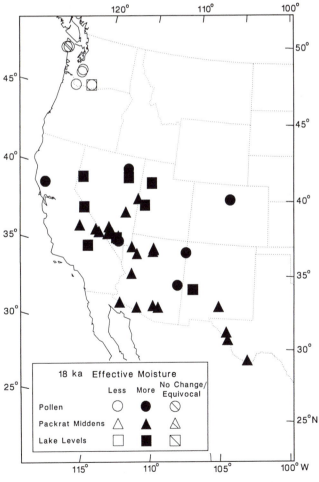

Fig. 18.5. Map of effective moisture anomalies (difference from present) at 18 ka (see also Tables 18.2–18.4).

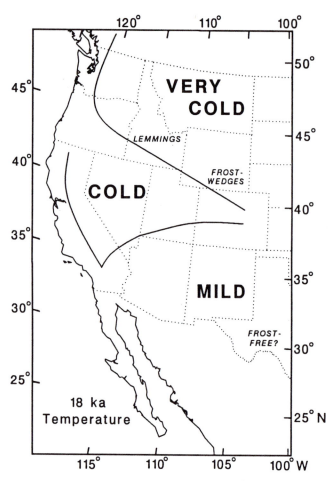

Fig. 18.6. Map of qualitative temperature anomalies (difference from present) at 18 ka. Comparison with the seasonality panel in Figure 18.2 indicates that the largest temperature anomalies were in the regions that today have highly seasonal temperatures.

Hay Lake on the Mogollon Rim of Arizona, which registered low water levels at what appears to be full-glacial time (Fine Jacobs, 1983).

Southwest

Packrat midden data from the Chihuahuan, Sonoran, and Mojave deserts indicate that many of the modern hot-desert plants were absent or displaced southward at 18 ka. In their place juniper, piñon pine, oak, and other woodland species descended to low elevations. In the northern Chihuahuan Desert of New Mexico, Douglas fir, Rocky Mountain juniper, oak, Colorado piñon pine, and woodland juniper grew in areas that today support desert scrub (Van Devender *et al.*, 1984). The entire modern elevational range of the Chihuahuan Desert in west Texas, southern New Mexico, and northern Mexico supported piñon pine-juniper-oak woodland during the full glacial (Van Devender, 1990a).

Fauna in Dry Cave in southeastern New Mexico in-

cluded pygmy rabbit and sagebrush vole, which indicates environmental conditions similar to those of today in southeastern Idaho (Harris, 1989, 1990), with cooler summers and more cool-season precipitation. At this time mammals that now live only on the Great Plains (prairie vole and least shrew) or in the Southeast (eastern woodrat) also lived in southeastern New Mexico (Harris, 1990).

Woodlands also occupied the area of modern desert scrub in the Sonoran Desert of southern Arizona, with single-needle piñon pine descending as low as 460 m (Van Devender, 1990b) and junipers growing at even lower altitudes. This woodland community extended as far up in elevation as 1500 m, where it mingled with Douglas fir, Arizona cypress, and other montane conifers (Van Devender and Spaulding, 1979; Van Devender, 1990b). Ponderosa pine, which today is the dominant tree of middle elevations in the Southwest, was present as a minor element in the full-glacial vegetation.

Rocky Mountain juniper, sagebrush, shadscale, Joshua tree, and other plants common today in the Rocky Mountains, Great Basin, and/or Mojave Desert grew at the Mexican border around 18 ka (Van Devender, 1987). The occurrence of these plants has been interpreted as reflecting 55-70% more precipitation than today and July temperatures 8-11°C lower than today (Van Devender, 1987). Border piñon pine and Arizona cypress reach their northern limits in southern Arizona today, and the presence of these plants in the full-glacial vegetation of southern Arizona may indicate that winter temperatures were not much below modern levels (Van Devender and Spaulding, 1979; Van Devender, 1990b).

Woodland and Great Basin steppe taxa descended 1000 m into the now arid landscape of the Mojave Desert (Wells and Jorgensen, 1964; Mehringer, 1967; Wells and Berger, 1967; Phillips, 1977; Van Devender and Spaulding, 1979; Spaulding, 1980, 1981, 1990a,b; Spaulding *et al.*, 1983). In the mountains bristlecone pine and limber pine grew in the modern woodland zone (Spaulding, 1981). Conversely, the upper limit of xerophytic shadscale moved upslope during the full glacial (Spaulding *et al.*, 1983; Spaulding, 1990b). In Death Valley juniper grew nearly 1500 m below its modern range (Wells and Woodcock, 1985; Woodcock, 1986). In contrast to data from desert regions farther south, the Mojave packrat midden data have been interpreted as indicating a mean annual temperature 6°C or more below the modern mean (Spaulding, 1990b).

Searles Lake in California once was thought to have overflowed during the full glacial, but new evidence bearing on the chronology of the system (Benson *et al.*, 1990; Dorn *et al.*, 1990) suggests that this lake, like those farther north in the Great Basin, was well below its maximum Late Wisconsin level until about 16 ka. Marsh complexes in southern Nevada were greatly expanded during the interval from 30 to 15 ka, with no marked maximum during the full glacial (Quade, 1986).

California

The southern Sierra Nevada supported a mixed conifer forest at 18 ka, as shown by a midden assemblage from Kings Canyon that contains red fir, western juniper, incense cedar, sugar pine, ponderosa pine, California nutmeg, and single-needle piñon pine (Cole, 1983). The past occurrence of these plants in areas of modern oak-chaparral vegetation has been interpreted as indicating a climate colder than today with near-modern precipitation levels. Dorn *et al.* (1987, 1990) obtained AMS radiocarbon dates on rock varnish on the outermost Tioga moraine on the east slope of the Sierra Nevada that indicate glaciers slightly predating 19-18 ka. This suggests an alpine glacial maximum apparently contemporaneous with the continental glacial maximum in this region.

Pollen data from Clear Lake in northern California (Adam *et al.,* 1981; Adam and West, 1983) indicate that the modern oak-dominated vegetation was absent at 18 ka and that instead junipers (and/or other members of the Taxaceae, Cupressaceae, or Taxodiaceae) and pines grew at the site. Transfer functions based on modern regional surface samples (Adam and West, 1983) suggest that full-glacial temperatures were 7-8°C cooler than present levels and precipitation was 300-350% greater.

12 KA: LATE-GLACIAL ENVIRONMENTS
(Figs. 18.7-18.9)

Northwest

By 12 ka the Cordilleran ice sheet and alpine glaciers had retreated from their full-glacial maxima. The Puget and Juan de Fuca lobes of the Cordilleran ice sheet melted rapidly after about 14.5 ka (Booth, 1987), and between 14 and 11 ka ice lobes both west and east of the Cascade Range retreated north of the U.S.-Canadian boundary (Waitt and Thorson, 1983). By 12 ka alpine glaciers in northwestern Wyoming, western Montana, and Colorado had melted back to near-modern size (Porter *et al.,* 1983).

Pollen data from the Olympic Peninsula record an increase in red alder, spruce, and western hemlock after 16 ka, suggesting warmer conditions than before (Heusser, 1973; Barnosky *et al.,* 1987). By 12 ka mixed communities of subalpine and lowland taxa grew throughout western Washington in areas where subalpine parkland had been at 18 ka (Heusser, 1977, 1983; Barnosky, 1981, 1984, 1985a; Tsukada *et al.,* 1981; Cwynar, 1987). This vegetation association commonly includes montane taxa, such as mountain hemlock, subalpine fir (perhaps), Engelmann spruce, and lodgepole pine, along with low-elevation species such as western hemlock, red alder, grand fir, and Sitka spruce. Macrofossils of hornwort, a temperate aquatic plant,

are recorded from the northeastern Olympic Peninsula and provide evidence that summer water temperatures at 12 ka may have been as warm as at present (Petersen *et al.,* 1983). In eastern Washington steppe vegetation continued to dominate the southern Columbia Basin, but warming is inferred from shallowing and increased spruce pollen at Carp Lake (Barnosky, 1985b). The earliest pollen spectra available from the Okanogan highlands date to this period and indicate the presence of tundra or steppe vegetation and thus cool climatic conditions (Mack *et al.,* 1976, 1978b,c).

Rocky Mountains

Many palynologic sites in the Rocky Mountain region have their basal dates near 12 ka. Tundra is generally the earliest recorded vegetation, with whitebark pine, spruce, and fir arriving at or soon after 12 ka and parkland developing by 11 ka (Baker, 1976, 1983; Mehringer *et al.,* 1977; Davis *et al.,* 1986; Gennett and Baker, 1986). A pollen record from Guardipee Lake (Barnosky *et al.,* 1987; Barnosky, 1989) on the plains of northeastern Montana reveals that the late-glacial veg-

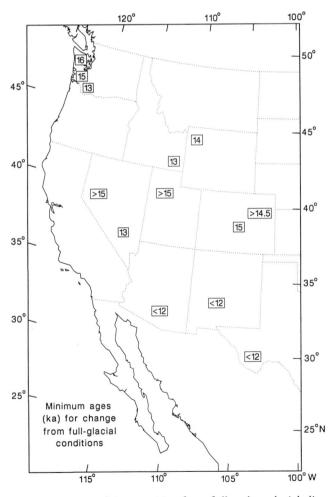

Fig. 18.7. Timing of the transition from full- to late-glacial climatic conditions.

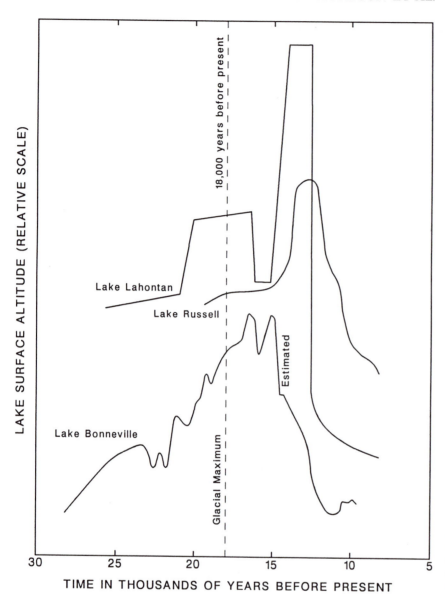

Fig. 18.8. Chronology of lake-level changes in three Great Basin lakes: Lake Lahontan, Nevada; Lake Russell, California; and Lake Bonneville, Utah. Modified from Benson and Thompson (1987).

etation was dominated by grasses, *Artemisia,* and chenopods. This record, together with evidence of the rapid retreat of Cordilleran ice by about 12 ka (Carrara *et al.,* 1986), suggests that the northern plains had warmed substantially by this time.

At Grays Lake in southeastern Idaho increased abundances of spruce, *Artemisia,* and other taxa reflect increased moisture between about 11.5 and 11.2 ka. Temperatures, while warmer than during the full glacial, remained cooler than today (Beiswenger, 1987, 1991). Moisture levels rose greatly near 13 ka, while temperatures remained cool (Beiswenger, 1987). Pollen data from a playa lake in northeastern Wyoming reveal the presence of treeless steppe vegetation from 13.2 ka to the present (Markgraf and Lennon, 1986).

Fossil insect assemblages on the High Plains near Denver indicate that the cold-dry prairie conditions of

the full glacial gave way to cold-moist tundra conditions by 14.5 ka (Elias and Toolin, 1990). In the San Juan Mountains of Colorado, deglaciation began by 15 ka (Carrara *et al.,* 1984), while pollen spectra dominated by *Artemisia,* grasses, composites, and juniper in central Colorado suggest that the upper timberline remained 500-700 m below modern levels from about 15 to 11 ka (Fall, 1988). Temperatures 3-4.5°C below modern temperatures and a predominance of winter precipitation are inferred from these spectra.

Great Basin

Alpine glaciers in the Great Basin retreated from their maximum Late Wisconsin extensions before 13 ka (Madsen and Currey, 1979; Wayne, 1984; Elliott-Fisk, 1987). Lacustrine records from across the central Great Basin (Fig. 18.8) (Currey and Oviatt, 1985; Benson and

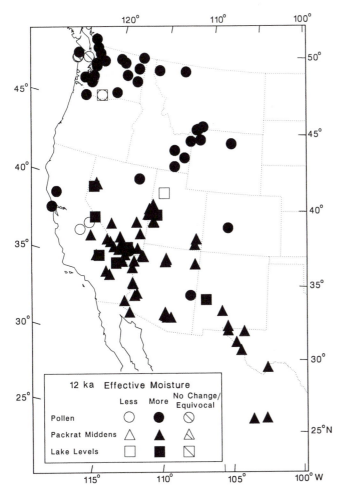

Fig. 18.9. Map of effective moisture anomalies (difference from present) at 12 ka (see also Tables 18.2-18.4).

Thompson, 1987; Benson *et al.,* 1990; Thompson *et al.,* 1990) indicate that lake levels were highest between about 16 and 13 ka. Physically based models of lake temperature and evaporation (Hostetler and Bartlein, 1990) suggest that Lake Lahontan could have reached and maintained its high level under climatic conditions of 3.8 times more runoff, 1.8 times more precipitation, and 42% less evaporation than at present (Hostetler and Benson, 1990).

Pleistocene lakes in the Great Basin appear to have desiccated rapidly, and Lake Bonneville, at least, was below modern levels between 12 and 11 ka (Currey and Oviatt, 1985). The abrupt end of the deep-lake phase in this basin led to the deposition of red beds and mirabilite (Currey, 1990), the latter indicating that temperatures were cool during the low-lake period. In contrast, the Sevier Lake basin, a southern arm of the Bonneville basin, held a freshwater lake (Lake Gunnison) (Oviatt, 1988) that flowed northward into the Great Salt Lake basin between 12 and 10 ka. Oviatt (1988) interpreted the greater moisture in the south as a reflection of intensified monsoonal activity at this early date.

However, Beiswenger (1987, 1991) argued that the pollen record from Grays Lake, on the northern edge of the Bonneville basin, indicates increasing (but still cool) temperatures with high levels of effective moisture. The juxtaposition of high moisture levels both north and south of a very dry (and apparently cold) Bonneville basin is difficult to attribute to a simple basinwide mechanism. The situation is further complicated by the apparent occurrence of moderately large lakes across the Great Basin between 11.1 and 10.3 ka (Benson *et al.,* 1990; Currey, 1990; Thompson, 1992).

Palynologic data from the Wasatch Range of Utah (Madsen and Currey, 1979) indicate that pine-spruce forest was established at an elevation of 2470 m by 12.3 ka. In contrast, regional pollen diagrams from 2835 m in the Raft River Mountains of Utah (Mehringer, 1977) and from 2519 m in the Albion Mountains of Idaho (Davis, 1981) reveal treeless vegetation until about 11 ka. The lower-elevation record from Swan Lake (Bright, 1966) indicates that the lower limits of pine and other montane species at 12 ka were below their modern limits. These taxa moved upslope after 11 ka, as the upper limits of subalpine conifers reached the Raft River and Albion Mountain sites.

Packrat middens from the western periphery of the Bonneville basin indicate that bristlecone pine, limber pine, and other subalpine and montane species grew in the modern woodland zone from the full glacial until after 11 ka (Thompson and Mead, 1982; Wells, 1983; Thompson, 1984, 1990). Woodland junipers and other relatively thermophilic plants began to appear by 12 ka, but the major change in floristic composition postdated 11 ka. Hydrogen isotope ratios from cellulose from these middens suggest that a major warming and/or a shift in the source of precipitation toward more southerly origins occurred between the full glacial and 14 ka (Siegal, 1983; Long *et al.,* 1990). Plant macrofossil assemblages changed relatively little between 14 and 11 ka, but hydrogen isotopes from cellulose from some of these middens indicate significant shifts (Siegal, 1983), perhaps reflecting a return to cooler temperatures and more dominant winter precipitation. In contrast to the full- to late-glacial changes in eastern Nevada described above, hydrogen-deuterium ratios from packrat middens in the Lahontan Basin of western Nevada record a large depletion of deuterium between about 16 ka and about 12 ka, suggesting a shift toward cooler temperatures and/or more winter rainfall (Flynn and Buchanan, 1990).

On the southern periphery of the Great Basin, plant associations and elevational ranges changed significantly between 13 and 11 ka. In the Eleana Range, limber pine and mountain mahogany, which dominated the vegetation at 13.2 ka, were replaced by piñon pine-juniper woodland by 11.7 ka (Spaulding, 1985). In the western Great Basin, western junipers invaded the recently vacated slopes of desiccating Lake

Lahontan by 12 ka (Thompson *et al.,* 1986), indicating conditions effectively moister than today.

Colorado Plateau

At 12 ka subalpine and montane conifers continued to dominate the modern woodland zone across much of the Colorado Plateau. Packrat middens from southeastern Utah provide evidence that blue spruce, limber pine, and prostrate juniper grew at least 850 m below their modern limits, and Engelmann spruce and subalpine fir were depressed at least 700 m (Betancourt, 1984, 1990). Colorado piñon pine grew with Rocky Mountain juniper in modern desert scrub south of the Grand Canyon between 13.7 and 10.8 ka (Cinnamon and Hevly, 1988). The changes in the elevational limits of subalpine taxa have been interpreted as reflecting temperatures 3–5°C cooler than today and precipitation levels 35–120% higher (Betancourt, 1984). The 1070-m depression of blue spruce may reflect summer temperatures 6°C cooler than today, while the absence of ponderosa pine and the rarity of summer herbs and annuals may indicate summers drier than at present (Betancourt, 1990). In this region the transition from full- to late-glacial climates was marked by more mesic conditions at some sites and more xeric conditions at others (Betancourt, 1990).

In other parts of southeastern Utah, spruce and Douglas fir were recovered from cave sediments of this age at Cowboy Cave (Spaulding and Van Devender, 1977), and blue spruce remains were recovered from a dung blanket dated around 12 ka at Bechan Cave (Mead *et al.,* 1987). In Canyon de Chelly in northwest Arizona, a packrat midden assemblage records blue spruce, limber pine, Rocky Mountain and prostrate junipers, and Douglas fir in association with high sagebrush pollen percentages (Betancourt and Davis, 1984). All of these finds indicate much more effective moisture than today in the central Colorado Plateau region. Farther west in the Grand Canyon region, vegetation apparently changed little between 18 and 12 ka (Cole, 1981, 1982, 1990a,b), but the rate of local extinctions between 12 and 10 ka was the highest for any period from 18 ka to the present.

Southwest

Packrat middens from the northern Chihuahuan Desert in Texas and New Mexico indicate that the full-glacial piñon pine-juniper-oak woodlands had changed little by late-glacial time (Wells, 1966; Van Devender and Spaulding, 1979; Lanner and Van Devender, 1981; Van Devender, 1990a). The eastward expansion of desert tortoise relative to its modern range (Van Devender *et al.,* 1976; Moodie and Van Devender, 1979) and the northward spread of cold-intolerant plants (Van Devender, 1990a) indicate lack of hard freezes in west Texas and southern New Mexico. Fossil insect remains from packrat middens indicate that high levels of effective moisture persisted until 12 ka

and were then followed by a trend toward warmer and drier conditions (Elias and Van Devender, 1990). In general, the vegetation data parallel this trend, although vegetation apparently reacted more slowly, with woodland plants disappearing between about 11.5 ka and about 10 ka (Elias and Van Devender, 1990).

In the southern Chihuahuan Desert two packrat middens from Bolson de Mapimi radiocarbon-dated at 12.3 and 12.7 ka demonstrate the southward expansion of papershell piñon pine-juniper or juniper woodland into Coahuila and Durango (approximately 26°N latitude) (Van Devender and Burgess, 1985). Many modern Chihuahuan Desert endemics grew with these woodland plants, suggesting mild winters without frequent hard freezes.

Woodland plants mixed with modern Mojave and Great Basin shrubs grew in the Arizona upland portion of the Sonoran Desert (Van Devender *et al.,* 1987; Van Devender, 1990b), whereas desert scrub was present at lower elevations (240–300 m) near the Colorado River at the U.S.-Mexican border by 13.4 ka (Cole, 1986, 1990a,b; Van Devender, 1990b). Packrat middens dated between 13.4 and 11 ka are dominated by creosote bush, Mormon-tea, brittlebush, and pygmy cedar. These assemblages do not reflect the modern Sonoran Desert vegetation but rather contain many Mojave Desert species such as Joshua tree, whipple yucca, blackbrush, and Mojave sage. These Mojavean elements disappeared from these sites by 11–10 ka.

In the Mojave Desert of southern Nevada and adjacent California, packrat middens indicate treeless desert scrub (presumably relatively arid conditions) as early as 17.5 ka (Spaulding, 1983, 1990a), although juniper woodland persisted at relatively low elevations in some habitats. Spaulding and Graumlich (1986) interpreted the decline of steppe shrubs before 12 ka and the abundance of succulents and grasses at that time in southern Nevada packrat middens as evidence of a shift from winter to summer precipitation dominance between the full and late glacial.

By 12 ka lakes in the San Agustin (New Mexico), Wilcox (Arizona), Manix (California), and Searles (California) basins had fallen below their deepest Late Wisconsin levels (Forester, 1987; Dorn *et al.,* 1989; Waters, 1989; Benson *et al.,* 1990), and marsh complexes in southern Nevada were in decline (Quade, 1986). In contrast, evidence from the Lake Mojave (California) system (Wells *et al.,* 1987; Enzel *et al.,* 1988; Dorn *et al.,* 1989; Meek, 1989) indicates that relatively deep waters persisted from 15.5 to 10.5 ka.

California

Packrat midden data from 12.5 ka from Kings Canyon in the southern Sierra Nevada indicate the persistence of western juniper and ponderosa pine from the full glacial (Cole, 1983). Palynologic data for 13.5–10.7 ka from a nearby site in a modern forest of mixed conifers (lodgepole pine, Jeffrey pine, red fir) are dominated by *Artemisia* and grasses, reflecting a relatively dry and cold climate. The presence of giant sequoia pollen in these assemblages suggests that although the climate was cooler than today it could not have been too extreme. Other sites from farther north indicate that deglaciation occurred by about 12.5 ka on the east side of the Sierra Nevada and by around 11 ka on the west side and that the upper-elevation localities supported few trees before 10 ka (Anderson, 1987). AMS radiocarbon dates on rock varnish from the innermost recessional Tioga moraines on the eastern flank of the Sierra Nevada predate 13 ka (Dorn *et al.,* 1987).

Oak pollen increased dramatically at Clear Lake (northwestern California) apparently near 12 ka, and Adam and West (1983) estimated near-modern climatic conditions for this time. Unfortunately, radiocarbon dating problems limit the usefulness of this record: depending on which Clear Lake core is used, the 12-ka spectra suggest that temperatures either were near their modern values or were considerably cooler

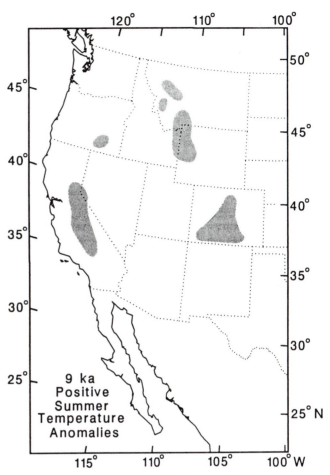

Fig. 18.11. Map of qualitative temperature anomalies (difference from present) at 9 ka.

(Adam *et al.,* 1981; Adam, 1988). Along the north-central coast, pollen assemblages from Coast Trail Pond dated at 12.3 ka indicate a closed Douglas fir-fir forest with inferred climatic conditions that were moister than today (Rypins *et al.,* 1989).

9 KA: EARLY-HOLOCENE ENVIRONMENTS
(Figs. 18.10 and 18.11)

Northwest

On the Olympic Peninsula Douglas fir was more prominent at 9 ka than before or after, reflecting the driest conditions of the postglacial period, although spruce-hemlock forest remained the dominant vegetation (Heusser, 1977). In western Washington and adjacent British Columbia, palynologic evidence of an expansion of Douglas fir, alder, and open-forest habitats implies drier summers (Mathewes, 1973; Hansen and Easterbrook, 1974; Barnosky, 1981). In eastern Washington, sediment cores from the Okanogan highlands indicate the desiccation of many sites (Mack *et al.,* 1978a,c, 1979). Where sediments are preserved, the

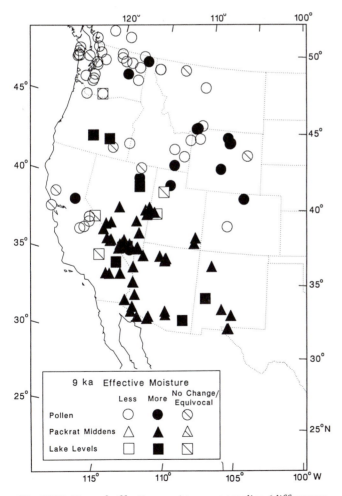

Fig. 18.10. Map of effective moisture anomalies (difference from present) at 9 ka (see also Tables 18.2–18.4).

pollen data indicate the expansion of *Artemisia* steppe (Mack *et al.*, 1978a,c, 1979; Nickmann and Leopold, 1984) and yellow pine parkland (probably lodgepole pine, but perhaps ponderosa pine [Mack *et al.*, 1978b; Nickmann, 1979]). Palynologic data from Carp Lake in the southwestern Columbia Basin provide evidence of a temperate grassland associated with low lake levels, implying very dry conditions (Barnosky, 1984).

Rocky Mountains

As in the Northwest, pollen data from the northern Rocky Mountains indicate expansion of Douglas fir-yellow pine parklands at some sites (Mack *et al.*, 1979, 1983; Baker, 1983). In the Yellowstone region in northwestern Wyoming, Douglas fir and lodgepole pine expanded to high-elevation sites during what appears to have been the period of maximum Holocene drought (Baker, 1976, 1983; C. Whitlock, unpublished data). At high-elevation sites in western Montana and southern Idaho, palynologic studies indicate the presence of spruce-whitebark pine or limber pine-fir parkland or forest (Mehringer *et al.*, 1977; Davis *et al.*, 1986). Lakes at upper elevations in southern Idaho and adjacent extreme northeastern Nevada (Davis, 1981; Thompson, 1984) were at low levels during the early Holocene. At the lower forest border in Yellowstone National Park, palynologic evidence suggests parkland of *Artemisia*, lodgepole pine, and limber pine, indicating that conditions were not as dry as at 6 ka (Gennett and Baker, 1986). However, high-elevation records from across the western interior (e.g., Andrews *et al.*, 1975; Petersen and Mehringer, 1976; Mehringer *et al.*, 1977; Madsen and Currey, 1979; Brant 1980, 1982) suggest that summer temperatures were warmer than at present (Fig. 18.11).

Beiswenger (1987, 1991) interpreted the pollen stratigraphy at the Snake River Plain of southeastern Idaho as indicating increasing aridity from 10.2 to 10 ka and a warm, xeric Holocene peaking at around 7.3 ka. At nearby Swan Lake, an *Artemisia* steppe replaced a forest of lodgepole pine, limber or whitebark pine, and spruce (Bright, 1966). In contrast, a pollen record from Ice Slough in central Wyoming suggests that moist early-Holocene climates supported a dense *Artemisia* stand until 8.2 ka, when conditions became drier (Beiswenger, 1987). Similarly, conditions in the Powder River Basin of Wyoming were apparently moister at 9 ka than during the middle Holocene (Markgraf and Lennon, 1986).

In the northern Great Plains at Guardipee Lake, pollen data provide evidence that a temperate grassland was replaced by a xerophytic grassland with more chenopods and less sagebrush near 9.3 ka (Barnosky, 1989). The nearby Lost Lake record indicates that a xerophytic grassland under conditions drier than modern was present from about 9.4 ka to about 6 ka (Barnosky, 1989).

Pollen spectra for 9 ka from Alkali Lake in central Colorado may reflect a downward expansion of treelines resulting from enhanced summer monsoons (Markgraf and Scott, 1981; Fall, 1985, 1988). The subalpine forest in this region was 300 m or more above modern limits between 9 and 4 ka (Fall, 1988). If lapse rates were the same as today, then July temperatures were about 2.1°C warmer than at present and mean annual temperatures were about 1.8°C warmer. Elsewhere in Colorado, timberlines higher than modern between 9 and 3 ka are inferred from the presence of coniferous macrofossils in alpine lake sediments as well as from palynologic evidence (Maher, 1961; Andrews *et al.*, 1975; Davis *et al.*, 1979; Jodry *et al.*, 1989). In the San Juan Mountains of southeastern Colorado, treeline was more than 80 m above modern elevations at 9 ka (Carrara *et al.*, 1984), and hydrogen isotope ratios indicate summer temperatures 1°C above today's, with a much higher proportion of summer rainfall (Friedman *et al.*, 1988). Fossil insect assemblages from high-elevation lakes in the Front Range indicate "a positive thermal anomaly" between about 9 and 6 ka (Elias, 1985). In the La Plata Mountains of southwestern Colorado, pollen data suggest that timberlines were low from 10 to 8.6 ka but higher than today from 8.6 to 8.3 ka (Petersen, 1988).

Great Basin

The few early-Holocene packrat midden assemblages available from the Great Basin for this time indicate a mixture of montane and woodland taxa in the elevational range of the modern piñon pine-juniper woodland. The absence of piñon pine and the abundance of Rocky Mountain juniper and other montane species imply conditions cooler and/or moister than today, a view supported by hydrogen isotope data (Siegal, 1983). Pollen records from the Ruby Marshes (Thompson, 1992) and Hidden Cave (Wigand and Mehringer, 1985) indicate more abundant sagebrush and conditions cooler and/or moister than today through the early Holocene until 7-6.5 ka.

In contrast to the Rocky Mountains of Colorado, in the Great Basin in the same latitudinal range there are no clear records of treelines higher than modern elevations during the early Holocene (even though a clear record exists for the middle Holocene, as discussed below). This lack of evidence may simply result from the history of preservation, or it may indicate that Great Basin treelines were not elevated during this period because the warming seen in Colorado did not extend this far west. However, pollen evidence from Steens Mountain in the far northeastern Great Basin, while not indicating higher treelines, does suggest conditions warmer than at present near 9 ka (Mehringer, 1985, p. 181). In addition, hydrogen-deuterium ratios from cellulose preserved in packrat middens in the Lahontan Basin (Flynn and Buchanan, 1990) indicate a major enrichment before 9 ka, which could reflect a major rise in temperature and/or an increase in summer rainfall from southerly sources.

Colorado Plateau

Packrat middens from across the Colorado Plateau (Betancourt, 1984, 1990) indicate that montane conifers persisted at relatively low elevations into the early Holocene, although major biotic changes took place through this time. Montane conifers and other mesophytic taxa present at low-elevation sites in southeastern Utah during the Late Wisconsin largely disappeared between 10.4 and 9.5 ka, as Utah juniper, little-leaf mountain mahogany, single-leaf ash, hackberry, and other modern woodland plants became established.

In contrast, an indurated midden layer from Natural Bridges National Monument indicates that Douglas fir, limber pine, blue spruce, Rocky Mountain juniper, prostrate juniper, and Engelmann spruce persisted below modern limits into the early Holocene (Mead et al., 1987). Packrat middens from extreme southeastern Utah dated at 9.5–9.4 ka contain woodland and steppe plants (Utah juniper, snowberry, barberry, and sagebrush) at elevations below their modern limits (Betancourt, 1990). The expansion of Gambell oak and ponderosa pine beyond their modern limits on the northern Colorado Plateau between about 10 ka and about 6 ka apparently indicates summer rainfall greater than modern levels (Betancourt, 1990).

Packrat midden data from northwestern New Mexico indicate that Douglas fir, Rocky Mountain juniper, limber pine, and (rarely) blue spruce were present in modern desert scrub-steppe habitat from 10.6 to 9.5 ka (Betancourt, 1990). The presence of sagebrush, saltbush, yucca, and prickly pear in these assemblages suggests an open woodland with the most mesophytic taxa restricted to alcoves. By 8.3 ka spruce and limber pine had apparently died out, and Rocky Mountain juniper and Douglas fir remained below their modern limits, growing with ponderosa pine, piñon pine, and single-seed juniper. Farther south in New Mexico, at the San Agustin Plains, freshwater conditions suggest that there was more moisture between 10 and 8.5 ka than before or after (Markgraf et al., 1983, 1984).

In the eastern Grand Canyon, packrat midden records indicate a major turnover in vegetation composition between 12 and 9 ka (Cole, 1990a,b), as subalpine conifers moved upslope from their full-glacial position. Juniper and other woodland plants persisted at relatively low elevations through the early Holocene, reflecting greater effective moisture than today. Cole (1990b) concluded that climatic conditions were slightly warmer (about 1°C) than today, with more summer rainfall.

Southwest

In the Sacramento Mountains of New Mexico, samples dated at 9.8 and 9.6 ka indicate that montane species (Douglas fir and Rocky Mountain juniper) present during the Late Wisconsin had departed, piñon pine-juniper-oak woodlands persisted, and the modern Chihuahuan Desert plants now living at the site had not yet arrived (Van Devender et al., 1984). In the Chihuahuan Desert in west Texas, woodland was present from before 10 ka to around 8.2 ka, when it was replaced by succulent desert scrub (Elias and Van Devender, 1990). Papershell piñon pine disappeared by 10.5 ka, while juniper departed and Hinckley oak decreased by 9.9 ka. Most temperate taxa in the insect faunas had disappeared 3000 yr earlier. The early-Holocene vegetation changed character soon after 9 ka, with the demise of the last remaining temperate taxa.

In the Sonoran Desert in southwest Arizona, piñon pine was gone from modern desert habitats around 11 ka, while xerophytic California juniper persisted until 9.1–7.9 ka (Van Devender, 1977, 1987). Modern desert plants are present in early-Holocene packrat midden assemblages from the Puerto Blanco Range. Some of these species (brittlebush, saguaro, and teddy bear cholla) still grow at the midden localities, whereas others (catclaw, mesquite, and blue palo verde) in this region are now restricted to washes. In the nearby Ajo Mountains, middens dated between 10.6 and 8.1 ka contain California juniper, redberry juniper, and Ajo oak, all of which indicate greater-than-present levels of effective moisture. Saguaro, a giant cactus intolerant of prolonged freezes, was present near its upper modern limits in the Ajo Mountains and nearby Hornaday Mountains of the northern Sonoran Desert by 10–9 ka (Van Devender et al., 1985, 1990). Van Devender (1977, 1987) interpreted the early-Holocene midden records from the Sonoran Desert as indicating 30% more rainfall than today, with greater summer rainfall. The presence of brittlebush and saguaros indicates little freezing weather.

In the Mojave Desert juniper remained below its modern limits into the early Holocene (Van Devender and Spaulding, 1979; Spaulding, 1990a,b), but a general trend toward aridification and warmer temperatures is evident. The continued presence of junipers and other woodland species in the Sonoran Desert and portions of the southern Mojave Desert led Spaulding and Graumlich (1986) and Spaulding (1990b) to postulate an enhanced monsoonal rainfall regime until at least 9 ka.

Lakes on the Mogollon Rim in Arizona appear to have been shallow during the early Holocene (Fine Jacobs, 1983). In contrast, a relatively deep lake stand occurred in the Wilcox Basin of southeastern Arizona near 8.9 ka (Waters, 1989), following several thousand years of lower water levels. Similarly, relatively deep (but not overflowing) lake levels are recorded in the Lake Mojave basin (California) between 9.5 and 8 ka (Wells *et al.*, 1987; Enzel *et al.*, 1988; Dorn *et al.*, 1989). In southern Nevada a trend toward declining marsh environments that began around 14 ka continued through the early Holocene, although water tables remained more than 25 m above modern levels until about 7.2 ka (Quade, 1986).

California

Pollen spectra from Balsam Meadows on the west slope of the Sierra Nevada are dominated by *Artemisia*, indicating that conditions were much drier than at present until about 7 ka (Davis *et al.*, 1985). Similarly, the pollen record from Exchequer Meadow (Davis and Moratto, 1988) has a xeric *Artemisia*-oak subzone dated at 10.7–7.1 ka. Studies of pollen cores in a transect across the central Sierra Nevada indicate that lake levels were lower than at present, that montane taxa grew at higher elevations, and that chaparral species were particularly abundant during the early Holocene (Anderson, 1987). Along the California coast, forest vegetation present during the Late Wisconsin gave way to coastal scrub and grassland by about 10–9.4 ka (Rypins *et al.*, 1989), indicating increasing aridity.

6 KA: MID-HOLOCENE ENVIRONMENTS
(Fig. 18.12)

Northwest

In western Washington State, the persistence of Douglas fir and alder in the modern hemlock-Sitka spruce forest indicates continued summer drought during the middle Holocene (Heusser, 1973, 1977). However, increasing amounts of western red cedar in some pollen records imply that the drought was less severe than at 9 ka (Mathewes, 1973; Barnosky, 1981; Tsukada *et al.*, 1981; Cwynar, 1987). Many records from the Okanogan highlands and Columbia Basin show the greatest development of steppe vegetation and lowest lake levels at this time, indicating xeric conditions (e.g., Mehringer, 1985). Other sites show development of ponderosa pine parkland following an expansion of steppe (Mack *et al.*, 1978c; Nickmann, 1979; Nickmann and Leopold, 1984), indicating slightly wetter conditions at 6 ka than at 9 ka. The arrival of ponderosa pine in the Columbia Basin around 8.3 ka reflects increased effective moisture during the transition from the early to middle Holocene (Barnosky, 1985b).

Rocky Mountains

Many records from Idaho, western Montana, and northwestern Wyoming suggest that the climate was warmer and drier at 6 ka than at 9 ka (Mehringer *et al.*, 1977; Mack *et al.*, 1979, 1983; Davis *et al.*, 1986; Gennett and Baker, 1986; Beiswenger, 1987). In contrast, upper-

elevation sites on the Yellowstone Plateau and in the Bighorn Mountains (Baker, 1983; C. Whitlock, unpublished data) and the lower-elevation Lost Lake, Montana (Barnosky, 1989) indicate xeric conditions at 6 ka differing little from those at 9 ka. At Lost Lake conditions became more mesic after 6 ka, as chenopods declined and sagebrush, conifer, and mesic shrub abundances increased in the pollen record.

In southern Idaho, pollen data from the western Snake River Plain (Henry, 1984) and from a high-elevation site (Davis, 1981) suggest that conditions were moister than at 9 ka although more arid than today. The eastern Idaho Snake River Plain and adjacent western Wyoming were more arid at 6 ka than at 9 ka (Bright, 1966; Davis, 1981; Bright and Davis, 1982).

Data from across Colorado provide a mixture of signals on the amplitude and timing of maximum summer warmth and monsoonal precipitation during the Holocene. The Alkali Basin pollen record (Markgraf and Scott, 1981) suggests that the monsoonal precipitation anomaly was greatest during the middle Holocene, whereas nearby high-elevation sites in cen-

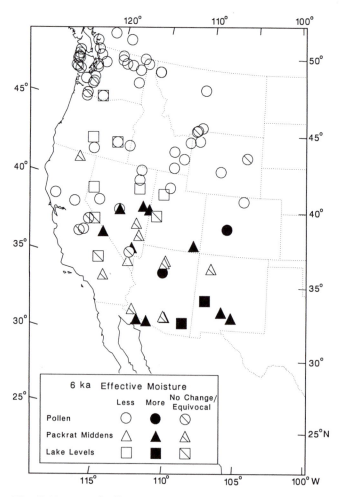

Fig. 18.12. Map of effective moisture anomalies (difference from present) at 6 ka (see also Tables 18.2–18.4).

tral Colorado indicate little change in the elevated summer warmth of the early and middle Holocene (Fall, 1988). Maher (1972) interpreted palynologic evidence from the Front Range as indicating that summer climates were cooler than at present at 10–7.6 and 6.7–3.0 ka, which is in opposition to most other interpretations. In the La Plata Mountains of southwestern Colorado, timberlines were as high from 6.8 to 5.6 ka as they were at 8.6 ka (Petersen, 1988), whereas in the San Juan Mountains of southeastern Colorado, the upper timberlines had apparently receded from their early-Holocene positions and summers were cooler (Elias, 1985; Friedman et al., 1988).

Great Basin

Plant macrofossil assemblages from packrat middens reveal major changes in the vegetation of the Great Basin between 9 and 6 ka. Montane plants indicative of cool conditions persisted at relatively low elevations through the early Holocene but retreated upslope by 6 ka (Thompson and Mead, 1982; Wells, 1983; Thompson, 1990). Piñon pine dispersed northward and approached its modern northern limit between 7.4 and 6 ka (Thompson and Hattori, 1983; Madsen and Rhode, 1990; Thompson, 1990). Boreal mammals, including pika and heather vole, survived outside their modern ranges until the middle Holocene (Grayson, 1982, 1987; Thompson and Mead, 1982; Mead, 1987), which suggests that maximum summer temperatures remained below current levels. In contrast, standing snags of subalpine conifers well above modern limits (LaMarche and Mooney, 1972; LaMarche, 1973) suggest that mean summer temperatures were nearly 2°C above those of today. Summer "monsoonal" rainfall higher than modern levels during the middle Holocene in the central Great Basin may explain the dispersal of piñon pine and the persistence of Rocky Mountain juniper outside its modern geographic and elevational limits (Thompson and Hattori, 1983; Thompson, 1990) as well as excursions in hydrogen isotope ratios (Siegal, 1983).

In Nevada palynologic data from valley-bottom settings at the Ruby Marshes (Thompson, 1992) and Hidden Cave (Wigand and Mehringer, 1985) indicate expanded sagebrush steppe until 7–6.5 ka, when xerophytic chenopod steppe expanded. Mehringer (1985) found that chenopod-sagebrush pollen ratios in a core from Great Salt Lake were highest between about 6.7 and 5.2 ka, reflecting the driest portion of the Holocene in this area. Similar pollen ratios from sites on Steens Mountain (Oregon) in the far northwestern Great Basin indicate low levels of effective moisture from about 8.3 to 5.4 ka and temperatures above modern levels from about 7 to 3.5 ka (Mehringer, 1985). Paleohydrologic indicators from Diamond Pond, near the base of Steens Mountain, show that water levels were more than 17 m below present levels before 5.4 ka (Wigand, 1987).

Colorado Plateau

Packrat midden evidence indicates that by 6 ka the last of the montane mesophytes present on the Colorado Plateau during the Late Wisconsin and early Holocene had retreated upslope to near-modern positions (Betancourt, 1984, 1990). Piñon pines dispersed northward on the Colorado Plateau between 9 and 6 ka (Van Devender et al., 1984), perhaps reflecting abundant summer rainfall.

Southwest

In the northern Chihuahuan Desert, piñon pine-juniper woodland plants mixed with new desert arrivals, such as mesquite and sotol (Van Devender et al., 1984). Van Devender et al. (1984) hypothesized that more frequent hard freezes may have prevented the growth of subtropical Chihuahuan Desert plants now present at the site. In west Texas, temperate grassland taxa were replaced by desert and desert-grassland plants by the middle Holocene (Elias and Van Devender, 1990; Van Devender, 1990a). Severe aridity is first detected in materials dated to about 6 ka, although some mesophytic species persisted until 2.5 ka.

In the Sonoran Desert the vegetation around 6 ka was dominated by plants characteristic of this desert today, but in different compositional mixtures (Van Devender, 1987). Climatic conditions approached those of today, but with more precipitation and a higher proportion of summer rainfall. Winter conditions are inferred to have been more severe than today, with freezes frequent enough to inhibit the northward advance of subtropical plants now in the Arizona desert but not so frequent or severe as to eliminate saguaro and brittlebush (Van Devender, 1987). The Mojave Desert was much drier at 6 ka than at 9 ka (Spaulding, 1990b, 1991), and many of the modern thermophiles arrived between 9.5 and 8 ka. Unlike the Sonoran Desert, the Mojave Desert apparently received less summer rainfall during the middle Holocene than earlier (Spaulding, 1990b, 1991).

Lakes in west Texas, New Mexico, and Arizona apparently were at higher levels than today during at least portions of the middle Holocene. In the Wilcox Basin of southeastern Arizona, a relatively deep lake stand has been dated to about 5.4 ka (Waters, 1989). In contrast, sites farther west, such as the marsh complexes in southern Nevada that were present through the Late Wisconsin and into the early Holocene, desiccated before 6 ka (Quade, 1986), as did early-Holocene lakes in the Mojave Desert of California (Wells et al., 1987).

California

Pollen and lake-level data from the Sierra Nevada record large increases in effective moisture between 9 and 6 ka (Anderson, 1987, 1990; Davis and Moratto, 1988). At Clear Lake, northwest of the Sierra Nevada sites, growth-increment widths from fossil tule perch scales indicate that summer temperatures were warmer than at present during the middle Holocene (Casteel *et al.*, 1977), and quantitative analysis of the pollen record suggests temperatures 1.4–2.1°C warmer than today (Adam and West, 1983).

ENVIRONMENTAL CHANGES DURING THE LATE HOLOCENE

Although the period from 6 ka to the present was not characterized by a monotonic trend, the northwestern, central, and western portions of the American West were generally cooler than during the early and middle Holocene. Indeed, portions of the period since 6 ka have been labeled the "Neoglacial" (Porter and Denton, 1967; Denton and Karlén, 1973) or the "Neopluvial" (Currey, 1990). The timing of the onset of cooler or moister conditions ranged from around 7.1–5.8 ka at Grays Lake (Beiswenger, 1987, 1991) to about 4.7 ka in the Great Basin (Bradbury *et al.*, 1989; Thompson, 1992), 3–2 ka in the Sierra Nevada (Anderson, 1987), and about 1.6 ka at Yellowstone (Gennett and Baker, 1986) and varied with elevation within each region. Post-Pleistocene moisture levels across the Great Basin appear to have been highest during the interval between about 3.8 and 2 ka (Mehringer, 1985; Wigand, 1987; Oviatt, 1988; Bradbury *et al.*, 1989; Currey, 1990; Stine, 1990). A regional drought of major proportions occurred across this region near 2 ka (Mehringer, 1985; Wigand, 1987; Bradbury *et al.*, 1989; Stine, 1990).

Cooler and drier conditions in Colorado began at about 4 ka, and aridity became more intense after 2.6 ka (Fall, 1988), presumably because of a reduction in growing-season temperature and monsoonal rainfall. The demise of Rocky Mountain juniper in the central Great Basin during the late Holocene may also be related to the latter phenomenon (Thompson, 1990). In the Chihuahuan Desert drier conditions after 4.7 ka were accompanied by warmer temperatures, and the modern environment represents the driest part of the Holocene (Elias and Van Devender, 1990; Van Devender, 1990a). In the Sonoran Desert a shift to warmer and drier conditions began at about 5.2 ka, and hard freezes became less frequent after 3.4 ka (Van Devender, 1987). The present Sonoran climate is the most arid in the late Quaternary. Although the Southwest appears to have been quite dry through most of the late Holocene, short intervals of relatively mesic conditions have been detected in paleolacustrine records (Enzel *et al.*, 1988; Waters, 1989) and in vegetation data (Van Devender, 1987).

SYNOPTIC-SCALE PALEOCLIMATIC INFERENCES

When the paleoclimatic inferences discussed above for individual sites or regions are mapped, large-scale coherent patterns emerge (Figs. 18.13 and 18.14). These patterns, and the inferred synoptic-scale paleoclimatic variations responsible for them, are summarized below.

18 ka

Alpine glaciers across the American West were at or near their maximal extents around this time, whereas in the Great Basin "pluvial" lakes were filling but were still well below their deepest Late Wisconsin levels. At 18 ka the elevational ranges of plants were depressed 600–1200 m or more below modern levels across the central interior of the western United States, indicating that summer temperatures were as much as 10°C below present levels. Winters were much colder than today in the northwestern sector but only slightly cooler near the Mexican border (Fig. 18.6).

Precipitation was lower than at present in the Northwest and the northern Rocky Mountain region but was greatly augmented in the southwestern deserts. This geographic pattern in effective moisture suggests that the band of westerlies that at present retreats north into Canada in the warm months was displaced southward into the southwestern deserts at 18 ka. Barnosky *et al.* (1987) proposed that the northwestern sector received stronger easterly flow off the cold and arid midcontinent at that time, while migratory low-pressure cells following the westerlies brought greatly increased precipitation to the Southwest. Antevs (1952, p. 101) offered this explanation 40 yr ago in an attempt to explain the rise of Pleistocene lakes in the Great Basin:

> During the glacial maxima with permanent high pressure over the cold northern part of the continent, the Aleutian Low persisted through the summer, and the Pacific Anticyclone remained relatively weak and far to the south, permitting rain-bearing cyclones to move over California and the Great Basin throughout the year.

12 ka

The scale of climatic variations from 15 to 11 ka dwarfs any before or after in the middle and northern portions of the American West. Postglacial warming began as early as 16 ka in the Pacific Northwest and by 14 ka in the southern Great Basin and Mojave Desert regions (Fig. 18.7). In the Great Basin region a series of major paleohydrologic fluctuations occurred as the high effective moisture levels of around 15–13 ka were abruptly replaced before 12 ka by apparently cold and dry conditions, which in turn gave way after 11 ka to moisture levels that were moderate but still above modern levels. Ice sheets and alpine glaciers retreated throughout the West by 12 ka.

In contrast to the dramatic changes farther north, the southern deserts changed little until 12–11 ka, and effective moisture was apparently above modern levels across the region. This combination permitted

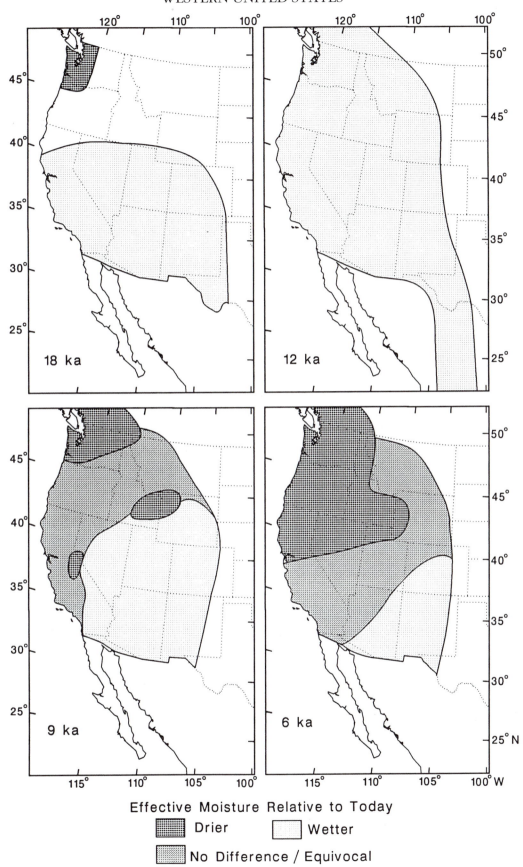

Fig. 18.13. Summary of regional-scale differences in effective moisture between 18, 12, 9, and 6 ka and the present.

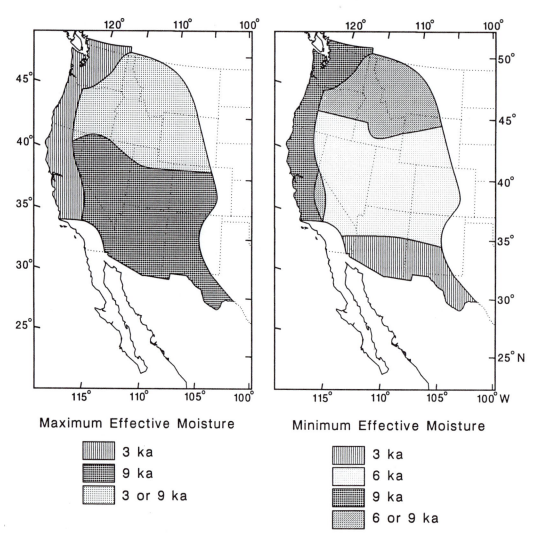

Fig. 18.14. Timing of maximum (left) and minimum (right) levels of effective moisture during the Holocene in the western United States.

many plants to expand their ranges northward and to higher elevations while still maintaining at least portions of their full-glacial southern and low-elevation ranges.

We propose that during the transition from full-glacial through late-glacial conditions, parts of the western interior experienced the following sequence of paleoclimatic events: (1) during the full glacial, easterly winds off the midcontinent and the Laurentide ice sheet led to the dominance of cold, dry Arctic air; (2) as the continental ice sheet diminished in extent and height, the core of the jet stream migrated northward and provided cool, wet air; (3) with further contraction of the ice sheet, the jet core moved north of the region, and cool, dry conditions ensued; and (4) finally, as insolation increased on the midcontinent in the latest Pleistocene and early Holocene, summer monsoonal rainfall from southern sources reached (at least) the southern and eastern portions of the region.

9 ka

Pollen and lake-level data from the Pacific Northwest indicate that conditions at 9 ka were the driest during the Holocene. The record is less clear in the northern Rocky Mountain region, with some sites indicating maximum Holocene dryness, while others were not as dry as at 6 ka. Pollen and plant macrofossil evidence for upper treelines above modern levels in the mountains of Colorado and California indicates that summer temperatures were warmer than at present. Downslope expansions of forest trees in Colorado and the presence of relatively mesophytic plants on south-facing desert slopes in Arizona both indicate increased summer "monsoonal" rainfall, whereas the generally arid conditions in the Sierra Nevada may illustrate the effects of warmer summers without an augmentation of monsoons in this generally summer-dry region.

The apparently cool and moist summer conditions

in the Great Basin, Colorado Plateau, and (to some degree) southwestern regions during the early Holocene have been the focus of many debates, as these regions lie nestled among areas of warmer-than-present temperatures. Recent studies from the Sierra Nevada indicating aridity during the early Holocene effectively eliminate increased rainfall from the North Pacific as the source of the moisture in the interior. Increased monsoonal rainfall from southerly sources appears to be the most plausible explanation for the cool-moist summer conditions in the southern portion of the American West around 9 ka. However, vertebrate paleontologists and other researchers in the Great Basin find it difficult to associate the cool temperatures seen in several paleobiotic data sets with increased southerly flow.

Whether increased monsoonal circulations were in place before 9 ka, as suggested by Spaulding and Graumlich (1986), remains unclear given the evidence for persistent cold temperatures through much of the interior until at least 12 ka. Climatic circulation patterns related to the El Niño southern oscillation (ENSO) have been suggested as the cause of Pleistocene marine varves off the California coast (Anderson et al., 1990) and cannot be ruled out as potential sources of moisture for the western interior.

6 ka

By the middle Holocene effective moisture levels were generally increasing in the Pacific Northwest, although they remained below present levels. The signal is mixed across the northern Rocky Mountain region, with some sites wetter than at 9 ka and others drier, apparently independently of elevation. Upper treelines remained above current levels in the Colorado Rockies and westward to the Great Basin and California. The southwestern deserts were warmer than at 9 ka and apparently received more summer monsoonal rainfall but also may have had more severe winters. Sites in California had more effective moisture than during the early Holocene but also apparently were warmer than today. The Great Basin and California regions both appear to have experienced major climatic change between 7 and 6 ka.

The pattern of climatic change between 9 and 6 ka may reflect a reduction in the strength of the subtropical Pacific high-pressure cell, which allowed more westerly moisture into the coastal regions, coupled with a change in the tracks of monsoonal storms, which brought more moisture from the Gulf of Mexico.

Holocene Chronology

Antevs (1948, 1952) proposed a division of the climatic history of the interior West into three periods: the "Anathermal," a cool-moist period from about 9 ka to about 7 ka; the "Altithermal," a warm-dry period from about 7 ka to around 4.5 ka; and the "Medither-

mal," a period with varying conditions, though generally cooler and moister than the preceding interval. This model was adopted by many researchers (e.g., Baumhoff and Heizer, 1965; Morrison, 1965), although some authors questioned its application to a region as climatically diverse as the western United States (e.g., Mehringer, 1977). Forty years of additional research and the development and widespread application of radiocarbon dating have shown that Antevs's model is invalid. The periods of maximum warmth and moisture were time-transgressive, and the range of climatic conditions was much broader than the cool-wet versus warm-dry opposition seen by Antevs.

As illustrated in Figure 18.14, effective moisture reached its maximum during the early Holocene in the Southwest and Great Basin but during the late Holocene in the Northwest. Effective moisture was at a minimum in the early Holocene in the Northwest and Sierra Nevada, during the middle Holocene in the Great Basin and Colorado Plateau, and during the late Holocene in the southwestern deserts. The northern Rocky Mountain region is difficult to characterize in this framework because of its patchwork of sites that apparently experienced their maximum (and minimum) levels of effective moisture at different times.

Comparison of Climate Model Simulations with Inferred Paleoenvironmental Changes

The spatial and temporal coherence evident in the inferred regional paleoclimatic patterns described above suggests that these patterns were produced by large-scale controls. To elucidate the nature of these controls, we examine the results of paleoclimatic simulations conducted with the Community Climate Model (CCM) of the National Center for Atmospheric Research (NCAR) by Kutzbach, Guetter, et al. (this vol.) and other selected simulations. These simulations reveal the large-scale spatial patterns in specific climate variables that are (physically) consistent with a particular set of controls or boundary conditions. These boundary conditions include, for example, the height of the ice sheet, sea-surface temperature, and the seasonal distribution of insolation (Kutzbach and Ruddiman, this vol.). The results of a particular experiment show the influence of the boundary conditions on large-scale atmospheric circulation features and in turn on continental-scale patterns of individual climate variables. The climate models therefore help us to identify the ultimate controls of observed regional paleoclimatic patterns, to the extent that the large-scale patterns in the simulations resemble those apparent in the paleoclimatic data.

Direct comparison of paleoclimatic simulations with paleoclimatic inferences derived from fossil data is complicated by several factors. Kutzbach, Guetter, et

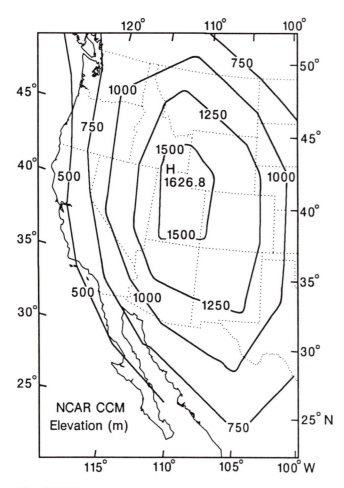

Fig. 18.15. Elevations in the western United States as portrayed in the Community Climate Model (CCM) of the National Center for Atmospheric Research (NCAR). H indicates the highest elevation in the western United States in the CCM portrayal of physiography. Modified from Dickinson *et al.* (1987).

al. (this vol.) discuss some of the general limitations of the model and the experimental design. For the western United States, the principal sources of uncertainty are the coarse spatial resolution of the model (4.4° latitude by 7.5° longitude) and the accompanying great smoothing of topography (Fig. 18.15). In the model the western Cordillera is represented by a broad dome, a little over 1.5 km high, centered roughly over northern Utah. Large-scale physiographic features such as the Sierra Nevada, Cascade Range, and Snake River Plain are therefore not represented, nor are the individual ranges of the mountainous West that produce much of the observed spatial variability of climate in the region.

A by-product of the combination of poor geographic and topographic resolution is the absence or mislocation of certain geographic features that influence the modern pattern of climate in the West. For example, at the location in the model corresponding to the Pacific Northwest, the model's topography is a broad plain, rising from 500 to 1250 m across a distance of about 1000 km. The orographic effects of the Cascade Range and northern Rocky Mountains are thus absent in the simulations. Similarly, in the Southwest, the present summer monsoonal precipitation is enhanced as moisture flows onshore from the Gulf of California and is forced to rise at the edges of the western plateaus, as along the Mogollon Rim in Arizona (Tang and Reiter, 1984). In the model, this configuration of a topographic barrier to the north of a moisture source occurs not at the location of Arizona but rather at grid cells corresponding to the location of central Mexico.

Comparisons of paleoclimatic simulations and inferences are further complicated by the different ways in which the simulations and observations are located in time. The simulations are spaced 3000 yr apart and are "snapshots" of conditions consistent with the "settings" of the large-scale controls at those times (which are themselves not perfectly known). This experimental design thus does not allow inferences about the timing of particular climatic changes, apart from establishing that they occurred between the experimental dates.

Specific point-by-point comparisons between the simulations and inferences from paleodata for a particular time are therefore almost certain to fail. The appropriate way to compare the two is to concentrate on the larger-scale features of the simulated climate—circulation features, such as the jet stream and "centers of action" (e.g., the East Pacific subtropical high-pressure system or the Aleutian low-pressure system), and continental-scale patterns of temperature and moisture—which are better simulated in the model than smaller-scale features.

To link the broad features in the simulations with the paleoclimatic data, we must consider variations in the large-scale features controlling modern regional climate patterns. For example, in the Pacific Northwest at present, weaker westerly flow would likely reduce orographic precipitation in that region. The unrealistic topography for this region in the model, however, would greatly diminish the response of simulated precipitation to such a circulation change. In a sense, the interpretation of the large-scale features of the simulated climate in light of present regional controls involves using the large-scale simulations as the boundary conditions of a *conceptual* mesoscale climate model.

In the following descriptions we make frequent reference to the maps in Chapter 4 of this volume by Kutzbach, Guetter, *et al,* henceforth referred to by number (e.g., Fig. 4.1). We also describe some of the larger-scale patterns of the simulated climate variables expressed in terms of area averages of the differences between paleoclimatic simulations and the modern "control" simulation (Fig. 18.16). A negative tempera-

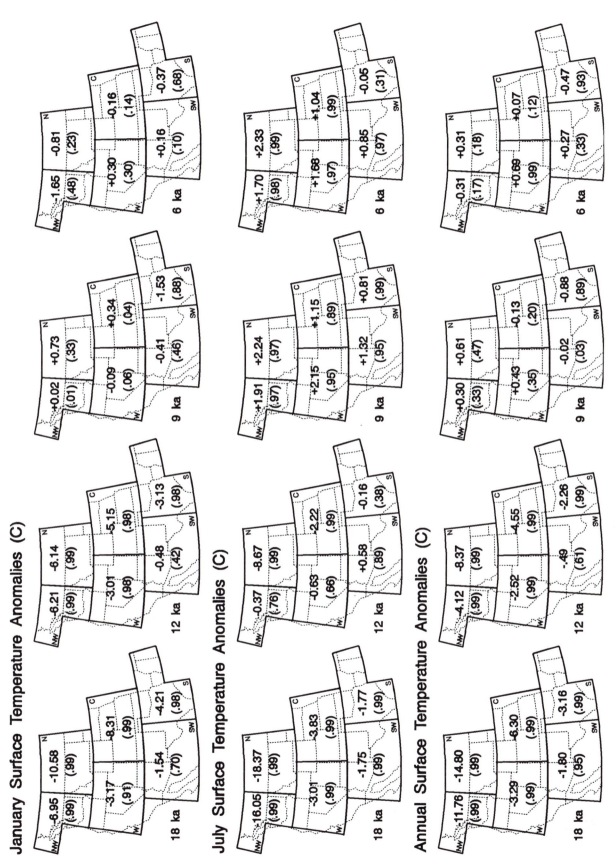

Fig. 18.16. Estimates of seasonal and annual surface temperature, precipitation, and precipitation-minus-evaporation (P-E) anomalies (difference from present) for 18, 12, 9, and 6 ka generated by the Community Climate Model of the National Center for Atmospheric Research for the northwest (NW), north (N), west (W), central (C), southwest (SW), and south (S) area-average regions of the western United States. The statistical significance of the estimates is shown in parentheses.

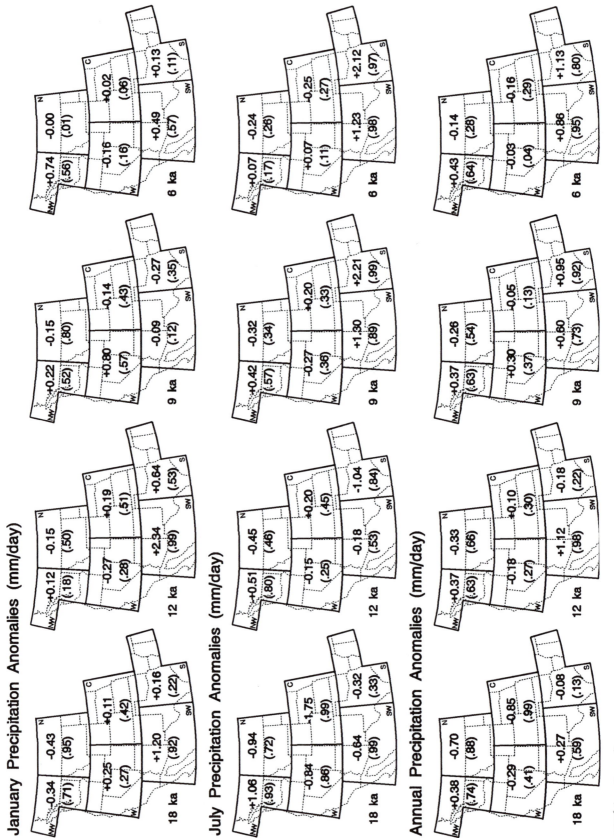

Fig. 18.16 (continued)

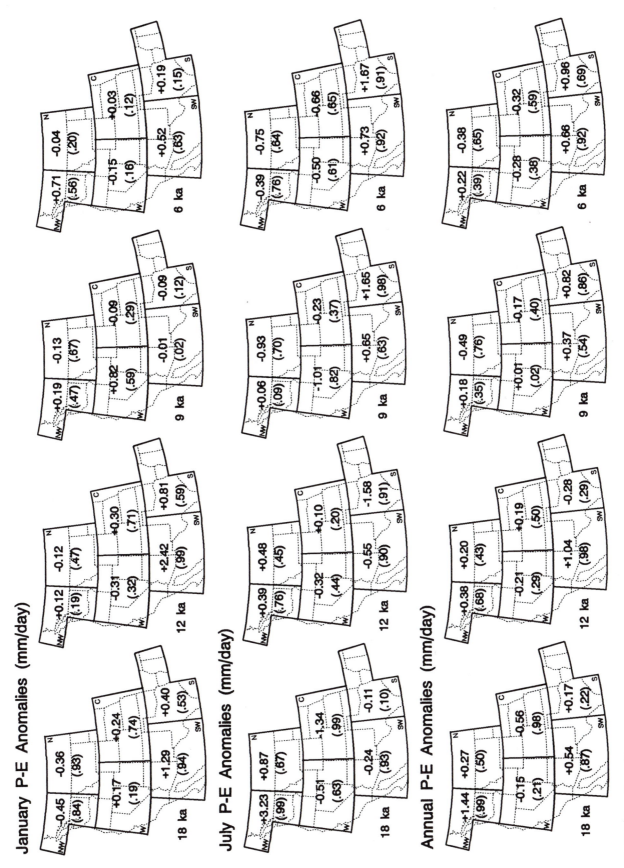

Fig. 18.16 (continued)

ture anomaly thus signifies conditions colder than at present, and a negative precipitation anomaly signifies conditions drier than at present. The statistical significance of these anomalies was assessed by the methods described by Kutzbach and Guetter (1986).

CLIMATE SIMULATIONS

General Features

The set of simulations described by Kutzbach, Guetter, *et al.* (this vol.) shows two major large-scale responses of the simulated climate over North America to the changing boundary conditions over the past 18,000 yr: (1) atmospheric circulation patterns were strongly influenced by the imposition of the large Laurentide ice sheet in the model, which split the jet stream into two branches and generated strong anticyclonic circulation at the surface; and (2) surface temperatures rose higher than at present in summer during the early to middle Holocene in response to increased summer insolation (Barnosky *et al.*, 1987; Kutzbach, 1987; Webb *et al.*, 1987; Kutzbach, Guetter, *et al.*, this vol.). The circulation response to the ice sheet and other glacial boundary conditions (i.e., lower sea-surface temperatures and more extensive sea ice) was most pronounced in the simulations for 18 ka and was still important in eastern North America at 12 and 9 ka. The response to the amplification of the seasonal cycle of insolation was generally greatest at 9 ka, but regions farther from the retreating ice sheet responded earlier.

18 ka

In the western United States the imposition of the glacial-age boundary conditions at 18 ka (Kutzbach and Ruddiman, this vol.) produced significant changes in atmospheric circulation in the simulations. The southern branch of the split jet stream was located at about 30°N in winter (January), approximately 20° south of its modern winter position (Figs. 4.18 and 4.22, this vol.). Simulated winds at the 500-mb level were also stronger than at present as they crossed the West Coast (Fig. 18.17). At the surface, circulation around the glacial anticyclone produced prevailing easterlies in the Pacific Northwest. Over the northern Pacific Ocean, the Aleutian low-pressure system was better developed than at present in both seasons, while over the continent higher pressure prevailed than at present. The southward displacement of the jet stream was still evident, though less prominent, in the summer (July) simulations, as was the glacial anticyclone, and the East Pacific subtropical high-pressure system was weaker than at present (Figs. 4.23 and 4.27, this vol.).

Simulated temperatures in both seasons were lower than at present, and the size of the anomaly increased toward the center of the continent. In the north, central, and northwest area-average regions (Fig. 18.16), these negative temperature anomalies were associated with increased easterly flow from the cold center of the continent. Smaller negative anomalies in the west and southwest regions resulted from

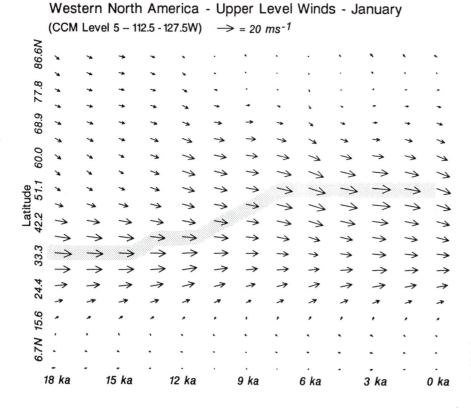

Western North America - Upper Level Winds - January
(CCM Level 5 – 112.5 - 127.5W)　　→ = 20 ms⁻¹

Fig. 18.17. Changes in the latitudinal placement of the maximum January upper-level winds over the last 18,000 yr.

the stronger onshore flow (winter) or weaker flow from the center of the continent (summer).

The storm track associated with the southward-displaced jet stream brought greater-than-present precipitation to the Southwest in winter. Relatively warm Pacific sea-surface temperatures (CLIMAP Project Members, 1981) and increased rising motions in the southwest quadrant of the jet stream core (Kutzbach and Wright, 1986) may also have contributed to this increase in precipitation. In summer in the Southwest, simulated precipitation was lower than at present, consistent with the development of high pressure over the continent and the consequent decrease in onshore flow there. The prevailing easterlies suppressed precipitation in the Northwest in winter, while in summer the strengthened Aleutian low and the location of the storm track along the southern margin of the ice sheet increased precipitation in the northwest area-average region, though mainly at the grid cells that were covered with ice.

In the area averages as a whole, the precipitation anomalies do not consistently reach statistically significant levels, probably because of the unrealistic orography in the model and the generally high spatial variability of the simulated precipitation (Kutzbach, Guetter, *et al.,* this vol.). However, if the simulated circulation changes for 18 ka were superimposed on a *realistic* landscape, annual precipitation would be greater than at present in the Southwest and less than at present in the Northwest.

The simulated circulation features (split jet and glacial anticyclone) and the attendant surface temperature and precipitation changes for 18 ka are robust features of paleoclimatic simulations for "full-glacial" conditions. For example, the simulation for the last glacial maximum by Broccoli and Manabe (1987a,b) clearly shows the development of the large negative temperature anomaly over the ice sheet and adjacent areas (Fig. 5 in Broccoli and Manabe [1987a]), the southward displacement of the southern branch of the split jet, and the development of the glacial anticyclone (Figs. 3 and 4 in Broccoli and Manabe [1987b]). These simulated circulation features are also robust with respect to the size of the ice sheet imposed in the model, provided the size exceeds some threshold value. Simulations with a smaller ice sheet than that included in the "standard" 18-ka experiment still show the split jet and glacial anticyclone (Figs. 4.28 and 4.29, this vol.).

12 ka

The interval around 12 ka was a time of transition between "glacial" and "interglacial" regional paleoclimates (e.g., Barnosky *et al.,* 1987; Webb *et al.,* 1987). The simulations for 12 ka, viewed in the context of those preceding and following, can therefore provide insight into the nature of that transition in the western United States.

In the simulations for 12 ka (Kutzbach, Guetter, *et al.,* this vol.), the height of the ice sheet was reduced to 40% of its 18-ka height, the area of the ice sheet was correspondingly reduced, and sea-surface temperatures and sea-ice limits were placed between their 18-ka and post-9-ka settings. The positive summer insolation anomaly was near its maximum, and the negative winter insolation anomaly was near its minimum (Kutzbach and Ruddiman, this vol.). The smaller ice sheet no longer split the jet stream into two branches, and the glacial anticyclone was smaller and was clearly evident only during summer (Figs. 4.19, 4.22, 4.24, and 4.27, this vol.) (COHMAP Members, 1988). The simulated January jet stream over North America took on the "West Coast ridge, East Coast trough" pattern of today, but wind speeds remained nearly as great as at 18 ka (Kutzbach, 1987; see also Fig. 18.17). In regions distant from the ice sheet, simulated July temperatures approached or even exceeded present values in response to the increased summer insolation (Kutzbach, 1987; Webb *et al.,* 1987).

In the western United States the jet stream and associated surface storm track moved northward toward their present positions in both seasons (Fig. 18.17), but the upper-level wind speeds remained greater than at present in winter across the region. In summer the jet stream and storm track were located along the southern edge of the ice sheet in the simulations, still south of their present positions. At the surface the prevailing easterlies in the Northwest in the 18-ka simulations were replaced by prevailing westerlies. The Aleutian low-pressure system approached its present strength in winter, while the East Pacific subtropical high-pressure system remained a little weaker than at present in the simulations for summer. The combined effect of these simulated circulation changes was to maintain greater-than-present onshore flow across the western United States at 12 ka.

Surface temperatures rose in both seasons compared to 18 ka and approached present values in the southwest area-average region in response to greater summer insolation. The spatial pattern of precipitation was not as well organized in the 12-ka simulations as in the 18-ka simulations, but it does suggest a continuation of wetter-than-present conditions in the southwest area-average region in winter. Elsewhere at 12 ka the simulated precipitation anomalies are mixed in both seasons.

Again, the simulated circulation patterns at 12 ka, if superimposed on realistic topography, would likely produce significant changes in orographic precipitation. In the Northwest the replacement of prevailing easterlies with prevailing westerly winds at the surface would probably greatly increase orographic precipitation in this region, while the general northward movement of the jet stream and storm track, particularly in winter, would likely reduce orographic precipitation in the Southwest but increase it in the Great

Basin. Overall, the continuation of stronger-than-present westerly winds in winter from 18 to 12 ka (Figs. 4.18 and 4.19, this vol.) would most likely have kept precipitation above present levels throughout the West at 12 ka.

9 ka

The size of the ice sheet was further reduced for the 9-ka simulations, and other "glacial" boundary conditions were set at their modern values (Kutzbach and Ruddiman, this vol.). The seasonal cycle of insolation was amplified, however, with summer insolation about 8% greater than at present and winter insolation correspondingly less than at present.

The large-scale circulation patterns over the whole of North America in winter were essentially similar to those at present, with only a slight reinforcement of the anticyclone by the ice sheet (Figs. 4.20 and 4.25, this vol.). In summer the residual ice sheet lowered temperatures in the region adjacent to it, while the center of the continent reached temperatures 2°C above present levels because of the greater summer insolation.

In the western United States the differences between the simulated circulations at 9 and 0 ka were less striking than earlier, particularly in winter, but were consistent with the first-order effects of the insolation anomalies. In summer the stronger heating of the center of the continent (by the enhanced insolation) intensified slightly the "normal" land-ocean temperature and surface-pressure contrasts; pressure lower than at present was simulated over the continent, and higher pressure was simulated over the oceans. This pattern resulted in a slight augmentation of the normal onshore flow of air in summer in the Southwest and strengthened the subtropical high-pressure system, which at present suppresses precipitation in the Northwest. The simulated summer (July) temperatures were greater than at present everywhere in the western area-average regions, while winter (January) temperatures were below or close to their present values.

Simulated winter precipitation at 9 ka differed little from the present. Simulated summer precipitation was greater than at present in the southwest and south area-average regions as a result of the slightly greater onshore flow there and was close to or below present levels elsewhere (Fig. 18.16). The greater summer insolation at 9 ka also resulted in the simulation of greater net radiation than at present and hence greater evaporation. Precipitation minus evaporation nevertheless was still greater than at present in the Southwest. Overall, the impression gained from the simulations is of generally wetter-than-present conditions in the Southwest at 9 ka, contrasting with drier conditions elsewhere.

As at 18 ka, the temperature, circulation, and attendant precipitation patterns for 9 ka are robust features in simulations performed with different models. For example, Mitchell et al. (1988) used the United Kingdom Meteorological Organization (UKMO) general circulation model to perform a set of experiments with 9-ka insolation inputs. They presented more detailed results for a simulation without an ice sheet but also presented results for a simulation that included a remnant ice sheet. Their simulation (without ice) shows a decrease in surface pressure over the continent and an accompanying expansion of the Pacific subtropical high-pressure system in summer (their Fig. 12) similar to those simulated by Kutzbach, Guetter, et al. (this vol., Fig. 4.25). Similarly, they simulated slightly wetter conditions than at present in the Southwest in summer and drier conditions than at present in the Northwest, also consistent with Kutzbach, Guetter, et al. (this vol.). (Again, keep in mind that the precipitation simulations can be interpreted only in the broadest fashion, because of the poor geographic resolution of the UKMO model, like that of the NCAR CCM.) The addition of an ice sheet reduced the simulated summer temperature anomaly over the center of the continent and consequently weakened some of the circulation responses, a result similar to that obtained by Kutzbach and Guetter (1984) with the NCAR CCM.

6 ka

In the simulations for 6 ka, all surface boundary conditions (ice sheets, sea ice, etc.) were set at their present values (Kutzbach and Ruddiman, this vol.). The simulations thus show the regional paleoclimatic responses to the amplification of the seasonal cycle of insolation alone. Insolation at 6 ka was still greater than at present in summer and less than at present in winter, although the anomalies were smaller than at 9 ka. At 6 ka the principal difference from the present circulation again was the development of low pressure over the center of the continent in summer in response to the enhanced summer insolation, accompanied by the expansion of the Pacific subtropical high-pressure system (Fig. 4.26, this vol.).

The area of the summer temperature anomaly (+2°C) evident at 9 ka expanded to include the area formerly occupied by ice, and the center of the positive anomaly therefore shifted northeast (Kutzbach, Guetter, et al., this vol.). As at 9 ka, onshore flow into the Southwest was slightly stronger than at present in summer, and simulated precipitation was again greater than at present in the southwest and south area-average regions.

The 6-ka simulations (without ice) can be viewed in sequence with the 9-ka set of simulations (with and without ice) to examine the impact of the final retreat of the ice sheet on regional climates at a time when summer insolation was greater than at present (Thompson, 1990). In regions distant from the residual ice, summer temperatures reached their greatest positive anomaly at 9 ka, while the response in regions close to the ice was still attenuated. In the interior and northeastern part of the continent, positive summer temperature anomalies were greatest at 6 ka, after the ice was gone (see also Webb, Bartlein, *et al.,* this vol.). In the distant regions the 6-ka anomalies were smaller than those observed at 9 ka because of the decrease in summer insolation at 6 ka compared to 9 ka.

At both 9 and 6 ka a similar summer surface-pressure pattern was maintained over the western part of the continent, with low pressure over the continent, high pressure over the Pacific, and consequently stronger-than-present onshore flow into the Southwest. At the same time, the expanded subtropical high-pressure system suppressed precipitation in the Northwest. The simulations (Figs. 4.25 and 4.26, this vol.) give a hint of a change in the circulation pattern from 9 to 6 ka that shifted the location of greater-than-present onshore flow into the eastern part of North America at 6 ka (see also Webb *et al.,* 1987). These simulated shifts in the location of the maximum positive temperature anomalies and the axis of onshore flow occurred in response to the shrinkage of the ice sheet and may explain the diachrony in many of the Holocene paleoclimatic records.

Summary of the Simulations

The simulated paleoclimates of the western United States reflect the replacement of one major control on atmospheric circulation, and hence on regional climates, by another. From 18 to 9 ka the influence of the ice sheet waned, while that of the amplified seasonal cycle of insolation grew. In the western United States the response to these changes consisted of a general increase in simulated temperature, along with regional-scale responses to the changing circulation (Fig. 18.18). As the ice sheet retracted, its influence on atmospheric circulation diminished, and the southern branch of the split jet in winter moved northward from about 30°N at 18 ka to its present position at about 50°N (Fig. 18.17). At the same time the glacial anticyclone that had brought prevailing easterlies to the Pacific Northwest at 18 ka contracted, so that the simulated winter atmospheric circulation essentially reached its modern configuration between 12 and 9 ka. Meanwhile, regional precipitation patterns changed: the precipitation maximum associated with the jet stream likely progressed northward across the West during this interval, and conditions in the Pacific Northwest became moister as prevailing westerlies became established.

At 9 ka summer insolation was close to its maximum, and it remained greater than at present at 6 ka. The main simulated response to this control was the development of lower surface pressure over the continent and higher pressure over the Pacific. This pattern resulted in the simulation of stronger onshore flow and consequently greater precipitation in the southwestern part of the continent at both 9 and 6 ka.

These general simulated changes probably are relatively robust, because they appear in several different general circulation models and simulation experiments. The specific regional or local effects of these changes in atmospheric circulation, temperature, and precipitation patterns are conditioned by the poor spatial resolution of the model. At the temporal and spatial scale we have examined here, most of the simulated climatic changes are progressive rather than abrupt. Even the most dramatic change, the disappearance of the split jet stream, was tempered somewhat by the maintenance of greater wind speeds after the single jet reappeared and by the apparently progressive northward migration of the location where the jet stream crosses the West Coast.

COMPARISON OF SIMULATED AND INFERRED PALEOCLIMATES

The simulated regional paleoclimatic changes described above appear to explain many of the inferred paleoclimatic changes summarized in Figures 18.7, 18.13, and 18.14, provided the several sources of uncertainty inherent in comparisons of the simulations and observations are taken into account. These sources include disparities in the temporal and spatial scales of the two sources of information and the resultant potential of the models to simulate the right change but at the wrong location or time.

The very cold "glacial" conditions at 18 ka indicated by the depression of vegetation zones and snowline across the West (along with the position of alpine glaciers at or near their maximum extent) seem to be well explained by the general role of the ice sheet in depressing temperatures across the continent. The paleoclimatic evidence of progressively colder conditions toward the center of the continent (Fig. 18.6) also is consistent with the temperature effect of the ice. Inferences of conditions drier than at present in the Northwest and moister than at present in the Southwest (Fig. 18.13) are consistent with the atmospheric circulation changes induced by the ice sheet. The replacement of prevailing westerlies by the prevailing easterlies associated with the glacial anticyclone in the simulations seems able to account for dry conditions in the Northwest, while the southward displacement of the southern branch of the split jet stream likely accounts for the moister-than-present conditions across the Southwest. The pluvial lakes of the Great Basin, while not at their highest levels, were larger than at present. The main focus of the enhanced precipitation associated with the jet stream therefore probably was located to the south of the Great Basin.

The interval between 16 and 11 ka saw great changes in the inferred paleoclimates across the West, and the 12-ka simulation can provide insight into the nature of the controls of these changes only if we compare it with the earlier (18-ka) and later (9-ka) simulations. The most striking change during this interval is that documented by the paleoclimatic evidence of the temporally coherent rise and subsequent fall of pluvial lake levels across the Great Basin between 15 and 13 ka. The progressive passage of the jet stream and its associated storm track across the Great Basin as circulation adjusted to the diminishing ice sheet (Fig. 18.17) could account for this sharp rise and fall (see also Hostetler and Benson, 1990).

Inferred conditions across the West were generally wetter at 12 ka than at present (Fig. 18.9). This general pattern is probably related to different controls in different regions: in the Southwest, to the maintenance of greater-than-present onshore flow in winter (Fig. 4.19, this vol.); in the Great Basin, to the northward progression of the jet (Fig. 18.17); and in the Northwest, to the resumption of prevailing westerly winds in winter (Figs. 4.18 and 4.19, this vol.).

The paleoclimatic evidence records temperature

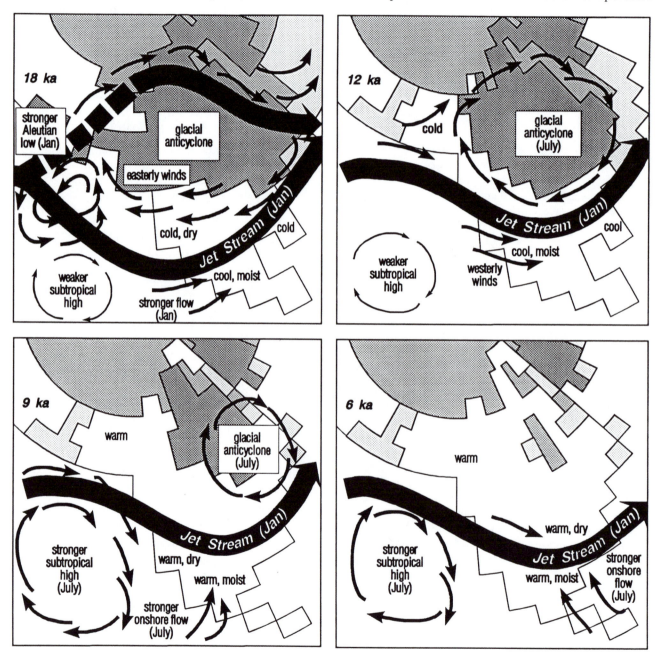

Fig. 18.18. Summary of the estimates of major changes in atmospheric circulation features over the last 18,000 yr for western North America as simulated by the Community Climate Model of the National Center for Atmospheric Research (see also Barnosky *et al.* [1987]).

increases across the West but at different times and of varying magnitude (Fig. 18.7). The simulations for both 12 and 9 ka suggest that temperatures should have risen first in regions distant from the ice sheet, but this is inconsistent with the data, which indicate little change in temperature in the southwestern deserts until after 12 ka.

The paleoclimatic evidence for 9 ka shows a fairly coherent pattern of inferred effective moisture (Fig. 18.10), drier than at present in the Northwest and far West and wetter than at present in much of the Southwest and Rocky Mountains. In the Southwest the downslope extension of forest trees and the persistence of mesophytic plants suggest increased summer "monsoonal" precipitation. The inferred summer temperature anomalies (Fig. 18.11) are less coherent but show generally warmer-than-present conditions in the drier-than-present region and a mixture of warmer and cooler conditions in the wetter-than-present region.

These patterns are generally consistent with the circulation changes simulated for summer at 9 ka (Fig. 4.25, this vol.). Drier conditions than at present in the Northwest are consistent with the suppression of precipitation by the expanded subtropical high (and with increased evaporation resulting from the direct effect of greater summer insolation). The wetter-than-present conditions inferred for the large region that includes the Southwest and southern Rocky Mountains (Fig. 18.10) are consistent with the simulation of slightly stronger-than-present onshore flow into the Southwest, which would enhance monsoonal precipitation. In the simulations this strengthened onshore flow does not seem extensive enough to influence regions as far from the moisture source as Colorado or Wyoming, but those regions are included in the general area affected by the summer monsoon at present (Tang and Reiter, 1984). The lower-than-present temperatures inferred for this wet region could be the result of increased cloudiness.

The basic pattern of inferred effective moisture at 6 ka (Fig. 18.12) is similar to that at 9 ka but with an expansion of the region of drier-than-present conditions in the Northwest and a contraction of the region of wetter-than-present conditions in the Southwest and southern Rocky Mountains. The Northwest, although drier than at present at 6 ka, was probably wetter than earlier (Fig. 18.14). This pattern of change in effective moisture across the West is consistent with both the decrease in summer insolation from 9 to 6 ka and the subtle shift in the onshore flow in summer noted in the simulations (Figs. 4.25 and 4.26, this vol.). As summer insolation decreased from 9 to 6 ka, its direct influence on effective moisture and indirect effects through the expansion of the subtropical high probably diminished in the Northwest. Although that region would remain drier than at present, it would not be as dry as at 9 ka, which is consistent with the evidence for the timing of maximum effective moisture during the Holocene summarized in Figure 18.14.

In the Southwest and Rocky Mountains a shift of the strongest onshore flow toward the east and the consequent reduction of summer monsoonal precipitation (except in the eastern part of this region), as suggested by the simulations for 6 ka, could produce changes in effective moisture that are also consistent with those described in Figures 18.13 and 18.14. The decrease in summer precipitation, coming at a time when summer insolation was still greater than at present, should have produced the driest conditions during the Holocene. These patterns may be summarized as follows: (1) regions that did not receive enhanced summer precipitation at 9 ka would appear to become moister from 9 to 6 ka because summer insolation decreased; (2) regions that did get enhanced precipitation at 9 ka but not at 6 ka would appear to become drier, because the drop in precipitation occurred while summer insolation was still greater than at present; and (3) regions that received enhanced summer precipitation at both 9 and 6 ka would appear to become moister, because summer insolation and the evaporation associated with it decreased over that time.

Summary and Conclusions

When viewed at the subcontinental or regional scale, the paleoclimatic evidence from the western United States shows relatively coherent patterns of variation. At 18 ka inferred temperatures were below present levels everywhere, with increasingly colder conditions toward the center of the continent. The Southwest was moister and the Northwest drier than at present. This large-scale pattern seems well explained by the influence of the ice sheet on climate over North America. The ice sheet had a direct cooling effect, lowering temperatures over the whole of the continent, and moreover had a profound effect on atmospheric circulation. The moistness of the Southwest appears to be tied to the splitting of the jet stream by the ice sheet and the consequent deflection of the southern branch of the jet to a latitude south of its present position. Dryness prevailed in the Northwest as the anticyclone generated by the ice sheet brought prevailing easterlies to the region instead of moisture-bearing prevailing westerlies.

Contraction of the ice sheet and increasing summer insolation produced large-scale adjustments in the circulation, accompanied by general warming. The paleoclimatic record for the West suggests that the region as a whole was wetter than at present at 12 ka. This pattern seems consistent with the adjustments in atmospheric circulation that accompanied the changes in the controls. Precipitation remained greater than at present in the Southwest in response to the continuation of stronger-than-present onshore flow there in winter. Precipitation in the Great Basin at 12 ka was also greater than at present because of the northward migration of the jet stream, and rainfall also increased in the Northwest as a result of the reestablishment of prevailing westerly winds.

When the ice sheet finally disappeared between 9 and 6 ka, circulation patterns reflected only the influence of the enhanced summer insolation. Summer temperatures greater than at present were generated over North America, and the land-sea temperature contrast produced lower surface pressure over the continent accompanied by intensification of the Pacific subtropical high-pressure system. The moister-than-present conditions recorded in the Southwest by the paleoclimatic evidence appear largely consistent with an enhancement of summertime "monsoonal" circulation during this interval resulting from the enhanced land-sea temperature contrast, while drier-than-present conditions in the Northwest reflect the enhancement of the subtropical high-pressure system. As the insolation anomaly gradually diminished toward the present, subtle circulation changes accompanied the reduction in the heating of the center of the continent, and these too seem apparent in the paleoclimatic data.

Considerable variation is evident in the paleoenvironments of the West when the climatic history is viewed at smaller spatial and shorter temporal scales than those considered in this chapter. The explanation for these variations must reside in the way in which mesoscale climatic and landscape features mediate the influence of the large-scale circulation controls. Better understanding of the finer-scale variations must await the development of both data sets and climate models of higher resolution. In general, however, much of the inferred paleoclimatic variation in the West does appear coherent at a large spatial scale and seems to be attributable to changes in large-scale atmospheric circulation controls.

Acknowledgments

Kathy Anderson and Beverly Lipsitz provided invaluable help in the preparation of graphics for this chapter. J. E. Kutzbach and P. J. Behling provided output from the Community Climate Model and help in interpreting these data. T. Webb III, H. E. Wright, Jr., R. F. Fleming, and D. J. Nichols provided useful reviews of the manuscript text.

References

Adam, D. P. (1967). Late-Pleistocene and Recent palynology in the central Sierra Nevada, California. In "Quaternary Paleoecology" (E. J. Cushing and H. E. Wright, Jr., Eds.), pp. 275-301. Yale University Press, New Haven.

——. (1988). "Palynology of Two Upper Quaternary Cores from Clear Lake, Lake County, California." Professional Paper 1363. U.S. Geological Survey, Washington, D.C.

Adam, D. P., and West, G. J. (1983). Temperature and precipitation estimates through the last glacial cycle from Clear Lake, California, pollen data. Science 219, 168-170.

Adam, D. P., Sims, J. D., and Throckmorton, C. K. (1981). 130,000-yr continuous pollen record from Clear Lake, Lake County, California. Geology 9, 373-377.

Alley, N. F. (1976). The palynology and paleoclimatic significance of a dated core of Holocene peat, Okanogan Valley, southern British Columbia. Canadian Journal of Earth Sciences 13, 1131-1141.

Anderson, R. S. (1987). "Late-Quaternary Environments of the Sierra Nevada, California." Unpublished Ph.D. thesis, University of Arizona, Tucson.

——. (1990). Holocene forest development and paleoclimates within the central Sierra Nevada, California. Journal of Ecology 78, 470-489.

Anderson, R. S., and Van Devender, T. R. (1991). Comparison of pollen and macrofossils in packrat (Neotoma) middens: A chronological sequence from the Waterman Mountains of southern Arizona, U.S.A. Review of Palaeobotany and Palynology 68, 1-28.

Anderson, R. Y., Linsley, B. K., and Gardner, J. V. (1990). Expression of seasonal and ENSO forcing in climatic variability at lower than ENSO frequencies: Evidence from Pleistocene marine varves off California. Palaeogeography, Palaeoclimatology, Palaeoecology 78, 287-300.

Andrews, J. T., Carrara, P. E., King, F. B., and Stuckenrath, R. (1975). Holocene environmental changes in the alpine zone, northern San Juan Mountains, Colorado: Evidence from bog stratigraphy and palynology. Quaternary Research 5, 173-197.

Antevs, E. (1948). The Great Basin, with emphasis on glacial and post-glacial times: Climatic changes and Pre-White man. Bulletin of the University of Utah 38, 168-191.

——. (1952). Cenozoic climates of the Great Basin. Geologische Rundschau 40, 94-108.

Baker, R. G. (1976). "Late Quaternary Vegetation History of the Yellowstone Basin, Wyoming." Professional Paper 729-E. U.S. Geological Survey, Washington, D.C.

——. (1983). Holocene vegetational history of the western United States. In "Late Quaternary Environments of the United States, Vol. 2: The Holocene" (H. E. Wright, Jr., Ed.), pp. 109-126. University of Minnesota Press, Minneapolis.

Barnosky, C. W. (1981). A record of late Quaternary vegetation from Davis Lake, southern Puget lowland, Washington. Quaternary Research 16, 221-239.

——. (1984). Late Pleistocene and early Holocene environmental history of southwestern Washington State, U.S.A. Canadian Journal of Earth Sciences 21, 619-629.

——. (1985a). Late Quaternary vegetation near Battle Ground Lake, southern Puget Trough, Washington. Geological Society of America Bulletin 96, 263-271.

——. (1985b). Late Quaternary vegetation in the southwestern Columbia Basin, Washington. Quaternary Research 23, 109-122.

——. (1989). Postglacial vegetation and climate in the northwestern Great Plains of Montana. Quaternary Research 31, 57-73.

Barnosky, C. W., Anderson, P. M., and Bartlein, P. J. (1987). The northwestern U.S. during deglaciation; vegetational history and paleoclimatic implications. In "North America and Adjacent Oceans during the Last Deglaciation" (W. F. Ruddiman and H. E. Wright, Jr., Eds.), pp. 289-321. Geology of

North America, Vol. K-3. Geological Society of America, Boulder, Colo.

Baumhoff, M. A., and Heizer, R. F. (1965). Postglacial climate and archaeology in the desert West. *In* "The Quaternary of the United States" (H. E. Wright, Jr., and D. G. Frey, Eds.), pp. 697-708. Princeton University Press, Princeton, N.J.

Beiswenger, J. M. (1987). "Late Quaternary Vegetational History of Grays Lake, Idaho and the Ice Slough, Wyoming." Unpublished Ph.D. thesis, University of Wyoming, Laramie.

——. (1991). Late Quaternary vegetational history of Grays Lake, Idaho. *Ecological Monographs* 61, 165-182.

Benson, L., and Thompson, R. S. (1987). The physical record of lakes in the Great Basin. *In* "North America and Adjacent Oceans during the Last Deglaciation" (W. F. Ruddiman and H. E. Wright, Jr., Eds.), pp. 241-260. Geology of North America, Vol. K-3. Geological Society of America, Boulder, Colo.

Benson, L. V., Currey, D. R., Dorn, R. I., Lajoie, K. R., Oviatt, C. G., Robinson, S. W., Smith, G. I., and Stine, S. (1990). Chronology of expansion and contraction of four Great Basin lake systems during the past 35,000 years. *Palaeogeography, Palaeoclimatology, Palaeoecology* 78, 241-286.

Betancourt, J. L. (1984). Late Quaternary plant zonation and climate in southeastern Utah. *Great Basin Naturalist* 44, 1-35.

——. (1990). Late Quaternary biogeography of the Colorado Plateau. *In* "Packrat Middens—The Last 40,000 Years of Biotic Change" (J. L. Betancourt, T. R. Van Devender, and P. S. Martin, Eds.), pp. 259-292. University of Arizona Press, Tucson.

Betancourt, J. L., and Davis, O. K. (1984). Packrat middens from Canyon de Chelly, northeastern Arizona: Paleoecological and archaeological implications. *Quaternary Research* 21, 56-64.

Betancourt, J. L., and Van Devender, T. R. (1981). Holocene vegetation in Chaco Canyon, New Mexico. *Science* 214, 656-658.

Betancourt, J. L., Van Devender, T. R., and Martin, P. S., Eds. (1990). "Packrat Middens—The Last 40,000 Years of Biotic Change." University of Arizona Press, Tucson.

Booth, D. K. (1987). Timing and processes of deglaciation along the southern margin of the Cordilleran ice sheet. *In* "North America and Adjacent Oceans during the Last Deglaciation" (W. F. Ruddiman and H. E. Wright, Jr., Eds.), pp. 71-90. Geology of North America, Vol. K-3. Geological Society of America, Boulder, Colo.

Bradbury, J. P., Forester, R. M., and Thompson, R. S. (1989). Late Quaternary paleolimnology of Walker Lake, Nevada. *Journal of Paleolimnology* 1, 249-267.

Brant, L. A. (1980). "A Palynological Investigation of Postglacial Sediments at Two Locations along the Continental Divide near Helena, Montana." Unpublished Ph.D. thesis, Pennsylvania State University.

——. (1982). A trail back through time. *Montana Outdoors* 13, 20-22.

Bright, R. C. (1966). Pollen and seed stratigraphy of Swan Lake, southeastern Idaho: Its relation to regional vegetational history and to Lake Bonneville history. *Tebiwa* 9, 1-47.

Bright, R. C., and Davis, O. K. (1982). Quaternary paleoecology of the Idaho National Engineering Laboratory, Snake River Plain, Idaho. *American Midland Naturalist* 108, 21-33.

Broccoli, A. J., and Manabe, S. (1987a). The influence of continental ice, atmospheric CO_2, and land albedo on the climate of the last glacial maximum. *Climate Dynamics* 1, 87-99.

——. (1987b). The effects of the Laurentide ice sheet on North American climate during the last glacial maximum. *Géographie physique et Quaternaire* 41, 291-299.

Bryson, R. A., and Hare, F. K. (1974). The climates of North America. *In* "Climates of North America" (R. A. Bryson and F. K. Hare, Eds.), pp. 1-47. World Survey of Climatology (H. E. Landsberg, Ed.), Vol. 11. Elsevier Press, New York.

Bryson, R. A., and Lowry, W. D. (1955). The synoptic climatology of the Arizona summer precipitation singularity. *American Meteorological Society Bulletin* 36, 329-339.

Burkart, M. R. (1976). "Pollen Stratigraphy and Late Quaternary Vegetational History of the Bighorn Mountains, Wyoming." Unpublished Ph.D. dissertation, University of Iowa.

Carrara, P. E., Mode, W. N., Meyer, R., and Robinson, S. W. (1984). Deglaciation and postglacial timberline in the San Juan Mountains, Colorado. *Quaternary Research* 21, 42-56.

Carrara, P. E., Short, S. K., and Wilcox, R. E. (1986). Deglaciation of the mountainous region of northwestern Montana, U.S.A., as indicated by late Pleistocene ashes. *Arctic and Alpine Research* 18, 317-325.

Casteel, R. W., Adam, D. P., and Sims, J. D. (1977). Late-Pleistocene and Holocene remains of *Hysterocarpus traski* (tule perch) from Clear Lake, California, and inferred Holocene temperature fluctuations. *Quaternary Research* 7, 133-143.

Cinnamon, S. K., and Hevly, R. H. (1988). Late Wisconsin macroscopic remains of pinyon pine on the southern Colorado Plateau, Arizona. *Current Research in the Pleistocene* 5, 47-48.

CLIMAP Project Members (1981). Seasonal reconstructions of the earth's surface at the last glacial maximum. *Geological Society of America Map and Chart Series* MC-36.

COHMAP Members (1988). Climatic changes of the last 18,000 years: Observations and model simulations. *Science* 241, 1043-1052.

Cole, K. L. (1981). "Late Quaternary Environments in the Eastern Grand Canyon: Vegetational Gradients over the Last 25,000 Years." Unpublished Ph.D. thesis, University of Arizona, Tucson.

——. (1982). Late Quaternary zonation of vegetation in the eastern Grand Canyon. *Science* 217, 1142-1145.

——. (1983). Late Pleistocene vegetation of Kings Canyon, Sierra Nevada, California. *Quaternary Research* 19, 117-129.

——. (1985). Past rates of change, species richness, and a model of vegetational inertia in the Grand Canyon, Arizona. *The American Naturalist* 125, 289-303.

——. (1986). The lower Colorado River Valley: A Pleistocene desert. *Quaternary Research* 25, 392-400.

——. (1990a). Late Quaternary vegetation gradients through the Grand Canyon. *In* "Packrat Middens—The Last 40,000 Years of Biotic Change" (J. L. Betancourt, T. R. Van Devender, and P. S. Martin, Eds.), pp. 240-258. University of Arizona Press, Tucson.

——. (1990b). Reconstruction of past desert vegetation along the Colorado River using packrat middens. *Palaeogeography, Palaeoclimatology, Palaeoecology* 76, 349-366.

Currey, D. R. (1990). Quaternary palaeolakes in the evolution of semidesert basins, with special emphasis on Lake Bonneville and the Great Basin, U.S.A. *Palaeogeography, Palaeoclimatology, Palaeoecology* 76, 189–214.

Currey, D. R., and Oviatt, C. G. (1985). Durations, average rates, and probable cause of Lake Bonneville expansions, stillstands, and contractions during the last deep-lake cycle, 32,000 to 10,000 years ago. *In* "Problems of and Prospects for Predicting Great Salt Lake Levels" (P. A. Kay and H. F. Diaz, Eds.), pp. 1–9. Center for Public Affairs and Administration, University of Utah, Salt Lake City.

Cwynar, L. C. (1987). Fire and the forest history of the North Cascade Range. *Ecology* 68, 791–802.

Davis, O. K. (1981). "Vegetation Migration in Southern Idaho during the Late Quaternary and Holocene." Unpublished Ph.D. thesis, University of Minnesota.

Davis, O. K., and Moratto, M. J. (1988). Evidence for a warm dry early Holocene in the western Sierra Nevada of California: Pollen and plant macrofossil analysis of Dinkey and Exchequer Meadows. *Madroño* 35, 132–149.

Davis, O. K., Anderson, R. S., Fall, P. L., O'Rourke, M. K., and Thompson, R. S. (1985). Palynological evidence for early Holocene aridity in the southern Sierra Nevada, California. *Quaternary Research* 24, 322–332.

Davis, O. K., Sheppard, J. C., and Robertson, S. (1986). Contrasting climatic histories for the Snake River Plain, Idaho, resulting from multiple thermal maxima. *Quaternary Research* 26, 321–339.

Davis, P. T., Upson, S., and Waterman, S. E. (1979). Lacustrine sediment variation as an indicator of late Holocene climatic fluctuations, Arapaho Cirque, Colorado Front Range. *Geological Society of America, Abstracts with Programs* 11, 410.

Denton, G. H., and Karlén, W. (1973). Holocene climatic variations—Their pattern and possible cause. *Quaternary Research* 3, 155–205.

Dickinson, R. E., Errico, R. M., Giorgi, F., and Bates, G. T. (1987). "Modeling of Historic, Prehistoric, and Future Climates of the Great Basin." Unpublished report to the U.S. Geological Survey Nevada Nuclear Waste Investigations Project.

Dohrenwend, J. C. (1984). Nivation landforms in the western Great Basin and their paleoclimatic significance. *Quaternary Research* 22, 275–288.

Dorn, R. I., Turrin, B. D., Jull, A. J. T., Linick, T. W., and Donahue, D. J. (1987). Radiocarbon and cation-ratio ages for rock varnish on Tioga and Tahoe morainal boulders of Pine Creek, eastern Sierra Nevada, California, and their paleoclimatic implications. *Quaternary Research* 28, 38–49.

Dorn, R. I., Jull, A. J. T., Donahue, D. J., Linick, T. W., and Toolin, L. J. (1989). Accelerator mass spectrometry radiocarbon dating of rock varnish. *Geological Society of America Bulletin* 101, 1363–1372.

——. (1990). Latest Pleistocene lake shorelines and glacial chronology in the western basin and range province, U.S.A.: Insights from AMS radiocarbon dating of rock varnish and paleoclimatic implications. *Palaeogeography, Palaeoclimatology, Palaeoecology* 78, 315–331.

Elias, S. A. (1985). Paleoenvironmental interpretations of Holocene insect fossil assemblages from four high-altitude sites in the Front Range, Colorado, U.S.A. *Arctic and Alpine Research* 17, 31–48.

Elias, S. A., and Toolin, L. J. (1990). Accelerator dating of a mixed assemblage of late Pleistocene insect fossils from the Lamb Spring site, Colorado. *Quaternary Research* 33, 122–126.

Elias, S. A., and Van Devender, T. R. (1990). Fossil insect evidence for late Quaternary climatic change in the Big Bend region, Chihuahuan Desert, Texas. *Quaternary Research* 34, 249–261.

Elliott-Fisk, D. L. (1987). Glacial geomorphology of the White Mountains, California and Nevada: Establishment of a glacial chronology. *Physical Geography* 8, 299–323.

Enzel, Y., Brown, W. J., Anderson, R. Y., and Wells, S. G. (1988). Late Pleistocene-early Holocene lake stand events recorded in cored lake deposits and in shore features, Silver Lake playa, eastern Mojave Desert, southern California. *Geological Society of America, Abstracts with Programs* 20, 158.

Fall, P. L. (1985). Holocene dynamics of the subalpine forest in central Colorado. *In* "Late Quaternary Vegetation and Climates of the American Southwest" (B. F. Jacobs, P. L. Fall, and O. K. Davis, Eds.), pp. 31–46. Contributions Series 16. American Association of Stratigraphic Palynologists Foundation.

——. (1988). "Vegetation Dynamics in the Southern Rocky Mountains: Late Pleistocene and Holocene Timberline Fluctuations." Unpublished Ph.D. thesis, University of Arizona, Tucson.

Fine Jacobs, B. (1983). "Past Vegetation and Climate of the Mogollon Rim Area, Arizona." Unpublished Ph.D. thesis, University of Arizona, Tucson.

Flynn, T., and Buchanan, P. K. (1990). "Geothermal Fluid Genesis in the Great Basin." Environmental Research Center Report 90R1. University of Nevada, Las Vegas.

Forester, R. M. (1987). Late Quaternary paleoclimate records from lacustrine ostracodes. *In* "North America and Adjacent Oceans during the Last Deglaciation" (W. F. Ruddiman and H. E. Wright, Jr., Eds.), pp. 261–276. Geology of North America, Vol. K-3. Geological Society of America, Boulder, Colo.

Friedman, I., Carrara, P., and Gleason, J. (1988). Isotopic evidence of Holocene climatic change in the San Juan Mountains, Colorado. *Quaternary Research* 30, 350–353.

Gehr, K. D. (1980). "Late Pleistocene and Recent Archaeology and Geomorphology of the South Shore of Harney Lake, Oregon." Unpublished M.A. thesis, Portland State University.

Gennett, J. A. (1977). "Palynology and Paleoecology of Sediments from Blacktail Pond, Northern Yellowstone Park, Wyoming." Unpublished M.S. thesis, University of Iowa, Iowa City.

Gennett, J. A., and Baker, R. G. (1986). A late Quaternary pollen sequence from Blacktail Pond, Yellowstone National Park, Wyoming, U.S.A. *Palynology* 10, 61–71.

Gilbert, G. K. (1890). "Lake Bonneville." Monograph 1. U.S. Geological Survey, Washington, D.C.

Graham, R. W., and Mead, J. I. (1987). Environmental fluctuations and evolution of mammalian faunas during the last deglaciation in North America. *In* "North America and Adjacent Oceans during the Last Deglaciation" (W. F. Ruddiman and H. E. Wright, Jr., Eds.), pp. 371–402. Geology of North America, Vol. K-3. Geological Society of America, Boulder, Colo.

Grayson, D. K. (1982). Toward a history of Great Basin mammals during the past 15,000 years. *In* "Man and Environment in the Great Basin" (D. B. Madsen and J. F. O'Connell, Eds.), pp. 82–101. Paper 2, Society of American Archaeology, Washington, D.C.

——. (1987). The biogeographic history of small mammals in the Great Basin: Observations on the last 20,000 years. *Journal of Mammalogy* 68, 359-375.

Hales, J. E., Jr. (1974). Southwestern United States monsoon source—Gulf of Mexico or Pacific Ocean? *Journal of Applied Meteorology* 13, 331-342.

Hansen, B. S., and Easterbrook, D. J. (1974). Stratigraphy and palynology of late Quaternary sediments in the Puget lowland, Washington. *Geological Society of America Bulletin* 85, 587-602.

Hansen, H. P. (1947). "Postglacial Forest Succession, Climate, and Chronology in the Pacific Northwest." American Philosophical Society, Philadelphia.

Harris, A. H. (1989). The New Mexican Late Wisconsin—East versus West. *National Geographic Research* 5, 205-217.

——. (1990). Fossil evidence bearing on southwestern mammalian biogeography. *Journal of Mammalogy* 71, 219-229.

Henry, C. (1984). "Holocene Paleoecology of the Western Snake River Plain, Idaho." Unpublished M.S. thesis, University of Michigan.

Heusser, C. J. (1972). Palynology and phytogeographical significance of a late-Pleistocene refugium near Kalaloch, Washington. *Quaternary Research* 2, 189-201.

——. (1973). Environmental sequence following the Fraser advance of the Juan de Fuca lobe, Washington. *Quaternary Research* 3, 284-304.

——. (1974). Quaternary vegetation, climate, and glaciation of the Hoh River Valley, Washington. *Geological Society of America Bulletin* 85, 1547-1560.

——. (1977). Quaternary paleoecology of the Pacific slope of Washington. *Quaternary Research* 8, 282-306.

——. (1978). Palynology of Quaternary deposits of the lower Bogachiel River area, Olympic Peninsula, Washington. *Canadian Journal of Earth Sciences* 15, 1568-1578.

——. (1983). Vegetational history of the northwestern United States, including Alaska. *In* "Late Quaternary Environments of the United States, Vol. 1: The Late Pleistocene" (S. C. Porter, Ed.), pp. 239-258. University of Minnesota Press, Minneapolis.

Hevly, R. H. (1985). A 50,000 year record of Quaternary environments; Walker Lake, Coconino Co., Arizona. *In* "Late Quaternary Vegetation and Climates of the American Southwest" (B. F. Jacobs, P. L. Fall, and O. K. Davis, Eds.), pp. 141-154. Contributions Series 16. American Association of Stratigraphic Palynologists Foundation.

Hibbert, D. M. (1979). "Pollen Analysis of Late-Quaternary Sediments from Two Lakes in the Southern Puget Lowland, Washington." Unpublished M.S. thesis, University of Washington, Seattle.

Hostetler, S. W., and Bartlein, P. J. (1990). Simulation of lake evaporation with application to modeling lake level variations of Harney-Malheur Lake, Oregon. *Water Resources Research* 26, 2603-2612.

Hostetler, S., and Benson, L. V. (1990). Paleoclimatic implications of highstand Lake Lahontan derived from models of evaporation and lake level. *Climate Dynamics* 4, 207-217.

Huning, J. R. (1978). "A Characterization of the Climate of the California Desert." Riverside Desert Planning Staff, California Bureau of Land Management, Riverside.

Jodry, M. A., Shafer, D. S., Stanford, D. J., and Davis, O. K. (1989). Late Quaternary environments and human adaptation in San Luis Valley, south-central Colorado. *In* "Water in the Valley, a 1989 Perspective on Water Supplies, Issues, and Solutions in the San Luis Valley, Colorado," pp. 189-200.

Eighth Annual Field Trip, Colorado Ground-Water Association.

King, T. J. (1976). Late Pleistocene-early Holocene history of coniferous woodlands in the Lucerne Valley region, Mojave Desert, California. *Great Basin Naturalist* 36, 227-238.

Kutzbach, J. E. (1987). Model simulations of the climatic patterns during the deglaciation of North America. *In* "North America and Adjacent Oceans during the Last Deglaciation" (W. F. Ruddiman and H. E. Wright, Jr., Eds.), pp. 425-446. Geology of North America, Vol. K-3. Geological Society of America, Boulder, Colo.

Kutzbach, J. E., and Guetter, P. J. (1984). The sensitivity of monsoon climates to orbital parameter changes for 9,000 years B.P.: Experiments with the NCAR general circulation model. *In* "Milankovitch and Climate," Part 2 (A. L. Berger, J. Imbrie, J. Hays, G. Kukla, and B. Salzman, Eds.), pp. 801-820. D. Reidel, Dordrecht, The Netherlands.

——. (1986). The influence of changing orbital parameters and surface boundary conditions on climate simulations for the past 18,000 years. *Journal of the Atmospheric Sciences* 43, 1726-1759.

Kutzbach, J. E., and Wright, H. E., Jr. (1986). Simulation of the climate of 18,000 yr B.P.: Results for the North American/North Atlantic/European sector and comparison with the geologic record of North America. *Quaternary Science Reviews* 4, 147-187.

Lajoie, K. P. (1968). "Late Quaternary Stratigraphy and Geologic History of Mono Basin, Eastern California." Unpublished Ph.D. thesis, University of California, Berkeley.

LaMarche, V. C., Jr. (1973). Holocene climatic variations inferred from treeline fluctuations in the White Mountains, California. *Quaternary Research* 3, 632-660.

LaMarche, V. C., Jr., and Fritts, H. C. (1971). Anomaly patterns of climate over the western United States, 1700-1930, derived from principal component analysis of tree-ring data. *Monthly Weather Review* 99, 138-142.

LaMarche, V. C., Jr., and Mooney, H. A. (1972). Recent climatic change and development of the bristlecone pine (*P. longaeva* Bailey) krummholz zone, Mt. Washington, Nevada. *Arctic and Alpine Research* 4, 61-72.

Lanner, R. M., and Van Devender, T. R. (1981). Late Pleistocene piñon pines in the Chihuahuan Desert. *Quaternary Research* 15, 278-290.

Legg, T. E., and Baker, R. G. (1980). Palynology of Pinedale sediments, Devlins Park, Boulder County, Colorado. *Arctic and Alpine Research* 12, 319-333.

Leopold, E. B., Nickmann, R. J., Hedges, J. I., and Ertel, J. R. (1982). Pollen and lignin records of late Quaternary vegetation, Lake Washington. *Science* 218, 1305-1307.

Leskinen, P. H. (1975). Occurrence of oaks in late Pleistocene vegetation in the Mojave Desert of Nevada. *Madroño* 23, 234-235.

Lipsitz, B. B. (1988). "Climatic Estimates for Locations between Weather Stations in the Pacific Northwest: Comparison and Application of Two Linear Regression Analysis Methods." Unpublished M.S. thesis, University of Oregon.

Long, A., and Rippeteau, B. (1974). Testing contemporaneity and averaging radiocarbon dates. *American Antiquity* 39, 205-215.

Long, A., Warneke, L. A., Betancourt, J. L., and Thompson, R. S. (1990). Deuterium variations in plant cellulose from fossil packrat middens. *In* "Packrat Middens—The Last 40,000 Years of Biotic Change" (J. L. Betancourt, T. R. Van Devender, and P. S. Martin, Eds.), pp. 380-396. University of Arizona Press, Tucson.

Mack, R. N., Bryant, V. M., Jr., and Fryxell, R. (1976). Pollen sequence from the Columbia Basin, Washington: Reappraisal of postglacial vegetation. *American Midland Naturalist* 95, 390-397.

Mack, R. N., Rutter, N. W., Bryant, V. M., Jr., and Valastro, S. (1978a). Late Quaternary pollen record from Big Meadow, Pend Oreille County, Washington. *Ecology* 59, 956-966.

——. (1978b). Reexamination of postglacial vegetation history in northern Idaho: Hager Pond, Bonner County. *Quaternary Research* 10, 244-255.

Mack, R. N., Rutter, N. W., and Valastro, S. (1978c). Late Quaternary pollen record from the Sanpoil River Valley, Washington. *Canadian Journal of Botany* 56, 1642-1650.

Mack, R. N., Rutter, N. W., Valastro, S., and Bryant, V. M., Jr. (1978d). Late Quaternary vegetation history at Waits Lake, Colville River Valley, Washington. *Botanical Gazette* 139, 499-506.

Mack, R. N., Rutter, N. W., and Valastro, S. (1979). Holocene vegetation history of the Okanogan Valley, Washington. *Quaternary Research* 12, 212-225.

——. (1983). Holocene vegetational history of the Kootenai River Valley, Montana. *Quaternary Research* 20, 177-193.

Madsen, D. B. (1973). "Late Quaternary Paleoecology in the Southeastern Great Basin." Unpublished Ph.D. thesis, University of Missouri.

Madsen, D. B., and Currey, D. R. (1979). Late Quaternary glacial and vegetation changes, Little Cottonwood Canyon area, Wasatch Mountains, Utah. *Quaternary Research* 12, 254-270.

Madsen, D. B., and Rhode, D. (1990). Early Holocene pinyon (*Pinus monophylla*) in the northeastern Great Basin. *Quaternary Research* 33, 94-101.

Maher, L. J., Jr. (1961). "Pollen Analysis and Postglacial Vegetation History in the Animas Valley Region, Southern San Juan Mountains, Colorado." Unpublished Ph.D. thesis, University of Minnesota.

——. (1972). Absolute pollen diagram of Redrock Lake, Boulder County, Colorado. *Quaternary Research* 2, 531-553.

Markgraf, V., and Lennon, T. (1986). Paleoenvironmental history of the last 13,000 years of the eastern Powder River Basin, Wyoming, and its implication for prehistoric cultural patterns. *Plains Anthropologist* 31, 1-12.

Markgraf, V., and Scott, L. (1981). Lower timberline in central Colorado during the past 15,000 yr. *Geology* 9, 231-234.

Markgraf, V., Bradbury, J. P., Forester, R. M., McCoy, W., Singh, G., and Sternberg, R. S. (1983). Paleoenvironmental reassessment of the 1.6 million year old record from San Agustin Basin, New Mexico. *In* "New Mexico Geological Society Guidebook, 34th Field Conference, Socorro Region II," pp. 291-297.

Markgraf, V., Bradbury, J. P., Forester, R. M., Singh, G., and Sternberg, R. S. (1984). San Agustin Plains, New Mexico: Age and paleoenvironmental potential reassessed. *Quaternary Research* 22, 336-343.

Martin, J. E., Barnosky, A. D., and Barnosky, C. W. (1982). Fauna and flora associated with the West Richland Mammoth from the Pleistocene Touchet Beds in South-Central Washington. *Research Report of the Thomas Burke Memorial Washington State Museum.*

Mathewes, R. W. (1973). A palynological study of postglacial vegetation changes in the University Research Forest, southwestern British Columbia. *Canadian Journal of Botany* 51, 2085-2103.

Mathewes, R. W., and Rouse, G. E. (1975). Palynology and paleoecology of postglacial sediments from the lower Fraser River canyon of British Columbia. *Canadian Journal of Earth Sciences* 12, 745-756.

Mead, E. M., and Mead, J. I. (1989). Quaternary zoogeography of the Nearctic *Dicrostonyx* lemmings. *Boreas* 18, 323-332.

Mead, J. I. (1987). Quaternary records of pika, *Ochotona*, in North America. *Boreas* 16, 165-171.

Mead, J. I., and Phillips, A. M. III (1981). The late Pleistocene and Holocene fauna and flora of Vulture Cave, Grand Canyon, Arizona. *Southwestern Naturalist* 26, 257-288.

Mead, J. I., Thompson, R. S., and Long, A. (1978). Arizona radiocarbon dates IX: Carbon isotope dating of packrat middens. *Radiocarbon* 20, 171-191.

Mead, J. I., Agenbroad, L. D., Phillips, A. M. III, and Middleton, L. T. (1987). Extinct mountain goat (*Oreamnos harringtoni*) in southeastern Utah. *Quaternary Research* 27, 323-331.

Mears, B., Jr. (1981). Periglacial wedges and the late Pleistocene environment of Wyoming's intermontane basins. *Quaternary Research* 15, 171-198.

Meek, N. (1989). Geomorphic and hydrologic implications of the rapid incision of Afton Canyon, Mojave Desert, California. *Geology* 17, 7-10.

Mehringer, P. J., Jr. (1967). Pollen analysis of the Tule Springs area, Nevada. *Nevada State Museum Anthropological Papers* 13, 129-200.

——. (1977). Great Basin late Quaternary environments and chronology. *Desert Research Institute Publications in the Social Sciences* 12, 113-167.

——. (1985). Late-Quaternary pollen records from the interior Pacific Northwest and northern Great Basin of the United States. *In* "Pollen Records of Late-Quaternary North American Sediments" (V. M. Bryant, Jr., and R. G. Holloway, Eds.), pp. 167-189. American Association of Stratigraphic Palynologists Foundation, Austin, Tex.

Mehringer, P. J., Jr., and Ferguson, C. W. (1969). Pluvial occurrence of bristlecone pine, *Pinus aristata*, in a Mohave Desert mountain range. *Journal of the Arizona Academy of Science* 5, 284-292.

Mehringer, P. J., Jr., and Wigand, P. E. (1986). Holocene history of Skull Creek dunes, Catlow Valley, Oregon, U.S.A. *Journal of Arid Environments* 11, 117-138.

——. (1987). Western juniper in the Holocene. *In* "Proceedings—Pinyon-Juniper Conference" (R. L. Everett, Compiler), pp. 109-119. General Technical Report INT-215. Intermountain Research Station, Ogden, Utah.

Mehringer, P. J., Jr., Arno, S. F., and Peterson, K. L. (1977). Postglacial history of Lost Trail Pass bog, Bitterroot Mountains, Montana. *Arctic and Alpine Research* 9, 345-368.

Mifflin, M. D., and Wheat, M. M. (1979). "Pluvial Lakes and Estimated Pluvial Climates of Nevada." Bulletin 94, Nevada Bureau of Mines and Geology, Reno.

Mitchell, J. F. B., Grahame, N. S., and Needham, K. J. (1988). Climate simulations for 9000 years before present: Seasonal variations and effect of the Laurentide ice sheet. *Journal of Geophysical Research* 93, 8283-8303.

Moodie, K. B., and Van Devender, T. R. (1979). Extinction and extirpation in the herpetofauna of the southern High Plains with emphasis on *Geochelone wilsoni* (Testudinidae). *Herpetologia* 35, 198-206.

Morrison, R. B. (1965). Quaternary geology of the Great Basin. *In* "The Quaternary of the United States" (H. E. Wright, Jr., and D. G. Frey, Eds.), pp. 265-286. Princeton University Press, Princeton, N.J.

Neilson, R. P. (1986). High-resolution climatic analysis and Southwest biogeography. *Science* 232, 27-34.

Nickmann, R. J. (1979). "The Palynology of Williams Lake Fen, Spokane County, Washington." Unpublished M.S. thesis, Eastern Washington State University, Cheney.

Nickmann, R. J., and Leopold, E. B. (1984). A postglacial pollen record from Goose Lake, Okanogan County, Washington; evidence for an early Holocene cooling. In "Chief Joseph Summary Report," pp. 131-148. Office of Public Archaeology, University of Washington, Seattle.

Oviatt, C. G. (1988). Late Pleistocene and Holocene lake fluctuations in the Sevier Lake basin, Utah, USA. Journal of Paleolimnology 1, 9-21.

Petersen, K. L. (1988). "Climate and the Dolores River Anasazi." Anthropological Papers 113. University of Utah, Salt Lake City.

Petersen, K. L., and Mehringer, P. J., Jr. (1976). Postglacial timberline fluctuations, La Plata Mountains, southwestern Colorado. Arctic and Alpine Research 8, 275-288.

Petersen, K. L., Mehringer, P. J., Jr., and Gustafson, C. E. (1983). Late-glacial vegetation and climate at the Manis Mastodon site, Olympic Peninsula, Washington. Quaternary Research 20, 215-231.

Péwé, T. L. (1983). The periglacial environment in North America during Wisconsin time. In "Late Quaternary Environments of the United States, Vol. 1: The Late Pleistocene" (S. C. Porter, Ed.), pp. 157-189. University of Minnesota Press, Minneapolis.

Phillips, A. M. (1977). "Packrats, Plants, and the Pleistocene of the Lower Grand Canyon, Arizona." Unpublished Ph.D. thesis, University of Arizona, Tucson.

Phillips, F. M., Peeters, L. A., and Tansey, M. K. (1986). Paleoclimatic inferences from an isotopic investigation of groundwater in the central San Juan basin, New Mexico. Quaternary Research 26, 179-193.

Porter, S. C., and Denton, G. H. (1967). Chronology of neoglaciation in the North America Cordillera. American Journal of Science 265, 177-210.

Porter, S. C., Pierce, K. L., and Hamilton, T. D. (1983). Late Wisconsin mountain glaciation in the western United States. In "Late Quaternary Environments of the United States, Vol. 1: The Late Pleistocene" (S. C. Porter, Ed.), pp. 71-114. University of Minnesota Press, Minneapolis.

Putnam, W. C. (1950). Moraine and shoreline relationships at Mono Lake, California. Geological Society of America Bulletin 61, 115-122.

Quade, J. (1986). Late Quaternary environmental changes in the upper Las Vegas Valley, Nevada. Quaternary Research 26, 340-357.

Rypins, S., Reneau, S. L., Bryne, R., and Montgomery, D. R. (1989). Palynologic and geomorphic evidence for environmental change during the Pleistocene-Holocene transition at Point Reyes Peninsula, central coastal California. Quaternary Research 32, 72-87.

Scott, W. E., McCoy, W. D., Shroba, R. R., and Rubin, M. (1983). Reinterpretation of the exposed record of the last two cycles of Lake Bonneville, western United States. Quaternary Research 20, 261-285.

Sellers, W. D. (1968). Climatology of monthly precipitation patterns in the western United States, 1931-1966. Monthly Weather Review 96, 585-595.

Siegal, R. D. (1983). "Paleoclimatic Significance of D/H and ¹³C/¹²C Ratios in Pleistocene and Holocene Wood." Unpublished M.S. thesis, University of Arizona, Tucson.

Smith, G. I., and Street-Perrott, F. A. (1983). Pluvial lakes of the western United States. In "Late Quaternary Environments

of the United States, Vol. 1: The Late Pleistocene" (S. C. Porter, Ed.), pp. 190-211. University of Minnesota Press, Minneapolis.

Snyder, C. T., Hardman, G., and Zdenek, F. F. (1964). "Pleistocene Lakes in the Great Basin." Miscellaneous Geologic Investigations Map I-416. U.S. Geological Survey.

Spaulding, W. G. (1980). "The Presettlement Vegetation of the California Desert." Riverside Desert Planning Staff, California Bureau of Land Management, Riverside.

——. (1981). "The Late Quaternary Vegetation of a Southern Nevada Mountain Range." Unpublished Ph.D. thesis, University of Arizona, Tucson.

——. (1983). Late Wisconsin macrofossil records of desert vegetation in the American Southwest. Quaternary Research 19, 256-264.

——. (1985). "Vegetation and Climates of the Last 45,000 Years in the Vicinity of the Nevada Test Site, South-Central Nevada." Professional Paper 1329, U.S. Geological Survey. U.S. Government Printing Office, Washington, D.C.

——. (1990a). Vegetational and climatic development of the Mojave Desert: The last glacial maximum to the present. In "Packrat Middens—The Last 40,000 Years of Biotic Change" (J. L. Betancourt, T. R. Van Devender, and P. S. Martin, Eds.), pp. 166-199. University of Arizona Press, Tucson.

——. (1990b). Vegetation dynamics during the last deglaciation, southeastern Great Basin, U.S.A. Quaternary Research 33, 188-203.

——. (1991). A middle Holocene vegetation record from the Mojave Desert of North America and its paleoclimatic significance. Quaternary Research 35, 427-437.

Spaulding, W. G., and Graumlich, L. J. (1986). The last pluvial climatic episodes in the deserts of southwestern North America. Nature 320, 441-444.

Spaulding, W. G., and Van Devender, T. R. (1977). Late Pleistocene montane conifers in southeastern Utah. Southwestern Naturalist 22, 269-286.

Spaulding, W. G., Leopold, E. B., and Van Devender, T. R. (1983). Late Wisconsin paleoecology of the American Southwest. In "Late Quaternary Environments of the United States, Vol. 1: The Late Pleistocene" (S. C. Porter, Ed.), pp. 259-293. University of Minnesota Press, Minneapolis.

Spaulding, W. G., Betancourt, J. B., Croft, L. K., and Cole, K. L. (1990). Packrat middens: Their composition and methods of analysis. In "Packrat Middens—The Last 40,000 Years of Biotic Change" (J. L. Betancourt, T. R. Van Devender, and P. S. Martin, Eds.), pp. 59-84. University of Arizona Press, Tucson.

Stine, S. (1990). Late Holocene fluctuations of Mono Lake, eastern California. Palaeogeography, Palaeoclimatology, Palaeoecology 78, 333-381.

Street-Perrott, F. A., and Harrison, S. P. (1985). Lake levels and climate reconstruction. In "Paleoclimate Analysis and Modeling" (A. D. Hecht, Ed.), pp. 291-340. John Wiley & Sons, New York.

Tang, M., and Reiter, E. R. (1984). Plateau monsoons of the Northern Hemisphere; a comparison between North America and Tibet. Monthly Weather Review 112, 617-637.

Thompson, R. S. (1984). "Late Pleistocene and Holocene Environments in the Great Basin." Unpublished Ph.D. thesis, University of Arizona, Tucson.

——. (1985). Palynology and Neotoma middens. In "Late Quaternary Vegetation and Climates of the American Southwest" (B. F. Jacobs, P. L. Fall, and O. K. Davis, Eds.), pp. 89-112. Contributions Series 16. American Association of Stratigraphic Palynologists Foundation.

——. (1988). Western North America—Vegetation dynamics in the western United States: Modes of response to climatic fluctuations. In "Vegetation History" (B. Huntley and T. Webb III, Eds.), pp. 415-458. Handbook of Vegetation Science, Vol. 7. Klüwer Academic Publishers, Dordrecht, The Netherlands.

——. (1990). Late Quaternary vegetation and climate in the Great Basin. In "Packrat Middens—The Last 40,000 Years of Biotic Change" (J. L. Betancourt, T. R. Van Devender, and P. S. Martin, Eds.), pp. 200-239. University of Arizona Press, Tucson.

——. (1992). Late Quaternary environments in Ruby Valley, Nevada. Quaternary Research 37, 1-15.

Thompson, R. S., and Hattori, E. M. (1983). Packrat (Neotoma) middens from Gatecliff Shelter and Holocene migrations of woodland plants. Anthropological Papers of the American Museum of Natural History 59, 157-167.

Thompson, R. S., and Kautz, R. R. (1983). Pollen analysis. Anthropological Papers of the American Museum of Natural History 59, 136-151.

Thompson, R. S., and Mead, J. I. (1982). Late Quaternary environments and biogeography in the Great Basin. Quaternary Research 17, 39-55.

Thompson, R. S., Van Devender, T. R., Martin, P. S., Foppe, T., and Long, A. (1980). Shasta ground sloth (Nothrotheriops shastense Hoffstetter) at Shelter Cave, New Mexico: Environment, diet, and extinction. Quaternary Research 14, 360-376.

Thompson, R. S., Benson, L., and Hattori, E. M. (1986). A revised chronology for the last Pleistocene lake cycle in the central Lahontan Basin. Quaternary Research 25, 1-9.

Thompson, R. S., Toolin, L. J., Forester, R. M., and Spencer, R. J. (1990). Accelerator-mass spectrometer (AMS) radiocarbon dating of Pleistocene lake sediments in the Great Basin. Palaeogeography, Palaeoclimatology, Palaeoecology 78, 301-313.

Tsukada, M., Sugita, S., and Hibbert, D. M. (1981). Paleoecology of the Pacific Northwest: I. Late Quaternary vegetation and climate. Verhandlungen der Internationalen Vereinung für theoretische und angewandte Limnologie 21, 730-737.

Van Devender, T. R. (1973). "Late Pleistocene Plants and Animals of the Sonoran Desert; a Survey of Ancient Packrat Middens in Southwestern Arizona." Unpublished Ph.D. thesis, University of Arizona, Tucson.

——. (1977). Holocene woodlands in the southwestern deserts. Science 198, 189-192.

——. (1987). Holocene vegetation and climate in the Puerto Blanco Mountains, southwestern Arizona. Quaternary Research 27, 51-72.

——. (1990a). Late Quaternary vegetation and climate of the Chihuahuan Desert, United States and Mexico. In "Packrat Middens—The Last 40,000 Years of Biotic Change" (J. L. Betancourt, T. R. Van Devender, and P. S. Martin, Eds.), pp. 104-133. University of Arizona Press, Tucson.

——. (1990b). Late Quaternary vegetation and climate of the Sonoran Desert, United States and Mexico. In "Packrat Middens—The Last 40,000 Years of Biotic Change" (J. L. Betancourt, T. R. Van Devender, and P. S. Martin, Eds.), pp. 134-165. University of Arizona Press, Tucson.

Van Devender, T. R., and Burgess, T. L. (1985). Late Pleistocene woodlands in the Bolson de Mapimi; a refugium for the Chihuahuan Desert biota? Quaternary Research 24, 346-353.

Van Devender, T. R., and Riskind, D. H. (1979). Late Pleistocene and early Holocene plant remains from Hueco Tanks State Historical Park: The development of a refugium. Southwestern Naturalist 24, 127-140.

Van Devender, T. R., and Spaulding, W. G. (1979). Development of vegetation and climate in the southwestern United States. Science 204, 701-710.

Van Devender, T. R., and Toolin, L. J. (1983). Late Quaternary vegetation of the San Andres Mountains, Sierra County, New Mexico. In "The Prehistory of Rhodes Canyon—Survey and Mitigation" (P. L. Eidenbach, Ed.), pp. 33-54. Human Systems Research, Inc., Tularosa, N.M.

Van Devender, T. R., and Wiseman, F. M. (1977). A preliminary chronology of bioenvironmental changes during the Paleoindian period in the monsoonal Southwest. In "Paleoindian Lifeways" (E. Johnson, Ed.), pp. 13-27. Museum Journal XVII. West Texas Museum Association, Lubbock.

Van Devender, T. R., Moodie, K. B., and Harris, A. H. (1976). The desert tortoise (Gopherus agassizi) in the Pleistocene of the northern Chihuahuan Desert. Herpetologia 32, 298-304.

Van Devender, T. R., Betancourt, J. B., and Wimberly, M. (1984). Biogeographic implications of a packrat midden sequence from the Sacramento Mountains, south-central New Mexico. Quaternary Research 22, 344-360.

Van Devender, T. R., Martin, P. S., Thompson, R. S., Cole, K. L., Jull, A. J. T., Long, A., Toolin, L. J., and Donahue, D. J. (1985). Fossil packrat middens and the tandem accelerator mass spectrometer. Nature 317, 610-613.

Van Devender, T. R., Thompson, R. S., and Betancourt, J. B. (1987). Vegetation history of the deserts of southwestern North America; the nature and timing of the late Wisconsin-Holocene transition. In "North America and Adjacent Oceans during the Last Deglaciation" (W. F. Ruddiman and H. E. Wright, Jr., Eds.), pp. 323-352. Geology of North America, Vol. K-3. Geological Society of America, Boulder, Colo.

Van Devender, T. R., Burgess, T. L., Felger, R. S., and Turner, R. M. (1990). Holocene vegetation of the Hornaday Mountains of northwestern Sonora, Mexico. Proceedings of the San Diego Society of Natural History 2, 1-19.

Waddington, J. C. B., and Wright, H. E., Jr. (1974). Late Quaternary vegetational changes on the east side of Yellowstone National Park. Quaternary Research 4, 175-184.

Waitt, R. B., Jr., and Thorson, R. M. (1983). The Cordilleran ice sheet in Washington, Idaho, and Montana. In "Late Quaternary Environments of the United States, Vol. 1: The Late Pleistocene" (S. C. Porter, Ed.), pp. 53-70. University of Minnesota Press, Minneapolis.

Waters, M. R. (1989). Late Quaternary lacustrine history and paleoclimatic significance of pluvial Lake Cochise, southeastern Arizona. Quaternary Research 32, 1-11.

Wayne, W. J. (1984). Glacial chronology of the Ruby Mountains-East Humboldt Range, Nevada. Quaternary Research 21, 286-303.

Webb, T. III, Bartlein, P. J., and Kutzbach, J. E. (1987). Climatic change in eastern North America during the past 18,000 years; comparisons of pollen data with model results. In "North America and Adjacent Oceans during the Last Deglaciation" (W. F. Ruddiman and H. E. Wright, Jr., Eds.), pp. 447-462. Geology of North America, Vol. K-3. Geological Society of America, Boulder, Colo.

Wells, P. V. (1966). Late Pleistocene vegetation and degree of pluvial climatic change in the Chihuahuan Desert. *Science* 153, 970-975.

——. (1983). Paleogeography of montane islands in the Great Basin since the last glaciopluvial. *Ecological Monographs* 53, 341-382.

Wells, P. V., and Berger, R. (1967). Late Pleistocene history of coniferous woodlands in the Mohave Desert. *Science* 155, 1640-1647.

Wells, P. V., and Hunziker, J. H. (1976). Origin of the creosote bush (*Larrea*) in the deserts of southwestern North America. *Annals of the Missouri Botanical Garden* 63, 843-861.

Wells, P. V., and Jorgensen, C. D. (1964). Pleistocene wood rat middens and climatic change in Mohave Desert: A record of juniper woodlands. *Science* 143, 1171-1174.

Wells, P. V., and Woodcock, D. (1985). Full-glacial vegetation of Death Valley, California: Juniper woodland opening to *Yucca* semidesert. *Madroño* 32, 11-23.

Wells, S. G., McFadden, L. D., and Dohrenwend, J. C. (1987). Influence of the Quaternary climatic changes on geomorphic and pedogenic processes on a desert piedmont, eastern Mojave Desert, California. *Quaternary Research* 27, 130-146.

Whitlock, C., Thompson, R. S., and Bartlein, P. J. (1990). Climatic assessment of the last deglaciation in the Pacific Northwest as inferred from paleobotanical data. *Geological Society of America, Abstracts with Programs* 1990, A354.

Wigand, P. E. (1987). Diamond Pond, Harney County, Oregon: Vegetation history and water table in the eastern Oregon desert. *Great Basin Naturalist* 47, 427-458.

Wigand, P. E., and Mehringer, P. J., Jr. (1985). Pollen and seed analyses. *In* "The Archaeology of Hidden Cave, Nevada" (D. H. Thomas, Ed.), pp. 108-124. Anthropological Papers 61, part 1. American Museum of Natural History, New York.

Woodcock, D. (1986). The late Pleistocene of Death Valley: A climatic reconstruction based on macrofossil data. *Palaeogeography, Palaeoclimatology, Palaeoecology* 57, 273-283.

Wright, H. E., Jr., Bent, A. M., Hansen, B. S., and Maher, L. J., Jr. (1973). Present and past vegetation of the Chuska Mountains, northwestern New Mexico. *Geological Society of America Bulletin* 84, 1155-1180.

CHAPTER 19

Climatic Changes during the Past 18,000 Years: Regional Syntheses, Mechanisms, and Causes

Thompson Webb III, William F. Ruddiman,
F. Alayne Street-Perrott, Vera Markgraf,
John E. Kutzbach, Patrick J. Bartlein,
H. E. Wright, Jr., and Warren L. Prell

Few periods of the earth's history have recorded climatic changes as large as those of the past 18,000 yr. At high to middle latitudes, large temperature changes dominated and were accompanied by major changes in atmospheric circulation and moisture balance. In the tropics, moisture variations dominated, resulting from variations in the monsoonal circulations. The data show that the timing, character, and patterns of climate changes have varied regionally. Understanding the relation of this regional variation to the changing global controls poses a fascinating problem in climatology, which COHMAP research (described in the preceding chapters) has been designed to help solve.

In our view, as the ice sheets melted and as seasonal insolation varied, nature performed a series of global-scale "experiments" on the climate system. We therefore have used climate models and a general understanding of the global controls (e.g., variations in insolation, ice volume, and related glacial-age boundary conditions) in an attempt to repeat nature's experiments and to illustrate the linkages between the varying regional climate patterns and the global controls. We also assembled large sets of data to describe the natural experiments and to evaluate the model results.

The large size of the climate changes during the late Quaternary presents a clear signal that is evident in many diverse sets of paleoclimatic data. The regional chapters describe the available data and the inferred climate patterns for each region. They also describe the model results and compare the data with the model results. In this chapter we combine these regional results into broad areal summaries, describe

how the data compare with the model results, and synthesize the results in terms of a general understanding of how the global controls have affected climate patterns during the past 18,000 yr. For these purposes we examined the data and model results for three sectors of the globe: (1) North America, the northern North Atlantic, and western Eurasia; (2) the monsoon sector of Africa, southern Asia, and the Indian Ocean; and (3) the midlatitudes of the Southern Hemisphere. To accommodate discussion of additional modeling results from mixed-layer ocean models, we include a section on oceans and describe the comparison of these model results with sea-surface temperature (SST) estimates.

For each sector, including the oceans, we describe the major similarities and differences between the model results and the climatic inferences derived from the data. These comparisons show how our experimental design can provide a fairly detailed check of the model simulations. We then summarize our understanding of the controls and large-scale circulation changes that led to the regional patterns of climate change. Similarities between data and models increase confidence in the models and encourage use of the model results in explaining the patterns in the data. Discrepancies indicate where improvements are needed in the boundary conditions, models, and data.

A key result of our study is the demonstration of how data and models can interact (Fig. 19.1) for the improvement of both. With continued improvements in the data and models (see Kutzbach, Bartlein, *et al.,* this vol.), we anticipate that certain of our current explanations for the observed changes will be modified.

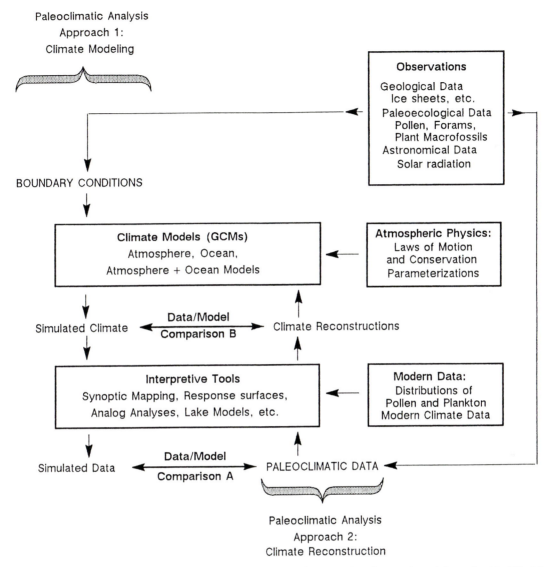

Fig. 19.1. Interactive system for designing modeling experiments and comparing data and model results. Modified from Webb and Bartlein (1988).

North American, Northern North Atlantic, and Western Eurasia Sector

PRESENT-DAY CONDITIONS

A key feature of northern midlatitude climates is the westerly jet stream aloft, which marks the boundary between warm and cold airmasses and governs the location and path of traveling storm systems (Fig. 19.2). In winter it crosses the west coast of North America at about 50°N. Inland east of the Rocky Mountains, it turns southeast and crosses the east coast at about 40°N. At the surface in summer, the Atlantic and Pacific subtropical high-pressure cells dominate the circulation, and in winter the Aleutian and Icelandic lows prevail over the northern oceans.

The Community Climate Model (CCM) of the National Center for Atmospheric Research (NCAR) simulates fairly accurately the modern location of these circulation features but is less accurate in representing their intensity. The simulation of the modern fields of temperature and precipitation also is far from perfect (Webb *et al.,* 1987; Kutzbach, Guetter, *et al.,* this vol.). Furthermore, the representation of topography and geography in the model is quite coarse. We therefore anticipate inaccuracies in the simulations of paleoclimates and accordingly focus on large-scale features.

The conditions at the earth's surface reflect the broad-scale climatic patterns. Permanent land ice is restricted mainly to Greenland. Sea ice covers much of the Arctic Ocean in summer as well as winter (Fig. 19.2). Spruce-dominated forest defines the boreal zone from Alaska to Labrador and across northern Eurasia

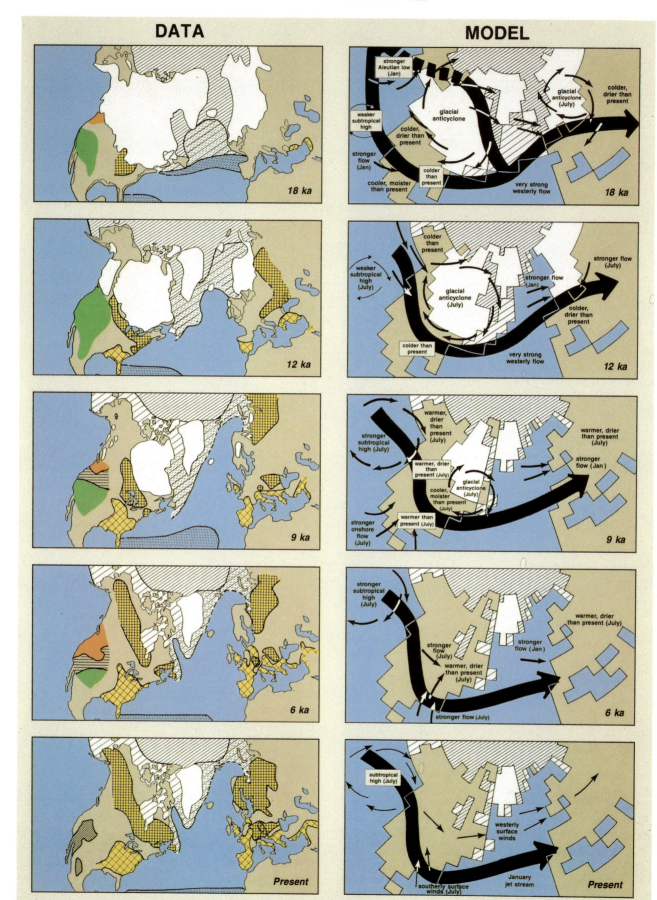

Fig. 19.2. Changes in the atmosphere, geosphere, and biosphere that accompanied the transition from glacial to interglacial conditions during the past 18,000 yr, as illustrated by selected geologic and paleoecologic evidence (left panel), and highlights of the paleoclimatic model simulations (right panel). (See also the key to Fig. 19.3.) The maps in the left panel show the extent of ice sheets and year-round and winter-only sea ice and the broadened land area from 18 to 9 ka that resulted from the lowered sea level. The distributions of oak and spruce populations as inferred from pollen data are shown for eastern North America and Europe. Moisture status relative to the present is shown for western North America from 18 to 6 ka. The present region where annual precipitation is less than 300 mm is shown in the southwestern United States. The maps in the right panel show the distributions of ice sheets and sea ice that served as boundary conditions for the simulations performed with the Community Climate Model (CCM) of the National Center for Atmospheric Research (NCAR). Surface winds and the position of the jet stream as simulated by the model are shown schematically by arrows (broken arrows if the jet is weak). Comparisons noted on the maps of model results (e.g., colder, wetter) are with reference to present conditions. From COHMAP Members (1988); reprinted by permission from *Science* vol. 241, pp. 1043–1052, copyright 1988 by the AAAS.

and covers high-elevation areas farther south on both continents. Temperate regions in eastern North America and western and southern Europe have broad-leaved deciduous forests, typically containing oak. (For illustrative purposes, only the distributions of oak and spruce are shown in Fig. 19.2, and the text therefore focuses on them.) The American Southwest is dry, with few lakes in the intermontane basins. Polar conditions occur in the northwestern North Atlantic, as shown by the abundance of the foraminifer *Neogloboquadrina pachyderma* (left-coiling) and by clastic detritus dropped from icebergs and sea ice. Subtropical conditions, indicated by *Globigerinoides ruber,* extend from Florida to northern Africa.

DYNAMICS OF THE MODEL SIMULATIONS

At 18 ka the boundary conditions included ice sheets at their maxima, with winter sea ice in the North Atlantic extending south to the coast of France (Fig. 19.2). South of the sea-ice border in both the North Atlantic and the North Pacific, prescribed SSTs were as much as 10°C lower than those at present (CLIMAP Project Members, 1981). Simulated July temperatures were generally 2–6°C lower than at present over the unglaciated land but 20–40°C lower than at present over the elevated and highly reflective ice sheets (Kutzbach, Guetter, *et al.,* this vol.). In winter the biggest differences in simulated temperatures were over the sea ice in the North Atlantic and just southeast of the Scandinavian ice sheet (Kutzbach, Guetter, *et al.,* this vol.). Strong anticyclonic circulation around the Laurentide ice sheet brought cold conditions to the North Atlantic and stronger-than-present easterly winds in regions near the southern flank of the ice sheet. The sharp temperature gradient at the southern edge of the continental ice sheets and the extensive sea-ice field were associated with a strengthened jet stream aloft, which extended across North America and east to Eurasia. The large Laurentide ice sheet used in the model (Kutzbach and Ruddiman, this vol.) was also responsible for splitting the flow of the jet stream in winter across North America, with one branch located over southern North America and the other along the northern edge of the ice sheet (Fig. 19.2).

By 12 ka the general warming of the climate associated with increased summertime solar radiation had begun in the model simulations (Fig. 3.1 in Kutzbach and Ruddiman, this vol.). The Laurentide ice sheet had decreased in size and thickness sufficiently so that the winter jet was no longer split, although it was still stronger than it is today. The simulated glacial anticyclone weakened in the northwestern United States, where westerlies replaced easterlies, and a high-pressure cell developed over the northeastern North Pacific. The glacial anticyclone over western Europe was also weaker than during the glacial maximum (Fig. 19.2), and the combination of increased summer insolation and still-large ice sheets created seasonally anomalous climates on both continents.

At 9 ka the increased insolation in summer played a much larger role, but the shrinking Laurentide ice sheet continued to influence the simulated climate. Only a small glacial anticyclone remained in eastern North America (Fig. 19.2). By then the Pacific subtropical high had strengthened off the west coast of North America, and northwesterly winds replaced westerly winds along the coast in the Northwest. The increased insolation produced simulated temperatures for summer higher than at present (by 2–4°C), especially in central Eurasia, Alaska, and west-central North America, but it was still colder than at present just south of the ice sheet in eastern North America (Kutzbach, Guetter, *et al.,* this vol.). A southerly summer monsoonal flow was evident along the Gulf Coast.

By 6 ka simulated summer temperatures were 2–4°C higher than at present throughout the continental interiors of North America and Eurasia, and the southerly flow into the eastern United States had strengthened at the surface. Strong westerlies were simulated near the surface in the Midwest and in Europe; they have weakened since 6 ka, and summer temperatures have declined with the reduction in summertime insolation.

PALEOCLIMATIC DATA

At 18 ka, with the ice sheets and North Atlantic sea ice at their maxima (Fig. 19.2), *Neogloboquadrina pachyderma* (left-coiling) was abundant far to the south of its current range, but the range of *Globigeri-*

noides ruber was largely unchanged in the subtropics. This pattern implies a steepened temperature gradient over the North Atlantic and eastern North America. Spruce and oak forests were absent from Europe because of the cold, dry conditions, and permafrost was widespread. The Mediterranean lowlands were treeless, although lake levels in Turkey and the Levant were high (Fig. 19.2) (COHMAP Members, 1988; Roberts and Wright, this vol.). In North America spruce trees grew relatively abundantly in the Midwest, and northern pines grew in the South except for Florida (Webb, Bartlein, *et al.*, this vol.). Subalpine parkland and periglacial steppe dominated at low elevations in the Pacific Northwest, while tundra prevailed in Alaska (Anderson and Brubaker, this vol.). In the Southwest conditions were moister than today, for lake levels were high and woodlands expanded (Thompson *et al.*, this vol.).

By 12 ka Northern Hemisphere insolation had increased in summer and decreased in winter, and the melting ice sheets had retreated somewhat and were lower (Fig. 19.2). The distribution of *Neogloboquadrina pachyderma* had contracted significantly in the northeastern North Atlantic as sea ice retreated north and SSTs rose. Air temperatures over land increased as oak forest spread across southeastern North America, and spruce trees grew along the southern edge of the retreating Laurentide ice sheet. Spruce forests also developed in eastern Europe, and oak woodlands expanded in the Mediterranean region, where conditions became wetter about 11 ka. In both Europe and North America, many pollen assemblages have no modern counterparts. Lakes in the southwestern United States reached their maximum extent between 15 and 12 ka. Conditions in the Pacific Northwest became wetter earlier than in the Southwest and were probably wetter than today. Birch populations became dominant over large areas in Alaska after 14 ka, as the climate warmed.

At 9 ka summer insolation was near a maximum. The Laurentide ice sheet had contracted further, and only fragments of the Cordilleran and Scandinavian ice sheets still persisted. SSTs continued to increase, the distribution of *Neogloboquadrina pachyderma* continued to decrease in the North Atlantic, and *Globigerinoides ruber* extended its range northward in the subtropics.

The most striking changes in vegetation patterns on both continents took place between 12 and 9 ka as temperatures apparently increased rapidly. In Europe the southern limit of boreal spruce forest was at its most northerly position of the Holocene, and in eastern North America the region with spruce trees continued to move north but was much diminished in north-south extent. Spruce trees grew north of their modern northern limit in northwestern Canada (Ritchie and Harrison, this vol.). Also indicative of warming was the northward extent of oak popula-

tions in eastern North America, while in southern Europe oak also extended its range northward. Lake levels were lower in Turkey and Iran and decreased in western North America, although summers remained wetter than today in the Southwest (Thompson *et al.*, this vol.). The northwest-southeast gradient in effective moisture in the West was not as steep as at present; the Pacific Northwest was effectively drier than today, and summer temperatures in northwestern Canada and Alaska were higher than at present.

The ice sheets in North America and Europe had disappeared by 6 ka, and foraminifera in the North Atlantic were distributed much as they are today. In the American Midwest the climate was drier than today, as evidenced by lower lake levels and the extension of prairie to the east. In Europe boreal spruce forest had expanded to the west and south as winters cooled, and oak populations reached their maximum distribution for the Holocene both to the north and to the south. Annual precipitation was at a maximum in central Europe. For the first time the majority of European pollen assemblages resembled those of today (Huntley and Prentice, this vol.). In northeastern North America spruce populations had moved farther north, and the spruce-dominated boreal forest grew in its modern location for the first time. Many deciduous forest trees were growing much farther north than earlier. Conditions drier than at present were more widespread in the western United States, and conditions remained wetter than at present only in Colorado, New Mexico, and eastern Arizona.

After 6 ka spruce dominance increased within the boreal forest of eastern North America, and the region of abundant spruce populations increased southward as summer temperatures decreased. Oaks declined in the southeastern United States as winter temperatures increased. In northwestern Europe, however, winter temperatures decreased, causing the southern and western limits of spruce populations to advance across Scandinavia and areas of maximum spruce dominance to move westward. The northeastern limit of oak in Europe contracted, and oak became less abundant over most of northern and upland central Europe where summer temperatures decreased. Effective moisture increased in much of the northwestern United States and remained greater than at present in parts of the Southwest.

Data-Model Comparisons

The patterns of change observed in the data are consistent with many of the changes in circulation, temperature, and moisture balance simulated by the model. In general, vegetation moved north and changed in composition as the ice sheets melted and temperatures increased (Ritchie, 1987; Anderson and Brubaker, this vol.; Huntley and Prentice, this vol.; Webb, Bartlein, *et al.*, this vol.). More specifically, western Canada experienced greatest summer warmth much earlier than central North America (Ritchie *et al.*, 1983; Kutzbach, Guetter, *et al.*, this vol.; Ritchie and Harrison, this vol.). At 18 ka both cooler-than-present temperatures and the southward displacement of the southern branch of the jet stream and associated storm tracks explain the higher-than-present lake levels and the expansion of woodlands in the American Southwest (Thompson *et al.*, this vol.). Cold and dry conditions in the Northwest can be attributed to the easterly winds along the southern side of the glacial anticyclone (Barnosky *et al.*, 1987).

Regional studies in North America (Barnosky *et al.*, 1987; Webb *et al.*, 1987; Harrison, 1989; Anderson and Brubaker, this vol.; Ritchie and Harrison, this vol.; Thompson *et al.*, this vol.; Webb, Bartlein, *et al.*, this vol.) and Europe (Huntley and Prentice, 1988, this vol.) have demonstrated that the NCAR CCM simulates the general geographic patterns of changes for selected climatic variables. In Alaska the model and data both show increasingly cold and dry conditions from eastern to western Beringia at 18 ka (Barnosky *et al.*, 1987). By 12 ka the simulated mean July temperature and annual precipitation are near their modern values, in good agreement with the pollen data. The period of maximum summer warmth was at 9 ka for both the data and model results, and the subsequent trends in temperature and moisture are similar in both the data and the model (Anderson and Brubaker, this vol.).

In northwestern Canada summer temperatures were higher than today at 9 ka both in the model results and in the data, which show the Arctic treeline north of its present position. At 6 ka black spruce was more abundant, indicating the development of conditions cooler and moister than those at 9 ka. The model also simulates such conditions. Since then the southern border of the boreal forest has moved south as summers have cooled (Ritchie and Harrison, this vol.).

In the western United States temperatures and moisture patterns at 18 and 9 ka agree well with the simulated patterns of atmospheric circulation there. The split in the winter jet stream and the consequent displacement of the southern branch and its associated storm tracks far to the south led to moister-than-present conditions in the Great Basin at 18 ka, but easterlies in the Northwest made conditions drier there than today. At 9 ka the subtropical high was expanded and moisture levels decreased (Thompson *et al.*, this vol.).

In eastern North America the data and model results agree that temperatures were lower than today at 18 ka, increased after that, increased most rapidly between 12 and 9 ka, and reached a maximum in the central region at 6 ka. The data and model results also agree that the modern gradients for July and January temperatures were established at 9 ka and that conditions were drier at 9 ka than at present in the northeastern United States (Webb, Bartlein, *et al.*, this vol.).

In Europe the patterns of simulated and inferred January temperatures are in qualitative agreement at 9 and 6 ka. The patterns of July temperatures show less agreement, but both the simulated and inferred temperatures at 6 ka were higher than at present in the continental interior. The simulations correctly indicate drier-than-present conditions over most of Europe at 9 ka (Huntley and Prentice, this vol.).

The regional studies also reveal several differences between the data and the NCAR CCM results. The model simulated January temperatures higher than at present in Alaska at 18 ka, in strong contradiction to the geologic evidence for winters colder than today (Hopkins, 1982). In the western United States the pattern and timing of temperature increases differed from those simulated by the model at 12 and 9 ka. The simulation showed an earlier temperature increase than the data appear to record. In southeastern North America the model correctly simulated conditions colder than at present from 18 to 12 ka but not nearly as cold as those inferred from the pollen data. In the Midwest the model correctly simulated dry conditions from 9 ka on but simulated maximum dryness at 9 ka instead of 6 ka as indicated by the pollen and lake-level data.

In Europe the simulated July temperatures at 9 ka were higher than at present across the whole continent, with the highest anomalies in the south. In contrast, the temperatures inferred from pollen data were lower than at present in all regions but the north. The data and model results agree that temperatures were generally higher at 6 ka than today, although the simulated values were not as high as the inferred values in the north and were higher rather than lower than at present in the south. Inferred and simulated moisture patterns also differ: the simulated values indicate conditions drier than at present in central Europe at 6 ka and in southern Europe at 9 and 6 ka, in contrast to the interpretations of wetter-than-present conditions inferred from data.

Overall the data agree with the model results more than they disagree, but improvements are needed in the model, boundary conditions, and data and are in progress (see Kutzbach, Bartlein, *et al.*, this vol.). Some additional experiments with the model (Kutzbach and Ruddiman, this vol.) illustrate that progress can be anticipated. Sensitivity experiments with the NCAR CCM for 18 ka showed improved agreement between the data and model results when a lower height for the Laurentide ice sheet was used (Kutzbach and Ruddiman, this vol.). Compared to the simulation with the higher ice sheet, the large region with January temperatures higher than at present in Alaska disappeared, although temperatures lower than at present were still not simulated there. The simulated temperatures in the southern United States were slightly lower and therefore were somewhat closer to those inferred from the pollen data. Such results illustrate the possibilities for combining new data with new model results to obtain improved paleoclimatic estimates. In some cases poor agreement between observations and simulations may be the result of incorrect choices of boundary conditions and may suggest revisions.

Monsoon Sector

PRESENT-DAY CONDITIONS

The present-day circulation of the northern summer monsoon is driven by the warmth of the African-Asian landmass compared with the surrounding ocean (Fein and Stephens, 1987). Hydrostatic considerations show that this thermal gradient results in low pressure over land and high pressure over the ocean, producing a southerly to southwesterly inflow of moist air and heavy monsoonal precipitation over western Africa and southern and eastern Asia. Much of northern Africa and central Asia lies beyond the penetration of the monsoon rains and is hot and dry (Fig. 19.3). Aloft, a large anticyclone develops, with a tropical easterly jet on its southern flank that stretches from the northern Indian Ocean to western Africa. During southern summer a similar land-ocean gradient brings monsoon rains to southern Africa and tropical Australia.

The NCAR CCM reproduces the main elements of the July summer monsoon circulation (Fig. 19.3). Some features of the January circulation are also well simulated, although the model overestimates the strength of the cross-equatorial flow from the Northern to the Southern Hemisphere in the Atlantic sector. A weak monsoon low develops in January over southern Africa; in Australia, on the other hand, the simulated summer monsoon winds and rains do not penetrate far enough inland (Kutzbach, Guetter, *et al.*, this vol.).

The principal controls on the modern distribution of vegetation in the monsoon regions are total annual rainfall and the timing, duration, and intensity of the dry seasons. Temperature also becomes an important influence at high altitudes and in the subtropics. In Africa a humid equatorial rainforest zone with a double rainy season is flanked on the north and south by broad belts of dry forest and savanna characterized by monsoonal climates with summer rains and winter drought. Rainforest, grading into dry forest toward the continental interiors and into subtropical and temperate forests along the Pacific seaboard in both hemispheres, also occurs in the monsoon regions of India, southeastern Asia, New Guinea, and northeastern Australia. Subtropical deserts are found at around 20–30° north and south latitude. In the north they stretch from northern Africa across Arabia to northwestern India and into central Asia (Fig. 19.3). In the south they occur in southwestern Africa and in the interior of Australia.

African lake levels are generally low today except for a few basins close to the equator (Street-Perrott *et al.*, 1989). Water levels are low in the Arabian Peninsula and low to intermediate in Afghanistan, northwestern India, and central Asia. In New Guinea and tropical Australia, water levels are intermediate to high, but they are low in the arid interior of Australia.

The modern distribution of the subpolar foraminifer *Globigerina bulloides* in the western Arabian Sea (Fig. 19.3) reflects coastal upwelling driven by the southwesterly summer monsoon winds. Tropical and subtropical water masses are illustrated by the distribution of *Globigerinoides ruber*.

DYNAMICS OF THE MODEL SIMULATIONS

At 18 ka the highly reflective ice sheets, generally cold oceans, and equatorward advance of sea ice in both hemispheres had a significant effect on the climate of the monsoon regions, even though the seasonal distribution of insolation at the top of the atmosphere was similar to that of today (Kutzbach and Street-Perrott, 1985). The low temperatures created by these glacial-age boundary conditions steepened the north-south temperature gradient over Eurasia (Kutzbach, Guetter, *et al.*, this vol.). As a result, the westerly jet stream aloft and its associated westerly storm track at the surface intensified and migrated equatorward in both summer and winter (Fig. 19.3).

Both summer and winter temperatures over tropical land areas were slightly colder than today. A west-east contrast in July temperatures was simulated between the African-Arabian sector, which was generally 0–2°C cooler than today, and southern, eastern, and southeastern Asia and Australia, where the decrease was 2–4°C (Kutzbach, Guetter, *et al.*, this vol.). A more zonal pattern of anomalies prevailed in the monsoon regions in January, when the cooling was 0–2°C in equatorial latitudes and southern Africa and 2–4°C elsewhere, except for eastern Asia, where advection of cold air from northeastern Siberia resulted in a lowering of more than 6°C.

Most of the tropics had rainfall similar to that of today or slightly less, with one or two exceptions related to the imposed SST pattern (CLIMAP, 1981); the contrast between the warmer-than-present prescribed SSTs in the western Indian Ocean and the cooler temperatures elsewhere resulted in the simulation of increased equatorial convergence and rainfall over eastern equatorial Africa, with a very marked weakening of the summer monsoon over southern Asia (Fig. 19.3). Precipitation also increased along the intertropical convergence zone (ITCZ) in southeastern Asia.

By 9 ka the remnant Laurentide ice sheet had a negligible effect on the climate of the monsoon regions (Kutzbach and Street-Perrott, 1985). All other surface boundary conditions were set to the same values as in the control cases. Hence the striking differences between the simulated climates of 9 ka and today can be

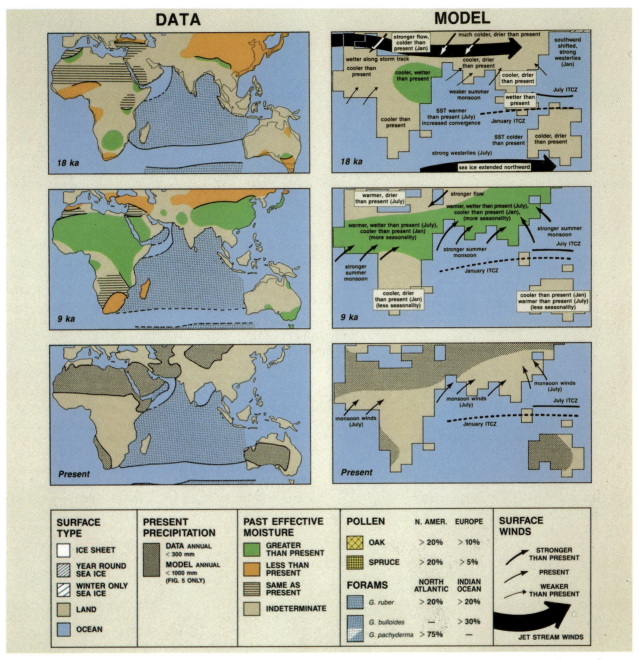

Fig. 19.3. Data and model results as in Figure 19.2 except for the regions around the Indian Ocean. In the simulation of present climate, regions where annual precipitation is less than 1000 mm are indicated. From COHMAP Members; reprinted by permission from *Science* vol. 241, pp. 1043-1052, copyright 1988 by the AAAS.

attributed almost exclusively to more July and less January radiation at the top of the atmosphere between 12 and 6 ka than at present (Kutzbach and Street-Perrott, 1985).

In July at 9 ka surface temperatures over the North Africa-Eurasian landmass were 2-4°C higher than today. The warmer conditions produced at the ground surface and aloft via convection intensified the summer monsoons, bringing enhanced rising mo-

tion, cloudiness, and precipitation, along with lower surface pressure, a stronger easterly jet aloft, and more vigorous southerly and southwesterly low-level winds over the Sahara Desert, Arabia, and southern and eastern Asia (Fig. 19.3). Consequently, wetter-than-present conditions developed over a large region. Northwestern Africa, which lay beyond the northern limit of the monsoon, received less precipitation than today. Summer rainfall increased slightly over the

Near East and central Asia but not enough to offset the high summer evaporation. The increased rising motion over land was balanced by increased subsidence and higher pressure over the northern subtropical oceans (Kutzbach, Guetter, *et al.*, this vol.).

In the simulation for January at 9 ka, the orbital changes resulted in lower land temperatures and less evaporation than at present, along with an intensified Siberian anticyclone with stronger outflowing northeasterly winds. The combined effect of the seasonal changes in precipitation and evaporation was a large annual water surplus, especially between 8.9 and 26.6°N (Kutzbach and Street-Perrott, 1985).

By 6 ka the magnitude of the radiation anomalies had declined slightly. The simulated climates for July and January were broadly similar to those for 9 ka, although the magnitude of the departures from the control cases was smaller. After 6 ka, as July insolation decreased, simulated land temperatures decreased, the monsoonal winds weakened, precipitation anomalies diminished over the northern continents, and the northern subtropical deserts expanded. Changes of opposite sign were simulated in the Southern Hemisphere.

PALEOCLIMATIC DATA

At 18 ka (Fig. 19.3) pollen data show that the extent of forests in the Atlas Mountains was greatly reduced, while dry steppe vegetation occupied the Mediterranean area, suggesting at first glance that northwestern Africa and the lands around the Mediterranean were cooler than today, with moisture levels similar to or slightly lower than those of today. But this interpretation conflicts with widespread evidence for high lake levels in the Near East (Roberts and Wright, this vol.) and less reliable indications of greater runoff from the Atlas into the northern Sahara (Street-Perrott and Perrott, this vol.). Most lakes in the southern Sahara were desiccated, implying conditions similar to or drier than today. Both pollen and lake-level data indicate a drier climate in western Africa, where the glacial-age cooling is estimated to have been at least 6°C (Maley, 1987). Eastern Africa is another area of conflict; here, most lakes were at levels similar to today's or even slightly higher in some cases, whereas arboreal vegetation was severely curtailed, and the lowering of the upper treeline suggests a cooling of 5-8°C (van Zinderen Bakker and Coetzee, 1972).

. Weaker southwesterly monsoonal winds and reduced upwelling are indicated by the diminution of eastern African montane pollen and by the presence of *Globigerina bulloides* in cores from the western Arabian Sea (Van Campo *et al.*, 1982; Prell, 1984; Prell and Van Campo, 1986). In southern Asia some evidence points to more extensive wet forests at 18 ka along the southern and southeastern flanks of the Himalayas and in Tibet, extending eastward into southwestern China (Yunnan). However, all of northern and eastern China, together with Japan, was much colder and drier than today, which suggests a weaker East Asian summer monsoon (Winkler and Wang, this vol.).

Between 12 and 6 ka the reconstructed moisture distribution was distinctly different from both 18 ka and today. At 9 ka (Fig. 19.3) the pollen and lake-level data concur that a belt of greatly enhanced effective moisture stretched from northern Africa eastward across Arabia, Afghanistan, northwestern India, Tibet, China, and Japan (Roberts and Wright, this vol.; Street-Perrott and Perrott, this vol.; Winkler and Wang, this vol.). The northern limit of this vast wet anomaly reached 30°N in Africa and southwestern Asia, more than 35°N in parts of western China, and 40°N in eastern Asia. In eastern Africa it extended across the equator down to at least 16°S.

Nowhere were the differences from today more marked than in the southern Sahara, where the vegetation belts shifted northward by 4-6° latitude (Ritchie *et al.*, 1985; Schultz, 1987; Lézine and Casanova, 1989), so that the surface albedo is estimated to have decreased by about 0.10-0.14 (Street-Perrott *et al.*, 1990). Remains of aquatic fauna (crocodile, hippopotamus, turtle, and Nile perch), with artifacts such as bone harpoons, testify to the widespread occurrence of permanent water, even in the central Sahara. The fossil record reveals that large grazing and browsing mammals, including elephant, rhinoceros, giraffe, and large antelope, all of which require great quantities of fodder, extended their range across areas today covered only by sparse shrubs and tufts of grass (Petit-Maire, 1986; Petit-Maire *et al.*, 1988). Lakes also developed during the early Holocene in many desert basins in Arabia, northwestern India, and southern Tibet (Xizang), while mesic temperate and subtropical forests migrated northward in eastern Asia in response to increased warmth and moisture. Paleohydrologic calculations have yielded estimates averaging 250 ± 130 mm/yr for the extra precipitation required to sustain the observed increase in lake area and vegetation cover between 5°S and 30°N in Africa and India (Street, 1979; Kutzbach, 1983; Swain *et al.*, 1983; Street-Perrott *et al.*, 1990).

The pattern of moisture anomalies at 9 ka has been attributed to an intensification and increased poleward penetration of the Northern Hemisphere summer monsoons during the early Holocene (Rossignol-Strick and Duzer, 1979; Street-Perrott and Roberts, 1983; Ritchie *et al.*, 1985; Street-Perrott *et al.*, 1990). This conclusion is supported by micropaleontological evidence from the Arabian Sea (Prell, 1984; Prell and Van Campo, 1986). The upwelling *Globigerina bulloides* fauna became more abundant in the west, and the tropical-subtropical *Globigerinoides ruber* fauna became more abundant in the east, although the location of the transition between them did not change. The influx of East African montane pollen to the sediments of the western Arabian Sea was higher than today (Van Campo *et al.*, 1982). In general, therefore, both terrestrial and marine data indicate stronger southwesterlies and more summer monsoon precipitation north of the equator.

Northwestern Africa, northern Greece, Anatolia, Iran, northwestern China, northeastern China, and northern Japan, all of which lay beyond the northern limits of the enhanced monsoons, were all drier than today (Roberts and Wright, this vol.; Street-Perrott and Perrott, this vol.; Winkler and Wang, this vol.). There is also evidence for reduced precipitation at 9 ka near the modern limits of the (southern) summer monsoon in southern Africa; pollen data from the Transvaal and the Orange Free State (24–31°S) suggest that semiarid thornbush had spread eastward into areas now occupied by subhumid grassland (Scott, 1989; Street-Perrott and Perrott, this vol.), while in central Madagascar (20°S) forest was less extensive and brushfires were more common than during the latter half of the Holocene (Burney, 1987a,b). In New Guinea the only lake-level record so far available suggests a low but rising water level (Garrett-Jones, 1979, 1980). The pollen and lake-level data from tropical Australia (Queensland) imply slowly increasing effective moisture, although the various sites studied are inconsistent in terms of the relative difference from modern conditions (Harrison and Dodson, this vol). In the southern Indian Ocean the *Globigerina bulloides* subpolar fauna had retreated slightly poleward. Note, however, that its position is poorly constrained by available data (COHMAP, 1988).

The pattern of moisture anomalies at 6 ka was broadly similar to that for 9 ka, apart from a few regional variations: the northeastern and southwestern Sahara, eastern equatorial Africa, and parts of northwestern and north-central China were slightly drier than at 9 ka, whereas northwestern Africa, Anatolia, Armenia, Iran, central Asia, northeastern China, and northern Japan were somewhat wetter (Roberts and Wright, this vol.; Street-Perrott and Perrott, this vol.; Winkler and Wang, this vol.). The trade winds off northwestern Africa had slackened (Sarnthein et al., 1981), causing a decrease in upwelling and a pronounced warming of the coastal waters (Petit-Maire, 1980; Mix et al., 1986). These changes are all compatible with an increased tendency for westerly airflow over the lower midlatitudes in summer compared with 9 ka (Street-Perrott et al., 1989).

DATA-MODEL COMPARISONS

18 ka

The simulated weakening of the South and East Asian monsoons at 18 ka agrees well with the geologic observations from the Arabian Sea, China, and Japan summarized in Figure 19.3. The equatorward displacement of the westerly storm tracks south of the ice sheets may help to explain the paradox of high lake levels in the Near East (and possibly also in the

Atlas piedmont) coexisting with pollen evidence for more open, steppelike vegetation. Increased cloud cover, occasional rains, and slightly cooler temperatures could have combined to produce a more positive water balance in summer, which is a season of intense evaporation at present. Winters could have been characterized by reduced precipitation (falling mainly as snow) and frequent outbreaks of severely cold air from central Asia, both of which are unfavorable for tree growth (Roberts and Wright, this vol.). Moreover, if the southward-shifted westerly jet stream steered depressions from the Mediterranean along the Himalayan mountain front all year round instead of just in winter as at present, this might explain the otherwise anomalous occurrence of wetter conditions at 18 ka on the slopes facing the Bay of Bengal (Winkler and Wang, this vol.) (Fig. 19.3).

An apparently serious disagreement between the geologic record for 18 ka and the simulations done with the CCM and other climate models (Manabe and Broccoli, 1985a,b; Rind and Peteet, 1985) centers around the temperature depression in the tropics. In the mountains of New Guinea, the lowering of the upper treeline and glacier equilibrium lines suggests a cooling of 5–9°C at high altitudes (Hope et al., 1976; Walker and Flenley, 1979), whereas adjacent SSTs were estimated to have been only around 2°C lower than today (CLIMAP, 1981). Only about one-third to one-half of the implied high-altitude terrestrial cooling is simulated by the CCM, in which the CLIMAP estimates of SSTs based on marine plankton were a prescribed surface boundary condition. In this situation the prescribed surface boundary condition with SSTs only 2°C lower than at present effectively prevents severe cooling over adjacent tropical lands. Faced with the discrepancy between terrestrial-based and marine-based paleoclimatic indicators, Webster and Streten (1978) and Rind and Peteet (1985) suggested that the CLIMAP estimates for the tropics were too warm, possibly by as much as 5°C. However, Prell (1982) argued that a substantial error in the marine data was unlikely given the stability of the faunal assemblages around New Guinea, and Broecker (1986) used the oxygen-isotope record to arrive at a similar conclusion.

The same problem arises in eastern Africa, where the glacial-age cooling reconstructed from the inferred descent of the upper treeline and glacier equilibrium lines is about 5–8°C (van Zinderen Bakker and Coetzee, 1972; Osmaston, 1975), whereas estimated SSTs offshore were within ±2°C of modern values (CLIMAP, 1981). Recent estimates of glacial-age cooling of 4 ± 2°C for eastern equatorial Africa, however, indicate that the difference may be less than the earlier estimate (Bonnefille et al., 1990).

Because modern meteorological data from the tropics tend to rule out a large change in environmental lapse rates (Webster and Streten, 1978) and because substantial errors in the CLIMAP reconstruction for the western equatorial oceans appear unlikely (Prell, 1982; Broecker, 1986), the most probable sources of conflict are the interpretation of the paleoclimatic data from the high mountains and the model simulations themselves (see below). The mass balance of tropical glaciers is highly sensitive to precipitation and net radiation (and hence to cloud cover and ice albedo) as well as to temperature (Hope *et al.,* 1976; Hastenrath, 1984). Although the maximum glacial advance in New Guinea and eastern Africa is very poorly dated, the occurrence of a large lake on the continental shelf between Australia and New Guinea at 18 ka (Torgerson *et al.,* 1988; McCullough *et al.,* 1989) makes it conceivable that more snow fell on the high mountains than today.

The lowering of the upper treeline was even greater in New Guinea and eastern Africa and has been more firmly dated to 18 ka than the equilibrium-line lowering (van Zinderen Bakker and Coetzee, 1972; Walker and Flenley, 1979). But the controls on the elevation of the forest limit are much more obscure: ultraviolet light intensity (J. R. Flenley, personal communication), atmospheric carbon dioxide content, wind stress, and the frequency of snowfall may have played important roles. Mountains are not merely tall masts projecting up into the free atmosphere. The extent of snow cover and the occurrence of cold-air drainage down mountain slopes act as important controls on the temperature regime on the valley floors, from which pollen samples are usually collected (Barry, 1981). For all these reasons, more research is needed before paleotemperature estimates for the tropical high mountains can be considered a reliable test of the midtropospheric temperatures simulated by general circulation models.

12 ka to the Present

In contrast to the serious disagreements that remain unresolved for 18 ka, the agreement between the CCM and the paleoclimatic evidence from the northern tropics for 12-6 ka is the outstanding feature of the whole period from 18 ka to the present. The simulation experiments indicate a pattern of intensified northern monsoon circulation and precipitation that agrees closely in space and time with the geologic data and explains the phenomenon in terms of orbitally induced changes in solar radiation.

To illustrate this broad agreement between observations and simulations, we compared the reconstructed record of changes in the status of closed-basin lakes with the model's simulation of precipitation-minus-evaporation (P-E) for the latitude belt 8.9-26.6°N (Fig. 19.4). Kutzbach and Street-Perrott (1985) compared these independent estimates of the surface hydrologic budget at 3000-yr intervals from 18 ka to the present. This period covers almost a whole precession cycle, ranging from near-present insolation

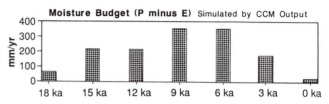

Fig. 19.4. Observed status of lake levels at 1000-yr intervals (upper row) and model-simulated moisture budget (precipitation minus evaporation) at 3000-yr intervals (lower row) from 18 ka to the present (0 ka) for the latitude belt 8.9-26.6°N. From COHMAP Members (1988); reprinted by permission from *Science* vol. 241, pp. 1043-1052, copyright 1988 by the AAAS.

values at 18 ka to a summer radiation maximum centered on 11-10 ka (Fig. 3.1 in Kutzbach and Ruddiman, this vol.). For the period 9-6 ka the various model runs simulated an increase in annual rainfall of 240-350 mm (20-30%) compared with present values and about a sevenfold increase in net moisture availability as measured by P-E (Fig. 19.4). By comparison, most northern tropical closed-basin lakes in our survey were at intermediate or high levels between 12 and 6 ka, whereas low levels prevail today. The simulated increase in rainfall at 9 ka agrees well with previous estimates of 250 ± 130 mm/yr based on biogeographic and lake-level evidence (Street, 1979; Kutzbach, 1983; Swain *et al.,* 1983; Street-Perrott *et al.,* 1990). This major conclusion strongly supports the importance of astronomic forcing as a direct control on tropical paleoclimates.

Figure 19.4, however, also reveals an interesting discrepancy between the model and the lake-level data for the late glacial. According to the NCAR CCM, the increased summer radiation would have already resulted in considerably strengthened summer monsoons over northern Africa and southern Asia by 15 ka, whereas in the real world most of the lakes in these areas remained low; the main exceptions were highland sites in the Tibesti Mountains and Tibet, where enhanced solar heating in summer may have been most effective in stimulating convection. Lakes in the southern Sahara, Arabia, and northwestern India did not generally start to rise until after 13 ka (Kutzbach and Street-Perrott, 1985; Street-Perrott *et al.,* 1989).

Part but not all of this discrepancy can be attributed to divergence of the radiocarbon time scale before 10 ka (Bard *et al.,* 1990). Kutzbach and Street-Perrott (1985) proposed other factors that may have contributed to this discrepancy. These factors include a delay in lake response caused by aquifer recharge after moister conditions developed and the impact of forcing factors not included in the modeling experiments; namely, the decreased levels of "greenhouse gases" and elevated tropospheric aerosol loadings in the glacial atmosphere (Thompson and Mosley-Thompson, 1981; Neftel *et al.*, 1982; Barnola *et al.*, 1987; Saigne and Legrand, 1987; Stauffer *et al.*, 1988). Street-Perrott and Perrott (1990) speculated that the onset of the monsoon was delayed by the impact on the tropical circulation of the suppression of deep-water formation in the northern North Atlantic before 13 ka (Boyle and Keigwin, 1987).

Southern Hemisphere Midlatitude Sector

PRESENT-DAY CONDITIONS

The principal atmospheric circulation features in the Southern Hemisphere are the same as in the Northern Hemisphere. They include the tropical easterlies, subtropical high-pressure belt, midlatitude westerlies, and circumpolar easterlies. They are much more intense in the Southern Hemisphere because of the far steeper pole-equator temperature gradient. Their seasonal latitudinal shift is less than in the Northern Hemisphere because the land-ocean ratio is much smaller and the Antarctic ice sheet helps keep the latitudinal temperature gradient strong in summer. Middle- to high-latitude westerly storm tracks and their seasonal latitudinal shifts are related in part to the seasonally contracting and expanding Antarctic sea ice. These seasonal shifts are responsible for the seasonal precipitation patterns characteristic of Australia, New Zealand, and southern South America and bring either winter rains (between 35 and 40°S) or seasonally uniform rains (between 40 and 50°S).

Additional geographic constraints such as high mountains and cool oceans accentuate the latitudinal climatic patterns and create regional or local orographic rainfall and rainshadows, which often result in steep environmental gradients. Depending on these factors, which determine the amount and duration of winter rains, vegetation in the southern latitudes ranges from sclerophyllous and temperate woodlands, rainforests, and subantarctic forests to moorlands, heath, and steppe.

The NCAR CCM reproduces the general features of the southern westerly storm tracks—their precipitation, pressure, and wind patterns (Kutzbach, Guetter, *et al.*, this vol.)—but underestimates their strength by 50% or more. The lack of elevational resolution in the model also leads to underestimation of the effects of the Andes on orographic precipitation and the rainshadow. The amplitudes of the waves in the midtro-posphere wind patterns are smaller in the simulations than the observed amplitudes.

DYNAMICS OF THE MODEL SIMULATIONS

Solar radiation at latitudes 40-60°S in July (winter) and January (summer) at 18 ka was similar to that of today (Fig. 3.1 in Kutzbach and Ruddiman, this vol.). After 15 ka insolation in winter rose, reaching a maximum of 3% more than at present between 12 and 9 ka, while insolation in summer fell to a minimum of 4-5% less than at present. The boundary of circum-Antarctic sea ice at 18 ka was estimated to lie about 7° latitude closer to the equator than today (CLIMAP, 1981). SSTs began to increase after 15 ka and were set at modern levels in the 9-ka experiment (Kutzbach and Ruddiman, this vol.). The ice caps that formed at 18 ka in southeastern Australia, New Zealand, and southern South America were too small to be included in the model grid. Although CLIMAP-based estimates of the changed height of the Antarctic ice sheet were included in the 18-ka boundary conditions, we have not attempted to analyze or isolate the effects of these changes on the overall simulation.

The paleoclimatic simulations for 18 ka for the southern middle and high latitudes show that temperatures over large areas were 2-5°C lower than today all year round, as a result of the lower prescribed SSTs. The seasonal temperature contrast was similar to that of today. Between 12 and 9 ka simulated temperatures increased substantially, and conditions were as warm as or warmer than today in winter (July), although they were still slightly cooler than today in summer (January). The seasonality of solar radiation and the resulting seasonal temperature contrast were at their lowest points for the past 18 ka.

Precipitation was generally lower than at present from 18 to 15 ka except for the eastern Pacific, eastern Indian, and Atlantic oceans, where SSTs were warmer than today according to CLIMAP (1981) estimates. This region of increased precipitation also included eastern Australia and the New Zealand region. (New Zealand is not represented as land on the model's grid.) Precipitation continued to be lower than today at 12-9 ka for much of South America, Australia, and the New Zealand region.

Simulated upper-level winds were stronger than today from 18 to 15 ka between latitudes 45 and 50°S but weaker between 35 and 45°S. Over New Zealand these winds had a stronger northwesterly component than at present. The subtropical high-pressure ridge was weaker than at present at 18-15 ka but stronger than at present at 9 ka.

DATA-MODEL COMPARISONS

The model results for 18-15 ka show lower temperatures and less precipitation than at present in this sector. Intensified southern westerlies were positioned farther poleward in Australia and South America but farther equatorward over New Zealand because of the northeasterly swing of the westerlies. The paleo-environmental data from southeastern Australia, New Zealand, and South America show depression of snowlines and upper-elevation vegetation by 600-800 m, dune field activity and loess deposition, and replacement of temperate forests by woodland, scrubland, and herb field, heath, or steppe. In contrast, expansion of cool-wet moorland in western Chile (Chiloé, latitude 43°S) and of forests in northern New Zealand indicates that precipitation was as high as or even higher than today in these regions and that the westerly storm tracks may have been located there. The relatively coarse resolution of the NCAR CCM, however, prevents detailed comparison over these small areas.

By 14 ka significant paleoenvironmental changes signal the end of full ice-age conditions, but full interglacial conditions were not established until between 12 and 9 ka. Throughout midlatitudes in Australia, New Zealand, and South America, forests expanded after 14 ka, whereas the high latitudes of South America (south of latitude 45°S) and South Island of New Zealand continued in a treeless glacial mode. These midlatitude forests were either more open than today or were dominated by drought- and frost-resistant taxa, indicating that although temperatures and precipitation had increased, the seasonal temperature and precipitation extremes still were more pronounced than today. These interpretations are consistent with the model simulations of increased temperature and precipitation in midlatitudes but diverge from the model results in terms of the degree and the timing of change. The model simulates climatic conditions for 12 ka that the data show were not reached until after 9 ka.

Data from lower midlatitudes (40-45°S) in New Zealand and South America indicate high temperatures and reduced precipitation between 9 and 6 ka relative to today. In southeastern Australia and between 45 and 50°S (i.e., higher midlatitudes) in South America, maximum precipitation for the Holocene is recorded and temperatures were lower than today. The high southern latitudes in South America and South Island of New Zealand received more moisture that led to reforestation, although precipitation was still somewhat below modern values. Moorland and rainforest expansion at latitude 50°S in South America indicates a relatively narrow band of high precipitation there. This evidence, combined with evidence for less-than-present precipitation in Tasmania and New Zealand, suggests that westerly storm tracks were confined mainly to a band from 45 to 50°S in winter.

As mentioned above, the data for 9-6 ka agree best with the simulated climate patterns for 12-9 ka. This temporal discrepancy could arise if poorly constrained boundary conditions such as sea-ice extent and SSTs were improperly specified for the 12- and 9-ka model experiments (Kutzbach and Ruddiman, this vol.). Only after 9 ka, when solar radiation changes became the principal climatic forcing, do the paleoenvironmental data and model predictions again agree.

For 6 and 3 ka both the data and the model predictions deviate from present values in the same direction and show an increased seasonal contrast relative to earlier times. Vegetation data from midlatitudes in southern Australia, New Zealand, and South America indicate expansion of forest types that are either drought- or frost-tolerant. In South America, lower latitudes as well as high southern latitudes record the most mesic vegetation of Holocene times. All of these changes are related to the increased seasonality of climate, which was far more pronounced during the late than the early Holocene. The changes also reflect the strengthening of the monsoonal patterns of summer rainfall that occurred as the seasonal contrast in insolation increased during the late Holocene.

The data also suggest that only in the late Holocene did the El Niño/Southern Oscillation (ENSO) pattern begin to affect climate patterns regularly throughout the southern latitudes (McGlone *et al.,* in press). Only during the late Holocene did environmental variability, linked today to ENSO events, become characteristic of records around the South Pacific. The version of the NCAR CCM used to simulate past climates, however, did not include an interactive dynamic ocean and therefore cannot be used to examine possible changes in the frequency or magnitude of ENSO events.

Oceans

Because SSTs were prescribed as fixed boundary conditions in the COHMAP deglacial experiments with the NCAR CCM, plankton-derived estimates of SSTs cannot be used to test the model results over the oceans. Such a test requires a coupled model that computes SST values and sea-ice limits in response to altered boundary conditions.

The ideal model for this purpose would include not just an atmospheric component, as in the NCAR CCM used by Kutzbach, Guetter, *et al.* (this vol.), but also several critical oceanic components, including (1) an interactive mixed layer with coastal and equatorial divergence and upwelling and a treatment of near-surface currents, (2) a treatment of the formation of deep and intermediate waters, and (3) an interactive sea-ice model. For the purposes of greatest interest to paleoclimatologists, we could then prescribe a few basic boundary conditions such as ice-sheet size, seasonal insolation, and atmospheric CO_2 concentrations for each model run and allow the interactive ocean, atmosphere, and sea-ice components of the model to simulate SSTs for regional comparisons with observed data.

Fully coupled models of this kind are not yet available, but climate modelers are working in this direction (see Kutzbach, Bartlein, *et al.,* this vol.). However, a number of paleoclimate experiments have used an atmospheric general circulation model coupled to a mixed-layer ocean and sea-ice model. Such models simulate mixed-layer SSTs and sea-ice limits but are far from accurate in regions where strong currents influence the temperature distribution. Comparison of the simulated SSTs with plankton-derived estimates of SST (and sea ice) is a first step in the process of assessing the reliability of both the SST estimates and the models and may also give some insights into the key mechanisms at work in the climate system.

In this section we briefly review the status of comparisons of marine data with results from models with mixed-layer oceans and interactive sea ice. Because most such experiments have focused on the glacial maximum at 18 ka (Manabe and Broccoli, 1985a,b; Broccoli and Manabe, 1987) and the interglacial maximum at 9 ka (Kutzbach and Gallimore, 1988; Mitchell *et al.,* 1988), our comparative data-model assessments are limited to these two periods.

18 KA

High-Latitude Northern Oceans

As noted by Kutzbach and Ruddiman (this vol.), Manabe and Broccoli (1985a,b) demonstrated that Northern Hemisphere ice sheets (particularly the Laurentide) can cause a decrease in SST and an increase in sea-ice extent comparable to those reconstructed by CLIMAP (1981) in high latitudes of the North Atlantic and North Pacific oceans. The primary mechanism altering North Atlantic SSTs in these models is the cold northwesterly winds that develop on the northeastern flanks of the Laurentide ice sheet and move cold airmasses across the subpolar North Atlantic (see also Keffer *et al.,* 1988). Lack of ocean dynamics and thermohaline circulation may explain why the Geophysical Fluid Dynamics Laboratory (GFDL) model underestimates cooling near the Gulf Stream and in the Norwegian Sea (Manabe and Stouffer, 1988). In contrast, Mitchell *et al.* (1988) used a Laurentide ice sheet much smaller than the maximum value and 9-ka insolation values as boundary conditions and simulated little cooling of the North Pacific and North Atlantic, except for the Labrador Sea near the residual ice.

The results of modeling experiments with large ice sheets support the finding from observations that ice-sheet effects transmitted through the atmosphere are an important control on sea-surface conditions at high northern latitudes. Additional evidence comes from the close relationship between SST values in the high-latitude North Atlantic and $\delta^{18}O$ values (a proxy for ice volume) over the full 2.5-million-year history of

Northern Hemisphere ice sheets (Ruddiman *et al.,* 1989; Raymo *et al.,* 1990). Ice sheets grew and decayed at a 41,000-yr rhythm late in the Pliocene and early in the Pleistocene and then shifted to a 100,000-yr rhythm late in the middle Pleistocene. These oscillations were mirrored by coherent, in-phase oscillations in North Atlantic SSTs at the same dominant periods.

Although northern ice sheets are thus thought to account for much of the overall change from glacial to Holocene SST values and sea-ice limits at high northern latitudes, other physical processes must be invoked to explain higher-frequency SST oscillations within glacial-maximum and deglacial times. The most prominent such oscillation is the mid-deglacial Younger Dryas cooling in the North Atlantic (Ruddiman *et al.,* 1977; Duplessy *et al.,* 1981; Ruddiman and McIntyre, 1981; Bard *et al.,* 1987; Broecker *et al.,* 1989). The size of ice sheets is unlikely to change rapidly enough to account for oscillations in ocean circulation modes from near-glacial to near-interglacial patterns within 1000 yr or less. The explanation appears to involve rapidly varying meltwater fluxes, but recently published data suggest that the problem is complex (Fairbanks, 1989; Broecker *et al.,* 1990).

High-Latitude Southern Ocean

The glacial-maximum simulation of Manabe and Broccoli (1985a,b) indicated that significant climatic effects from large Northern Hemisphere ice sheets are not transmitted through the atmosphere into the Southern Hemisphere and thus cannot account for the glacial-maximum SST decrease and sea-ice advance reconstructed from proxy indicators in the subantarctic by CLIMAP (1981). A later experiment (Broccoli and Manabe, 1987) that included lower glacial atmospheric CO_2 concentrations as an additional boundary condition did produce a subantarctic SST cooling very similar to the CLIMAP estimates. Although the ocean models used in these experiments omitted numerous dynamic processes, this agreement between data and model results implies that atmospheric CO_2, and presumably methane, are critical factors influencing the Southern Ocean.

Changes in North Atlantic deep-water (NADW) formation, with subsequent effects on the amount of NADW advected into the Southern Ocean, may also influence subantarctic surface-ocean temperatures and salinity (Crowley and Parkinson, 1988). This interesting hypothesis cannot yet be verified with existing ocean models because of the lack of sufficiently accurate treatment of ocean dynamics and the thermohaline circulation.

Tropical-Subtropical Oceans

As noted by Kutzbach and Ruddiman (this vol.) and in the monsoon sector section of this chapter, Rind and Peteet (1985) pointed out the fundamental mismatch between the modest CLIMAP SST cooling at low latitudes and the large equilibrium-line lowering on adjacent continents. Subsequent research has tended to support the original CLIMAP estimates (Prell, 1985; Broecker, 1986) and thus has not resolved this discrepancy.

The Broccoli and Manabe (1987) experiments with mixed-layer ocean models indicated that glacial-maximum ice sheets and CO_2 together produce a larger zonal-mean cooling of low-latitude oceans than that estimated by CLIMAP (1981). Even these models, however, do not treat wind-driven thermocline adjustments and upwelling processes in an interactive way and thus lack the physical processes that have the most important effects on surface-ocean structure at tropical and subtropical latitudes. Therefore, comparisons of data-derived estimates and model simulations are particularly premature at this time.

9 KA

Estimates of ocean temperature at 9 ka are available only for a few locations, and therefore only preliminary data-model comparisons are possible. Ruddiman and Mix (this vol.) surveyed the scattered observational evidence of ocean temperatures for the North Atlantic at 9 and 6 ka. The estimated ocean temperatures are lower than at present near Labrador and southern Greenland, along the northwest African coast, and in the equatorial Atlantic. They are higher than at present in the extreme western North Atlantic near Cape Hatteras. The coupled atmosphere/mixed-layer ocean model used by Kutzbach and Gallimore (1988) also simulates cooler conditions for 9 ka, compared to the present, throughout all but high northern latitudes. In general, however, the observed temperature differences are larger than those simulated by the model. The observed temperature differences may of course be the result of changes in ocean dynamics (currents and upwelling) as well as thermodynamics, whereas the model includes only thermodynamic effects. Moreover, the large heat capacity of the model ocean causes a seasonal lag in the response to insolation changes. In the simulations for 9 ka, the northern midlatitude ocean was warmest, relative to the present day, in fall (September-November, or several months after the peak in summer insolation) and coldest in spring (i.e., after the decreased winter insolation).

In high northern latitudes the model simulated significant reductions in the thickness and duration of Arctic sea ice for 9 ka. Some observations support a warm and more ice-free Arctic in the early Holocene (Haggblom, 1982; Salvigsen and Osterholm, 1982; Stewart and England, 1983). The simulated pattern of warmer annual-average conditions in very high latitudes and slightly cooler conditions in the middle and low latitudes at 9 ka (compared to the present) is a direct consequence of the latitudinal distribution of annual-average changes in solar radiation associated with increased axial tilt (Kutzbach and Gallimore, 1988).

Causes and Mechanisms

OVERVIEW

The results in this book point to two fundamental cause-and-effect relationships in the climate system during the last 18,000 yr (see Fig. 3.1 in Kutzbach and Ruddiman, this vol.; Kutzbach and Webb, this vol.). First, during glacial-maximum and early-deglacial times, when seasonal insolation was close to today's values, ice-related boundary conditions explain many of the climatic departures from modern values, particularly in middle and high latitudes of the Northern Hemisphere. Second, later in the deglaciation and into the early-middle Holocene, as ice-related climatic influences diminished and finally became negligible, the large differences in seasonal insolation became the major cause of climatic anomalies relative to modern conditions. After 6 ka both insolation and climatic anomalies slowly approached modern values.

The conceptual understanding of the orbitally induced changes in seasonal insolation and the variations in surface boundary conditions (mainly ice sheets) originated with Milankovitch, Spitaler, Köppen, Wegener, and Brooks (see Wright, this vol.). Initial modeling experiments by Kutzbach and Guetter (1986) provided the basic framework for evaluating these two fundamental factors. The updated modeling experiments outlined by Kutzbach and Ruddiman (this vol.) and Kutzbach, Guetter, et al. (this vol.) have summarized how these two factors have affected regional climates over the last 18 ka. The data-model comparisons in preceding chapters in this book show how specific regional climatic changes resulting from variations in these two factors can help to explain a wide array of deglacial climatic responses inferred from paleoclimatic data, including major shifts in vegetation, hydrologic budgets, and marine plankton.

The two major climatic controls—the seasonal distribution of insolation and the ice-related boundary conditions—represent external and internal boundary conditions, respectively. Boundary conditions related to ice sheets mainly involve the "slow physics" of the climate system; the size of ice sheets present at any point in geologic time is a function, at least in part, of the integrated effect of insolation changes over the preceding thousands (or tens of thousands) of years (Imbrie and Imbrie, 1980). Because ice sheets build and decay slowly, their lagged response, relative to insolation changes, allows for the possibility of complex interactions between ice-sheet forcing and seasonal insolation forcing.

The variation in the seasonal distribution of insolation also influences the "fast physics" of the climate system. Insolation levels at the top of the atmosphere for any season of any year are set entirely by the orbital configuration of that particular interval in time. In regions remote from the climatic influence of the ice sheets, local climatic responses (such as the peak intensity of summer heating of large landmasses) may track this "instantaneous" insolation forcing with no detectable lag other than the thermal inertia characteristic of seasonal thermal responses (typically one to three months, depending on whether the region is land or ocean).

Much of the uncertainty in our understanding of how the climate system has operated over the past 18 ka involves the estimation of the relative influence of these slow and fast processes and of the interactions between them. For example, changes in sea ice appear to be explained by a combination of ice-sheet and insolation influences (Manabe and Broccoli, 1985a,b; Kutzbach and Gallimore, 1988; Mitchell et al., 1988).

TRACKING CAUSE AND EFFECT DURING THE DEGLACIATION

The influence of the ice sheets was expressed in its purest form during the last glacial maximum near 18 ka, when summer and winter insolation values were almost identical to those of today. Model-simulated patterns of atmospheric temperature and precipitation caused mainly by the influence of the Laurentide ice sheet can account rather well for the changes in relative abundance patterns of several types of vegetation (e.g., spruce, southern pines, and prairie forbs) found in North American pollen records (Webb et al., 1987; Webb, Bartlein, et al., this vol.).

On the other hand, the data and the model results diverge significantly with regard to other vegetation types, such as oak and northern pines in the southeastern United States (Webb et al., 1987). Recent sensitivity experiments with lower-elevation ice sheets (Kutzbach and Ruddiman, this vol.) removed part of these discrepancies, but other refinements to the boundary conditions (e.g., reduced atmospheric CO_2 and methane levels, lower SSTs in the Gulf of Mexico) may also be necessary to bridge the gap between the data and model results.

The model simulations described by Kutzbach, Guetter, et al. (this vol.) show that the large Laurentide ice sheet split the jet stream and displaced the storm track to the south. This configuration may explain the late-glacial interval of high lake levels in the southwestern United States (Thompson et al., this vol.). Farther north, the model-simulated anticyclone over the ice sheet brought increased easterly flow along the ice margin in the northwestern United States, providing a possible explanation for the drier-than-present vegetation indicated by the pollen data there (Barnosky et al., 1987; Thompson et al., this vol.).

Other model results (Kutzbach and Ruddiman, this vol.; Ruddiman and Mix, this vol.) indicate that long-term, orbital-scale variations in North Atlantic SST and

sea-ice extent can be explained in large part by ice-sheet forcing. The ice-sheet signal is transmitted to the high-latitude ocean via cold northwesterly winds blowing from the northeastern flanks of the Laurentide ice sheet.

These climatic changes in the North Atlantic surface ocean can also be transmitted elsewhere in the climate system. Model experiments show that the North Atlantic has a large downstream effect on Europe, especially the western part (Rind et al., 1986). The bitterly cold late-glacial climate in Europe indicated by the absence of trees north of the Alps, by high-Arctic beetle fauna, and by permafrost features appears to be consistent with a North Atlantic Ocean covered by sea ice in winter (Atkinson et al., 1987; sensitivity tests in Kutzbach and Ruddiman, this vol.; Kutzbach, Guetter, et al., this vol.). Present dating of the European records does not fully resolve the timing of extreme cold conditions within the broad interval of the last glacial maximum and early deglaciation.

The North Atlantic may have even larger-scale effects on climate because of its key role in deep-water circulation. The formation of deep water in the North Atlantic was suppressed at the last glacial maximum (Boyle and Keigwin, 1985) and during early glacial maxima throughout the last 2.5 million years (Raymo et al., 1990). Changes in NADW can influence temperature, salinity, nutrient content, and alkalinity in the Southern Ocean and, ultimately, atmospheric CO_2 levels and thus global climate (Broecker and Peng, 1989). Discussion of these global-scale linkages is beyond the scope of this chapter, which focuses on the influence of Northern Hemisphere ice on proximal parts of the climate system via the atmosphere.

As deglaciation progressed, the influence of the shrinking ice sheets waned and that of the growing seasonal insolation anomalies became predominant. If no ice sheets had been present at middle and high latitudes of North America and Europe in the last 18 ka, the direct impact of the additional heating of landmasses by increased summer insolation would probably have been greatest between 11 and 10 ka. During this interval summer insolation reached maximum values nearly 8% higher than today, and winter insolation was almost 8% lower than today (Fig. 3.1 in Kutzbach and Ruddiman, this vol.). Without ice sheets, this configuration would have produced maximum seasonality of climate (hot summers, cold winters) at this time.

In actuality the ice sheets were still a sufficiently significant climatic factor even at mid-deglaciation that they delayed the maximum impact of insolation anomalies on heating and cooling over the ice-free land areas of the northern middle latitudes. Indeed, it is difficult to find records that are demonstrably free of ice-induced lags, although regions in the Northern Hemisphere most distant or upwind from the ice sheets apparently responded earlier to insolation forcing (McCulloch and Hopkins, 1966; Ritchie *et al.*, 1983).

Pollen distribution maps for North America remained in the characteristic glacial pattern of prominent north-south gradients between 15 and 12 ka in response to the continuing influence of the Laurentide ice sheet. Second-order changes in vegetation, however, began to be felt as early as 15 ka in the southeastern United States, where rising summer insolation had begun to have some climatic impact far from the influence of the ice sheet (Webb, Bartlein, *et al.*, this vol.). Relatively early melting of ice in northwestern North America also created a strong thermal gradient between the remaining ice sheet in the northeast and the ice-free, insolation-heated land in the southwest.

As a result of rising summer insolation and melting of the ice sheets, vegetation in east-central North America was substantially reorganized between 12 and 9 ka into a pattern of northeast-southwest and northwest-southeast gradients, which subsequently evolved slowly toward the modern distribution patterns (Webb, Bartlein, *et al.*, this vol.). Strong summer insolation heating of ice-free regions in the southwest and midcontinent produced somewhat warmer summer temperatures in the midcontinent by 9 ka, as evident in both the model simulations (Kutzbach and Ruddiman, this vol.) and the pollen data (Webb, Bartlein, *et al.*, this vol.). The absence of ice by 6 ka led to still warmer summer temperatures due to higher summer insolation values, and data in Europe also indicate higher-than-present temperatures (Huntley and Prentice, this vol.). Thus the warmest temperatures in North America and Europe appear to have been registered between 9 and 6 ka, even though summer insolation was well past its maximum by then. The effect of the ice sheets in delaying maximum summer insolation heating of the Northern Hemisphere midcontinents thus is in the range of 2000–5000 yr.

Monsoonal Climates and Links to Ice Sheets

Another extensive data set suitable for large-scale data-model comparisons is the collection of lake-level records in North Africa (Street and Grove, 1979; Street-Perrott *et al.*, 1989; Street-Perrott and Perrott, this vol.). In this region variable summer insolation heating of the African (and Asian) landmass is a major forcing factor determining summer monsoonal circulation (Kutzbach, 1981), and the distant Northern Hemisphere ice sheets are less influential.

The primary mechanism by which insolation forces monsoonal circulations is the differential response of land and sea to orbital radiation (Spitaler, 1921; Kutzbach, 1981). The relatively small heat capacity over land (see Fig. 2.3 in Kutzbach and Webb, this vol.) permits a much larger heating response of large landmasses to excess early-Holocene summer insolation than over the ocean, where insolation effects are attenuated by penetration into and mixing within the upper-ocean mixed layer. The excess summer heating over land (relative to the ocean) enhances rising motion over land, outflow of air from land to ocean at upper levels, and sinking motion over the oceans and strengthens land-sea pressure gradients that drive the enhanced summer monsoonal inflow. In contrast, during winters in the early to middle Holocene, the reduction in insolation heating caused net cooling of the land surface compared to today. The operation of the monsoon system is discussed in more detail in the monsoon sector section of this chapter and in Street-Perrott and Perrott (this vol.) and Kutzbach and Webb (this vol.).

The dominance of the insolation control is clearly demonstrated by two observations: (1) African lake levels at the last glacial maximum were similar to those of today (very low), even though ice sheets were large; and (2) lake levels were highest early in the Holocene, when summer insolation considerably above modern values drove a strong monsoonal response. Comparison of the modeling experiments of Kutzbach and Guetter (1986) and tropical lake-level data shows that much of the African lake-level response over the last 18 ka is consistent with the simple model of direct seasonal insolation forcing (Kutzbach and Street-Perrott, 1985; Kutzbach and Webb, this vol.).

The lake-level response, however, may not be entirely governed by direct insolation forcing, because it appears to lag somewhat behind the radiative signal. The strongest summer insolation forcing can be positioned anywhere between about 11 ka (if perihelion at the June 21 summer-solstice radiation maximum is used as the reference season) and about 8 ka (if perihelion coincident with the mid-August monsoonal maximum is invoked as the reference season). If the lake-level maximum is assumed to have been between 9 and 8 ka, then depending on the choice of critical insolation season, the lake response could have lagged behind insolation forcing by as much as 3000 yr.

More diagnostic still is the delay in the perceptible rise in African lake levels until 12.5 ka (Fig. 19.4), even though summer insolation values had been rising for several thousand years. Lags of this magnitude are probably too large to be attributable entirely to delayed aquifer responses. However, if the radiocarbon time scale is in error (Kutzbach and Ruddiman, this vol.), then much of this lag may disappear (see discussion in monsoon sector section).

Sensitivity experiments with and without Northern Hemisphere ice sheets at 9 ka indicated that moderate-sized ice sheets in the Northern Hemisphere have little direct influence on African aridity (Kutzbach and Guetter, 1984). This result implies that the ice sheets are not a likely explanation of the lagged responses in Africa, although that possibility cannot be ruled out entirely. Sensitivity tests conducted by Rind (1987) similarly suggested that Northern Hemisphere ice sheets have little direct climatic influence in Africa but also indicated that North Atlantic SST anomalies at high latitudes have a larger impact (see also Rind *et al.,* 1986). That is, ice sheets may affect African climate via their indirect effects on the North Atlantic.

Prell and Kutzbach (1987) proposed that "glacial boundary conditions" influence African climatic responses, including aridity. They inferred that low-latitude SST was one of the most important glacial boundary conditions because of its impact on evaporation over the oceans and on precipitation over adjacent land areas. This analysis suggested that the slow rise in oceanic SST values to postglacial warmth may have delayed the onset of full monsoonal precipitation until several thousand years after the insolation forcing. Prell and Kutzbach also noted that the persistence of lower atmospheric CO_2 levels early in the deglaciation, a factor not included in these initial experiments, may have weakened the monsoon by cooling the African interior. The subsequent fall in lake levels to much lower levels late in the Holocene is in accord with the slow decline of summer insolation toward modern values. Higher-frequency (about 1000-yr) oscillations in the North African monsoon evident during the Holocene (Street-Perrott and Perrott, this vol.) are not explained by the slow changes in insolation values or glacial boundary conditions discussed here.

This analysis of causes and mechanisms has necessarily been limited to the two key factors (insolation and ice sheets) used as upper and lower boundary conditions in the sequence of glacial-to-Holocene modeling experiments. As summarized here, these two factors can explain a substantial part of the observed climatic responses of the last 18 ka. As resolution and understanding of the sequence of events during the last deglaciation improve, other climatic variables, such as atmospheric CO_2, methane, and aerosol loadings, will no doubt emerge as additional sources of regional or global climatic forcing and feedback. The causal role of changes in deep-water circulation (and links to the CO_2 system) also remains unclear. The uncertainty surrounding the height of the ice sheet during the deglaciation is another problem that needs to be addressed by further sensitivity tests like the one reported in Kutzbach and Ruddiman (this vol.) and Kutzbach, Guetter, *et al.* (this vol).

SOUTHERN HEMISPHERE

Causes and mechanisms are as yet difficult to decipher in the Southern Hemisphere. The predominance of ocean and the sparse coverage of terrestrial data restrict the kind of detailed data-model comparisons on which the COHMAP strategy relies. Data from ocean cores are also generally sparse (Morley and Dworetzky, this vol.).

The potential link between precessional and monsoonal variations suggests that the precessional insolation minimum in the Southern Hemisphere during the early and middle Holocene might have caused a minimum in the seasonality of temperature and in monsoon strength at that time over tropical and subtropical latitudes of South America, Africa, and Australia. The sparse terrestrial data available both agree and disagree with this model (Harrison and Dodson, this vol.; Markgraf, this vol.; Street-Perrott and Perrott, this vol.), and the data base is far from adequate for a full evaluation (see discussion in the monsoon sector section). In addition the small size of the Southern Hemisphere continents might be expected to result in rather weak thermal and monsoonal signals.

At higher latitudes the Antarctic ice sheet was relatively stable in size through the deglaciation and thus is not likely to have been a strong source of slow-response forcing analogous to that of the Northern Hemisphere ice sheets. Nevertheless, both terrestrial and marine climatic signals from the Southern Ocean (e.g., McGlone *et al.,* this vol.; Morley and Dworetzky, this vol.) are surprisingly similar to those in the Northern Hemisphere, with a well-marked deglacial warming that is roughly in phase with that in the Northern Hemisphere. This out-of-phase timing with respect to direct insolation suggests that Northern Hemisphere ice sheets exert an indirect but basic control over Southern Ocean circulation. Deep-water changes are a plausible mechanism for the transference of climatic signals from northern to southern polar regions (Broecker and Peng, 1989). Indications of a small lead in Southern Ocean responses compared to changes in Northern Hemisphere ice could mean that other interactive factors are also involved, possibly including control of atmospheric CO_2 levels by deep-ocean circulation.

Evaluation and Conclusions

Much COHMAP research has been focused on identifying the underlying causes of the timing, pattern, and magnitude of the long-term regional- to global-scale climatic changes of the past 18 ka. This research has benefited from the availability of (1) a general knowledge of past global climatic controls, (2) large sets of well-dated data that can be calibrated or interpreted in climatic terms, and (3) general circulation models that can simulate climatic patterns. This framework has enabled us to check the results of a series of paleoclimatic experiments and thus to advance this method of paleoclimatic research introduced by Williams *et al.* (1974), CLIMAP (1976), and Gates (1976).

Our main conclusion from comparisons of the data and model results is that two major cause-and-effect relationships appear to stand out: (1) the heightened seasonality of Northern Hemisphere solar radiation around 12–6 ka increased the seasonality of Northern Hemisphere climates (greater temperature and moisture extremes, enhanced monsoon circulations), and

(2) the large ice sheets and more extensive sea ice during glacial times altered atmospheric flow patterns and associated temperature and precipitation patterns, especially in middle to high latitudes.

Despite these relatively clear results, we recognize that the limitations of the data, models, and research design leave much room for improvement in the estimates of the sensitivity of climate to these changes at the upper and lower boundaries of the atmosphere. For example, in the interior of a large continent at 9 ka, was the summer temperature 1-2°C or 2-4°C higher than at present? We are fairly certain that the warming was in this general range, but significant uncertainty remains concerning the magnitude of the anomaly. Questions of magnitude are critical both for understanding the past and for making accurate predictions concerning future climates.

Many vexing questions about regional and local climates remain unanswered. How much colder were the tropics at the glacial maximum? What seasonal values of temperature and rainfall are consistent with the vegetation associations that grew south of the Laurentide ice sheet and have no modern analogues? What refinements in regional-scale modeling and data analysis are needed to resolve the fine-scale structure of paleoclimates in the basin and range province of the American West, a huge tract that is represented in current models by only a few grid squares with a much smoothed topography? Here we recognize that the limitations of the current model (version 0 of the NCAR CCM), current choice of boundary conditions (Kutzbach and Ruddiman, this vol.), and current estimates from the data have led to many regional discrepancies between the simulations and the data.

Firm conclusions about whether the data or the model is "correct" or "incorrect" in a particular region are therefore premature. We also recognize that it would be misleading to conclude that both the data and the model results are correct because they agree! As we look forward to the fresh insights and greater accuracy that will come with improved data, models, and research designs (see Kutzbach, Bartlein, *et al.*, this vol.), we are confident that both orbitally induced insolation changes and the slowly changing glacial-age boundary conditions will be essential elements in the explanation of climatic history since glacial times. The importance of the orbitally induced insolation changes in explaining postglacial climates was not widely recognized when we started this work in the 1970s (Wright, this vol.). The importance of glacial-age features has long been recognized, but climate models provide, for the first time, quantitative estimates of the magnitude and spatial scale of both of these effects. Our results are just a beginning, however. Kutzbach, Bartlein, *et al.* (this vol.) outline potential opportunities for an even more thorough examination of the large climatic changes of the past 18 ka.

References

Atkinson, T. C., Briffa, K. R., and Coope, G. R. (1987). Seasonal temperatures in Britain during the past 22,000 years, reconstructed using beetle remains. *Nature* 325, 587-592.

Bard, E., Arnold, M., Moyes, J., and Duplessy, J.-C. (1987). Reconstruction of the last deglaciation: Deconvolved records of δO-18 profiles, micropaleontological variations, and accelerator mass spectrometric C-14 dating. *Climate Dynamics* 1, 101-112.

Bard, E., Hamelin, B., Fairbanks, R. G., and Zinder, A. (1990). Calibration of the ^{14}C time scale over the past 30,000 years using mass-spectrometric U-Th ages from Barbados corals. *Nature* 345, 405-410.

Barnola, J. M., Raynaud, Y. S., Korotkevitch, Y. S., and Lorius, C. (1987). Vostok ice core provides 160,000-year record of atmospheric CO_2. *Nature* 329, 408-414.

Barnosky, C. W., Anderson, P. M., and Bartlein, P. J. (1987). The northwestern U.S. during deglaciation: Vegetational history and paleoclimatic implications. *In* "North America and Adjacent Oceans during the Last Deglaciation" (W. F. Ruddiman and H. E. Wright, Jr., Eds.), pp. 289-321. The Geology of North America, Vol. K-3. The Geological Society of America, Boulder, Colo.

Barry, R. G. (1981). "Mountain Weather and Climate." Methuen, London.

Bonnefille, R., Roeland, J. C., and Guiot, J. (1990). Temperature and rainfall estimates for the past 40,000 years in equatorial Africa. *Nature* 346, 347-349.

Boyle, E. A., and Keigwin, L. (1985). Comparison of Atlantic and Pacific paleochemical records for the last 215,000 years; changes in deep ocean circulation and chemical inventories. *Earth and Planetary Science Letters* 76, 135-150.

——. (1987). North Atlantic thermohaline circulation during the past 20,000 years linked to high-latitude surface temperature. *Nature* 330, 35-40.

Broccoli, A. J., and Manabe, S. (1987). The influence of continental ice, atmospheric CO_2, and land albedo on the climate of the last glacial maximum. *Climate Dynamics* 1, 87-99.

Broecker, W. S. (1986). Oxygen isotope constraints on surface ocean temperatures. *Quaternary Research* 26, 121-134.

Broecker, W. S., and Peng, T.-H. (1989). The cause of the glacial to interglacial atmospheric CO_2 change: A polar alkalinity hypothesis. *Global Biogeochemical Cycles* 3, 215-240.

Broecker, W. S., Kennett, J. P., Flower, B. P., Teller, J., Trumbore, S., Bonani, G., and Wolfli, W. (1989). The routing of Laurentide ice-sheet meltwater during the Younger Dryas cold event. *Nature* 341, 318-321.

Broecker, W. S., Bond, G., Klas, M., Bonani, G., and Wölfli, W. (1990). A salt oscillator in the glacial Atlantic? 1. The concept. *Paleoceanography* 5, 469-477.

Burney, D. A. (1987a). Late Quaternary stratigraphic charcoal records from Madagascar. *Quaternary Research* 28, 274-280.

——. (1987b). Pre-settlement vegetation changes at Lake Tritivakely, Madagascar. *Palaeoecology of Africa* 18, 357-381.

CLIMAP Project Members (1976). The surface of the ice age earth. *Science* 191, 1131-1136.

——. (1981). Seasonal reconstructions of the earth's surface at the last glacial maximum. *Geological Society of America Map and Chart Series* MC-36.

COHMAP Members (1988). Climatic changes of the last 18,000 years: Observations and model simulations. *Science* 241, 1043-1052.

Crowley, T. J., and Parkinson, C. L. (1988). Late Pleistocene variations in Antarctic sea ice. II: Effect of interhemispheric deep-ocean heat exchange. *Climate Dynamics* 3, 93-103.

Duplessy, J.-C., Delibrias, G., Turon, J. L., Pujol, C., and Duprat, J. (1981). Deglacial warming of the northeastern Atlantic Ocean: Correlation with the paleoclimatic evolution of the European continent. *Palaeogeography, Palaeoclimatology, Palaeoecology* 35, 121-144.

Fairbanks, R. G. (1989). A 17,000-year glacio-eustatic sea level record: Influence of glacial melting rates on the Younger Dryas event and deep-ocean circulation. *Nature* 342, 637-642.

Fein, J. S., and Stephens, P. L. (1987). "Monsoons." Wiley, New York.

Garrett-Jones, S. E. (1979). "Holocene Vegetation and Lake Sedimentation in the Markham Valley, Papua New Guinea." Unpublished Ph.D. dissertation, Australian National University, Canberra.

——. (1980). Holocene vegetation change in lowland Papua New Guinea. *In* "Abstracts," p. 149. Fifth International Palynological Conference, Cambridge, U.K.

Gates, W. L. (1976). Modeling the ice age climate. *Science* 191, 1138-1144.

Haggblom, A. (1982). Driftwood in Svalbard as an indicator of sea-ice conditions. *Geografiska Annaler* 64A, 81-84.

Harrison, S. P. (1989). Lake levels and climatic change in eastern North America. *Climate Dynamics* 3, 157-167.

Hastenrath, S. (1984). "The Glaciers of Equatorial East Africa." D. Reidel, Dordrecht, The Netherlands.

Hope, G. S., Peterson, J. A., Radok, U., and Allison, I., Eds. (1976). "The Equatorial Glaciers of New Guinea." Balkema, Rotterdam, The Netherlands.

Hopkins, D. M. (1982). Aspects of the paleogeography of Beringia during the late Pleistocene. *In* "Paleoecology of Beringia" (D. M. Hopkins, J. V. Matthews, Jr., C. E. Schweger, and S. B. Young, Eds.), pp. 3-28. Academic Press, New York.

Huntley, B., and Prentice, I. C. (1988). July temperatures in Europe from pollen data, 6000 years before present. *Science* 241, 687-690.

Imbrie, J., and Imbrie, J. Z. (1980). Modeling the climatic response to orbital variations. *Science* 207, 943-953.

Keffer, T., Martinson, D. G., and Corliss, B. H. (1988). The position of the Gulf Stream during Quaternary glaciations. *Science* 241, 440-442.

Kutzbach, J. E. (1981). Monsoon climate of the early Holocene: Climate experiment using the earth's orbital parameters for 9000 years ago. *Science* 214, 59-61.

——. (1983). Monsoon rains of the late Pleistocene and early Holocene: Patterns, intensity and possible causes of changes. *In* "Variations in the Global Water Budget" (A. Street-Perrott, M. Beran, and R. Ratcliffe, Eds.), pp. 371-389. D. Reidel, Dordrecht, The Netherlands.

Kutzbach, J. E., and Gallimore, R. G. (1988). Sensitivity of a coupled atmosphere/mixed-layer ocean model to changes in orbital forcing at 9000 years B.P. *Journal of Geophysical Research* 93, 803-821.

Kutzbach, J. E., and Guetter, P. J. (1984). Sensitivity of late-glacial and Holocene climates to the combined effects of orbital parameter changes and lower boundary condition changes: "Snapshot" simulations with a general circulation model for 18-, 9-, and 6-ka B.P. *Annals of Glaciology* 5, 85-87.

——. (1986). The influence of changing orbital parameters and surface boundary conditions on climate simulations for the past 18,000 years. *Journal of the Atmospheric Sciences* 43, 1726-1759.

Kutzbach, J. E., and Street-Perrott, F. A. (1985). Milankovitch forcing of fluctuations in the level of tropical lakes from 18 to 0 kyr B.P. *Nature* 317, 130-134.

Lézine, A.-M., and Casanova, J. (1989). Pollen and hydrological evidence for the interpretation of past climates in tropical West Africa. *Quaternary Science Reviews* 8, 45-55.

McCulloch, D. S., and Hopkins, D. M. (1966). Evidence for an early recent warm interval in northwestern Alaska. *Geological Society of America Bulletin* 77, 1089-1108.

McCullough, M. T., De Deckker, P., and Chivas, A. R. (1989). Strontium isotope variations in single ostracod valves from the Gulf of Carpentaria, Australia: A palaeoenvironmental indicator. *Geochimica et Cosmochimica Acta* 53, 1703-1710.

McGlone, M. S., Kershaw, A. P., and Markgraf, V. (1993). El Niño/Southern Oscillation climatic variability in Australasian and South American paleoenvironmental records. *In* "Paleoclimate Aspects of El Niño/Southern Oscillation" (H. F. Diaz and V. Markgraf, Eds.), pp. 435-462. Cambridge University Press, Cambridge.

Maley, J. (1987). Fragmentation de la forêt dense humide africaine et extension des biotopes montagnards au Quaternaire récent: Nouvelles données polliniques et chronologiques. Implications paléoclimatiques et biogéographiques. *Palaeoecology of Africa* 18, 307-334.

Manabe, S., and Broccoli, A. J. (1985a). A comparison of climate model sensitivity with data from the last glacial maximum. *Journal of the Atmospheric Sciences* 42, 2643-2651.

——. (1985b). The influence of continental ice sheets on the climate of an ice age. *Journal of Geophysical Research* 90, 2167-2190.

Manabe, S., and Stouffer, R. J. (1988). Two stable equilibria of a coupled ocean-atmospheric model. *Journal of Climate* 1, 841-866.

Mitchell, J. F. B., Grahame, N. S., and Needham, K. J. (1988). Climate simulations for 9000 years before present: Seasonal variations and effect of the Laurentide ice sheet. *Journal of Geophysical Research* 93, 8283-8303.

Mix, A. C., Ruddiman, W. F., and McIntyre, A. (1986). Late Quaternary paleoceanography of the tropical Atlantic. 1: Spatial variability of annual mean sea-surface temperatures, 0-20,000 years B.P. *Paleoceanography* 1, 43-66.

Neftel, A., Oeschger, H., Schwander, J., Stauffer, B., and Zumbrunn, R. (1982). Ice core sample measurements give atmospheric CO_2 content during the past 40,000 yr. *Nature* 295, 220-223.

Osmaston, H. A. (1975). Models for the estimation of firnlines of present and Pleistocene glaciers. *In* "Processes in Physical and Human Geography" (R. Peel, M. Chisholm, and P. Haggett, Eds.), pp. 218-245. Heineman, London.

Petit-Maire, N. (1980). Holocene biogeographical variations along the northwestern African coast (28°-19°N): Paleoclimatic implications. *Palaeoecology of Africa* 12, 365-377.

——. (1986). Palaeoclimates in the Sahara of Mali: A multidisciplinary study. *Episodes* 9, 7-16.

Petit-Maire, N., Riser, J., Sabre, J., and Commelin, D. (1988). "Le Sahara à l'Holocène: Mali (carte 1:1,000,000)." Centre national de la recherche scientifique, Marseille-Luminy, France.

Prell, W. L. (1982). Reply to comments by P. J. Webster and N. A. Streten regarding "Surface circulation of the Indian

Ocean during the last glacial maximum, approximately 18,000 YBP." *Quaternary Research* 17, 128-131.

——. (1984). Variation of monsoonal upwelling: A response to changing solar radiation. *In* "Climate Processes and Climate Sensitivity" (J. E. Hansen and T. Takahashi, Eds.), pp. 48-57. Geophysical Monograph 29. American Geophysical Union, Washington, D.C.

——. (1985). "The Stability of Low-Latitude Sea-Surface Temperatures: An Evaluation of the CLIMAP Reconstruction with Emphasis on the Positive SST Anomalies." Contract Report, Contract No. DE-AC02-83ER60167. U.S Department of Energy, Washington, D.C.

Prell, W. L., and Kutzbach, J. E. (1987). Monsoon variability over the past 150,000 years: Comparison of observed and simulated paleoclimatic time series. *Journal of Geophysical Research* 92, 8411-8425.

Prell, W. L., and Van Campo, E. (1986). Coherent response of Arabian Sea upwelling and pollen transport to late Quaternary monsoonal winds. *Nature* 323, 526-528.

Raymo, M. E., Ruddiman, W. F., Shackleton, N. J., and Oppo, D. W. (1990). Evolution of Atlantic-Pacific $\delta^{13}C$ gradients over the last 2.5 m.y. *Earth and Planetary Science Letters* 97, 353-368.

Rind, D. (1987). Components of the ice-age circulation. *Journal of Geophysical Research* 92, 4241-4281.

Rind, D., and Peteet, D. (1985). Terrestrial conditions at the last glacial maximum and CLIMAP sea surface temperature estimates: Are they consistent? *Quaternary Research* 24, 1-22.

Rind, D., Peteet, D., Broecker, W. S., McIntyre, A., and Ruddiman, W. F. (1986). The impact of cold North Atlantic sea-surface temperatures on climate: Implications for the Younger Dryas cooling (11-10 k). *Climate Dynamics* 1, 3-33.

Ritchie, J. C. (1987). "Postglacial Vegetation of Canada." Cambridge University Press, Cambridge.

Ritchie, J. C., Cwynar, L. C., and Spear, R. W. (1983). Evidence from north-west Canada for an early Holocene Milankovitch thermal maximum. *Nature* 305, 126-128.

Ritchie, J. C., Eyles, C. H., and Haynes, C. V. Jr. (1985). Sediment and pollen evidence for an early to mid-Holocene humid period in the eastern Sahara. *Nature* 314, 352-355.

Rossignol-Strick, M., and Duzer, D. (1979). West African vegetation and climate since 22,500 B.P. from deep-sea cores palynology. *Pollen et spores* 21, 105-134.

Ruddiman, W. F., and McIntyre, A. (1981). The North Atlantic Ocean during the last deglaciation. *Palaeogeography, Palaeoclimatology, Palaeoecology* 35, 145-214.

Ruddiman, W. F., Sancetta, C. D., and McIntyre, A. (1977). Glacial/interglacial response rate of subpolar North Atlantic surface waters to climatic change: The record in oceanic sediments. *Philosophical Transactions of the Royal Society of London Ser. B* 280, 119-142.

Ruddiman, W. F., Raymo, M. E., Martinson, D. G., Clement, B. M., and Backman, J. (1989). Pleistocene evolution: Northern Hemisphere ice sheets and North Atlantic Ocean. *Paleoceanography* 4, 353-412.

Saigne, C., and Legrade, M. (1987). Measurements of methane-sulphonic acid in Antarctic ice. *Nature* 330, 240-242.

Salvigsen, O., and Osterholm, H. (1982). Radiocarbon-dated raised beaches and glacial history of the northern coast of Spitsbergen, Svalbard. *Polar Research* 1, 97-115.

Sarnthein, M., Tetzlaff, G., Koopmann, B., Wolter, K., and Pflaumann, U. (1981). Glacial and interglacial wind regimes over the eastern subtropical Atlantic and north-west Africa. *Nature* 293, 193-196.

Schultz, E. (1987). Die holozäne Vegetation der zentralen Sahara (N-Mali, N-Niger, SW-Libyen). *Palaeoecology of Africa* 18, 143-161.

Scott, L. (1989). Climatic conditions in southern Africa since the last glacial maximum inferred from pollen analysis. *Palaeogeography, Palaeoclimatology, Palaeoecology* 70, 340-353.

Spitaler, R. (1921). "Das Klima der Eiszeitalters." R. Spitaler, Prague.

Stauffer, B., Lochbronner, E., Oeschger, H., and Schwander, J. (1988). Methane concentration in the glacial atmosphere was only half of the pre-industrial Holocene. *Nature* 332, 812-814.

Stewart, T. G., and England, J. (1983). Holocene sea-ice variations and paleoenvironmental change, northernmost Ellesmere Island, N.W.T., Canada. *Arctic and Alpine Research* 15, 1-17.

Street, F. A. (1979). Late Quaternary precipitation estimates for the Ziway-Shala Basin, southern Ethiopia. *Palaeoecology of Africa* 11, 135-143.

Street, F. A., and Grove, A. T. (1979). Global maps of lake-level fluctuations since 30,000 yr B.P. *Quaternary Research* 12, 83-118.

Street-Perrott, F. A., and Perrott, R. A. (1990). Abrupt climatic changes in the tropics—An Atlantic Ocean feedback mechanism. *Nature* 343, 607-612.

Street-Perrott, F. A., and Roberts, N. (1983). Fluctuations in closed-basin lakes as an indicator of past atmospheric circulation patterns. *In* "Variations in the Global Water Budget" (A. Street-Perrott, M. Beran, and R. Ratcliffe, Eds.), pp. 331-345. D. Reidel, Dordrecht, The Netherlands.

Street-Perrott, F. A., Marchand, D. S., Roberts, N., and Harrison, S. P. (1989). "Global Lake-Level Variations from 18,000 to 0 Years Ago: A Palaeoclimatic Analysis." Technical Report TR046. U.S. Department of Energy, Washington, D.C.

Street-Perrott, F. A., Mitchell, J. F. B., Marchand, D. S., and Brunner, J. S. (1990). Milankovitch and albedo forcing of the tropical monsoons: A comparison of geological evidence and numerical simulations for 9,000 y BP. *Transactions of the Royal Society of Edinburgh, Earth Sciences* 81, 407-427.

Swain, A. M., Kutzbach, J. E., and Hastenrath, S. (1983). Estimates of Holocene precipitation for Rajasthan, India, based on pollen and lake-level data. *Quaternary Research* 19, 1-17.

Thompson, L. G., and Mosley-Thompson, E. (1981). Temporal variability of microparticle properties in polar ice sheets. *Journal of Volcanic and Geothermal Research* 11, 11-27.

Torgerson, T., Luly, J., De Deckker, P., Jones, M. R., Searle, D. E., Chivas, A. R., and Ullman, W. J. (1988). Late Quaternary environments of the Carpentaria Basin, Australia. *Palaeogeography, Palaeoclimatology, Palaeoecology* 67, 245-261.

Van Campo, E., Duplessy, J.-C., and Rossignol-Strick, M. (1982). Climatic conditions deduced from a 150-kyr oxygen isotope-pollen record from the Arabian Sea. *Nature* 296, 56-59.

van Zinderen Bakker, E. M., Sr., and Coetzee, J. A. (1972). A reappraisal of late Quaternary climatic evidence from tropical Africa. *Palaeoecology of Africa* 7, 151-181.

Walker, D., and Flenley, J. R. (1979). Late Quaternary vegetational history of the Enga District of upland Papua New

Guinea. *Philosophical Transactions of the Royal Society of London Ser. B* 286, 265-344.

Webb, T. III, and Bartlein, P. J. (1988). Late Quaternary climatic change in eastern North America: The role of modeling experiments and empirical studies. *Bulletin of the Buffalo Society of Natural Sciences* 33, 3-13.

Webb, T. III, Bartlein, P. J., and Kutzbach, J. E. (1987). Climatic change in eastern North America during the past 18,000 years: Comparisons of pollen data with model results. *In* "North America and Adjacent Oceans during the Last Deglaciation" (W. F. Ruddiman and H. E. Wright, Jr., Eds.), pp. 447-462. The Geology of North America, Vol. K-3. The Geological Society of America, Boulder, Colo.

Webster, P. J., and Streten, N. A. (1978). Late Quaternary ice age climates of tropical Australasia: Interpretations and reconstructions. *Quaternary Research* 10, 279-309.

Williams, J. R. G., Barry, R. G., and Washington, W. M. (1974). Simulation of the atmospheric circulation using the NCAR global circulation model with ice age boundary conditions. *Journal of Applied Meteorology* 13, 305-317.

CHAPTER 20

Epilogue

J. E. Kutzbach, P. J. Bartlein, I. C. Prentice, W. F. Ruddiman, F. A. Street-Perrott, T. Webb III, and H. E. Wright, Jr.

This book presents the results of a sequence of paleoclimatic modeling experiments and compares them with paleoenvironmental data. We used the modeling experiments to illustrate how slowly changing external and internal boundary conditions have induced major changes in global climates since the last glacial maximum. We also compiled and described the available data sets and introduced methods to allow clear comparisons of the data and the model results. New developments now make it important to repeat the experiments with improved boundary conditions and upgraded, higher-resolution climate models, to expand the data sets, and to improve the methods for data-model comparisons.

Fairbanks (1989) and Bard *et al.* (1990) recently published results that may require key modifications to the boundary conditions used in the COHMAP-CLIMAP experiments (Fig. 20.1). Fairbanks (1989) used dated cores from submerged coral reefs off Barbados to establish a new sea-level curve that helps constrain estimates for the size of the ice sheets used as boundary conditions in the climate-model experiments. Using high-resolution uranium-series (U) dates from the same Barbados cores, Bard *et al.* (1990) found evidence that radiocarbon (^{14}C) dates are anomalously young by amounts ranging from less than 1000 yr at 9000 ^{14}C yr B.P. to more than 3000 yr at and beyond 15,000 ^{14}C yr B.P. (Fig. 20.1). If these new U dates are correct, then the original COHMAP climate-model experiments described in Kutzbach and Guetter (1986), Kutzbach and Ruddiman (this vol.), and Kutzbach, Guetter, *et al.* (this vol.) involved boundary conditions (both ice-sheet size and astronomically determined insolation) that became increasingly mismatched back through the deglaciation (Fig. 20.2).

These findings, combined with other modeling and data developments, open up several major research opportunities. The first opportunity is to revise the previous set of boundary conditions and to create a new standard set. The revised ice-sheet sizes and ^{14}C corrections can be used to bring the geologically determined boundary conditions in line with astronomically determined insolation values for each date modeled (Table 20.1 and Fig. 20.2). Some revision is also needed for the other boundary conditions, including the original sea-surface temperatures (SSTs) (CLIMAP Project Members, 1981) for the glacial maximum.

A second opportunity is to rerun the original experiments with the revised boundary conditions and new climate models, such as those being developed at the National Center for Atmospheric Research (NCAR). For example, the new NCAR Community Climate Model 2 (CCM 2) will have significantly improved spatial resolution compared to previous models and better treatment of the physics of atmospheric circulation and of surface processes such as vegetation, soil moisture, and runoff. The added spatial resolution of the new model simulations will offer many new opportunities for comparisons with the geologic data.

A third opportunity is to use the newly revised "standard" set of boundary conditions to promote comparisons of paleoclimatic simulations generated by different climate models for the same paleoclimatic boundary conditions. Paleoclimatic simulations from different models differ significantly at a broad scale, and comparisons of the model results with paleoclimatic data will permit evaluation of the relative sensitivity and biases of the models. Comparative analyses can also be done of the feedback mechanisms associated with forcing from orbital factors and glacial-age boundary conditions relative to the forcing associated with future greenhouse-gas scenarios.

A fourth opportunity is to combine regional paleoenvironmental data sets, such as those described in

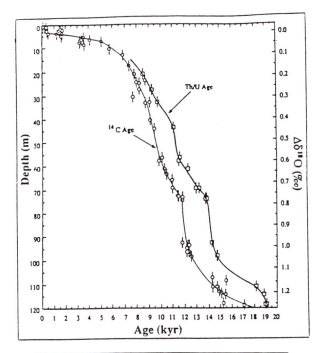

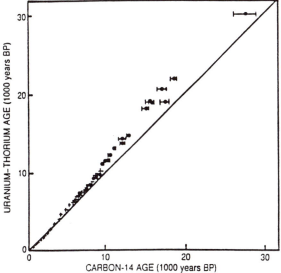

Fig. 20.1. New deglacial sea-level chronology (a) and plot of age offsets (b). Circles in part a show a sea-level curve in meters based on radiocarbon dating (^{14}C age) of submerged corals from sites near Barbados (Fairbanks, 1989). Data have been corrected for the mean Barbados uplift rate of 34 cm/ka. Squares indicate uranium-series (^{230}Th/^{234}U) redating of the Barbados sea-level curve by Bard *et al.* (1990). Error bars on dates are smaller than the symbols shown. Estimated oxygen-isotope changes equivalent to sea-level changes are shown at right. The chronologies partially converge at the bottom of the sequence because the oldest ^{14}C dates are ignored. This convergence is not apparent in plot b (Bard *et al.,* 1990), where extra dates earlier than 20,000 yr B.P. are shown. Part a from Fairbanks (1989); reprinted by permission from *Nature* vol. 342, p. 639, copyright 1989 Macmillan Magazines Ltd. Part b from Bard *et al.* (1990); reprinted by permission from *Nature* vol. 345, p. 407, copyright 1990 Macmillan Magazines Ltd.

this book, to form continental and hemisphere-wide data bases. Such broad-scale data bases would allow more efficient and extensive data-model and model-model comparisons, facilitate broad searches for modern analogues for fossil assemblages, and aid in comparative studies of vegetation dynamics in different regions.

With these improvements in the boundary conditions, data sets, and models, increasingly accurate studies of regional-scale climate dynamics and the climatic impacts on ecosystems can be anticipated. The detailed model output from such new experiments may permit analysis of how systems controlled by climate (e.g., lakes, soils, terrestrial and marine ecosystems, and even human populations) have responded to the climatic changes since the last glacial maximum.

The model output can also be used as input to physiologically based biome models (Prentice *et al.,* 1989) to simulate past vegetation on a global scale. The availability of biome models will significantly extend the scope of data-model comparisons, and eventually asynchronous coupling of a biome model to a climate model should allow researchers to incorporate vegetation feedback into models of the climate system.

New Boundary Conditions

The new sea-level data from Fairbanks (1989) and Bard *et al.* (1990) provide a strong impetus for revision of the boundary conditions used for the experiments discussed in this book (Fig. 20.1). In the decade that has passed since CLIMAP (1981) published its glacial-maximum reconstruction, numerous groups have used the estimates in simulations and sensitivity tests of the last glacial maximum. A substantial revision of glacial and deglacial boundary conditions is now warranted for several reasons. First, Quaternary researchers need more realistic model-based climate simulations for comparison with data-based climatic reconstructions (Webb, Ruddiman, *et al.,* this vol.). Second, the development of higher-resolution climate models permits (and requires) more accurate specification of boundary conditions. And third, the proliferation of modeling groups using glacial and deglacial boundary conditions has expanded opportunities for comparisons among models, which are useful for validating model performance.

The revised boundary conditions could become the definitive input to climate-model experiments conducted by several international modeling groups over the next 5 yr or so. Table 20.1 lists five deglacial time levels that are prime for study. The series of experiments envisioned in Table 20.1 would focus mainly on the definition of (1) the smooth envelope of climatic forcing provided by insolation and by changes in ice-sheet volume, atmospheric CO_2, and SST at time steps of several thousand years and (2) the effects of these two kinds of forcing on regional climatic

A) PREVIOUS EXPERIMENTS:
    ASSUME ^{14}C yr BP = ORBITAL yr BP

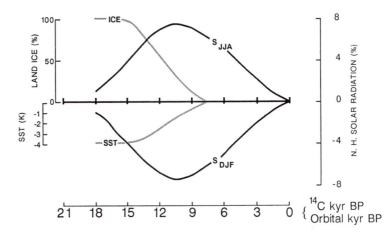

B) PLANNED EXPERIMENTS:
    ASSUME U yr BP = ORBITAL yr BP

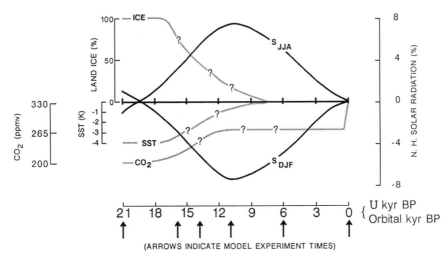

(ARROWS INDICATE MODEL EXPERIMENT TIMES)

Fig. 20.2. Schematic diagram of major changes since 21,000 U yr B.P. in external forcing (Northern Hemisphere solar radiation in June–August [S$_{JJA}$] and December–February [S$_{DJF}$], as percentage difference from the present) and internal boundary conditions (land ice [ICE] as a percentage of the glacial maximum, global mean annual sea-surface temperature (SST) as departure from the present, and atmospheric CO$_2$ concentration [ppmv]). The top figure (a) assumes that ^{14}C years, used to specify internal boundary conditions, are equivalent to orbital years (absolute, astronomic time scale), as assumed in the initial set of COHMAP experiments. The bottom figure (b) corrects ^{14}C years to the new calibration based on the U time scale (see also Table 20.1 and Fig. 20.1). This new relationship can be the basis for new experiments. The internal boundary conditions sketched in part b are based on the values shown in part a, but age-corrected. The question marks indicate that the actual values should be reviewed. The main initial set of COHMAP experiments (a) did not include CO$_2$ variations, so the schematic CO$_2$ curve is shown in part b only. The times of possible new model experiments and data-model comparisons are indicated with arrows at the bottom.

changes. This approach directly addresses two key climate-system components (external and internal) that were responsible for moving regional climatic responses from a glacial to an interglacial mode.

By focusing on the broad envelope of climatic forcing and regional responses, one avoids certain complications inherent in the study of abrupt (less than 1,000-yr) climatic changes and meltwater-influx episodes (e.g., Boyle and Keigwin, 1987; Jones and Keigwin, 1988; Broecker *et al.,* 1989; Jansen and Veum,

1990). These abrupt changes were particularly significant during the time of rapid deglaciation.

Once the "envelope" experiments are completed (Table 20.1), researchers will be in a better position to design sensitivity tests to elucidate the still unexplained Younger Dryas cooling at 11,000–10,000 ^{14}C yr B.P. (Rind *et al.,* 1986; Oglesby *et al.,* 1989; Overpeck *et al.,* 1989). Further sensitivity tests should also be run on the climatic effects of various ice-sheet heights and shapes that existed during deglaciation, especially for times when the

Table 20.1. Choice of time levels for new modeling experiments (ages approximate)

U yr B.P.	¹⁴C yr B.P.	Reasons for selection of level
6000	6000	Ice sheets mostly melted; insolation 6% higher (lower) in northern summer (winter) compared to present; test of "pure" insolation effect.
11,000	9000	Ice sheets much smaller (about 35% of glacial-age volume); insolation near maximum difference (about 8% higher [lower] in northern summer [winter] compared to present); monsoonal responses at a maximum.
14,000	12,000	Ice sheets smaller (about 60% of glacial-age volume); period immediately preceding Younger Dryas; insolation near maximum difference (about 6% higher [lower] in northern summer [winter] compared to present).
16,000	13,500	Ice sheets melting (about 85% of glacial-age volume); insolation changing rapidly (4% higher [lower] in northern summer [winter] compared to present).
21,000	18,000	Ice sheets at maximum size; insolation near modern values (at solstices), with the phase of precession almost identical to present and tilt slightly less than present; test of "pure" ice-sheet effect.

estimates are uncertain (Fairbanks, 1989). Ultimately, a full treatment of climate change over this time span will incorporate both gradual and abrupt processes, which will require fully coupled ice-ocean-atmosphere models that can handle transient responses.

New Models and Planned Experiments

The original set of COHMAP experiments using CCM 0 was completed in 1985 (Kutzbach and Guetter, 1986; COHMAP Members, 1988). CCM 0 contained many simplifications compared to current models. It simulated "perpetual" January and July conditions rather than the complete seasonal cycle, and it used prescribed values for soil moisture, snow cover, sea ice, and SST.

Significantly improved models, including CCM 1 and CCM 2, are now available or are being developed. CCM 1 simulates the full seasonal cycle, has interactive soil moisture and snow cover, and is coupled to a mixed-layer ocean with interactive temperature and (thermodynamic) sea ice (Covey and Thompson, 1989). A newer version of this model will have improved land-surface process parameterizations.

CCM 2 has significantly better horizontal resolution than CCM 1 and CCM 0. It uses spectral wave-form representations allowing as many as 42 waves around a latitude circle and corresponding to a resolution of approximately 2.8° latitude by 2.8° longitude, or about 280 km by 280 km in middle latitudes. CCM 0 and CCM 1 employ only 15 waves, corresponding to a resolution of approximately 4.4° latitude by 7.5° longitude, or about 440 km by 750 km in middle latitudes. CCM 2 also has 15 vertical levels, improved cloud and convective schemes, an improved boundary-layer scheme, semi-Lagrangian transport of water vapor, and a diurnal-cycle capability. One version of CCM 2 will have an improved land-surface process package, a version of the biosphere-atmosphere transfer scheme (BATS) of Dickinson et al. (1986).

The availability of these or other new models with improved treatments of vegetation and hydrology is an important development for paleoclimate studies because paleovegetation and paleohydrologic indicators are used in data-model comparisons. With these advances CCM 1 and CCM 2 should provide significantly better simulations and more opportunities to explore internal climate feedback mechanisms than CCM 0.

The following subsections highlight some of the research topics and issues that can be explored in more depth with more sophisticated climate models, revised boundary conditions, and improved observational data sets.

ICE SHEETS

CCM 0 was used to simulate the glacial-maximum climate with CLIMAP's (1981) minimum and maximum estimates of glacial-age ice; the two sets of ice-sheet heights differed by about 25% (Kutzbach and Ruddiman, this vol.). Different specifications of ice-sheet height produced significant differences in the simulated location of jet streams, storm tracks, and planetary waves. Land-surface climate and precipitation-minus-evaporation (P-E) fluxes over the ocean also changed significantly. Shinn and Barron (1989) found a similar sensitivity of the simulated climate to ice-sheet size and shape.

Revised ice-sheet heights will be a critical new component of glacial-age simulations. The higher horizontal and vertical resolution of CCM 2 compared to CCM 0 and CCM 1 should allow more realistic prescriptions of ice-sheet topography and provide more accurate simulation of the atmospheric dynamics associated with flow over and around the ice sheets.

DEGLACIATION PHASING

Introduction of the revised time scale (U vs. ¹⁴C) (Fig. 20.2) will produce different phase relations for orbital forcing compared to both the lower boundary forcing and the observed climatic response in various regions of the globe. These differences, amounting to several thousand years in the early deglaciation period, can now be addressed.

ADDITIONAL VARIABLES

The availability of additional variables simulated in CCM 1 and CCM 2 (but not in CCM 0)—the seasonal cycle, soil moisture, snow cover, sea ice, SST—will encourage the development of new process models for interpreting paleoenvironmental evidence. In studies of midcontinental aridity and tropical wetness, the effects of changes in soil moisture, runoff, P-E, and potential evapotranspiration can now be considered (Delworth and Manabe, 1988; Rind *et al.,* 1990). At high latitudes, new kinds of comparisons of the observations with the simulations of climatic features such as snow cover and Arctic sea ice will also be possible.

FEEDBACK MECHANISMS

The period around 6000 yr B.P. provides the opportunity to study some of the types of feedback that may accompany increases in atmospheric CO_2. Of course doubled CO_2 induces a change in downward longwave radiation (about 4 W/m²) and is therefore different from the initial orbitally induced change in downward shortwave radiation (about 10-20 W/m²), not only in its magnitude but also in its seasonal and latitudinal distribution (Gallimore and Kutzbach, 1989; Mitchell, 1990). On the other hand, the effects of both the CO_2 changes and the orbital changes are either amplified or damped according to the actions of internal feedback mechanisms associated with water vapor, clouds, soil moisture, snow cover, and sea ice. Feedback mechanisms associated with vegetation changes can also be examined (Street-Perrott *et al.,* 1990). Moreover, land and ocean respond differentially to both orbital changes (Kutzbach and Gallimore, 1988; Mitchell *et al.,* 1988) and CO_2 changes (Washington and Meehl, 1989; Manabe *et al.,* 1990). The role of feedback mechanisms and responses can be studied in new experiments for 6000 yr B.P. Because the observed climate at 6000 yr B.P. has already been described from paleoenvironmental records, the accuracy of the model's treatment of various processes can be assessed.

New Model-Model Comparisons

We see a major opportunity to promote the use of a standard set of boundary conditions for times like 6000 and 18,000 yr B.P. and to compare paleoclimatic simulations from different models that use the identical boundary conditions. The need for comparisons among models is growing now that atmospheric models are being coupled to ocean, land-surface, and ice-sheet models.

This need arises because differences in the atmospheric simulations directly influence the simulations of the other components of the climate system. For example, surface fluxes from the atmospheric simulations at the glacial maximum influence the forcing of the glacial ocean (wind stress, atmospheric temperature and heat flux, freshwater flux). Glacial simulations generated by atmospheric general circulation models are probably not yet accurate enough to provide accurate surface fluxes to a glacial-age ocean model. By studying the one-way forcing (atmosphere to ocean) before attempting full two-way coupling of atmosphere and ocean, researchers can learn how sensitive the dynamic ocean models are to the input forcing at glacial times.

New Data Sets and Techniques for Data-Model Comparisons

Pollen, macrofossils, and lake-level fluctuations are the primary data sets described in this book and used for paleoclimatic reconstructions and comparisons with model results, primarily because these data sets are large and well-dated and can yield quantitative estimates of climate variables. Many paleoecologists around the world have contributed to the efforts toward a global synthesis. The distribution of data is highly variable, however. Few studies have been completed in the tropical rainforests, for example, largely because of the paucity of suitable sites for pollen analysis. Recent work in Amazonia suggests that the commonly accepted picture of dry conditions during the last glacial maximum may be less important than reduced temperatures in modifying the vegetation cover (Bush *et al.,* 1990).

In areas where the primary data are thin or do not provide quantitative estimates of climatic variables, other approaches need to be considered. For example, stratigraphic oxygen-isotope analysis of lake sediments can lead to estimates of late-glacial changes in atmospheric temperature in lakes with a short residence time (and thus minimal evaporation) (Siegenthaler and Eicher, 1986). Or in closed-basin lakes with high evaporation, it can reveal changes in salinity (Stuiver, 1970). Salinity can also be reconstructed from diatom analyses (Fritz, 1991), ostracod analyses (Smith, 1991), and trace-metal analyses of ostracod shells (Engstrom and Nelson, 1991).

Another data set that needs to be enlarged and evaluated pertains to the persistent problem of the disparity during the last glacial maximum between the relatively stable SST in tropical areas and the inferred deep depression of the snowline and treeline in tropical mountains (Rind and Peteet, 1985). Recent work in East Africa (Bonnefille *et al.,* 1990) based on pollen transfer functions implies that the discrepancy may not be as large as previous estimates (see discussion in Webb, Ruddiman, *et al.,* this vol.). The same conclusion is reached for snowline depressions in the Bolivian Andes, where precipitation is believed to be more important than temperature (Seltzer, 1990). More studies in tropical and subtropical mountains are needed to address this question.

Huntley and Prentice (this vol.) and Webb, Bartlein, *et al.* (this vol.) describe the general procedure for comparing paleoclimatic simulations with the fossil pollen record. The need is apparent (1) to develop further the methods for illustrating comparisons between data and models; (2) to integrate the major sets of regional pollen, macrofossil, and lake-level data to create continental-scale data sets for data-model comparisons; (3) to augment existing data sets and techniques to allow the simulation of vegetation patterns at the global scale and their comparison with the fossil record; and (4) to develop data sets of paleolimnological indicators and of peatland distributions for paleoclimatic analysis.

In addition to paleoenvironmental data, the comparison of paleoclimatic simulations with the fossil record requires methods for establishing the link between the present climate and the various environmental systems with fossil records (e.g., vegetation, lake hydrology). In practice, this link is applied to climate-model output to translate it into estimates of, say, the distribution or abundance of plants or the status of lakes. The predicted patterns can then be compared with those distilled from the fossil record. Alternatively, the link can be applied to the fossil record in order to interpret it in climatic terms. Climate reconstructions can then be directly compared with paleoclimatic simulations.

Two classes of vegetation models can be used for this purpose: static and dynamic (Prentice and Solomon, 1991). Dynamic vegetation models, those that describe the time-dependent variations in plant composition or structure, are not yet available for global-scale simulations, although efforts to build them are under way (Prentice *et al.,* 1989).

The class of static vegetation models includes the response surface approach, which we have used to compare paleoclimatic simulations with paleoecological records in eastern North America and Europe, as well as biome models (see review by Prentice and Solomon [1991]). When calibrated with data on relative abundance, response surfaces show how the relative abundance of a particular taxon varies as a function of a few climatic variables. When applied to mapped vegetation range data, response surfaces can show how the probability of occurrence, or incidence, of a taxon varies with a few climatic variables. Although we have not yet done so, it is technically feasible to develop response surfaces with a global set of climate and vegetation data. Because the response surfaces must be fitted with modern climate and vegetation data, their application under climatic conditions without modern analogues is subject to some uncertainty.

The same uncertainty surrounds the application of biome models, such as the Holdridge scheme, that rely on map comparison or screening of climate data to determine the limiting values of certain climatic variables. Prentice *et al.* (in press) describe a biome model that is based on ecophysiological limits determined for different plant functional forms. Although vegetation dynamics are still not represented in this model, its physiological basis implies that its application to past climate systems without modern analogues would be more appropriate than would the application of response surfaces.

Data-model comparisons using either response surfaces or a biome model would be implemented by applying either model to a particular paleoclimatic simulation to examine the global-scale pattern of individual taxa, plant forms, or biomes. In regions with extensive sets of fossil data, such as North America, the simulated vegetation distributions could be compared quantitatively with the fossil distributions. In other regions where only the broad patterns of past vegetation can be estimated from the fossil data, qualitative comparisons would still be possible.

Conclusions

The continuing improvements in climate models, boundary conditions, observational data sets, and techniques for interpreting the data and model output present a rich array of new opportunities for continued research into the causes and processes of climatic change. Both the new data and the refined model results will increase our understanding of how changes in large-scale controls of the climate system govern the regional-scale climate response.

References

Bard, E., Hamelin, B., and Fairbanks, R. G. (1990). U/Th ages obtained by mass spectrometry in corals from Barbados: Sea level during the past 130,000 years. *Nature* 346, 456-458.

Bonnefille, R., Roeland, J. C., and Guiot, J. (1990). Temperature and rainfall estimates for the past 40,000 years in equatorial Africa. *Nature* 346, 347-349.

Boyle, E. A., and Keigwin, L. (1987). North Atlantic thermohaline circulation during the past 20,000 years linked to high-latitude surface temperature. *Nature* 330, 35-40.

Broecker, W. S., Kennett, J. P., Flowers, B. P., Teller, J., Trumbore, S., Bonani, G., and Wolfli, W. (1989). The routing of Laurentide ice-sheet meltwater during the Younger Dryas cold event. *Nature* 341, 318-321.

Bush, M. B., Colinvaux, P. A., Wiemann, H. C., Piperno, D. R., and Liu, K.-B. (1990). Late Pleistocene temperature depression and vegetation change in Ecuadorian Amazonia. *Quaternary Research* 34, 330-345.

CLIMAP Project Members (1981). Seasonal reconstructions of the earth's surface at the last glacial maximum. *Geological Society of America Map and Chart Series* MC-36.

COHMAP Members (1988). Climatic changes of the last 18,000 years: Observations and model simulations. *Science* 241, 1043-1052.

Covey, C., and Thompson, S. L. (1989). Testing the effects of ocean heat transport on climate. *Palaeogeography, Palaeoclimatology, Palaeoecology* 75, 331-341.

Delworth, T. L., and Manabe, S. (1988). The influence of potential evaporation on the variabilities of simulated soil wetness and climate. *Journal of Climate* 1, 523-547.

Dickinson, R. E., Henderson-Sellers, A., Kennedy, P. J., and Wilson, M. F. (1986). "Biosphere-Atmosphere Transfer Scheme (BATS) for the NCAR Community Climate Model." NCAR Technical Note 275+STR. National Center for Atmospheric Research, Boulder, Colo.

Engstrom, D. R., and Nelson, S. (1991). Paleosalinity from trace metals in fossil ostracodes compared with observational records at Devils Lake, North Dakota. *Palaeogeography, Palaeoclimatology, Palaeoecology* 83, 295-312.

Fairbanks, R. G. (1989). A 17,000-year glacio-eustatic sea level record: Influence of glacial melting rates on the Younger Dryas event and deep-ocean circulation. *Nature* 342, 637-642.

Fritz, S. C. (1991). Twentieth century water-level and salinity fluctuations in Devils Lake, N.D.: A test of diatom-based transfer functions. *Limnology and Oceanography* 35, 1771-1781.

Gallimore, R. G., and Kutzbach, J. E. (1989). Effects of soil moisture on the sensitivity of a climate model to earth orbital forcing at 9000 yr B.P. *Climatic Change* 14, 175-205.

Jansen, E., and Veum, T. (1990). Evidence for two-step deglaciation and its impact on North Atlantic deep-water circulation. *Nature* 343, 612-616.

Jones, G. A., and Keigwin, L. D. (1988). Evidence from Fram Strait (78°N) for early deglaciation. *Nature* 336, 56-59.

Kutzbach, J. E., and Gallimore, R. G. (1988). Sensitivity of a coupled atmosphere/mixed-layer ocean model to changes in orbital forcing at 9000 years B.P. *Journal of Geophysical Research* 93, 803-821.

Kutzbach, J. E., and Guetter, P. J. (1986). The influence of changing orbital parameters and surface boundary conditions on climate simulations for the past 18,000 years. *Journal of the Atmospheric Sciences* 43, 1726-1759.

Manabe, S., Bryan, K., and Spelman, M. J. (1990). Transient response of a global ocean-atmosphere model to a doubling of atmospheric carbon dioxide. *Journal of Physical Oceanography* 20, 722-749.

Mitchell, J. F. B. (1990). Greenhouse warming: Is the mid-Holocene a good analogue? *Journal of Climate* 3, 1177-1192.

Mitchell, J. F. B., Grahame, N. S., and Needham, K. J. (1988). Climate simulations for 9000 years before present: Seasonal variations and effect of the Laurentide ice sheet. *Journal of Geophysical Research* 93, 8283-8303.

Oglesby, R. J., Maasch, K. A., and Saltzman, B. (1989). Glacial meltwater cooling of the Gulf of Mexico: GCM implications for Holocene and present day climates. *Climate Dynamics* 3, 115-133.

Overpeck, J. T., Peterson, L. C., Kipp, N., Imbrie, J., and Rind, D. (1989). Climatic change in the low-latitude North Atlantic region during the last deglaciation. *Nature* 338, 553-557.

Prentice, I. C., and Solomon, A. M. (1991). Vegetation models and global change. *In* "Global Changes of the Past" (R. S. Bradley, Ed.), pp. 365-383. University Corporation for Atmospheric Research, Office of Interdisciplinary Earth Sciences, Boulder, Colo.

Prentice, I. C., Webb, R. S., Ter-Mikhaelian, M. T., Solomon, A. M., Smith, T. M., Pitovranov, S. E., Nikolov, N. T., Minin, A. A., Leemans, R., Lavorel, S., Korzukhin, M. D., Hrabovszky, F. P., Helmisaari, H. O., Harrison, S. P., Emanuel, W. R., and Bonan, G. B. (1989). "Developing a Global Vegetation Dynamics Model: Results of an IIASA Summer Workshop." International Institute of Applied Systems Analysis (IIASA), Laxenburg, Austria.

Prentice, I. C., Cramer, W., Harrison, S. P., Leemans, R., Mouserud, R. A., and Solomon, A. M. (in press). Predicting global vegetation patterns from plant physiology and dominance, soil properties and climate. *Journal of Biogeography.*

Rind, D., and Peteet, D. (1985). Terrestrial conditions at the last glacial maximum and CLIMAP sea-surface temperature estimates: Are they consistent? *Quaternary Research* 24, 1-22.

Rind, D., Peteet, D., Broecker, W., McIntyre, A., and Ruddiman, W. F. (1986). The impact of cold North Atlantic sea-surface temperatures on climate: Implications for the Younger Dryas cooling (11-10 k). *Climate Dynamics* 1, 3-33.

Rind, D., Goldberg, R., Hansen, J., Rosenzweig, C., and Ruedy, R. (1990). Potential evapotranspiration and the likelihood of future drought. *Journal of Geophysical Research* 95, 9983-10,004.

Seltzer, G. O. (1990). Recent glacial history and paleoclimate of the Peruvian-Bolivian Andes. *Quaternary Science Reviews* 9, 137-152.

Shinn, R. A., and Barron, E. J. (1989). Climate sensitivity to continental ice sheet size and configuration. *Journal of Climate* 2, 1517-1537.

Siegenthaler, U., and Eicher, U. (1986). Stable oxygen and carbon isotope analyses. *In* "Handbook of Holocene Palaeoecology and Palaeohydrology" (B. Berglund, Ed.), pp. 407-423. Wiley, New York.

Smith, A. J. (1991). "Lacustrine Ostracodes as Paleohydrological Indicators in Holocene Lake Records of the North-Central United States." Unpublished Ph.D. thesis, Brown University, Providence, R.I.

Street-Perrott, F. A., Mitchell, J. F. B., Marchand, D. S., and Brunner, J. S. (1990). Milankovitch and albedo forcing of the tropical monsoons: A comparison of geologic evidence and numerical simulations for 9,000 yr B.P. *Transactions of the Royal Society of Edinburgh, Earth Sciences* 81, 407-427.

Stuiver, M. (1970). Oxygen and carbon isotope ratios of freshwater carbonates as climatic indicators. *Journal of Geophysical Research* 75, 5247-5257.

Washington, W. M., and Meehl, G. A. (1989). Climate sensitivity due to increased CO_2: Experiments with a coupled atmosphere and ocean general circulation model. *Climate Dynamics* 4, 1-38.

Notes on Contributors

About the Editors

Patrick J. Bartlein is associate professor in the Department of Geography the University of Oregon. He completed his Ph.D. at the University of Wisconsin-Madison in 1978. His current research interests focus on the controls of regional paleoclimatic variations and methods for analyzing paleoecological data.

John E. Kutzbach is director of the Center for Climatic Research, professor in the Department of Atmospheric and Oceanic Studies, and Environmental Studies, and Plaenart-Bascom Professor of Liberal Arts at the University of Wisconsin-Madison. He is also a summer visiting scientist at the National Center for Atmospheric Research in Boulder, Colorado. He obtained his Ph.D. in meteorology from the University of Wisconsin in 1966.

William F. Ruddiman received a B.A. in geology from Williams College in 1964 and a Ph.D. in marine geology from Columbia University in 1969. He has worked at the Lamont-Doherty Geological Observatory of Columbia University and the U.S. Naval Oceanographic Office and is currently at the University of Virginia. His research interests have focused on ice ages and the reasons for them and on glacial-interglacial cycles.

F. Alayne Street-Perrott is project leader in paleoclimatology at the Environmental Change Unit of the University of Oxford. She was educated at Cambridge University and the University of Colorado. Her main research interests are tropical paleolimnology and paleoclimatology.

Thompson Webb III obtained his Ph.D. in meteorology from the University of Wisconsin-Madison in 1971 and currently is a professor of geological sciences at Brown University. He has been active in COHMAP research since 1974. His research has focused on the climatic and vegetational interpretation of the pollen record on scales ranging from a square kilometer to global.

H. E. Wright, Jr. holds B.A. and Ph.D. degrees from Harvard University. He has been a member of the faculty of the Department of Geology and Geophysics at the University of Minnesota since 1947, most recently as Regents' Professor of Geology, Ecology, and Botany and director of the Limnological Research Center. Currently retired, he continues to teach part-time and advise graduate students. His research has focused on the natural and cultural history of landscapes in North America, the Andes, and the Near East during and since the last glacial period. He has edited or coedited several books on the Quaternary history of North America, most recently *The Patterned Peatlands of Minnesota.*

About the Other Contributors

Katherine H. Anderson obtained her master's degree from Moss Landing Marine Laboratories in 1985. For the last six years she has been developing analysis techniques and analyzing pollen data for use in paleoclimatic reconstructions and data-model comparisons. She is currently associated with Brown University.

Patricia M. Anderson received her M.A. and Ph.D. degrees from Brown University, where she pursued an interdisciplinary program in palynology and archaeology. Her primary research interests are the late-Quaternary vegetation and climates of Beringia. She is currently on the staff of the Quaternary Research Center at the University of Washington.

Pat J. Behling received her B.S. in sociology and psychology from the University of Wisconsin in 1976. She worked with the Climate Food Group at the Institute for Environmental Studies at the University of Wisconsin-Madison for 10 years, then joined the Center for Climatic Research. Her research interests include the analysis of large data sets and the development of scientific visualization tools for research applications.

Linda B. Brubaker is professor of forest ecology at the College of Forest Resources at the University of Washington. Her research focuses on the late-Quaternary vegetation history of Alaska and the responses of conifer forests in the Pacific Northwest to short-term climate variation.

John Dodson is a graduate of Monash University and the Australian National University and is currently associate professor and head of the School of Geography at the University of New South Wales. His research interests lie in the areas of Quaternary environmental change, biogeography, and human impact on the environments of Australia, the Pacific, and Ireland.

Beth Ann Dworetzky completed her studies at Lamont-Doherty Geological Observatory in 1986, then worked for private industry for three years. She now writes and works as an advocate for children with special health care needs.

Peter J. Guetter is an associate scientist at the Center for Climatic Research at the University of Wisconsin-Madison. He received an A.B. degree from Augustana College in 1954. His research interests include using computer general circulation models to simulate past climates.

Sandy P. Harrison is a paleoclimatologist with a special interest in the use of lake-level data for reconstructing regional climates. She has worked in Australia, Great Britain, the United States, and Sweden. Currently she teaches at Lund University in Sweden.

Brian Huntley received B.A., M.A., and Ph.D. degrees from the University of Cambridge. He has worked as a research assistant and associate at the University of Cambridge and as lecturer in biological sciences at the University of Durham, where he is currently a core staff member at the Environmental Research Centre.

Matt S. McGlone is a paleoecologist at DSIR Land Resources, Christchurch. He has specialized in pollen-analytical studies of the New Zealand Quaternary, with particular emphasis on climate change over the last glacial-interglacial cycle.

Vera Markgraf is research professor at the Institute of Arctic and Alpine Research and the Department of Geography at the University of Colorado. She studies late-Quaternary paleoclimatic changes interpreted from pollen-assemblage changes in sediment sections from Europe, the southwestern United States, and southern South America.

Alan C. Mix received his Ph.D. in geology in 1986 from Columbia University. He studies paleoceanography

using species of fossil plankton (foraminifera) and stable isotopes of oxygen and carbon. He has contributed to the paleoclimate projects CLIMAP, COHMAP, and SPECMAP. He teaches at Oregon State University in Corvallis.

Neville T. Moar is a research associate at DSIR Land Resources, Christchurch. He has published extensively on late-Quaternary palynology, vegetation, and climate change in New Zealand. He is currently completing an atlas of New Zealand angiosperm pollen.

Joseph J. Morley is a micropaleontologist and paleoclimatologist. He has concentrated his research efforts on describing variations in the earth's climate over the past several million years as recorded by variations in radiolarians (single-celled siliceous microfauna) preserved in marine sediments from the Atlantic, Pacific, and subantarctic oceans. He is on the staff of the Lamont-Doherty Geological Observatory.

Richard A. Perrott is a lecturer in geography at Wadham College, Oxford. After a first career as an architect, he obtained a B.Sc. in geography at the New University of Ulster and became a tropical palynologist. His advanced studies were on the late-Quaternary environmental history of East Africa. He has also done research in Mexico, Jamaica, Morocco, Nigeria, and Scotland.

Gilbert M. Peterson completed a Ph.D. in land resources at the University of Wisconsin-Madison. After working as a freelance Russian technical translator, he studied data processing at Madison Area Technical College. He is a programmer-analyst for the state of Wisconsin.

Warren L. Prell obtained his Ph.D. from Lamont Geological Observatory of Columbia University in 1974 and is currently the Henry L. Doherty Professor of Oceanography in the Department of Geological Sciences at Brown University in Rhode Island. His research has centered on the evolution of climate as recorded in marine sediments and especially the evolution and variability of monsoonal climates in the Arabian Sea sector of the Indian Ocean. He has been chief scientist of several cruises to the Indian Ocean and has edited several volumes presenting the results of the Ocean Drilling Program in this area.

I. Colin Prentice is an ecologist with current research interests in forest dynamics, vegetation models, and climate-biosphere interactions. He obtained his Ph.D. in Cambridge in 1977 and since then has worked as a research scientist in several European countries. He was based in Uppsala during 1983-1991 and in 1991 was appointed professor of plant ecology at Lund University.

James C. Ritchie, Ph.D., D.Sc., F.R.S.C., is retired professor of botany at the University of Toronto, Scarborough College. His research interests include modern and late-Quaternary plant ecology of arctoboreal regions of North America and of Mediterranean and North African ecosystems.

Neil Roberts is senior lecturer in physical geography at Loughborough University of Technology (U.K.). His research interests include environmental change in low- and middle-latitude regions, notably the circum-Mediterranean lands and East Africa. He is author of *The Holocene: An Environmental History.*

M. Jim Salinger is a principal scientist at the National Climate Centre of the New Zealand Meteorological Service. He specializes in climate change studies and has done extensive research on New Zealand climate history on geological, historical, and instrumental time scales.

Richard D. Selin is an associate information processing consultant at the Center for Climatic Research, University of Wisconsin-Madison. He received his B.S. in meteorology from the University of Wisconsin in 1977. He has worked as an applications programmer at the Forest Products Laboratory of the U.S. Department of Agriculture and the Space Science and Engineering Center of the University of Wisconsin. His area of expertise is the analysis and display of large meteorological data sets.

W. Geoffrey Spaulding manages Dames & Moore's Cultural Resources Center in Las Vegas, Nevada. He specializes in the paleoecology of the Mojave Desert and for the last 20 years has pursued archaeobotanical and paleoecological studies throughout the Southwest and California. He holds degrees in geosciences and anthropology from the University of Arizona, Tucson.

Robert S. Thompson is a geologist with the U.S. Geological Survey in Denver. He studies fossil pollen and plant macrofossils from the western United States. His research interests include the exploration of methods of reconstructing past climates from vegetation data and investigations of the nature and causes of climatic variations through the late Cenozoic. He is currently investigating climatic conditions during the period of sustained global warmth around 3 Ma.

Pao K. Wang received his Ph.D. in atmospheric sciences in 1978 from the University of California, Los Angeles, and is now professor of meteorology at the University of Wisconsin-Madison. His research interests include cloud and aerosol physics, cloud dynamics, and climate studies.

Cathy Whitlock is an associate professor of geography at the University of Oregon. Her research interests focus on the Quaternary vegetation and climate history of the western United States. In addition to her research in the Pacific Northwest and the Rocky Mountains, she has conducted research in Minnesota, Pennsylvania, Ireland, and China.

Marjorie Green Winkler is an associate scientist at the Center for Climatic Research, University of Wisconsin-Madison. She received her A.B. in zoology from Cornell University in 1956 and spent subsequent years raising four children and working in field biology and environmental education. She obtained her M.S. and Ph.D. degrees from the University of Wisconsin-Madison in 1982 and 1985, respectively. Her research interests include the evolution of wetlands and the reconstruction of past environments from pollen, diatom, charcoal, and isotopic analyses of lake sediments and peat.

Index

Compiled by Robert Grogan